# Packings and Stationary Phases in Chromatographic Techniques

## CHROMATOGRAPHIC SCIENCE

*A Series of Monographs*

Editor: JACK CAZES
*Sanki Laboratories, Inc.*
*Sharon Hill, Pennsylvania*

Volume 1: Dynamics of Chromatography
*J. Calvin Giddings*

Volume 2: Gas Chromatographic Analysis of Drugs and Pesticides
*Benjamin J. Gudzinowicz*

Volume 3: Principles of Adsorption Chromatography: The Separation of Nonionic Organic Compounds (out of print)
*Lloyd R. Snyder*

Volume 4: Multicomponent Chromatography: Theory of Interference (out of print)
*Friedrich Helfferich and Gerhard Klein*

Volume 5: Quantitative Analysis by Gas Chromatography
*Joseph Novák*

Volume 6: High-Speed Liquid Chromatography
*Peter M. Rajcsanyi and Elisabeth Rajcsanyi*

Volume 7: Fundamentals of Integrated GC-MS (in three parts)
*Benjamin J. Gudzinowicz, Michael J. Gudzinowicz, and Horace F. Martin*

Volume 8: Liquid Chromatography of Polymers and Related Materials
*Jack Cazes*

Volume 9: GLC and HPLC Determination of Therapeutic Agents (in three parts)
*Part 1 edited by Kiyoshi Tsuji and Walter Morozowich*
*Parts 2 and 3 edited by Kiyoshi Tsuji*

Volume 10: Biological/Biomedical Applications of Liquid Chromatography
*Edited by Gerald L. Hawk*

Volume 11: Chromatography in Petroleum Analysis
*Edited by Klaus H. Altgelt and T. H. Gouw*

Volume 12: Biological/Biomedical Applications of Liquid Chromatography II
*Edited by Gerald L. Hawk*

Volume 13: Liquid Chromatography of Polymers and Related Materials II
*Edited by Jack Cazes and Xavier Delamare*

Volume 14: Introduction to Analytical Gas Chromatography: History, Principles, and Practice
*John A. Perry*

Volume 15: Applications of Glass Capillary Gas Chromatography
*Edited by Walter G. Jennings*

Volume 16: Steroid Analysis by HPLC: Recent Applications
*Edited by Marie P. Kautsky*

Volume 17: Thin-Layer Chromatography: Techniques and Applications
*Bernard Fried and Joseph Sherma*

Volume 18: Biological/Biomedical Applications of Liquid Chromatography III
*Edited by Gerald L. Hawk*

Volume 19: Liquid Chromatography of Polymers and Related Materials III
*Edited by Jack Cazes*

Volume 20: Biological/Biomedical Applications of Liquid Chromatography IV
*Edited by Gerald L. Hawk*

Volume 21: Chromatographic Separation and Extraction with Foamed Plastics and Rubbers
*G. J. Moody and J. D. R. Thomas*

Volume 22: Analytical Pyrolysis: A Comprehensive Guide
*William J. Irwin*

Volume 23: Liquid Chromatography Detectors
*Edited by Thomas M. Vickrey*

Volume 24: High-Performance Liquid Chromatography in Forensic Chemistry
*Edited by Ira S. Lurie and John D. Wittwer, Jr.*

Volume 25: Steric Exclusion Liquid Chromatography of Polymers
*Edited by Josef Janča*

Volume 26: HPLC Analysis of Biological Compounds: A Laboratory Guide
*William S. Hancock and James T. Sparrow*

Volume 27: Affinity Chromatography: Template Chromatography of Nucleic Acids and Proteins
*Herbert Schott*

Volume 28: HPLC in Nucleic Acid Research: Methods and Applications
*Edited by Phyllis R. Brown*

Volume 29: Pyrolysis and GC in Polymer Analysis
*Edited by S. A. Liebman and E. J. Levy*

Volume 30: Modern Chromatographic Analysis of the Vitamins
*Edited by Andre P. De Leenheer, Willy E. Lambert, and Marcel G. M. De Ruyter*

Volume 31: Ion-Pair Chromatography
*Edited by Milton T. W. Hearn*

Volume 32: Therapeutic Drug Monitoring and Toxicology by Liquid Chromatography
*Edited by Steven H. Y. Wong*

Volume 33: Affinity Chromatography: Practical and Theoretical Aspects
*Peter Mohr and Klaus Pommerening*

Volume 34: Reaction Detection in Liquid Chromatography
*Edited by Ira S. Krull*

Volume 35: Thin-Layer Chromatography: Techniques and Applications, Second Edition, Revised and Expanded
*Bernard Fried and Joseph Sherma*

Volume 36: Quantitative Thin-Layer Chromatography and Its Industrial Applications
*Edited by Laszlo R. Treiber*

Volume 37: Ion Chromatography
*Edited by James G. Tarter*

Volume 38: Chromatographic Theory and Basic Principles
*Edited by Jan Åke Jönsson*

Volume 39: Field-Flow Fractionation: Analysis of Macromolecules and Particles
*Josef Janča*

Volume 40: Chromatographic Chiral Separations
*Edited by Morris Zief and Laura J. Crane*

Volume 41: Quantitative Analysis by Gas Chromatography, Second Edition, Revised and Expanded
*Josef Novak*

Volume 42: Flow Perturbation Gas Chromatography
*N. A. Katsanos*

Volume 43: Ion-Exchange Chromatography of Proteins
*Shuichi Yamamoto, Kazuhiro Nakanishi, and Ryuichi Matsuno*

Volume 44: Countercurrent Chromatography: Theory and Practice
*Edited by N. Bhushan Mandava and Yoichiro Ito*

Volume 45: Microbore Column Chromatography: A Unified Approach to Chromatography
*Edited by Frank J. Yang*

Volume 46: Preparative-Scale Chromatography
*Edited by Eli Grushka*

Volume 47: Packings and Stationary Phases in Chromatographic Techniques
*Edited by Klaus K. Unger*

*Additional Volumes in Preparation*

HPLC of Biological Macromolecules: Methods and Applications
*Edited by Karen M. Gooding and Fred E. Regnier*

Modern Thin-Layer Chromatography
*Edited by Nelu Grinberg*

# Packings and Stationary Phases in Chromatographic Techniques

*edited by*

**Klaus K. Unger**

*Institute of Inorganic Chemistry*
*and Analytical Chemistry*
*Johannes Gutenberg University*
*Mainz, Federal Republic of Germany*

**MARCEL DEKKER, INC.** **New York and Basel**

Library of Congress Cataloging-in-Publication Data

Packings and stationary phases in chromatographic
techniques.
(Chromatographic science series ; v. 47)
Includes index.
1. Packings (Chromatography) 2. Stationary phase
(Chromatography) 3. Chromatographic analysis.
I. Unger, K. K. (Klaus K.). II. Series:
Chromatographic science ; v. 47.
QD79.C4P33 1990 543'.090 90-16869
ISBN 0-8247-7940-1 (alk. paper)

This book is printed on acid-free paper.



MARCEL DEKKER, INC.
270 Madison Avenue, New York, New York 10016

Current printing (last digit):
10 9 8 7 6 5 4 3 2 1

PRINTED IN THE UNITED STATES OF AMERICA

# Preface

Although chromatography with all its branches has become a powerful and mature technique in separation and analysis, the important role of packings and stationary phases has not yet been thoroughly reviewed and critically examined. This is particularly valid for high performance liquid chromatography (HPLC) with its bewildering number of packings and stationary phases.

Modern packings and stationary phases are designed, modeled, and tailored for their specific application in chromatography. Having the desired properties, they serve as chemical sensors, able to recognize and discriminate between analytes of closely related chemical and physical structures in chromatographic separation, isolation, and purification procedures.

This book provides the first comprehensive review on packings and stationary phases in chromatography written by highly experienced and internationally recognized chromatographers. It begins with a brief historical introduction and surveys the types of packings and their role in separation processes. The major part deals with the manufacture, structural properties and chromatographic behavior, and use of stationary phases in gas, thin-layer, and column liquid chromatography. The book provides valuable information that will enable the user to choose, handle, and evaluate stationary phases for a given separation problem and to develop separation strategies. It is addressed to analytical chemists, chemists in synthetic laboratories, and biochemists confronted with analytical problems in environmental, pharmaceutical, bio- and polymer analysis, in purity control of industrial products and chemicals, and in structure elucidation of substances.

It also serves as a textbook for graduate students in analytical chemistry and for those who want to update their knowledge in chromatography.

*Klaus K. Unger*

# Contents

# Contributors

**V. A. Davankov** Nesmeyanov Institute of Organo-Element Compounds, Academy of Sciences, Moscow, USSR

**J. V. Dawkins** Department of Chemistry, Loughborough University of Technology, Loughborough, Leicestershire, England

**W. Engewald** Department of Chemistry, Analytical Center, Karl Marx University, Leipzig, German Democratic Republic

**H. E. Hauck** R & D Chromatography, E. Merck, Darmstadt, Federal Republic of Germany

**Helfried Hemetsberger** Institute of Organic Chemistry II, Ruhr University, Bochum, Federal Republic of Germany

**Jan-Christer Janson** Biochemical Separation Center, Uppsala University Biomedical Center, Uppsala, Sweden

**Willi Jost** R & D Chromatography, E. Merck, Darmstadt, Federal Republic of Germany

**Tore Kristiansen** Pharmacy LKB Biotechnology AB, Uppsala, Sweden

**Per-Olof Lagerström** Department of Bioanalytical Chemistry, AB Hässle, Mölndal, Sweden

**B. A. Persson** Department of Bioanalytical Chemistry, AB Hässle, Mölndal, Sweden

**Donald J. Pietrzyk** Department of Chemistry, The University of Iowa, Iowa City, Iowa

**William H. Pirkle** School of Chemical Sciences, University of Illinois, Urbana, Illinois

**Thomas C. Pochapsky** School of Chemical Sciences, University of Illinois, Urbana, Illinois

**Jürgen Pörschmann** Department of Biotechnology, Chemical Plant Construction Industry, Leipzig, German Democratic Republic

**Klaus K. Unger** Institute of Inorganic Chemistry and Analytical Chemistry, Johannes Gutenberg University, Mainz, Federal Republic of Germany

**Ursula Wintermeyer** Public Relations Department, E. Merck, Darmstadt, Federal Republic of Germany

# 1

# Historical Review

Ursula Wintermeyer / *E. Merck, Darmstadt, Federal Republic of Germany*

## ETYMOLOGICAL INTRODUCTION

One hundred and fifty years ago "chromatography" meant something very different from what it means today. A textbook on color theory was published under this title by G. Field in 1835. Chroma (χρωμα = color) and graphein (γραφειν = writing) occur very frequently in modern languages in scientific and technological terminology. Besides "graphics," now a word in its own right, the suffix "-graph" is used for all kinds of instruments that record measurements. In his 1873 publication, "On Comparative Vegetable Chromatology," H. C. Sorby discussed the identification of vegetable dyes by spectroscopic examination [1].

The word "chromatographic" was first used by M. S. Tswett in 1906 in connection with his new technique for separating natural chlorophyll with the the aid of adsorbents. In his publication in 1903 [2], Tswett cited Sorby's work and in 1906 [3]–referring to this publication–he compared the color zones of the separated components of a dye mixture on an adsorption column to the "rays of light in the spectrum." Tswett's neologism "chromatogram" was obviously influenced by Sorby's "chromatology."

## PREVIOUS HISTORY

Today chromatography denotes analytical and preparative separation of substance mixtures by distribution processes between a mobile and a stationary phase. While the mobile phase is gaseous or liquid, the stationary phases are solids or liquids fixed in gels or on solid surfaces. Since the properties of the stationary phase are decisive for the success of the chromatographic separation, a wide range of supports for carrying stationary phases has been developed in chromatographic technology, working by various adsorption principles. Many of the physicochemical processes taking place at stationary phase boundaries were known as regards their adsorption effect long before the advent of chromatography as such, and were utilized for separating or purifying solutions and mixtures of gases. An historical review of the development of adsorbents and stationary phases in chromatography therefore inevitably leads first of all to a consideration of their uses before chromatography as we know it today had evolved.

## Adsorption

In 1657 Glauber reported that dirty water became potable after it has been filtered through fine earth and that red wine lost its color [4]:

> Die beste ▽ion, (so aber etwas kostbar fället) geschiehet durch einen sonderlichen ∵/ den mir nicht beliebt zu offenbaren/ welcher alles ⊖/Schleim/Gestanck und Unreinigkeit aus dem ▽/ Mistlachen/ die angebohrne Röthe im rothen Wein/ Bier/ 2c. 2c. in wenig Stunden fället/ daß alles weiß/ klar und hell als Brunnen▽ wird/ daß also alles gut zu trincken/ und aus rothem ein weisser Wein wird. Wird doch das Meer-Wasser süß/ wann es durch gemeinen ∵ lauffet/ und allda sein Saltz hinter sich lässet: geschieht nun solches natürlich, warum nicht auch durch die Kunst?

(The best precipitation takes place through especially fine soil, soil that I will not disclose, which precipitates out all salts, slime, stench and impurities in a few hours from water or pools of manure, or the redness out of red wine, beer, etc., so that all becomes white, clear and pure as spring water and as good to drink, and white wine from red. Will sea-water too become sweet when it runs through common, fine soil leaving all its salts behind: If such things occur naturally, then why could they not be made to happen?)

Adsorptive filtration for the purification of liquids, or for removing a component from a solution, goes back a long way. The oldest application, and one still used today, is for obtaining drinking water from drainage water. Beds of soil and sand were the first adsorbents. Desalination of seawater by adsorptive filtration–to use the modern term–is a problem that has engaged man from the earliest times. Pliny the Elder, for example, reported on the property of alumina, which desalinates seawater trickling through it [5], "nam in terra marina aqua argilla percolata: dulcescit."

Whether the practical application of the technique was always successful must remain debatable. The report by Hales from 1739 sound very convincing [6]: "sea water being filtered through stone cisterns, the first pint that runs through will be like pure water, having no taste of the salt, but the next pint will be as salt as usual."

In his *Handbuch der theoretischen Chemie*, Gmelin described in 1819 the adsorptive properties of organic charcoal [7]: "The nature of charcoal differs depending on the composition of the organic material turned into charcoal." Many experiments were performed to find out which gases, vapors, or liquids were absorbed by charcoal. According to Gmelin, individual gases were "consumed" by the charcoal in varying amounts, but they could sometimes be driven out again by water or by heat.

"Charcoal has an affinity for very many odorifous substances." Vapors of prussic acid, tobacco, or fumigating powders are consumed by charcoal. Foul water and rotten meat "lose their evil smell and taste with charcoal powder." Red wine and acidic indigo solution were decolorized by shaking with charcoal [7].

In the middle of the nineteenth century, as the demand for food grew with increasing population, agricultural scientists turned to adsorptive filtration to establish which components of fertilizers are held in different types of soil and which are washed out by rain. Liebig reported as early as 1840 [8] that "soils containing iron oxide and burnt clay . . . are therefore true absorbers of ammonia, which they keep from evaporating by a chemical attraction. . . . Every time it rains, the absorbed ammonia passes into the water and is supplied to the soil in solution."

In accordance with the aim and purpose of their work, the agricultural scientists investigated various types of soil for their adsorptive activity with respect to dissolved fertilizers. Prompted by the work of Thompson [9], who investigated arable soils in 1845 with regard to the best use of fertilizers, Way started systematic experiments in 1848. He investigated the adsorptive effects of different soils, ground brick, ignited soil substances, and powdered clay from clay pipes. Thompson and Way let the solution under examination run through a column packed with the adsorption material (between 6 and 20 in.) through a filter. Way also observed that the first portion of the filtrate was free from dissolved substances; the salts and the ammonia were at first retained by column packing. In order to investigate the action of rainwater in soil, Thompson allowed a quantity of water corresponding to 3 in. of rain to trickle through the filter column.

In numerous investigations–Way described about 100 experiments the two English agricultural scientists ascertained the adsorptivity of various soils and of the other materials mentioned with respect to ammonia, a range of salts, and organic fertilizers. Way did not consider the phenomena that he observed to be "surface attraction to which the name capillarity is given," and he continued [6a]: "Furthermore, the former property is only the resultant of two opposite forces, that of the surface attraction of the sand and of water for the salt. It can only therefore operate a condensation of the salt in relation to the strength of the solution, the salt continually shared in given proportions between the sand and the water, so that eventually the whole is washed away. Such, however, is not the case with the compounds which are formed in the soil." On the basis of these observations, Way suggested [6b] "that there are compounds of alumina with silica, having in some respects the same chemical properties as alumina itself."

The experiments on adsorption filtration were not restricted to agricultural problems. Already in 1850 Way was able to report on the basis of his work that solutions of various dyes such as logwood, sandalwood, cochineal, litmus, and other substances were completely decolorized by filtration through earth, but that sugar solution passed through the soil filter unchanged [6b].

## Capillary Analysis

Around the middle of the last century another type of adsorption on stationary phases increased in importance, which can be regarded as the precursor of Tswett's chromatography. The carrier of the stationary phase was the cellulose in paper, whose adsorptive effect has been described by Cramer [10] in 1953 as follows: "Roughly speaking, cellulose surrounds itself with a hydrate shell that acts as the aqueous phase of the distribution chromatogram, in which hydrophilic molecules dissolve and are retained, and come into equilibrium with the moist organic phase flowing past."

In Roman times papyrus was used to carry chemical information. In his work *Historia Naturalis*, Caius Plinius Secundus described the detection of iron sulfate additions to verdigris by means of papyrus impregnated with gallnut extract [11]: "Experimentum in batillo fereo. Nam quae sincera est suum colorem retinet: quae mixta atramento rubescit. Depraehenditur et papyro galla prius macerato." [The test is carried out on a (hot) iron shovel; the pure substance retains its color, the substance mixed with ferrous sulfate turns red. The detection can also be done by means of paper impregnated with gallnut extract.]

If to Pliny paper served as the matrix of a chemical reaction, another property of paper—called "capillarity" by Runge in 1850—was later used for the separation of individual substances from dye mixtures [12]: "By means of its capillarity, paper separates a drop placed upon it into its component parts and, depending on the characteristics of the liquid, forms a pattern with a dark spot in the middle and light or even colorless rings or haloes."

Another paper chromatographic procedure was demonstrated in 1861 by Schönbein in his lecture, "On some separation effects arising out of the capillary attraction of paper," which he presented before the Natural Science Society of Basle and in the subsequent publication in *Poggendorff's Annalen.* This work was to stimulate many subsequent experiments. Schönbein [13a] let "8-inch long 1-inch wide strips of white, unsized paper . . . hang vertically with their lower ends in the test liquid until they were wetted to a height of 1" by capillary action. The test liquids used were dilute aqueous solutions of alkalis, acids, salts, and dyes."

Since the result of his experiments indicating that [13b] "with few exceptions, water migrates faster than the substances dissolved along the capillary route" is of fundamental importance to the separation of substance mixtures at phase boundaries, his experiments with dyes should be considered more closely [13c]: "If one lets a strip of paper dip into water that has been colored so deeply with tincture of indigo that linen or suchlike dipped into it is turned quite a dark blue, until the paper is wetted to a distance of one inch by capillary action, then only the lower half of the wetted part appears blue; the upper half is completely colorless."

Schönbein observed that individual dyes differed in their migration distance, as did inorganic salts. Table 1 has been constructed from his data on organic dyes. The concentration distribution shown in Fig. 1 can be drawn up for the clear dye boundaries described by Schönbein with even colorations, i.e., concentrations. In experiments with acidic litmus solution, Schönbein distinguished three different regions: the lower third of the wetted field was colored litmus red, the middle colorless part gave an acid reaction, and only "the small uppermost part" was pure water. In 1861, Schönbein recognized and described what is known today as frontal analysis, which is shown in Fig. 2.

**Table 1** Migration Distance of the Dyes

| Dye | Migration distance of dye / Migration distance of water |
|---|---|
| Indigo | 1:2 |
| Hematoxylin | 1:3 |
| Logwood | 1:5 |
| Pernambuco | 1:10 |
| Litmus | 11:12 |
| Acid litmus | 1:3 |

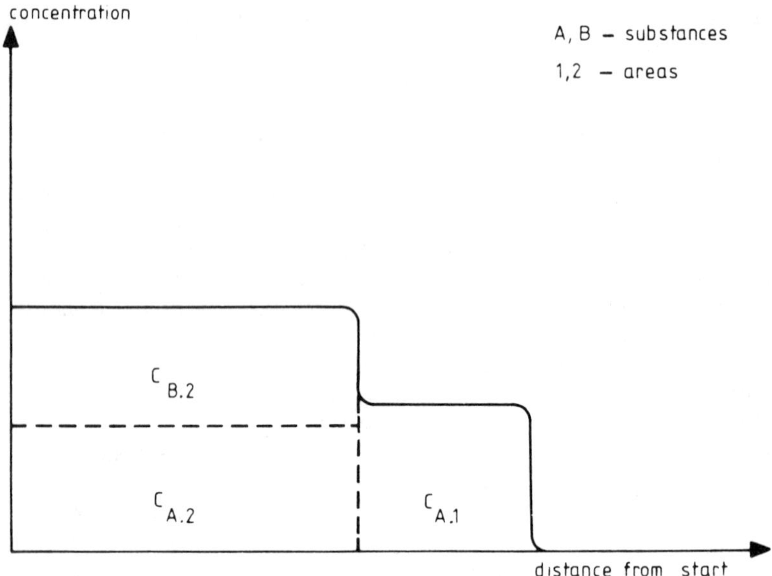


Fig. 1 Course of the concentration in Schönbein's separation experiments in "capillary attraction" (according to the theoretical concepts of Claesson) [14].

Goppelsröder chose for his field of study "capillary analysis based on capillarity and adsorption phenomena" after the lecture in 1861 by his tutor Schönbein. In his extensive series of experiments Goppelsröder used, apart from paper, diatomaceous earth and sand, various types of soil and rock, glass capillaries, wood fibers, and a large number of textile fibers [15b]. In 1906 he summarized the results of his many experiments on the use of other carrier substances [16]: "Already at the start of my experiments genuine Swedish filter paper for quantitative analysis proved to be the only medium suitable for capillary-analytical purposes, although later the highest-purity filter paper from Schleicher und Schüll in Düren, used for analytical purposes, proved to be suitable."

The method used by Goppelsröder for separating mixtures corresponded in principle to Schönbein's experiments: "Naturally, the dyes do not separate completely from one another in the first such capillary experiment," he wrote in 1910 [15a]. "The lower layers contain ever-decreasing amounts of those dyes, which by and large ascent." He recommends eluting the different zones and separating them "sharply from each other" by repeated "capillarization." This does not, however, bring with it any decisive progress, for with a zone ratio of

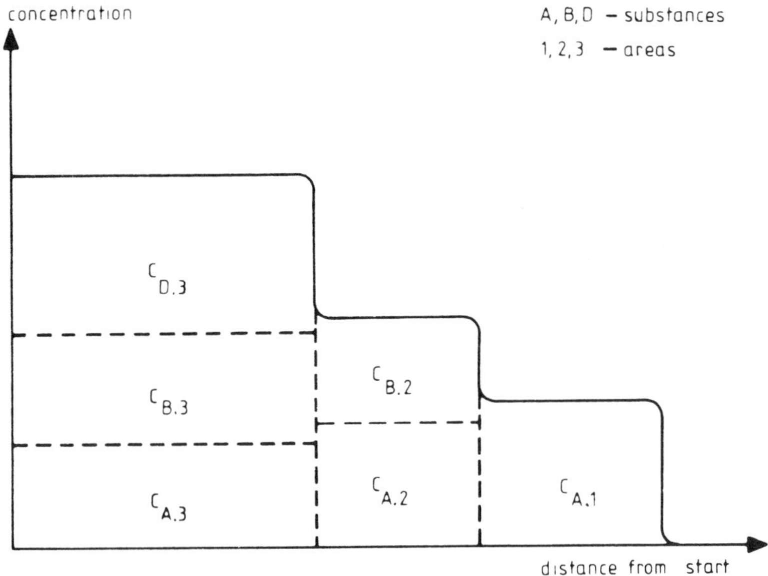


Fig. 2 Schematic substance distribution with acidic litmus tincture (after Claesson) [14a].

1:2 still more than 10% of the target substance with the greater migration distance is present in the lower zone after three separations.

This deficiency, which affected all adsorption analytical work in the second half of the nineteenth century, prevented a breakthrough in this method. From the point of view of scientific history, the works that contributed to further clarification of the separation mechanism and meant progress in the theory are important. The publication by Bayley in 1878 should be mentioned in this context. Bayley noticed that, on a piece of filter paper, the silver from a drop of silver nitrate solution remained concentrated at the center. He attempted to establish experimentally [17] "whether filter paper has any power of withdrawing silver salts from solution." His experiments showed that solutions of metal salts absorbed in rolls of filter paper or applied as drops to filter paper were retained in different ways by the paper, depending on the concentration and nature of the metal [17a]: "The salts of silver, lead and persalts of mercury, when moderately concentrated, give a wide waterring, while the salts of copper, nickel, and cobalt must be much more dilute to present the same appearance. Cadmium seems especially able to pass through filter-paper."

Ostwald saw in the different degrees of adsorption of substances by moist paper the root cause of the separation of substances and "effects of a similar nature, such as those described for animal charcoal. It is the adsorptive processes that cause the separation—capillarity only asserts itself in the transport of the separated components" [18].

Fischer, whose name is linked with pioneering work in sugar chemistry, questioned this interpretation of the separation effect in a joint paper with Schmidmer [19]:

> With many of the dyes investigated by Schönbein and Goppelsröder a relationship to cellulose is undeniable, so that Oswald's explanation applied. But simple inorganic substances like acids, bases, and salts are retained to such a small extent by pure paper that the reason for the change in the solution surely cannot be sought in this relationship alone. We are more inclined to believe that the separation is caused in this case mainly by the different diffusion of the dissolved substances. The following experiments, which were performed for different reasons, actually show that out of two salts the one whose diffusion rate is faster will rise up the paper more quickly.

However, the time had not yet come for a theoretical consideration of the separation processes at phase boundaries by diffusion processes. Half a century was to elapse before Fischer and Schmidmer's observations made within the framework of other research work found a satisfactory explanation.

## Gas Adsorption

The development of adsorption on stationary phases considered above had one feature in common: the substances to be separated were all in solution. The adsorption of gases on solid supports developed separately, an adsorption method which later evolved into gas chromatography. In his 1833 textbook, *Pharmakologischen Tabellen*, Schwartze reported that animal charcoal has [20] "the capacity to absorb significant amounts of gases in contact with it in its pores, and to concentrate them to a certain extent (a property that charcoal has in common with other porous materials)."

These adsorption properties of charcoal, which were also used to decolorize and to purify liquids, were discovered and described almost simultaneously in the second half of the eighteenth century by three researchers: Scheele in Uppsala, Fontana in Florence, and Lowitz in St. Petersburg.

About a century later, in 1881, Chappuis investigated the dependence of the amount of gas adsorbed on temperature and pressure. He found that at a constant temperature the amount of carbon dioxide adsorbed depended on the $CO_2$ pressure and increased "more and more rapidly" with increasing gas pressure. Chappuis hence decided that at 0°C carbon dioxide forms "a layer the thickness of one molecule"; since he observed "two or three times stronger adsorption" at lower temperatures, he concluded that "the layer must comprise two or three layers of molecules" [21].

In 1905 Ramsay [22] was the first to make use of the adsorbent properties of charcoal for analysis of gas mixtures. Following the work of Dewar, who investigated the adsorptive effect of gases on chilled "coconut charcoal," Ramsay passed air over wood charcoal at two levels. Nitrogen, oxygen, and argon were completely adsorbed in the lower level, which was cooled to -100°C. He then conducted the remaining gases over charcoal cooled with liquid air and separated helium from neon: the helium was evaporated off and the neon "expelled by heat" [22a].

In 1907, Freundlich [23] attempted to organize known experimental results mathematically, using an empirical approach. For a gas concentration he found the relationship between the amount of gas adsorbed and the amount of adsorbent [23a].

A decade later Langmuir laid down the basis for a theoretical consideration of gases at adsorbent surfaces. On the assumption that the surface forces are chemical in nature and only act over short distances (0.1 nm), Langmuir concluded that the adsorbing surface was only covered by a monomolecular layer. The amount of gas adsorbed is then proportional to the gas pressure and dependent on the gas and the adsorbent [24].

In 1922, Langmuir's monomolecular layers at the adsorbent surface were confirmed by Carver [25]: "Adsorption will not produce layers of toluene greater than 1 molecule deep on glass. If the thicker layers obtained by other investigators cannot be accounted for by capillary condensation, some other effects must be sought."

The theoretical considerations were not increasingly replaced by attempts to make the differential adsorption of individual gases of practical use. Before coming to the most interesting gas chromatographic investigation in the latent phase of chromatography, namely, the adsorption and desorption of gases and vapors on activated charcoal carried out by Berl's research group at the Technical University of Darmstadt with original ideas, we shall discuss the further developments toward the end of the last century, which can be regarded as the precursors of liquid chromatography.

## Frontal Chromatography

All separation experiments on adsorbent-packed columns during the second half of the last century were afflicted by the problem of incomplete separation of all the components after the first component, but the testing of different adsorbents as column packings and other progress in application techniques paved the way for chromatography. Apart from the agricultural scientists, mention should be made of Reed, who in 1893 separated dye solutions not only on paper but also with "satisfactory results . . . by using tubes containing powdered kaolin lightly rammed down" [26]. His hope [26] "that this method of separation, or some modification of it, might have proved available for the separation of alkaloids from organic matters of different nature" was only to be fulfilled decades later.

In 1886 Engler of the Technical University Karlsruhe published a paper in which he reported jointly with Böhm on the separation of the "oxygen-containing constituents" from the "hydrogen-poor unsaturated hydrocarbons" [27b] in crude oil fractions used in the production of Vaseline by "pressing them through a small battery of filters like the bone-ash filters used in sugar mills" [27a]. Although the Engler and Böhm paper is on the practical side of the utilization of the adsorbent properties of charcoal, in its technical form it influenced the petroleum chemist Day. Day attempted to confirm experimentally the so-called filtration hypothesis, according to which the different types of oil were produced by primary petroleum deep underground filtering upward through different ground layers. Powdered limestone, Fuller's earth, and Florida earth proved in the experiments to be best at separating crude oil "into fractions of different consistencies, differing specific gravities, and different boiling ranges" [28].

In 1901 Engler and Albrecht resumed these investigations and added a drawing of their experimental arrangement to the description of their work with Florida earth. Under a small static pressure the crude oil was forced from below through a "filter pipe" 4 cm in diameter and containing an 80- to 90-cm column of adsorbent (see Fig. 3a) [29].

Gilpin and Bransky used the arrangement shown in Fig. 3b for their experiments on "the diffusion of crude petroleum through Fuller's earth" [30]. The experiments had been prompted by Day, whose apparatus had probably been similar. In 1914 silica gel was introduced successfully into the petroleum industry as a new adsorbent for spatial separation by selective adsorption [31]. This development ran parallel to the first industrial manufacture of a specific adsorbent.

## First Industrially Manufactured Adsorbent for Frontal Chromatography (Fibrous Alumina as per Wislicenus)

The development of aluminum oxide for adsorption analytical purposes in frontal chromatography was taken up by Wislicenus, then in charge of the chemical laboratories of the Royal Saxon Forestry Academy in Tharandt near Dresden. In 1904 he approached the Merck Company in Darmstadt with the request to manufacture a "swelled" alumina that he had already made in the laboratory for the adsorption analysis of natural tanning principles [32]. Merck presented this special "aluminum

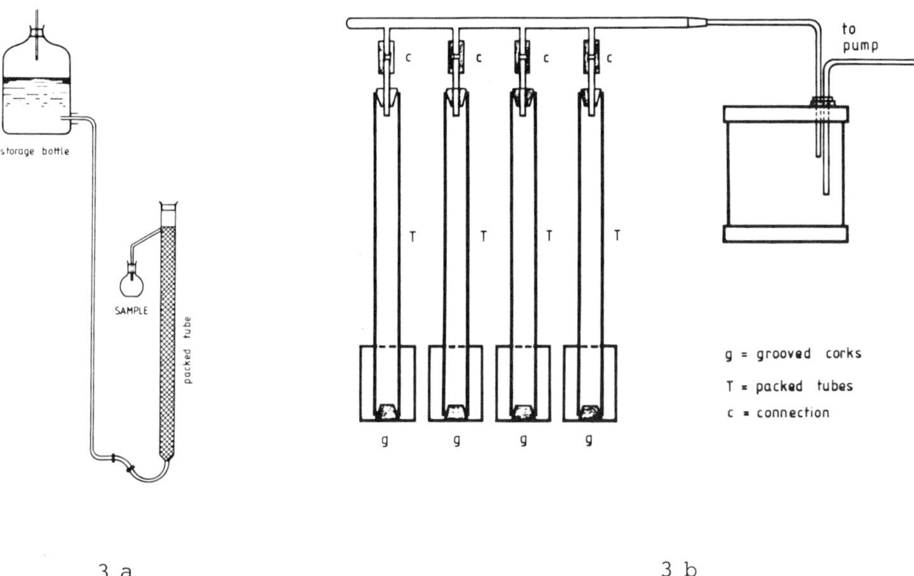


3 a

3 b

Fig. 3 Experimental procedure according to (a) Engler and Albrecht [29], (b) Gilpin and Bransky [30], for the adsorptive purification of crude petroleum.

oxide acc. to Wislicenus" in the same year, at the 7th Conference of the International Association of Leather Industry Chemists in Turin, September 18-23, 1904.

Until that time hide powder made from animal skins had been used exclusively for the determination of the "total amount of tanning substances" in natural tanning principles–information that was important for the manufacture of leather. The variations in the measurement results [32a]–due essentially to the variable nature of the hide powder as a natural product–prompted the search for standardized adsorbents and for the development of a new method of testing specific to the purpose. After the first personal contacts, in correspondence lasting several years Wislicenus conveyed his concepts and improvements on the original method of making aluminum oxide to achieve a constant quality of the "swelled alumina." He placed particular emphasis on the surface quality of the aluminum oxide [33]: "If one increases the surface area of dry aluminum oxide (hydrate) powder to a great extent, its similarity to the hide substance will not remain restricted to its chemical nature, but adsorption effects must also emerge." Wislicenus achieved the increase in the surface area of the aluminum oxide by compounding the purified coarse aluminum powder several times alternately with NaOH and mercury or a mercury salt.

> It is necessary to spread the material out as thinly as possible, on flat trays or on packing paper, and best of all in shallow dishes with an edge a few cm high. Letting it stand overnight is advantageous, as the productive swelling still needs a little time even if the main part of the effect takes only a few minutes. Now, of course, comes the main difficulty: separating the flakes from the contaminated and somewhat less well swelled fractions and from aluminum residues [34].

Wislicenus recommended suction or blowing for the separation of the aluminum oxide flakes, in apparatus which he drew in a letter to Merck. It is critical "that it always be constructed in the same way" (see Fig. 4). The method developed by Wislicenus and named adsorption analysis was intended to detect

1. Total amount of tanning substance
2. Total amount of nontanning substance
3. Amount of water
4. Amount of insoluble (interfering) substances

in natural tanning substances [32b].

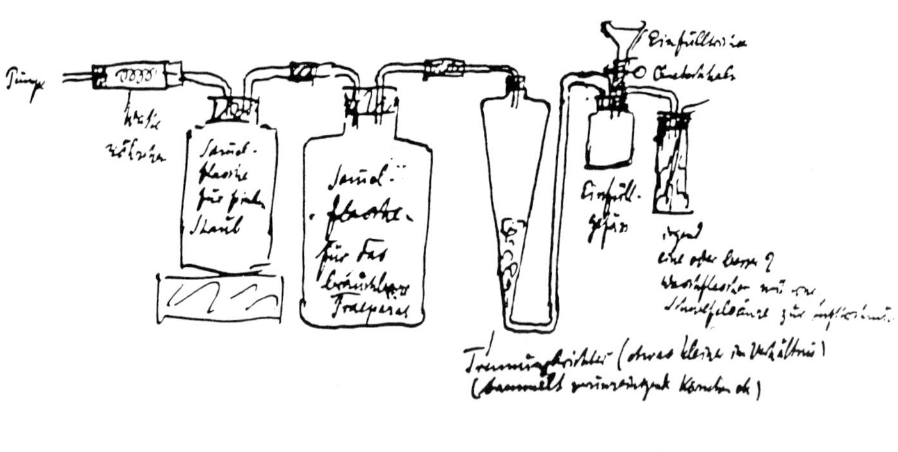


Fig. 4 Sketch of apparatus by Wislicenus [34].

This is how Wislicenus described the adsorbent-filled columns that he developed [35]: One uses

> tubes 15 mm in diameter and 10 cm long, with a narrow 2 cm long tube inserted in the top. Into these tubes, which are also provided with a cotton-wool plug and are tared, one now places 2.50 g of swelled alumina which has been freshly ignited in a porcelain crucible and kept in a desiccator, taps it down, and covers it with a second cotton-wool plug.

As early as 1904 Wislicenus [36] was investigating "dye extracts and artifical dyes" by an "analytical principle in which adsorption plays at the very least the leading role." In the course of the subsequent years Wislicenus considerably broadened the range of the substances he investigated and tried to classify them into a series with selective adsorptive action of the swelled alumina.

## MIKHAIL SEMENOVICH TSWETT

Until the turn of the century the works of Wislicenus and all other investigations into the separation of substances that were related to adsorption–some with other support materials–had the disadvantage that the substances to be separated were not obtained in isolation from each other. Each substance contained in the solution spread evenly from its chromatographic front to the base (frontal chromatography); only the constituent with the lowest adsorption could be isolated in pure form as far as the front of the next substance. The other substances were always overlapped by the ones ahead. Repeated separations often met with only limited success, and in any event they were laborious and time consuming. This was the state of chromatography shortly after the turn of the century, until one man, Mikhail Tswett, decisively improved the adsorption analytical method and 30 years later led the triumphant march of analytical and preparative chromatography.

In an ingenious combination of the familiar and the new, Tswett engineered the critical methodological improvement of the separation technique by developing the chromatograms with pure solvent. Many Nobel prize winners of this century based their scientific knowledge on Tswett's chromatography.

Tswett was a botanist, and his interest lay in the isolation and identification of vegetable pigments. He started his first adsorption experiments in 1901 on filter paper. He had also carried out physicochemical studies in Geneva, and now turned completely to adsorption techniques of separation in his research on pigment mixtures. He explained these techniques comprehensively in his famous paper read before the biological section of the Warsaw Society for Natural Science on March 21, 1903. (According to the old Russian calendar, the date would have been March 8, 1903. The entry of both dates in the original publication shows that there were contacts with other Western research institutions.) He referred to adsorption from organic solvents [2a] as "terra incognita," but "it is precisely investigations of adsorption from organic media that could lead to very interesting results."

Tswett started out from the consideration that "different powdered substances" exhibit different adsorption behavior toward chlorophyll solutions. Hence he investigated "over 100 substances from a wide range of groups of the chemical system" for their adsorption capacity. He achieved his "clearest and fastest" results for chlorophyll solutions with inulin [2b]: "Its more or less rounded particles have a diameter of about 2 μ[m]. The total surface area of inulin can be easily calculated as 2.22 $m^2/g$ from the density of 1.35 on the assumption of spherical particles with a diameter of 2 μ[m]." Tswett recognized the importance of the choice of adsorbent. He reported the closest agreement between the inulin experiments and the observations made with calcium carbonate, aluminum oxide, sugar, and other adsorbents. However, many of the substances examined tended to "decompose carotin and other pigments chemically," and also "the chemical composition of the pigments is changed by many adsorbents" [2d].

In his first paper on chromatography in 1903, Tswett described three different separation techniques based on adsorption:

1. Filtration in a glass tube drawn out at the bottom, packed with a finely powdered adsorbent and plugged with filter paper.
2. Addition of the adsorbent to chlorophyll solution in a test tube, which was then shaken and centrifuged.
3. Addition of coarse adsorbent to the solution in a mortar for subsequent pulverization (with hygroscopic sorption media).

Tswett described his observations in the column experiment very graphically [2c]:

> First of all colorless liquid runs out of the tube, then yellow liquid (carotin), while a light green ring forms in the upper part of the inulin column, which, after a short time, acquires a yellow edge at the bottom. After letting pure ligroin flow through the inulin column, both rings—the green and the yellow—broaden considerably and are carried down to a certain extent. . . . If the adsorption column in which the filtration is carried out is not long enough to hold all the pigments by adsorption, the yellow ring can reach the open end as it migrates down and yellow ligroin solution flows out.

Whereas Tswett had listed three possible ways of carrying out the experiments in 1903, in the reports of the German Botanical Society in 1906 column separation is given as the best method. Tswett [3] postulated an adsorption series corresponding to that of the pigments "analyzed from top to bottom in different colored zones, the more strongly adsorbed dyes displacing the less strongly retained ones further down. . . . I call such a preparation a chromatogram, and the corresponding method the chromatographic method."

In contrast to the adsorptive filtrations carried out hitherto, a different distribution of the adsorbed substances on the column is obtained by elution with pure solvent. Figure 5 shows the product distribution in the direction of flow corresponding to the adsorption and desorption behavior of the substances. Tswett's two-part publication in 1906 showed that after countless experiments with different solvents and adsorbents he had succeeded in making use of the very slight differences in the adsorbability of the individual components in the pigment mixtures for their separation. Figure 6 shows "a chromatographic apparatus for rapid separation of small amounts of pigment solutions" [3a].

Although Tswett still recommended in this publication calcium carbonate as well as inulin or powdered sugar as the "most suitable adsorption medium" for the separation of chlorophyll solutions, in his Russian dissertation of 1910—considered to be his magnum opus—he indicated that "aggressive" adsorbents decompose the vegetable pigments and that only powdered sugar and inulin can be recommended for the separation of chlorophyll [37].

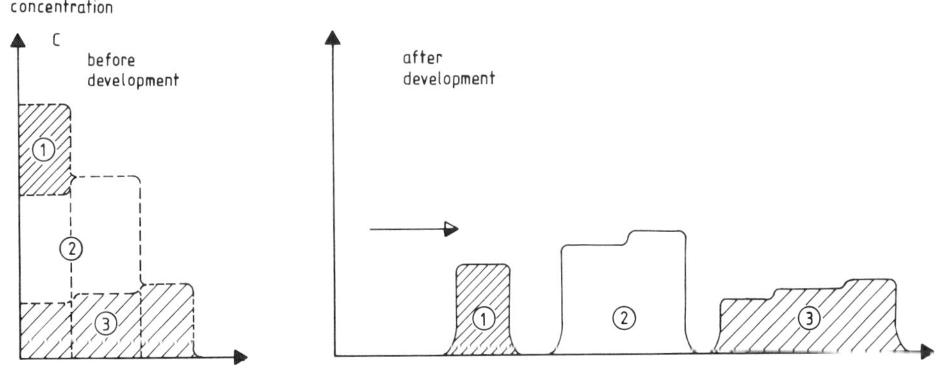


Fig. 5 Distribution of a constituent in Tswett's column before and after development.

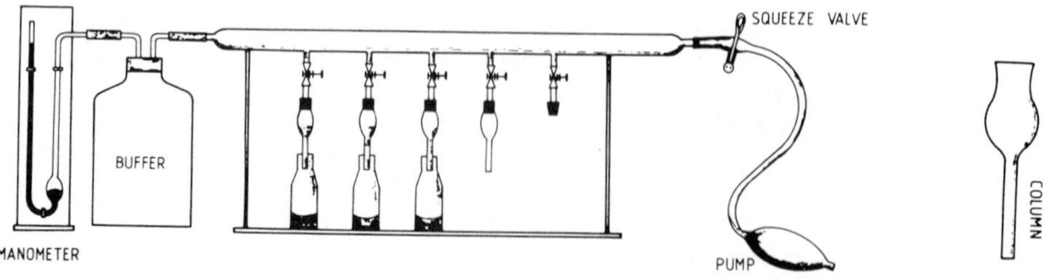


Fig. 6 Chromatography apparatus according to Tswett [3b].

## THE DORMANT PERIOD OF CHROMATOGRAPHY

Accidents and seemingly trivial details can decide the success or failure of new discoveries. Willstätter, the "pope" of chlorophyll research [38], had the only German translation of Tswett's much cited book, *Chromophylls in the Plant and Animal Kingdoms*, which had been written in Russian (1910). In 1913 Willstätter wrote in a treatise that Tswett's adsorption analysis "has so far been applied very little," and that it was "an original method" [39a] but "unsuitable for preparative work" [39b]. Willstätter's skeptical attitude about chromatography could have been due to the fact that in their experiments his associates used unsuitable adsorbents [37] because he had proved that the dye undergoes changes when adsorption of chlorophyll on $CaCO_3$ is done using the Tswett method [43]. Presumably, it had been overlooked that for chlorophyll Tswett had particularly recommended powdered sugar and inulin.

The events of World War I and Tswett's early death, and also the attacks made by some researchers in the field of vegetable pigments (e.g., Kohl, Molisch, and Marchlewski) on Tswett and the results of his work, which went far beyond objective criticism, did not exactly make Tswett's chromatography seem to be an attractive proposition. It is therefore not surprising that not only this method of separation but also other chromatographic processes were used only sporadically during the next 25 years.

The work of Lehmann [40] falls within this dormant period of chromatography as an interesting variation. For the purpose of separation, Lehmann used the differences in the diffusion rates of the test substances in gels rather than their differences in adsorption. He dipped a mixture of 5-10% gelatin or 2% agar solution set in a tube 1 cm in diameter into an equimolar solution of eosin and tartrazine. Above a lower 3- to 4-cm layer the color of the solution, a pure yellow ring about 3 cm broad is formed. Lehmann wrote about the separation of the pigments in the solution [40a]:

> One naturally obtains one constituent pure, the one that diffuses fastest. . . . Further separation can be achieved, even if it is somewhat more laborious, if one suspends the tube containing the gelatin colored by the pigment mixture in pure water. The second pigment in the lowest layer also diffuses out of this in pure form.

This publication from 1907 found as little response in its time as did a remarkable work on gas chromatography from the Chemical Engineering and Electrochemistry Institute of the Technical University of Darmstadt. The apparatus sketched in Fig. 7 looks modern, but it was used by Berl and Schmidt in 1923 to separate various gas mixtures in a column by their adsorption and subsequent desorption on active charcoal. As the experimental arrangement shows, the refractive index "of gases coming out of the last adsorption tube is monitored against the original gas mixture" using a gas interferometer [41]. The change in the interferometer reading is plotted as a function of the amount of gas passed through the instrument; in this way, Berl and Schmidt may have constructed the first gas chromatograms (see p. 19). By raising the temperature in the adsorption column, the gases are desorbed and then displaced with water or benzene vapor (Fig. 8). Table 2 shows the quality of the separations achieved. Like Tswett's liquid chromatography, Berl's gas chromatographic work failed to strike any perceptible response in the scientific world in the subsequent years. See Figs. 9 and 10.

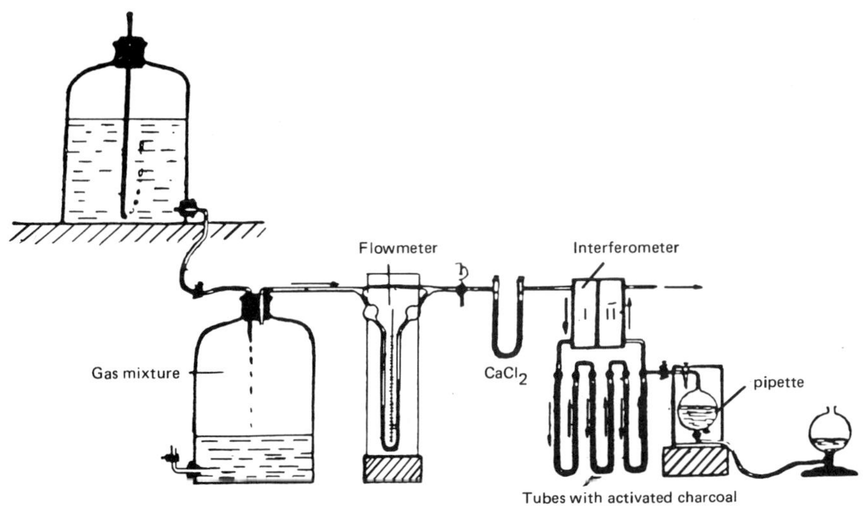


Fig. 7 Berl and Schmidt [41b]: experimental arrangement for partial adsorption. (By permission of VCH Verlagsgesellschaft.)

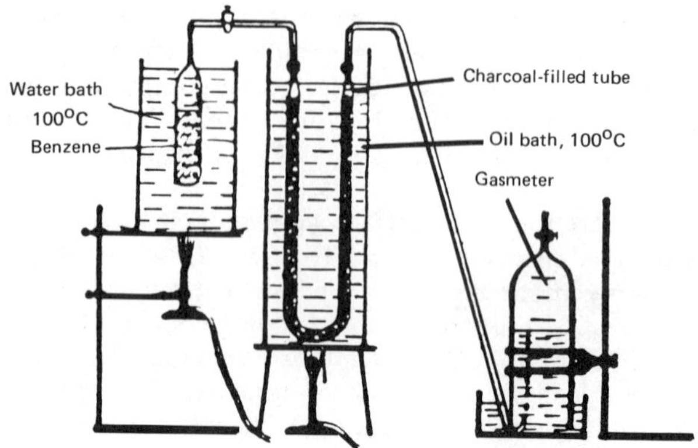


Fig. 8 Berl and Schmidt [41c]: experimental arrangement for desorption of gases and vapors. (By permission of VCH Verlagsgesellschaft.)

Some works on liquid chromatography by biologists and plant physiologists, in all of which reference is made to Tswett's work, fall within the dormant period of chromatography. Palmer deserves a special mention, as after initial publications in collaboration with his mentor Eckles and other publications in this field he exerted a positive influence on the use of chromatography with his 1922 book, *Carotinoids and Related Pigments* [42]. Dhéré and his associates (between 1911 and 1916), Coward (1924), and Lippmaa (1926) were also inspired to isolate pigments from animal and plant tissues by chromatographic analysis through Tswett's adsorption. The column packing was in all cases calcium carbonate, as preferred by Tswett. Although Palmer only mentioned $CaCO_3$ in the descriptions of his experiments, he certainly investigated other substances for their adsorption capacity, for one can read in his book of 1922 [42a]: "The characteristic properties which may be employed for this purpose include . . . adsorption affinity towards finely divided agents like $CaCO_3$."

Table 2 Experimental Results Obtained by Berl and Schmidt (% by volume)

| Phase | $C_2H_4$ | $CH_4$ | $O_2$ | $N_2$ |
|---|---|---|---|---|
| Unadsorbed gas | 0 | 90 | 2 | 8 |
| Temperature desorbate | 74 | 20 | 1 | 5 |
| Water desorbate | 93 | 7 | 0 | 0 |

*Source*: Ref. 41a.

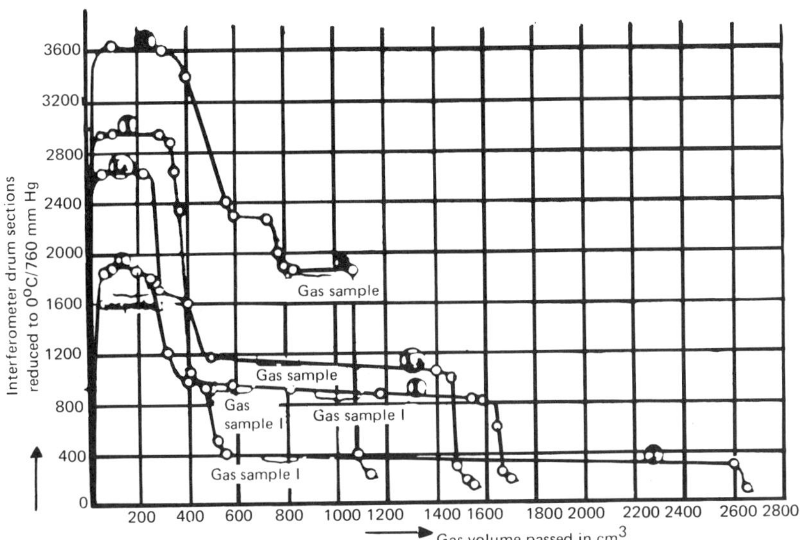


Fig. 9 Gas chromatograms by Berl and Schmidt [41d], 1923. (By permission of VCH Verlagsgesellschaft.)

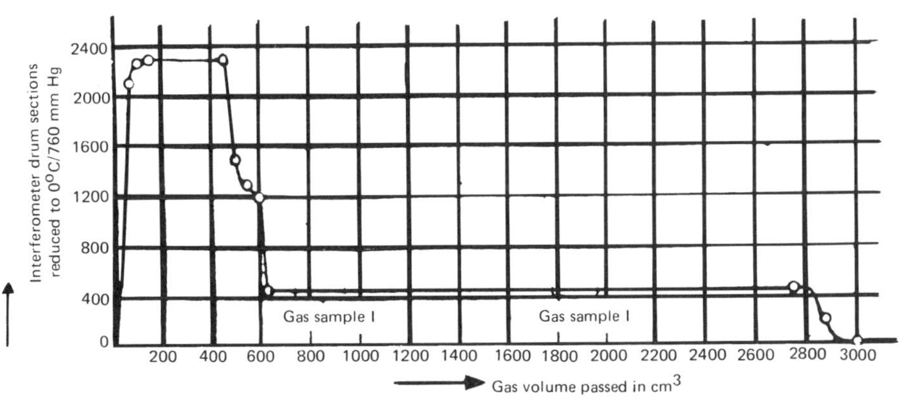


Fig. 10 Gas chromatogram by Berl and Schmidt [41b], 1923. (By permission of VCH Verlagsgesellschaft.)

## THE REBIRTH OF CHROMATOGRAPHY

The breakthrough in chromatography occurred some years later, at the beginning of the 1930s, at the Kaiser-Wilhelm Institute for Medical Research in Heidelberg.

### Richard Kuhn's Team

Edgar Lederer came to Heidelberg in September 1930 to start postdoctoral work with a young professor named Richard Kuhn. His first task was to check experimentally Kuhn and Winterstein's hypothesis that lutein was a mixture of xanthophyll, which had been isolated 20 years earlier by Willstätter, and zeaxanthin, which had been isolated by Karrer one year previously. Lederer thought that with two such similar substances, separation by the hitherto customary chemical methods was impossible. In the course of literature searches he came across Palmer's 1922 book [42] and the short description of adsorption chromatographic separation as per Tswett.

In 1972 Lederer reminisced [37a]: "J'en parlais à Kuhn, qui par bonheur possédait une traduction allemande, manuscrite, du livre de Tswett que Willstätter avait fait faire. C'est dans ce manuscrit que j'ai pu lire tous les détails nécéssaires." [I was talking about it to Kuhn, who fortunately had a German translation, a manuscript, of Tswett's book, which Willstätter had had made. I was able to read all the necessary details in that manuscript.]

Thanks to the excellent results obtained in isolating plant pigments, chromatography spread rapidly wherever closely related compounds or isomers had to be separated. Brockmann, one of Kuhn's team in Heidelberg, said later looking back [43]:

> It was lucky that the first preparative application of Tswett's method was in the field of carotenoids, which demonstrated the efficiency of the method particularly impressively, as this provided the stimulus to try chromatographic adsorption with other classes of substances. Winterstein in particular was of great service in this respect.

Kuhn's team first of all used calcium carbonate for their chromatographic separations, the adsorbent favored by Tswett. Nevertheless, other adsorbents already known at that time were also checked for their efficiency. Carotene, which was hardly adsorbed at all by calcium carbonate, was

> drawn through a column (diameter 1 cm) of fibrous alumina (after Wislicenus, E. Merck). It was then washed with petroleum ether until the colored zone was 2-3 cm wide. The uppermost and bottom

layers of this zone (3 mm each) were skimmed off and extracted by shaking with 90% methanol + petroleum ether. The absorption spectra in carbon disulfide showed that extensive separation of α- and β-carotene had been achieved [44].

Fuller's earth, a form of aluminum silicate used in oil processing before the turn of the century, had also proved unsuitable for separating carotenes, since, according to Brockmann, it "destroyed" carotene [43a]. Kuhn and Brockmann wrote in 1932 [45] that "the use of suitable adsorbents occasionally allows us to separate isomeric pigments such as lycopene and carotene easily and quantitatively. In other cases, however, specific adsorbents that would allow a quantitative separation of say carotene into α- and β-carotene, or of lutein and zeaxanthin, are not yet known."

Finding suitable adsorbents was therefore the major problem in the early years of chromatography for all those using this method of separation.

Winterstein, then Kuhn's assistant, contributed to the rapid growth of this separation technique with his detailed working instructions for "fractionation and purification of substances of vegetable origin according to the principle of chromatographic adsorption analysis" in Klein's 1933 *Handbuch der Pflanzenanalyse* [46] and with his numerous lectures accompanied by experimental demonstrations. It was through one of Winterstein's lectures, in which he demonstrated the separation of crude carotene on a chalk column, that Martin received the first incentive for his later pioneering theoretical and practical works on chromatography [38a,47].

## Early Industrial Manufacture of Specific Chromatographic Adsorbents; Aluminum Oxide as per Brockmann; the Brockmann Scale

In his publication of 1933, Winterstein wrote that in particular aluminum oxide, Wislicenus' fibrous alumina, calcium carbonate, and sugar were used as adsorbents in Heidelberg, but that "in theory all salts, oxides, etc. can be considered as potential adsorbents" [46a]. Examination of the papers on chromatography in the 1930s, whose number increased from year to year, with regard to the choice of column packing material, shows that adsorption analysis was at first always undertaken by Tswett's method and with the adsorbents he had suggested. Only when some kind of difficulty arose, or if another adsorbent had been discovered by chance, was the range of adsorbents extended. In 1938 Zechmeister and v. Cholnoky wrote in the second edition of their book on the chromatographic adsorption method [48]: "One chooses an adsorbent on an empirical basis, although there is already much information in the literature that can be used as a guide within certain substance classes."

In 1933, Karrer and Walker separated α- and β-carotene on calcium oxide and calcium hydroxide. In 1934, Strain was the first to use magnesium oxide for the separation of carotenoids. "Fibrous alumina, however," remarked Strain in retrospect [49], "proved difficult to prepare, handle and pack into columns. . . . One day, largely by accident, I discovered an unusual adsorbent. This material was powdered magnesium oxide." In his publication of 1934 on the separation of carotenes, Strain wrote [50]: "Of the numerous adsorbents utilized for the separation of carotenes, magnesium oxide possesses the greatest number of desirable properties." If new reagents for chromatographic work were described, it was not only their chemical properties with respect to the substances to be separated that were decisive, for the quality of the adsorption properties was determined by the method of manufacture of the reagent. Thus, Strain remarked in a footnote [50a] that

> the adsorption capacity of magnesium oxide depends, to a large extent, upon its method of preparation. Thus, oxide prepared by rapid calcination of the basic carbonate is only a moderately good adsorbent, while the oxide prepared from magnesium hydroxide under suitable conditions is a most satisfactory adsorbent.

At the end of his contribution to Klein's 1933 *Handbuch der Pflanzenanalyse*, Winterstein mentioned a special development in adsorbents for chromatographic purposes on an industrial scale [46a]: "E. Merck, Darmstadt, have recently marketed an aluminum oxide which has been activated and standardized as per H. Brockmann, and which has proved itself very well in our experiments in place of the expensive fibrous alumina."

Brockmann reported that in these initial years of the industrial manufacture of special chromatographic adsorbents aluminum oxide was used most often "for the preparation of adsorption columns, because at particle sizes allowing a good filtration rate in the adsorption column it has a pronounced adsorption capacity for many substance classes" [51]. In the case of aluminum oxide, it transpired during those first few years in Heidelberg that because of the inhomogeneity of the material, it was difficult to obtain reproducible conditions [52].

Through the scientific contacts between the Heidelberg research group and the Merck Company in Darmstadt, Brockmann found suitable support for his idea that chromatography, then in its infancy, would be greatly helped by the availability of suitable adsorbents with a standard adsorption capacity. Merck was receptive to this idea, and with the product "Brockmann's standardized aluminum oxydatum anhydricum" was already offering two special adsorbents for chromatographic purposes in 1934, i.e., in the developmental period of chromatography. In a personal communication to the present author [53], Brockmann mentioned that this aluminum oxide preparation was "defined as guaranteed with respect to its adsorption affinity according to criteria I had suggested."

In a joint paper with Schodder in 1940, Brockmann described a method for the manufacture of the standardized aluminum oxide with graded adsorption capacities. The activity of the adsorbent was modified by the adsorption of water, leading to classification into five grades of activity, which subsequently entered the literature as the "Brockmann scale." The grades of activity were determined by the chromatographic distribution of five azo dye test mixtures on a chromatographic column (Table 3).

> In this scheme the Roman numerals signify the activity grade of the aluminum oxide, starting with the most effective preparation, and the two lines present the upper and lower edges of the adsorption column. The dyes between these lines show where their zones are in the column after the development of the chromatogram. If a dye passes into the solvent during development, its name appears below the lower line. The Arabic numerals above the individual columns indicate the dye mixture used [51a].

The possibility of reproducible representation of differences in the activities of other adsorbents used in chromatography was in part realized [54]:

> For if it is possible to manufacture several preparations with graded and defined activities from each of the common adsorbents, there is a prospect of having a series of adsorbents running from the weakest to the most active, which would permit a continuous variation in the adsorption medium and therefore optimal separation performance.

In 1947 Brockmann published an extended table of sorbents and test mixtures for the classification of adsorbent activity (Table 4). The table shows the range for $SiO_2$ to be comparable with the activity of aluminum oxide, whereas other adsorbents at the most cover three grades, or in some cases only one grade, of the Brockmann scale.

## Silica

By $SiO_2$ Brockmann understood precipitated silicic acid or silica gel with a particle size of 60-90 μm. van Bemmelen [55] had already observed in 1897 the adsorptive properties of dried silicic acid on vapors, gases, and liquids. In 1898 he found [56] that "the effect of heat on the colloids generally causes a lowering—and finally the total loss—of the absorption capacity. If the gel has become porous during dehy-

Table 3 Scheme for Activity Testing

| I | II | | III | | IV | | V |
|---|---|---|---|---|---|---|---|
| 1 | 1 | 2 | 2 | 3 | 3 | 4 | 5 |
| Methoxy-azobenzene | | Sudan yellow | | Sudan red | | Amino-azobenzene | Oxyazo-benzene |
| Azobenzene | Methoxy-azobenzene | Methoxy-azobenzene | Sudan yellow | Sudan yellow | Sudan red | Sudan red | Amino azobenzene |
| | Azobenzene | | Methoxy-azobenzene | | Sudan yellow | | |

*Source*: From Ref. 51a, by permission of VCH Verlagsgesellschaft.

**Table 4** Adsorbent Activities According to Brockmann

| | | | | |
|---|---|---|---|---|
| *p*-Methoxyazobenzene<br>Azobenzene | $Al_2O_3$ I | $SiO_2$ I | MgO I | |
| Sudan yellow<br>*p*-Methoxyazobenzene | $Al_2O_3$ II | $SiO_2$ II | MgO II | $CaSO_4$ I |
| Sudan red<br>Sudan yellow | $Al_2O_3$ III | $SiO_2$ III | MgO III | $CaSO_4$ II |
| *p*-Aminoazobenzene<br>Sudan red | $Al_2O_3$ IV | $SiO_2$ IV | | |
| *p*-Hydroxyazobenzene<br>*p*-Aminoazobenzene | $Al_2O_3$ V | | $MgCO_3$ I | |
| *p*-Hydroxyazobenzene<br>*p*-Aminoazobenzene<br>in filtrate | | | $CaCO_3$ I | |
| Fat orange<br>Fat blue | | | $CaCO_3$ II | $CaSO_4$ III |

*Source*: Ref. 54a by permission of VCH Verlagsgesellschaft.

dration, like the hydrogel of $SiO_2$, the volume of these pores or cavities becomes smaller under the influence of heat."

Marcus of Frankfurt patented the adsorption of unwanted components in gases and liquids on silica gel in Germany in 1911 [57]. Numerous patents followed, including in other countries. From 1914 onward he improved the methods of manufacturing activated silica gel [57a]. From the Institute of Therapy, Brussels University, Zunz reported in 1913 on investigations on the adsorption of toxins, lysines, and their antibodies by silicic acid [58]. In the same year the American Patrick obtained his doctorate with a thesis on "the uptake of gases by the gel of silicic acid" [59]. In subsequent years he dedicated himself to the clarification of the adsorption behavior of silica gel and to the improvement of its methods of manufacture, which resulted in numerous publications and patents. After silica had proved itself as an adsorbent in gas mask filters on the Allied side in the First World War, the Silica Gel Corporation and the Davison Corporation in the United States started producing silica gel in 1919 for industrial adsorption purposes [60]. With production in the United States, the research innovations were at first also predominantly made in the United States, and only in the 1920s were inroads made into the German research institutions.

In 1925 Holmes and Anderson reported at the 65th Meeting of the American Chemical Scoiety, New Haven, Connecticut, on "the preparation of highly adsorbent gels" [61]. Silica gel and other metal oxide gels, made with various porosities, exhibited strong adsorptive effects with respect to certain substances. Thus, the sulfur components in crude oil could be reduced considerably by adsorptive filtration with a "gel from iron" as well as with silica gel. The authors comment [61a] that "the loss of water molecules in the drying process gives capillarity to the usual silica gel. . . . The porosity of these gels is greatly affected by the rate of drying–the slower the better."

By 1927 Ruff and Mautner, who had been working on the adsorption capacity of activated charcoal, referred to the technical importance of silica gel after a comparative study of charcoal and silica gel [62]: "Oil refining is almost exclusively the domain of silica gel and related substances."

A new and–in the context of the development of chromatography–remarkable application of silica took place one year later at the University of Würzburg. Referring to Patrick's report on the adsorption of dissolved substances on silica and the separation of vapors, Grimm and Wolff stated [63]: "However, to our knowledge it is not known that the adsorptive properties of silica gel can be used for the separation of liquid mixtures, particularly in cases where fractional distillation is of no use."

The separation technique described by the authors as the "dripping method" is frontal chromatography with silica as the adsorbent.

> The simplest and, as will be shown, the most effective method is to let the mixture of liquids for separation drip slowly through a layer of silica gel of a defined size. The most suitable dimensions of the layer and the best particle size are determined by trial and error. The liquid dripping from the layer is collected in small fractions [63].

As with all frontal chromatograms, even with silica, the authors only obtained some of the first component of the liquid mixture dripping from the layer in a pure form. Other separations of mixtures of saturated hydrocarbons and aromatics followed the first separation of chloroform and alcohol. The authors established from their experiments that the separation effect increases in proportion to the heat of wetting. Variation of the particle size and the pore diameter led to the result "that the separation effect within the ranges examined improves with decreasing pore size and decreasing particle size [63a].

Whitehorn's use of silica as the adsorbent in chromatographic columns in 1934 followed its use in frontal chromatography [64]. It was chance that led Whitehorn to the first use of this substance as a stationary phase [64a]: "In the course of attempts to prepare aluminosilicates containing as little iron as possible . . . I found silicic acid itself a very good adsorbent for epinephrine, and not destructive." The commercial material was specially processed to obtain the adsorptive properties required by the author.

The "selective adsorption by silicic acid" [64c] mentioned by Whitehorn as a method of determining epinephrine in blood is based on effective distribution chromatography with buffered solutions in a 20 × 200 mm column [64b]:

> Pour the neutralized blood filtrate into the adsorption tube and suck it through the silicic acid at the rate of about 3 drops per second. Rinse the tube and silicic acid with three successive 8 cc portions of distilled water recently boiled and cooled. The epinephrine is adsorbed by the silicic acid, the glutathione and other non-basic or relatively feebly basic substances pass through.

The epinephrine was subsequently washed out of the chromatographic column with 2/3 N sulfuric acid. The effect on which the separation is based was explained in 1941 by Martin and Synge in the paper, "A new form of chromtogram employing two liquid phases" [65]. The aimed use of silica gel in chromatography really started with this publication.

Martin had designed apparatus for countercurrent extraction to separate substances with similar partition coefficients between two solvents. In 1940, together with Synge, Martin tried to move two liquids in an experimental setup against each other, the equilibrium being established as quickly as possible so as not to prolong unduly the duration of the experiment.

In the course of these experiments the authors came to the following conclusions [65]:

> In most forms of countercurrent extraction column the very small drop required for the rapid attainment of equilibrium, and hence for high efficiencies, cannot be used owing to the difficulty of preventing it moving in the wrong direction. In the case of a solid, however, for any reasonable size of particle a filter will prevent movement in any undesired direction. Consideration of such facts led us to try absorbing water in silica gel, etc. and then using the water-saturated solid as one phase of a chromatogram, the other being some fluid immiscible with water, the silica acting merely as mechanical support.

In retrospect, Martin wrote [47a],

> Then I suddenly realized that it was not necessary to move both liquids; if I just moved one of them the required conditions were fulfilled. I was able to devise a suitable apparatus the very next

> day, and a modification of this eventually became the partition chromatograph with which we are now familiar. Synge and I took silica gel intended as a drying agent from a balance case, ground it up, sieved it and added water to it. We found that we could add almost its own weight of water to the gel before it became noticeably wet. We put this mixture of silica gel and water into a column. . . . One foot of tubing in this apparatus could do substantially better separations than all the machinery we had constructed until then.

In terms of historical importance, the Martin and Synge publication dated November 19, 1941 is extraordinary: a practical application and the related theory of new research results appear together in the first publication. In addition to the outstanding experimental work, the authors translated their extensive theoretical knowledge of countercurrent extraction into a theory of chromatographic separation and introduced the concept of an equivalent theoretical plate (HETP) into chromatography.

## Gas Chromatography

Martin and Synge in their 1941 paper had also given consideration to gas phase separation [65a]. However, in a paper published a year earlier (December 6, 1940) [66], Hesse et al. had already reported on adsorption chromatographic experiments in the gas phase. A carrier gas was to replace the solvent in "ordinary adsorption analysis" for the separation of $C_6$ acids [66a].

In 1942, Hesse and Tschachotin [67] jointly published the results of investigations of adsorptive separation of gases and vapors carried out as part of Tschachotin's project for his diploma dissertation. With these low molecular weight substances, "adsorption analysis is necessarily transferred from the liquid into the gas phase" [67]. The main component of the experimental apparatus was similar to the equipment used for liquid chromatography: the adsorbent (silica or activated charcoal) was packed into a 66-cm-long glass tube with an internal diameter of 1.55 cm, which was positioned vertically and surrounded by a heating mantle. The weighed quantity of the substance mixture (about 20 g) was added to the stream of carrier gas from a small evaporation flask and condensed in fractions at the outlet of the experimental apparatus. The authors achieved partial separations or enrichment of azeotropic and isomeric mixtures with various carrier gases.

In 1939, Wicke [68] in Göttingen investigated the theorectical relationships in adsorption and desorption processes of gases and vapors on activated charcoal. He ascertained that the course of the concentration distribution and the "breakthrough curves" is determined by two factors, namely, the curvature of the adsorption isotherms and the diffusion resistance in the interior of adsorbent particles [68a].

Damköhler and Thiele [69] called their adsorption chromatographic method for the separation of gases (published in 1943) the "overflow method." "The principle of the method has already long been known in chemistry as chromatography," wrote the authors [69a]. The substances were determined at the end of the adsorbent layer with an interferometer or a thermal conductivity cell. "The overflow method is the only technique capable of delivering both components of a mixture in a completely pure state" [69a].

In addition to an adsorption gas chromatogram [70], Damköhler and Theile in 1943 published the first gas distribution chromatograms (see Fig. 11). By charing tempered clay with glycerol as the stationary phase, they separated various mixtures (e.g., ethanol-methanol, benzene-cyclohexane) with nitrogen or hydrogen as the carrier gas. For optimal separation the authors required an apparatus that avoided turbulence during flow and enabled complete equilibrium to be set up at every point in the adsorbent column.

In the mid-1940s, Cremer and her team in Innsbruck developed (with no knowledge of the work by Damköhler and Thiele) an apparatus for the microanalytical separation of ethylene and acetylene on the basis of their different adsorption energies [17]. Stimulated by the detailed description of column chromatography by Hesse [72], theoretical considerations led Cremer to the conviction that separation of two substances in a carrier gas is possible even if there is only a very slight difference between their adsorption energies [73]. Her brief theoretical note "On the migration rates of zones in chromatographic analysis" for *Naturwissenschaften*, in which it was announced that the experimental results supporting these statements were to be published, could no longer appear in 1944 owing to the chaos of war and consequent relocation of the publishing house. The practical investigations for corroborating Cremer's theory were likewise only carried out in 1946-1947 by Prior. The first gas adsorption chromatogram for the analytical separation of two substances (air and carbon dioxide) was published in 1947 in Prior's dissertation [74] (see Fig. 12).

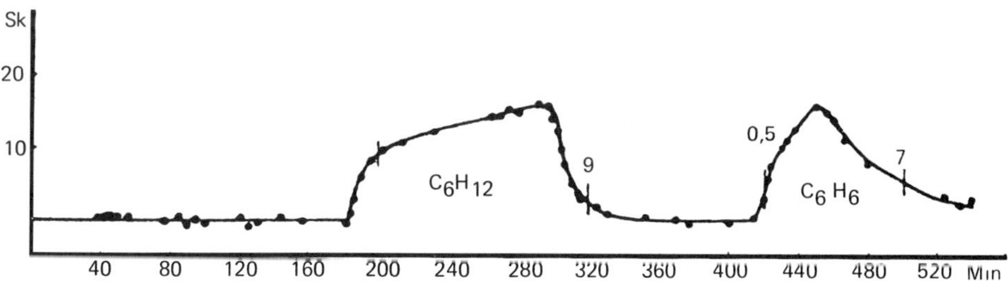


Fig. 11 Separation of benzene cyclohexane in a column of glycerinated potter's clay by atmospheric pressure and -40°C in a carrier gas. (From Ref. 69 by permission of VCH Verlagsgesellschaft.)

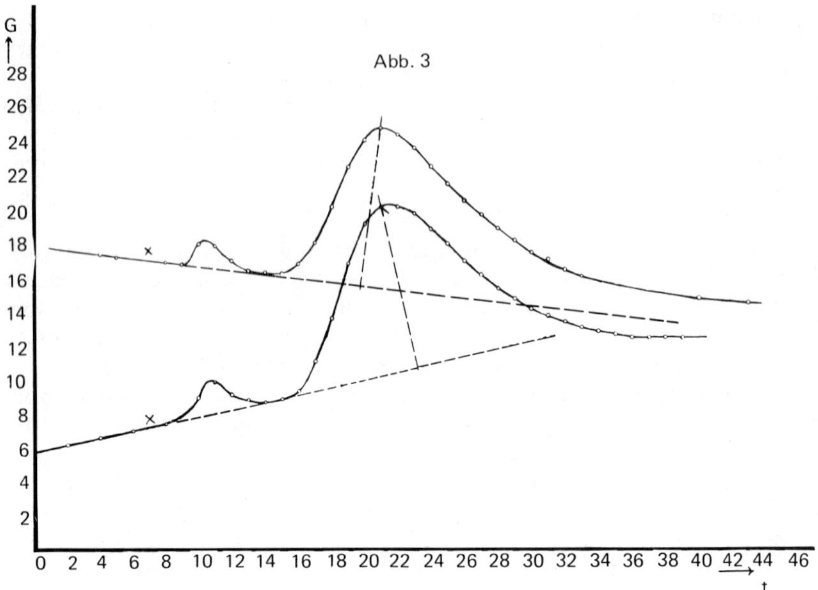


Fig. 12 The first analytical gas chromatogram; original figure from the thesis of Prior [74]. (By permission of Cremer.)

Although difficulties with the baseline and the tailing of peaks were still apparent here [73a], these problems were overcome in the subsequent dissertation project by Müller [75,76]. Figure 13 shows the apparatus used by Cremer and her associates.

"A brand new analytical method for qualitative and quantitative analysis" had been found according to Cremer in a historical review [73a], even if there was still doubt among organic chemists in 1949 about "using many grams of impure adsorbents . . . to quantitatively determine $10^{-5}$ g substance" [73b].

"Chromathermography" was the name given by Turkeltaub [77] to the procedure by which he and his group separated hydrocarbons in Moscow in 1950-1951. For the thermal desorption of gases while air was being passed over, a hot zone was conducted from one end of the silica column to the other with the aid of a cylindrical furnace.

At the beginning of the 1950s, Janak, prompted by the works of Cremer and Turkeltaub [78], constructed in Brno (Czechoslovakia), at the Institute of Petroleum Research, a gas chromatograph for the investigation of natural gas and light hydrocarbons, using carbon dioxide as the carrier gas. Janak is the holder of the first patent for a gas chromatograph worldwide [78], for which he applied in 1952.

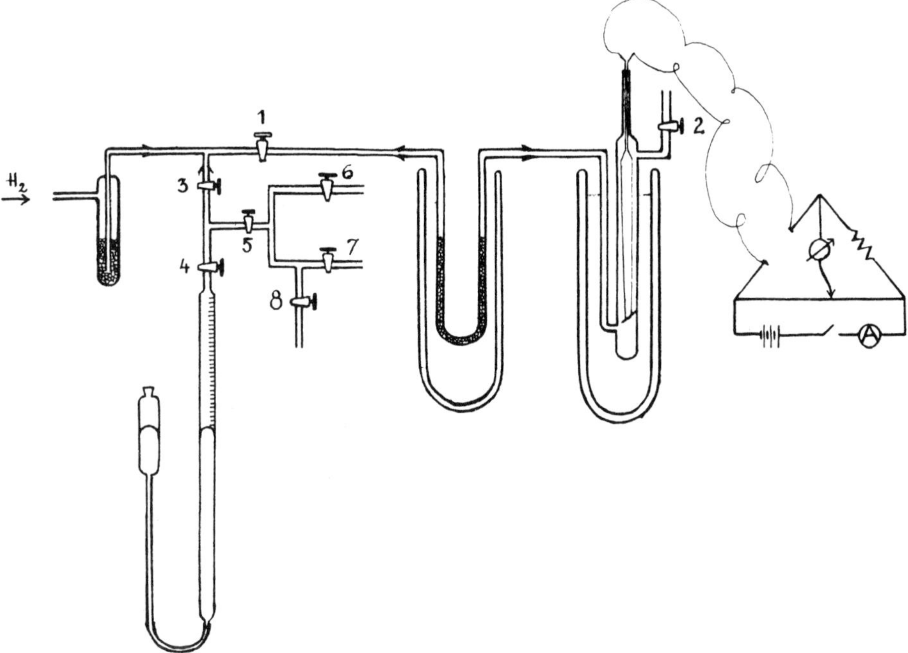


Fig. 13 Apparatus used by Cremer, Prior, and Müller, from the thesis of Prior [74]. (By permission of Cremer.)

Even if today this instrument can only be admired as a museum piece, one must not forget that it contained the key elements of a modern gas chromatograph.

## ION EXCHANGE CHROMATOGRAPHY

There are allusions to the action mechanism of another type of stationary phase in chromatography in the works of the agricultural scientists of the middle of the last century, quoted in the introduction (Liebig, Thompson, Way). The observed adsorption and buffering behavior led Lemberg back to ion exchange in the soil at the end of the last century [80]:

> A study [79] of the conversion of alkali metal silicates with alkali metal carbonates led me to investigate the mass actions in more detail, and it emerged that they exert the greatest influence in

> the case of silicates. . . . All naturally occurring stretches of water contain a certain amount of dissolved salts, and it is important to know how the acids and bases in question are bound, since the electro-negative component of a dissolved salt can often exert a considerable influence on the latter's effect on a mineral.

From the results of many experiments on the interactions between soils from various sources and ions from different salt solutions, Lemberg concluded [80a] that "some soil silicates . . . enter into chemical interactions with dissolved substances very rapidly; if a soil is treated for a long time with salt solutions, all the strong bases will be substituted by others."

The first commercial use of natural zeolites as ion exchangers was reported by the sugar industry. In 1896 Harm patented a process for removing potassium and sodium ions from sugarbeet juice by using a natural silicate [81].

On the occasion of the International Congress on Applied Chemistry [82] in 1903, Rümpler reported on the first technical application of synthetic zeolitic compounds in the processing of molasses. When filtering the molasses through these zeolites, the potassium content of the molasses is reduced in exchange with the calcium of the zeolites and its ability to crystallize is improved.

Shortly after the turn of the century, in the second half of the first decade, Gans synthesized zeolites of the $Na_2Al_2Si_3O_{10}$ type from clay, sand, and sodium corbonate for use as cation exchangers; several inorganic silicates of a similar type had already been synthesized by Way [83,84]. Besides the technical application of synthetic aluminum silicates in the improvement of molasses [84a], Gans points out that various salts of an acid can be produced [84b] "by filtering an easily obtainable salt of this acid through an aluminum silicate containing the base desired for the acid bound in exchangeable form." The constantly equivalent conditions, independent of the method of presenting these water-containing synthetic aluminum silicates, as well as their behavior in the dissociation of water and in the exchange for neutral salt solutions induced Gans in 1913 to regard permutites as chemical compounds [85]. However, like the later aluminum sulfate and sodium silicate synthetic zeolites, they were only of limited use owing to their sensitivity to acids.

A rapid increase in the use of ion exchangers came about in 1935 with the discovery of synthetic resins by Adams and Holmes. These resinous products from polyhydric phenols and formaldehyde and from aromatic amines and formaldehyde opened up many possibilities for their use owing to their resistance [86]. The authors described the following possible applications in their first publication: adsorption of alkalis, removal of anions, adsorption of gaseous ammonia, and bactericidal activities [86a].

In 1938 Taylor and Urey published a paper on the separation of lithium and potassium isotopes on columns packed with synthetic zeolites [87]. In the authors' opinion these experiments fell far short of answering all the questions about the possibilities and the effectiveness of the separation, and some thought should be given "to the nature of the fractionation process" [87a]. Three columns were constructed for the experiment, in different sizes and from different materials (Table 5). The authors stated that these first series of experiments were still unable to yield definite information, as only two forms of the same zeolite type were tested without changing the solvent medium [87b]:

> Any change in the system affecting the nature of the solvent medium will affect the binding of the two isotopic ions, and if this is not in the same ratio for the solvation and binding of the ions to the zeolite, there will be a change in the effectiveness of the fractionation. We cannot predict whether this would improve or decrease the desired fractionation, nor can calculations be made for such a complicated system.

Only one year later Samuelson of Helsingborg described in several publications the preparation of sulfonic acid permutites and their potential uses in analytical chemistry [88]. The author emphasized as one important application the quick and simple removal of anions or even cations that could interfere with an analytical determination, suggesting the funnel shown in Fig. 14.

The synthetic ion-exchange resins achieved worldwide importance through their use in the separation of the rare earths and the transuranic elements and their fission products. In the November 1947 issue of their journal [89], the American Chemical Society published a series of papers on this subject written by various authors from the Clinton National Laboratory in Oak Ridge, Tennessee; from the Institute for Atomic Research and the Department of Chemistry, Iowa State College; as well as a contribution from the Dow Chemical Co., Midland, Michigan on the "fundamental properties of a synthetic cation exchange resin."

**Table 5** Columns According to Taylor and Urey in 1938

| Material | Diameter (in.) | Length (ft) | Sodium zeolite |
|---|---|---|---|
| 1. Stainless steel | 0.75 | 35 | 2 kg |
| 2. Hard rubber | 1.25 | 30 | 4 kg |
| 3. Hard rubber | 1.25 | 100 | 13 kg |

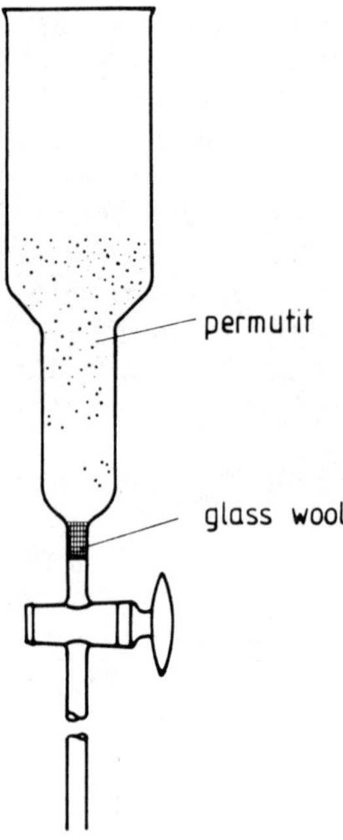


Fig. 14 Apparatus according to Samuelson [88a].

In 1980 Ettre wrote on the subject [90]: "It is an excellent demonstration of the wide-ranging capability of chromatography, involving the separation of from kilogram quantities to only a few atoms."

Between January and March 1945 Spedding's team succeeded for the first time in isolating large amounts of cerium and yttrium of spectrographic purity by using an ion exchange column packed with Amerlite resin [91]. This was a method whose further development made it possible to obtain larger amounts of the rare earths and rare-earth-like metals, whose isolation by conventional methods is laborious and time consuming owing to their chemical similarity in states of very high purity.

This new method of separating chemically similar substances by ion exchange was not only employed in laboratories, but also found braod application on a large industrial scale. In the book *Ion-Exchange Technology* of 1956, in which experienced users discuss the important technological and engineering aspects of ion exchange processes, Spedding and Powell also presented a pilot plant for the "isolation in quantity of individual rare earths of high purity" [92]. Figures 15 and 16 show a schematic diagram of the apparatus and a diagram of the separation of heavy rare earths.

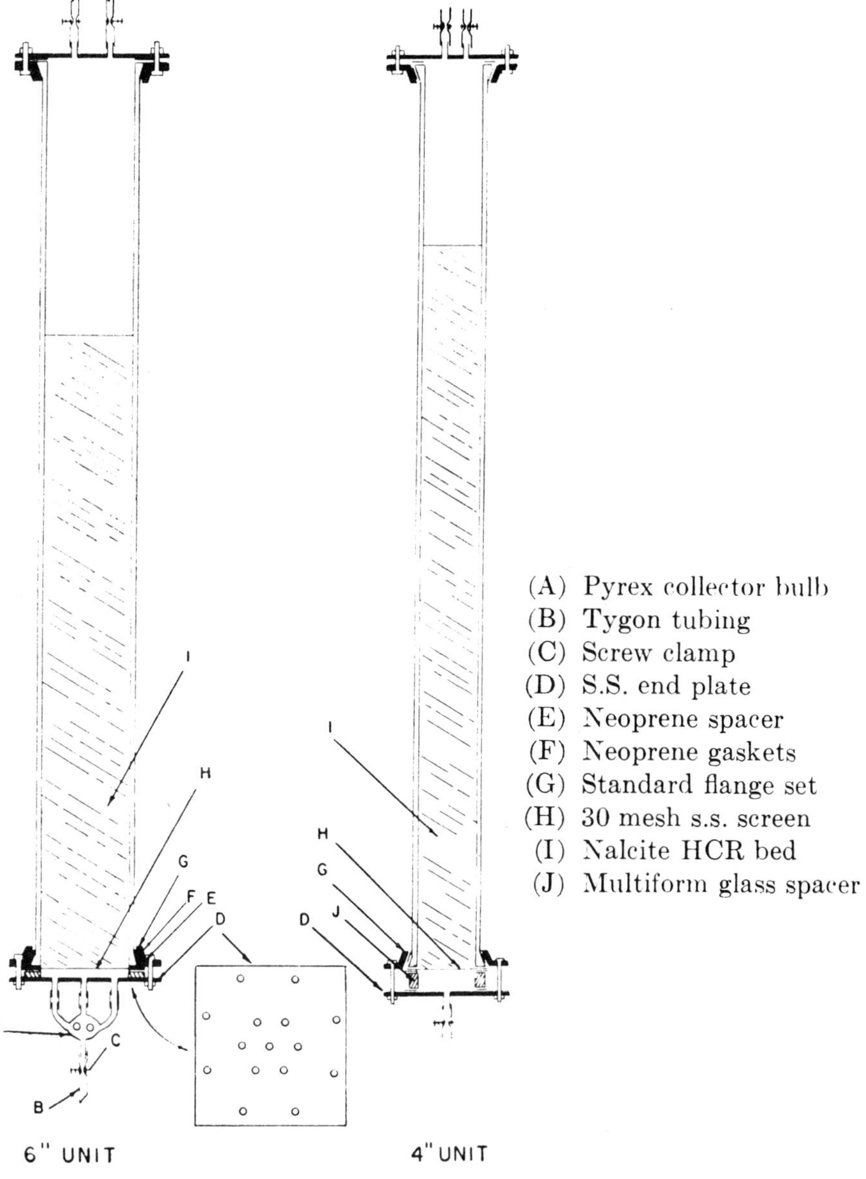


**Fig. 15** Schematic diagrams of the individual 6-in. and 4-in. units used in the pilot plant operations by Spedding and Powell [92a]. (By permission of Academic Press.)

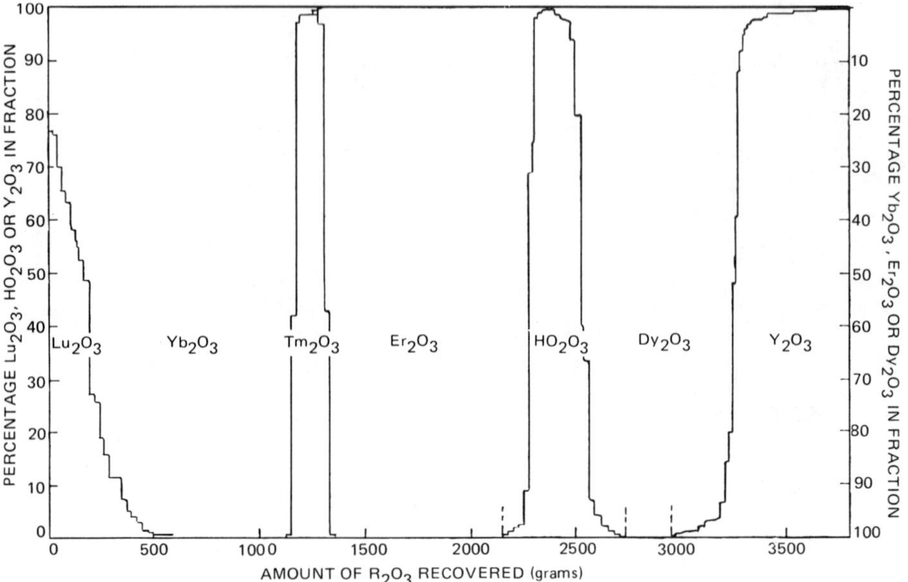


Fig. 16 Composite curve showing the separation achieved by secondary elution [92b]. (By permission of Academic Press.)

Further applications of ion exchangers in various other fields, e.g., the metals industry, analytical and preparative chromatography, biochemistry, and medicine, will not be discussed here in any detail. The synthetic manufacture of the exchangers made it possible to tailor the properties of these resins to particular purposes. The three-dimensional network structure with cavities and pores causes a selection of the molecules according to size, quite apart from the ion exchange capacities. Gel chromatography, also known as gel filtration, makes use of this size exclusion effect.

## GEL FILTRATION

In the course of his studies with Tiselius at the Department of Biochemistry in Uppsala at the beginning of the 1950s, Flodin observed that proteins and small molecules behave differently if starch is used in electrophoresis instead of filter paper. He did not follow up this phenomenon, but was reminded of it a few years later when, as a researcher at AB Pharmacia in Uppsala, he was working on the development of a new dextran-based expander. Porath, who was looking for inert supports for column electrophoresis, tested a polymerizate synthesized

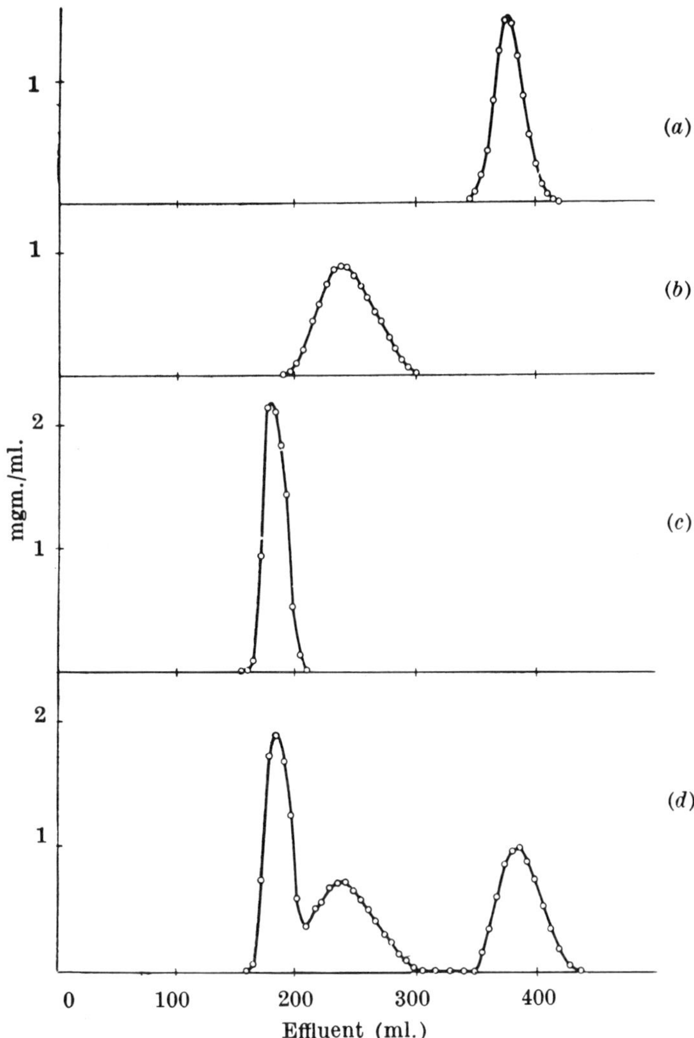


Fig. 17 First separation of oligosaccharides by Flodin and Porath, 1959 [94a]. (By permission from *Nature*, Macmillan Journals Ltd.)

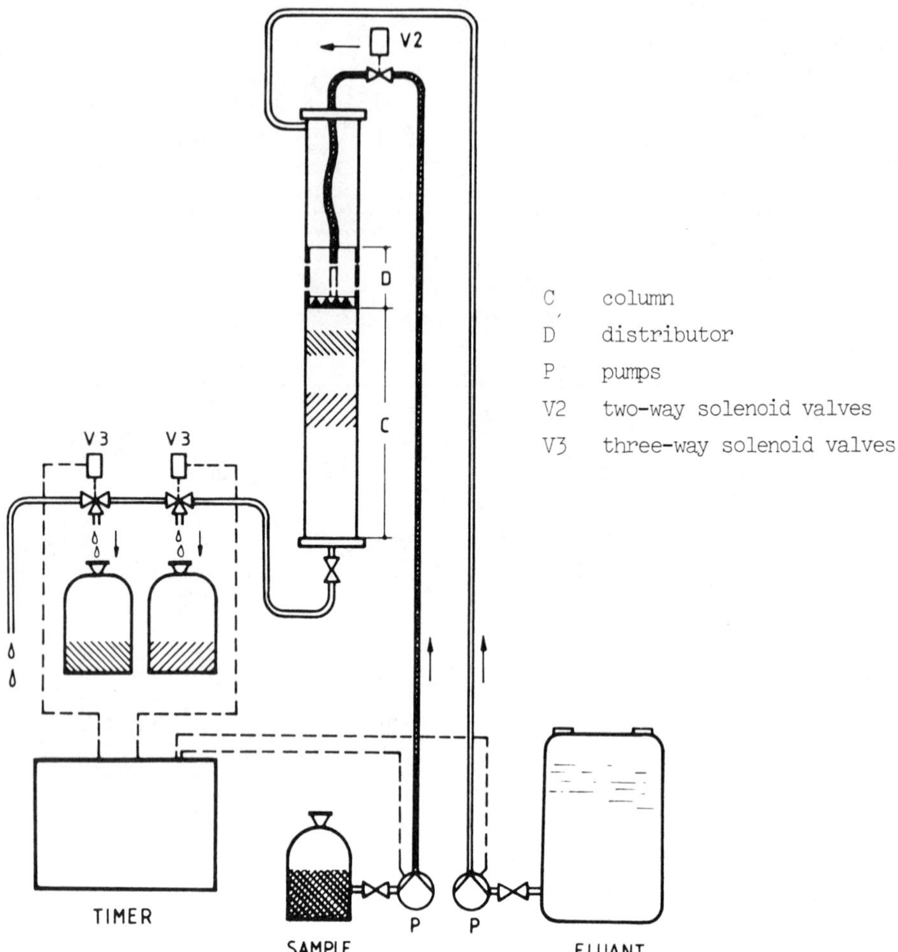


Fig. 18 Apparatus used by Flodin in 1962 for gel filtration to remove salts and other low molecular weight solutes from macromolecules [93b].

by Flodin from dextran and epichlorohydrin. Flodin recalled in 1975 [93] that "he found it to be an excellent support with very little adsorption of proteins; it behaved like starch and not like cellulose when large and small molecules were passed through the column."

The researchers pointed out that chromatography could not be carried out with these gels in the traditional sense, but that they could replace dialysis [93a]. In May 1957 Flodin presented to the company management dextran gels for column electrophoresis, for ion exchangers, as molecular weight-selective membranes for electrophoresis, and for separation by restricted fiffusion in gels [93a]. At Tiselius' suggestion the method was called gel filtration and the dextran gels were given the name Sephadex, an abbreviation of SEParation, PHArmacia, and DEXtran [93a].

Flodin and Porath presented their results with numerous experiments at the Gordon Conference on Protein at New Hampton in January 1959, at the same time the well-known paper [94], "Gel filtration: A method for desalting and group separation," was published in *Nature*. The researchers described their product, which was to be followed in the course of time by other Sephadex developments intended for specific areas of application and having defined pore diameters, with the words [94]: "These gels consist of hydrophilic chains which are cross-linked. They are devoid of ionic groups, the polar character being almost entirely due to the high content of hydroxyl groups. . . . The fractionation depends primarily on differences in molecular size although phenomena have been observed which indicate the influence of other factors." The number of possible applications was as great in industry as it was in research. Figure 17 and 18 show the graphs of the first separations of oligosaccharides in 1959 and an automatic apparatus for gel filtration, dating back to 1962. Further progress was made in 1964 by Moore of the Dow Chemical Company with the manufacture of synthetic gels based on macroporous, crosslinked polystyrenes for separating high molecular weight polymers with neighboring molecular weights [95]: "It is apparent that polystyrene gels, crosslinked in the presence of appropriate diluents, can be used in column chromatography to make molecular size fractionations over an extremely wide range. The term 'gel permeation chromatography' is proposed for this technique."

## REFERENCES

1. H. C. Sorby, *Proc. Roy. Soc.*, *21*:442 (1873).
2. M. S. Tswett, *Tr. Protok. Varshav. Obshch. Estestvoispyt. Otd. Biol. 14*:20 (1903, publ. 1905); (transl. in Michael Tswett's erste chromatographische Schrift, Woelm-Eschwege, (a) 8, (b) 14, (c) 15; (d) 16) (1954).
3. M. S. Tswett, *Ber. Deut. Botan. Ges.*, *24*:322 (a) 386 (b) Tafel XVIII (1906).
4. J. Glauber, Glauberus Concentratus, Leipzig und Breßlau 1715, 507 (reprint Ulm 1961).
5. C. Plinius Secundus, *Historia Naturalis*, Venedig 1491, lib. XXXI, Cap. VI.
6. Th. Way, *J. Roy. Agr. Soc. Engl.*, *11*:315 (a) 316; (b) 365 (1850).

7. L. Gmelin, *Handbuch der theoretischen Chemie III*, F. Varrentrapp, Frankfurt/Main, 1819, 1462-1469.
8. J. Liebig, *Die organ. Chemie in ihrer Anwendung auf Agricultur u. Physiologie*, F. Vieweg Braunschweig, 1840, p. 83.
9. H. S. Thompson, *J. Roy. Agr. Soc. Engl.*, *11*:68 (1850).
10. F. Cramer, *Papierchromatographie*, VCH, Weinheim, 1953, p. 21.
11. C. Plinius Secundus, *Historia Naturalis*, Venedig 1491, lib. XXXIIII, Cap. XI.
12. F. F. Runge, *Farbenchemie* III, E. S. Mittler, Berlin, 1950, p. 15.
13. C. F. Schönbein, *Pogg. Ann.*, *24*:(a) 275; (b) 280; (c) 278 (1861).
14. S. Claesson, *Ark. Kemi, 23A*1, 59; (a) 66 (1946).
15. F. Goppelsröder, *Kapillaranalyse*, Th. Steinkopff, Dresden, 1910, (a) 5; (b) 7.
16. F. Goppelsröder, Anregung zum Studium der auf Capillaritäts- und Adsorptionserscheinungen beruhen den Capillaranalyse, Helbing und Lichtenhahn, Basel, 1906, p. 24.
17. Th. Bayley, *J. Chem. Soc.*, *33*:304; (a) 305 (1878).
18. W. Ostwald, *Lehrb. d. allgem. Chemie I*, W. Engelmann, Leipzig, 1885, p. 789.
19. E. Fischer, and E. Schmidmer, *Liebigs Ann.*, *272*:157 (1893).
20. G. W. Schwartze, *Pharmakologische Tabellen*, Leipzig, 1833, p. 803.
21. P. Chappuis, *Ann. Phys. Chem.*, *12*:179 (1881).
22. W. Ramsay, *Proc. Roy. Soc. A*, *76*:111; (a) 113 (1905).
23. H. Freundlich, *Z. Physik. Chem.*, *57*:385; (a) 451 (1907).
24. I. Langmuir, *J. Am. Chem. Soc.*, *40*:1363 (1918).
25. E. K. Carver, *J. Am. Chem. Soc.*, *45*:66 (1923).
26. L. Reed, *Proc. Chem. Soc.*, *9*:126 (1893).
27. C. Engler and M. Böhm, *Dingler's Polytechn. J.*, *262*:(a) 470; (b) 472 (1886).
28. D. Holde, *Angew. Chem.*, *13*:1201 (1900).
29. C. Engler and Albrecht, E., *Angew. Chem.*, *14*:889 (1901).
30. J. E. Gilpin and O. E. Bransky, *Am. Chem. J.*, *44*:262 (1910).
31. V. Heines, *Chem. Tech.*, *1*:281 (1971).
32. H. Wislicenus, *Z. Anal. Chem.*, *44*:97; (a) 99; (b) 98 (1904).
33. H. Wislicenus, *Angew. Chem.*, *17*:802 (1904).
34. H. Wislicenus, Letter of June 15, 1905 (Firmena archiv Merck, Darmstadt).
35. H. Wislicenus, *Collegium*, *245*:62, 63 (1907).
36. H. Wislicenus, *Kolloid-Z.*, *2*:IX (1907).
37. E. Lederer, *J. Chromatog.*, *73*:361; (a) 362 (1972).
38. L. S. Ettre, *Anal. Chem.*, *47*:422 A; (a) 435 A (1975).
39. R. Willstätter and A. Stoll, *Untersuchungen über Chlorophyll*, J. Springer, Berlin, 1913, (a) 16; (b) 157.

40. E. Lehmann, *Z. Physik. Chem.*, *57*:718; (a) 720 (1907).
41. E. Berl, and O. Schmidt, *Z. Angew. Chem.*, *36*:247; (a) 250; (b) 248; (c) 252; (d) 251 (1923).
42. L. S. Palmer, *Carotinoids and Related Pigments*, Chemical Catalog Company, Inc., New York, 1922, p. 247.
43. H. Brockmann, *Angew. Chem.*, *53*:384; (a) 385 (1940).
44. R. Kuhn and E. Lederer, *Chem. Ber.*, *64*:1355 (1931).
45. R. Kuhn and H. Brockmann, *Z. Physiol. Chem.*, *206*:42 (1932).
46. A. Winterstein, Fraktionierung und Reindarstellung von Pflanzenstoffen nach dem Prinzip der chromatographischen Adsorptionsanalyse in: *Handbuch der Pflanzenanalyse* (G. Klein, ed.), III, J. Springer, Wien, 1933, pp. 1403-1437; (a) p. 1436.
47. A. J. P. Martin, in *75 Years of Chromatography: A Historical Dialogue* (L. S. Ettre and A. Zlatkis, eds.), Elsevier, Amsterdam, 1979, p. 286; (a) pp. 288, 291.
48. L. Zechmeister, and L. v. Cholnoky, *Die chromatographische Adsorptionsmethode*, 2. Aufl., J. Springer, Wien, 1938, p. 43.
49. H. H. Strain, Ref. 47, p. 438.
50. H. H. Strain, *J. Biol. Chem.*, *105*:524; (a) 524 footnote (1934).
51. H. Brockmann and H. Schodder, *Chem. Ber.*, *74*:73; (a) 74 (1941).
52. H. Brockmann, *Angew. Chem.*, *53*:389 (1940).
53. H. Brockmann, Letter of June 18, 1983 (*Archiv Merck*, Darmstadt).
54. H. Brockmann, *Angew. Chem. A*, *59*:200; (a) 201 (1947).
55. J. M. van Bemmelen, *Z. Anorg. Chem.*, *13*:296 (1897).
56. J. M. van Bemmelen, *Z. Anorg. Chem.*, *18*:122 (1898).
57. R. Marcus, DP 263 388; (a) DP 279 075.
58. E. Zunz, *Z. Imm. Forsch.*, *19*:326 (1913).
59. W. A. Patrick, Die Aufnahme von Gasen durch das Gel der Kieselsäure. Inaugural Dissertation Göttingen, 1914; *Kolloid-Z.*, *16*:118 (1915).
60. Ullmann (ed.) *Encyklopädie der Technischen Chemie*, *15*:716 (1964), *21*:459 (1976).
61. H. N. Holmes and J. A. Anderson, *Ind. Eng. Chem.*, *17*:280; (a) 282 (1925).
62. O. Ruff, and P. Mautner, *Z. Angew. Chem.*, *40*:433 (1927).
63. H. G. Grimm, and H. Wolff, *Z. Angew. Chem.*, *41*:98; (a) 103 (1928).
64. J. C. Whitehorn, *J. Biol. Chem.*, *108*:633; (a) 634; (b) 639; (c) 643 (1934).
65. A. J. P. Martin and R. L. M. Synge, *Biochem. J.*, *35*:1358; (a) 1359 (1941).
66. G. Hesse, H. Eilbracht and F. Reicheneder, *Liebigs Ann. Chem.*, *546*:233 (a) 251 (1941).
67. G. Hesse and B. Tschachotin, *Naturwissenschaften*, *30*:387 (1942).

68. E. Wicke, *Oel Kohle, 37*:405; (a) 410 (1941).
69. G. Damköhler and H. Thiele, *Angew. Chem., 56*:353; (a) 355 (1943).
70. G. Damköhler and H. Thiele, *Beiheft z. Z. deut. Chemiker, 49* Verlag Chemie, Berlin, S. 18, 1944.
71. E. Cremer, Ref. 47, p. 22.
72. G. Hesse, *Adsorptionsmethoden im chemischen Laboratorium*, de Gruyter, Berlin, 1943.
73. E. Cremer, *Chromatographia, 9*:363; (a) 365; (b) 366 (1976).
74. F. Prior, Dissertation, Universität Innsbruck, 1947.
75. E. Cremer and F. Prior, *Z. Elektrochem., 55*:66 (1951).
76. E. Cremer and R. Müller, *Mikrochem., 36*:553 (1951).
77. A. A. Zhukhovitskii, O. V. Zolotareva, V. A. Sokolov and N. M. Turkeltaub, *Chem. Zentr. (1951)* II, 1932.
78. J. Janak, Ref. 47, p. 177; Czech. Patent 83991.
79. J. Lemberg, *Z. Deut. Geol. Ges., 22*:803 (1870).
80. J. Lemberg, *Z. Deut. Geol. Ges., 28*:526-527. (a) 590 (1876).
81. F. Harm, DP 95.447.
82. A. Rümpler, *Fifth Int. Kongr. Angew. Chem.*, III:59 (1903).
83. J. Schubert and F. C. Nachod, in *Ion Exchange Technology* (F. C. Nachod and J. Schubert, eds.), New York, 1956, p. 3.
84. R. Gans, *Jahrb. Preuss. Geol. Landesanst., 27*:63; (a) 92; (b) 93 (1906).
85. R. Gans, *Jahrb. Preuss. Geol. Landesanst., 34*:281 (1913). *Zentralbl. Mineral. Geol.* 741 (1913).
86. B. A. Adams and E. L. Holmes, *J. Soc. Chem. Ind., Trans. Commun., 54*:1; (a) 4 (1935).
87. T. I. Taylor and H. C. Urey, *J. Chem. Phys., 6*:429; (a) 434; (b) 437 (1938).
88. O. Samuelson, *Z. Anal. Chem., 116*:328: (a) 331 (1939).
89. *J. Am. Chem. Soc., 69*:2769-2881 (1947).
90. L. S. Ettre, Evolution of liquid chromatography: A historical overview, in *High-Performance Liquid Chromatography* (C. Horvàth, ed.), Academic Press, New York, 1980, p. 55.
91. F. H. Spedding, et al. Ref. 89, p. 2777.
92. F. H. Spedding and J. E. Powell, Ref. 83, p. 359; (a) 380; (b) 385.
93. P. Flodin, Ref. 47, p. 69; (a) 70; (b) 74.
94. J. Porath and P. Flodin, *Nature, 183*:1657; (a) 1658 (1959).
95. J. C. Moore, *J. Polym. Sci. A, 2*:842 (1964).

# 2

# Survey of Types of Chromatographic Packings and Stationary Phases and Their Role in Separation Processes

Klaus K. Unger / *Johannes Gutenberg University, Mainz, Federal Republic of Germany*

## DEFINITIONS AND PRINCIPLES OF CLASSIFICATION

Generally, chromatography is understood as a zone migration method that results in a selective dilution of individual sample components contained in the starting mixture [1]. More precisely, chromatography is a physical separation method whereby the components are distributed between two phases, namely, the stationary phase and the mobile phase. Separation is primarily governed by the differences in the distribution coefficients of the sample components that are involved in distribution processes during migration along the chromatographic bed. Chromatograpic methods are divided into gas and liquid, depending on the type of mobile phase. A third method is termed supercritical fluid chromatography whereby the mobile phase is a supercritical fluid. In gas chromatography (GC), the stationary phase is the active surface of an adsorbent, or a liquid film either coated on an inert support packed in a column or layered on the wall of an open tube. Liquid chromatography (LC) offers two choices depending on the arrangement of the stationary phase: (a) Thin-layer chromatography (TLC), as an open-bed technique, uses a thin layer of adsorbent particles which constitutes the stationary phase. (b) In column liquid chromatography (CLC), the chromatographic bed consists of a dense packing of support or adsorbent particles. As usual, the support merely functions as a carrier for the liquid stationary phase, while the adsorbent provides an extended stationary phase for interaction. In addition to packed columns, open tubular columns are employed in CLC, with a thin layer of stationary phase on their walls.

Since the terms support, packing, adsorbent, and stationary phase are often used quite loosely, specific definitions are called for. A support is understood as an inert solid carrying the stationary phase. A packing is simply a solid which makes up the chromatographic column and can be either an adsorbent or a support. The term adsorbent refers to solid particles, either totally porous or carrying a porous layer, possessing a high ratio of surface area to volume and capable of sorption interactions with solute components. "Stationary phase" stands for that part of the chromatographic phase system which is in equilibrium with the mobile phase. The stationary phase might be an interfacial layer, a mono- or multilayer, or a bulk liquid.

In chromatographic literature, supports, packings, and stationary phases are treated and categorized under widely differing aspects. In CLC, the grouping of packings into rigid, semirigid, and soft is related to the mechanical strength of the materials. Porous and pellicular packings differ in the way the porosity is distributed across the particles. Pellicular packings carry a thin porous layer on an inert core. The chemical bulk composition serves as another criterion whereby supports and packings are divided into inorganic and organic (listed in Table 1). Among these, crosslinked organic polymers and silica and their functionalized derivatives are by far the most widely used materials.

Since the major objective in chromatographic separation is a gain in selectivity arising from specific interactions of sample components in the phase system, it is useful to classify packings and stationary phases according to the type of interactions and the resulting selectivities. The operative interactions are mainly of a physical nature, but might also include chemical equilibria. In GC the mobile phase possesses an inert character, and thus retention is controlled via direct solute-stationary phase interactions. These range from the nonspecific van der Waal's type to specific dipole-dipole orientation and donor-acceptor interactions [2]. Kiselev and his associates [3] established a classification of adsorbents in gas-solid chromatography based on nonspecific and specific molecular interactions. Similar concepts have been proposed to characterize liquid stationary phases in GLC [2]. A more practical method for discriminating between stationary liquids in GC is to utilize the retention of solutes under defined conditions and to establish a polarity scale [4] as discussed in Chap. 3. In LC we meet a more complicated situation since the mobile phase plays an active role in the distribution process by participating in the interactions. Liquid stationary phases in liquid-liquid chromatography are, for example, classified on the basis of the solubility parameter ($\delta$), derived from the theory of regular solutions [5].

**Table 1** Survey of Basic Types of Supports and Packings According to Their Chemical Bulk Composition

| | |
|---|---|
| *Inorganic* | |
| Carbonates | $CaCO_3$, $MgCO_3$ |
| Oxides | $SiO_2$, $Al_2O_3$, $TiO_2$, $ZrO_2$, $CeO_2$, MgO |
| Phosphates | Zirconium phosphate, hydroxyapatite |
| Silicates | Magnesium silicate, aluminosilicates |
| Carbon | |
| Porous glasses | |
| Nitrides | BN |
| *Organic* | |
| Cellulose | |
| Dextran | |
| Agarose | |
| Polyamide | |
| Polyacrylamide | |
| Styrene-divinylbenzene copolymers | |
| Polyvinylalcohols | |
| Vinylacetate copolymers | |
| Polymethylacrylates | |
| Hydroxylated polyether | |
| Divinyl-*N*-vinylpyrrolidones | |
| *Composites* | |
| Silica/carbon | |
| Silica/polymers | |

Up to now, no generally valid and reliable classification of adsorbents and stationary phases in LC has been established. One major cause arises from the diversity of interactions taking place in LC as opposed to GC. In addition to those mentioned previously, there are hydrogen-bonding, acid-base, charge-transfer, complexation, and coloumbic interactions operating in the various modes of LC. As a result, the retention of solutes is often the sum of several contributions. Another serious hindrance to adsorbent classification is the lack of well-defined adsorbents which might serve as reference materials.

The most rational way of treating adsorbents and stationary phases in chromatography is to group them according to the underlying dominant mechanism which controls retention and selectivity, whereby the term mechanism does not refer to the kinetics of chromatographic processes but rather to the specific type of distribution equilibria by which the solutes are retained. This convention is used throughout the volume.

## PHYSICAL AND CHEMICAL STRUCTURE OF SUPPORTS, PACKINGS, AND STATIONARY PHASES

Often chromatographers are unaware that for different chromatographic techniques they are employing the same (with slight variations) type of packing. Moreover, packings of differing chemical composition can be described under commonly valid aspects. For this reason we should first discuss the physical and chemical properties in general before going on to specific structure-solute relationships. When we follow the consecutive stages of a packing–from the manufacture up to its ultimate chromatographic application–we can derive distinct structural properties at three levels.

Level 1 corresponds to a stage where the supports and packings are manufactured as bulk powders and physical and chemical methods can be applied in the characterization of these solids in order to determine their macroscopic and molecular properties. However, any one method used individually would permit a rather limited insight into the structure; only the combined application of modern surface analysis provides an all-over reliable characterization. On the other hand, the pretreatment conditions imposed on the samples during these physical-chemical measurements are often very different from those applied in chromatographic use. The question has therefore been put whether a comprehensive characterization of bulk powder is actually useful; the entire characterization should be derived from chromatographic data instead. There are a number of arguments against this statement. First, a knowledge of the physical and chemical parameters of bulk packings largely enables the chromatographer to understand and interpret the retention mechanisms of chromatographic processes through an appropriate correlation. The obtained measurements provide useful guidelines for choosing a suitable packing for a given separation problem and enable the chromatographer to differentiate between various types of packings on a rational rather than purely empirical basis. Second, since packings are in a continuous process of improvement and refinement, a knowledge of structural, as well as chromatographic, properties will help to produce tailor-made packings for future use.

Level 2 refers to the confection stage, where supports or packings are ready-made for application. Typical confection stages are the coating of capillary columns with a stationary phase, the layering of plates in TLC, and the packing of columns in GC and LC. The quality control of the confection is almost entirely in the hands of the manufacturer and is performed more on an empirical basis than on the basis of sophisticated physical measurements. Nevertheless, the parameters of the confection are essential for the flow dynamics and they ultimately control the performance of the chromatographic system.

Level 3 describes the situation where the packing is in operation, i.e., equilibrated with the mobile phase. The data collected at this stage are representative for a specific type of sample component; in other words, in order to obtain an overall characterization of the system, a series of test solutes must be applied under various conditions. The chromatographic data reflect the behavior of the whole phase system and, as in LC, associated mobile phase effects must be considered in order to determine any correlation to stationary phase properties. The chromatographic evaluation provides the most useful and reliable characterization since it is closest to the real situation. However, with the passage of time chromatographic values of a system may deviate from those measured initially, since the stationary phase is highly reactive and has a tendency to approach a more stable state.

## Physical Structure Properties of Supports and Packings at the Bulk Powder or Granulate Stage

The primary goal in the manufacture of a given support or packing is to produce a solid of defined physical structure (e.g., particle morphology, particle size, particle porosity, and pore size). Adjustment of a defined surface chemical structure is often achieved by employing secondary processes during production, e.g., by appropriate thermal and hydrothermal treatment, chemical surface modification, etc. In liquid chromatography the stationary phase composition is finely controlled by the eluent composition and by mobile phase additives. The physical structure is defined by a series of parameters (listed in Table 2).

### *Particle Shape and Size*

Chromatographic packings are employed as irregularly shaped and spherical particles with average particle diameters ranging from 2 to 200 μm (see Table 3). Angular particles are manufactured from larger granules by milling and grinding processes with intermediate size classification. This technology is highly developed, e.g., in the production of silica grades for TLC. Shaping into beads is accomplished via specific procedures applied during the formation of the packing, e.g., emulsification-polycondensation or polymerization, sol-gel transformation in a two-phase system, agglutination, spray drying, and others. In the most favorable case, the process conditions are optimized in such a way that beads of predetermined size with narrow distribution and high yield are formed. However, generally it is necessary to size the beads by consecutive cutting of the lower and upper ends of the

Table 2 Physical Structure Parameters of Supports and Packings

| Term | Symbol | Definition |
|---|---|---|
| Particle shape | | Angular, spherical |
| Particle size | dp | Particle diameter defined according to the method of determination (see Table 4) |
| Average particle size | $dp_{50}$<br>$dp_m$ | Average particle diameter at 50% of the cumulative distribution (median), $dp_{50}$, or the most frequent average particle diameter of the relative distribution (mode), $dp_m$ |
| Particle size distribution | dpsd | Number, volume, weight, or surface area distribution |
| Specific pore volume | vp | Uptake in ml of liquid per unit mass of unit volume of packing to fill the internal pores |
| Micropore volume | vp (micro) | vp of pores of pd < 2 nm |
| Mesopore volume | vp (meso) | vp of pores of 2 < pd < 50 nm |
| Macropore volume | vp (macro) | vp of pores of pd > 50 nm |
| Specific surface area | $a_s$ | Internal and external surface area per unit mass or volume of packing |
| Pore shape | | Assumed to be cylindrical in most cases |
| Pore diameter<br>Micropores<br>Mesopores<br>Macropores | pd | Width of the pore of a given shape<br>pd < 2 nm<br>2 < pd < 50 nm<br>pd > 50 nm |
| Hydraulic pore diameter | $pd_n$ | Ratio of four times the specific pore volume divided by the specific surface area |
| Kelvin pore diameter | $pd_k$ | Pore diameter according to the Kelvin equation [see Eq. (6)] |
| Washburn pore diameter | $pd_w$ | Pore diameter according to the Washburn equation [see Eq. (2)] |
| Average pore diameter | $pd_{50}$<br>$pd_m$ | Average pore diameter at 50% of the cumulative distribution (median), $pd_{50}$, or the most frequent average |

**Table 2** (continued)

| Term | Symbol | Definition |
|---|---|---|
| | | pore diameter of the relative distribution (mode), $pd_m$ |
| Pore diameter distribution | pdd | Distribution of vp or $a_S$ as a function of the average pore diameter |
| Interstitial column porosity | $\varepsilon_0$ | Ratio of interparticle column volume to total geometrical column volume |
| Internal column porosity | $\varepsilon_p$ | Ratio of intraparticle column volume to total geometrical column volume |
| Total column porosity | $\varepsilon_{(total)}$ | $\varepsilon_{(total)} = \varepsilon_0 + \varepsilon_p$ |

distribution; the process is controlled by measuring the size distribution. Sizing is accomplished by fluid classification employing two systems: counterflow equilibrium and transverse flow separation [6].

Surprisingly, the correct assignment of the particle size in chromatography is less carefully evaluated. It is elementary that (a) particle size is uniquely defined by the diameter of spherical particles, (b) the diameter of irregularly shaped particles is expressed as an equivalent

**Table 3** Typical Particle Size Ranges of Supports and Packings Applied in Various Chromatographic Techniques

| Technique | Range of average particle diameter (dp/μm) |
|---|---|
| Gas chromatography (GC) | 100-200 |
| High pressure GC | 50-100 |
| Liquid chromatography (LC) | |
| Thin-layer chromatography (TLC) | 10-30 |
| High-performance TLC | 5 |
| Column liquid chromatography | |
| Large-scale | 50-200 |
| Preparative | 10-60 |
| Analytical, including the high-performance mode | 2-40 |

sphere diameter, and (c) the particle diameter given depends largely on the method evaluation (see Table 4). Size analysis is carried out variously by sieving, microscopy, the coulter counter, laser diffraction, and so on, depending on the average diameter and the width of distribution. The particle size distribution is then expressed as number, mass, surface or volume distribution according to the weighing factor. From the distribution the mode (the average value at the peak maximum of the relative distribution, $dp_m$) or the median (the average value at 50% of the cumulative distribution, $dp_{50}$) are derived. The arithmetic mean particle diameter is the sum of the diameters of individual particles divided by the number of particles. There are also a geometrical mean and a harmonic mean. Mean, median, and mode coincide only for a symmetrical distribution, whereas for asymmetrical distributions one obtains the sequence dp (mean, arithmetic) > median > mode. For more details, see Ref. 8.

The following additional information is essential in order to reliably compare the dp values of different chromatographic packings:

1. The method of evaluation of dp
2. The weighing factor of averaging, i.e., number, surface area, etc.
3. The type of average calculated (mean, mode, median)
4. Quantities to describe the width of the distribution, e.g., the standard deviation or the values $dp_{10}$ and $dp_{90}$.

*Porosity*

The porosity of supports and packings originates from internal voids and cavities, termed pores. By definition, open pores are those communicating with the surface of the particle while closed pores are inaccessible [9]. Porosity relates to a single particle or fractions of the particle, and also to bulk powders and granular masses. The term "particle porosity" describes the ratio of the volume of open pores to the total volume of the particle. Powder porosity is the ratio of the volume of voids between the particles plus the volume of open pores occupied by the powder to the total volume. The specific pore volume of the packing corresponds to the uptake of milliliters of liquid per unit mass or unit volume to fill the internal pores. Porosity can be distributed across the whole particle, as in the case of totally porous particles, or confined to a porous layer encasing an inert core or coated on the wall of a tube.

Also, there are packings possessing a permanent porosity, or a so-called swelling porosity, or features of both. Rigid packings, such as i.a. silicas, aluminas, zeolites, exhibit a permanent porosity, i.e., when these materials are exposed to vapors or immersed in liquids, the porosity remains constant. In contrast to most inorganic packings,

Table 4 Definitions of Particle Size

| Symbol | Name | Definition |
|---|---|---|
| $dp_v$ | Volume diameter | Diameter of a sphere having the same volume as the particle |
| $dp_s$ | Surface diameter | Diameter of a sphere having the same surface as the particle |
| $dp_{sv}$ | Surface volume diameter | Diameter of a sphere having the same external surface to volume ratio as a sphere |
| $dp_d$ | Drag diameter | Diameter of a sphere having the same resistance to motion as the particle in a fluid of the same viscosity and at the same velocity ($dp_d$ approximates to $dp_s$ when Reynold's number is small) |
| $dp_f$ | Free-falling diameter | Diameter of a sphere having the same density and the same free-falling speed as the particle in a fluid of the same density and viscosity |
| $dp_{stk}$ | Stokes' diameter | Free-falling diameter of a particle in the laminar flow region (Reynold's number < 0.2) |
| $dp_a$ | Projected area diameter | Diameter of a circle having the same area as the projected area of the particle resting in a stable position |
| $dp_p$ | Projected area diameter | Diameter of a circle having the same area as the projected area of the particle in random orientation |
| $dp_c$ | Perimeter diameter | Diameter of a circle having the same perimeter as the projected outline of the particle |
| $dp_A$ | Sieve diameter | Width of the minimum square aperture through which the particle will pass |
| $dp_F$ | Feret's diameter | Mean value of the distance between pairs of parallel tangents to the projected outline of the particle |
| $dp_M$ | Martin's diameter | Mean chord length of the projected outline of the particle |
| $dp_R$ | Unrolled diameter | Mean chord length through the center of gravity of the particle |

*Source*: From Ref. 7, reprinted by permission of the publisher.

organic-based packings except macroporous are generally nonporous in the dry state and swell by taking up the solvent, the extent depending on the degree of crosslinking: organic packings with a low degree of crosslinking swell considerably when immersed in a liquid. The extent of swelling is also a function of the polarity of the solvent relative to the surface polarity of the packing [10]. Highly crosslinked organic polymers possess a negligibly small swelling porosity; their porosity is predominantly permanent. Swelling is associated with the strength of the three-dimensional linkage of polymer chains or polymer networks. Chains, which are bound by noncovalent links, show a flexible structure affected by solvation and other phenomena. Swelling is also observed for inorganic packings with a layer structure, e.g., layer silicates, graphite, etc. Application of external pressure may lead to a change in the porosity, depending on the compressibility of the packing. There are few data on this phenomenon.

Inorganic and organic packings are often treated as different structures with regard to pore space and porosity [11]. Irrespective of whether a packing is composed of an assembly of three-dimensionally linked solid particles (e.g., spheres) or covalently linked chains in a space network, the structure is collectively described by the term "gel." As stated by Stauff [12], a gel is defined as a system in which the dispersed substance and the dispersion medium form a mutually penetrating coherent system. Thus, coherence is the overall common criterion applicable to both networks and to coordinated particles of colloidal size, and hence all packings can be treated under common aspects.

In the assessment of particle porosity, a further distinguishing feature is based on the size of the pores. Pores are split into three categories according to their width (w): micropores of $w < 2$ nm, mesopores of $2 < w < 50$ nm, and macropores of $w > 50$ nm [9]. Each category of pores then has a pore volume assigned to it. The sum makes up the total pore volume [vp(total)]. The total specific pore volume is estimated by fluid displacement methods, measuring the apparent density due to helium and mercury [13], $\rho(\mathrm{He})$ and $\rho(\mathrm{Hg})$, according to the equation

$$\mathrm{vp(total)} = \frac{1}{\rho(\mathrm{Hg})} - \frac{1}{\rho(\mathrm{He})} \tag{1}$$

Since, according to the Washburn Eq. (2), mercury at atmospheric pressure is able to penetrate pores below 15 $\mu$m in diameter, and helium with a molecular diameter of $\sim 0.3$ nm enters pores larger than 0.3 nm, the total pore volume calculated this way is only a relative quantity. The micropore volume is usually assessed by gas adsorption measurements, converting the amount adsorbed ($n^a$) at a given temperature into milliliters of liquid adsorbate [9,14]. Purely mesoporous solids exhibit a type IV isotherm in sorption measurements, with a more or less pronounced hysteresis between the adsorption and desorption branch, and a flat part parallel to the relative pressure abscissa at $p/p^0 > 0.95$. The uptake at saturation ($n^a_{satur}$), corresponding to the flat portion of the isotherm converted into milliliters of liquid adsorbate, yields the mesopore volume [15].

The macropore volume is measured by mercury porosimetry, whereby mercury is intruded into the porous solid with increasing pressure [16]. The Washburn equation provides a relationship between the equilibrium pressure (p) and the pore diameter of those pores which are filled with liquid mercury. Assuming a surface tension of mercury of $\gamma = 0.48\ N\ m^{-2}$ and a contact angle ($\Theta$) of 140° at 293 K, one obtains

$$dp_w\ (nm) = \frac{1470}{p(MPa)} \tag{2}$$

This means that at 29 MPa those pores with a diameter equal to or larger than 50 nm are filled with mercury. Depending on the maximum pressure generated by the porosimeter, the pore volume of a large portion of mesopores can thus be determined.

### *The Surface Area* [17]

Supports and packings are characterized by a high surface area to mass or volume ratio. The surface area originates from the external surface area of small particles and from the internal surface area made up of the area of pore walls. The extent to which the external and internal surface areas contribute to the total surface area depends on the particle size and on the pore size. By definition the external surface area ($As_{(ext)}$) is regarded as the envelope surrounding the discrete particles and includes all the prominences as well as the surface of those cracks and fissures that are wider than they are deep [9]. The internal surface area ($As_{(int)}$) refers to the walls of pores and cavities which are accessible to adsorption. In the simplest case, the external surface area relates to the particle size (assuming spherical particles of uniform diameter) thus:

$$a_{s(ext)} = \frac{6}{\rho \cdot dp} \tag{3}$$

where $\rho$ is the true solid density [18]. Setting dp = 100, 10, and 1 μm and $\rho = 2\ g\ ml^{-1}$, $a_{s(ext)}$ is calculated at 0.03, 0.3, and 3 $m^2\ g^{-1}$. When A refers to unit mass or unit volume of packing, it is termed 'specific surface area' ($a_s$).

The internal surface area is inversely proportional to the pore size. As a rule of thumb, macroporous solids develop specific internal surface area ($a_{s(int)}$ of <50, and microporous solids $a_{s(int)}$ values of

>500. Thus, for porous supports of average particle diameter (dp > 10 μm), the external surface plays a subordinate role compared to the internal surface.

By far the most widely applied method for estimating the specific surface area is nitrogen adsorption at 77 K and applying the BET equation [19]. The solid sample, outgassed under defined conditions, is thermostatted at the temperature of liquid nitrogen and the adsorption isotherm is measured in a relative pressure range $p/p^0$ between 0.05 and 0.30. On rearranging the data in terms of the BET equation, a straight line is obtained with a slope (s) and an intercept (i), from which the monolayer capacity $n_m^a$ in moles of adsorbate per mass adsorbent) is calculated. Avogadro's constant, $N_A$, multiplied by $n_m^a$, and the molecular cross-sectional area of the nitrogen molecule $a_m(N_2) = 0.162\ nm^2$, express the specific surface area:

$$a_s = n_m^a \cdot N_A \cdot a_m \tag{4}$$

There are other calculation procedures for assessing the specific surface area from sorption measurements; these are described in Ref. 20. All methods, including the BET method, provide relative $a_s$ values, as each method carries its own systematic error. Under carefully standardized conditions with calibrated devices, the reproducibility of the specific surface area according to BET lies around ±2-5% [21]. The best way to obtain reliable data involves calibrating the equipment using reference substances of similar chemical composition and surface area (if available).

*Pore Size, Pore Shape, and Pore Size Distribution*

A pore is defined as a cavity that is deeper than it is wide. The sizes of pores cover several orders of magnitude, starting from very small pores of molecular dimensions to large pores that can be seen through a microscope. Models are used to indicate the pore shape, e.g., cylindrical, slit-shaped, platelike, inkpot-shaped, etc. A hydraulic pore diameter, $pd_h$, is defined as follows:

$$pd_h = \frac{4}{vp/a_s} \tag{5}$$

where vp is the specific pore volume and $a_s$ the specific surface area. The model of a cylindrically shaped pore is the most common; in other words, the pore size is expressed in terms of the diameter of open cylindrical pores. Various experimental methods are available to determine the pore size distribution of a given packing: sorption, mercury intrusion, small-angle X-ray scattering, neutron scattering, electron microscopy, flow methods, light microscopy, and touch methods. How-

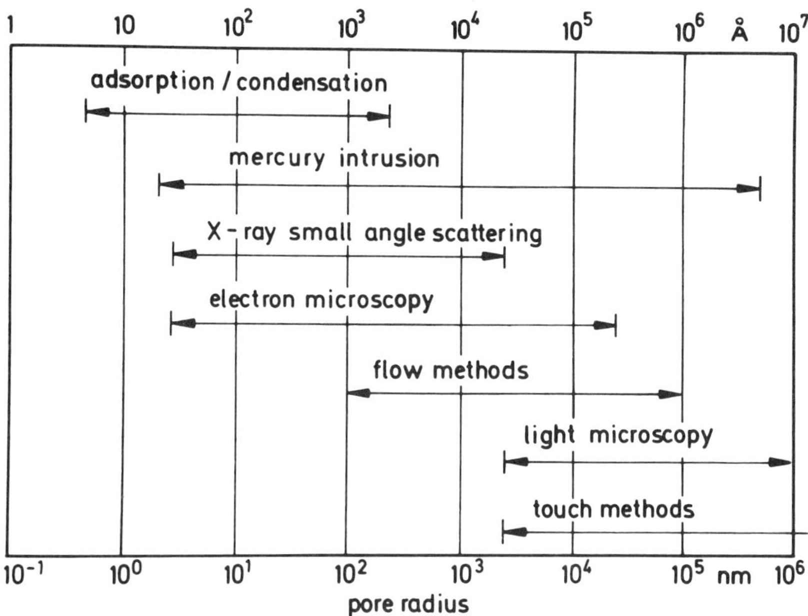


Fig. 1 Survey on pore size determination methods. (By permission of Ref. 22.)

ever, application of the different methods is confined to certain pore size range, as is illustrated in Fig. 1.

Gas adsorption, employing nitrogen as adsorbate, is the method of choice for determining the distribution of micropores and mesopores for a porous material. After outgassing the sample at the appropriate temperature and thermostatting at the temperature of liquid nitrogen (77 K), the adsorption and desorption branches are measured over the whole range of the isotherm.

In micropore analysis, the approach of Brunauer and coworkers is applied to the isotherm data, using the t plot of a nonporous reference material of the same surface chemical composition [23,24]. The calculation gives the distribution of the specific pore volume or specific surface area of the microporous material as a function of the hydraulic diameter $pd_h$.

Mesopore analysis uses the hysteresis range of the nitrogen isotherm and is based on the Kelvin equation which describes the condensation of the adsorbate to a liquid in pores of a certain diameter $pd_K$, termed Kelvin pore diameter [25]. Assuming a cylindrical pore shape, the Kelvin equation is given by

$$pd_k = \frac{4 \cdot \gamma \cdot V_L \cdot \cos\theta}{RT \ln(p^0/p)} \qquad (6)$$

where $\gamma$ is the surface tension of the nitrogen wetting the surface, $V_L$ the molar volume of liquid nitrogen, $\theta$ the wetting angle of nitrogen between the liquid nitrogen and the surface, R the universal gas constant, and T the Kelvin temperature. Since capillary condensation sets on at an already adsorbed multilayer by exceeding a relative pressure ration $p/p^0 > 0.4$, the Kelvin pore diameter must be corrected for the thickness of this multilayer (t). t is a function of $p/p^0$ and is taken from a nitrogen adsorption isotherm measured on a nonporous reference solid under identical conditions. The corrected pore diameter is then expressed as

$$pd = pd_k + 2t \qquad (7)$$

Numerous calculation methods have been developed to assess the mesopore distribution. They differ in the choice of adsorption or desorption branch of the isotherm, and in the assumption of a variable or constant thickness of the adsorbed multilayer. The distribution of specific pore volume or specific surface area as a function of the average pore diameter is thus obtained; from this value both the $pd_{50}$ (the median of the distribution) and the $pd_m$ (the mode) values are derived. The relative distribution is preferentially plotted as $\Delta vp/(\log \Delta pd) \cdot vp$ versus $\overline{pd}$ or log $\overline{pd}$ [21].

As in particle size analysis, in mesopore size analysis it is equally essential to take into account the following factors:

1. Type of sorption branch used for calculation
2. Calculation method
3. Type of t curve employed for correcting the data
4. Type of average pore diameter, i.e., median, mode, etc.

For several reasons, pores of pd > 50 nm cannot be assessed from nitrogen isotherms with sufficiently high precision (see Ref. 21); that is to say that the application of the Kelvin equation to nitrogen isotherms is experimentally limited to the mesopore size range.

Data on macropore distribution is obtained by means of mercury porosimetry, utilizing the phenomenon of capillary depression [16]. Since mercury is a nonwetting liquid, it is forced (intruded) into the pores of the solid by applying external pressure. According to the Washburn Eq. (2), an inverse relation holds between the equilibrium pressure (p) and the pore diameter $pd_w$, assuming a cylindrically shaped pore:

$$pd_w = -\frac{4 \cdot \gamma \cdot \cos\theta}{p} \tag{8}$$

where $pd_w$ is the pore diameter according to the Washburn equation, $\gamma$ the surface tension of mercury, and $\theta$ the contact angle of mercury with the solid. Usually, values of $\gamma = 0.48$ N $m^{-2}$ and $\theta = 140°$ at 298 K are adopted. The pore volume distribution is calculated, on the basis of the Washburn equation, by measuring the volume of intruded mercury as a function of p. The dependence of $\gamma$ and $\theta$ on the surface chemical composition, on the pressure, and on the surface curvature are neglected in the calculation. Aside from this and other systematic errors inherent in the method, reproducibility is largely a function of accurate pressure and volume measurements. On insertion of the corresponding values of $\gamma$ and $\theta$ into the Washburn equation [see Eq. (2)], it becomes obvious that mercury porosimetry is capable of covering also the mesopore size range when sufficiently high pressure is applied; p = 450 MPa corresponds to a pore diameter of pd = 4 nm. It is often claimed that pore size distributions of porous materials falling in the overlap regime and measured by nitrogen sorption and mercury porosimetry coincide very well. However, since both calculations are based on different assumptions, this is purely accidental or, in other words, the parameters have been chosen to fit both results. Like nitrogen sorption, mercury porosimetry provides relative, not absolute, data on the pore volume distribution of packings.

Size exclusion chromatography of standard polymers on porous packings has been developed as an alternative tool for assessing pore size distribution [26]. The pore size of the material and the size of the solvated polymer are approximated by appropriate models; these then enable the chromatographer to correlate the elution volume of a polymer solute with the pore size and to calculate the pore volume distribution. Although this method seems easy to apply to pore size analysis, it contains two major drawbacks with regard to resolution. Calculations have shown [27] that even in the case of a packing with uniform pores of equal size, the calibration curve, log molecular size vs. elution volume, spans almost 1.5 decades of the molecular weight. Moreover, the whole spectrum of pore size, from 1 nm to 10 μm, is matched by 10 to 15 data points, corresponding to the number of narrow molecular weight standards employed. As a consequence, the distributions calculated from elution volumes on this basis do not provide a high resolution.

## Chemical Structure Properties of Supports and Packings in the Form of Bulk Powders and Granulates

In the assessment of chemical structure properties there must be a clear distinction between bulk and surface characteristics, the latter being most decisive in chromatographic separations. The bulk composition is generally given by the type and content of constituents, determined by elemental analysis, X-ray fluorescence, atomic absorption spectroscopy, etc., whereas more detailed information on the structure (with regard to phase composition, lattice constants, etc.) is obtained by X-ray and neutron diffraction spectroscopy. The surface composition differs from the bulk composition since the surface is coordinatively saturated by linkage of surface functional groups, the type of which can be varied over a wide range. There is an enormous amount of literature on the nature and characterization of surface functional groups on solid surfaces, particularly on the assessment methods applied. Recently, an informative survey was published by Boehm and Knözinger [28].

Broadly speaking, there are physical and chemical procedures; these are listed in Table 5. Physical methods provide detailed insight into the surface structure by determination of the types of surface functional group and their coordination, as well as quantitation. However, in order to ensure clean surfaces, in many cases the materials to be measured must be pretreated under ultrahigh-vacuum conditions and high temperatures. This is a long way from the practical conditions to which the materials are subjected in chromatography. Furthermore, some of the techniques (e.g., infrared nuclear magnetic resonance spectroscopy, etc.) include internal, i.e., bulk, as well as surface functional groups, and hence combined methods should be applied in order to assess the surface properties. In other words, since individual methods reflect quite specific molecular properties, an overall reliable estimate of the surface structure is only possible by using several techniques in combination and to correlate the results.

There are fewer drawbacks when chemical methods, e.g., isotopic exchange, chemical reaction, and specific adsorption, are employed. However, these techniques are not without constraints either, arising from slow kinetics or pore size effects occurring in porous systems during reaction, the complexity of the stoichiometry of reactions due to multifunctional surface sites and the reaction sites of probe molecules, the formation of byproducts, and other phenomena.

All in all, detailed knowledge, experience, and critical evaluation of the data are prerequisites in the successful application of techniques for assessing the type, concentration, and reactivity of surface functional groups. The surface chemistry of packings is treated in detail in the single chapters.

### Characterization of the Packing Structure in the Chromatographic Bed

A reproducible and homogeneous flow pattern is essential for the high performance of chromatographic separations. The flow pattern in turn

Table 5 Survey on Methods to Characterize Surface Functional Groups at Solid Surfaces

| |
|---|
| *Physical methods* |
| Spectroscopic |
| Vibrational spectroscopies |
| Infrared transmission-absorption spectroscopy |
| Infrared reflection spectroscopy |
| Raman spectroscopy |
| Infrared photoacoustic spectroscopy |
| Electron vibrational spectroscopies |
| Inelastic neutron scattering |
| Optical spectroscopies |
| Absorption spectroscopy |
| Luminescence spectroscopy |
| Magnetic resonance |
| Electron spin resonance (ESR) |
| Nuclear magnetic resonance (NMR) |
| Photoelectron spectroscopy (XPS) |
| Auger electron spectroscopy (AES) |
| Secondary ion mass spectrometry (SIMS) |
| Thermal desorption spectroscopy (TDS) |
| *Chemical methods* |
| Isotopic exchange |
| Chemical reaction |
| Specific adsorption |

*Source*: Ref. 28.

depends on the structure of the chromatographic bed. By assembling porous particles in a layer or in a column bed, a certain type of space structure is generated. This structure is characterized by decisive parameters: the shape, size, and size distribution of the particles, and their average contact number in the bed. Models of packed beds exist for uniform spheres of equal diameters, packed with a defined coordination number. The volume of the packed column can be divided into three compartments:

Interstitial volume, i.e., the volume of the interstices between the packed particles

Internal volume, i.e., the volume attributed to the pore volume of the particles in the column
Volume originating from the true solid

It is common practice to relate these volumes to the geometrical column volume. The dimensionless quantities obtained are given by

Interstitial column porosity $\varepsilon_0$
Internal column porosity $\varepsilon_p$
Total column porosity $\varepsilon(\text{total})$ $\qquad \varepsilon(\text{total}) = \varepsilon_0 + \varepsilon_p$

The packing structures resulting from model packings of nonporous spheres of uniform size are listed in Table 6. The interstitial porosities of packed columns reported in the literature [30] are in the range of $0.35 < \varepsilon_0 < 0.55$. Although there is a small number of packings with monodisperse particle size distribution available, values of $\varepsilon_0 < 0.3$, theoretically feasible for hexagonally densest packings, have never been obtained. In the case of a synthetic opal, consisting of spheres 220 nm in diameter, a porosity of $\varepsilon_0 = 0.26$ has been measured, which is identical to a hexagonally densest packing [21].

The packing density ($\rho_p$), being the mass of packing per unit volume of column, and the specific pore volume (vp), are used to calculate the internal column porosity:

$$\varepsilon_p = vp \cdot \rho_p \tag{9}$$

The $\rho_p$ value is 0.2 g $ml^{-1}$, for highly porous materials and 0.8 g $ml^{-1}$ for materials of low porosity [31]. For nonporous spheres, a packing density $\rho_p$ of 1.6 g $ml^{-1}$ has been achieved [32]. This implies that the high specific pore volume of a packing cannot be utilized to the full on account of its low packing density with respect to $\rho_p$. Internal porosities commonly range from 0.3 to 0.5; the maximum total column porosities reported in the literature amounted to $\varepsilon(\text{total}) \sim 0.8$ [33].

The homogeneity of the packing structure is not only governed by particle shape and particle size distribution; it is also affected by the ratio of column to particle diameter. At values of $d_c/dp < 100$ (where $d_c$ is the column bore), wall effects become noticeable [34], whereby this phenomenon depends also on the flexibility of the column walls. While stainless steel columns exhibit minimum flexibility and glass columns some, the highest degree of flexibility under pressure is observed for columns made of plastic.

The quality of the bed structure and the flow pattern of a column or layer can be made visible by running colored samples. It is measured quantitatively by the chromatographic permeability, the column resistance factor, the plate height, and the peak shape; these will be discussed in the next section.

Table 6 Characteristics of Bodies of Packed Spheres of Radius dp

| Type of packing | n | $\varepsilon_0$ | Pore volume per $cm^3$ of spheres ($cm^3/cm^3$) | Radius of sphere inscribed in the cavities, dp" | Radius inscribed in the throats connecting cavities, dp' |
|---|---|---|---|---|---|
| Hexagonal close-packed | 12 | 0.260 | 0.350 | 0.225 dp octahedral<br>0.414 dp tetrahedral | 0.155 dp |
| Body-centered tetragonal | 10 | 0.302 | 0.432 | 0.291 dp | 0.265 dp<br>0.155 dp |
| Primitive hexagonal | 8 | 0.395 | 0.654 | 0.527 dp | 0.414 dp<br>0.155 dp |
| Primitive cubic | 6 | 0.476 | 0.910 | 0.732 dp | 0.414 dp |
| Tetrahedral | 4 | 0.660 | 1.94 | 1.00 dp | 0.732 dp |

*Source*: From Ref. 29, reprinted by permission of the publisher.

## Chromatographic Characterization of Packings and Stationary Phases

### *Retention Characteristics* [1,35]

In order to define the differential migration of a sample component in a chromatographic system, and to assess reliable retention parameters which characterize the packing and stationary phase, we shall start with a formal description of the structure of a packed column. In the foregoing section we saw how a packed column divides into the following volume constituents (see Fig. 2): the volume of the truly solid phase, the volume of the stationary phase, the volume occupied by the stagnant eluent in the pores of the packing, and the volume of the mobile eluent in the interstitial voids of the packing. A further distinction should now be made between the stationary zone (the chromatographic medium within the particles) and the mobile zone (the fluid outside the particles). The mobile zone moves with an average linear velocity $u_0$. In chromatographic separation, the sample molecule is distributed between the mobile and the stationary zone. Because of the distribution of the solute in the stationary zone, the sample i migrates with an average linear velocity $u_i$. At equilibrium, the fraction of the solute in the mobile zone is $u_i/u_0$. This fraction is equal to the mass of solute in the mobile zone $Q_{mz}$ divided by the total mass of solute in the mobile zone and the stationary zone $Q_{mz} + Q_{sz}$

$$\frac{u_i}{u_0} = \frac{Q_{mz}}{Q_{mz} + Q_{sz}} = \frac{1}{1 + Q_{sz}/Q_{mz}} = \frac{1}{1 + k''} \tag{10}$$

The ratio of $Q_{mz}/Q_{sz}$ is termed zone capacity factor of k". This scheme describes solely a specific mode of chromatography, namely, size exclusion, where polymer solutes are distributed between the two zones. In all other modes of chromatography, the stagnant fluid within the particles does not contribute to retention, which essentially occurs in the stationary phase (being an active surface) or a stationary liquid. Hence, the migration of a solute in these other modes is related to the average linear velocity of the mobile phase (u). Equation (10) is then converted to

$$\frac{u_i}{u} = \frac{Q_m}{Q_m + Q_s} = \frac{1}{1 + Q_s/Q_m} = \frac{1}{1 + k'} \tag{11}$$

where $Q_m$ and $Q_s$ are the mass of the solute in the mobile and stationary phase, and k' is the capacity factor of the solute. Assuming equilibrium conditions, $Q_m$ and $Q_s$ are defined as

<table>
<tr><td>volume of truly solid phase</td><td>volume of stationary phase</td><td>volume of stagnant eluent in the pores</td><td>volume of mobile eluent in the interstitial voids of the column</td></tr>
<tr><td></td><td colspan="2">stationary zone</td><td>mobile zone</td></tr>
<tr><td></td><td>stationary phase</td><td colspan="2">mobile phase</td></tr>
</table>

Fig. 2 Structure of a packed column (volumes, zones, and phases).

$$Q_m = C_m \cdot V_m \quad \text{and} \quad Q_s = C_s \cdot V_s \tag{12}$$

where $C_m$ and $C_s$ are the concentration of the solute at equilibrium in the mobile phase and stationary phase, and $V_m$ and $V_s$ are the volume of the mobile phase and the stationary phase, respectively.

In linear elution chromatography, the ratio of $C_s/C_m$ is equal to the distribution coefficient $^cK$, which is constant at infinite dilution of the solute in the mobile phase. The capacity factor k' is then related to $^cK$ by

$$k' = \frac{V_s}{V_m} \cdot {}^cK \tag{13}$$

Since

$$\frac{u_i}{u} = \frac{V_m}{V_m + {}^cK \cdot V_s} \tag{14}$$

and the retention volume of the solute $V_R$ is inversely proportional to the average linear velocity $u_i$, one obtains

$$V_R = V_m (1 + k') = V_m + {}^cK \cdot V_s \tag{15}$$

or

$$t_R = t_m (1 + k') \tag{16}$$

where $t_R$ is the retention time of the solute and $t_m$ the retention time of an unretained solute.

The ratio of the rate of movement of solute band to the rate of movement of mobile phase can be expressed as

$$\frac{u_i}{u} = \frac{t_m}{t_m + t_s} = \frac{1}{1 + k'} \tag{17}$$

where $t_m$ and $t_s$ are the average residence time of the solute molecule in the mobile phase and stationary phase. Rearrangement of Eq. (17) gives

$$k' = \frac{t_s}{t_m} \tag{18}$$

The retention of a solute in column chromatography is characterized by the capacity factor k'. The assessment of k' requires the measurement of $V_R$ and $V_m$ or $t_R$ and $t_m$. In the case of k' = 0, the solute is unretained. An increase in k' marks an increase also in retention.

The selectivity of a phase system is usually defined by the selectivity coefficient $\alpha$:

$$\alpha = \frac{k_j'}{k_i'} \quad (k_j' > k_i') \tag{19}$$

where $k_j'$ and $k_i'$ are the capacity factors of solutes j and i. In thin-layer chromatography, retention is characterized by the $R_f$ value, whereby $R_f$ is the ratio of the average linear velocity of the solute $u_i$ to the average linear velocity of the solvent front, $u_{\text{solvent front}}$:

$$R_f = \frac{u_i}{u_{\text{solvent front}}} \tag{20}$$

Since the average linear velocity of the solvent front is larger than that of the following mobile phase, i.e.,

$$\frac{u_i}{u} > \frac{u_i}{u_{\text{solvent front}}} \tag{21}$$

the $R_f$ value in TLC varies from zero to unity. Low $R_f$ values indicate a strong retention and vice versa:

$$R_f \sim \frac{1}{1 + k'} \tag{22}$$

*Dispersion Characteristics* [*1,35*]

The dispersion of a solute zone in a chromatographic bed is described by the plate height H, which is defined for a uniform column, free from velocity and concentration gradients:

$$H = \frac{\sigma_L^2}{L} \tag{23}$$

where $\sigma_L^2$ is the square of the standard deviation of the peak in length units and L the column length. A quantity related to plate height is the number of theoretical plates per column length N given by

$$N = \frac{L}{H} \tag{24}$$

N is measured by the statistical moment method. The zero (area), first (mean), and second (variance) statistical moments are calculated from the chromatogram, applying the equation:

$$\mu_0^\circ = \int f(t)\, dt \tag{25}$$

$$\mu_1^\circ = \frac{1}{\mu_0^\circ} \int t\, f(t)\, dt \tag{26}$$

$$\mu_2^\circ = \frac{1}{\mu_0^\circ} \int (t - \mu_1^\circ)\, f(t)\, dt \tag{27}$$

N is then defined by

$$N = \frac{\text{mode}^2}{\mu_2^\circ} \tag{28}$$

The peak symmetry or peak skewness $\tau$ is calculated by

$$\tau = (\mu_1^\circ - \text{mode}) \cdot \mu_2^{\circ\,-1/2} \tag{29}$$

The plate height is a function of the linear velocity of the eluent and of the capacity factor of the solute. At optimum flow rates, H corresponds to about two times the particle diameter of the packing for an unretained solute.

In the optimization of a chromatographic system the analysis time is a decisive quantity. It is related to the pressure drop along the column. At a laminar flow, the fluid velocity (u) in fluid flux per unit area is given by [37]

$$u = -\frac{K_0}{\eta \cdot \varepsilon} \frac{dp}{dz} \tag{30}$$

where $K_0$ is the specific permeability with a dimension of length squared, $\varepsilon$ the porosity, $\eta$ the fluid viscosity, and dp/dz the pressure drop per unit length.

The specific permeability is related to the average particle diameter (dp) of the column packing as follows:

$$K_0 = \frac{dp^2}{\Phi} \tag{31}$$

where $\Phi$ is the flow resistance parameter. The dependence of the average linear velocity and the specific permeability of a packed column on other parameters is expressed by the Kozeny-Carman equation:

$$u_0 = \frac{\Delta p \cdot dp^2}{180 \cdot \eta \cdot L} \cdot \left(\frac{\varepsilon_0}{1-\varepsilon_0}\right)^2 = \frac{\Delta p \cdot dp^2}{\Phi \cdot \eta \cdot L} = \frac{\Delta p \cdot K_0}{\eta \cdot L} \tag{32}$$

where $u_0$ is the average linear velocity of the mobile zone for a column packed with nonporous regular uniform spheres of the interstitial column porosity $\varepsilon_0$.

$\Phi$ is then calculated by

$$\Phi = 180 \left(\frac{1-\varepsilon_0}{\varepsilon_0}\right)^2 \tag{33}$$

Typical values are $\varepsilon_0 = 0.35$, $\Phi = 620$; $\varepsilon_0 = 0.40$, $\Phi = 405$; $\varepsilon_0 = 0.45$, $\Phi = 268$.

The extension to columns packed with porous particles gives

$$u = \frac{\varepsilon_0}{\varepsilon(\text{total})} \cdot u_0 = \frac{\Delta p \cdot dp^2}{\Phi \cdot \eta \cdot L} \; \frac{\varepsilon_0}{\varepsilon(\text{total})} = \frac{\Delta p \cdot dp^2}{\Phi' \cdot \eta \cdot L} \tag{34}$$

where u is the average linear velocity of the mobile phase and

$$\Phi' = \Phi \cdot \frac{\varepsilon(\text{total})}{\varepsilon_0} \tag{35}$$

At $\varepsilon_0 = 0.4$, $\varepsilon(\text{total}) = 0.8$ and $\Phi = 405$, $\Phi'$ equals 810. Although the flow patterns of gases and liquids on columns with defined porosity and laminar flow are identical, the velocity and pressure distribution in a gas chromatographic column deviates from the ideal case on account of gas compressibility. Although the flow velocity varies along the column, the product $(p \cdot v)$ remains constant; v can thus be substituted by

$$v = \frac{p_i \cdot v_i}{p} \tag{36}$$

where the index i refers to the inlet pressure; this leads to

$$p \cdot dp = \frac{2 \cdot \Phi \cdot \eta \cdot p_i \cdot v_i}{dp^2} = I \cdot dz \tag{37}$$

Integration within the limits $p \to p_0$, $Z \to L$ gives

$$p^2 - p_0^2 = I(L - Z) \tag{38}$$

This equation represents the pressure distribution as a function of the distance (L-Z) from the outlet.

In thin-layer chromatography [37,38], the flow is driven by capillary forces. The flow velocity, i.e., the rate of advance of the front at $z_f$, is proportional to the driving force and inversely proportional to the viscosity $\eta$ multiplied by the length $z_f$ of capillary.

$$\frac{dz_f}{dt} = \frac{\text{const} \cdot \gamma}{\eta \cdot z_f} \tag{39}$$

Integration within the limits $0 \to t$ and $0 \to z_f$ gives

$$z_f^2 = 2 \text{ const } \left(\frac{\gamma}{\eta}\right) \cdot t = k \cdot t \tag{40}$$

## ROLE OF PACKING AND STATIONARY PHASE PROPERTIES IN CHROMATOGRAPHIC PROCESSES

The separation process in ideal elution chromatography is best divided into two aspects: the thermodynamic and the kinetics. Peak retention, expressed by the solute capacity factor k" or k', is essentially determined by the thermodynamic equilibrium constant which is associated with the distribution of the solute between the mobile and stationary zones or phases. Peak dispersion, however, characterized by the plate height, originates from kinetic processes such as flow pattern effects, molecular diffusion, and mass transfer. Therefore, the role of packings and stationary phases should likewise be viewed under these two distinct aspects.

### Influence of Packings and Stationary Phases on Retention

Solute retention on a given phase system and at constant temperature is described by the phase capacity factor k' or the zone capacity factor k", which relate to the distribution coefficient as follows:

$$k' = \frac{v_s}{v_m} \cdot {}^cK \quad \text{or} \quad k'' = \frac{v_s}{v_m} \cdot {}^cK \tag{41}$$

where $v_s/v_m$ is the ratio of the volume of stationary phase (zone) to the volume of mobile phase (zone), and $^cK$ is the solute distribution coefficient given in concentration units at infinite dilution. Equation (41) is formally valid for solute retention in all modes of elution chromatography, e.g., adsorption, partition, ion exchange, etc. On a given type of phase system, the capacity factors k' and k" are modulated by $v_s/v_m$ at constant distribution coefficient. In adsorption chromatography, $v_s$ is proportional to the specific surface area per unit volume of adsorbent. In gas-liquid and liquid-liquid partition chromatography, $v_s$ is equal to the volume of the liquid stationary phase per unit of column volume. In ion exchange, $v_s$ is determined by the effective capacity per volume unit of the ion exchanger, and in size exclusion k" is controlled by the volume of the stationary zone, i.e., the volume of stagnant eluent in the pores of the packing. Since the specific surface area of adsorbents covers two orders of magnitude (from 5 to 500 $m^2$ $ml^{-1}$), the phase ratio can be varied systematically within these limits. In partitioning systems, the lower limit of loadability ($v_s$) is given by the volume corresponding to an adsorbed monolayer of liquid stationary phase, the upper limit of $v_s$ being the specific pore volume per unit volume of support, which is then completely filled with stationary liquid. A linear relationship results when the retention volume $V_R$ is plotted against $v_s$, with the distribution coefficient $^cK$ as slope and $v_m$ as intercept. Deviations from linearity indicate that more than one mechanism is involved in retention, e.g., adsorption, partitioning.

$^cK$ is related to the thermodynamic distribution constant $K^0$ via the activity coefficients $\gamma_s$ and $\gamma_m$ of the solute in the stationary and the mobile phase, which depend on the standard and reference state chosen in the two phases. This is discussed in depth in Ref. 39. As usual, $K^0$ reflects the changes in partial molar enthalpy ($\Delta h^0$) and partial molar entropy ($\Delta s^0$) of the solute in the given phase system:

$$RT \ln K^0 = -\Delta\mu° = \Delta h^0 - T\,\Delta s^0 \tag{42}$$

where R is the gas constant, T the Kelvin temperature, and $\Delta\mu°$ the change in the chemical potential for the solute in the chosen standard state. The dependence of $K^0$ on temperature is given by the van't Hoff equation:

$$\frac{d \ln K^0}{d(1/T)} = -\frac{\Delta h^0}{R} \tag{43}$$

The selectivity coefficient $\alpha$ is equal to the ratio of the thermodynamic distribution constant of two solutes i and j:

$$\alpha = \frac{K_j^0}{K_i^0} \quad (K_j^0 > K_i^0) \tag{44}$$

With a dilute solution as reference state, one obtains

$$RT \ln \alpha = -(\Delta\mu_j^0 - \Delta\mu_i^0) \tag{45}$$

$\Delta\mu_j^0$ and $\Delta\mu_i^0$ are the difference in standard chemical potential of solutes j and i between the two distributing phases. This way, changes in the selectivity coefficient can be expressed quantitatively in terms of $\Delta(\Delta\mu^0)$ given in J mol$^{-1}$. For example, $\alpha = 1.01$ gives -23 J mol$^{-1}$ at 298 K.

Size exclusion is predominantly an entropy-controlled process, but the majority of chromatographic retention processes are enthalpy-controlled. Types of enthalpic interactions cover a wide range, from nonspecific van der Waals to highly stereospecific, and biospecific. Stereospecificity results from the simultaneous operation of three interactions, forming a three-point attachment to the solute. A similar situation is met in biospecific recognition. Intermolecular forces are coulombic, dipole-dipole, hydrogen bonding, $\pi$-$\pi$ (i.e., charge transfer), and others. Retention is often the sum of contributions by several types of interactions, where one is dominant and the others play a lesser role. In liquid chromatography, the individual contributions can be manipulated through changes in the eluent composition.

In gas chromatography, solute retention is essentially governed by the interaciton between the solute and the stationary phase, and the column temperature. The distribution equilibrium is described by the corresponding isotherm, which is the solute concentration in the stationary phase at equilibrium ($c_s$) plotted against the solute concentration in the mobile phase ($c_m$) at constant temperature. The distribution coefficient is then given by the slope of the isotherm and changes with $c_m$, except in the low concentration range, where $^cK$ is constant. The linear part of the distribution isotherm for gas-liquid equilibria is described by Henry's law:

$$p = K^0 \gamma \cdot x \tag{46}$$

where p is the partial pressure of the solute in the gas phase, x the mole fraction, and $\gamma$ the activity coefficient of the solute in the liquid phase.

There is an enormous amount of literature on gas adsorption phenomena (cf. Refs. 40 and 41). The measured experimental isotherms are plotted as the amount adsorbed ($n^a$) in mole per unit mass or unit surface area of adsorbent against the relative pressure ratio $p/p^0$, p being the equilibrium pressure at adsorption and $p^0$ the saturation pressure of the pure adsorbate at the given temperature. The course of the experimental isotherms is approximated by equations derived from a theoretical treatment of the adsorption process, assuming a localized or nonlocalized adsorption, distinct adsorption sites, a homogeneous surface, etc. Table 7 lists the most common equations describing adsorption isotherms. In the linear range, the slope of the isoterms toward the low-pressure range is characterized by Henry's constant or by an adsorption coefficient ($K_{ads}$), given in ml $g^{-1}$ or ml $m^{-2}$. Although the treatment of the thermodynamics of adsorption phenomena has greatly advanced (e.g., by use of the Gibbs equation employing excess quantities and by the potential theory of Polanyi), the application to real adsorbents with more or less heterogeneous surfaces remains fairly limited.

A more straightforward approach proposed by Ross and Olivier [43] divides adsorbents into Gaussian, whereby the surface is described by a Gaussian distribution of the adsorption potential, and non-Gaussian, whereby a multimodal distribution is exhibited. From experimental isotherms of gases measured at low temperature, a number of adsorbents were identified as Gaussian, e.g., graphitized carbon black, boron

Table 7 A Selection of Equations Describing Adsorption Isotherms

| Name | Equation | Comment |
|---|---|---|
| Langmuir | $\frac{n^a}{n^a_m} = \frac{b \cdot p}{1 + b \cdot p}$ | b = constant of the Langmuir equation |
| Freundlich | $n^a = k \cdot p^{1/n}$ | (n > 1) |
| Henry | $n^a = k'p$ | |
| Brunauer, Emmet, and Teller | $\frac{p}{n^a(p^0-p)} = \frac{1}{n^a_m \cdot C} + \frac{C-1}{n^a_m \cdot C} \cdot p/p^0$ | BET two-parameter equation |

*Source*: From Ref. 42, reprinted by permission of the publisher.

nitride, diamond, synthetic zeolite (Linde 13X), rutile, and anatase. The most thorough and valuable attempt to classify adsorbents and their interactions in gas chromatography, made by Kiselev and coworkers [3,44], categorizes absorbents and solutes into groups according to their intermolecular interactions, and relates retention data to the physicochemical properties of adsorbents. Vice versa, gas chromatography has been developed as a tool to characterize adsorbents and to elucidate molecular structure parameters of solutes; the latter approach is termed chromatoscopy [44].

Liquid-liquid equilibria are characterized by the Nernst distribution law:

$$^{x}K = K^{0} \left(\frac{\gamma_s}{\gamma_m}\right) \tag{47}$$

where $^{x}K$ is the distribution coefficient expressed as the ratio of mole fractions of solute in the two phases, $K^0$ the thermodynamic distribution constant, and $\gamma_s$ and $\gamma_m$ the activity coefficients of the solute in the stationary phase and the mobile phase. For dilute solutions, $\gamma_s$ and $\gamma_m$ remain constant; $^{x}K$ is a constant at a given temperature [39].

The principal question arising in the discussion on liquid-solid adsorption in chromatography concerns the types of interactions and distribution equilibria that can be collectively described by the term adsorption. Apparently, adsorption is involved in all those cases where adsorbents exhibiting a high-phase boundary for interactions are applied and bulk phenomena are excluded. From this point of view, chromatography on organic- and inorganic-based adsorbents, including their chemically modified derivatives, is considered as adsorption. Consequently, the solvophobic theory applied to reversed phase chromatography by Horvath and his associates [45] is not in contradiction to this concept. Further specification begins when chemical equilibria, such as acid-base, charge transfer, ion pair formation, and ion exchange, are involved.

The major criterion characterizing liquid-solid adsorption systems, and making them more specific than gas-solid systems, is the active role of the mobile phase, i.e., the eluent. Thus, retention in liquid-solid adsorption stems from three competing interactions:

Solute-mobile phase
Solute-stationary phase
Mobile phase-stationary phase

Therefore, slight changes in the composition of the phase system (mobile and stationary) have a remarkable effect on the solute capacity factor. It is worth mentioning that the energetic differences–in terms of molar enthalpies and entropies–between the solutes to be separated do not need to be very large. As shown by Eq. (45), which relates the selectivity coefficient to the difference in chemical potential of two solutes, $\Delta(\Delta\mu^\circ)$, such values are small at $\alpha \sim 1.01$.

When we consider the types of adsorbents and mobile phases that can be applied in liquid-solid systems, the enormous variability becomes immediately evident. On the one hand, this provides great flexibility for the separation system but has the drawback that retention data are difficult to predict because of the complexity of the phase system.

Consequently, it seems irrelevant to divide adsorbents into polar and unpolar, and subsequently into straight phase and reversed phase systems. It must be emphasized that all adsorbents have both a hydrophobic and a hydrophilic nature; it is through the choice of the mobile phase that the relative effectivity of the two characters is either suppressed or enhanced. Strictly speaking, there is no known adsorbent with an entirely hydrophobic or hydrophilic surface. For instance, pure graphite behaves in a predominantly hydrophobic way–which is associated with the aromatic layer structure–but it still possesses polar surface sites to a small extent. The majority of adsorbents is bi- and even multifunctional in terms of surface sites and surface chemical character. From this viewpoint, chromatography on silica and reversed phase silicas differs insofar as on silica retention is governed by specific solute-surface interactions, whereas retention in reversed phase systems is predetermined by the strength of solvation of the solute in the mobile phase as a precursor state, followed by nonspecific van der Waals interactions between the bonded $n$-alkyl chain of the silica and the hydrophobic domains of the solute molecule.

The stationary phase composition of an adsorbent, characterized under clean conditions by physical and chemical methods, is not the same as that in operation under liquid chromatographic conditions. The actual surface functionality is controlled by the composition of the adsorbed layer and the phase boundary between the layer and the stagnant eluent. In general terms, the surface functionality covers a wide polarity range since the mobile phases range from organic solvents to mixtures of organic solvents with water to neat buffered solutions. Furthermore, the surface functionality of an adsorbent in any system is known to be highly sensitive to so-called modifiers, which are adsorbed at low concentration levels. The term modifier encompasses, for example, polar solvents as additives, bases, acids, and detergents. Their role in chromatographic processes will be discussed in detail later.

This variability in surface functionality inherent in liquid adsorption systems is best demonstrated by the so-called solvent-generated, liquid-solid adsorption systems [46]. The solubility equilibrium of a ternary solvent system, comprising three solvents (A, B, and C) that

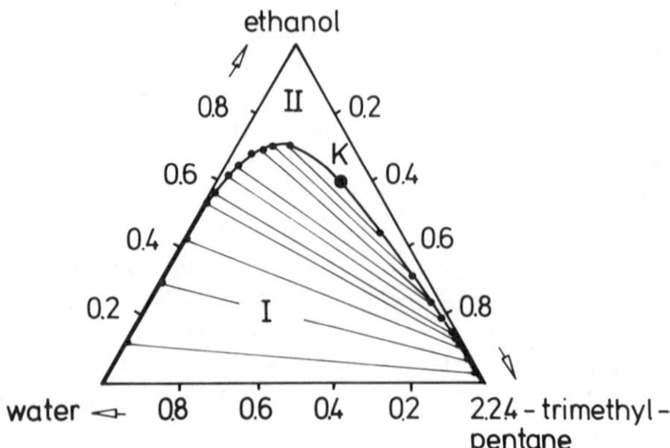


Fig. 3 Liquid-liquid equilibria curve of a ternary solvent system (water, ethanol, and 2.2.4-trimethylpentane). (From Ref. 62, reprinted by permission of the publisher.)

differ in polarity, is described graphically by a Gibbs triangle, as shown in Fig. 3. The system separates into two domains, I and II. In domain I the system splits into two immiscible phases when a composition within the domain is chosen, except for point K where complete miscibility is observed. In domain II there is complete miscibility between the three solvents, i.e., one phase is obtained. When an appropriate support is employed, selective liquid-liquid partitioning systems can be produced with the mobile phase composition maintained within domain I, whereas within domain II, liquid adsorption systems with a distinctly different selectivity are generated, using the same adsorbent. Another advantage of this systematic approach is that the corresponding isotherms of the solutes can be measured under static conditions; also, the distribution coefficients can be assessed from the isotherm data, and the capacity factors can be predicted for the system when adapted to chromatographic operation. Furthermore, the measured adsorption coefficient $K_{ads}$ can be related to the structural parameters of the solute, and this to a large extent facilitates the prediction of solute retention on a given phase system on the basis of the molecular structure.

The application of secondary chemical equilibria through ionization, solvation, ion pair formation, complexation, etc., has considerably expanded the scope of CLC [47]. When secondary equilibria are involved, the solute may exist in several distinct forms in the phase system. As a result, the distribution coefficient is replaced by a distribution ratio (D) defined by

$$D = \frac{\text{stoichiometric amount of solute in the stationary phase}}{\text{stoichiometric amount of solute in the mobile phase}} \tag{48}$$

The solute capacity factor then becomes

$$k' = \frac{V_m}{V_m} \cdot D \tag{49}$$

Depending on the chemical quilibrium, D contains constants of the equilibria involved. This approach has become very useful for the separation of ionic substances, e.g., bases, acids, and polyelectrolytes [48].

To conclude, the surface chemistry of the stationary phase in gas-solid chromatography is directly expressed by solute-surface interactions. In partitioning systems, e.g., gas-liquid and liquid-liquid, the molecular properties of the bulk liquid govern the retention through the equilibrium constant. For simple systems, a relation between the molecular structural properties of the partitioning phases and the solute molecular structure might be established, provided adsorption effects do not operate. However, this would be extremely complicated for binary, ternary, and quaternary liquid systems. In liquid-solid chromatography, the mobile phase controls the surface chemistry and induces significant changes in retention through solvation, ionization, complexation, and other phenomena. At present, the surface chemistry of an adsorbent and its effect on retention can be rationalized only on a phenomenological basis.

### Influence of Packing and Stationary Phase Characteristics on Dispersion

The concepts and mathematical description of band dispersion or zone spreading in chromatography originate from the pioneering work of Wilson, Martin and Synge, De Vault, Weiss, and others in the 1940s [49]. In the following decade, the treatment of dispersion phenomena was thoroughly extended and substantial progress was made by the contribution of Lapidus and Amundson, van Deemter, Zuiderweg, Klinkenberg, Glueckauf, and Golay. The subsequent application of the theory led to the development of highly efficient columns in gas chromatography. Concurrently, Giddings developed the generalized nonequilibrium theory, this being a straightforward and profound approach applicable to a wide range of kinetic situations in chromatography [50]. Based on Giddings' work, Knox and associates [51-54] emphasized the reduced parameter concept, with the reduced plate height reduced velocity dependence as a main feature to characterize dispersion. Simultaneous examinations of Huber [55] on mass transfer and equilibria in chromatography resulted in empirical mass transport functions, clearly demonstrating the importance of the packing particle diameter as a parameter governing performance in column liquid chromatography.

These results initiated research on highly efficient columns packed with microparticulate silicas. Further in-depth studies by Stout et al. [56] and Knox and Scott [57] provided more detailed understanding of the kinetic phenomena in columns and showed that the basic kinetic theory of dispersion in liquid chromatography describes the experimental results qualitatively as well as quantitatively.

It is generally accepted that the dispersion of the solute band in a chromatographic bed arises from three independent kinetic processes: the flow pattern and eddy diffusion, the axial molecular diffusion, and slow mass transfer in the stationary zone [50]. Each of these processes is characterized by a specific term and contributes independently to dispersion in the plate height-linear velocity dependence. In gas chromatography, the plate height equation is known as the Van Deemter equation, written as

$$H = A + \frac{B}{u} + C \cdot u \tag{50}$$

where A is the eddy diffusion term, assumed to be independent of the linear velocity u. B/u, the axial diffusion term, becomes dominant at low flow velocities, and the mass transfer term C·u is a linear function of the velocity. A more common approach uses reduced dimensionless parameters, such as the reduced plate height h:

$$h = \frac{H}{dp} \tag{51}$$

and the reduced linear velocity:

$$\nu_0 = \frac{u_0 \cdot dp}{D_{mz}} \qquad \nu_m = \frac{u_m \cdot dp}{D_{mp}} \tag{52}$$

where $\nu_0$ and $\nu_m$ are the reduced linear velocities of the mobile zone and the mobile phase [57], dp is the average particle dimaeter of the packing, and $D_{mz}$ and $D_{mp}$ are the solute diffusion coefficients in the mobile zone and mobile phase. In column liquid chromatography, the reduced plate height-reduced velocity equation becomes [57]:

$$h = A \cdot \nu_0^{1/3} + \frac{B}{\nu_m} + C \cdot \nu_0 \tag{53}$$

A, B, and C are constants for a particular column and phase system. Typical values are A = 1, B = 2, and C = 0.02-0.05 [52-54,56]. For an open tubular column, the reduced plate height h is defined as

$$h = \frac{H}{d_c} \tag{54}$$

where dc is the column bore and

$$\nu_0 = \frac{u \cdot d_c}{D_{mz}} \tag{55}$$

The h vs. $\nu_0$ dependence, known as the Golay equation, then becomes [58]

$$h = \frac{B}{\nu_0} + C_m \cdot \nu_0 + C_s \cdot \nu_0 \tag{56}$$

where $C_m$ and $C_s$ are the mass transfer coefficients of the solute in the mobile zone and the stationary zone.

In his generalized nonequilibrium theory, Giddings [50] discussed in depth the contributions of packing structure and stationary phase to total dispersion within a theoretical frameowrk. In order to elucidate what type of properties are involved and the extent to which they are weighted according to their importance for dispersion, the individual terms in the plate height equation must be analyzed and substituted by more explicit terms, and the coincidence between theoretical predictions and experimental values must be established.

The term A represents contributions to the plate height from flow anisotropy of the mobile zone and hence serves as a measure of the quality of the geometrical structure of the chromatographic bed. Van Deemter [59] divided A into a velocity-independent eddy diffusion terms A and a velocity-dependent term $C \cdot \nu_0$, which accounts for the slow mass transfer in the mobile zone. Giddings [50] coupled the two terms, assuming a cooperative effect of velocity variations, to

$$\left(\frac{1}{A} + \frac{1}{C_m} \cdot \nu_0\right)^{-1} \tag{57}$$

where A refers to the dispersion by flow and $C_m$ to the mass transfer in the mobile zone. Kennedy and Knox [51] showed that the expression $A \cdot \nu_0^{1/3}$ provided a reasonably practical compromise to fit in with the experimental results. Huber [55] devised an A term in the nonreduced form:

$$\left[2 \frac{\lambda_1 \cdot dp}{1 + \lambda_z \frac{D_{mz}}{u_0 \cdot dp}^{1/2}}\right] \tag{58}$$

where $\lambda_1$ and $\lambda_2$ are coefficients describing the packing structure with respect to eddy diffusion. Whichever expression is adopted, the term responds to the regularity of the packing through $\lambda$, and includes the diffusion coefficient of the solute in the mobile zone as well as the particle diameter, the latter being of minor importance as seen from $A \cdot \nu_0^{1/3}$ and Eq. (53). For packed columns with uniform spherical particles, A was found to lie between 0.5 and 0.8 [56]. The term A has no relevance in open tubular columns [60].

The term $B/\nu_m$ represents the contributions from axial molecular diffusion, where B equals $2 \cdot \gamma$. $\gamma$ is the tortuosity or obstructive factor and characterizes the constricted pathway when the solute molecules diffuse through the interstitial and internal voids of the column. $\gamma$ amounts to 0.65 for an unretained solute in a bed packed with non-porous particles and $\varepsilon_0 = 0.40$, while the value approaches unity in the case of porous particles [56,57]. More detailed expressions have been derived to account for the dependence of B on solute retention. Stout et al. [56] proposed the term

$$B = 1.28\left[ x + (1 - x) \frac{dp}{D_{mz}} + k' \frac{D_s}{D_{mz}}\right] \tag{59}$$

where x is the fraction of mobile phase in the column, $D_{mz}$ the solute diffusion coefficient in the mobile zone, and $D_s$ the solute diffusion coefficient within the stationary phase, i.e., the surface diffusion along the pore walls. The apparent surface diffusion coefficient $D_s$ is roughly half that of the bulk liquid diffusion coefficient.

The term $C \cdot \nu_0$ reflects the contributions from slow mass transfer within the mobile and stationary zones. A common term to express C is

$$C = \left[\frac{1}{30} \gamma (1 - x)\right] \left[\frac{(1 + k' - x)}{(1 + k')}\right]^2 \tag{60}$$

where $\gamma$ is the obstructive factor and x is the fraction of mobile zone in the column [56]. Typical values measured for C were 0.02 and 0.13 for $x = \gamma = 0.64$, and $k' = 0$ and $k' = 10$. It is obvious that C decreases with k' when $k' \gtrsim 1$. The examination of experimentally measured h vs. $\nu$ plots on columns with various packings indicated a remarkable dependence on the C value, calculated at constant k', for different packings [56]. These variations were rationalized by introducing a so-called restricted diffusion of solutes, due to steric interaction and/or pore size effects, using a restriction factor $\rho$ [56]. Equation (59) was then expanded by multiplication with the factor $D_{mz}/\bar{D}_p \cdot \rho$, where $\bar{D}_p$ is the effective solute diffusion coefficient in the stationary zone.

Huber [55] provided an in-depth treatment of mass transfer and mass distribution phenomena in a chromatographic column in CLC and suggested that the total plate height is the sum of four independent contributions:

$$H = H_{Md} + H_{Mc} + H_{Ef} + H_{Eb} \tag{61}$$

where $H_{Md}$ represents the contribution of mixing by diffusion $H_{Mc}$ the contribution of eddy diffusion, $H_{Ef}$ the contribution of mass transfer in the mobile zone, and $H_{Eb}$ the contribution of mass transfer in the stationary zone.

The individual terms are related to column and solute parameters and the operation conditions as follows:

$$H_{Md} = 2\chi \cdot \frac{D_{im}}{u} \tag{62}$$

$$H_{Mc} = 2\left[\frac{\lambda_1 \cdot dp}{1 + \lambda_2 \, (D_{im}/u \cdot dp)^{1/2}}\right] \tag{63}$$

$$H_{Ef} = \frac{1}{3}\Psi_f\left[\frac{\varepsilon_f}{1-\varepsilon_f}\right]\left[\frac{dp^{3/2} \cdot \nu_m^{1/2} \cdot u^{1/2}}{D_{im}^{2/3}}\right]\left[\frac{k_i'}{1+k_i'}\right]^2 \tag{64}$$

$$H_{Eb} = \frac{1}{30}\Psi_b\left[\frac{dp^2 \cdot u}{D_{ip}}\right]\left[\frac{k_i'}{(1+k_i')^2}\right] \tag{65}$$

where $\chi$ is the tortuosity or obstructive factor, u is the linear velocity of the eluent. $\lambda_1$ and $\lambda_2$ are dimensionless factors describing the geometry of the column bed with regard to eddy diffusion, $\psi_f$ and $\psi_b$ are dimensionless factors characterizing the geometry of the interstitial voids of the packing, k' is the capacity factor of solute i, $\nu_m$ the kinematic viscosity of the eluent, and $\varepsilon_f$ the interstitial porosity.

The extent to which each term contributes to the total plate height (as a function of u) can be estimated by plotting the partial curves of each term in the form of H against u (see Fig. 4). For a given

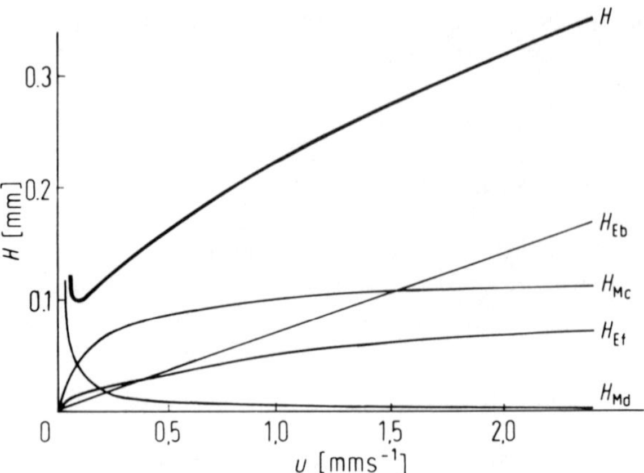


Fig. 4 Dependence of the total plate height H and the single terms $H_{Mc}$, $H_{Hd}$, $H_{Eb}$, and $H_{Ef}$ on the linear velocity u of the eluent. (From Ref. 55, reprinted by permission of the publisher.)

column, the packing and solute capacity factor $H_{Eb}$ increases linearly with u, while $H_{Ef}$ asymptotically approaches a limiting value. At low linear velocity and at the optimum where the total plate height is at a minimum. $H_{Ef}$ is larger than $H_{Eb}$ by a factor of 3. With increasing flow velocity $H_{Eb}$ dominates over $H_{Ef}$. In other words, at high flow rates mass transfer phenomena in the stationary zone govern peak dispersion.

An analysis of the individual terms of the plate height equation provides insight into the relative importance of the properties of the packing and the stationary phase. The A term expresses the uniformity of the column bed and is related to particle shape, particle size distribution, and packing density by the parameter $\lambda$. The most uniform bed is achieved by employing monodisperse spheres as packing materials. A slight dependence of A on the solute diffusion coefficient in the mobile zone has been observed.

On converting the plate height equation into its nonreduced form it becomes clear that A is proportional to dp. In order to obtain a column with excellent performance characteristics, it is of primary importance to employ a regular and uniform packing. The value of A determines the basic level of h, to which mobile and stationary phase contributions must then be added.

The term $B/\nu_m$ becomes dominant at low linear velocities and is dependent on the uniformity of the packing structure and pore structure through the obstructive factor $\gamma$. As shown by Stout et al. [56], the dependence of B on the capacity factor varies significantly for different stationary phases, even when the mean pore diameter of the packing is constant. This implies that the structure of the stationary phase also plays a role in determining the value of B. One should emphasize that the plate height H remains unaffected by the particle diameter with respect to B.

The term $C \cdot \nu_0$ increases with $\nu_0$. On rearranging the term into its nonreduced form it becomes apparent that $C \cdot u_0 \cdot dp^2 \cdot D_{mz}^{-1}$ is proportional to the plate height H, i.e., with regard to slow mass transfer H equals the average particle diameter of the packing squared (compare also Ref. 55). Again, the packing structure and the pore structure affect the value of C through the obstructive factor and through the effective diffusion coefficient of the solute.

On combining the terms $A \cdot \nu_0^{1/3}$ and $B/\nu_m$ for a given packing and taking an unretained solute, the total plate height measured allows an estimation of the order of the mobile phase contribution to h. Since A amounts to 0.5-0.8 at reduced velocity, in the optimum of the h vs. $\nu$ plot h rises to about 1.75 at $\nu \sim 1.75$ and beyond that through mobile phase contributions. A further increase in h to about 2-3 at $\nu \sim 1.75$ is achieved by adding the slow mass transfer effects.

For open tubular columns, the reduced plate height equation takes the form [58,59]:

$$h = \frac{B}{\nu_0} + C_m \cdot \nu_0 + C_s \nu_0 \tag{66}$$

and

$$h = \frac{2}{\nu_0} + \left[\frac{1+6k'' + 11(k'')^2}{96(1+k'')^2}\right] \nu_0 + \frac{2}{3}\left[\frac{k''}{(1+k'')^2}\left(\frac{d_f}{dc}\right)^2\right] \cdot \left[\frac{D_{mz}}{D_{sz} \cdot \nu_0}\right] \tag{67}$$

where k" is the zone capacity factor (i.e., the quantity of solute in the stationary zone divided by the quantity of solute in the mobile zone), $d_f$ is the thickness of the stationary zone, and $D_{mz}$ and $D_{sz}$ are the solute diffusion coefficients in the mobile and stationary zones, respectively. The minimum of the reduced plate height/reduced velocity curve for open tubular columns is attained when $C_s$ equals zero. Depending on k", the following values are achieved: $0.3 < h < 1.0$ at $4 < \nu < 14$ [60]. The C values estimated from experimental data were found to lie between $C_m$ and $2C_m$.

The minimum reduced plate height of packed columns in gas and liquid chromatography amounts to roughly 2-3 at reduced velocities of about 2-5 [60]. A value of $\nu = 1$ corresponds to a linear flow rate of 100 mm $sec^{-1}$ in gas chromatography, and $u \sim 0.01$ mm $sec^{-1}$ in column liquid chromatography, the difference reflecting the ratio of the diffusion coefficients $D_m$ in gases compared to liquids.

The following three types of packing and stationary phase properties qualitatively determine the solute band dispersion through flow pattern and various diffusion processes:

Structure of the stationary phase, i.e., its uniformity and homogeneity in geometrical aspects
Pore structure of the particles composing the bed, i.e., uniformity of the pores in terms of openings, cavities, etc., across the particles
Packing structure, i.e., uniformity of interstitial voids

In column liquid chromatography, the diffusivities are specific for solutes according to their molecular weight, and dependent on the characteristics of the eluent employed as mobile pahse.

So far, major emphasis has been placed on flow and diffusion as the dominant parameters affecting dispersion. The kinetics of thermodynamic processes were assumed to be fast and hence considered of lesser importance. This, however, is not the case for several reasons. The packing, pore, and stationary phase structures cannot be assumed to be homogeneous and uniform. The stationary phases of real packings possess a considerable surface heterogeneity in energetic terms, attributed to multiple functional groups, to remaining support effects, and so on, this causing slow mass transfer kinetics, e.g., in adsorption-desorption processes. Because of the multiple surface functionality of stationary phases, several competing equilibria participate in retention, each having its specific kinetic characteristics. Secondary chemical equilibria, e.g., acid-base interactions, ion pair formation, complexation, and the like, are often applied in CLC to increase selectivity. Dispersion on such phase systems is known to be largely controlled by the kinetics of these processes.

Dispersion has been discussed in terms of the plate height, i.e., the variance of the peak relative to the column length. In practice, the peak symmetry is equally important. Tailing or fronting of peaks may arise from exceeding the linear part of the isotherms. Tailing can be associated with extremely slow kinetics caused by micropores present in the packing, or by several distribution processes in which the solutes are involved.

## CONCLUSION

In this chapter the various aspects of packings and stationary phases in the chromatographic separation processes have been critically examined. A variety of methods are available to characterize the physical and chemical structure of packings and stationary phases as bulk material as well as in situ under chromatographic operation. However, a cursory literature search reveals that these methods were not consistently applied, the properties derived were poorly described in terms of standardized conditions and statistical evaluations, and thus establishing a reliable comparison of packings and stationary phases remains a crucial task. Profound models of the chromatographic retention and dispersion processes have been elaborated that allow us to estimate and predict the role of physical and chemical properties of packings and stationary phases on the selectivity and efficiency of systems in both gas and liquid chromatography.

Up to now major emphasis has been placed on the separation of low molecular weight compounds, whereas for high molecular weight substances, e.g., biopolymers, novel theoretical concepts are needed.

## REFERENCES

1. J. C. Giddings, *Dynamics of Chromatography*, Part I, Marcel Dekker, New York, 1965, pp. 1-11.
2. E. Leibniz and H. G. Struppe (eds.), *Handbuch der Gas-Chromatographie*, Vol. 2, Auflage, Verlag Chemie, Weinheim, 1970, pp. 366-368.
3. A. V. Kiselev and Y. I. Yashin (eds.), *Gas-Adsorption Chromatography*, Plenum Press, New York, 1969, pp. 10-67.
4. E. Leibniz and H. G. Struppe (eds.), *Handbuch der Gas-Chromatographie*, Vol. 2. Auflage, Verlag Chemie, Weinheim, 1970, pp. 377-378.
5. B. L. Karger, L. R. Snyder, and C. Horvath, *An Introduction to Separation Science*, John Wiley and Sons, New York, 1973, pp. 268-276.
6. T. Allen (ed.), *Particle Size Measurements*, Chapman and Hall, London, 1981, pp. 325-347.
7. T. Allen (ed.), *Particle Size Measurements*, Chapman and Hall, London, 1981, p. 104.
8. K. K. Unger, *Porous Silica*, J. Chromatogr. Libr. Vol. 16, Elsevier, Amsterdam, 1979, pp. 150-151.
9. K. S. W. Sing, D. H. Everett, R. A. W. Haul, L. Moscow, R. A. Pierotti, J. Rouquerol, and T. Siemieniewska, *Pure Appl. Chem.*, 57:603-619 (1985).
10. J. Seidler, J. Malinsky, K. Dussek, and W. Heitz, *Adv. Polym. Sci.*, 5:113-213 (1967).
11. R. Epton (ed.), *Chromatography of Synthetic and Biological Polymers*, Vol. 1, E. Horwood, Chichester, UK, 1978, pp. 1-6.
12. J. Stauff, *Kolloidchemie*, Springer-Verlag, Göttingen, 1960, pp. 665-673.
13. S. J. Gregg and K. S. W. Sing, *Adsorption, Surface Area and Porosity*, Academic Press, London, 1982, p. 187.

14. S. J. Gregg and K. S. W. Sing, *Adsorption, Surface Area and Porosity*, Academic Press, London, 1982, pp. 209-228.
15. S. J. Gregg and K. S. W. Sing, *Adsorption, Surface Area and Porosity*, Academic Press, London, 1982, pp. 111-115.
16. S. J. Gregg and K. S. W. Sing, *Adsorption, Surface Area and Porosity*, Academic Press, London, 1982, pp. 173-194.
17. S. J. Gregg and K. S. W. Sing, *Adsorption, Surface Area and Porosity*, Academic Press, London, 1982, pp. 41-73.
18. S. J. Gregg and K. S. W. Sing, *Adsorption, Surface Area and Porosity*, Academic Press, London, 1982, pp. 30-36.
19. S. J. Gregg and K. S. W. Sing, *Adsorption, Surface Area and Porosity*, Academic Press, London, 1982, pp. 41-84.
20. K. K. Unger, *Porous Silica*, J. Chromatogr. Libr., Vol. 16, Elsevier, Amsterdam, 1979, pp. 27-31.
21. B. Straube, Ph.D. thesis, Johannes Gutenberg-Universität, Mainz, FRG, 1985.
22. R. Sh. Mikhail and E. Robens, *Microstructure and Thermal Analysis of Solid Surfaces*, John Wiley and Sons, Chichester, UK, 1983, p. 29.
23. K. K. Unger, *Porous Silica*, J. Chromatogr. Libr., Vol. 16, Elsevier, Amsterdam, 1979, pp. 32-35.
24. R. Sh. Mikhail and E. Robens, *Microstructure and Thermal Analysis of Solid Surfaces*, John Wiley and Sons, Chichester, UK, 1983, pp. 175-182.
25. S. J. Gregg and K. S. W. Sing, *Adsorption, Surface Area and Porosity*, Academic Press, London, 1982, pp. 111-166.
26. H. J. Knox and H. P. Scott, *J. Chromatogr., 316*:311 (1984).
27. W. W. Yau, C. R. Ginnard, and J. J. Kirkland, *J. Chromatogr., 149*:465 (1979).
28. H. P. Boehm and H. Knözinger, in *Catalysis*, Vol. 4, (J. R. Anderson and M. Boudart, eds.), Springer-Verlag, Berlin, 1983, pp. 40-189.
29. R. G. Avery and J. D. F. Ramsay, *J. Colloid Interface Sci., 42*:597 (1973).
30. R. Ohnmacht and I. Halasz, *Chromatographia, 14*:155 (1981).
31. K. K. Unger, J. N. Kinkel, B. Anspach, and H. Giesche, *J. Chromatogr., 296*:3 (1984).
32. H. Giesche, K. K. Unger, U. Esser, B. Eray, U. Trüdinger, and J. Kinkel, *J. Chromatogr.*, in print.
33. K. Unger and R. Kern, *J. Chromatogr., 122*:345 (1976).
34. J. H. Knox, *J. Chromatogr. Sci., 15*:352 (1977).
35. J. H. Knox, *High Performance Liquid Chromatography*, Edinburgh University Press, 1978.
36. J. J. Kirkland, W. W. Yau, H. J. Stoklosa, and C. H. Dilks, Jr., *J. Chromatogr. Sci., 15*:303 (1977).

37. J. C. Giddings, *Dynamics of Chromatography*, Part I, Marcel Dekker, New York, 1965, pp. 205-215.
38. F. Geiss, *Die Parameter der Dünnschichtchromatographie*, F. Vieweg, Braunschweig, FRG, 1972, pp. 6-12.
39. B. L. Karger, L. R. Snyder, and C. Horvath, *An Introduction to Separation Science*, John Wiley and Sons, New York, 1973, pp. 12-33.
40. E. A. Flood, *The Solid-Gas Interface*, Vols. 1 and 2, Marcel Dekker, New York, 1967.
41. J. M. Thomas and W. J. Thomas, *An Introduction to the Principles of Heterogeneous Catalysis*, Academic Press, New York, 1967.
42. Ref. 41, p. 33.
43. S. Ross and J. P. Olivier, *On Physical Adsorption*, Interscience, New York, 1964.
44. A. V. Kiselev and J. A. Jasin, *Gas- und Flüssigkeits-Adsorptionschromatographie*, A. Hüthig Verlag, Heidelberg, 1985.
45. C. Horvath, W. Melander, and I. Molnar, *J. Chromatogr.*, *125*: 129 (1976).
46. J. F. K. Huber, M. Pawlowska, and P. Rarkl, *Chromatographia*, *17*:653 (1983).
47. B. L. Karger, J. N. LePage, and N. Tanaka, in *High Performance Liquid Chromatography, Advances and Perspectives*, Vol. 1 (C. Horvath, ed.), Academic Press, New York, 1980, pp. 113-200.
48. M. T. W. Hearn, *Ion-Pair Chromatography*, Chromatographic Sciences Series, Vol. 31, Marcel Dekker, New York, 1985.
49. J. C. Giddings, *Dynamics of Chromatography*, Part I, Marcel Dekker, New York, 1965, pp. 13-19.
50. J. C. Giddings, *Dynamics of Chromatography*, Part I, Marcel Dekker, New York, 1965.
51. J. H. Knox and M. Saleem, *J. Chromatogr. Sci.*, *7*:614 (1969).
52. G. J. Kennedy and J. H. Knox, *J. Chromatogr. Sci.*, *10*:549 (1972).
53. J. H. Knox and A. Pryde, *J. Chromatogr.*, *112*:171 (1975).
54. J. H. Knox, *J. Chromatogr. Sci.*, *15*352 (1977).
55. J. F. K. Huber, *Ber. Bunsenges. physik. Chem.*, *77*:179 (1973).
56. R. W. Stout, J. J. de Stefano, and L. R. Snyder, *J. Chromatogr.*, *282*:263 (1983).
57. J. H. Knox and H. P. Scott, *J. Chromatogr.*, *282*:297 (1983).
58. M. Golay, *Gas Chromatography* (D. H. Desty, ed.), Butterworths, London, 1959, p. 36.
59. J. J. van Reemter, F. J. Zuiderweg, and A. Klinkenberg, *Chem. Eng. Sci.*, *5*:271 (1958).
60. J. H. Knox, *J. Chromatogr. Sci.*, *18*:453 (1980).
61. A. W. J. de Jong, *J. Chromatogr.*, *193*:181 (1980).
62. J. F. K. Huber, *J. Chromatogr. Sci.*, *9*:72 (1971).

# 3

# Solid and Liquid Stationary Phases in Gas Chromatography

Jürgen Pörschmann / *Chemical Plant Construction Industry, Leipzig, German Democratic Republic*

W. Engewald / *Karl Marx University, Leipzig, German Democratic Republic*

## PRINCIPLES OF GAS CHROMATOGRAPHY

Gas chromatography (GC) has been very much improved in recent years, mostly in terms of instrumentation and novel stationary phases [1,2]. In GC, the separation arises from the frequent transition of the components between a continuously flowing gas phase and the stationary phase. The stationary phase may be an adsorbent or a liquid coated on a support which is assumed to be inactive. The variants of GC are therefore divided into gas-liquid partition chromatography and gas adsorption chromatography. Since it is often difficult to distinguish exactly between adsorption and solution phenomena, it appears reasonable to classify the principal techniques according to the state of aggregation of the stationary phase used into gas-solid chromatography (GSC) and gas-liquid chromatography (GLC).

### Basic Relationships

In what follows the nomenclature is based on recommendations given by Ettre [3] in a detailed overview. Fortunately, a uniform nomenclature is more advanced in GC than in high-performance liquid chromatography (HPLC), where confusion prevails.

The partition coefficient K of a solute between the two phases is defined as follows:

$$K = \frac{\text{mass of vapor / unit volume of stationary phase}}{\text{mass of vapor / unit volume of gas phase}} = \frac{c_s}{c_m} \tag{1}$$

For a given combination of solute-stationary phase K is a constant at constant temperature. The net retention volume $V'_R$ is directly related to the partition coefficient by

$$V'_R = K_L \cdot V_L \tag{2}$$

where $K_L$ is the partition coefficient in GLC and $V_L$ the volume of the liquid stationary phase.

In GSC, $K_L$ is replaced by the Henry constant of adsorption $K_H$, which is equal to the net retention volume for small sample size per unit surface area (cf. [4,5]). The partition coefficients $K_L$ and $K_H$ are constant in the range of low concentrations of analytes. At high concentrations, peaks with different asymmetries result according to the curvature of the sorption and partition isotherm (Fig. 1). Figure 1 shows the concentration of analytes in the stationary phase vs. the

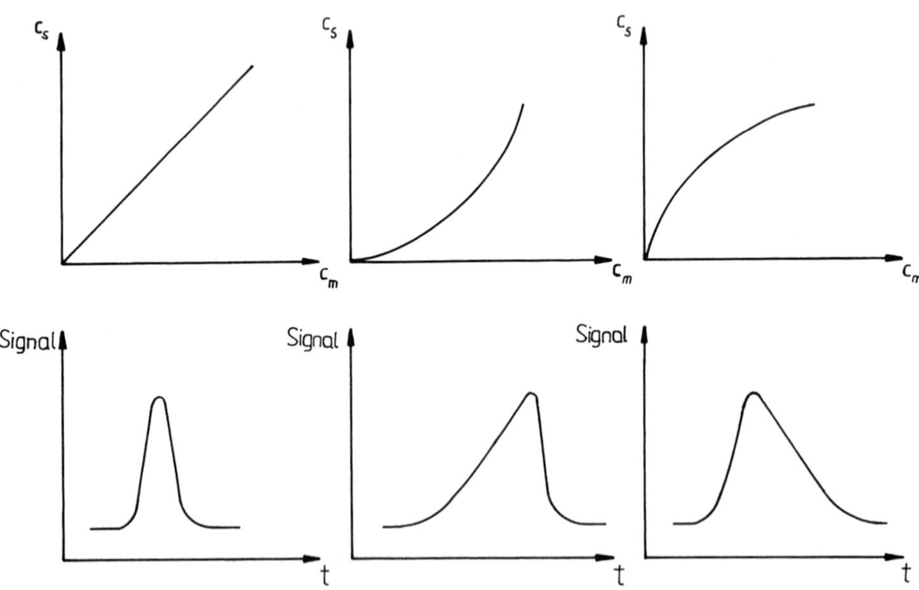


Fig. 1 Dependence of isotherm curvature on peak shape.

concentration of analytes in the mobile phase. In this respect, an important difference between GSC and GLC will be mentioned: tailed peaks are observed on overloading the adsorption column (Fig. 1, right), while an overloading in GLC results in leading peaks (Fig. 1, middle). Further sources of tailed peaks are of kinetic origin (cf. below).

Equation (1) can be rewritten by including the column phase ratio ($\beta$) and the solute capacity ratio (k'):

$$K_L = \frac{V_G}{W_G} \cdot \frac{W_L}{V_L} = k' \cdot \frac{V_G}{V_L} = k' \cdot \beta \tag{3}$$

where $V_G$ is the volume of mobile phase and $V_L$ the weight of liquid stationary phase.

The conclusions derived from Eq. (3) are very useful in selecting any proper column type as well as film thickness (cf. [6] and below). The thermodynamic partition coefficient K is related to the change of free enthalpy $\Delta G$ of a given solute or adsorbate:

$$K = e^{-\Delta G/RT} \tag{4}$$

$$\ln K = \frac{-\Delta G}{RT} = \frac{-\Delta H}{RT} + \frac{\Delta S}{R} \tag{5}$$

The free enthalpy of solution or adsorption is a quantity that reflects the strength of interaction between the stationary phase and the solute or adsorbate. Thus, by means of Eq. (5) it becomes possible to determine differential changes in enthalpy or entropy by plotting ln K against the reciprocal of the absolute temperature [1,7,8]. Since the energies of interaction in GSC are greater than those in GLC, the analysis temperature in GSC must be raised by about 100 K. Consequently, GSC should be preferred for the analysis of compounds having low K values, e.g., gases and low-boiling substances.

During migration along the column the solute(s) are subjected to several dispersion mechanisms that result in an additional broadening of the eluted peaks. This kinetic dispersion is characterized by the height equivalent to one theoretical plate H. The smaller H, the narrower the peaks. The total column dispersion is assumed to be the sum of four independent contributions:

$$H = A + \frac{B}{\bar{u}} + C_L \cdot \bar{u} + C_G \cdot \bar{u} \tag{6}$$

where $\bar{u}$ is the average gas velocity. The coefficients A, B, $C_L$, and $C_G$ represent the Eddy diffusion, longitudinal diffusion, mass transfer in the liquid phase, and mass transfer in the gas phase. Considering the GSC technique, we have to substitute the $C_k$ mass transfer term (k means kinetic) for the $C_L$ mass transfer term. In the case of GSC capillary columns the $C_k$ value is assumed to be in the range of $10^{-4}$-$10^{-6}$ sec only, provided no active sites occur on the adsorbent surface. Obviously, in the case of capillary GC, in general, the A term in Eq. (6) is neglected (cf.[10]).

From the following Eq. (7) it becomes apparent that peak dispersion depends on a variety of chromatographic factors:

$$H = 2\lambda dp + \frac{2\gamma D_G}{\bar{u}} + \frac{2}{3}\frac{k'}{(k'+1)^2}\frac{d_f^2}{D_L}\bar{u} + \frac{dp^2}{D_G}\frac{k'^2}{(k'+1)^2}\bar{u} \tag{7}$$

where dp is the particle diameter
- $d_f$ the film thickness
- $D_{G,L}$ the diffusion coefficient of the solute in the mobile and liquid stationary phase
- $\lambda$ the packing irregularity factor
- $\gamma$ the tortuosity factor

Besides these factors, the temperature (via $D_G$, $D_L$, k', $\bar{u}$) and the pressure drop (via $D_G$, $\bar{u}$) are "hidden" in Eq. (7) (for detailed information cf. [11] and references cited therein). To decrease H it is advantageous to work with small particles, provided the column can be properly packed. Equation (7) also indicates that H will decrease with decreasing film thickness. At very low phase loadings, however, the support surfaces or the tube walls in the capillaries are often not completely coated, resulting in undesirable adsorption effects, especially in analyzing polar solutes. It follows from Eq. (3) that the film thickness (or the capacity ratio) is dependent on the analysis temperature. Thus, in order to attain convenient capacity ratios, it appears reasonable to select a low film thicknesses (this is associated with high phase ratios) when analyzing polynuclear hydrocarbons with large partition coefficients. Usually an average partition ratio of k' = 4-8 is preferred. When k' is less than 2, the efficiency is substantially decreased; cf. Eq. (8). On the other hand, when k' is larger than 10, peaks become broad and flat in the isothermal mode.

Consequently, as can be taken from these considerations, thick film columns are useful for the separation of lower molecular weight compounds (cf. [12]). Thick films give an increase in the capacity factor, which is important for early eluting peaks. Under these conditions, the number of theoretical plates needed for a given solution is reduced (cf. [13] and Eq. (8)). The practical limit of the film thickness in capillary GC is approximately 8 μm. The ideal phase for thick coatings in capillary GC appears to be a nonpolar phase such as PS-255

(Petrarch System, Pennsylvania) due to the gum character and easy immobilization. There is evidence from the literature that capillary columns with thick films of polar phases are presently being developed. A further advantage of very thick films of immobilized stationary phases is their utilization for repetitive collection of volatile organic compounds (preparative capillary GC).

In a manner similar to that of thick films, Goretti et al. [14] prepared thick-layer open tubular columns with graphitized carbon black to separate low-boiling substances. Thus, in summary, a careful selection or better optimization of the film thickness is very important, particularly in capillary GC. A narrow range of film thickness would reduce the performance of GC needlessly.

In order to reduce the coefficient associated with the mass transfer in the gas phase, a carrier gas should be employed that has a high diffusion coefficient. Consequently, the use of hydrogen as carrier gas should be preferred at high velocities as opposed to heavier gases such as $CO_2$ and nitrogen. In the past hydrogen was frequently avoided due to explosion hazards. Presently, many chromatographers use hydrogen with capillary columns where the flow rates are lower and the hazard is less significant [16]. Beside these considerations, the choice of a proper carrier gas is also dependent on the type of detectors, e.g., thermal conductivity detector, electron capture detector, gas chromatography/mass spectrometry (GC/MS) coupling.

Obviously, in practice there is a need to compromise on the selection of chromatographic parameters. It must be emphasized that H represents an average value. As discussed by Giddings and Chovin (cf. [17]), the gradient of the pressure drop within the column causes differences in the carrier speed from the beginning to the end of the column, which influences the H value. Thus, the number of theoretical plates is proportional to the length of the column only to a first approximation (for more detailed information, cf. [11]).

However, the analyst is not in the enviable position to estimate the quality of any separation by means of the H value only. The chromatographic resolution ($R_S$) is defined as the distance between the two peak maxima divided by the mean peak width at the base and expresses how well two consecutive peaks 1 and 2 are separated. The resolution is dependent on the column characteristics, the capacity factor, and the relative retention [Eq. (8)]. Since for two peaks with a small relative retention the column efficiency is nearly the same for both solutes, we may rewrite [6,10]:

$$R_s = \frac{\sqrt{N}}{4}\left(\frac{\alpha - 1}{\alpha}\right)\left(\frac{k'_2}{k'_2 + 1}\right) = \frac{1}{4}\sqrt{\frac{L}{H}}\left(\frac{\alpha - 1}{\alpha}\right)\left(\frac{k'_2}{k'_2 + 1}\right) \tag{8}$$

where $\alpha$ is the selectivity coefficient defined by $\alpha = k'_2/k'_1$.

Utilizing this well-known relationship derived by Purnell it becomes possible to calculate the number of theoretical plates required ($N_{req}$) or the column length required ($L_{req}$) for two adjacent peaks:

$$\frac{L_{req}}{H} = N_{req} = 16R_s^2 \left(\frac{\alpha}{\alpha - 1}\right)^2 \left(\frac{k'_2 + 1}{k'_2}\right)^2 \quad (9)$$

Equation (8) reflects the quantification of the chromatographic separation in a simplified manner. As an example, the required resolution also depends on the concentration ratio of the two analytes, especially if the major analyte is eluted first.

As can be inferred from Eqs. (8) and (9), $R_s$ increases with the square root of $N_{req}$. Hence, the column length must be increased almost four times to double the resolution. From this point of view it becomes evident that the separation of closely eluting compounds (e.g., isomeric mixtures) cannot be performed via efficiency alone because the use of long columns entails large analysis times. Therefore, we must also take into account selectivity. Obviously, the greater the selectivity, the shorter the analysis time. The dependence of $N_{req}$ on the relative retention is depicted in Fig. 2, which indicates that small changes in relative retentions significantly reduce $N_{req}$. It therefore appears advantageous to select relative retentions near the largest bend of the curve. Since slight improvements of stationary phase selectivity have a great influence on resolution, many efforts have been made on tuning phase selectivity (see below).

A comparison of the resolution capability in GC and HPLC shows that polarity and selectivity of the separation system in GC cannot be varied to the same extent as in HPLC. Hence, GC requires higher separation efficiencies due to the limited possibilities for selectivity changes of the GC stationary phase.

## Development of Gas-Solid and Gas-Liquid Chromatography

GSC was a well-established technique prior to the introduction of GLC by James and Martin in 1952. The research groups of Cremer in Austria, Turkeltaub and Zhukovitskii in the Soviet Union, and Janak in Czechoslovakia pioneered the analysis of gases and physical-chemical investigations by means of GSC [19,20]. With the rapid development of GLC in the 1960s, traditional GSC technique lost its importance and found application mainly in the analysis of mixtures of hydrogen isotopes, inorganic gases, and low molecular weight hydrocarbons. There were important reasons for this development:

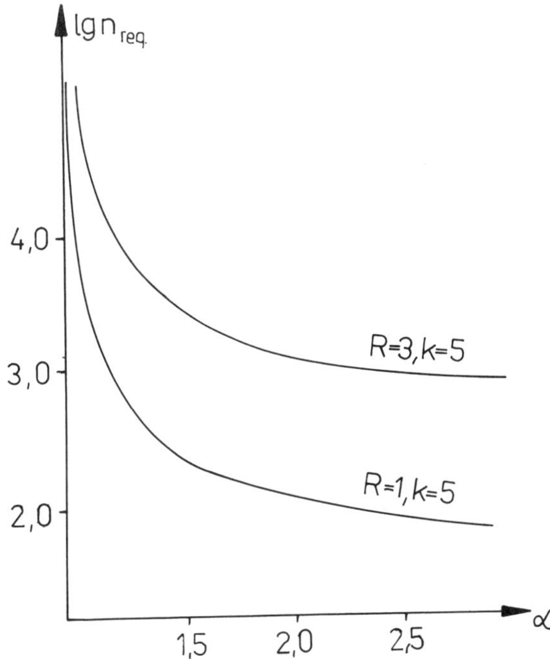


**Fig. 2** Dependence of relative retention on required theoretical plate numbers.

1. The number of commercially available adsorbents suitable for GC purposes was limited compared with the huge number of stationary liquids.
2. The surface of the adsorbents employed, e.g., activated carbon, silica, alumina, was heterogeneous, often causing peak tailing and incomplete recovery.
3. The adsorbents were not standardized with regard to their properties and suffered from poor reproducibility.

From the late 1960s on GSC has seen a renaissance due to the availability of adsorbents with uniform and homogeneous surfaces, which tend to promote the linearity of the isotherm. By definition, a surface is considered to be homogeneous and uniform if the adsorption energy is merely a function of the distance between the adsorbate and the adsorbent surface. Further developments that substantially influenced the use of GSC were the highly sensitive detectors that avoided column overloading and the improved methods in adsorbent deactivation (see below). Porous polymers, graphitized thermal carbon black, and several types of silicas (e.g., the Porasil series) belong to those adsorbents specifically developed for GSC (see below). With the advent and the maturity of capillary GC, there has also been an ever-growing interest in extending the use of those adsorbents well suited for GSC purposes to capillary GC. This demand has been met by Chrompack (Middelburg, The Netherlands) in particular. Chrompack markets high-performance GSC capillaries, with molecular sieve types 5 Å and 13X, alumina and silica to accomplish specific and difficult separation problems (see below).

Apart from its analytical application, GSC has become a suitable tool in physicochemical examinations of the surface chemistry of solids and in investigations of molecular interactions (cf. [21] and references cited therein). For these applications GSC has some advantages over the traditional static methods, i.e., methods based on GSC do not require expensive equipment, need only tiny amounts of adsorbates, and are time saving. GLC was pioneered by Martin and Synge in 1941, whose work dealt with liquid-liquid chromatography. Their original paper (cf. [19]) contains the often quoted prediction: "The mobile phase need not be a liquid but may be a vapour." Eventually, in 1951, Martin and James put this idea into practice while analyzing fatty acids by means of a classical packed column. This work stimulated the expansion of GC.

Considering peak dispersion phenomena, GLC and GSC have in common the sum of contributions from Eddy diffusion [A term in Eq. (6)], longitudinal diffusion in the gas phase (B term), and mass transfer from gas to stationary pahse ($C_G$ term). However, as already mentioned, the $C_L$ term is different from the $C_k$ term. Because the latter is smaller than the former (provided that homogeneous surfaces are given), GSC offers a great challenge with respect to fast and effective separations.

The efficiency of GSC columns is to a large extent influenced by the specific surface area, the specific pore volume, and the distribution of the pore sizes of the support [22]. On shortening the inner diffusion paths by using superficially porous adsorbents instead of totally porous ones, the resolution is increased, but the sample capacity is reduced. The term "sample capacity" denotes the amount of a solute to be analyzed without dramatic reduction (10% are usually assumed) in peak resolution.

The advantages of GSC over GLC consist of providing a higher column efficiency and a markedly reduced column bleeding, the latter being particularly useful in electron capture detection (e.g., in analyzing traces of pesticides or their metabolites in biological samples) and in GC/MS coupling. As it will be outlined in a later section, much work has been done in the field of reducing column bleeding with liquid phases both by developing low-bleeding liquid phases and appropriate immobilizing procedures.

The advantages of GLC over GSC include the multitude of liquid phases commercially available and the larger part of linearity of the distribution isotherm compared to that of the adsorption isotherm. In GLC the amount of liquid phase can be varied easily, which allows the use of the same phase both in high-efficiency, open tubular columns and in preparative columns. In the authors' opinion, both variants should supplement one another rather than compete.

### Column Types

The columns in GC are classified into open tubular and packed. The latter comprises classical packed columns and packed capillaries [23,24]. In the authors' opinion, the term "packed capillaries" is wrong. These columns, produced by drawing out glass tubes filled with particles, are semipacked at best.

The classical packed type of column is usually further divided into types with a diameter larger than 1 mm and the so-called micropacked columns, the latter having particle-to-inner-column diameters (dp/dc) of less than 0.3 mm, inner column diameters of less than 0.8 mm, and high packing densities. Figure 3 illustrates the arrangement of particles of graphitized thermal carbon black (see below) in a micropacked glass column. The performance of micropacked columns was scrutinized by Cramers and Rijks [26] and Welsch et al. [25,28] (cf. also the reference [27]). The results obtained show that micropacked columns represent a good compromise between separation efficiency, column capacity, and analysis time. The reduced particle size in micropacked columns entails a forcing up of the required inlet pressure [29]. Consequently, it would appear impossible to lengthen the column as desired. On the other hand, it would be worthwhile to construct a high-pressure gas chromatograph for high-performance micropacked columns with an overall efficiency of more than $10^6$ theoretical plates. Recently, Golay reported the advantages derived from the use of high inlet pressures [30].

The introduction of open tubular columns by Golay in 1956 (cf. [10,19]) represented the real breakthrough in terms of higher column performance and permeability [12]. Additional stimulation was derived from the glass capillary drawing machine of Desty because it allowed scientists to make their own columns [31]. Considering efficiency, speed of analysis, and sensitivity, it is clear that most of the topical separation problems can be solved better using capillary columns. This is mainly due to the fact that the limiting factor of the separation capability, the slow diffusion of the solute molecules within the pores of the support, is eliminated. However, increased efficiency is gained at the expense of sample capacity. The main column characteristics of the two types are summarized in Table 1.

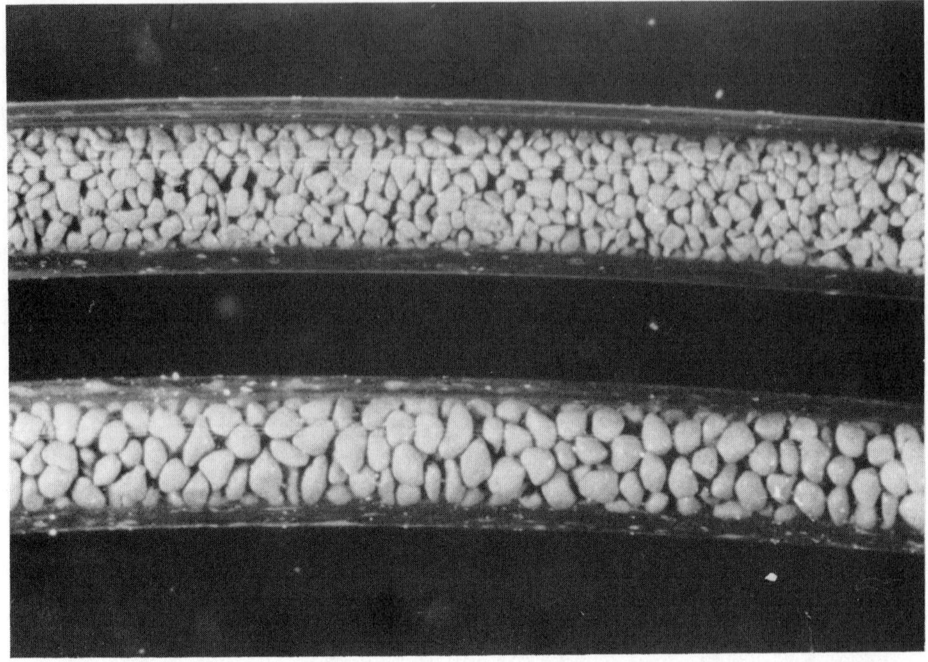

**Fig.** 3 Microphotographs of a 0.45-mm-i.d. column filled with GTCB (Sterling MT). (top) dp = 0.09-0.12 mm; (bottom) dp = 0.16-0.20 mm.

Open tubular columns comprise wall-coated open tubular columns (WCOT), where the stationary liquid is coated as a thin film on the tube wall, and support-coated open tubular columns (SCOT), where the fine support particles coated with a liquid are deposited on the tube wall. WCOT columns produce the highest efficiency, while SCOT columns, which tend to produce lower efficiencies, offer the advantage of higher sample capacity. Pioneering work in the preparation of SCOT columns was carried out by Halasz and Horvath [37]. Ettre et al. [10] developed these columns further and made them commercially available (cf. also [38,39]).

There is a tendency to replace SCOT columns by WCOT columns with thick coatings. By using narrow-bore WCOT columns with an inner diameter of 60 μm, a further improvement of efficiency and speed of analysis can be achieved [32], but, at the expense of sample capacity. Columns whose stationary layer is a porous adsorbent are termed porous layer open tubular columns (PLOT). They were examined in

Table 1 Column Characteristics

| Characteristic | Classical packed columns | Capillary columns | |
|---|---|---|---|
| | | WCOT | SCOT |
| Internal diameter d(mm) | 2-5 | 0.1-0.6 | 0.3-1.0 |
| Length L (m) | 1-8 | 10-80 | 10-30 |
| Average HETP (mm) | 1 | 0.3 | 0.6 |
| $N_{eff.}$/meter (k' = 5) | 400-650 | 2000-3000 | 800-1400 |
| Permeability ($cm^2 \cdot 10^{-7}$) | 1 | 100 | Intermediate values corresponding to inner diameter |
| Preferred applications | Simple mixtures | High efficiency is essential for complex mixtures | |
| Sample capacity | High | Low | Moderate |
| Cost | Low | Higher | Moderate |

[33-36]. Advantages and disadvantages of the basically different types of capillary columns were summarized by Ettre [6,10]. The problems facing the chromatographer engaged in work with WCOT columns are detailed in a later section.

Today the cardinal limitations of capillary columns are the small capacity and the specific demands on the instrument. Recently, attention has been focused on larger bore open tubular columns with a large sample capacity to overcome these drawbacks. Using a high carrier gas flow, the wide-bore "capillaries" still provide better separations than those of the packed columns, but in a fraction of time and at higher sensitivities. Furthermore, on using fused silica as column material, better peak shapes in analyzing polar solutes can be attained because the fused silica surface is less adsorptive than the diatomaceous earth commonly applied in packed columns (see below). A further advantage of wide-bore capillaries is that old-fashioned instruments, not designed for capillary GC, can be run with these columns. In summary, wide-bore capillaries combine high capacity and ease of use with high efficiency. In [41] the separation efficiencies of a packed column, a narrow-bore, and a wide-bore capillaries are compared.

At present capillary columns with inner diameters ranging from 0.1 to 0.8 mm and film thicknesses of 0.1 to 5 μm can be purchased from several sources (e.g., the 530-μm series from Hewlett-Packard, the 750-μm series from Supelco). Thus, the analyst is in the enviable position to choose the column best suited for the solution of his problem. However, despite its limited efficiency and poor permeability, the classical packed column type has been the commonly used one since the original work of James and Martin. As pointed out by Jennings [31], at least 60% of all applications in the United States and probably 50-60 of those in Europe are performed on packed columns. In view of the necessity to apply high-performance columns, GC is becoming more and more capillary (cf. [42]). This tendency receives support from recent developments done in the field of stationary phases, deactivation procedures with glass and fused silica, variation of capillary column inner diameter and film thickness, injection modes, etc.

## Classification of Stationary Phases

### *Solute/Adsorbate-Stationary Phase Interactions*

The retention of a solute in GLC or of an adsorbate in GSC results from several types of interactions.

*Dispersion Forces* The universal London dispersive forces arise from fluctuating dipoles and are mainly responsible for the interaction between nonpolar molecules. In contrast to orientation forces (see below), neither dispersive nor inductive forces are temperature-dependent. Dispersive forces decrease rapidly with increasing distance between the interacting centers. As soon as the molecules approach each other, the attraction switches to repulsion.

*Orientation Forces* Orientation forces exist between two molecules having permanent dipole moments. According to Keesom (1921), the interaction energy varies directly with the square of the dipole products and inversely with both the thermal energy per molecule and the sixth power of the distance between the dipoles. From the two latter points it can be concluded that the analysis temperature should be set as low as possible and that small molecules represent highly selective phases. Since low temperatures ensure highly selective separations, the further extension of supercritical fluid chromatography (SFC) at about 40°C becomes important in this manner, provided a high column performance is generated. Besides Keesom forces, the hydrogen bond is a further type of orientation force. The more electronegative the atom to which the small hydrogen atom is bonded, the greater the development of partial positive charge on the hydrogen. A literature search proved that there is a similarity between hydrogen bonds and charge-transfer interactions [43].

*Induction Forces* This type of intermolecular force results from an interaction between a permanent dipole in either solute or stationary phase and an induced dipole in the other. Induction or Debye forces are largely dependent on polarizability. Induction forces are weaker than orientation forces.

*Charge-Transfer Interactions* If there are, on the one hand, molecules having high electron affinity and, on the other hand, molecules having $\pi$-electron systems with low ionization potentials, the development of charge-transfer systems is encountered, e.g., in analyzing aromatics on cyanosilicones (see Chap. 8).

*Coulombic Forces* This kind of interaction occurs between ions, e.g., when molten salts are used. As a rule they do not affect the solubility in GC.

### *Classification of Adsorbents*

In contrast to the solution of solute molecules in a stationary liquid, the adsorption of molecules on a surface offers a considerably simpler mechanism. It seems reasonable to divide adsorbents according to the nature of adsorption interactions and to their specific surface area. The first is determined by the chemical structure of the adsorbents, whereas the latter is one of the terms determining the geometrical structure of adsorbents.

*Classification According to the Chemical Structure* The chemical structure of the adsorbent surface governs the energy and the nature of the interactions occurring between interacting molecules. Classification as suggested by Kiselev [45] warrants special mention. Since adsorption occurs through several forces of physical attraction ranging from unspecific dispersion forces to specific ones, Kiselev subdivided the adsorbents into three types according to their capacity to enter into any kind of intermolecular interaction with the adsorbate molecules:

Adsorbents of the first type are unspecific ones such as graphitized thermal carbon black and boron nitride, carrying neither functional groups nor exchangeable ions.

Adsorbents of the second type bear localized positive charges as active sites, shifted to the outside and concentrated in a region of small radius (acidic hydroxyl groups at silica). The second type also includes adsorbents with aprotic acid centers or small-radius cations with the compensating negative charges disturbed over the inner bonds of a large complex anion (zeolites).

Adsorbents of the third type have concentrated negative charges bulging out on their surfaces, including functional groups such as nitriles.

It will be useful to consider the adsorbates from the same point of view as the corresponding adsorbents. Therefore, Kiselev [45] subdivided adsorbates into four groups according to the different spherical electron density distribution of their bonds and linkages:

Group A includes molecules with a spherically symmetrical electron shell, i.e., there is no locally concentrated electron density on their peripheries (rare gases, saturated hydrocarbons).

Group B involves molecules with concentrated electron densities on the periphery (molecules with $\pi$ electrons and lone electron pairs such as unsaturated hydrocarbons, tertiary amines, ketones, ethers). Molecules of group B, in contrast to those of group D (see below), are not capable of associating.

Group C substances possess locally concentrated positive charges in small-radius linkages (organometallic compounds).

Group D molecules possess neighboring links of small radius with a positive charge concentrated in one of them and the electron density concentrated on the periphery of the other (alcohols, primary and secondary amines).

The concept suggested by Kiselev has gained wide acceptance due to its usefulness in selecting appropriate stationary solid phases and estimating retention behavior of the adsorbates. Attention should be paid to the fact that the molecules of group A can interact with any other molecules only nonspecifically through universal dispersive forces. The same is valid for adsorbents of the first type. Clearly, specific adsorbents (second and third type) are capable of undergoing specific interactions with adsorbate molecules of the B, C, and D groups.

*Classification According to Geometrical Structure* (*cf. Chap. 2*) The pore structure of adsorbents is governed by the mean pore diameter (pd); the specific surface area ($a_s$), which expresses the extension of the pore area per gram adsorbent; the specific pore volume (vp), which expresses the total volume of the pores per gram adsorbent; and the pore size distribution (psd) [46]. Based on the ideas of Dubinin [48], Kiselev and Yashin divided adsorbents into four basic structural types [47]:

Type I, nonporous adsorbents. Although these adsorbents are called "nonporous," they dispose of macropores formed by interstices between the primary particles (average pore diameter >200 nm). The secondary pores are sufficiently large to ensure fast mass exchange.

Type II, uniformly wide-pore adsorbents. This type includes, in particular, ordinary wide-pore silicas and large-pore porous glasses with specific surface areas of less than 300-400 $m^2$ $g^{-1}$ and pore sizes of about pd > 10 nm.

Uniformly fine-pore adsorbents. Type III includes popular adsorbents such as zeolites, carbon molecular sieves, fine-pore glasses with a mean pore diameter of about 2-4 nm and a specific surface area of more than 500 $m^2$ $g^{-1}$. Therefore, the mean pore diameter is of the same order as the diameters of large adsorbate molecules.

Type IV, nonuniformly porous adsorbents. This type is not in common use in GSC.

The pore size has an important influence on the speed of the mass transport to the active sites at the surface, where the adsorption and desorption process takes place. In order to establish a fast mass transfer, the adsorbent employed should exhibit only macropores. This is especially valid in analyzing medium- and high-boiling substances. Therefore, in this case adsorbents with low specific surface areas and large pores are employed [49]. In contrast, adsorbents having a high surface area provide high-capacity factors necessary in analyzing low-boiling substances such as light hydrocarbons. According to Guillemin [22], it is possible to compensate the loss of efficiency caused by shorter columns by using adsorbents with higher specific surface areas (see also [50]).

There is an urgent need in analytical practice to compromise between speed and separation efficiency and to adapt the specific surface area to the separation problem at hand. Several attempts have been made to improve the relation between speed and resolution (cf. the "brush"-type phases detailed below).

*Characterization of Stationary Liquids*

According to Herington, the distribution coefficient K (or net retention volume $V'_R$), the vapor pressure $p^0$, and the activity coefficient $f^0$ at infinite dilution of two solutes 1 and 2 are connected in the following manner:

$$\lg \frac{V'_{R,2}}{V'_{R,1}} = \frac{K_2}{K_1} = \lg \frac{p_1^0}{p_2^0} + \lg \frac{f_1^0}{f_2^0} \tag{10}$$

Clearly, the greater the vapor pressure of the solute at a given column temperature, the smaller the partition coefficient. If two solutes have the same vapor pressure, the difference in the partition coefficients must be governed by the difference in their activity coefficients in the stationary phase. The activity coefficient expresses the interactions between solute molecules and stationary liquid molecules.

As GC has matured, the need has developed for a convenient method to establish reproducibly and precisely whether one stationary liquid is different from another and in what way [51]. At first, characterization of stationary phases was performed in terms of polarity, with the term polarity referring to the molecule under chromatographic conditions and not in the isolated state. With the passing of time, several empirical parameters were recommended for use as polarity indices. It is evident that it would be fairly useless to classify a given phase by means of only one parameter such as the dipole moment.

Real breakthroughs in the classification of stationary liquids were achieved by Kovàts and by Rohrschneider. Kovàts introduced the retention index, which was rapidly accepted as the method of choice for reporting chromatographic data. In this retention index system the logarithmic retention of a solute is interpolated between those of two standard components, namely, *n*-alkanes:

$$I = 100 \cdot z + 100 \frac{\lg t'_{R,i} - \lg t'_{R,z}}{\lg t'_{R,z+1} - \lg t'_{R,z}} \tag{11}$$

where $t'_{R,z} < t'_{R,i} < t'_{R,z+1}$ and $t'_{R,i}$ is the adjusted net retention time of solute i and $t'_{R,z}$ is the adjusted net retention time of *n*-alkane with z carbon atoms.

Hence, the retention index is theoretically based on the linear relationship between the logarithm of the net retention and the number of carbon atoms in a homologous series of *n*-alkanes under isothermal conditions. Strictly speaking, this general linear relationship, which requires the linearity of the saturation vapor pressures within a homologous series as well as the constancy of the activity coefficients of the reference *n*-alkanes at infinite dilution, does not exist.

The determination of retention indices can be made with great precision and accuracy. As will be shown below, the presentation of retention data in the retention index form provides a great deal of information. In the analysis of fatty acids and steroids the introduction of the equivalent chain length (ECL) and the steroid number (SN), respectively, has proved useful (cf. [52]). However, it is difficult to transform ECL and SN values into thermodynamic quantities. A promising attempt has been made to extend the retention index concept to other chromatographic techniques such as HPLC and TLC [53].

By definition, the retention index of any *n*-paraffin is always 100z at a given temperature and on a given column. The retention index difference of a given solute i on a polar and a less polar phase, the latter serving as reference, characterizes the proportion of specific interactions:

$$\Delta I = I_i^{polar} - I_i^{nonpolar} \tag{12}$$

The more polar the phase, the higher the retention index of a polar solute on it. A special case of $\Delta I$ values are the well-known homomorphic factors (abbreviated H) [54].

By using Eq. (12) it is possible to evaluate the index contributions of different structural units of a molecule or a functional group. In order to estimate the polarity of any stationary liquid it is necessary to obtain information on its capacity to enter into various kinds of intermolecular interactions. The question therefore arises, how many independent intermolecular attractions are operating? Rohrschneider [55] found five. A given solute was selected for each independent kind of polar force in order to define that force, e.g., in each case five polarity factors were used to characterize the stationary phases (x, y, z, u, s) as well as the solute (a, b, c, d, e):

$$\Delta I = ax + by + cz + du + es \tag{13}$$

By analyzing each of m solutes on each of n liquid phases, mn values are obtained, i.e., a system of mn equations with 5(m + n) unknowns. Rohrschneider solved the undetermined system by fixing arbitrary values for x, y, z, u, and s: $x = \Delta I_{benzene}/100$, $y = \Delta I_{ethanol}/100$, $z = \Delta I_{methylethylketone}/100$, $u = \Delta I_{nitromethane}/100$, $s = \Delta I_{pyridine}/100$. Thus, the greater the Rohrschneider constant, the greater the retention for that particular compound. Rohrschneider himself characterized 22 stationary liquids by the five index differences of benzene (x), ethanol (y), methylethylketone (z), nitromethane (u), and pyridine (s) using squalane as nonpolar reference at 100°C. Incidentally, a truly "nonpolar" phase does not exist. In addition to the requirements of a nonpolar reference the stationary phase can itself present problems due to variation in the retention behavior owing to thermal instability, oxidative degradation, and impurities in the phase itself (cf. references cited in [56]).

Rohrschneider interpreted the five polarities of the solutes as a measure of hydrogen bonding (b, H donor; c, H acceptor), orientation forces (e), and charge-transfer forces (a and d). Consequently, the retention indices of the five test compounds serve to define the polarity of a given stationary phase, whereas the $\Delta I$ values define its selectivity. Thus, selectivity characterizes selective interactions. In this manner, selectivity describes a minor part of the overall interaction because the retention index of a given solute on a nonpolar phase is mostly higher than the index shift on a polar phase.

In the course of time it turned out that the standard solutes ethanol, methylethylketone, and nitromethane were less suitable in analytical practice due to their low retention at 100°C. McReynolds [57] suggested substitution by the higher members $n$-butanol, 2-pentanone, and nitropropane, respectively. The index differences of the five McReynolds solutes (squalane as reference) have been designated as X, Y, Z, U, and S; the indices were measured at 20% loading at 120°C. Another suggestion by McReynolds, to extend the scope of test substances to 10, has not gained complete acceptance [58].

Although the five constants X, Y, Z, U, and S correspond to the original Rohrschneider constants, there are two differences: First, instead of relying on y, z, and u, the higher boiling homologues are used. Second, the $\Delta I$ values themselves serve to define the McReynolds constants instead of the $\Delta I/100$ values as in the case of Rohrschneider constants. Table 2 lists the five substances suggested by McReynolds and the classes of compounds characterized by them.

The usefulness of the Rohrschneider concept and its modified form by McReynolds lies in the fact that over 200 stationary phases could be characterized. The McReynolds constants are presented in the form of tabulations of $\Delta I$ for the test probes (see below). By means of these constants it is possible to arrange and compare the separation ability of stationary phases for various classes of solutes. They serve as a useful tool in selecting an appropriate stationary liquid and also in predicting elution orders. Furthermore, McReynolds constants are helpful in choosing a substitute phase of similar selectivity. To the authors' minds, in the case of polar stationary phases with relatively poor reproducibility the retention indices should be given together with the column manufacturer.

The limitations of the concept are as follows:

1. McReynolds constants are determined at a fixed temperature of 120°C.
2. The retention behavior of the solute depends on the type of the support, its deactivation, etc.
3. Phases can be exposed to thermal or catalytic degradation, oxidation, loss of phase, etc., which influences the separation characteristics.
4. The use of $n$-alkane standards implies errors in determining retention indices on strongly polar phases.
5. McReynolds constants are determined with an unusual high-phase loading of 20%. As has long been known, the retention indices can vary with the phase loading.
6. The Rohrschneider or McReynolds concept is not capable of giving a complete reflection of the types of selectivities of various stationary liquids. In many separations it is not a single force which dominates, but the physical forces discussed above act in a very complex manner.

Table 2 McReynolds Substances

| Symbol | Substance | Characterized classes |
|---|---|---|
| X | Benzene | Aromates, olefines |
| Y | Butanol-1 | Alcohols, nitriles, organic acids |
| Z | 2-Pentanone | Ketones, ethers, aldehydes |
| U | 1-Nitropropane | Nitro compounds, nitriles |
| S | Pyridine | *N*-heterocyclic compounds |

Since the consideration of the solid support is outside the scope of this chapter, only some brief comments should be made: The effect of the solid support on the retention indices of McReynolds test solutes was detailed by Berezkin [59,60]. These studies were extended by several workers; cf. for example, [61]. Excessive surface activity of the support is undesirable because this frequently leads to irreversible adsorption phenomena and loss of solutes. An essential part of column technology is to minimize these undesirable phenomena. From this point of view emphasis should be placed on the tailor-made supports recently developed by Merck (Volaspher series).

Additional attempts to characterize polarity and selectivity were based on a strongly thermodynamic treatment. Golovnya and Misharina [62-65] suggested the definition of the polarity by means of the capacity of a stationary liquid for various intermolecular interactions that are governed by the partial molar free energy of solution in GLC or by the partial molar free energy of adsorption in GSC. Since there are no solutes capable of simultaneously undergoing all of the possible kinds of intermolecular interactions with the stationary phase, six parameters were proposed for evaluating polarity: the partial molar free energy of solution of the *n*-alkane methylene group ($\Delta G^{CH_2}$) representing dispersive interactions, and the partial molar free energy of solution of the McReynolds five test substances. The higher the value of $\Delta G$, the more polar the phase. The most nonpolar stationary phase exhibits the five lowest values of $\Delta G$ compared to other phases. Thus, polarity in chromatography is a measure of the occurrence of intermolecular interactions calculated in terms of $\Delta G$. The proposed system of evaluation of polarity is universal and does not require the selection of a stationary phase with "zero polarity." Its usefulness has been demonstrated in establishing structure-retention relationships (cf. references in [64,66]). A similar concept has been developed by Sevcik and Loewentap [67].

For the sake of completeness some comments should be made on the structural characterization of stationary liquids by means of popular analysis techniques. These include size exclusion chromatography, infrared spectroscopy, nuclear magnetic resonance spectroscopy, and thermal analysis (thermogravimetry) [68]. A useful tool in analyzing stationary phases is provided by their thermal decomposition patterns. By means of pyrolysis or hydrolysis, oligomers and/or monomers are obtained [69] that can be converted into volatile derivatives and then analyzed by GC/MS.

As commented by Grob [70], GC itself can be favorably used to analyze stationary liquids. However, the full power of capillaries employed for separation is only guaranteed in combination with an appropriate sampling technique (on-column injector).

## PACKINGS IN GAS-SOLID CHROMATOGRAPHY

### Carbon Adsorbents (cf. Chap. 6)

#### *Graphitized Thermal Carbon Blacks (GTCB)*

This widely used type of adsorbent is produced by graphitization of thermal carbon black at ∿3000°C in the absence of oxygen. On graphitization the content of hydrogen and oxygen decreases to less than 0.4% [71]. By this treatment the small carbon black particles are converted into graphitic polyhedra.

Adsorption on GTCB is caused primarily by nonspecific dispersive forces. Therefore, GTCB can be categorized as a nonspecific adsorbent of the first type according to Kiselev's classification [4,45]. Due to its pronounced chemically and geometrically homogeneous surface, the retention of adsorbate molecules is especially sensitive to their particular geometry. This in turn means that the retention index of the analyte is largely controlled by the number of points of contact with the plane surface. These properties make GTCB highly suitable for the separation of both structural and spatial isomers. In this respect GTCB is a selective adsorbent.

The energy of adsorption is high because of the considerable surface activity caused by the large concentration of carbon atoms at the homogeneous surface of graphite. Consequently, relatively high temperatures are required for the elution of adsorbates with medium and high boiling points.

The different separation mechanisms in GSC with GTCB and GLC with commonly used nonpolar phases are shown by chromatograms of a mixture containing substances which belong to the A, B, and D groups according to Kiselev's classification (cf. Fig. 4). Some of these substances bearing polar functional groups are capable of undergoing both specific and nonspecific interactions, whereas substances of group

A are capable only of nonspecific interactions. The sequence of elution of the substances analyzed differs substantially on OV-1 and GTCB columns. In the latter case, all substances, even those with a large dipole moment, were eluted from the GTCB column before *n*-decane (group A substance), which has the most preferential orientation on the GTCB surface.

The outstanding selectivity of GTCB toward isomeric solutes compared to commonly used stationary liquids is well demonstrated by the separation of isomeric dimethyl cyclohexanes. As shown in Fig. 5, all of the possible isomers can be entirely separated on GTCB [72], while in GLC the pairs 1,3-*cis*/1,4-*trans* as well as 1,3-*trans*/1,4-*cis*-dimethylcyclohexane were coeluted. The elution order of the isomers on GTCB is in accord with the preferential orientation of the molecules on the surface of GTCB.

An essential advantage of using GTCB consists of the possibility of predicting the elution sequences by calculating the potential energies. The values obtained from this approximate calculation are in sufficient agreement with experimental data measured at low surface coverages on GTCB (see below and [4,73]).

GTCB can be easily modified with liquid and solid phases, resulting in many selectivities to be set. Pure and modified GTCB (see below) was used for the separation and the investigation of the retention behavior of many classes of substances such as saturated aliphatic hydrocarbons [74], unsaturated aliphatic hydrocarbons [75-78], alkylbenzenes, alkylnaphthalenes and polynuclear aromatic hydrocarbons [79-83], cyclic saturated hydrocarbons [72,74,84], aldehydes, ketones, and alcohols [21,85,86], ethers [87], amines [88,89], mercaptans [90], phenols [91], and free fatty acids [92,93]. In the reviews [4] and [94], the application of GTCB is detailed further.

A survey of commercially available GTCBs is given in Table 3. As already mentioned, the higher the specific surface area, the higher the required elution temperature. Therefore, the use of Carbopack B with a specific surface area of about 100 $m^2\ g^{-1}$ proved to be favorable to gas analysis. In the case of higher boiling substances, preference should be given to the use of Carbopack C, Sterling MT, etc. The Carbopack C adsorbent exhibits a higher degree of graphitization compared to Carbopack B, the latter having a more developed surface. Recently, a new graphitized carbon with a low surface area (about 6 $m^2\ g^{-1}$) and good mechanical properties was introduced in GSC [95]. The mechanical strength of GTCB particles can be improved by treatment with an adhesive polymer, e.g., Apiezon L (0.01% by weight). This highly efficient adsorbent was called Carbochrom by Kiselev [45].

The satisfactory tailing-free elution from pure GTCB is hindered by traces of oxygen and sulfur polar compounds on the adsorbent surface, which are capable of undergoing undesirable interactions with the functional polar groups of the adsorbates. Such chemical inhomo-

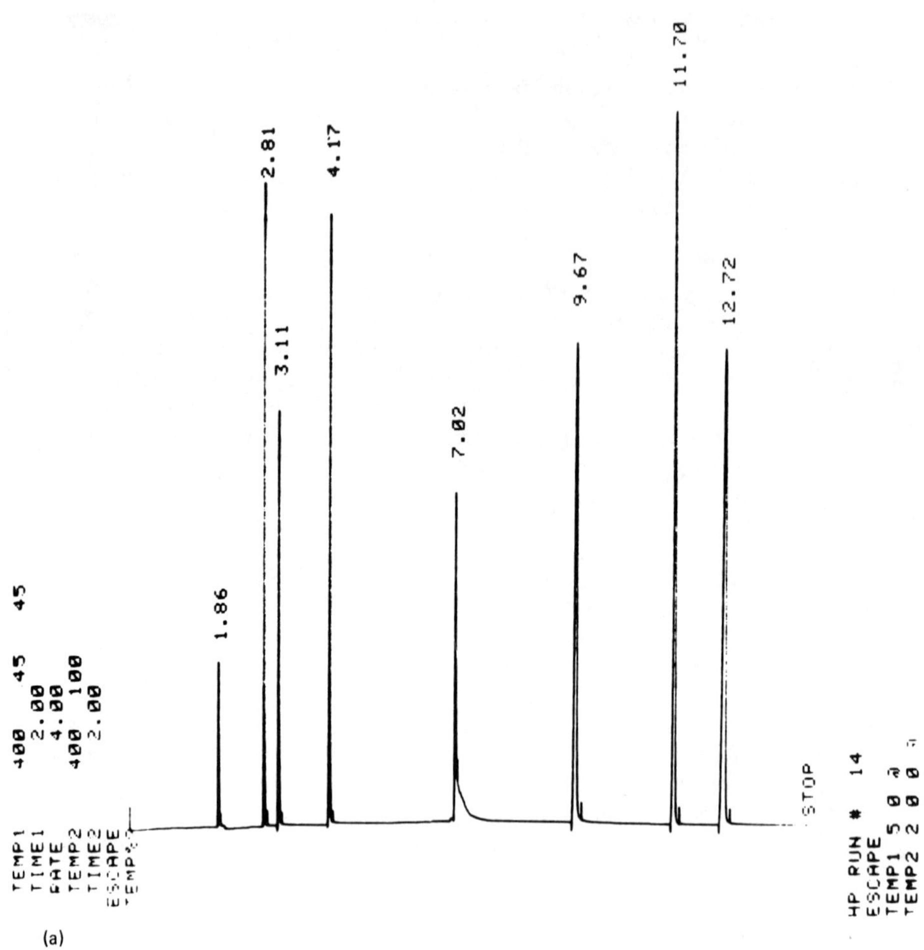


Fig. 4 Test mixture on OV-1 fused silica capillary column (a) and GTCB micropacked column (b). Column temperature, see figure.

Retention time (min)

| OV-1 | GTCB | |
|---|---|---|
| 2.81 | 1.50 | cyclohexane |
| 3.11 | 1.30 | dioxane |
| 4.17 | 3.63 | toluene |
| 7.02 | 2.23 | cyclohexanol |
| 9.67 | 4.35 | aniline |
| 11.70 | 14.87 | *n*-decane |
| 12.72 | 10.15 | acetophenon |

(From Pörschmann, Engewald, unpublished results.)

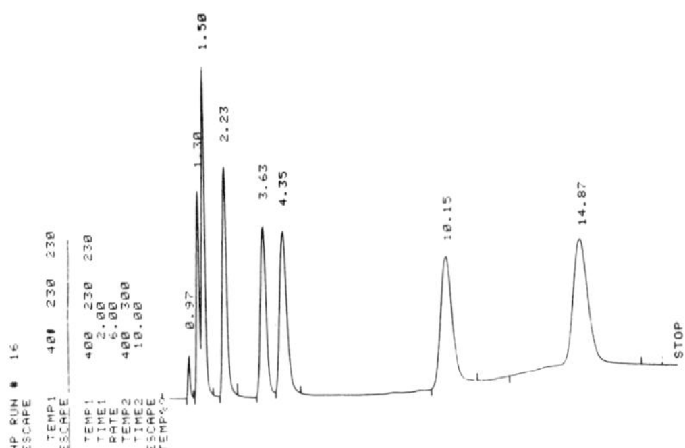


Fig. 4 (b)

geneities provoke an initial pronounced "knee-bend" in the adsorption isotherm to yield asymmetrical peaks, loss of adsorbates, ghosting phenomena, etc. As suggested by Di Corcia and Bruner [85], a hydrogen treatment of the adsorbent at 1100°C results in removal of polar functional groups, leading to a greater homogeneity. It has thus become possible to extend the availability of GTCB to the analysis of polar compounds (cf. references mentioned above and also see below).

However, there is evidence from other studies that the contaminating sulfur and oxygen complexes could not be completely removed. Di Corcia et al. [75,96] extracted hydrogen-treated GTCB (abbreviated GTCB HT) with water and measured a pH value of 10.5 of the water extract. Hence, GTCB with basic properties obviously creates difficulties in the analysis of acidic compounds. It is important to note that an addition of an acidic deactivating agent such as phosphoric acid is only partly effective because of the deterioration of the column due to the neutralization of phosphoric acid by basic surface groups and because of the reduced temperature stability (see below). After washing GTCB with an aqueous solution of phosphoric acid or a low-boiling acid such as acetic acid, the undesirable contaminants were removed completely resulting in the tailing-free elution of low free fatty acids and phenols even at a ppm level [86,96].

Recent studies by Petsev et al. [97] showed that the chromatographic properties of thermal carbon black can be improved by treatment with a high-frequency plasma in the presence of benzene vapor, which leads to a real reorganization of the adsorbent structure. The procedure yields uniform and nonspecific surfaces which approximate those of GTCB.

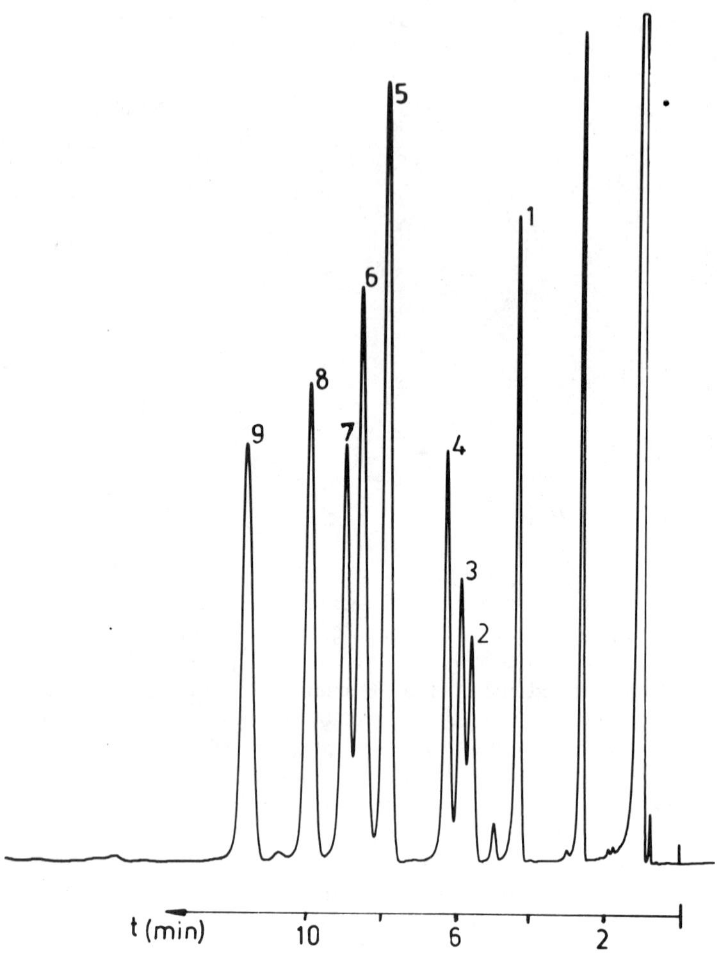


Fig. 5 Separation of isomeric dimethyl cyclohexanes (DMCH) on GTCB. Column: 2.5 m × 0.45 mm i.d.; dp = 0.09-0.12 mm; column temperature; 140°C; inlet pressure, 0.6 MPa. Peaks: 1, 1.1-DMCH; 2, 1r.4c-DMCH; 3, 1r.2c-DMCH; 4, 1r.3t-DMCH; 5, $nC_7$; 6, ethyl cyclohexane; 7, 1r.2t-DMCH; 8, 1r-3c-DMCH; 9, 1r.4t-DMCH. (From Ref. 72 with permission.)

Table 3 Properties and Use of Graphitized Thermal Carbon Black (GTCB)

| Trade name | Supplier | $a_s$ ($m^2$ $g^{-1}$) | Preferred use |
|---|---|---|---|
| Carbopack B HT[a] | Supelco Inc. (Bellefonte, PA, USA) | 100 | Gas analysis, low-boiling substances |
| Spheron-6 | Phase Separations Ltd. (Queensferry, UK) | 87 | Adsorbents to trap organics from water or air |
| Carbopack C HT | Supelco | 9 | Efficient separation of structure and spatial isomers (trapping method) |
| Sterling MT | Cabot Corp. (Billerica, MA, USA) | 9,6 | |
| Sterling FT | Cabot Corp. | 11,5 | |

[a]HT = hydrogen-treated.

A novel approach to prepare activated carbon similar to Carbopack B was taken by Gilbert et al. [98]. The so-called porous glassy carbon was produced by impregnation of silica or porous glass with a phenol-formaldehyde resin mixture. After polymerization within the pores of the template material, the polymer was carbonized by heating in an inert atmosphere to about 1000°C. The silica template was then removed by an alkaline treatment and the remaining material calcined at about 2500°C to form glassy carbon. This adsorbent can also be applied to HPLC owing to its good mechanical strength. Further carbon-surface adsorbents tailored for HPLC use (and having all the advantages of GTCB) are the so-called pyrocarbons produced by depositing thin pyrocarbon layers on either carbon black or silicas. The use of these adsorbents in GC, which were produced by pyrolysis of alcohols, was detailed by Leboda (see below).

*Carbon Molecular Sieves*

Based on the work of Gvozdovich et al. [99], Kaiser [100,101] used synthetic polymers such as polyvinylidene chloride as precursors of porous carbon. By means of a controlled pyrolytic degradation of the starting polymer HCl was removed, leaving a fine-pore, inert, geometrically and chemically homogeneous carbon adsorbent of the first type according to Kiselev's classification. The resulting adsorbent has a mean pore radius of about 1.5 nm and a specific surface area of about 500-1200 $m^2 g^{-1}$, both depending on the temperature of carbonization [102]. These adsorbents are called carbon molecular sieves (CMS) because of their pore sizes in the molecular range.

According to the exceedingly high adsorption energies appearing on CMS, the permanent gases, light hydrocarbons, and low molecular weight organic compounds with hetero atoms are the preferred substances for analysis [103,104]. The nature of CMS is hydrophobic. Thus, water is eluted ahead of methane with a sharp, symmetrical peak, and hence CMS are extremely useful in the trace analysis of water. The water content in solid matrices, e.g., coal, can also be determined by means of the reaction of calcium carbide with water to form acetylene, the latter being analyzed on CMS [105]. The extremely low polarity of CMS is expressed by retention indices of Rohrschneider substances (cf. Table 4).

Similar to the retention behavior of GTCB, the adsorbates are eluted from CMS according to their degree of saturation: The higher the saturation, the higher is the retention. This behavior proves to be useful in determining traces of unsaturated substances in more saturated ones, e.g., the trace analysis of methane and acetylene in ethylene. When using CMS, a carrier gas that is free of oxygen should be employed in order to avoid surface oxidation. In addition, CMS should not be exposed to air for prolonged periods of time because of their ability to adsorb contaminants. Therefore, the producers–among them Analabs (Spherocarb), Supelco (Carbosieve S-II, Carbosieve G, the latter earlier called Carbosieve B with a specific surface area of

**Table 4** Retention Indices of Rohrschneider Substances on Carbon Molecular Sieve (CMS) and Some Porous Polymers

| | Retention indices on | | | |
|---|---|---|---|---|
| Substance | CMS (320°C) | Tenax GC (175°C) | Chromosorb 101 (200°C) | Squalane (100°C) |
| Benzene | 546 | 645 | 745 | 649 |
| Ethanol | 300 | 395 | 495 | 384 |
| Methylethylketone | 469 | 560 | 645 | 531 |
| Nitromethane | 263 | 540 | 650 | 457 |
| Pyridine | 544 | 740 | 850 | 695 |

*Source*: Data from [101,106] with permission.

only 100 $m^2$ $g^{-1}$), Chrompack (Carbosphere, Carbosieves)–supply CMS in sealed ampules.

A literature search has shown that further attention is focused on developing novel CMS. Activated carbon prepared by the reduction of polytetrafluoroethylene by lithium amalgam has been introduced as an adsorbent in GSC [107]. Using the starting polymer polyacrylonitrile, "activated carbon fibres" with acceptable chromatographic properties are obtained after several pyrolytic processing steps.

## Porous Polymers

One of the most creative papers in the field of stationary phases was published by Hollis in 1966 [109], who introduced porous polymers into GC. These adsorbents are prepared by copolymerization of a monomer with a vinyl group such as styrene (STY) or ethylvinylbenzene (EVB) and a crosslinking agent (about 8-12%), the most frequently used of which is *p*-divinylbenzene (DVB), in the presence of a diluent (cf. Fig. 6).

By using a polar starting monomer such as acrylonitrile (ACN), vinylpyrrolidone, acrylic esters, etc., or by changing the conditions of polymerization, it becomes possible to obtain adsorbents modified

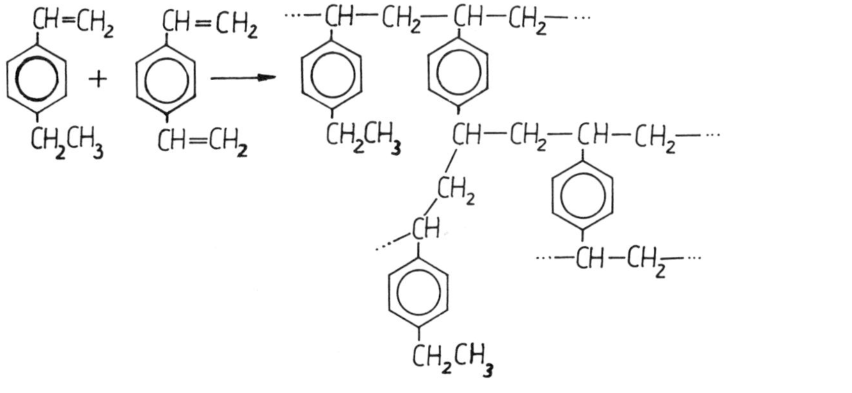

Fig. 6 Preparation of porous polymers (scheme).

with respect to polarity and specific surface area. These polar monomers are capable of facilitating desirable interactions between the substances being separated and the stationary phase. The higher the amount of the polar monomer used, the higher is the retention of polar substances.

It is self-evident that the porosity can be controlled by the selection of appropriate crosslinking agents. An increase in the proportion of the latter causes a decrease in specific surface area. Beyond this, different kinds and amounts of solvents also influence the crosslinked skeleton [102]. Among the various kinds of polymerization, suspension polymerization yielding spherical beads is preferred. Spherical beads, in turn, lead to reduced A terms in the van Deemter Eq. (6) via the superior column-packing quality compared to angular particles.

To sum up, porous polymers offer several advantages such as the possibility of regulating both geometrical and chemical structure (see above), high thermal stability, adequately homogeneous surface, spherical bead form, ruggedness to withstand packing into the column, and sharp elution of polar components such as water ([110,111], and references cited therein). Therefore, the application of porous polymers covers a wide range of compounds with different polarities, the separation of which is difficult to achieve with conventional column packings. Moreover, porous polymers are increasingly employed in analyzing gas mixtures, thus replacing the "classical" packings such as silica, zeolites, and alumina. Further possibilities of improving resolution consists in mixing porous polymers of different polarity or in coupling different Porapak columns. The optimization of mixed columns, e.g., by window diagrams, will be discussed in more detail in a later section.

The extraordinary separation power of porous polymers has prompted many companies to develop these adsorbents, the most important being the Chromosorb Century Series (Manville, USA) consisting of Chromosorb 101-108, and the Porapak Series (Millipore, USA). These packings and some of their most important properties are listed in Table 5. Recently, Chrompack (The Netherlands) offered the so-called Haye Sep porous polymers which are claimed to have comparable or better performance than the Porapak equivalents [113]. In addition to the Chromosorb and Porapak series, several other polymers have been marketed which have gained local importance, among them Cekachrom 1-6 in the GDR [114], polyfunctional polymer sorbent Polysorb in the Soviet Union [115,116], GDX 1-5 in China [102], polar Spherons in Czechoslovakia [117-119], and Polichrom A in Poland [120], the last derived from dimethacryloyloxymethyl naphthalenes. Further selective porous polymers, e.g., with imidazole or pyrazole and others as well as their derivatives, are conceivable.

Among the commercially available porous polymers the 2,6-diphenyl-*p*-phenyleneoxide shows an outstanding maximum operating temperature up to 375°C [121]. This adsorbent, designated as Tenax GC, with a specific surface area of about 20 $m^2$ $g^{-1}$, is marketed by Enka N.V. (The Netherlands) and distributed by Applied Science Lab. (USA). Tenax is capable of weakly specific interactions with appropriate solutes [122]. Owing to its low adsorption activity, Tenax GC is well suited to separate high-boiling polar compounds such as alcohols, diols, phenols, mono- and diamines, ketones, and aldehydes without undergoing derivatization [106,123,124]. Tenax has been successfully applied to aqueous solutions of polyamines, $C_1$-$C_4$ alcohols, $C_2$-$C_6$ monocarboxylic acids, and phenols [125]. A further field of application is the adsorption of pollutants in environmental matrices (see below). One drawback of Tenax is related to its relatively low column efficiency according to its low geometrical homogeneity. Besides this, Tenax GC is mainly available with 60/80 and 35/60 mesh sizes, resulting in a small pressure drop over the column and low efficiency. Conditioning of Tenax GC is detailed in some manufacturers catalogues (cf. Phase Separation catalogue 1985).

The fields of application of porous polymers and the temperature range available can be extended by chemical modifications of the adsorbents such as bromination, acylation, and nitration [126,127]. Further reactions such as amination, sulfation, or phosphorylation are conceivable. These modified adsorbents are more suitable for the separation of polar compounds. Recently, this approach of adsorbent modification was rediscovered by Sakodynski et al. [128]. As shown in Table 5, the analyst may choose among a variety of tailor-made columns providing a unique functionality and porosity for a given separation problem. In manufacturers catalogues silanized products, e.g., Porapak P-S and Porapak Q-S, are often listed. The silanization serves for removing residual phenolic hydroxyl groups and undesirable active centers as well, which leads to improved peak shape for polar compounds.

The interactions between adsorbates and porous polymers have been studied intensively. Gvozdovich [129] classified porous polymers carrying no functional groups as weakly specific adsorbents of the third type according to Kiselev's classification, whereas packings containing functional groups are assigned to strongly specific ones. Therefore, molecules of group D are retained on these porous polymers longer than the corresponding molecules of groups A and B. Further, the retention on porous polymers has been interpreted with respect to electron polarizability, dipole moment, and carbon number of adsorbate molecules. The discussed retention mechanisms include both adsorption and solution phenomena [112]. The higher the temperature, the higher the share of solution phenomena due to the increased mobility of the individual chains of the adsorbent.

The selectivity of porous polymers was assessed by the method of Rohrschneider and McReynolds (see below). The popular reference squalane was generally replaced by more suitable references

Table 5 Properties and Use of Porous Polymers

| Phase | Polymer | Maximum working temperature (°C) | Specific surface area $a_s$ ($m^2\ g^{-1}$) | Pore diameter pd (nm) | Retention indices at 200°C | | | | $\Sigma I$ | Use |
|---|---|---|---|---|---|---|---|---|---|---|
| | | | | | $I_{Benzene}$ | $I_{t\text{-}Butanol}$ | $I_{2\text{-}Butanone}$ | $I_{Acetonitrile}$ | | |
| Chromosorb 101 | Sty-DVB | 275 | 30-40 | 300-400 | 745 | 565 | 645 | 580 | 2535 | Alkanes, ethers, glycoles, free fatty acids, alcohols, ketones |
| Chromosorb 102 | Sty-DVB | 250 | 300-400 | 8.5 | 650 | 525 | 570 | 450 | 2205 | Permanent gases, polar low boilers, adsorbent to trap organics from water or air |
| Chromosorb 103 | Crosslinked poly-styrene | 275 | 15-25 | 300-400 | 720 | 575 | 640 | 565 | 2500 | Amines, amides, hydrazides, (glycoles are adsorbed) |
| Chromosorb 104 | ACN-DVB | 250 | 100-200 | 60-80 | 845 | 735 | 860 | 885 | 3325 | Nitroparaffins, nitriles, traces of $H_2O$, $CO_2$, $NH_3$, etc. |
| Chromosorb 105 | Polyaromatic | 250 | 600-700 | 40-60 | 635 | 545 | 580 | 480 | 2240 | Light hydrocarbons, permanent gases, formaldehyde in water |
| Chromosorb 106 | Crosslinked poly-styrene | 250 | 700-800 | 5 | 605 | 505 | 540 | 405 | 2055 | As above, low alcohols and fatty acids |
| Chromosorb 107 | Crosslinked acrylic ester | 250 | 400-500 | 8-9 | 660 | 620 | 650 | 550 | 2480 | Efficient separation of various classes of compounds in general and of formaldehyde, sulfur gases in particular |

| | | | | | | | | | | |
|---|---|---|---|---|---|---|---|---|---|---|
| Chromosorb 108 | Crosslinked acrylic | 250 | 100-200 | 23-25 | 710 | 645 | 675 | 605 | 2635 | Gases, alcohols, aldehydes, ketones, glycoles, etc. |
| Porapak N | Vinylpyrol-lidone | 250 | 250-350 | 7-9 | 735 | 605 | 705 | 595 | 2640 | HCHO and $NH_3$ in aqueous solutions |
| Porapak P | Sty-DVB | 250 | 100-200 | 9-10 | 765 | 560 | 650 | 590 | 2565 | Separation of wide variety of carbonyl compounds |
| Porapak Q | EVB-DVB | 250 | 500-600 | 7.5-10 | 630 | 538 | 580 | 450 | 2448 | Most widely used, use as above, organics in water, light hydrocarbons |
| Porapak R | Sty-DVB (vinylpyrol-lidone) | 250 | 450-600 | 7-10 | 645 | 545 | 580 | 455 | 2225 | Ethers, esters, nitro compounds, separation of $H_2O$ from HCl |
| Porapak S | Sty-DVB (vinyl-pyridine) | 250 | 300-450 | 7-9 | 645 | 550 | 575 | 465 | 2235 | Carbonyl compounds, alcohols |
| Porapak T | Ethylene glycol dimetha-crylate | | 250-350 | 7-9 | 675 | 675 | 700 | 635 | 2685 | Highest polarity and greatest water retention, formaldehyde in aqueous solutions |

*Abbreviations*: Sty, styrene; DVB, divinylbenzene; ACN, acrylonitrile; EVB, ethylvinylbenzene.
*Source*: Manufacturer catalogues and [112], with permission.

such as GTCB and the so-called Kovats hydrocarbon $C_{87}H_{176}$ (Apolane) (see below) [126,130,131] due to drawbacks of squalane such as negative $\Delta I$ values, fast elution of low-boiling hydrocarbons, scatter of retention data with different supports, and high volatility preventing its use above 140°C.

Castello and D'Amato [130,131] estimated the order of polarity of porous polymers by summing up the $\Delta I$ values of butanol, methylpropylketone, benzene, and pyridine related to the least polar Chromosorb 106: Chromosorb 106 $<$ Porapak Q $<$ Chromosorb 102 $<$ Chromosorb 105, Porapak R $<$ Porapak N $<$ Chromosorb 101 $<$ Porapak P $<$ Chromosorb 103 $<$ Porapak T $<$ Chromosorb 104. In comparison with liquid phases, Porapak T and Chromosorb 104 exhibit an average polarity close to that of trifluoropropylsilicones (OV-210, QF-1), while the polarity of Chromosorb 105 and Porapak Q almost equals that of methylsilicones. Porapak T exhibits strong interactions with the oxygen atom in ethers, alcohols, and ketones [132]. Some effort has yet to be made to compare in terms of overall polarity available liquid phases with available porous polymers. On this basis, a porous polymer with very low bleeding properties may then replace a corresponding liquid packing in cases where the use of highly sensitive detectors at high temperatures requires low bleeding.

It may be seen from Table 5 that the order of polarity according to the sum of retention indices of the four compounds listed in Table 5 and the order of polarities determined by Castello and D'Amato disagrees in parts. This is attributed to two main reasons: Firstly, many workers (cf. [131]) have scrutinized the influence of rate, sample size, previous aging, etc., on retention behavior. In the authors' opinion, differences in retention behavior on such packings originate largely from differences in the employed pretreatment procedure. Aging, which must be carried out in a stream of oxygen-free gas, is required for the removal of low molecular monomers and impurities. Hollis [133] described an expensive cleaning procedure for porous polymer beads involving extraction with various solvents followed by vacuum drying at elevated temperatures. Since the aging procedure involves a certain shrinkage of the porous polymers, it should be carried out separately and not in a chromatographic column. Second, in addition to differences in the separation properties caused by different aging procedures, an unsatisfactory batch-to-batch reproducibility was observed [134] that could be ascribed partly to the presence of unreacted vinyl groups in the polymers.

### Silica (cf. Chp. 6)

On account of the presence of acidic hydroxyl groups on the surface, silica is classified as an adsorbent of the second type to enter into hydrogen bonding with molecules of the B and D groups. Among silanol groups one distinguishes between isolated silanol groups and those that are hydrogen-bonded to adjacent ones. The concentration of silanol groups may be evaluated by means of the deuterium exchange method or infrared spectroscopy (cf. [49]), the latter method also being capable of differentiating between the two different silanol group types. Molecules possessing spherical electron shells or only σ bonds are only adsorbed by nonspecific dispersive forces.

In accordance with the varying water content of silica (between 2 and 10% w/w), its universal formula is written as $(SiO_2)_n \cdot xH_2O$. The total water content of porous silica originates from bulk water through the condensation of internal hydroxyl groups, chemisorbed water through the condensation of surface hydroxyl groups and physisorbed water [49]. The latter can be removed by thermal treatment at 100-200°C under vacuum, leading to an increased surface activity. More elevated temperatures cause a dehydroxylation of hydroxyl groups into siloxane groups, the latter being incapable of strong specific interactions. Thus, the concentration of hydroxyl groups on the surface and within the bulk of the particles depends on the nature and the crystalline structure of the silica and on its pretreatment [46,135,136]. By hydrothermal treatment a widening of the pores accompanied by a reduction in total surface area is achieved, which results in a reduced nonlinearity of the adsorption isotherm [137].

A further reason for the heterogeneity of silica is given by Lewis acid impurities such as aluminum and boron, which cause an increase in the energy of bonds formed with electron donor molecules. The influence of such centers on the increase of interaction energy becomes more important as a result of dehydroxylation when these centers are being exposed [45]. Consequently, strong organic bases such as amines are chemisorbed on the silica surface. However, pure silica does not behave like a chemisorbent toward molecules of the B and D groups, but merely like a specific molecular adsorbent of the second type.

Hence, there is a need for preparing silica of higher purity than technical silica or for blocking these active centers by modification. Such an adsorbent with higher purity has been prepared by Bebris et al. [138] (Silochrom C 80; Stavropol Plant for Chemicals, Stavropol I, USSR) and by Guillemin et al. [139]. The product of the latter authors is called porous silica beads, marketed under the trade name of Porasil or Spherosil and available in a wide range of surface areas and particle diameters (see Table 6). Porasil beads are very rigid and incompressible despite their moderate to high porosity. Owing to their graduated porosity, the applicability extends from gas analysis (pore size not over 2 nm) to the analysis of relatively high-boiling hydrocarbons (pore size larger than 20 nm). In analyzing polar compounds it is advantageous both to add slight amounts of a selected modifier and to perform a hydrothermal treatment [137] in order to avoid tailed peaks and chemisorption. The amount of the modifier re-

Table 6 Properties and Use of Commercial Silicas

| Trade name | | | Specific surface area $a_s$ ($m^2$ $g^{-1}$) | Mean pore diameter pd (nm) | Preferred use |
|---|---|---|---|---|---|
| Porasil[a] | A Spherosil[b] | XOA 400 | 350-500 | 8-10 | Gas analysis |
| | B | XOA 200 | 125-200 | 10-20 | |
| | C | XOA 075 | 50-100 | 20-40 | Organic compounds belonging to A and B groups with medium boiling points |
| | D | XOB 030 | 25-45 | 40-80 | |
| | E | XOB 015 | 6-30 | 80-150 | Compounds with higher boiling points |
| | F | | 2-6 | 150 | |
| | | XOC 005 | 5-15 | 300 | |

[a]Millipore, Waters Chromatography, Milford, MA, USA.
[b]Pechiney, St. Gobain, France.
*Source*: Manufacturers catalogues.

quired to achieve symmetrical peaks depends on the type of Porasil and the components to be separated. As already mentioned, the lower the specific surface area, the less modifier is needed. Analogously, Porasil can be used as support in GLC as well.

Usually, modification of silica adsorbents is performed by chemical reactions and by coating with organic or inorganic compounds. The two former methods are described in a later section. The commonly used inorganic modifiers to provide symmetrical peaks are salts such as LiBr, $Na_3PO_4$, $Al_2(SO_4)_3$, $Na_2MoO_4$, [140,141], CsCl, CKl, LiCl [142], or lanthanum chloride [143]. The salt-modified silica is capable of undergoing specific interactions with the analytes, e.g., charge-transfer. Varying anions or cations permits a setting of different selectivities [144].

In order to obtain adsorbents with surface properties like those of GTCB and a good mechanical resistance like silica, Leboda [145-148] prepared carbon-silica adsorbents by means of pyrolysis of several aliphatic and aromatic alcohols on the surface of silica. The properties of the so-called Carbosils are governed by the conditions of pyrolysis and the kind of alcohol used. Rayss [149] investigated the properties of an octadecanol film on a carbon-silica adsorbent prepared by thermal decomposition of $CH_2Cl_2$ at 450°C.

Special attention has been focused on preparing capillary columns with silica. Pioneering work on this subject was done by Halasz and Gerlach [150] as well as by Mohnke and Saffert [151], the latter using an etching procedure for the internal surface of the glass capillary by means of hydroxide solutions. A silica layer of about 20 μm thickness (depending on the strength of the alkali treatment) behaves as an absorbent for the separation of hydrogen isotopes at low temperatures (77.6 K) (see Fig. 7). Incidentally, the ideas of Mohnke and Saffert stimulated the development of capillary GC [152].

Mathews et al. [153-155] prepared glass capillaries by means of surface-modified porous silica and used it for hydrocarbon analysis. In addition, Ober et al. [156] described the preparation of GSC columns containing a porous silica layer produced by the procedure of Mohnke and Saffert, which was modified by adsorption of rubidium chloride. Such columns are characterized by an outstanding thermal stability to allow the separation of polynuclear hydrocarbons. In this way, the concept of salt-modified supports successfully applied with packed columns [157] was extended to capillary columns.

A further material of silicic origin used in GSC is provided by porous glass. Contrary to silica, the porous glasses exhibit a narrower pore size distribution. Their preparation and utilization in chroma-

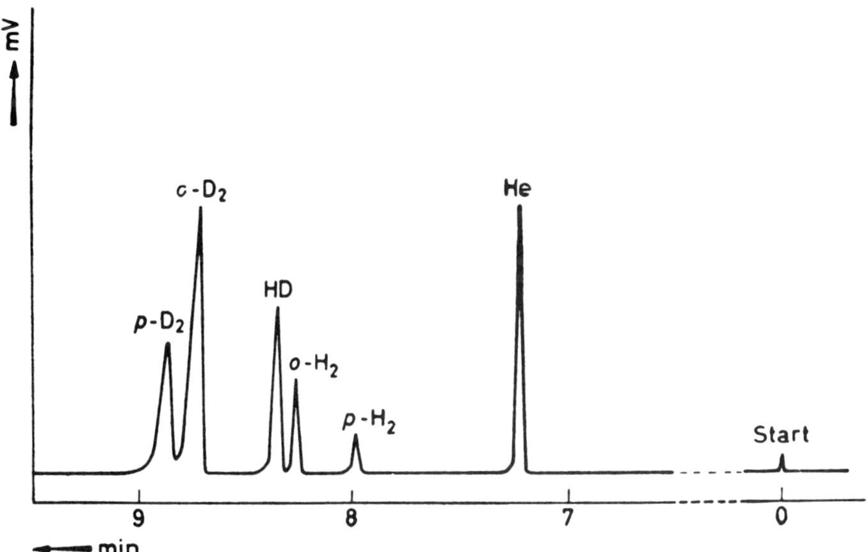


Fig. 7 Separation of hydrogen isotopes and their nuclear spin isomers: column length = 80 m; temperature, 77.6 K; carrier flow, 2 ml Ne/min. (From Ref. 151, with permission.)

tography has been comprehensively reviewed by Janowski and Heyer [158,159]. The separation of mixtures consisting of CO, $H_2$, $CH_4$, $C_2H_4$, $C_2H_6$, etc., as well as hydrocarbons of up to 8-10 carbon atoms, is successfully achieved by using porous glasses possessing a pore size suited to the respective separation problem. Generally speaking, porous glasses with narrow pores in the range of 1 nm are applicable as molecular sieves. According to their narrow pore size distribution, porous glasses such as the commercially available Corning controlled-pore glass packings (Corning Glass Work, USA) are used in size exclusion chromatography of proteins, enzymes, nucleic acids, and synthetic polymers as well [160,161].

### Alumina

In the early period of GC the use of alumina with a highly specific surface area was relatively widespread for the analysis of gases and low-boiling organic compounds. Today its place is mostly taken by other stationary phases because the chromatographic peaks resulting on alumina adsorbents are often asymmetrical. The cause for this is twofold: First, the surface of alumina adsorbents is heteropolar due to Lewis acid (electron accepting) sites, oxide ions, and hydroxyl groups, the most active of which are the first [162]. These Lewis acid centers are able to adsorb polar molecules. For example, ketones form stable complexes via the oxygen atom of the carbonyl group. Second, in addition to the chemical inhomogeneity the distribution of pores is very inhomogeneous.

The asymmetrical peaks can be eliminated by deactivation with water, inorganic salts, or a thin film of a high-boiling liquid [163,164]. The selectivity of alumina depends very much on the amount of water adsorbed, as is the case with silicas or zeolites. Thus, water-free carrier gas is necessary for highly reproducible analysis. Modification of the surface properties of alumina by coating with an inorganic salt is advantageous because salt-modified adsorbents show high thermal stability and bring about selective separations, e.g., of cis and trans isomers. Moriguchi et al. (cf. [165]) prepared salt-modified alumina by treatment with potassium fluoride-hydrofluoric acid solution or by coating with alkali metal fluoride and ammonium fluoride. A surface layer was found to form on the alumina surface by reaction with the alkali metal fluoride and the alumina. After converting the original surface into a uniform and less adsorptive one, sharp and symmetrical peaks were obtained. Naito et al. [166-168] extended these investigations using dipotassium hydrogen phosphate and potassium phosphate. The chromatographic behavior of these modified adsorbents has proved to be dependent on the nature of the modifying agent, the salt loading, and the temperature of preheating and postheating.

Separations of hydrocarbons are the preferred field of application for alumina. The saturated hydrocarbons are eluted ahead of the olefins with the same carbon number due to the interaction of electron-rich unsaturated bonds of the olefins with the electropositive sites on the alumina lattice. On the other hand, the strong interaction with all polar analytes such as alcohols or ketones may lead to their complete adsorption. Then these compounds are not eluted from the column even at high temperatures approaching 250°C. This phenomenon results in altered retention behavior of the adsorbent.

Highly efficient PLOT columns for GSC of $C_1$-$C_{10}$ hydrocarbons are described in [169,170]. According to these authors, a wetting of the carrier gas is superfluous. Besides this, such columns are presently being developed by Chrompack.

## Zeolites

Natural and synthetic zeolites are crystalline metal-alumosilicates with the composition $M_x^+(AlO_2)_x^-(SiO_2)_y zH_2O$. Their structures consist of a three-dimensional framework of $AlO_4$ and $SiO_4$ tetrahedra capable of exchanging alkali for calcium, magnesium, etc. Zeolites show a large inner surface of the order of 700-800 $m^2\ g^{-1}$, which is heteropolar.

The vast potential for generating synthetic zeolites is due to the multitude of possibilities for exchanging cations and for changing the Si/Al ratio, as well as to the differences in the three-dimensional framework. In the Linde Research Laboratory, a variety of different species were prepared, of which those mostly used in GC are designated as type A and type X zeolites. The description of the structure of zeolites is outside the scope of this book, and the reader is referred to the literature [171]. The pore openings are of molecular dimensions, hence their name zeolitic molecular sieves. Thus the principal types, zeolite 3A (KA), 4A (NaA), 5A (CaNaA), 13X (NaX), and 10X (CaNaX), differ in the pore openings.

The chromatographic behavior of adsorbates on zeolites is based on various phenomena:

1. The steric or geometrical effect. Only molecules with a critical molecular diameter equal to or smaller than the pore opening of the molecular sieve are capable of entering the cavities. In this way, the separation of straight-chain n-alkanes with a critical molecular diameter of about 0.49 nm from branched-chain iso-alkanes and cyclo-alkanes becomes possible.
2. The thermodynamic or energetic effect. Since positive charges are located at the surface (second type adsorbent according to Kiselev), molecules with high electron density are preferentially adsorbed. Based on the energetic effect, ethane is eluted earlier than ethylene. Specific interactions also occur with

molecules having permanent quadrupole moments such as $N_2$, CO, and $CO_2$. In summary, the selectivity of zeolites can be changed by charge, size, and position of the cation in the framework [172-174].

3. The kinetic effect. The separation can be accomplished by different velocities of the adsorbates, e.g., the separation of *cis*-butene-2 from the other butenes on zeolite 5A.

The entire selectivity of zeolites is governed by all of the above effects. Because of high adsorption activity, GC employment of zeolites is preferred for the analysis of low-boiling substances. Typical applications are the separation of hydrogen isotopes and of gaseous mixtures containing $H_2$, $O_2$, $N_2$, He, Ne, Ar, and low hydrocarbons such as $CH_4$, $C_2H_2$, $C_2H_4$, $C_2H_6$ up to benzene and thiophene (cf. references in [94,173,175]). The separation of mixtures of organic compounds may create difficulties due to strong adsorption onto zeolites or the catalytic transformations on these adsorbents.

A further field of application is the selective removal of one component or a group of components from the test mixture (some sort of substraction method). Generally, the subtraction method is known to be a rapid method for functional group assessment and is therefore a useful aid in analyzing complex mixtures. Like an agent, which selectively retains components by formation of nonvolatile derivatives, the zeolites only adsorb and retain components which can pass through the molecular sieve openings due to their molecular shape and size. It was shown (cf. [176]) that *n*-parafins $C_3$-$C_{11}$ were adsorbed quantitatively on a column with molecular sieve 5A, whereas aromatic hydrocarbons, napththenes, and branched alkanes pass through the column unaffected. Molecular sieves are also used in the subtraction method for analyzing relatively high-boiling compounds (cf. [176]). Thus, rapid methods by means of molecular sieve 5A were developed for the determination of *n*-alkanes in kerosenes and gas oil [177].

Recently, De Zeeuw and De Nijs of the Chrompack Company and Mohnke jointly described fused silica PLOT columns coated with molecular sieve, illustrating the possibility of a fast and efficient separation of permanent gases as shown in Fig. 8 (cf. [178]). With the introduction of a capillary column coated with molecular sieve, all the advantages of capillary columns have been combined with molecular sieve adsorption chromatography.

On the other hand, Andronikashvili et al. [172,179] (based on the guiding ideas of Kaiser [180]) prepared highly efficient columns as follows: A fine zeolite powder is shaken with a large-pore solid support such as Chromosorb W. The adsorbent enters the macropores of the support and is retained there by adhesive forces. The surface layer sorbents make it possible to increase the column efficiency without increasing the resistance to the carrier gas flow. However, the

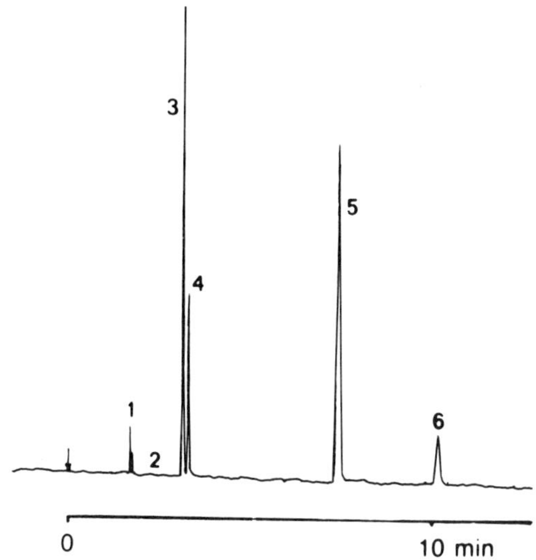


Fig. 8 Separation of permanent gases on Molsieve 5 Å (30 μm); column, 25 m × 0.32 mm fused silica PLOT; column temperature, 30°C; carrier gas, $H_2$, 60 kPa; detector, thermal conductivity detector; sample size, 30 μl. Peaks: 1, He; 2, Ne; 3, Ar; 4, $O_2$; 5, $N_2$; 6, $CH_4$. (From Ref. 178, with permission.)

poor capacity of such packings will presumably prevent a wide acceptance in GC.

In the experimental use of zeolites, attention should be focused on dry carrier gases because of their ability to adsorb water. Regeneration is simply achieved by heating the column for a few hours at about 300°C under a stream of dry carrier gas. Inert, poorly adsorbed gases such as He, $H_2$, and $N_2$ are common carrier gases in zeolite applications. However, the use of an "active" carrier gas such as carbon dioxide, which occupies some of the active centers of the adsorbent, may improve the separation ability of zeolites without completely suppressing the effect of the cation nature on the separation of mixtures, thereby widening the range of use of zeolites in GC. Besides the use of an active carrier gas and the packing of columns in the surface layer mode, a further possibility for expanding the applicability of zeolites consists of reducing the cation density, which diminishes the strong adsorption of certain analytes.

### Clathrates

Like zeolites, clathrates are able to distinguish between molecular structures. In separating isomers, the most highly branched structure is eluted first and the *n* isomer last. The first clathrate used in GSC was thiourea [181], followed by potassium, rubidium, and sodium benzenesulfonate and some so-called Werner complexes (cf. [182]). The retention on the three salts is governed by clathration, hydrogen bonding, and interactions with the lone pairs of electrons on the metal ion.

Today cyclodextrins have found wide application as clathrates. Cyclodextrins are torus-shaped oligosaccharides composed of six, seven, and eight D(+)-glycopyranose units with α-1,4 linkages. With cavities of 0.5-0.8 nm, cyclodextrins are capable of including molecules with the appropriate dimensions. The better the guest molecules fit into the cavity, the more stable is the inclusion complex (host-guest chromatography). Thus, cyclodextrins exhibit many interesting features such as stereospecificity (e.g., the separation of xylene positional isomers) and enantiomeric specificity (cf. [183,184]). Besides chromatographic separations, GC with cyclodextrins may also serve to investigate inclusion phenomena.

In analyzing polar adsorbates, the retention is controlled to a high degree by interactions with the cyclodextrin hydroxyl groups. On decreasing the cyclodextrin polarity by means of methylation of hydroxyl groups, stereoslectivity is evident even in the separation of alcohols [185]. By polymerization of cyclodextrins with several diisocyanates, cyclodextrin polyurethane resins were obtained with an upper temperature limit of about 230°C.

A key advantage of cyclodextrins over zeolites is the capability of forming inclusion compounds in the solid state as well as in solution. The use of cyclodextrins in high-efficiency capillary GC has been neglected up to now, a situation that will soon change.

In addition to their employment in GC, cyclodextrins and chemically modified versions of them are of increasing importance in HPLC.

### Salts and Saltlike Adsorbents

Of these compounds boron nitride, molybdenum and tungsten sulfide, and barium sulfate have gained moderate acceptance as stationary phases in GC. The crystalline structure of BN, $MoS_2$, and $WS_2$ is of the lamellar type and resembles that of graphite. In the case of the nonpolar, nonspecific $MoS_2$, for instance, the lamellar structure is made up of S-Mo-S units. With BN, there are alternate boron and nitrogen atoms instead of carbon atoms in the hexagonal basal plane. If the lattices are free of defects (e.g., crystal imperfections, chemical admixtures), the basal faces are homogeneous and nonpolar. Therefore, adsorption on the three adsorbents mentioned is governed mainly by ubiquitous nonpolar dispersive forces [45,186,187]. On account of the two latter aspects–lamellar structure and predominance of nonspecific forces–the retention order is largely dependent on the geometry of the molecules to be analyzed, in conformity with the elution of GTCB. According to the lower adsorption activity of $MoS_2$ to GTCB, the temperature of analysis will be lower when $MoS_2$ is used.

Barium sulfate as a type II-specific adsorbent with a low specific surface area of approximately 2.5 $m^2$ $g^{-1}$ is capable of undergoing strong specific interactions with aromatic hydrocarbons. For example, the isomeric xylenes can be separated in the sequence para < meta < ortho [188]. This retention order agrees with the increase in asymmetry of the electron density distribution in the xylene molecules.

The use of nonporous ionic crystals such as alkali metal halides, which are mainly sensitive to the electron structure of the separated compounds, as stationary phases in GSC dates back to 1964 (cf. [94]). Using different salts and treatment techniques, many specificities of intermolecular interactions with the adsorbates may be established. The most promising ionic adsorbents are LiCl, CsCl, and $CaCl_2$ as well as copper (I and II) chloride, commonly coated on Chromosorb P with 10-25% loading. The preferred field of application is the separation of aromatic hydrocarbons, whereas more polar adsorbates produce broad and/or tailed peaks. Systematic investigations showed anion-cation effects to be significant (e.g., elution on sulfates or nitrates occurs at higher temperatures than on chlorides). Some further generalizations are given in [189,190]. Several authors (cf. [191,192]) have expanded the scope of salts used as stationary phases in GSC by introducing compounds with different numbers of available 3d electrons (e.g., V(II)-$3d^3$, Co(II)-$3d^7$) to achieve efficient separations of various unsaturated and more polar compounds. Being Lewis acids, these transition metal salts (e.g., chlorides) are capable of coordinatively interacting with olefins or aromatic compounds. The strength of interaction is dependent on both the number of free d orbitals and the $\pi$-electron density of the adsorbate molecule. However, in considering chromatograms in literature obtained with salts deposited on support particles or in the form of pure salts, poor column efficiency becomes evident. Improvements in this respect are expected in the future with metal complexes chemically bonded to the support (e.g., to formed ligand surfaces).

In this section, the ion exchange resins should be considered only briefly (see Chap. 10). Organic cation exchangers (as found on slightly sulfonated Porapak Q or Tenax GC) substituted with metals such as $Ag^{+}$, $Cr^{3+}$, $Zn^{2+}$, $Cd^{2+}$, $Ni^{2+}$, $Cu^{2+}$, etc., have been applied in GSC (cf. [193-195]). Highly specific donor-acceptor interactions are observed with these adsorbents. In some cases they can be used as agents in the subtraction methods, e.g., for removing amines, nitriles, or alcohols. By means of modern methods of data processing such as factor analysis and topological analysis, the major influences on the behavior of saturated and unsaturated hydrocarbons on some ion exchangers in different cationic forms could be identified. In the case of alkenes the retention characteristics are broken down into electronic (charge transfer) and steric effects. Thus, a proper ion exchanger packing provides the necessary separation of alkanes with columns having only low plate numbers.

## Gas Chromatography on Modified Adsorbents

Despite the considerable progress achieved in GSC, some disadvantages remain:

1. Because of the high adsorption energy compared to the solution energy, the retention times of high-boiling compounds are too high, although the working temperature range of some adsorbents (CMS, GTCB) has been raised to 400°C and higher.
2. Unlike GLC with its immense number of available liquids of different polarity, in GSC there are only a few homogeneous adsorbents.
3. The so-called uniform surface may be contaminated by small residual surface heterogeneities, which may be geometrical or chemical in nature, or both, resulting in a nonconstant adsorption energy over the entire surface of an adsorbent [196].

The geometrical irregularities at the surface (cracks, edges, etc.) may affect the symmetry of chromatographic peaks, even those of *n*-alkanes. The chemical inhomogeneity, e.g., caused by traces of metal salts used as catalysts in the production of porous polymers, may influence the adsorption of polar compounds. The polar sites are few in number but they can establish specific, strong interactions with polar molecules [196]. This leads to undesirable effects such as tailed peaks, ghosting phenomena, and incomplete recovery of the sample. Therefore, efforts are dedicated to eliminating such surface heterogeneities.

### *Vapor Phase Modification*

The blocking of chemical surface impurities by mixing the carrier gas with various strongly adsorbed vapors such as water and ammonia is a well-established method in GSC (cf. [20]). With surface coverages of more than a monolayer of vapor the retention time of solutes is diminished and in most cases the peak shape is improved [197]. Besides the effect of deactivation of adsorbents, the main advantage of using vapors that adsorb in or on the stationary phase is the ability to alter and control the separation power in situ by external manipulation of the type and pressure of the volatile modifier. The higher the activity of the adsorbents, the higher are the changes in retention behavior of the adsorbates.

Vapors such as *n*-pentane and ethanol [198-200], several amines [201], water [202,203], formic acid [204,205], etc., have been used to influence the retention time and peak shape of different solutes. Hence it has become possible to extend the availability of silica gels and alumina to analyze high-boiling and polar organics. The schematic diagram of a steam generator, which can be adapted to any commercially available chromatograph, is presented in [203].

By using the so-called steam GC with a composite mobile phase (e.g., steam/nitrogen = 90:10) Guillemin et al. [206] succeeded in analyzing aqueous mistures. The success is attributed to the fact that sample and mobile phase have the same matrix, namely, water. Hence, sample treatment can be avoided.

As pointed out by Parcher [197], vapor phase chromatography may be able to replace the use of liquid-modified adsorbents, although the former technique has not yet gained wide acceptance, as it is not always convenient and can be impracticable, e.g., if sensitive detectors are used. The modifier and the detector must be compatible.

The acceleration of analysis and the production of linear isotherms by means of vapor mobile phases is another important aspect in GLC [197,198].

### *Modification by Coating with Liquid and Solid Organic Stationary Phases*

Eggertsen et al. [207] were the first to add a nonvolatile liquid phase (squalane) to carbon. Since the modifier influences the separation occurring on the adsorbent surface, the coating of adsorbents with liquid stationary phases is usually called gas-liquid-solid chromatography (GLSC), though this term is not quite appropriate to describe the features of a modified solid [196,208].

This technique combines some favorable features of GSC, such as high separation efficiency, with those of GLC, such as the extremely wide range of selectivities available. In particular, GLSC is characterized as follows:

1. In the submonolayer region, molecules of the liquid will first saturate the most active sites of the heterogeneous surface,

which results in a more linear adsorption isotherm and, in turn, improved peak symmetry. The choice of the best deactivating liquid should be made on its capability of being adsorbed more strongly on heterogeneous surface sites than the particular solute.

2. On using polymers or large modifier molecules with various functional groups, quite homogeneous specific adsorbents can be obtained. Varying the liquid/solid ratio will yield a whole range of selectivities. Thus, GLSC columns can be tailored for any separation problem difficult to solve. When the modifier is coated in quantities that do not exceed the monolayer range, a modification of the adsorbent characteristics will result without markedly altering the surface geometry.
3. The adsorbed liquid may cause increased retention of the sample solutes due to strong lateral interactions, or it may cause diminished retention by blocking the active sites or the entire surface, when there is sufficient liquid to form a complete monolayer [209].
4. The mass exchange on the surface of dense layers takes place much more quickly than in the bulk of liquids [210]. The higher the liquid load, the higher is the mass transfer term in the van Deemter Eq. (6).
5. The vapor pressure of the liquids held on a large surface area of an active adsorbent is lower than that on the small and weakly adsorbing surfaces used in GLC because the stationary liquid phase is located in the strong adsorption field of the active adsorbent. Therefore, GLSC columns can be employed at temperatures higher than the upper limit of temperatures at which the corresponding GLC columns can be used, resulting in the availability of GLSC columns for programmed temperature chromatography, for trace analysis, and for coupling GC/MS [211].

Modification in the chromatographic process arises from the combined effect of the force centers of both the liquid and the solid phase. Hence the retention is composed of interactions at the stationary liquid-adsorbent interface plus the gas-stationary liquid interface plus the partition between the gas and liquid phases (cf. [212]). Studies of the chromatographic mechanism of Chromosorb 102, coated with different loads of squalane, show that the behavior of Chromosorb 102 and an inactive support is comparable above 15% of squalane, but the adsorption effect is predominant below 15% [213].

The modified adsorbents most widely used are the GTCBs and, to a certain extent, wide-pore silicas and porous polymers. Di Corcia and coworkers developed packings using Carbopack graphitized carbons that are commercially available from the Supelco Company. The outstanding ability to separate organics in the range $C_1$-$C_{10}$, such as alcohols, free acids, phenols, amines, etc., is shown in Table 7.

Table 7 Coated Carbopack Applications

| Modified Carbopack | Maximum working temperature (°C) | Preferred application |
|---|---|---|
| Carbopack C/0.3% Carbowax 20M/0.1% phosphoric acid | 200 | $C_2$-$C_5$ acids |
| Carbopack B AW/5.0% Carbowax 20M [96, Supelco Bull. 784] | 225 | Alcoholic beverages, blood |
| Carbopack B AW/6.6% Carbowax 20M [86] | 225 | |
| Carbopack C/0.2% Carbowax 1500 [196, Supelco Bull. 744C] | 175 | |
| Carbopack C/0.3% Carbowax 20M [214] | 225 | Alcohols, esters, ketones, vinylchloride in air, general-purpose |
| Carbopack C/0.2% Oronite N/W | 200 | |
| Carbopack B/2.5% Oronite N/W | | |
| Carbopack B/4% Carbowax 20M/0.8% KOH [215], see [88] | 220 | Amines |
| Carbopack C/0.8% THEED [216] | 125 | Ethylene oxide |
| Carbopack B/1% SP-1000 | 225 | Halogenated organics |
| Carbopack C/0.1% SP-1000 [217, Supelco Bull. 747C] | 225 | Phenols, solvents |
| Carbopack B/3% SP-1500 | 230 | Solvents, industrial |
| Carbopack B/1% SP-1510 | 230 | |
| Carbopack B/1.5% XE 60/1% phosphoric acid [218, Supolco Bull. 722F] | 150 | Sulfur gases |
| Carbopack C/0.19% picric acid [219] | 120 | Unsaturated $C_4$ hydrocarbons |

*Source*: Supelco catalog 21, 1983.

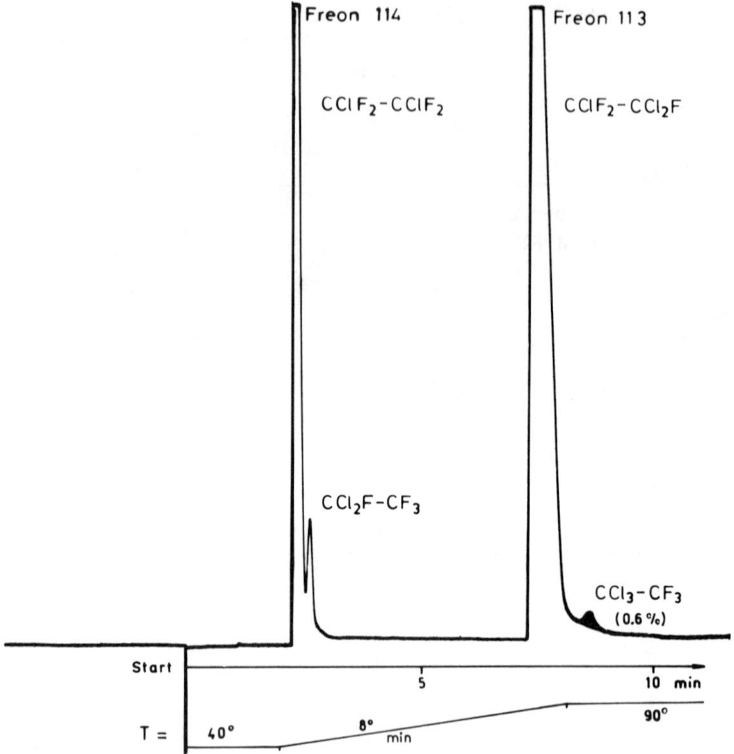


Fig. 9 Separation of halocarbons on modified Sterling MT (from Engewald, Pörschmann, Welsch, unpublished results). Column: 1.1 m × 0.6 mm i.d. packed with 0.4% (w/w) DBTCP on Sterling MT, dp = 0.07-0.09 mm.

The separation of freons on Sterling MT modified with 0.4% dibutyl tetrachlorobenzene (DBTCB) is depicted in Fig. 9. Compared to non-halogenated compounds, the halogenated ones have a greater distance from the flat surface. Thus, they are eluted before the corresponding hydrocarbons. The use of a GTCB column allows separation of halogenated hydrocarbons without requiring very low or subambient temperatures.

As noted above, Carbopack C and Carbopack B have specific surface areas of approximately 12 $m^2 g^{-1}$ (similar to Sterling FT) and approximately 100 $m^2 g^{-1}$ (similar to Graphon), respectively. Hence, the amount necessary to block the heterogeneous centers of the adsorbent will be considerably increased for Carbopack B compared to Carbopack C (Table 7).

In selecting modifying liquids special attention should be given to polyphenylether (PPE 20) and Dexsil 400 [211]. By means of these highly thermostable phases it becomes possible to separate alkanes ranging in size from $C_4$ to $C_{40}$ as well as steranes. The decision on the best amount of any modifying liquid should be based on a detailed evaluation of the influence of the liquid load on retention time with regard to optimizing the analysis. Thus, for example, the separation of the unsaturated $C_4$ hydrocarbons is performed quite well with Carbopack C/0.19% picric acid [219]. Increasing or decreasing the amount of picric acid will lower the resolution. In this case the most appropriate amount of modifier is found by the minimum amount necessary to block the surface inhomogeneties. Incidentally, the coating of unmodified GTCB with picric acid or 2,4,5,7-tetranitrofluorenone [75] (the two latter have a marked capability of undergoing charge-transfer interactions) offers an example of conversion of a nonspecific adsorbent into a specific one.

Proceeding along the same lines as the workers mentioned, Vidal-Madjar and Guiochon [94,221] modified some GTCB with solid phthalocyanines. The latter adsorbents (third type according to Kiselev) contain different functional groups such as -N= and HN=, which can result in specific interactions with analytes. The prospect of controlling the specific nature of the adsorbent has been widened by introducing metallic complex-forming ions such as Co, Ni, Zn, and Cu. These packings are suitable for the separation of alcohols, phenols, aromatic bases, etc. (cf. [94]). Modified GTCB with monolayers of other nonvaporizable derivatives having different functional groups (further porphyrins, crown ethers, etc.) are expected to be used in the future.

Copper and nickel phthalocyanates coated on silica (Porasil C) exhibit a high selectivity for nitrogen-containing substances [222,223]. The advantages of using phthalocyanines and phthalocyanates include a reduction in analysis time, an improvement in peak symmetry, and a high thermal stability. Guillemin et al. [139,224] coated Porasil silicas with polar liquids in order to analyze polar compounds in a short time. On account of the high purity of these adsorbents, it was necessary to load only small amounts of liquid phases.

### *Modification by Chemical Reactions of the Adsorbent*

Utilizing the reactivity of the hydroxyl groups in silica, a tremendous number of organic radicals can be bonded to the silica surface, giving selective sorbents extensively used in GC and HPLC. The substitution of the acidic hydrogen atom having a small radius reduces the specific interactions with molecules containing polar functional groups such as OH, NH, etc. By means of chemical reactions, the major obstacle to good performance in GSC of untreated silicas–the surface heterogeneity–is reduced or eliminated. In addition, these chemically bonded stationary phases allow fast mass exchange due to the monolayer phenomenon and show low bleeding compared to physically adsorbed phases.

The formation of chemically bonded adsorbents can be established by the following reactions (see also Chap. 6):

1. Phases with ≡Si-O-C≡ bonds can be prepared by esterification of the acidic hydroxyl groups with appropriate alcohols such as 3-hydroxypropionitrile (OPN) [225]. Halasz called these materials "brush-type packings" because attaching the liquid molecules onto the adsorbent provides a structure attached chemically to the silica surface at one end, with the rest of the molecule freely available to the sample components (cf. [226]). These adsorbents are marketed under the trade name of Durapak (see Table 8). The poor hydrolysis stability of the ≡Si-O-C≡ bonds calls for carefully dried carrier gas and water-free samples. As mentioned by Little et al. [227], the polarity of the *n*-octane packings is higher than that of Carbowax 400 and OPN. This is caused by some residual hydroxyl groups remaining on the silica surface.

2. Phases with ≡Si-O-Si-C≡ bonds prepared mainly by reaction with organochlorosilanes or organoalkoxysilanes. Depending on the organochlorosilane used and the experimental conditions, monolayers as well as polymer layers are formed. Kirkland and De Stefano (cf. [230]) have grafted silane compounds on a support with a thin porous layer on a hard inert core (Permaphase). Abel et al. (cf. [228,229]) prepared modified supports of polymer type by polymerization of hexadecyltrichlorosilane with Celite 545 as support. Such a bonded phase is stable against hydrolysis and offers a high-temperature stability up to 300-400°C [230].

3. Phases with a ≡Si-C≡ bond, which are thermally and solvolytically stable, were prepared by means of the Grignard reaction [231]. For example, halogenated Porasil C was reacted with naphthyl magnesium bromide to yield polynaphthyl Porasil. This method has the disadvantage that magnesium occlusion salts are formed. Later, Sebestian and Halasz [232] synthesized such ≡Si-C≡ bonded phases with a brush structure. A further horizon might be opened by using the chlorination Grignard method with agents containing vinyl or allyl groups. These groups provide opportunities for further modifications (e.g., the introduction of enantioselective groups). The resulting adsorbents may also be of interest in HPLC.

## Typical Applications of Gas-Solid Chromatography

### *Analysis of Permanent Gases and Light Hydrocarbons*

The determination of permanent gases, e.g., $N_2$, $O_2$, $H_2$, Ar, $CO_2$, and light hydrocarbons $C_1$-$C_4$, in the chemical industry [233,234], in

Table 8 Chemically Modified Silicas

| Trade name | Type | Maximum working temperature (°C) | Preferred use |
|---|---|---|---|
| Durapak *n*-Octane/Porasil C | $\equiv Si-O-(CH_2)_nCH_3$ | 180 | Hydrocarbons |
| Durapak Carbowax 400/Porasil C | $\equiv Si-O-(CH_2-CH_2-O)_nH$ | 180 | Alcohols, ketones, aldehydes, hydrocarbons |
| Durapak Carbowax 400/Porasil F | See above | 230 | As above |
| Durapak OPN/Porasil C | $\equiv Si-O-CH_2-CH_2-CN$ | 140 | Low-boiling hydrocarbons |
| Durapak Phenylisocyanate/Porasil C | $\equiv Si-O-CO-NH-C_6H_5$ | 120 | Unsaturated and saturated hydrocarbons |

*Source*: Manufacturer catalogues.

medicine [235-238], in pollution control [239-241], in investigations of thermal fragmentation [242], in isotope research and in astronautics [243] is an important task.

GC has almost completely displaced the classical methods of gas analysis. The aim of GC is to separate simultaneously as many gases as possible on a single column in a short time. Since the heats of adsorption are larger than those of partition (see above), GSC should preferably be used for analysis of such low-boiling compounds. Using GLC with small and moderate film thicknesses for gas analysis would result in operating in unrealistically low temperatures. As already outlined (see above), that problem may be circumvented by thick-film capillaries (cf. [244]).

For special aspects the reader is directed to the "atlas" of gas analysis by Leibrand [245] and Mindrup [246] as well as to the books devoted to this subject [247,248]. Many separations have been performed by means of the most important adsorbents employed in separating permanent gases and light hydrocarbons such as GTCB, porous polymers, and zeolites. Drawbacks of these adsorbents may consist in the insufficient resolution of certain gases and in the tight bonding of some gases to the adsorbent used. Thus, $CO_2$ is adsorbed irreversibly on zeolites. The separation of $CO_2$ in mixtures with $H_2$, $O_2$, $N_2$, $CH_4$, and CO with CMS or some porous polymers was successfully accomplished. Using a dual system containing Porapak Q and zeolite 5A, a sufficient separation of gas mixtures containing $H_2$, $O_2$, $N_2$, $CH_4$, CO, and $CO_2$ is possible [249]. Further arrangements of columns or mixtures of packings in a single column in order to achieve sufficient separations are described in [250-257]. A contribution to the theory of the efficiency of serially connected columns has been given by Purnell [258].

Recently, Riederer (cf. [259]) introduced polymer Schiff base metal complexes in GSC. With these molecular sieves many separations (including $CO_2$) can be performed in a single run. The merits of GSC and GLSC are expressed best in trace analysis of gases, e.g., in the determination of impurities in the ppm range in pyrolysis gases. By modifying GTCB with picric acid, trace impurities in 1,3-butadiene can be estimated [260]. Modification with electron acceptors has proved necessary to ensure a larger retention of the matrix 1,3-butadiene compared to that of the impurities (Fig. 10a). As known, olefins form molecular complexes with $\pi$ acceptors. The extraordinary selectivity for isomeric compounds favors GTCB for the resolution of the $C_4$ hydrocarbons, as can be inferred from Fig. 10b.

Gases containing reactive sulfur such as $H_2S$, $SO_2$, COS, $CH_3SH$, and $(CH_3)_2SH$ have been separated on modified or unmodified porous polymers [254,261-263], modified GTCB [218], OPN-Porasil C [264], specially pretreated silicas [265] (Chromosils, distributed by Supelco; see Supelco Bull. 722F), and CMS (see Supelco Bull. 712).

### *Analysis of Organic Compounds with Polar Functional Groups*

*Free Short-Chain Fatty Acids and Alcohols* The accurate determination of free short-chain fatty acids, mainly occurring in aqueous solutions, is of interest in the study of metabolic processes of microorganisms. They also occur in various other biological sources (cf. references cited in [266]).

There are difficulties in analyzing highly polar solutes by means of GLC due to their strong interactions with the supports, giving rise to badly tailed peaks, loss of sample, or ghosting phenomena [267]. As is customary in GLC, the undesired interactions with the support are reduced by adding small amounts of a deactivating agent such as orthophosphoric acid to the stationary phase. Such deactivation implies several disadvantages: thermal instability of $H_3PO_4$ at a temperature above 190°C, partial deterioration of the stationary phase caused by the acid at high temperatures, anomalous interactions between the acid and alcohols to be analyzed. Generally, the mentioned problems may be circumvented by derivatization procedures, but in this context these techniques are outside the scope of this chapter. The same goes for amines. The reader is directed to the *Handbook of Derivatization* by King and Blau (Heyden and Son Ltd., 1978).

An alternative to an inert support is the use of high-purity adsorbents with homogeneous and uniform surfaces such as GTCB or porous polymers. Generally speaking, the separation of polar molecules with very slight differences in electron density distribution frequently involves problems in GLC. However, since there are differences in either the geometrical structure or polarizability, a properly modifed GTCB can be employed. As mentioned in a previous section, the presence of small amounts of oxygen and sulfur chemisorbed on the graphite surface can lead to undesirable adsorption effects with acidic compounds. These centers can be blocked by deactivation of GTCB with $H_3PO_4$ and subsequent coating (cf. [268] and Supelco Bull. 751A). However, GTCB modified in such a manner offers the same drawbacks as mentioned above. Washing the surface of GTCB is an effective means to remove the active complexes reasonable for chemisorption. On acid washing Carbopack B modified with 6.6% of PEG 20M, small amounts of acids, alcohols, and aldehydes are eluted as symmetrical peaks as shown in [86]. Di Corcia et al. [92,93,214] extended the scope of acid-washed GTCB to the determination of underivatized hydroxy acids such as 3-hydroxybutyric acid and drugs (e.g., phenobarbitals), which are important in therapeutic monitoring, drug overdose determination, recognition of diabetes, etc.

Besides GTCB, porous polymers have been used for the determination of low free fatty acids with and without modification of the stationary phase (cf. [120,269-271]). As pointed out [272,273], the removal

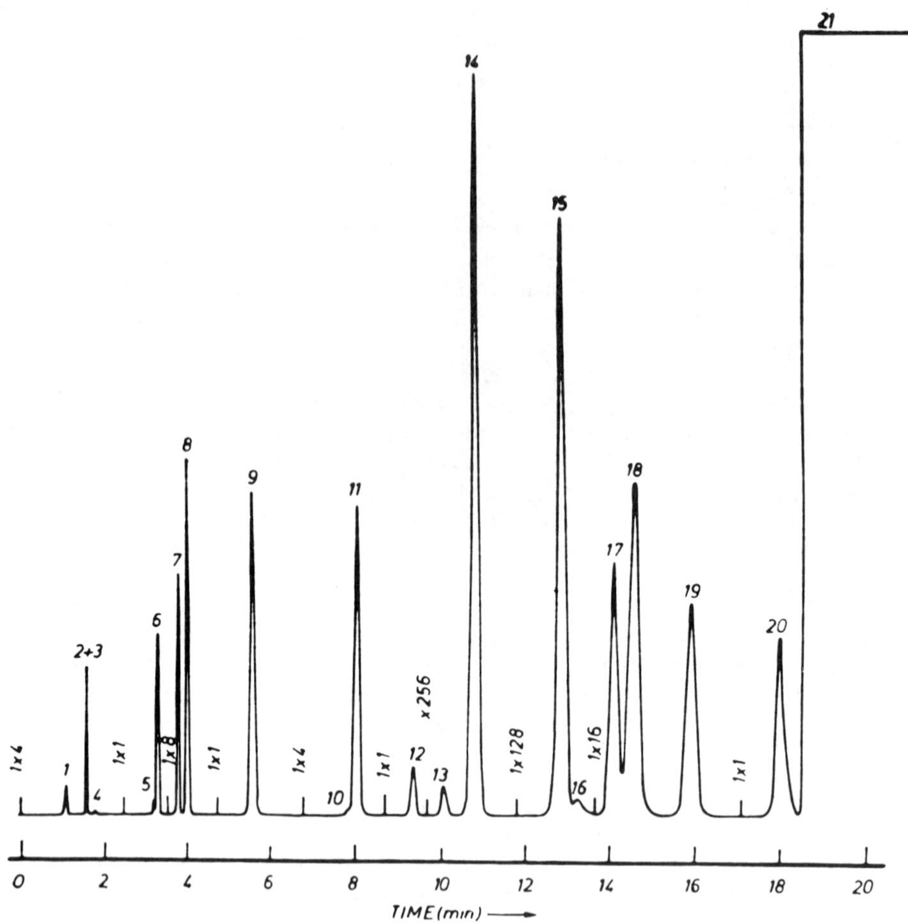


Fig. 10 (a) Separation of impurities in 1,3-butadiene. Column, 5 m × 1.5 mm i.d. packed with Carbopack B + 4.94% (w/w) picric acid; column temperature, 46°C. Peaks: 1, methane; 2, ethane; 3, ethylene; 4, acetylene; 5, cyclopropane; 6, propane; 7, propene; 8, propadiene; 9, propine; 10, unknown; 11, isobutane; 12, neopentane; 13, butane; 14, butene-1; 15, isobutene; 16, butadiene-1.2; 17, butene; 18, butene-2-trans; 19, butine-1; 20, 1-buten-3-ine; 21, butadiene-1.3. (From Ref. 260, with permission.)

(b) Analysis of a mixture of gaseous hydrocarbons on a micropacked column with GTCB. Column: 1.7 m × 0.28 mm i.d.; GTCB, Sterling MT, particle size 0.06-0.08 mm: column temperature, 40°; carrier gas, hydrogen; inlet pressure 7.2 bar. Peaks: 1, methane; 2, ethane; 3, propene; 4, propane; 5, isobutane; 6, 1-butene; 7, isobutene; 8, *n*-butane; 9, *cis*-2-butene; 10, *trans*-2-butene. (From Ref. 25, with permission.)

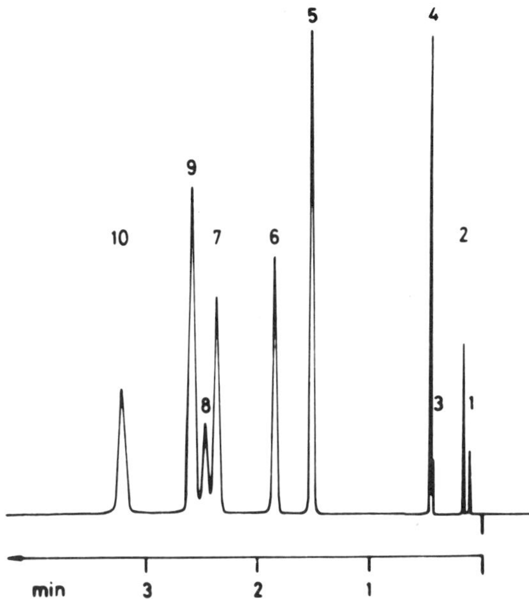


Fig. 10 (b)

of monomeric material, e.g., by thorough extraction with acetone, as well as the conditioning of the porous polymers is essential for achieving reproducible results. Acidifying the aqueous solution with oxalic acid (cf. [274]) allows direct injection on untreated porous polymers with virtually no ghosting problems. In many cases some isomeric pairs are coeluted (e.g., 2 M butyric acid, 3 M butyric acid). In this event coating with a proper liquid has proved to be favorable [275]. Recently, Goretti et al. [33,376] employed kaolin-pretreated glass capillaries coated with FFAP (see below) for the analysis of acidic compounds.

GC determination of alcohols has become increasingly important in the food industry [277]. For example, the relative concentration of methanol, ethanol, and acetaldehyde characterizes the quality of citrus products and is used to a suspected falsification. Preferred adsorbents in determining alcohols are porous polymers [277-279], modified carbon blacks (Supelco Bull. 738 and [86,280]), modified Porasils [281], and Durapak [226], the last allowing fast analysis due to small mass exchange resistance.

Conversely, GSC with polymers such as Porapak Q is a very suitable method for analyzing the water content in both the liquid and the vapor phases [282]. Incidentally, water can be determined easily by reaction GC and subsequent GSC [263].

*Free Amines* The analysis of amines is important because they are commonly found in foodstuffs, tobacco leaves, human urine and blood, as well as in microorganisms [283,284].

Problems similar to those faced by chromatographers concerned with the analysis of free fatty acids also occur in the determination of free amines. Both the deactivation of supports by treatment with alkali and the employment of appropriate adsorbents have proved to be successful approaches to reducing undesirable interactions with compounds to be analyzed. The major disadvantage of alkali-pretreated packings lies in the insufficient thermal stability, preventing temperature-programmed analysis within a wide range. Liquid-modified GTCB as well as modified and tailor-made porous polymers are frequently applied to the analysis of amines. Di Corcia et al. [88] (see also [264, 285,286]) developed the packing Carbopack B/0.3% KOH/4.8% Carbowax 20M in determining amines up to 12 carbon atoms even at the nanogram level. For analyzing higher boiling substances Tenax GC may be used.

In employing the above-modified GTCB packing attention should be paid to conditioning, including heating at 220°C with a nitrogen stream and subsequent 30 injections of freshly boiled water. With porous polymers tailed peaks are often observed, caused by the occurrence of simple acidic centers and metal ions. Peak tailing can be reduced by coating the polymer with polyethyleneimine, tetraethyleneamine, or potassium hydroxide [287].

A tailor-made porous polymer for basic compounds is Chromosorb 103 (Table 5), which totally adsorbs acidic components. Having a hydrophobic character, Chromosorb 103 can be used in analyzing aqueous solutions. In Fig. 11 the tailing-free elution of some low amines and diamines on Chromosorb 103 is depicted. Even in microbiology the determination of diamines is becoming increasingly important [284,288].

### Analysis of Inorganic Compounds

A comprehensive survey of inorganic GC was given by Uden [289] (see also [506]). Here we shall confine our discussion to anions such as halides or cyanides, and metals.

In inorganic GC chemisorption takes place. Thus, in order to determine halides, derivatization procedures have been employed using, for example, tetra-*n*-butylammonium succinimide [290] as well as dimethyl sulfate as methylation agent [291]. The resulting alkyl halides have been separated on the porous polymers Porapak Q and Porapak P and subsequently detected by specific detectors. Reference [292] reports that the volatile metal bromides (Zr, Nb, Mo, Sn) were separated on quartz by using nitrogen as carrier gas, to which boron tribromide was added. Cyanides and thiocyanates in water have been analyzed by reaction with acetaldehyde to yield cyanohydrins [294] as well as by reaction with bromine to yield BrCN [293], with the resulting compounds chromatographed on porous polymers.

006

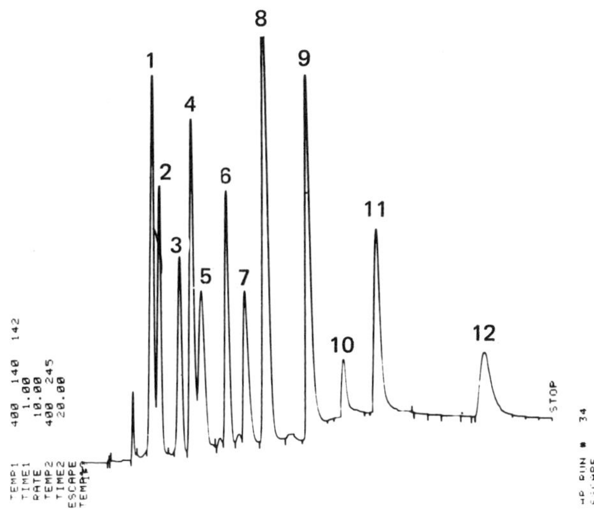


Fig. 11 Separation of free amines on Chromosorb 103; column, 1.5m x 0.60 mm i.d.; dp = 100/120 mesh; column temperature: hold 1 min at 140°C, then to 245°C at 10°/min. Peaks: 1, *n*-propylamine; 2, diethylamine; 3, *i*-butylamine; 4, *n*-butylamine; 5, triethylamine; 6, *n*-pentylamine; 7, 1,2-diaminopropane; 8, *n*-hexylamine; 9, putrescine; 10, cadaverine; 11, hexamethylenediamine; 12, octamethylenediamine. (From Porschmann, Liebetrau, unpublished results.)

Free metals, which are virtually nonvolatile, may be separated at temperatures above 500°C on appropriate adsorbents such as silica, alumina, and graphite (cf. [295]). The metals As, Se, Ge, Sn have been sequentially determined by means of hydride generation followed by GSC on Chromosorb 102 with an atomic adsorption detector in the single-channel mode [296]. A review of metal determination by chromatography including the use of selective detection and derivatization procedures was recently given by Nickless [297].

*Adsorbents in Preconcentration Techniques*

In the face of an increasing scale of environmental pollution there is an urgent need for the quantitative and qualitative determination of pollutants. Volatile compounds causing odor nuisance or hazardous ones such as chlorinated organic compounds, including pesticides, occur in concentrations ranging from 1 mg $m^{-3}$ to far below 1 μg $m^{-3}$. Since the detection limit of GC is poorer, even though selective detectors are used to increase the response to compounds of interest, concentration techniques become essential. Therefore, the chromatographer of today must be sufficiently familiar with the methods of enrichment. In the case of gaseous samples, the following methods for reaching the detection limit of the GC system are popular:

Solid adsorption. The sample stream is conducted through a tube filled with an appropriate solid adsorbent ("trap") on which the volatile compounds are adsorbed.

Liquid adsorption. The sample stream flows through a solvent to dissolve the volatile compounds.

Cryogenic method. The sample stream flows through a cooled vessel condensing the volatile compounds.

Cryogenic trapping can lead to analytical difficulties because freezing out of water vapor occurs along with the organic content in air, whereas solvent scrubbing is insufficiently sensitive for the analysis of organics below the ppm range [298].

Solid adsorbents should satisfy the following demands:

1. The volatile compounds are effectively concentrated from a large volume of air without influence of relative humidity on the adsorbing capacity of the adsorbent.
2. The desorption of adsorbates should be complete and without losses due to degradation. Desorption can be performed by displacement of small volumes of organic solvents or by thermal means. The latter method yields unsatisfactory results for thermally labile substances.

3. The adsorbent should provide an inert, highly pure surface to omit background peaks. The latter can be considerably reduced by a thorough cleaning of the adsorbent [299].
4. The adsorbent should exhibit reproducible adsorption properties after repeated cooling and heating in the presence of large amounts of humidity and oxygen.

The application of adsorbents as trapping media is surveyed in [300-302]. Both the synthetic porous polymers, including Tenax GC, Porapak series, and Chromosorb series, and GTCB have proved to be more advantageous than charcoal.

Regarding the tube filled with adsorbent as a short chromatographic column, the chromatographer is able to determine the breakthrough volume of a component to be enriched. The breakthrough volume is defined as the gas volume to be passed through a sorbent before the component under study begins to be eluted from the cartridge containing the adsorbent. The breakthrough volume is an important criterion in selecting the most effective adsorbent for collecting air samples for quantitative analysis. It depends mainly on the adsorption capacity of the adsorbent as well as the size of the trap tube (cf. [303-305]). For instance, Tenax GC offers a low capacity for the preconcentration of highly volatile organics owing to its low adsorption activity. The same is partly true of Carbopack C and Chromosorb 101. In this case the use of strongly adsorbing media such as Carbopack B is favored. Tenax GC exhibits the best retention properties in the collection and thermal desorption of hazardous chlorinated aromatic hydrocarbons since the specific retention volumes of these compounds are relatively large at ambient temperatures and small at temperatures of about 300°C [306]. More recently, a hydrophobic inorganic adsorbent with high thermal stability (up to 500°C), designated as Thermosorb, was employed for trace analysis with aqueous solutions or air [315]. Thermosorb may be a supplement or even an alternative to Tenax GC.

In summary, the choice of adsorbent merits great care as changes in the composition of the sample mixture engendered by selection of inappropriate adsorbents cannot be corrected at any later step of the analysis.

Highly reactive pollutants such as formaldehyde cause problems on the sampling media, which can be circumvented by using solutes coated with *N*-benzylethanolamine to form oxazolidines with aldehydes [307,308]. In a similar manner, the strongly polar amines can be trapped as coordination complexes on columns containing copper(II) salts [309], whereas lower fatty acids in air can be collected by using $Sr(OH)_2$ [310]. A substraction technique using alkaline precolumns has been employed to identify fatty acids in complexly composed matrices [311].

Contaminants in water can be sampled by "purging," i.e., bubbling a pure gas through the solution and conducting that gas through the sorbent trap. XAD resins [300,312] and Carbopack adsorbents [31] are widely used in the enrichment of water pollutants. However, compounds having high water solubility (low aldehydes and alcohols) offer a low purge efficiency [314]. A novel technique containing adsorption from aqueous solutions onto an adsorbent such as Tenax GC followed by thermal desorption of the dried adsorbent has been introduced for nonpolar organics of intermediate molecular weight [316].

In addition to the use in environmental analysis, enriching and isolating procedures using adsorbents can also be applied in clinical chemistry. As an example, long-chain fatty acids in blood serum or urine may be isolated by adsorption on Carbopack columns. After desorption and vaporization of the eluting liquid, the isolated fatty acids are methylated as usual.

### *Structure-Retention Relationships*

Over the last 30 years GC has become the most powerful separation method available for vapors. By means of this method our knowledge about the quantitative and qualitative composition of complex natural and synthetic mixtures has been extended.

GLC is mainly employed to investigate mixtures of organic compounds. By application of highly efficient capillary columns, considerable advances in the analysis of individual components of complex mixtures and in the separation of spatial and structure isomers have been attained. Further potential for the solution of difficult separation problems is provided by the development of adsorbents with specific separation properties.

The facilities for identification of the separated compounds have not kept abreast of the improved separation power. Because of the small loading capacity of highly efficient thin-film capillaries and the fast elution of peaks, only two powerful spectroscopic methods, mass spectroscopy (MS) and Fourier transform infrared (FTIR) spectroscopy, can be used in direct on-line coupling with the capillaries. These powerful but very expensive coupling techniques are often not sufficient for the unambiguous identification of separated compounds. Thus, the increased importance of chromatographic methods as an aid to identification cannot be overlooked. This approach is based on measuring relative retention data by means of the retention indices on capillaries of well-defined polarity and offers a sensible complement to the structural information obtained by GC/MS and GC/FTIR combination.

At the present level of the theory of solution it is impossible to predict the volatility fo solutes, and with it the retention, by means of molecular parameters. The only way to calculate the interaction of solutes with the given stationary phase consists in solving the Schrödinger equation, considering each electron-electron interaction separately. However, the complexity of the calculations grows exponentially with the size of the atoms. As a result of numerous and expensive investigations, many quantitiative and semiquantitative relationships between molecular structure of solutes/adsorbates and their retention on stationary phases could be found empirically [317-319]. These findings constitute a valuable help in the identification of substances.

In comparison with GLC the relationships between structure and retention are theoretically more surveyable in GSC with adsorbents having unspecific surfaces. In GLC, molecules dissolved in the bulk of a liquid film are mobile and the molecules surrounded by other molecules on all sides. In GSC molecules interact only with the nearest force centers of the adsorbent. In this respect, GTCB is the adsorbent of choice because it consists only of atoms of one sort in the same electronic configuration and its surface is homogeneous, flat, and chemically well defined. In what follows the structure-retention relationships will therefore be restricted to this adsorbent, though on using the comprehensive data obtained with other adsorbents, e.g., the retention values obtained by Castello and D'Amato (see above) with porous polymers, especially nonpolar ones, presumably would also allow the establishment of such relations. Likewise, coated GTCB will be excluded from these considerations because the more complex separation mechanism involving partition phenomena would complicate matters. Thus, to predict elution sequences it appears convenient to use an atom-atom approximation for the potential energy of paired interactions. The calculated potential energies are then compared with the experimentally measured ones. The potential energy $\phi$ is approximated as the sum of the atom-atom potential functions of each atom A of the molecule adsorbed (M) with each carbon atom of the adsorbent:

$$\phi = \Sigma\, \pi_{A(M)\,.\,.\,.\,C(GTCB)} \tag{14}$$

$$\pi = -C_1 \cdot r^{-6} - C_2 \cdot r^{-10} + B \cdot \exp(-r/\rho) \tag{15}$$

The equation for $\pi$ can be written in form of the Buckingham-Corner potential (6, 8, exp) as well. The constants of attraction $C_1$ and $C_2$ are usually calculated by means of the Kirkwood-Müller equation. The repulsion constant B is determined from equilibrium conditions (with $\rho$ = 0.028 nm). Poshkus and Kiselev [320,321] estimated the potential functions $\pi$ for carbon atoms having the electronic configuration $sp^3$, $sp^2$, and sp, and for hydrogen atoms. Thus, to compute the energy of paired interactions of saturated hydrocarbons, the use of only the two atom-atom potentials $^{\pi}C(sp^3)\ .\ .\ .\ C(Gr.)$ and $^{\pi}H\ .\ .\ .\ C(Gr.)$ is essential (Gr. means graphite).

As proved, the prediction of retention by means of nonbound interactions is not sufficient for an exact description of retention behavior. First, in addition to the isosteric (or differential) heat of adsorption, which is close to potential energy, the entropy must also be considered, because the retention volume depends on the *free* differential energy of adsorption. Hence, it is necessary to introduce the potential energy into the molecular-statistical expressions for Henry's constant $K_1$. In the case of nonadsorbing carrier gas this constant equals the retention volume per unit surface for zero sample size (see above). Several research groups developed semiempirical molecular statistical methods for the calculation of retention volumes of hydrocarbons on GTCB (cf. [73,322,323]). Poshkus and Kiselev [321,324] as well as Guiochon et al. [79] computed retention volumes of aromatic hydrocarbons by means of the molecular statistical theory of adsorption. Second, with unsaturated hydrocarbons it becomes essential to take into account differences of the electronic structure of the adsorbate molecules, e.g., in the case of progressive ring condensation or progressive methyl substitution of alkyl aromatics. The possibility of development of charge-transfer interactions between adsorbate molecules and carbon atoms of the graphite lattice cannot be ruled out [325].

In conclusion, calculated values of retention volumes can be used to establish the elution sequence of substances with known molecular structure and/or to identify unknown components in a given mixture. Moreover, Kiselev [326] has suggested a solution to the reverse problem: the investigation of the structure of molecules on the basis of chromatography. By analogy with spectroscopy this method is termed chromatoscopy or chromatostructural analysis.

Employing chromatoscopy, several Soviet workers (mainly from Kiselev's group) [7,320,326,327] reported on the determination of diverse structure parameters such as potential barriers to internal rotation of alkyl groups with respect to the benzene ring in alkyl aromatics and of the benzene rings with respect to each other in biphenyl. With the aid of chromatoscopy Poshkus and Grumadas [321] found that the equilibrium angle between the planes of the benzene rings in 2,6-dimethylbiphenyl is 68°. Kalashnikova et al. [327] examined some alkyl derivatives of biphenyl. They found, for example, that the weakest adsorption occurs for 2,6,2',6'-tetramethylbiphenyl with the most inhibited internal rotation of the molecule (Form. 1).

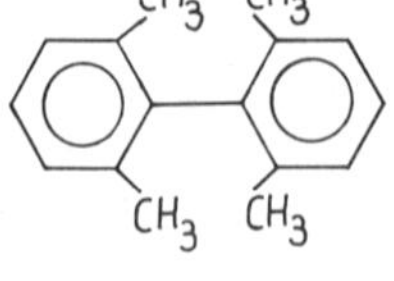

$CH_3$ $CH_3$ / $CH_3$ $CH_3$

Formula 1: 2,6,2',6'-tetromethylbiphenyl

Generally, the findings obtained by chromatoscopy are in good agreement with those obtained by expensive analytical tools such as the electron diffraction method. Chromatoscopy can also be extended to specific adsorbents of known structure (e.g., zeolites) to solve problems concerning the electronic configuration of molecules.

In recent years there has been much progress in the identification of unknown substances using the retention index concept both in GLC and GSC [328-330]. Comparing retention indices and the derived values such as $\Delta I$ value or homomorphic factor measured on GTCB and liquid phases, it becomes possible to draw important conclusions, e.g.,

With progressive branching of the carbon chain the retention decreases both on GTCB and on liquid phases. Thus, isomeric butylbenzenes exhibit the following elution order: $I_n > I_{iso} > I_{sec} > I_{tert}$. As expected, the differences of retention indices for the isomers mentioned are more pronounced in GSC with GTCB (Table 9). It is possible to achieve complete and fast separation of all (10) isomeric *p*-dibutylbenzenes on GTCB, whereas it is impossible to separate all isomers on liquid stationary phases. As shown in Table 9, *p*-di-*n*-butylbenzene exhibits the largest retention on GTCB due to its most favorable orientation on the adsorbent surface resulting in a high number of contact points. The index difference from the first eluted isomer (tert-tert) to the last one (n-n) amounts to 131 index units with squalane and 339 index units with GTCB. This finding demonstrates the outstanding selectivity of GTCB toward structural isomers.

The retention data of polymethylbenzenes are always considerably higher as compared to isomeric alkyl- and dialkylbenzenes.

In contrast to GLC, all of the octaines and octadiines are eluted faster from GTCB columns as the homomorphic *n*-octane according to larger distances between some of the force centers to the graphite lattice.

## STATIONARY PHASES IN GAS-LIQUID CHROMATOGRAPHY

Several hundred stationary liquids have been reported in the chromatographic literature. This multitude of phases is most likely attributed to the fact that they have probably been devised in order to compensate for the lack of efficiency that occurs with the packed column type. The question facing today's chromatographer is either to restrict himself to a few liquid stationary phases, or to exploit the overall scope of the phases available. There are many examples showing clearly that by restriction to a few liquids the extraordinary power of GC would be needlessly reduced. However, when a desired separation

**Table 9** Retention Indices of *p*-Dibutylbenzenes on Squalane(s) and Graphitized Thermal Carbon Black (Sterling MT)

| Isomer | $I_{100}^{S}$ | $I_{310}^{GTCB}$ |
|---|---|---|
| *n-n* | 1412.5 | 1334.5 |
| *n-iso* | 1359.6 | 1308.9 |
| *iso-iso* | 1307.7 | 1285.8 |
| *n-sec* | 1358.4 | 1248.8 |
| *iso-sec* | 1305.8 | 1235.9 |
| *n-tert* | 1345.3 | 1219.5 |
| *tert-iso* | 1290.6 | 1206.9 |
| *sec-sec* | 1304.3 | 1185.6 |
| *tert-sec* | 1290.7 | 1142.5 |
| *tert-tert* | 1281.7 | 1095.6 |

*Source*: From [80], with permission.

cannot be performed with a given liquid stationary phase, e.g., Carbowax 1540, then the chance of success with a similar phase, e.g., Carbowax 1500, will be small as well. As a further example, the same is true for Emulphor ON-870, Triton X-100, and Ucon 50-HB-2000, all of which are similar to one another.

In terms of material availability, today's chromatographer must often select from industrial products developed for other than chromatographic purposes, e.g., plasticizers such as polyesters, surfactants, and lubricants (Carbowaxes, Ucons). Such products often exhibit poor batch-to-batch reproducibility and low thermal stability. They contain catalytically active impurities and additives, and often possess a wide molecular weight distribution. Thus, purification is frequently needed to make them suitable for chromatographic purposes. Nevertheless, an increasing number of liquid stationary phases has been specifically manufactured for pure chromatographic applications, e.g., the OV-silicones (Ohio Valley), Silars (Applied Sci.), and CP phases (Chrompack).

### Requirements for Liquid Stationary Phases

A stationary phase in gas-liquid chromatography should satisfy a number of requirements:

1. It should not react irreversibly with the support and the analytes. For example, the tailing-free elution of free fatty acids is frequently achieved by adding phosphoric acid, which promotes the dehydration of alcohols releasing additional peaks.
2. It should not be affected by the carrier gas. Generally, problems in applying liquid phases have arisen with oxygen and water in the carrier gas. As examples, it was found that polysiloxanes are more stable against oxidation than $C_{87}$ hydrocarbons or squalane (see below). Among the polysiloxanes, the phase OV-17 is more resistant than SE-30 or OV-1. In contrast, polyesters are quite susceptible to oxidation by trace air in the carrier gas. The presence of water in the carrier gas has a profound harmful influence on polyesters and polyalkylene oxides. Acids and bases frequently have harmful effects on liquid phases, in that polyesters are preferentially degraded by bases and polyalkylene oxides by acids.
3. It should be sufficiently thermally stable. Thermal stability is related to two phenomena: the decomposition of the stationary phase by increasing the temperature (chemical thermal stability) and the formation of droplets at the support surface by increasing the temperature (physical thermal stability). The applicability of liquid stationary phases in the low-temperature range is limited by the melting point and by the fact that viscosity increases with decreasing temperature. Some efforts have been made to extend the temperature operating range of phases (see below).
4. The synthesis of the stationary phase must be highly reproducible. It seems that this aim can be fully achieved with nonpolar to moderately polar phases only.
5. It should be chemically stable and not contribute to undesired adsorptive activities. Impurities in the liquid stationary phase originated from the manufacturing process should be chromatographically inactive.
6. The liquid stationary phase should maintain its properties over a prolonged time of usage, particularly in situations where the phase is subjected to various forms of stress such as high temperature, injection of aqueous solutions, etc.
7. To ensure a high separation efficiency the stationary liquid should form a homogeneous film at the surface of the support. A prerequisite for this is a small contact angle between the support and the liquid interface.

8. The stationary liquid should be capable of sufficiently dissolving the analytes, otherwise the retention becomes inacceptably low. As a rule of thumb, solubility is dictated by the chemical similarity between the stationary liquid and the analyte.
9. It should be immobilized in situ for purposes of capillary GC.

## Classes of Liquid Stationary Phases

### *Nonaromatic Hydrocarbons*

The retention of analytes on hydrocarbon stationary phases containing neither polar nor polarizable groups is determined solely by nonspecific dispersive forces. Thus, the apolar hydrocarbon solutes can be selectively retarded as opposed to polar solutes, both having similar boiling points. This retention behavior is useful in analyzing traces of polar compounds in an apolar matrix.

Due to their chemical inertness, apolar hydrocarbons are applicable to the resolution of nearly all of the volatile samples, even reactive ones. In operating these phases care should be taken to remove air because of the susceptibility of hydrocarbon liquids to oxidation, which results in the formation of hydroperoxides with tertiary hydrogen atoms. Representatives of strongly apolar hydrocarbons are squalane, the so-called Kovats $C_{87}$ hydrocarbon [331], poly-α-olefins [332], paraffin waxes and paraffin oils, and apiezones.

Squalane, produced by hydrogenation of squalene, ranks as the nonpolar reference phase to assess the McReynolds constants. However, squalane has not gained wide acceptance as the nonpolar standard due to its low temperature stability limit of 140°C and the possible presence of squalene. The latter falsifies the retention indices of the test substances dramatically (see Table 10). Impurities in squalane can be removed by a treatment on activated charcoal or alumina columns. In the course of time the place of squalane has been taken by either some apiezones [333] (usable up to 250°C) or polydimethylsiloxanes such as SE-30 (usable up to 350°C).

$$(CH_3)_2CH-(CH_2)_3-CH(CH_3)-(CH_2)_3-CH(CH_3)-(CH_2)_4-CH(CH_3)-(CH_2)_3-CH(CH_3)-(CH_2)_3-CH(CH_3)_2$$

Formula 2: Squalane

The $C_{87}$ hydrocarbon (available from Applied Science; see Formula 3) is an example of a liquid stationary phase specifically designed for GC. This tailor-made, well-characterized, and well-defined phase without large batch-to-batch variation has also been devised to substitute for squalane as nonpolar standard. However, its high price as well as difficulties encountered in the coating of $C_{87}$ hydrocarbon

**Table 10** McReynolds Constants of Hydrocarbon Liquids

| Phase | Temperature[a] min./max. (°C) | X (benzene) | Y (1-butanol) | Z (2-pentanone) | U (1-nitropropane) | S (pyridine) | Σ(X+Y+Z+U+S) |
|---|---|---|---|---|---|---|---|
| Nujol | 0/150 | 9 | 5 | 2 | 6 | 11 | 33 |
| $C_{87}$ hydrocarbon | /280 | 21 | 10 | 3 | 12 | 25 | 71 |
| Apiezon MH[b] | | 21 | 10 | 7 | 18 | 30 | 86 |
| Apiezon M (grease) | 50/300 | 31 | 22 | 15 | 30 | 40 | 138 |
| Apiezon L (grease) | 50/300 | 32 | 22 | 15 | 32 | 42 | 143 |
| Apiezon I (oil) | 20/250 | 38 | 36 | 27 | 49 | 57 | 207 |
| Apiezon N (grease) | 50/300 | 38 | 40 | 28 | 52 | 58 | 216 |
| Apiezon H (grease) | 20/250 | 59 | 86 | 81 | 151 | 129 | 506 |
| Squalene | | 152 | 341 | 238 | 329 | 344 | 1404 |
| Absolute squalane indices at 120°C | | 653 | 590 | 627 | 652 | 699 | |

[a]The statement of upper temperature limits is approximate, dependent on column technology, solid support modification, carrier gas purity, column type.
[b]H, hydrogenated.
*Source*: Manufacturer catalogues.

$$(H_{37}C_{18})_2CH-(CH_2)_4-C(C_2H_5)_2-(CH_2)_4-CH(C_{18}H_{37})_2$$

Formula 3: KOVATS' "$C_{87}$ HYDROCARBON"
24, 24 Diethyl-19,29-dioctadecylheptatetracontane

even on untreated fused capillary columns with high surface tension (quite apart from the fact that this phase does not wet high-temperature silylated surfaces) restrict its application. As will be shown later, this phenomenon is due to the low viscosity of the hydrocarbon.

The series of poly-α-olefins has been found to have advantageous properties as stationary phases in GC, offering McReynolds values lower than those of squalane.

Paraffin waxes and oils (e.g., Nujol) are mixtures of normal and branched alkanes, partly including naphthenes. They are less common due to their changeable composition and limiting temperature range.

The commercially available apiezones are branched aliphatic hydrocarbon mixtures obtained by subjecting lubricating oils to a high-temperature treatment. The residue is then purified. As shown by NMR and IR spectroscopy, some impurities such as low aromatics and carbonyl compounds are retained in the product, leading to a poor batch-to-batch reproducibility. The purification of apiezones is usually performed by passage through an alumina column using hexane as eluent [335]. Table 10 lists the McReynolds constants of some hydrocarbon stationary phases commercially available. Incidentally, it should be noted that the list is by no means complete. Many more phases can be found in commercial catalogues. The same holds true for the tables that follow.

A promising approach to the preparation of novel hydrocarbon phases was taken by the group of Sandra [343] by synthesizing an apolar gum phase named RSL-110 with a molecular weight of about 100,000. Today the hydrocarbon liquids are extensively replaced by polydimethylsiloxanes, the latter having better thermal stability, batch-to-batch reproducibility, and film-forming characteristics. The aromatic hydrocarbon liquids such as benzyldiphenyl, alkalnaphthalenes (available under the trade name of Fluhyzon), or trimeric stilbene are less commonly used.

### *Alcohols, Polyglycols, and Polyalkylene Oxides*

These phases, the chemistry of which was detailed in [338], were among the most popular for a long time [336,337]. They are capable of undergoing hydrogen bonding with appropriate analytes such as alcohols, acids, phenols, primary and secondary amines, resulting in a larger retention time compared to the corresponding nonpolar hydrocarbon analogs. Thus, the separation of apolar substances in polar matrices is easy to achieve by means of polar alcoholic phases. Typical members of low molecular alcohols are the multivalent alcohols such as glycerol, diglycerol, and the derivatives of carbohydrates such as inosit, erythrit, and sorbit. All of them, however, exhibit rather limited operating temperature ranges, e.g., erythrit can be operated in the temperature range 120-150°C only.

The high molecular weight alcohols have become more widely used than the monomers. The simultaneous presence of acceptor atoms (hydroxyl and ether oxygen) and donor atoms (hydroxyl hydrogen) in polyglycols and polyalkylene oxides enables their low molecular members in particular to enter into strong interactions with solutes having not only hydroxyl groups and primary amino groups but also carbonyl and secondary or tertiary amino groups. The lower the molecular weight of the stationary phase, the stronger the interaction with the polar analytes. As an example, the retention of butanol-1 decreases in the sequence polyethylene glycol PEG 400 > PEG 1000 > PEG 4000.

Given these considerations, the main application of these phases is to all substances containing oxygen or nitrogen as well as sulfur and halogens, e.g., solvent mixtures, flavor ingredients, fatty acid methyl esters. Although the solubility of hydrocarbons in these phases is insufficiently low, there is a reasonable selectivity toward branched and nonbranched saturated analytes. Using Carbowax 20M below its lower temperature limit, positional and spatial isomers of hydrocarbons were successfully resolved [339]. Figure 12 shows the separation of fatty acid methyl esters of rapeseed oil on a Carbowax 20M capillary column. The methyl esters with the same carbon number are eluted according to the degree of saturation.

When applying polyglycols and polyalkylene oxides, careful attention should be paid to thermal conditioning and the use of very dry, oxygen-free carrier gas [340]. For example, oxygen oxidizes polyethylene oxide to acetaldehyde and/or acetic acid, especially at higher temperatures. Acetaldehyde in turn can react with amines, anilines, and other classes of compounds. Consequently, polyglycols should not be applied for the determination of trace amines. In addition, the decomposition of polyglycols is substantially influenced by the nature of the support. For instance, Carbowax 20M is known to be much more stable on fused silica capillaries or borosilicate glass capillaries than on soda glass [341]. Polyglycols may contain high amounts of catalyst residues, which might catalyze phase decomposition.

Sandra et al. [342,343] prepared Superox phases (available from RSL, Belgium and Alltech, USA)–a refined version of Carbowax phases, especially devised for capillary GC–with low contents of catalyst residues. Superox-4 (average molecular weight 4,000,000) as a typical

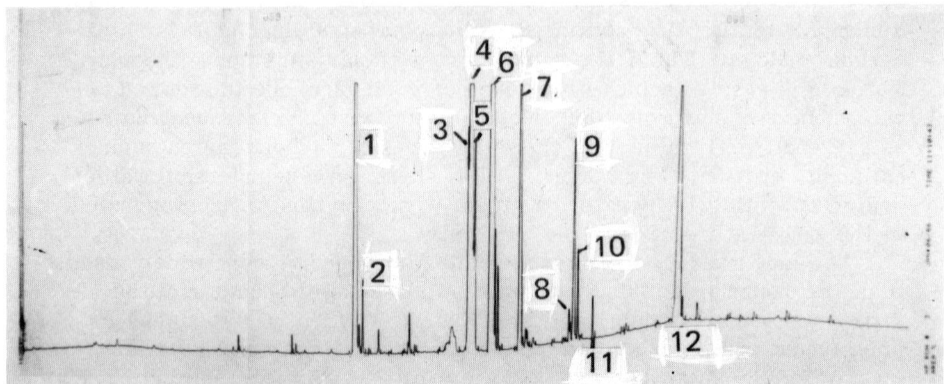


Fig. 12 Separation of rapeseed oil fatty acid methyl esters on Carbowax 20M. Glass column, 20 m × 0.25 mm i.d.; column temperature, hold 2 min at 60°C, then to 225°C at 4°/min. Peaks (abbr.): 1, 16:0; 2, 16:1; 3, 18:0; 4, 18:1 (oleic); 5, 18:1; 6, 18:2 (linoleic); 7, 18:3 (α-linolenic); 8, 20:0; 9, 20:1; 10, 20:1; 11, 20:2; 12, 22:1 (erucic). (From Pörschmann, Hommel unpublished results.)

member of that series has two favorable features: it can withstand high temperatures up to 300°C in continuous use and as a gum with high viscosity it adheres well to capillary surfaces (see below).

Tables 11 and 12 summarize the McReynolds constants for relevant polyethylene glycols and polyoxyethyleneoxypropylenes, respectively. It should be emphasized that the tables contain only a selection of stationary phases. For example, Chrompack offers a series of Carbowaxes including Carbowax 200, 300, 400, 550, 600, 750, 1000, 1500, 1540, 4000, 6000, 20M. The lower the molecular weight of the phases, the lower the operating temperatures as compared to Carbowax 20M.

For a better understanding of the structure of diverse representative stationary phases, their chemical compositions are listed below:

1. Polyethylene glycol

   $HO{-}CH_2{-}CH_2{-}({-}O{-}CH_2{-}CH_2{-})_n{-}O{-}CH_2{-}CH_2{-}OH$

   Trade name: Polyethylene glycol 400, 1000, . . ., Carbowax 400, 1500, . . . (Union Carbide).

2. Polyoxyethyleneoxypropylene

   $-({-}O{-}CH_2)_m\,({-}O{-}\underset{\displaystyle CH_3}{\underset{|}{C}}H{-}CH_2)_n{-}$

   Trade name: Pluronic P-84

**Table 11** McReynolds Constants for Selected Polyethylene Glycols, Polyethylene Glycol Derivatives, and Diglycerol

| Compounds | Temperature min./max. (°C) | X | Y | Z | U | S | Σ(X+Y+Z +U+S) |
|---|---|---|---|---|---|---|---|
| Ethofat 60/25 (polyethylene-glycolmonostearate) | 50/125 | 191 | 382 | 244 | 380 | 333 | 1530 |
| Emulphor ON-870 (polyethyleneglycoloctadecylether) | 0/200 | 202 | 395 | 251 | 395 | 344 | 1587 |
| Triton X-100 [polyethyleneglycolmono (tetramethylbutyl) phenylether] (M = 600) | 0/200 | 203 | 399 | 268 | 402 | 362 | 1634 |
| Tween 80/polyoxyethylene sorbitan monoleate | 20/125 | 227 | 430 | 283 | 438 | 396 | 1774 |
| Igepal CO-880 [nonylphenoxypoly-(ethyleneoxy)ethanol), N = 30; identical to Antarox CO-880 | 100/200 | 259 | 461 | 311 | 482 | 426 | 1939 |
| Igepal CO-900, as above, N = 100, identical to Antarox CO-990 | 60/220 | 298 | 508 | 345 | 540 | 475 | 2166 |
| Triton X-305 (as above, M = 1500) | 0/200 | 262 | 467 | 314 | 488 | 430 | 1961 |
| Carbowax 4000 | 60/220 | 282 | 496 | 331 | 517 | 467 | 2093 |
| Carbowax 20 M terephthalic acid | 60/225 | 321 | 537 | 367 | 573 | 520 | 2318 |
| Carbowax 4000 | 60/200 | 317 | 545 | 378 | 578 | 521 | 2339 |
| FFAP | 50/250 | 340 | 580 | 397 | 602 | 627 | 2546 |
| Carbowax 600 | 30/125 | 350 | 631 | 428 | 632 | 605 | 2646 |
| Carbowax 1540 | 50/175 | 371 | 639 | 453 | 666 | 641 | 2770 |

Table 11 (continued)

| Compounds | Temperature min./max. (°C) | X | Y | Z | U | S | Σ(X+Y+Z +U+S) |
|---|---|---|---|---|---|---|---|
| Diglycerol | 20/100 | 547 | 977 | 708 | 828 | 1001 | 4061 |
| CP Wax 51 40M ($M_w$ = 40,000) | 60/280 | | | | | | |
| CP Wax 600M ($M_w$ = 600,000) | 60/300 | | | | | | |
| CP Wax 4000M ($M_w$ = 4,000,000) | 60/325 | | | | | | |
| Superox-4[a] ($M_w$ = 4,000,000) | 140/300 | | | | | | |

[a]Retention indices for five McReynolds test compounds at 66.5°C are given in [342].
*Source*: Manufacturer catalogues and [57] with permission.

3. Polyoxyalkylene derivatives

$$R-(O-CH_2-CH_2)_m-(O-\underset{\displaystyle CH_3}{\underset{|}{C}H}-CH_2-)_n-R'$$

3.1 n = O, R = H, R' = $C_{17}H_{35}COO$

Trade name: Carbowax 600, 1000, . . ., monostearate, Ethofat 60/25

3.2 n = O, R = H, R' = $C_8H_{17}-C_6H_4-O-$

Trade name: Triton X-100 (m = 10), Triton X-305 (m = 30)

3.3 n = O, R = H, R' = $C_9H_{19}-C_6H_4-O-$

Trade name: Igepal CO-880 (m = 30), Igepal CO-990 (m = 100)

3.4 n = O, R = $C_{18}H_{37}-$, R' = $C_{18}H_{37}O-$

Trade name: Emulphor ON-860

3.5 n = O, R = $HOOC-C_6H_3(NO_2)-CO-$, R' = $HOOC-C_6H_3(NO_2)-COO-$

Trade name: FFAP, SP-1000

Table 12 McReynolds Constants for Selected Polyoxyethyleneoxypropylene Stationary Phases

| Phase | Temperature min./max. (°C) | X | Y | Z | U | S | Σ(X+Y+Z+U+S) |
|---|---|---|---|---|---|---|---|
| UCON LB[a] 550 X | 0/200 | 118 | 271 | 158 | 243 | 206 | 996 |
| UCON LB+ 1751 | 0/200 | 132 | 297 | 180 | 275 | 235 | 1119 |
| UCON 50 HB 280X | 0/200 | 177 | 362 | 277 | 351 | 302 | 1469 |
| UCON 50 HB 2000 | 0/200 | 202 | 394 | 253 | 392 | 341 | 1582 |
| UCON 50 HB 5100 | 0/200 | 214 | 418 | 278 | 421 | 375 | 1706 |
| UCON 75 H 90000 | 0/200 | 255 | 452 | 299 | 740 | 406 | 1882 |
| Pluronic L-81 | 0/200 | 144 | 314 | 187 | 289 | 249 | 1183 |
| Pluronic P-85 | 0/200 | 201 | 380 | 247 | 388 | 335 | 1551 |
| Pluronic P-65 | 0/200 | 203 | 394 | 251 | 383 | 340 | 1581 |
| Pluronic F-35 | 0/200 | 206 | 406 | 257 | 398 | 349 | 1616 |
| Pluronic F-68 | 0/200 | 264 | 465 | 309 | 488 | 423 | 1949 |
| Pluronic F-88 | 0/200 | 262 | 461 | 306 | 483 | 419 | 3880 |

[a]LB: Propylene glycol units 50%, thus weakly water-soluble character.
HB: Ethylene glycol units 50%, thus water-soluble character.
*Source*: From manufacturer catalogs.

3.6 n = O, R = H, R' = O-alkyl
Trade name: Ucon LB 1200X, LB 1715, LB 1800X

3.7 m ≅ n, R = H, R' = alkyl
Trade name: Pluronic F-68, F-88, L-35, L-81, P-65, P-85, UCON 50-HB-280X, UCON 50-HB-2000, UCON 50-HB-5100

3.8 n = O, R = $C_5H_8$–, R' = $C_{17}H_{33}COO$–
Trade name: Tween 80

By significant reaction of the polyglycol Carbowax 20M with 2-nitroterephthalic acid one obtains FFAP (*free fatty acid phase*), which has become a useful liquid stationary phase in the analysis of a wide variety of organic compounds, except for aldehydes and triazines, which are irreversibly adsorbed [344]. The Supelco phase SP-1000 is similar to FFAP but shows a greater thermal stability. Another product similar to FFAP is Carbowax 20M-TPA, prepared by reaction of Carbowax 20M with terephthalic acid.

The Pluronics marketed by Fluka (Switzerland), listed in the lower part of Table 12, were occasionally used as liquid stationary phases in the early period of GC but did not become popular. They were rediscovered by Grob [345]. Pluronics have a narrower molecular distribution than Carbowax 20M [346]. The polarities of Pluronic and Ucon members are lower than those of Carbowax 20M.

As will be outlined later, with the exception of Carbowax 20M little attention has been paid to the polyglycols and polyalkylene oxides by renowned capillary manufacturers. The main reasons for this are poor film-forming properties and temperature stability. This situation might change profoundly in future if chromatographers succeed in bonding the stationary phase with an OH-terminated group to the leached and dehydrated glass capillary surface (see below). Surely, the bonding procedure would result in improved thermal stability, film stability (and thus in higher column efficiency), and inertness as opposed to capillaries produced by roughening methods, e.g., by means of barium carbonate.

*Polyesters*

In the early days of GC, polyesters were extremely popular. Today they are largely replaced by polar polysiloxanes (see below), which show better batch-to-batch reproducibility and film-forming characteristics, or, to a lesser extent, by polyglycols. Since esters possess carboxyl oxygen atoms with donor properties which can enter into hydrogen bonds, the retention of solutes with protonated functional groups is high on such stationary phases. Polyesters are among those phases that exhibit the highest polarity.

Members of the monomeric esters are phthalates, adipates, and sebacates such as dinonylphthalate, didecylphthalate, and dioctylsebacate. All of these have very poor thermal stability. In contrast to monomeric esters, the thermal stability of polyester phases ranges between 200 and 225°C. As the name suggests, polyesters are macromolecules formed by a multitude of carboxylic ester linkages, derived from the reaction of a polybasic acid and a polyhydric alcohol. The initially good thermal stability was notably improved by 40-60°C using a proprietary process (Analabs). Some other companies, among them Supelco and United Technologies, offer polyesters stabilized with polysiloxanes (see below and Table 16). Such phases show slightly improved temperature stability and decreased polarity compared to the unmodified polyesters. However, a literature search reveals that the use of organosilicone polymers has markedly decreased at the expense of polar polysiloxanes.

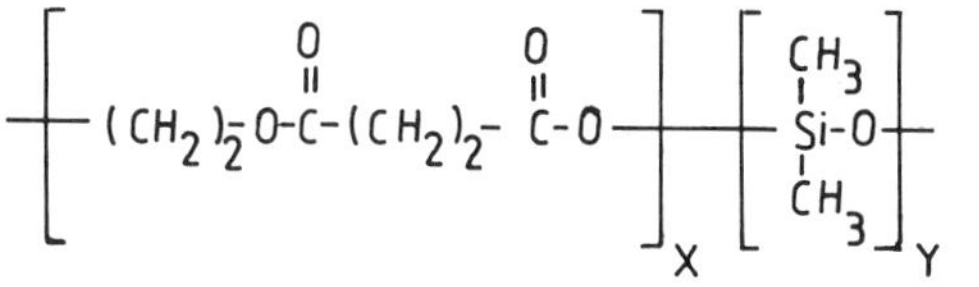

$$-\left[(CH_2)_2-O-\overset{O}{\overset{\|}{C}}-(CH_2)_2-\overset{O}{\overset{\|}{C}}-O\right]_X-\left[\underset{CH_3}{\overset{CH_3}{Si}}-O\right]_Y-$$

EGSS-X: Y/X < 1
EGSS-Y: Y/X > 1

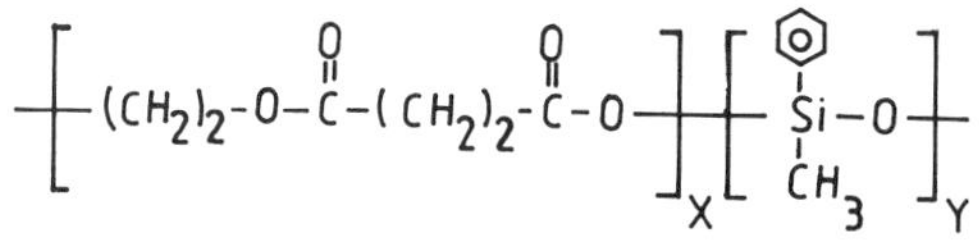

$$-\left[(CH_2)_2-O-\overset{O}{\overset{\|}{C}}-(CH_2)_2-\overset{O}{\overset{\|}{C}}-O\right]_X-\left[\underset{CH_3}{\overset{C_6H_5}{Si}}-O\right]_Y-$$

EGSP-A: Y/X < 1
EGSP-Z: Y/X > 1

$$-\left[(CH_2)_2-O-\overset{O}{\overset{\|}{C}}-(CH_2)_2-\overset{O}{\overset{\|}{C}}-O\right]_X-\left[\underset{CH_2CH_2CN}{\overset{CH_3}{Si}}-O\right]_Y-$$

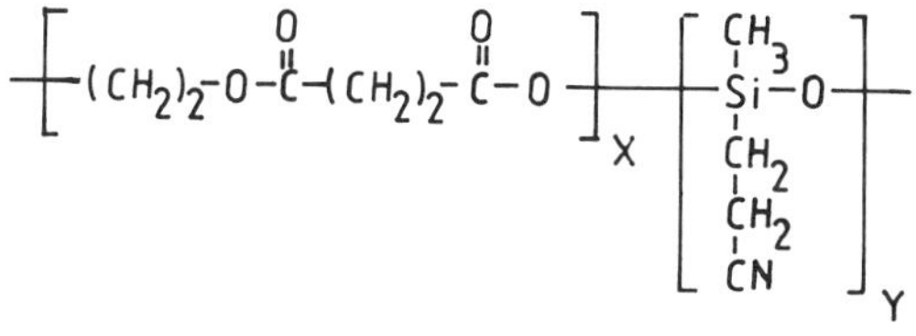

ECNSS-S: Y/X < 1
ECNSS-M: Y/X > 1

Formula 4

Typical polyesters are listed in Table 13. The most important one is polydiethylene glycol succinate (DEGS), frequently used to analyze methyl esters of saturated and unsaturated fatty acids as well as pesticides [348]. As indicated by the McReynolds constants listed in Table 13, the number of methylene groups of polyesters has an essential importance to the polarity of the phase.

Interesting polyesters are offered by Supelco, namely, SP-216-PS and SP-222-PS, which are stabilized with phosphoric acid for separating $C_{12}$-$C_{20}$ free fatty acids and their methyl esters. The less polar ester phases SP-1200 and SP-1220 are also intended for the separation of free fatty acids.

Table 13 McReynolds Constants for Polyester Stationary Phases[a]

| Polyester type | Symbol | Temperature min./max. (°C) | X | Y | Z | U | S | Σ(X+Y+Z+U+S) |
|---|---|---|---|---|---|---|---|---|
| Polyneopentylglycolsuccinate, identical to Lac 18-R-767 | NPGS | 50/200 | 275 | 472 | 367 | 543 | 489 | 2146 |
| Polybutanediol succinate, identical to Lac 6-R-860 | BDS | 50/225 | 367 | 548 | 436 | 645 | 564 | 2560 |
| Polyethylene glycol adipate, identical to Lac 13-R-741 | EGA | 100/200 | 378 | 564 | 448 | 641 | 606 | 2637 |
| Polyphenyldiethanolaminesuccinate | PDEAS | 15/210 | 387 | 543 | 470 | 673 | 625 | 2698 |
| Stabilized EGA | | 100/260 | 370 | 582 | 452 | 652 | 666 | 2722 |
| Polydiethylene glycol adipate, identical to Lac 1-R-296 | DEGA | 0/200 | 382 | 600 | 458 | 666 | 691 | 2797 |
| Stabilized DEGA | | 0/260 | 372 | 760 | 432 | 652 | 649 | 2875 |
| Polydiethylene glycol succinate, identical to Lac 3-R-728 | DEGS | 20/200 | 504 | 753 | 590 | 844 | 855 | 3546 |
| Stabilized DEGS | | 20/240 | 509 | 757 | 584 | 851 | 895 | 3596 |
| Polyethylene glycol succinate, identical to Lac 4-R-886 | EGS | 100/210 | 545 | 779 | 642 | 908 | 881 | 3755 |
| SP-216-PS | | 25/200 | 632 | 875 | 733 | 1000 | 680 | 3920 |

[a]McReynolds constants of polyesters offered by Supelco (Cat. 21, 1983) differ considerably from those offered by Analabs. A comprehensive survey on polyesters including Lac series is given by Packard Catal. (GC Systems).
*Source*: From [347], with permission

In the use of polyesters attention should be paid to the following:

1. Polyesters must be very well protected from contact with water and oxygen. The more hydrophobic the glycols and dicarboxylic acids, the higher the stability of polyesters toward hydrolysis. Oxidation leads to the formation of hydroperoxides, which decompose the polymer chain by means of free radicals.
2. Polyesters are degraded by strong acids and bases. In the presence of catalyst fragments, transesterification of alcohols, weak acids, and esters will occur in the polyester phase.
3. Compounds capable of reacting with residual carboxyl or hydroxyl groups, e.g., epoxides, isocyanates, anhydrides, halo acids, and acyl halides, should not be analyzed on these phases.
4. Since polyesters are hygroscopic, a special conditioning procedure is necessary [350].
5. Published McReynolds constants should be viewed with caution because they may be based on different batch lots or different manufacturers [336].

For a better understanding, some important polyesters will be given by their formulae:

- Polydiethylene glycol succinate (DEGS)

$$(-CH_2-CH_2-O-CH_2-CH_2-O-CO-CH_2-CH_2-CO-O-)_n$$

- Polydiethyleneglycol adipate (DEGA)

$$(-CH_2-CH_2-O-CH_2-CH_2-O-CO-(CH_2)_4-CO-O-)_n$$

- Polybutanediol succinate (BDS)

$$(-CH_2-CH_2-CH_2-CH_2-O-CO-(CH_2)_2-CO-O-)_n$$

### *Amines and Amides*

Aliphatic amines capable of forming hydrogen bonds provide a high selectivity toward alcohols, pyridines, glycols, and mercaptans. Owing to the high values of Y, Z, U, and S compared to X, aliphatic amines permit the separation of aromatics in the presence of oxygen-containing compounds.

Aromatic amines with $\pi$-electron systems may interact with aromatic hydrocarbons. Thus, they display a high selectivity in difficult separations of isomeric alkylbenzenes, e.g., xylenes [351], as found with some adsorbents (see above).

Among the amides, dimethylformamide and formamide were frequently employed in the analysis of low-boiling hydrocarbons in the early period of GC. As GC matured these low-boiling phases were

**Table 14** McReynolds Constants for Amine and Amide Stationary Phases

| Phase | Type | Description | Temperature min./max. (°C) | X | Y | Z | U | S | Σ(X+Y+Z+U+S) |
|---|---|---|---|---|---|---|---|---|---|
| Armeen SD | Amine | $C_{16}/C_{18}$ primary amines | 35/80 | 44 | | | | 78 | |
| Amine 220 | Amine | 1-Ethanol-2-(heptadecyl)-2-isoimidazole | 0/180 | 117 | 380 | 181 | 293 | 133 | 1105 |
| Polypropyleneimine | Amine | | 0/200 | 122 | 425 | 168 | 263 | 224 | 1202 |
| Quadrol | Amine | *N*,*N*,*N*,*N*-Tetrakis-(2-hydroxypropyl)-ethylene | 0/150 | 214 | 571 | 357 | 472 | 489 | 2103 |
| Polyethyleneimine | Amine | | 0/180 | 322 | 800 | | 573 | 524 | |
| THEED | Amine | Tetrahydroxyethyl-ethylene diamine | 0/150 | 463 | 942 | 626 | 801 | 893 | 3725 |
| Hallcomid M-18 | Amide | Dimethylstearamide | 30/150 | 79 | 268 | 130 | 222 | 146 | 845 |
| Hallcomid M-18-OL | Amide | Dimethyloleamide | 30/150 | 89 | 280 | 143 | 239 | 165 | 916 |
| Versamid 930 | Amide | Polyamide resin | 115/150 | 108 | 309 | 137 | 208 | 207 | 969 |
| Poly-A 103 | Amide | Polyamide | 70/275 | 115 | 331 | 149 | 263 | 214 | 1072 |
| Poly-I 110 | Amide | Polyamide | 90/275 | 115 | 357 | 151 | 262 | 214 | 1099 |
| Poly-A 135 | Amide | Polyamide | 70/250 | 163 | 389 | 168 | 340 | 269 | 1329 |
| Poly-A 101a | Amide | Polyamide | 50/275 | | | | | | |

*Source*: From manufacturer catalogues.

replaced by others. Elongation of the carbon chain (e.g., Hallcomid; see Table 14) leads to improved temperature stability but sharply decreased selectivity.

By condensation of piperidines with dicarboxylic acids, Mathews et al. [352] succeeded in preparing liquid phases that were stable up to 275°C (see Table 14, where other amines are also listed).

### *Nitriles and Nitrile Ethers*

Nitrile groups are known as strong electron acceptors due to their high electronegativity, and they therefore selectively retain compounds of low ionization energy. In addition to the charge-transfer interactions, strong orientation forces and hydrogen bonds are effective. Thus, favorable selective interactions may reinforce each other and are desirable.

Consequently, the separation of saturated from unsaturated hydrocarbons, in particular aromatics, may be successfully performed by means of nitriles and nitrile ethers. The first nitrile phases used in GLC were β,β'-oxydipropionitrile and 1,2,3-tris(2-cyanoethoxy)propane [353]. One of the most polar phases is N,N-bis(2-cyanoethyl)-formamide, which has not gained wide acceptance due to poor thermal stability.

Table 15 lists nitriles and nitrile ethers that offer the highest retention for organic compounds. Selectivities are similar to that of OV-225 (Table 16) but are more pronounced. As can be inferred from Table 15, all of these phases have an insufficient range of operating temperature.

### *Poly-m-phenylethers*

Poly-*m*-phenylethers such as OS-124 (five-ring polyphenylether) or OS-138 (six-ring polyphenylether) exhibit an excellent thermal stability up to 400°C and hence are preferred in high-temperature gas chromatographic work. Owing to this favorable feature, polyethers have proved very useful for the separation of aromatic hydrocarbons from olefins and alkanes [354].

Polyphenylethers have been modified in several ways: Mathews et al. [35]] synthesized sulfonated polyphenylethers such as Poly S-179 (temperature range 200-400°C) and Poly S-176 (150-400°C) by reacting dinuclear aromatic sulfonyl chlorides with five- or six-ring polyphenylethers using Friedel-Crafts catalysts. The structure of a polyoxyarylsulfonylarylene is depicted in formula 5 (from [355]).

Schomburg et al. [356] expanded the scope of Poly S-179 to capillary GC. Sufficient separations of high-boiling hydrocarbons and sterols were achieved. The polarity of Poly S-179 is similar to that of Carbowax 20M [357], i.c., unsaturated compounds are retained more strongly than the corresponding saturated ones.

**Table 15** McReynolds Constants for Nitriles and Nitrile Ethers

| Phase | Formula | Temperature (min./max.) (°C) | X | Y | Z | U | S | Σ(X+Y+Z+U+S) |
|---|---|---|---|---|---|---|---|---|
| β,β'-Oxydipropionitrile | $NC-CH_2-CH_2-O-CH_2-CH_2-CN$ | 0/75 | 588 | | | | 919 | |
| Tris(2-cyanoethyl)-nitromethane | $NO_2-C(-CH_2-CH_2-CN)_3$ | 20/140 | 635 | | | | | |
| Tetracyanoethylated pentaerythritol | $C(CH_2O-CH_2-CH_2-CN)_4$ | 30/175 | 562 | 782 | 677 | 920 | 837 | 3778 |
| 1,2,3,4,5,6-Hexakis-(2-cyanoethoxy) cyclohexane | $C_6H_6(-O-CH_2-CH_2-CN)_6$ | 125/150 | 567 | 825 | 713 | 978 | 901 | 3984 |
| 1,2,3-Tris(2-cyano-ethoxy)propane | $CH_2-O-CH_2-CH_2-CN$<br>$CH-O-CH_2-CH_2-CN$<br>$CH_2-O-CH_2-CH_2-CN$ | 0/125 | 594 | 857 | 759 | 1031 | 917 | 4158 |
| 1,2,3,4-Tetrakis-(2-cyanoethoxy) butane | $CH_2-O-(CH_2)_2-CN$<br>$CH-O-(CH_2)_2-CN$<br>$CH-O-(CH_2)_2-CN$<br>$CH_2-O-(CH_2)_2-CN$ | 110/200 | 617 | 860 | 773 | 1048 | 941 | 4239 |
| Cyanoethyl sucrose | | 20/200 | 647 | 919 | 797 | 1043 | 976 | 4382 |
| *N,N*-bis(2-cyanoethyl) formamide | $H-CO-N(-CH_2-CH_2-CN)_2$ | 0/125 | 690 | 991 | 853 | 1110 | 1000 | 4644 |

*Source*: From manufacturer catalogues and [57], with permission.

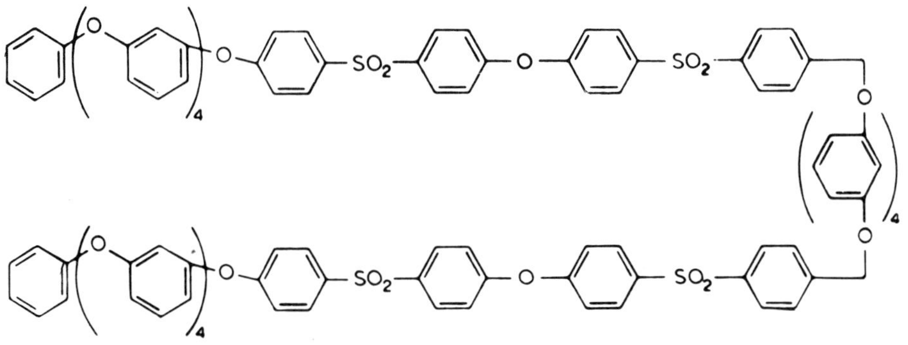

Formula 5

Polyphenylethers of different polarity were prepared by Poole et al. [358]; Hammann et al. [359] prepared polyphenylethers with a biphenyl central core.

*Polysiloxanes*

Polysiloxanes represent the major group of liquid stationary phases in GC and have been reviewed in several papers [2,334,360] which report a multitude of polysiloxanes available from the beginning (Ohio Valley Specialty Company) to the present. Generally, three types of polysiloxanes are commercially available: oils, gums, and rubbers. The first type comprises linear polysiloxanes. Silicone gums are linear, high molecular weight compounds. The term "rubber" denotes a cross-linked silicone gum. As shown later, rubbers yield higher efficiencies than oils in capillary GC (see below). Thus, in this case the prediction of diffusivity and hence efficiency [cf. Eq. (7)] on the basis of viscosity is completely misleading [361].

The wide application of polysiloxanes in GC can be attributed to several extraordinary features:

1. Polysiloxanes are chemically inert and cover nearly the entire range of chromatographic polarities by substituting various amounts of pendant phenyl, fluoroalkyl, and cyanopropyl groups for methyl in the "backbone" polymer structure.

2. They can be used over a wide temperature range extending from 0°C up to 250°C for cyano- and trifluoroalkylpolysiloxanes and up to 350°C for methyl- and phenyl-containing phases, provided the carrier gas is free of traces of oxygen and water. Thereby the influence of oxygen is less significant than for polyglycols. The high physical temperature stability can be related to the small influence of temperature on viscosity, which ensures stable films even at elevated

temperatures (see below). The slight viscosity-temperature dependence is assumed to be caused by the coiled helical structure of polysiloxanes, with the hydrocarbon moiety protruding outwards. This helical structure is disturbed in the presence of bulky side groups, which leads to a stronger dependence of viscosity on temperature.

In addition to their outstanding physical thermal stability, an excellent chemical thermal stability is observed due to the high binding energy of Si-O and Si-C bonds, the effective strength of these bonds being greatly influenced by neighboring atoms. Phenyl and methyl groups withstand higher thermal stress than polar groups. This can be attributed to the fact that the activation energy of decomposition is diminished in the presence of hetero atoms within the polysiloxane. Therefore, polysiloxanes having methyl and phenyl groups only are preferentially applied to GC/MS coupling owing to their low bleeding [362].

By the way, starting from the premise that the linkage between silicon atoms and phenyl groups is very strong and arylene groups reduce the chain flexibility of the polymer chain, a promising approach would be to replace certain oxygen atoms in the polysiloxane by phenyl groups, which would presumably lead to better selectivity toward aromatics and some other classes of compounds and even better thermal stability of the polysiloxane. By introducing enantioselective or liquid crystalline groups into the phenyl backbone, new horizons could be opened.

Polysiloxanes, in particular gum phases [363], exhibit an extraordinarily high diffusivity as distinct from other polymers, which results in high column efficiencies.

As is the case with all polymers, there is some batch-to-batch variation in the molecular weight distribution, leading to both slight differences in retention characteristics and reduced thermal stability (cf. [364]). Low molecular weight impurities are removed by conditioning the column above the working temperature. Conditioned columns can be operated below the aging temperature for a long time, provided that equilibration catalysts have been removed or at least neutralized. In the presence of catalyst traces, strong acids or bases injected onto silicone phases may cause depolymerization. Catalyst residues are usually removed by reprecipitation, conducted by the dropwise addition of ethanol/water (85:15 v/v) to a solution of the polymer in chloroform, acetone, or ethyl acetate. Then the polymer is washed twice with ethanol (85% v/v) and dried over phosphorpentoxide.

In addition to the decomposition pattern in the column, polysiloxanes might be decomposed during the coating procedure [365,366]. For instance, the molecular weight of SE-52 or SE-54 will be dramatically reduced if the solvent (e.g., chloroform) contains traces of hydrochloric acid. In order to block active hydroxyl groups, commercially available polysiloxanes are usually end-capped. Hydroxyl-terminated phases may be very useful in the future, as pointed out below.

*Polydimethylsiloxanes* Polydimethylsiloxanes are among the most popular phases in GLC. As mentioned, they have a nearly constant viscosity at different temperatures owing to the coiled helical structure. From the general formula (see below) it becomes evident that polydimethylsiloxanes exhibit low polarity (cf. [367]). Thus, they tend to separate solutes according to their boiling points.

$$CH_3-\underset{CH_3}{\overset{CH_3}{Si}}-O\left[Si-O\right]_n Si-CH_3$$

Formula 6

Table 16 lists some properties of polydimethylsiloxanes offered as chromatographic specialties by renowned companies. The table comprises only products prepared for chromatographic use. This is also true for the subsequently mentioned types of polysiloxanes. Industrial, miscellaneous, and obsolete polysiloxanes, recommended for use in GLC, are reviewed in [334,360]. The GC grades offer a narrower molecular weight distribution throughout, and thus provide better separations [367]. A very popular commercial grade phase is SE-30. Of eight different liquid phases, SE-30 was the only phase to elute a variety of drugs under study [368]. Polyalkylsiloxanes such as DC-730 (ethyl substitution) that exhibit a reduced polarity compared with methyl polysiloxanes are less extensively used in practice (possibly with the exception of the Soviet Union as indicated by [360]) owing largely to reduced thermal stability.

Highly viscous purified polydimethylsiloxanes are applicable up to 350°C in the absence of oxygen and water. This feature has proved favorable in analyzing high-boiling substances as polycyclic aromatic hydrocarbons, terpenes, or derivatized steroids. The separation of underivatized fats on mostly short OV-1 capillary columns can serve as an example for this (Fig. 13). Although the temperature-programmed run extends to 365°C, excessive bleeding does not occur. For these purposes, high temperatures of about 365°C have proved necessary to elute triglycerides with up to nearly 60 acyl carbon atoms, even when a capillary column with a low film thickness is used. For separating triglyceride isomers according to the degree of unsaturation, the mixed phase SE-30/$AgNO_3$ is recommended [369].

*Polymethylphenylsiloxanes* Historically, the phenyl-substituted polysiloxane DC-550 with 25% phenyl substitution was the first phase

Table 16 Specialty Polysiloxanes for GC

| Phase | Substituent | Supplier | Molecular weight | Nominal viscosity at 25°C (cS) | Maximum working temperature (°C) | McReynolds constants | | | | |
|---|---|---|---|---|---|---|---|---|---|---|
| | | | | | | X | Y | Z | U | S |
| OV-1 | 100% methyl | Ohio Valley | 300,000-400,000 | Gum | 350 | 16 | 55 | 44 | 65 | 42 |
| SP-2100 | 100% methyl | Supelco | | 600 | 350 | 17 | 57 | 45 | 67 | 43 |
| ASI-100 methyl | 100% methyl | Applied Sci. | | 12,500 | | 17 | 55 | 45 | 67 | 43 |
| OV-101 | 100% methyl | Ohio Valley | 30,000 | 1200-1500 | 350 | 17 | 57 | 45 | 67 | 43 |
| SE-30 GC | 100% methyl | General Electric | $1.0\text{-}2.5 \times 10^6$ | $9.5 \times 10^6$ | 300 | 15 | 53 | 44 | 64 | 41 |
| OD-1 | 100% methyl | Analabs | | Gum | | 16 | 53 | 45 | 66 | 42 |
| JXR | 100% methyl | Applied Sci. | | Gum | 300 | 15 | 53 | 45 | 64 | 41 |
| CP SiL-5 | 100% methyl | Chrompack | | | 350 | | | | | |
| CP SiL-8 | 5% phenyl | Chrompack | | | 350 | | | | | |
| OV-73 | 5.5% phenyl | Ohio Valley | 800,000 | Gum | 325 | 40 | 86 | 76 | 114 | 85 |
| OV-3 | 10% phenyl | Ohio Valley | 20,000 | 500 | 350 | 44 | 86 | 81 | 124 | 8 |
| OV-7 | 20% phenyl | Ohio Valley | 10,000 | 500 | 305 | 69 | 113 | 111 | 171 | 128 |
| OV-61 | 33% phenyl | Ohio Valley | 40,000 | 50,000 | 350 | 101 | 143 | 142 | 213 | 174 |
| OV-11 | 35% phenyl | Ohio Valley | 7,000 | 500 | 350 | 102 | 143 | 145 | 219 | 178 |

| | | | | | | | | | | |
|---|---|---|---|---|---|---|---|---|---|---|
| OV-17 | 50% phenyl | Ohio Valley | 4,000 | 1300-1500 | 375 | 119 | 158 | 162 | 243 | 202 |
| SP-2250 | 50% phenyl | Supleco | | | 375 | 119 | 158 | 162 | 243 | 202 |
| ASI-50[a] methyl | 50% phenyl | Applied Sci. | | | | 119 | 159 | 162 | 243 | 202 |
| OV-22 | 65% phenyl | Ohio Valley | 8,000 | 50,000 | 350 | 160 | 188 | 191 | 283 | 253 |
| OV-25 | 75% phenyl | Ohio Valley | 10,000 | 100,000 | 350 | 178 | 204 | 208 | 305 | 280 |
| SP-400 | 11% chlorophenyl (like DC 560) | Supelco | | | 350 | 32 | 72 | 70 | 100 | 68 |
| OV-202 | 50% trifluoropropyl | Ohio Valley | 10,000 | 500 | 275 | 146 | 238 | 358 | 468 | 310 |
| OV-215 | 50% trifluoropropyl | Ohio Valley | 200,000 | Gum | 275 | 149 | 240 | 363 | 478 | 315 |
| OV-210 | 50% trifluoropropyl | Ohio Valley | 18,600 | 10,000 | 275 | 146 | 238 | 358 | 468 | 310 |
| SP-2401 | 50% trifluoropropyl | Supelco | 2,600 | 700 | 275 | 146 | 238 | 358 | 468 | 310 |
| ASI-50 methyl[b] | 50% trifluoropropyl | | | | | 146 | 238 | 358 | 468 | 310 |
| OV-105 | 5% cyanopropyl | Ohio Valley | | 1,500 | 275 | 36 | 108 | 93 | 139 | 86 |
| AN-600 | 25% cyanoethyl | Analabs | | Gum | | 202 | 369 | 332 | 482 | 408 |
| OV-225 | 25% cyanopropyl<br>25% phenyl | Ohio Valley | 8,000 | 9,000 | 265 | 228 | 369 | 338 | 492 | 386 |
| ASI-50 methyl[c] | 25% cyanopropyl<br>25% phenyl | Applied Sci. | | | | 228 | 369 | 338 | 492 | 386 |
| SILAR 5 CP | 50% phenyl<br>50% cyanopropyl | Dist. by Applied Sci. | | | 250 | 319 | 495 | 446 | 637 | 530 |
| SP-2300 | 50% phenyl<br>50% cyanopropyl | Supelco | | | 275 | 319 | 495 | 446 | 637 | 530 |

Table 16 (continued)

| Phase | Substituent | Supplier | Molecular weight | Nominal viscosity at 25°C (cS) | Maximum working temperature (°C) | McReynolds constants | | | | |
|---|---|---|---|---|---|---|---|---|---|---|
| | | | | | | X | Y | Z | U | S |
| CP SiL-58 | 50% phenyl<br>50% cyanopropyl | Chrompack | | | 275 | | | | | |
| CP SiL-76 | 75% cyanopropyl<br>25% phenyl | | | | 275 | | | | | |
| SILAR 7CP | 75% cyanopropyl<br>25% phenyl | Dist. by Applied Sci. | | | 250 | 440 | 638 | 605 | 844 | 673 |
| SP-2310 | 75% cyanopropyl<br>25% phenyl | Supelco | | | 275 | 440 | 637 | 605 | 840 | 670 |
| CP SiL-84 | 90% cyanopropyl<br>10% phenyl | Chrompack | | | 275 | | | | | |
| SILAR 9CP | 90% cyanoproply<br>10% phenyl | Dist. by Applied Sci. | | | 275 | 489 | 725 | 631 | 913 | 778 |
| SP-2330 | 90% cyanopropyl<br>10% phenyl | Supelco | | | 275 | 490 | 725 | 630 | 913 | 778 |
| SILAR 10C | 100% cyanopropyl | Dist. by Applied Sci. | | | 250 | 523 | 755 | 659 | 942 | 801 |
| SP-2340 | 100% cyanopropyl | Supelco | 5,000 | 20,000 | 275 | 520 | 757 | 659 | 942 | 800 |
| CP SiL-88 | 100% cyanopropyl | Chrompack | | | 275 | | | | | |
| OV-275 | 100% cyanopropyl | Ohio Valley | | | 250 | 629 | 872 | 763 | 1106 | 849 |

| | | | | | | | | | | |
|---|---|---|---|---|---|---|---|---|---|---|
| OV-1701 | 7% cyanopropyl 7% phenyl | Ohio Valley | | Gumlike | 300 | | | | | |
| SP-2125 | 5% phenyl 2% cyanopropyl | Supelco | | Gum | | | | | | |
| Organcsilicone Polymers | | | | | | | | | | |
| OV-330 | Phenylsiloxane-Carbowax copolymer | Ohio Valley | 5,000 | 500 | 250 | 222 | 391 | 273 | 417 | 368 |
| EGSS-X | Polyethylenesuccinate-dimethylsiloxane, low methyl | Appied Sci. | | | 225 | 484 | 710 | 585 | 831 | 778 |
| EGSS-Y | Polyethylenesuccinate-dimethylsiloxane, medium methyl | Applied Sci. | | | 230 | 391 | 597 | 493 | 693 | 661 |
| EGSP-A | Polyethylenesuccinate-methylphenylsiloxane, low phenyl | Applied Sci. | | | 220 | 397 | 629 | 519 | 727 | 700 |
| EGSP-Z | Polyethylenesuccinate-methylphenylsiloxane, medium phenyl | Applied Sci. | | | 230 | 308 | 474 | 399 | 548 | 549 |
| ECNSS-S | Polyethylenesuccinate-methylcyanoethylsiloxane, low cyanoethyl | Applied Sci. | | | 210 | 439 | 659 | 566 | 820 | 722 |
| ECNSS-M | As above, medium cyanoethyl | Applied Sci. | | | 210 | 421 | 690 | 581 | 803 | 732 |

Table 16 (continued)

| Phase | Substituent | Supplier | Molecular weight | Nominal viscosity at 25°C (cS) | Maximum working temperature (°C) | McReynolds constants | | | | |
|---|---|---|---|---|---|---|---|---|---|---|
| | | | | | | X | Y | Z | U | S |
| Carborane-siloxane polymers | | | | | | | | | | |
| Dexsil 300 GC | Polymethylcarborane/methylsiloxane = 1:4 | Dist. by Analabs | 16,000-20,000 | | 400 | 47 | 80 | 103 | 148 | 96 |
| Dexsil 400 GC | Polymethylcarborane/methylphenylsiloxane = 1:5 | Dist. by Analabs | 12,000-16,000 | Viscous | 400 | 59 | 114 | 140 | 187 | 173 |
| Dexsil 410 GC | Polymethylcarborane/methylcyanoethylsiloxane = 1:5 | Dist. by Analabs | 9,000 | Viscous | 400 | 85 | 165 | 169 | 242 | 180 |

[a] ASI 50 methyl/50 phenyl.
[b] ASI 50 methyl/50 trifluoropropyl.
[c] ASI 50 methyl/25 cyanopropyl/25 phenyl.
*Source*: From Haken [334] and manufacturer catalogues.

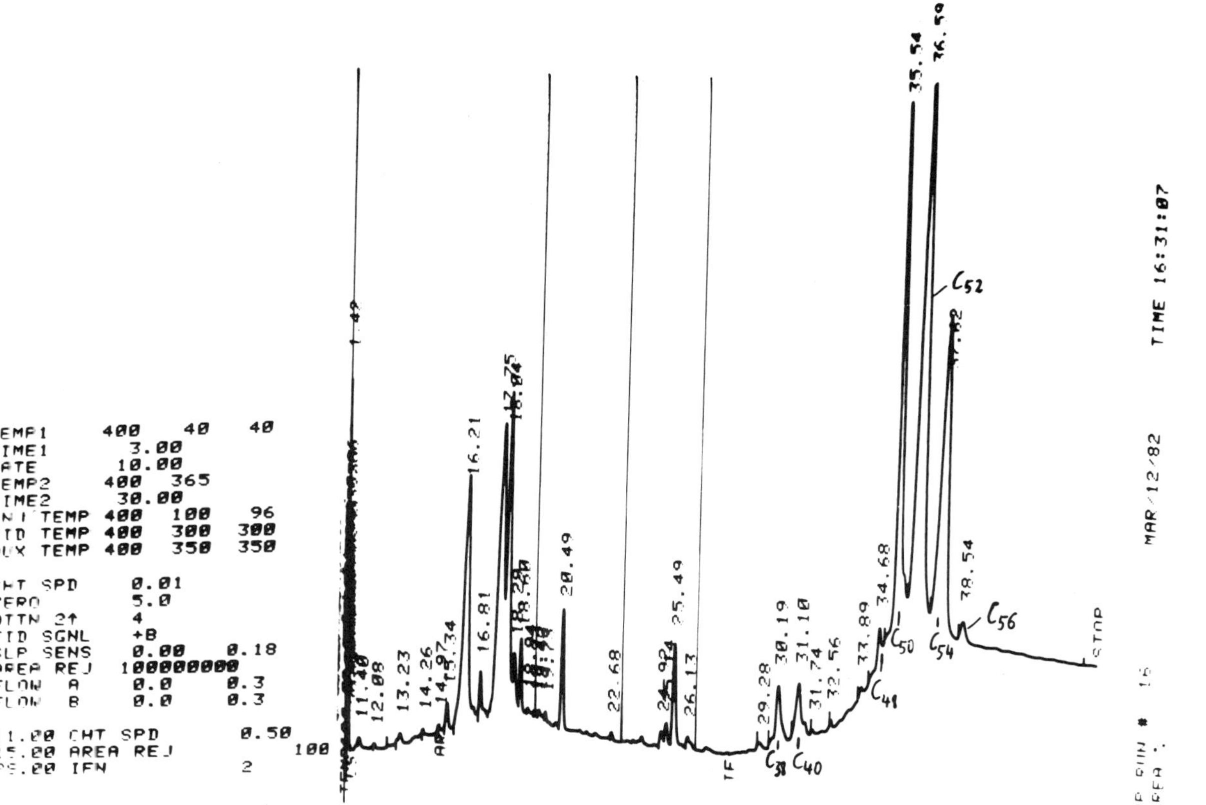


Fig. 13 Separation of an underivatized cacao fat on a short OV-1 glass capillary column; column, 5 m × 0.20 mm i.d.; film thickness, 0.15 μm; $C_{50}$, etc., indicate the number of acyl carbon atoms; column temperature, see Fig. 13. (From Pörschmann, Welsch, Püchner, unpublished results.)

ever used in GC by James and Martin. In the course of time the older phases were replaced by products specifically manufactured for GC purposes. The substitution of phenyl for methyl leads to an enhanced intrinsic thermal stability due to oxygen lone pair electrons being drawn into the bond, to a hindered oxidizing attack, as well as to reduced efficiency due to the slower diffusion of hydrocarbons into phenylsiloxanes [372]. As shown in [370], in using polymethylphenylsiloxanes, the optimum phenyl content seems to be about 65%. The oxidation attack toward polymethylphenylsiloxanes was detailed by Evans [373]. Reactions comprise the oxidation of the methyl groups to form carbonyl groups or the rupture of the silicon-carbon bond, the latter leading to the formation of a polymer gel (see also [374]). Recently, OV-17 (see below) was modified by heating to 340 and 380°C in oxygen, resulting in improved peak symmetry of polar solutes and a slight reduction of the polarity of the stationary phase [375]. In the authors' opinion, crosslinking phenomena are the most probable cause for these changes.

In terms of retention behavior, the substitution of a more polar moiety for an alkyl group commonly leads to more advantageous selectivity toward polar solutes. The polarities of several polymethylphenylsiloxanes have been detailed in [370]. The differential polarity of the phases under study depends on differential polarizability effects (see also [364]) and in addition, a low level of hydrogen-bonding acceptor strength is observed. The effect of increasing phenyl substitution on the Rohrschneider constants is shown in Fig. 14.

The relationship between selectivity and polarizability has prompted the research group of Lee to develop phases substituted with strongly polarizable groups such as biphenyl, naphthyl, or phenoxyphenyl functional groups [371]. With a 25% biphenyl-substituted phase the successful separation of, in particular, aromatic isomers with very similar properties has been accomplished without excessive retention of these isomers. Such outstanding separations could not be achieved with more polar cyanopropyl silicones (see below). Presumably, the "softer" dipole-induced dipole interaction between the polar solute and polarizable stationary phase was more sensitive than the dipole-dipole interaction between the polar solute and the polar cyano group, which leads to higher selectivities toward minor differences in molecular structure. Obviously, to meet the ever growing demand for polar and medium polar phases, further attempts are expected in the future from Lee's group and others to create unique selectivities by positioning a large number of polar order medium polar groups on the phenyl ring.

At present, one of the most widely used liquid stationary phases is OV-17, offered as 50% phenyl, with an upper temperature limit of 375°C after thorough conditioning. NMR and UV adsorption studies indicate a content of phenyl groups of 42% [364]. This finding coincides

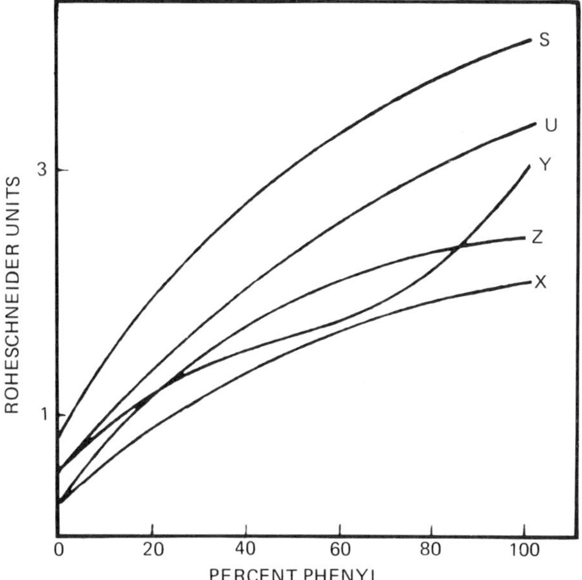


Fig. 14 Relationships between Rohrschneider constants and phenyl substitution. (From Ref. 334, with permission.)

with the data of [360], where a molecular weight of nearly 4000 is assumed, which would result in a phenyl content of about 45%, assuming the structure $(CH_3)_3Si{-}O{-}(\underset{\displaystyle C_6H_5}{\underset{|}{Si}}(CH_3){-}O)_{29}{-}Si(CH_3)_3$.

OV-17 has been successfully used with packed columns, especially in drug analysis [368]. Further fields of application are aromatic and heteroaromatic compounds, trimethylsilyl derivatives of steroids, and the like. However, because of the low film-forming capacity owing to its poor viscosity, OV-17 is not very common in capillary GC (see below). In this respect the gum OV-1701 has proved superior [376, 377]. The structure of OV-1701, available from Alltech (USA), has been elucidated by Temmermann et al. [68] using both spectroscopy (NMR, IR) and GC analysis of hydrolysis products (cf. Formula 7). As shown in Table 16, the polarity of OV-1701 is significantly lower than that of OV-17.

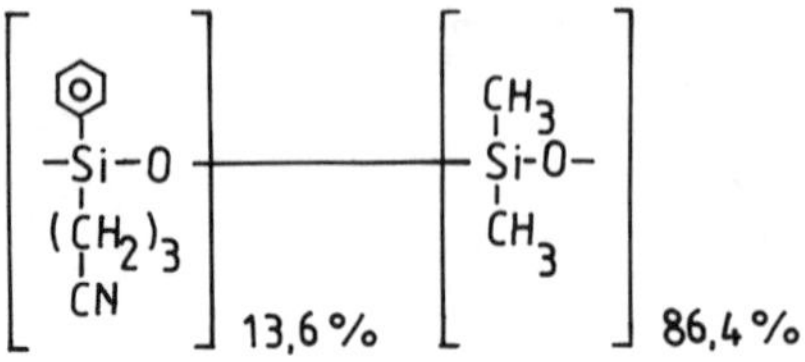

$$\left[ \begin{matrix} C_6H_5 \\ | \\ -Si-O- \\ | \\ (CH_2)_3 \\ | \\ CN \end{matrix} \right]_{13{,}6\,\%} \left[ \begin{matrix} CH_3 \\ | \\ -Si-O- \\ | \\ CH_3 \end{matrix} \right]_{86{,}4\,\%}$$

Formula 7

Several companies, among them Ohio Valley and Supelco, offer products ranging from 5.5% to 75% phenyl substitution. The general structure of polymethylphenylsiloxanes is presented in more detail in Table 17 and in Formula 8. Among corresponding 50% phenyl types, preference should be given to SP-2250 rather than OV-17 because of the narrower molecular weight distribution of the former [378,379]. Recently, vinylated products of OV-17 and OV-1701 have been marketed by several companies. As will be discussed later (see below), these phases are intended for immobilization.

The substantially distinctive retention mechanisms of the almost nonpolar polysiloxane SE-54 and the polar polyethylene glycol Carbowax 20M is exemplified by Fig. 15, which shows the separation of peppermint oil. While the polar terpene alcohols such as menthol or 1,8-cineol are strongly retained on Carbowax 20M, the less polar terpene hydrocarbons such as limonene, α- and γ-terpinene, or terpinolene have larger retention values on SE-54 compared to those on Carbowax 20M.

Analogous regularities are observed with fatty acid methyl esters (FAME). A comparison of Fig. 12 with Fig. 16 reveals the distinctive retention behavior on the Carbowax 20M and SE-54 phases. Unsaturated FAMEs, e.g., oleic acid, are eluted after the homomorphic ones (stearic acid) on Carbowax 20M, whereas homomorphic FAMEs are eluted after unsaturated ones on SE-54. The retention behavior of FAMEs on nonpolar phases appears to be less advantageous for identification purposes because unsaturated solutes (e.g., linol, linoleic, oleic acid) are coeluted with branched-chain saturated FAMEs [380]. In contrast, the retention sequence on polar phases is as follows: branched-chain saturated < nonbranched-chain saturated < monounsaturated < diunsaturated, etc. However, for analyzing strongly polar hydroxy FAME the use of nonpolar phases coated onto well-deactivated supports such as high-temperature silylated capillary columns proved to be better because the polar hydroxy FAMEs are adsorbed on roughened, insufficiently deactivated glass capillary surfaces. As shown in a later paragraph, these problems may be circumvented by using cyanopropylsilicones coated onto surfaces that have been modified with silylating agents containing cyano groups.

Table 17 Polymethylphenylsiloxanes Used as Liquid Stationary Phases

| GC grade | Commercial grade | Content of phenyl % (w/w) | Content of monomer(s) | | | |
|---|---|---|---|---|---|---|
| | | | a | b | c | d |
| OV-73 | – | 5.5 | 94.5 | – | 5.5 | – |
| OV-3 | – | 10 | 80 | 20 | – | – |
| OV-7 | – | 20 | 60 | 40 | – | – |
| OV-61 | – | 33 | 67 | – | 33 | – |
| OV-11 | – | 35 | 30 | 70 | – | – |
| OV-17 | – | 50 | – | 100 | – | – |
| SP-2250 | – | 50 | – | 100 | – | – |
| OV-22 | – | 65 | – | 70 | 30 | – |
| OV-25 | – | 75 | – | 50 | 50 | – |
| – | SE-52 | 5 | 95 | – | 5 | – |
| – | SE-54 | 5 | 93 | – | 5 | 2 |
| – | DC-550 | 25 | 50 | 50 | – | – |
| – | DC-710 | 50 | – | 100 | – | – |

*Source*: From [2], with permission.

$$CH_3-Si(CH_3)_2-O-\left[Si(CH_3)_2-O\right]_a\left[Si(CH_3)(C_6H_5)-O\right]_b\left[Si(C_6H_5)_2-O\right]_c\left[Si(CH_3)(CH{=}CH_2)-O\right]_d-H$$

Formula 8

*Chlorophenyl- and Fluoroalkyl-substituted Polysiloxanes* Among the siloxanes containing chlorophenyl groups such as DC-560, F-560, F-60, F-61, F-4050, the Lukooil and KhS series (the last two produced in Czechoslovakia and the Soviet Union, respectively) [334,360], and SP-400, preference should be given to the latter because the Supelco product is the only material specifically offered as a stationary phase. As can be inferred from Table 16, the polarity of SP-400 with approxi-

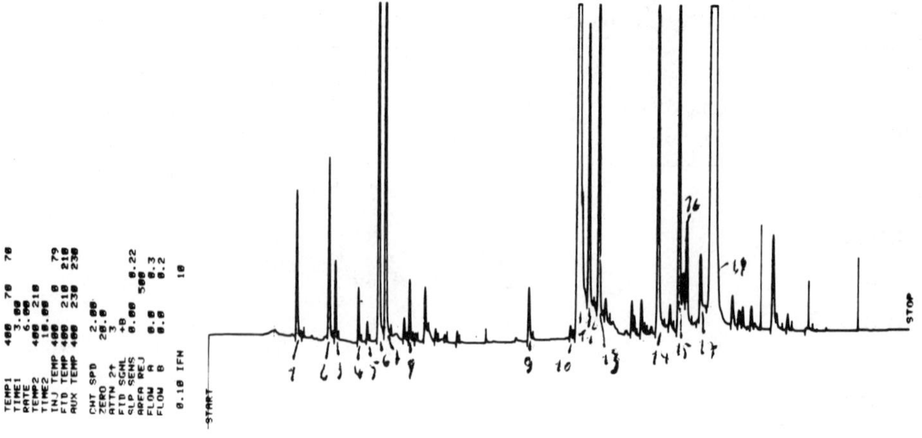

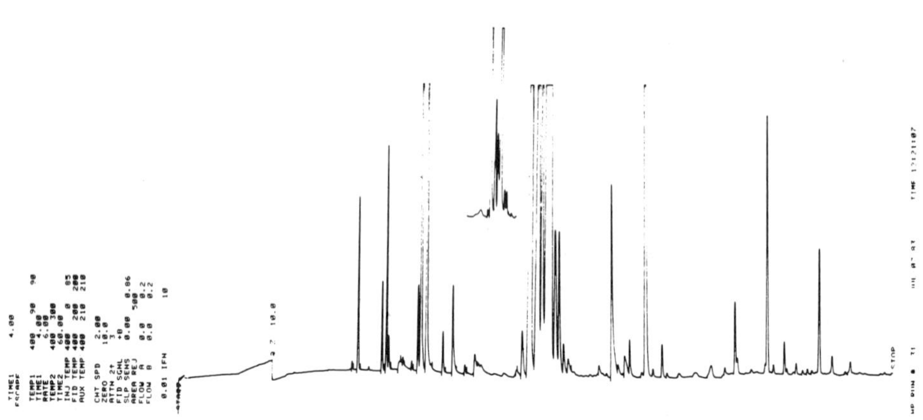

Fig. 15 Separation of a peppermint oil on Carbowax 20M (a) and SE-54 (b). Glass columns, (a) 40 m × 0.25 mm i.d.; (b) 40 m × 0.22 mm i.d.; film thickness, 0.22 μm; column temperatures, see figure. (From Pörschmann, Welsch, Görig, unpublished results.)

Symbols: 1 = α-pinene; 2 = β-pinene; 3 = sabinene; 4 = myrcene; 5 = α-terpinene; 6 = L-limonene; 7 = 7,8-cineole; 8 = 8-terpinene; 9 = octanol-3; 10 = trans-sabinene hydrate; 11 = L-methone; 12 = menthofurane; 13 = D-isomenthone; 14 = menthyl acetate; 15 = neomenthol; 16 = caryophyllene; 17 = neo-isomenthol; 18 = L-menthol.

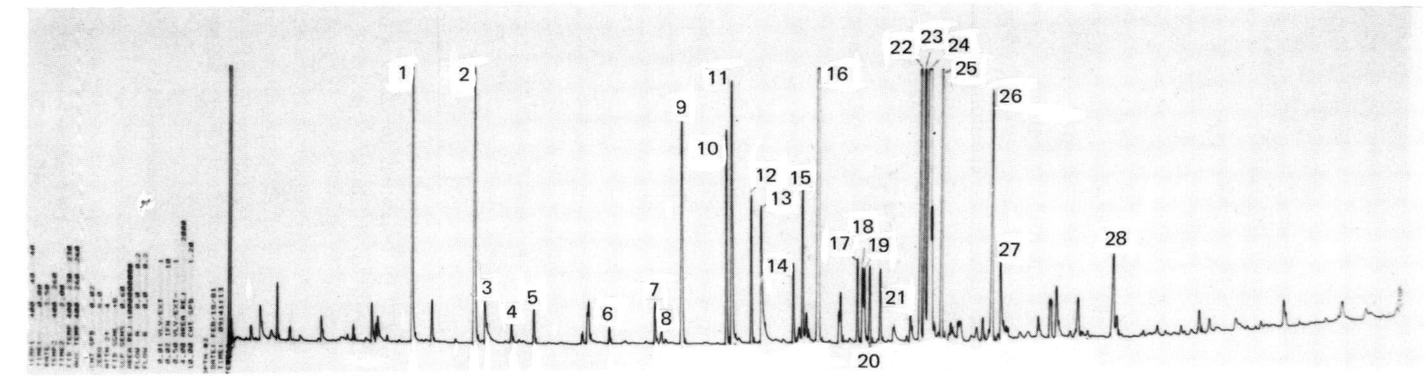


**Fig. 16** Separation of fatty acid methyl esters on SE-54. Glass column, 40 m × 0.25 mm; film thickness, 0.25 μm; deactivation by high-temperature silylation with di-*n*-butyltetramethyldisilizane. (From Pörschmann, Welsch, Engewald, unpublished results.)

Symbols (shorthand designation): 1 = 10:0; 2 = 11:0; 3 = 2-OH-10:0; 4 = i-12:0; 5 = 12:0; 6 = 13:0; 7 = i-14:0; 8 = 14:1; 9 = 14:0; 10 = i-15:0; 11 = a-15:0; 12 = 15:0; 13 = 2-OH-14:0; 14 = i-16:0; 15 = 16:1ω7c; 16 = 16:0; 17 = i-17:0; 18 = a-17:0; 19 = 17:1; 20 = 17:0▽(9,10c); 21 = 17:0; 22 = 18:2ω6,9cc; 23 = 18:1ω9c; 24 = 18:1ω7c; 25 = 18:0; 26 = 19:0▽(11,12c); 27 = 2-OH-18:1ω7c; 28 = 2-OH-19:0(11,12c).

mately 11% chlorophenyl substitution is similar to that of the methylphenylpolysiloxane OV-73 in terms of the sum of the McReynolds constants.

The currently available fluoroalkylsilicones are all homopolymers of 3,3,3-trifluoropropylmethylsiloxane produced by base-catalyzed polymerization of the cyclic trimer [334]. The general structure of poly-3,3,3-trifluoropropylsiloxanes is given in Formula 9. Trifluoropropylsilicones display medium polarity. The most widely used are OV-202, OV-205, OV-210, OV-215, QF-1, and SP-2401, the gumlike phase OV-215 yielding good results with capillary columns due to its sufficient film-forming properties [381]. Recently, novel polysiloxanes specifically intended for crosslinking (see below), such as PS-286, a methyl-3,3,3-trifluoropropylmethylvinylpolysiloxane gum from Petrarch Systems, or Silastic LS-420, a 50% 3,3,3-trifluoropropyl, 1% vinylpolysiloxane from Analabs, have been introduced in GC.

$$CH_3-\underset{CH_3}{\overset{CH_3}{|}}{Si}-O-\left[\underset{\underset{CF_3}{|}}{\underset{(CH_2)_2}{|}}{\overset{CH_3}{|}}{Si}-O\right]_n-\underset{CH_3}{\overset{CH_3}{|}}{Si}-CH_3$$

Formula 9

The unique selectivity of halogen-substituted polysiloxanes is based on the high dipole moment and the strong acceptor capability of the functional group. Since the X value of OV-210 is similar to that of OV-17 and OV-22, the employment of trifluoropropylpolysiloxanes allows sufficient separations of alkanes and cycloalkanes from alkenes and aromatics, all of them having similar boiling points. On considering the high Z and U values of OV-210 in contrast to those of OV-17, the acceptor character of the trifluoropropyl phases becomes evident. The Z values being higher than the Y values, trifluoroalkylpolysiloxanes are capable of selectively retarding ketones compared to alcohols because the $CF_3$ group interacts with free electron pairs of carbonyl oxygen. This feature favorably influences the separation of keto and hydroxy steroids. OV-210 also has a pronounced ability to separate heroin and structurally related compounds.

The temperature stability of trifluoropropyl-substituted phases is less than that of phases containing only methyl groups (Table 16 and [374]). When employing these phases one should be aware of the fact that the silicone bond in fluorosilicones is susceptible to cleavage by either trace water or a strong base and can undergo degradation by oxidation. Recently, the use of polymeric fluoroalkyl esters as stationary phases in GLC was reported [382].

*Polycyanoalkylsiloxanes* The cyano group represents one of the most stable polar substituents to be connected to a silicone backbone. Polycyanoalkylsiloxanes are commercially available in a range of cyanoalkyl contents from 5 to 100%. A specific advantage of these phases in comparison with low molecular nitriles is their high temperature stability (Tables 15 and 16). This feature is especially valid for cyano groups in the $\gamma$ position. A strong electron attraction in the $\alpha$ position would substantially weaken the Si-C bond [383]. A slight increase in terms of thermal stability is achieved by incorporating phenyl groups instead of methyl groups. Thus, the separation of natural glycerides has been attained with Silar 5-CP (50% phenyl, 50% nitrile; see Table 16) at operating temperatures above 250°C [384].

As mentioned above, stationary phases containing cyano groups with strong electron-accepting properties are common in GLC due to their pronounced selectivity toward polar and polarizable compounds. The selectivity of polycyanoalkylsiloxanes toward various classes of solutes changes depending on the content of nitrile groups in the stationary liquid. With increasing polarity of the polycyanoalkylsiloxane the selectivity increases for the separation of aromatics from aliphatics (cf. [385]), of aromatics from olefins, of alcohols from ethers, and of ketones from primary alcohols, but in the last case to a smaller extent. Polycyanoalkylsiloxanes with high cyano contents are not recommended for analyzing saturated hydrocarbons due to their low retention values. Polycyanoalkylphenylsiloxanes exhibit a polarity somewhat different from that of polycyanoalkylsiloxanes due to the ability of phenyl groups to attract polarizable groups more strongly. Thus, the choice of cyanosilicones such as Silar 10C, XE-60, OV-275, or the novel Supelco phase SP-2560 as stationary liquid offers many advantages for the analysis of complex fatty acid methyl ester mixtures including cis/ trans isomaer (trans isomers are eluted earlier than cis isomers), pharmaceuticals, as well as alditol acetates of monosaccharides [386-390]. Having investigated the selectivity of several cyanosilicones, Stark et al. [364] suggested the replacement of polyethylene glycols by silicones of 40-45% cyanopropyl and 20-25% phenyl substitution (cf. below).

Commercially available cyanosilicones are collected in Table 16. Detailed information on the structure of the SP-2300 series was given by Ottenstein et al. [391] (see also [2]). The general structure of some popular cyanosilicone members is given in Table 18 and Formula 10.

There is some confusion in the literature on the structure of OV-275, the most polar phase of the polycyanoalkylsiloxanes commercially available. Recently, Jones et al. [383] evaluated that OV-275 should have 38% pendant cyanoethyl groups, 57% pendant cyanopropyl groups, and 5% carboxyamide groups ($-CONH_2$). The last stems from the synthesis from dichlorosilane monomers.

There are some drawbacks in using polycyanoalkylsiloxanes:

**Table 18** Cyanosilicones Used as Liquid Stationary Phases

| Stationary phase | Substitute | Structure | Content of: Cyanopropyl % (w/w) | Content of: Nitrile[a] % (w/w) |
|---|---|---|---|---|
| SP-2300 | Silar 5CP | R = A | 50 | 12.4 |
| SP-2310 | Silar 7CP | R = AB | 75 | 20.0 |
| SP-2330 | Silar 9CP | R = A9B | 90 | 27.0 |
| SP-2340 | Silar 10C | R = B | 100 | 28.4 |
| OV-105 | (XF-1105) | R = D9C | 5 | |
| OV-225 | – | R = ED | 25 | |
| AN-600 | XE-60 | R = CF | 25 (ethyl) | |

[a]These values refer to the molecular weight of the cyano group divided by the molecular weight of the entire polymeric molecule. For instance, assuming a molecular weight of 5000 for SP-2340 [334], n will be approximately 27 (27 × 2 × 26/5000 = 28%).

*Source*: Manufacturer catalogues.

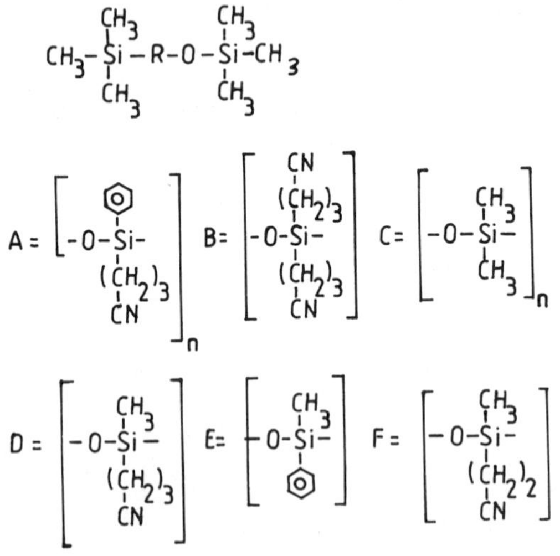

$$CH_3-Si(CH_3)_2-R-O-Si(CH_3)_2-CH_3$$

$$A = \left[-O-Si(C_6H_5)((CH_2)_3CN)-\right]_n \quad B = \left[-O-Si((CH_2)_3CN)((CH_2)_3CN)-\right] \quad C = \left[-O-Si(CH_3)(CH_3)-\right]_n$$

$$D = \left[-O-Si(CH_3)((CH_2)_3CN)-\right] \quad E = \left[-O-Si(CH_3)(C_6H_5)-\right] \quad F = \left[-O-Si(CH_3)((CH_2)_2CN)-\right]$$

Formula 10

1. The relatively poor batch-to-batch reproducibility [392], presumably caused partly by the presence of the aforementioned carboxyamide group.
2. The susceptibility of cyano polymers to oxidation and hydrolysis in the presence of air and water [393].
3. The much more strongly decreasing viscosity with increasing operating temperature compared with polydimethylsiloxanes [394,395].

To circumvent the last problem a stabilizaiton of the polar silicone phase by in-situ crosslinking has been recommended (see below).

*Carborane-Siloxane Polymers* In the early 1970s a high-temperature-stable carborane-siloxane polymer phase, designated as Dexsil 300, was marketed by Olin Corp. The chemical structure of the three currently available Dexsil phases is detailed in Fig. 17. The Dexsils consist of *meta*-carborane units connected by siloxane units. Covalent bonds link each hydrogen to a boron and a silicone to each of the two carbon atoms. The high-temperature stability is attributed to the fact that the carborane ring acts as an energy sink stabilizing the neighboring siloxane units in the entire Dexsil structure. As detailed in [397], Dexsil 300 yields lower bleeding than a number of other high-temperature phases.

Among the Dexsil phases Dexsil 410 offers the highest polarity owing to the incorporation of a 2-cyanoethyl group into the polymer chain. The polarity of Dexsil 410 is very similar to that of OV-17 (Table 16), while the polarities of Dexsil 300 and Dexsil 400 are similar to those of OV-3 and OV-7, respectively. Thus, Dexsil 410 exhibits an outstanding selectivity toward high-boiling compounds containing $\pi$ electrons (aromatics, esters, ketones). Dexsil 400 is a highly suitable phase for resolving polynuclear aromatics.

Although Dexsil phases were believed to be stable up to 450°C, only a few successful separations have been reported in the literature, among them the separation of cholesteryl esters (cf. [398]). In addition to that, they have not gained wide acceptance in capillary GC due to their only moderate film-forming properties (see below).

As pointed out recently [360], novel carborane-siloxane polymers have been prepared with properties similar to the Dexsils, but having higher molecular weights.

### *Specialty Stationary Liquid Phases*

When designing specific stationary phases, the resulting interactions between the stationary phase and the solute should be strong enough to ensure a selective retention, but also sufficiently labile to ensure elution from the column. Highly specific interactions can occur when transition metal complexes, enantioselective compounds, liquid crystals, and organic molten salts are employed as stationary phases.

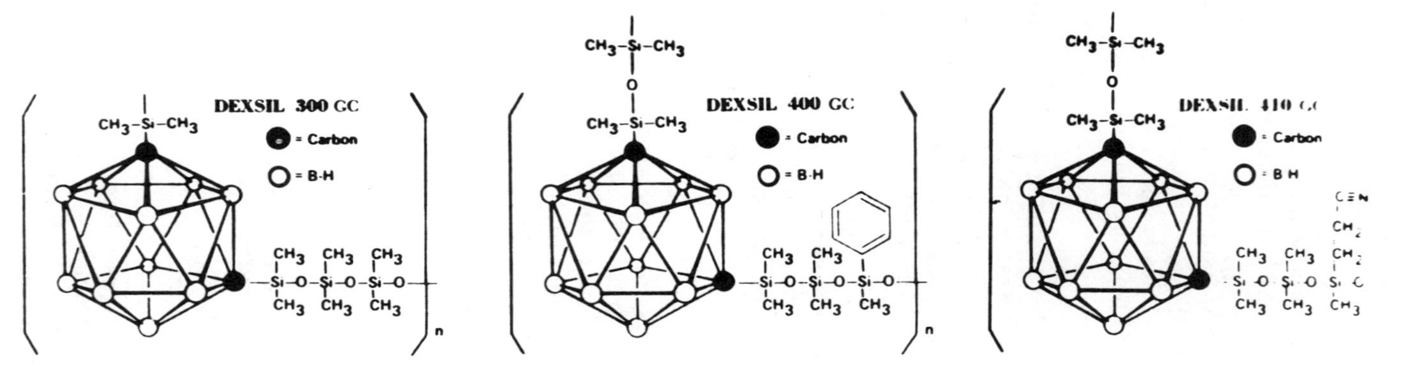


*DEXSIL 300 GC*, **a Carborane/Methyl Silicone**

*DEXSIL 400 GC*, **a Carborane/Methyl/Phenyl Silicone**

*DEXSIL 410 GC*, **a Carborane/Methyl/2-Cyanoethyl Silicone. (The polymer chain is end-capped with trimethylsilyl groups.)**

Fig. 17 Structures of Dexsil-carborane silicone polymers. (From Ref. 396, with permission.)

*Transition Metal Complexes* It is well known that transition metals possess d, p, and s orbitals interacting with ligand orbitals in a defined spatial orientation. The chromatographic ability of the transition metal complexes can be ascribed to properties of the central metal ion and the ligand, the latter being capable of forming steric or inductive forces, e.g., the trans effect for square-planar complexes. In turn, the separation ability of analytes is governed by electronic, steric, and polarizability factors. Thus, it becomes possible to resolve substances differing only slightly in their structure such as cis and trans isomers by means of chemical complexation (cf. [399,400] and references cited therein). The selectivity of the silver-containing phases, for instance, results form the high sensitivity of small structural changes in the olefins to the stability constants of the complexation. The same holds true for other organic molecules containing $\pi$ bonds or free electron pairs.

Using *N*-dodecyl salicylaldimines and methyl-*n*-octyl-glyoximes of Ni, Pd, Pt, and Cu, Cartoni et al. [401] succeeded in separating substances that possess ligands (amines, ketones, alcohols, molecules containing double bonds). Schurig (see below) examined the use of enantioselective transition metal complexes to separate enantiomers.

On account of their extraordinary separation ability transition metal complexes are frequently termed "superselective" packings. Crystal hydrates also rank as "superselective" packings [402,403]. Superselective phases are usually coated on the surface of a support or adsorbent (GTCB, pure silica). Using silver and cadmium forms of zeolites, strong bonding of olefins has been observed. A molecular sieve packing based on a polymer Schiff base-chromium complex has been introduced for GC separations [404].

Polymetalloorganopolysiloxanes, a new kind of polysiloxane possessing divalent or higher metal ions in the siloxane chain, have been examined as selective phases (cf. [405]). The introduction of a metal into a siloxane chain notably influences separation characteristics. Polymetalloorganopolysiloxanes are capable of interacting coordinatively with compounds having $\pi$ bonds or lone electron pairs.

In a similar manner, lanthanide metal chelates such as strong electron-withdrawing $\beta$-diketonates have been utilized as packings in preanalytical columns for the selective retention of nucleophilic species (cf. [144]). Substances incapable of entering into complex formation such as alkanes remain unaffected. After heating of the precolumn, the complexes are cleaved and the oxygenated compounds are driven into the analytical column. This subtraction technique allows both a simplification of the analytical procedure and a classification according to substance classes.

Recently, porous polymers containing various metal chelates bonded to nitrogen functionalities on the surface of the polymers have been synthesized and found to bind oxygen reversibly [406].

*Enantioselective Stationary Phases* The importance of assessing the configuration of organic compounds is of increasing interest as illustrated by the following points:

1. The configuration of drugs has a decisive effect on their biological activity (e.g., thalidomide, penicillamine, small peptides applied as therapeutic or sweetening agents).
2. The nutritional value of D-amino acids in special proteins (e.g., microbial-produced biomass) may be less than that of L-amino acids as is assumed for lysine and others [407,408].
3. The data of geological samples based on the ratio of enantiomeric amino acids demands unequivocal determination of the degree of racemization.
4. Control of assymmetrical synthesis, e.g., peptides.

There is a vast number of references concerned with the subject of enantioselective phases. Hence, the reader is referred to reviews only [236,409-412], by means of which it is possible to enter the field. Recently, the separation of optical isomers was treated by Souter [501].

There are two fundamental approaches to separating enantiomers: (1) the conversion of enantiomers into diastereomeric derivatives by reaction with absolutely optically pure chiral reagents and subsequent separation of the derivatives on commonly used phases. As an example, the separation of some diastereomeric *N*(*O*,*S*)-trifluoroacetyl-L-menthyl-amino acid esters on SE-54 is shown in Fig. 18. (2) Direct separation of enantiomers on enantioselective phases. Gil-Av et al. were the first to prepare an enantioselective stationary phase, namely *N*-trifluoroacetyl-L-isoleucine lauryl ester, to separate highly volatile amino acid derivatives (alanine, valine). Poor temperature stability (upper limit of about 100°C) and enantioselectivity lead to the development of dipeptide phases such as *N*-trifluoroacetyl-L-valyl-L-valine cyclohexyl ester (cf. Formula 11) or *N*-trifluoroacetyl-L-phenylalanine-L-leucine-cyclohexyl ester having two asymmetrical centers (upper temperature limit 110 and 140°C, respectively).

The substitution of tripeptide derivatives for dipeptides failed to improve both the temperature stability and enantioselectivity. However, optically active tetraamide stationary phases derived from L-phenylalanine were recently prepared [413].

As a result of systematic investigations it became apparent that the $-NH-CO-\overset{+}{C}H(R)-NH-CO-$ group plays the essential role in dipeptides, which has lead to diamide phases such as *N*-dodecanoyl-L-valine-*tert*-butylamide (cf. Formula 11) with an upper temperature limit of about 180°C. As shown in Fig. 19, the diamide phase *N*-stearoyl-L-valine-*tert*-butylamide exhibits a satisfactory enantioselectivity but an unsatisfactory temperature stability. The elution of derivatives that are less volatile than those of glutamic acid seems to be problematic.

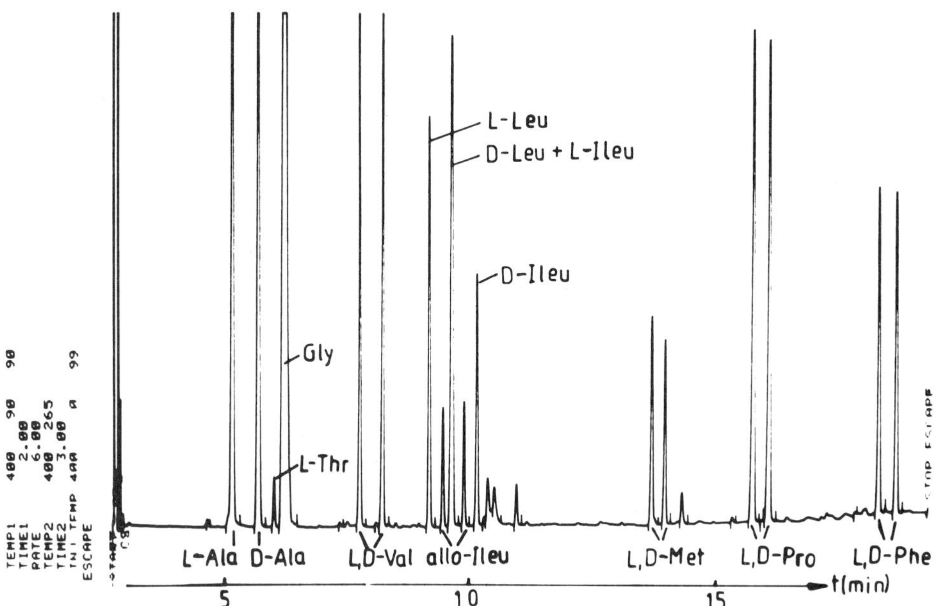


Fig. 18 Separation of *N*-(*O*,*S*)-trifluoroacetyl-L-methylamino acid esters on a SE-54 glass capillary column, 20 m × 0.25 mm i.d.; film thickness, 0.32 μm; deactivation by high-temperature silylation with diphenyltetramethyldisilizane. (From Pörschmann, Wettengel, unpublished results.)

$$CF_3\text{–CO–NH–CH}(CH(CH_3)_2)\text{–CO–NH–CH}(CH(CH_3)_2)\text{–C}(=O)\text{–O–}C_6H_{11}$$

$$CH_3\text{–}(CH_2)_{10}\text{–CO–NH–CH}(CH(CH_3)_2)\text{–C}(=O)\text{–NH–C}(CH_3)_3$$

Formula 11

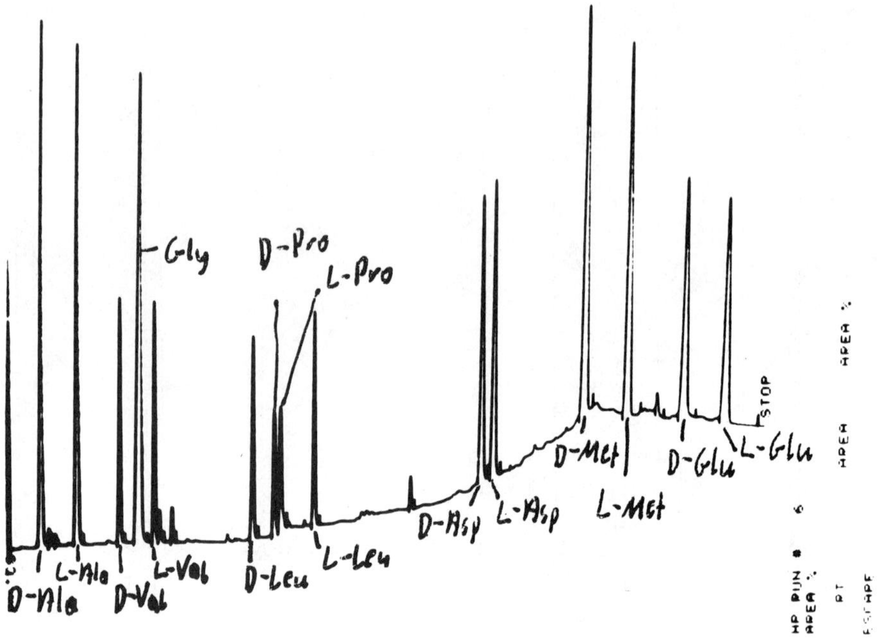


Fig. 19 Separation of *N*-(*O*,*S*)-trifluoroacetylisopropyl esters of racemic amino acids on *N*-stearoyl-L-valine-*tert*-butylamide; glass capillary column, 18 m × 0.25 mm i.d., dynamically coated; column temperature, hold 2 min at 80°C, then to 155°C at 4°/min. (From Pörschmann, Wettengel, Öhme, unpublished results.)

As can be inferred from bulletins and catalogues, Supelco offers the enantioselective phase *N*-lauroyl-L-valine-*tert*-butylamide under the trade name SP-300, while Serva offers *N*-behenoyl-L-valine-*tert*-butylamide.

A third class of low molecular enantioselective phases are the carbonylbisamino acid esters such as carbonylbis-L-valineisopropyl ester, which proved less suitable in analyzing amino acid derivatives but more suitable for enantiomeric amines.

The real breakthrough in terms of temperature stability and extension of operating temperature has been effected by the introduction of polymeric chiral phases, prepared in the following two ways:

Frank et al. synthesized polysiloxanes with L-valine-*tert*-butylamide covalently bound in the side chain. This chiral moiety possesses an extraordinary enantioselectivity as has already been shown (cf. [414]). In the polymer the chiral moieties were separated from each other by approximately seven dimethylsiloxane units. According to the inventors, this arrangement is an important factor for enantioselectivity and thermal stability as well. The phase designated as Chirasil-Val has an outstanding thermal stability up to at least 220°C to permit the separation of both high- and low-volatile amino acid derivatives in a single chromatographic run. Capillaries coated with Chirasil-Val are available from Applied Science Laboratories and Chrompack. A modified version of Chirasil-Val involving a 15% phenyl substitution, designated as Heliflex Chirasil-Val, is marketed by Applied Science (cf. also [415]). The polarity of Chirasil-Val as indicated by the McReynolds constants is relatively low [416]. In comparison, the sum of the $\Delta I$ values of the five test probes is similar to that of Dexsil 400 (cf. Table 16). However, the high value for *n*-butanol ($\Delta I = 264$) indicates the capability of Chirasil-Val to form hydrogen bonds.

The second approach involves the modification of commercially available cyanopolysiloxanes such as OV-225, XE-60, or Silar 10C by hydrolysis of the cyanoalkyl side chains and coupling L-valine-*tert*-butylamide via an amide linkage. The chiral moiety L-valine-(*S*)-α-phenylethylamide coupled to the hydrolyzed phase also shows an outstanding enantioselectivity. The conversion of XE-60 to a chiral polysiloxane with this moiety is shown in Fig. 20. Beside hydrolysis, the reduction of the cyano group to yield an aminoethyl group and further modification by reaction with carboxylic groups is a further possibility for polysiloxane modification. Intensive pioneering work in this field was carried out by Verzele [420] and König (cf. [409]). In order to improve the stability against solvent rinsing and deterioration of film stability, the enantioselective polymers have been crosslinked [419]. Recently, fused silica capillary columns coated with XE-60-L-valine (*S* or *R*)-α-phenylethylamide became available from Chrompack.

In addition to improved thermal stability, a decisive feature of the progress achieved by means of Chirasil-Val or XE-60-L-valine-α-phenyl-

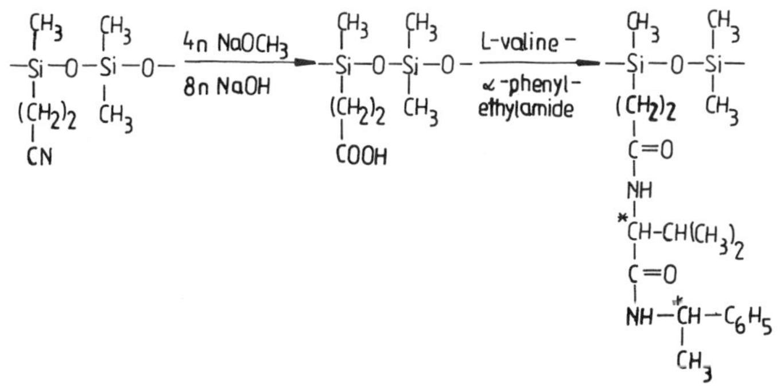

$$-Si(CH_3)[(CH_2)_2CN]-O-Si(CH_3)_2-O- \xrightarrow[8n\ NaOH]{4n\ NaOCH_3} -Si(CH_3)[(CH_2)_2COOH]-O-Si(CH_3)_2-O- \xrightarrow[\alpha\text{-phenylethylamide}]{\text{L-valine-}} -Si(CH_3)[(CH_2)_2-C(=O)-NH-{}^{*}CH(CH(CH_3)_2)-C(=O)-NH-{}^{*}CH(CH_3)-C_6H_5]-O-Si(CH_3)_2-$$

Fig. 20 Conversion of XE-60 to a chiral polysiloxane via hydrolysis.

ethylamide is the extension of the field of application to other classes of compounds, e.g., amino alcohols, hydroxy acids, haldgenocarboxylic acids, trifluoroacetylated carbohydrates, sulfur-containing compounds, ketones, etc. [421-425]. As an example, Fig. 21 shows the resolution of 2-hydroxyalkanoic acids on Chirasil-Val. By introducing the enantiomer-labeling technique using optical antipodes (D-amino acids, for instance) as internal standard compounds (first described in [417]), the quantitation of GC amino acid analysis has been vitalized, especially in determining biological fluids (cf. [418]). Enantiomer labeling based on the valid assumption that optical antipodes have identical chemical and physical properties, circumvents losses in sample preparation.

Several attempts have been made to explain the separation mechanisms. Generally speaking, the success of the separation of enantiomers on chiral phases depends on the formation of diastereomeric complexes between chiral phase and chiral analyte, termed selectand-selector interactions. For instance, hydrogen bondings are mainly responsible for the enantioselective resolution of amino acids via their *N*-acyl-esters. The separation will take place when the complexes are formed rapidly and reversibly. According to a theory established by Dalgliesh three independent selectand-selector interactions are necessary to separate enantiomers. In the future it will become possible to separate more and more classes of enantiomeric substances by careful choice of the type of selectand-selector association.

When highly efficient capillary columns are used, the difference in the free enthalpies of the diastereomeric association complexes must be at least 40 J/mol (at room temperature) in order to attain a sufficient separation. By using Chirasil-Val (with a L-valine-containing moiety as already mentioned) the L enantiomer of amino acid derivatives is eluted last owing to its more stable complex with the stationary phase. In an analogous manner, on applying Chirasil-Val with a D-valine moiety the D enantiomer is eluted last, which is especially useful in determining traces of L enantiomers in a D-enantiomer matrix.

Schurig et al. [426,427] have extended the field of application by introducing enantioselective phases containing β-ketoenolates. High-resolution complexation GC with chiral metal chelates has proved to be a valuable tool for underivatized oxygen-containing insect pheromones.

*Liquid Crystalline Stationary Phases* Liquid crystals have an intermediate position between isotropic liquids and crystalline solids. Although they are incapable of converting from the solid to the isotropic liquid phase, they do undergo discrete phase transformations including liquid crystal intermediate phases, so-called mesophases. This term indicates a liquid state with an ordered structure. Above the transition temperature the orientation is lost and isotropic liquids are formed.

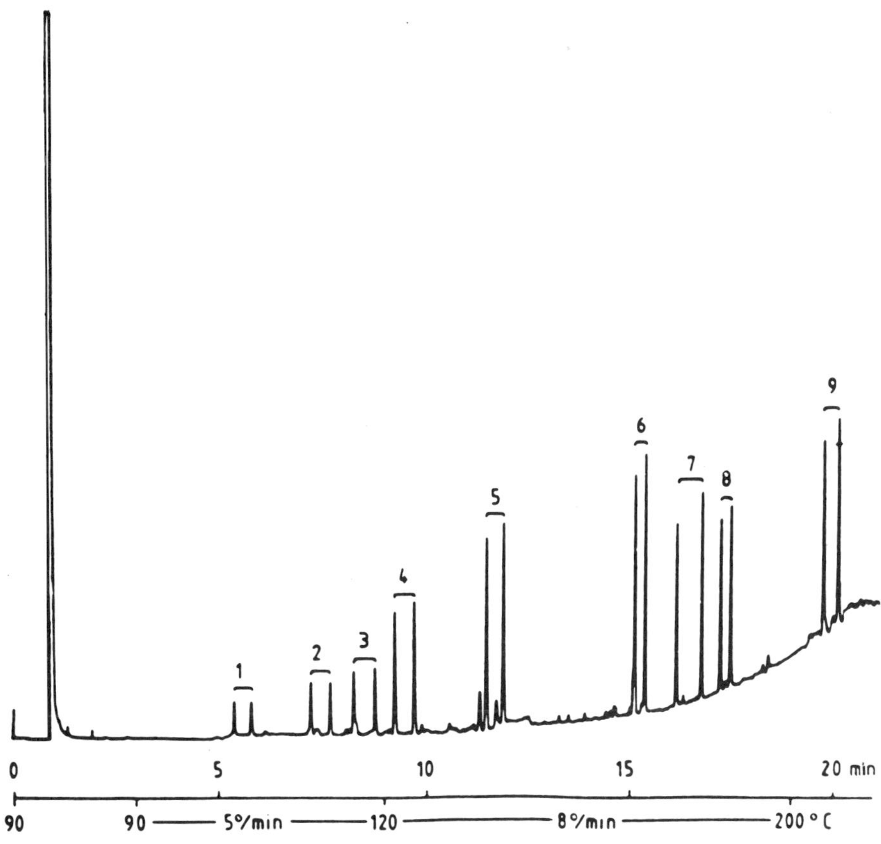


Fig. 21 Gas chromatographic resolution of 2-hydroxyalkanoic acids as 2-trimethylsiloxycarboxamides on 20 m × 0.28 mm glass capillary, coated with Chirasil-Val. Peaks: 1, lactic acid; 2, 2-hydroxybutyric acid; 3, 2-hydroxyisovaleric acid; 4, 2-hydroxyvaleric acid; 5, 2-hydroxycaproic acid; 6, 2-hydroxycaprylic acid; 7, mandelic acid; 8, β-phenyllactic acid; 9, *m*-hydroxymandelic acid. (From Ref. 423, with permission.)

There are two main types of mesophases: the smectic and the nematic. The molecular structure of nematic phases were examined recently [502]. The smectic phases, being the more effective in GC, are stabler at lower temperatures than the nematic. Smectic phases are characterized by a parallel orientation of molecules in layers, whereas the nematic phases are less ordered and the molecules are free to move within the limits of parallel configurations; layered structures were not observed. A variety of the nematic phase is the cholesterinic type, whereby the nematic layers are twisted to a helixlike structure. Cholesteryl benzoate and cholesteryl nonanoate are among the compounds showing the cholesterinic form.

The chemical structure of liquid crystals is commonly as follows:

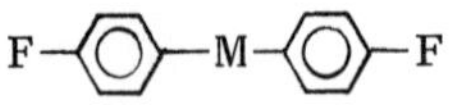

F—⟨◯⟩—M—⟨◯⟩—F

F - $C_nH_{2n+1}$

$C_nH_{2n+1}O$

$C_nH_{2n+1}OCO$

$C_nH_{2n+1}OCOO$

M - –N=N–

–CH=N–

–C≡C–

–CO–O–

–N=N–
↓
O

Since 1963, when Kelker first used a liquid crystal as stationary phase, the importance of liquid crystals in GC has constantly grown due to their high selectivity for geometrical and positional isomers, which frequently cannot be separated satisfactorily on conventional stationary phases (cf. [428-432,504] and references cited therein). Several kinds of interactions between the stationary phase and the solute such as dispersive, orientation, and inductive forces contribute to retention. But the success of separation on liquid crystalline phases is dependent to a high degree on the compatibility of the dimensions of the solute molecules and the structure of the stationary phase itself. Thus, the elution order of geometrical isomers is governed by the length-to-breadth ratio and planarity of the solute molecules. Rodlike molecules penetrate the orientated phase more readily than broad ones. For example, the *p* isomer is more elongated than the *m* isomer and fits between the layers of the mesophase more readily. Hence, the *p* isomer is eluted later.

Interesting separation features with liquid crystals were obtained in their so-called supercooled temperature range. Kraus [433] compared the selectivity of the crystalline, nematic, and supercooled ranges with respect to the separation of isomeric xylenes (Fig. 22). The superior separation in the supercooled state is a result of a higher degree of order, while the separation at the same temperature in the solid state is made possible by the parallel orientation of the surface molecules.

Previous results show the structure of liquid crystal films to depend strongly on the kind of support, i.e., the chromatographic properties of the film coated are different from those of the bulk liquid crystals.

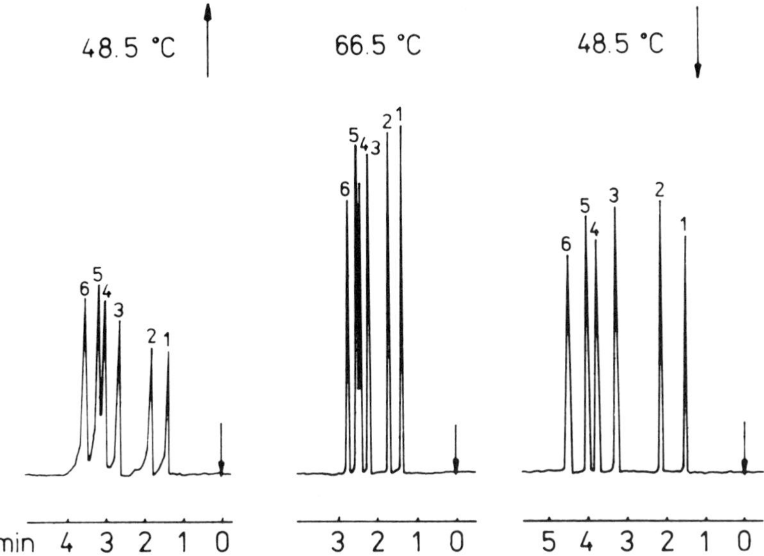


Fig. 22 Glass capillary, 6 m × 0.25 mm, coated with *p,n*-heptyloxyphenyl-4-(trans-4-*n*-propylcyclohexanecarbonyloxy)-2-methylbenzoate, sample 0.2 μl, split ratio 130:1, u = 8.4 cm $sec^{-1}$. Crystalline range (48.5°C), nematic range (66.5°C), and supercooled range (48.5°C). Peaks: 1, benzene; 2, toluene; 3, ethylbenzene; 4, *m*-xylene; 5, *p*-xylene; 6, *o*-xylene. (From Ref. 433, with permission.)

Such phenomena occur with packed [434] as well as capillary columns [435]. Capillaries coated with liquid crystals were prepared by Sojak and Kraus [436,437]. It has frequently been observed that liquid crystals have poor film-forming properties. In such cases efficient columns may be prepared by depositing liquid crystals into a matrix of silicone gum [438] (see below). A novel attempt to prepare capillary columns with liquid crystals is to synthesize the stationary phase directly on the surface [439,440].

Despite their merits, low molecular weight liquid crystals have not gained wide acceptance for several reasons:

1. The restricted mesomorphic temperature range in which the desirable high selectivity is displayed
2. The highly crystalline structure
3. Poor efficiency as a result of slow kinetics of mass transfer
4. Excessive column bleeding at high temperatures

Recently, progress in polymer chemistry has allowed polysiloxanes to be prepared that have side chains with liquid crystal properties. The mesomorphic side chains can have nematic or smectic features. These synthetic routes represent a useful means for overcoming the above limitations [441,442]. By using so-called Mepsil phases (mesomeric polysiloxanes), which have good column efficiencies and temperature stabilities up to 300°C, the entirely satisfactory separation of polycyclic hydrocarbons could be achieved in a temperature range of 140-280°C. The extraordinary separation power for the ubiquitous environmental pollutants, based on differences in solute molecular shape (in particular length-to-breadth ratio) and probably on the development of charge-transfer interactions, exceeds that of the frequently used "slightly polar" phases such as SE-54 or SE-52. Furthermore, liquid crystalline polymers are more suitable for preparing capillaries due to their sufficient film-forming capabilities. Obviously, the kind of mesomorphic moiety of the liquid crystalline polysiloxane may be selectively varied. Therefore, further liquid crystalline polymer phases are expected from the leading groups in this field in the near future.

*Liquid Organic Molten Salt Phases* In contrast to liquid crystalline phases, which have just been discussed, the liquid organic molten salt phases behave as stable isotropic liquids within a fixed temperature range. They were discovered for GC purposes, mainly by the group of Poole (cf. [44,507] and literature cited therein), in pursuit of high-polarity and high-selectivity separations. Among the ionic liquids there are three types: inorganic molten salts, inorganic hydrated melts, and organic molten salts. The last type is only of interest in GC because the organic salts are capable of sufficiently dissolving organic molecues, while inorganic salts and hydrated melts fail to do so. Typical members of liquid organic molten salt phases used in GC include quaternary ammonium salts with various anions such as tetra-*n*-butylammonium picrate, tetra-*n*-butylammonium tetrafluoroborate, or tetra-*n*-hexylammonium benzoate, thiocyanate salts such as alkylammonium thiocyanates, or miscellaneous salts such as ethylpyridinium chloride.

The interactions between the stationary phase and the solute are mainly governed by strong orientation and moderate to strong proton acceptor forces with the dispersive interactions being only weak. If one succeeded in extending the working temperature range and in enhancing viscosity to ensure stable films in capaillary GC (see below), these monomolecular liquid organic salt phases with a highly defined, reproducible structure might serve as standard reference polar phases in future.

## Stationary Phases in Capillary GC

### *General Considerations*

Although a huge number of liquid stationary phases have been applied in GC of packed columns, not all of them are suitable for capillary GC, mostly due to poor film-forming properties, poor reproducibility, and limited temperature range. Most of the limited number of useful phases are highly viscous polysiloxanes and polyglycols. Therefore, special emphasis has been placed on developing tailored phases for capillary columns.

In capillary GC the formation of a homogeneous film of stationary phase on an inert and nonadsorbing surface is essential for the success of a separation. In order to achieve a sufficient wettability, the surface tension of the supporting surface must be greater than the surface tension of the stationary phase (cf. [443-445]). Glass, which for many years was the predominant support material [15] for capillaries, is wettable by apolar liquids without prior surface modification. However, more polar phases tend to break up into droplets owing to the fact that the glass surface readily undergoes adsorption and hydration, resulting in low-energy surface tensions. Typical stationary phases have surface tensions in the range of 20-50 dyne/cm (e.g., OV-101: 20 dyne/cm; Carbowax 400: 44 dyne/cm), whereas the critical surface tension for smooth clean glasses in smaller than 30 dyne/cm. To form a homogeneous film with medium polar or polar phases, an increase of the surface tension of glass is required. Capillary surface roughening methods such as whisker formation, carbon black or barium carbonate deposition were found to be useful [443,446]. However, these roughening methods have been largely replaced by other techniques such as the high-temperature silylation with agents containing cyano groups (see below), because they fail to provide a film of constant thickness. This in turn results in losses of potential column efficiency.

As important as the homogeneous film formation is the deactivation of the glass surface. Active sites of the supporting capillary material are capable of causing adsorption of polar solutes and the formation of artefacts. The separation of underivatized solutes is frequently desirable because derivatization procedures may yield disadvantates in terms of reproducibility, recovery, time spent, etc. While polar or moderately polar liquids (particularly in the case of thick films) can deactivate the glass surface to some degree by masking active sites, films of apolar liquids fail to do so. Therefore, a number of methods have been devised to eliminate active sites originating from various glass ingredients or from the silica structure itself. Besides the Carbowax deactivation suggested by Schomburg (cf. [447]), a very common method is the combined leaching-silylation process (cf. [443,444,448, 449]). The leaching procedure using 20% (w/w) hydrochloric acid removes alkali and basic oxide components to yield a silica-rich surface layer and at the same time increases the density of surface hydroxyl groups. Since soda glass has a high content of metal ions which are able to catalyze silicone decomposition leading to a high column bleeding rate at elevated temperatures, leaching is a prerequisite for the successful use of silicone-coated soda glass capillary columns. In the authors' opinion, among the leaching procedures created by Grob, Schomburg, Dirkes, Neu (static methods), Lee, and Venema (dynamic methods), preference should be given to Grob's HCl method because of its simplicity and effectiveness.

After leaching, active silanol groups are replaced by silyl ether groups by silylation. Silanol groups may also be removed by esterification to yield stable, inert, and hydrophobic alkoxy linkages [454]. More recently, silanol groups on soft glass capillary column surfaces have been esterified in the presence of polyethylene glycol 400 [455]. In chromatographic practice three silylation procedures are popular:

1. High-temperature silylation ("persilylation" according to Grob [450]) with hexamethyldisilazane (Welsch, Grob)
2. Alkylpolysiloxane degradation (PSD method, Schomburg)
3. Reaction with cyclic siloxanes (e.g., octamethylcyclotetrasiloxane) according to Stark and Blomberg [448,449,451-453]

Owing to the strength of the Si-O-Si linkage, such deactivated surfaces display high stability. When hexamethyldisilazane is used as silylation agent, deactivation takes place at the expense of wettability because a methylated surface has a low resultant critical surface tension of about 21 dyne/cm, allowing efficient wettability by methylsilicone gums such as OV-1. However, the wettability of the modified layer may be adapted to more polar stationary liquids by adding more polar functionalities into the silylation agent (Welsch [456], Grob [457]). In a similar manner, Blomberg et al. [394,395,458] employed cyclic siloxanes that carry polar functional groups to yield surfaces that are deactivated and also wettable by polar phases. Schomburg [459] used phenyl- and trifluoropropylsilicones with the PSD method to improve wettability.

However, the energy criterion, which takes into account the surface tension only, does not suffice to clarify whether a stationary phase will wet or not. As an example, after deactivation with $C_6H_5(CH_3)_2$-Si-NH-Si-$(CH_3)_2C_6H_5$ the glass surface has a critical surface tension of about 33 dyne/cm [460]. Surprisingly, such a surface is not wettable by the low-viscous, phenyl-containing phase OV-17, which has a surface tension of about 31 dyne/cm. Furthermore, untreated fused silica (cf. below) with its sufficiently high surface energy should be wettable by almost all stationary phases. However, poor coating efficiencies are observed with low viscous phases.

The viscosity of the stationary phase presumably plays an essential role for physical film stability, especially in capillary GC. About 10 years ago Grob (cf. [461,462]) referred to the benefits of gum phases. Gum phases were shown to yield columns with more stable films than liquids. With highly viscous phases the resistance to droplet formation is enhanced. In principle, in the case of low-viscous phases the film stability might be enhanced by increasing the capillary column diameter and/or diminishing the film thickness. Such modifications, in turn, influence the phase ratio, capacity ratio, etc. (see above), frequently in an undesired way. From these considerations follows further that the range of film thickness to be formed with gum phases is wider than that of liquids (0.02-3 μm compared to 0.05-0.4 μm). This feature is important with thick-film capillaries.

On considering the viscosity criterion, the restriction of the number of stationary phases useful for capillary GC as mentioned at the beginning of this paragraph becomes evident: In comparison with packed columns, where the stationary phase is relatively firmly attached to the support, any film coated on an inner surface of a crylindrical tube is less stable, as has long been known. Hence, the low-viscosity phases frequently used in GC with packed columns are unsuitable for capillary GC because they show insufficiently low coating efficiencies [377,462]. On the other hand, the smaller number of applicable liquid phases at the experimenter's disposal is compensated by efficiency.

A milestone in the development of capillary GC was the introduction of flexible fused silica in 1979 [463]. Fused silica capillaries offer two major advantages: mechanical flexibility, which facilitates column handling and connecting to gas chromatographs or mass spectrometers, and extremely low metal ion content (<1 ppm), which results in low catalytic activity to the stationary phase. Fused silica exhibits a high surface energy which allows wetting with polar liquid phases, provided that the phase has a sufficiently high viscosity (see above). As emphasized by Grob [15], the introduction of fused silica was ideally timed because at this very moment the know-how for the production of inert columns had become fully available from the work with glass and could be applied directly to fused silica. Drawbacks of fused silica may be its high cost compared to glass, the present inability of manufacturers to ensure the constant quality which is necessary in coating with polar phases (cf. [449,464]), and to a certain extent the limitations in producing tailored columns, which can be made much more readily of glass. Although the concentration of silanol groups in fused silica is substantially lower than in glass [465], it is necessary to remove remaining active silanol and siloxane surface sites. As pointed out by Lee's group [466], purging the freshly drawn fused silica capillary with nitrogen at 300°C prior to silylation proved favorable.

Nondeactivated fused silica surfaces are silightly acidic in behavior, resulting in tailed peaks for basic solutes (Fig. 23a) but in remarkable

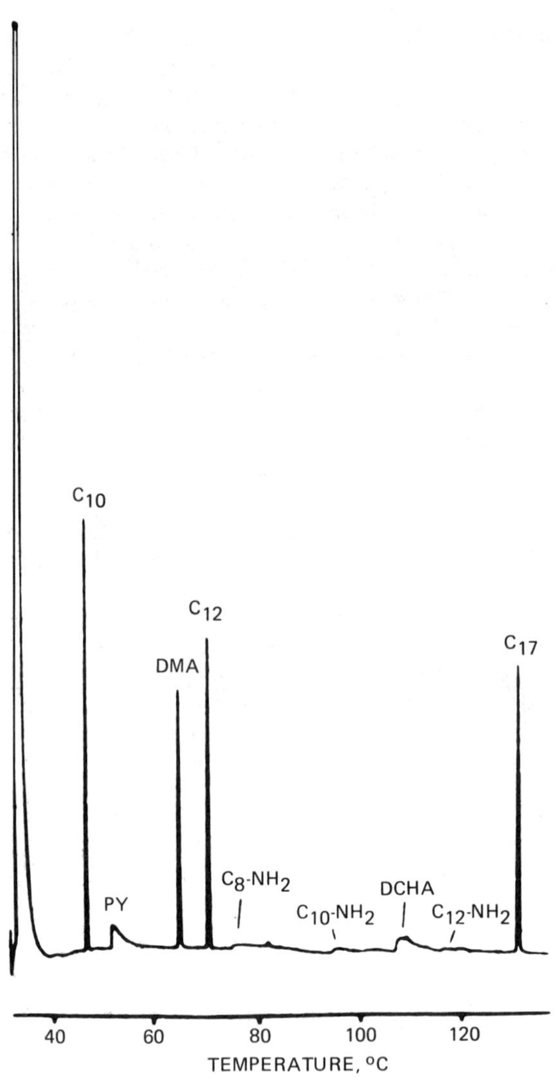


(a)

Fig. 23 Basic test mixture on nondeactivated fused silica (a) and on fused silica deactivated with polymethylhydrosiloxane (b) at 250°C. Abbreviations: PY, 3.5-dimethylpyrimidine; DMA, 2.6-dimethylaniline; $C_{8,10,12}$-$NH_2$, 1-aminooctane (decane, dodecane); DCHA, *N*,*N*-dicyclohexylamine. (From Ref. 467, with permission.)

$C_{10}$
PY
$C_8$-$NH_2$
$C_{10}$-$NH_2$
$C_{17}$
$C_{12}$-$NH_2$
$C_{12}$
DMA
DCHA
40
60
80
100
120
TEMPERATURE, °C

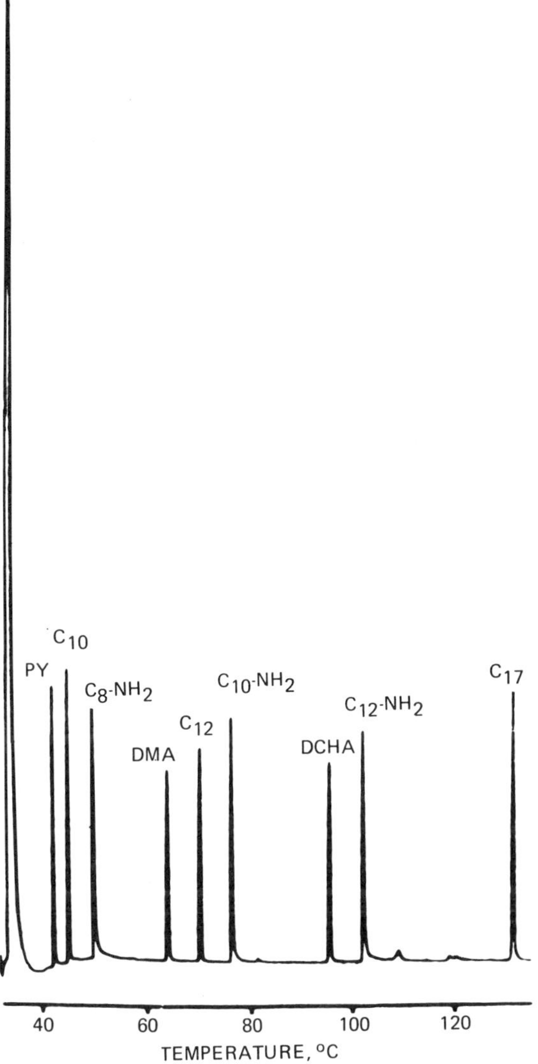

Fig. 23 (b)

separations of free fatty acids. In addition to surface silanol groups, which are slightly acidic in behavior, the acidic character of fused silica is presumed to stem partly from residues of HCl and $HNO_3$ originally present in the preforms of fused silica [454,468]. Fused silica is offered commercially with a polyimide outer coating, which cannot frequently be subjected to high-temperature silylation due to the limited thermal stability of the polyimide coating (in the vicinity of 350°C). An alternative may be the use of a polymethylhydrosiloxane containing silyl hydride groups which reacts effectively at lower temperatures [467]. The high inertness of capillaries deactivated in such a manner is shown in Fig. 23b.

By incorporating phenyl and cyanopropyl functional groups into the organosilicon hydride reagent, an efficient wettability for phases containing 50% phenyl and 50% cyanopropyl groups on well-deactivated, neutral fused silica surfaces could be achieved [452]. The upper temperature limit of the polyimide outer coating also restricts the use of fused silica capillary columns for separating high-boiling compounds such as triglycerides or polycyclic aromatic hydrocarbons with a high degree of condensation. In order to take advantage of stationary phases, which are temperature-resistent up to 400°C (poly-*m*-phenylethers or crosslinked polymethylsiloxanes), an interesting attempt has been made to use aluminum outer coatings (Quadrex, USA).

At the end of this paragraph a few commercial aspects will be briefly outlined: At present the analyst tends to buy ready-made fused silica columns from renowned companies rather than untreated fused silica columns because he is often not experienced enough in capillary preparation or cannot keep up with the technical development in this field. He can be sure of buying a column which shows a good reproducibility in separation properties. However, soft glass, which was the dominant material from the beginning of capillary GC up to the introduction of fused silica, represents still an important material for laboratory-made columns owing to its wide availability and low price. On the other hand, from the point of view of the producer of stationary phases, there is a tendency to combine manufacturing and marketing of these columns because there is no stationary market founded on liquid phases due to the very small amounts needed for the preparation of capillaries.

*Immobilized Phases*

Since the early 1980s the preparation of so-called immobilized phases is an important approach in capillary GC. On the basis of the investigations of Grob (cf. [449]) and Madani and Chambaz [469], Blomberg prepolymerized siloxanes prior to their coating inside the capillary, resulting in nonextractable stationary phases with high film stability. Immobilization can be attained through crosslinking of the polymer chains within the bulk liquid phase. A further possibility consists of chemical bonding of the stationary phase to surface groups such as silanol groups, double bonds (vinyl or allyl groups from the silylating agent), aminopropyl groups from the silylating agent, epoxy groups, etc.

Crosslinking has a number of advantages:

1. A crosslinked stationary phase can withstand various strains. As an example, they become insoluble in commonly used solvents (such as in the case of cold-on-column or splitless injection) including water.
2. Undesired contaminants from dirty samples deposited at the beginning of the column can be washed away with solvents to regenerate the column.
3. The operating temperature range is extended and the background is decreased due to reduced bleeding, the latter being especially relevant for thick-film capillary columns. The temperature tolerance extends not only toward higher operating temperatures, but also toward lower ones [369,470]. Lower bleeding properties and higher working temperatures are of special interest in GC/MS coupling because baseline shifts and additional artefact peaks are reduced. As a result of extensive work, research groups, among them those of Lee and Blomberg, succeeded in synthesizing cyanopropylpolysiloxanes and polyethylene glycols with a temperature stability approaching 300°C.
4. The column lifetime is prolonged and phase rearrangements or deterioration of column performance are minimal.
5. The chemical stability of the stationary phase is considerably improved. For instance, in the case of glass capillary columns there is a diffusion of metal ions to the surface. An attack of ions on the phase, which leads to its destruction, is largely hindered with immobilized phases.

At present, immobilization is mainly achieved by subsequent crosslinking of the coated phase. The immobilization procedure is initiated by a free radical reaction using peroxides (Grob, Sandra, Blomberg), azo compounds (Lee), $\gamma$ radiation (Schomburg, Hubball, Pretorius, Vigh), accelerated electrons (Blomberg), and ozone (Blomberg) (cf. [451,471,472 and references cited therein). The radiation curing method appears to be the most suitable because no foreign compounds have to be added to the stationary phase and the radiation is highly reproducible. In addition, a test prior to immobilization becomes feasible. However, very few laboratories have such expensive equipment at their disposal.

Immobilization by ozone [464] may offer a useful alternative, whereby only an ordinary equipment is required. The peroxide-initiated reaction is attractive because of its simplicity. However, the peroxide must not form decomposition products to catalyze silicone decomposition or cause adsorption activity. Dicumyl peroxide meets these requirements best. When polydimethylsiloxanes are subjected to peroxide reaction, the thermal homolysis of peroxides produces carbon-based radicals from the methyl groups of the dimethylsiloxane. Then a covalent linkage is formed (Fig. 24). After curing, the reaction products are extracted from the crosslinked phase.

The immobilization procedure initiated by peroxides is described in detail in [376,466,473]. Generally, the solid peroxides (dicumyl peroxide, benzoyl peroxide) are added to the static coating solution and vulcanization is achieved by heating the coated column at about 150°C. Finally, the column is flushed with convenient solvents leaving a stable, nonextractable film. The concentration of catalyst ranges from 0.1 to 1.0% (w/w) for the polysiloxanes provided the crosslinking occurs between methyl groups. The amount of catalyst depends on the molecular weight of the prepolymer. The higher its molecular weight, the lower the amount of initiator needed. In addition, the amount of catalyst needed also depends on functional groups present in the polysiloxane. Phenyl and cyanopropyl groups were observed to hinder crosslinking, since they act as a "free radical sink." However, as already outlined, crosslinking of cyanopropyl- and phenylsilicones was found to be especially advantageous because these phases exhibit a decrease in viscosity when raising the temperature attributed to the disturbance of the helical structure, which results in uneven films at elevated temperatures.

Peroxide applied in amounts up to 5% can lead to an appreciable adsorption activity of the stationary phase. Crosslinking of both phenyl- and cyanopropylpolysiloxanes has been greatly facilitated by

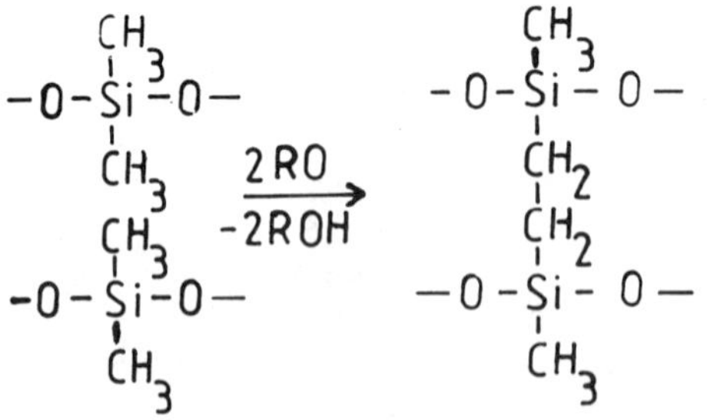

$$
\begin{array}{c}
CH_3 \\
| \\
-O-Si-O- \\
| \\
CH_3 \\
CH_3 \\
| \\
-O-Si-O- \\
| \\
CH_3
\end{array}
\quad
\xrightarrow[-2ROH]{2RO}
\quad
\begin{array}{c}
CH_3 \\
| \\
-O-Si-O- \\
| \\
CH_2 \\
| \\
CH_2 \\
| \\
-O-Si-O- \\
| \\
CH_3
\end{array}
$$

Fig. 24 Typical free radical crosslinking mechanism of polydimethylsiloxanes.

incorporating tolyl groups into the prepolymer [393,394,474]. As high percentages of peroxides may oxidize tolyl groups, the use of azo-*tert*-butane and other compounds as crosslinking initiators has been recommended [393,475,476]. Recently, the preparation of wide-bore (320 μm) and narrow-bore (50 μm) fused silica capillary columns has been described for immobilized self-made cyanopropylsilicones containing 60 and 88% cyanopropyl substitution [477].

The presence of vinyl groups in stationary phases aids crosslinking, in contrast to phenyl and cyanopropyl groups. In the case of polysiloxanes containing vinyl groups, the principal crosslinking reaction is not methyl-to-methyl but rather methyl-to-vinyl crosslinking [466]. Thus, high molecular silicones with vinyl groups such as PS-255 (a methylsilicone gum with 1% vinyl groups), PS-286 (see above), SE-31, SE-33, SE-54, and OV-1701-Vi are especially suitable for immobilization. These phases are synthesized by mixing an appropriate (mostly small) amount of polar to moderately polar monomers with a majority of dimethylsilane (e.g., dimethyldichlorosilane) and a small amount of vinylmethylsilanes. Modern synthetic methods (cf. [476]) enable circumvention of the formation of the harmful HCl during the polymerization process.

As already outlined at the beginning of this section, the immobilization ensures resistance against solvents. This holds especially true for nonpolar phases. More polar phases can be detached from the supporting surface by repeated injections with large sample volumes. This problem has been circumvented by depositing so-called anchor groups, e.g., vinyl, on the surface [478]. The vinyl group, introduced for instance by divinyltetramethyldisilazane as silylating agent, is capable of forming bonds with active groups of the stationary phase molecules (e.g., vinyl groups from PS-255) during the immobilization procedure.

For the sake of completeness it should be mentioned that several authors have succeeded in elaborating immobilization procedures with polyethylene glycols or polyollike phases (e.g., RSL-310) coated on capillary surfaces [458,479-481,505] as well as on such ordinary supports as Chromosorb W. Crosslinked and bonded PEG phases have a high stability toward aqueous samples and are stable against oxidative and pyrolytic attacks (up to 280°C). The success of such an immobilization depends greatly on the molecular weight of the polyethylene glycol. Carbowax 400 and Superox-4 are difficult to crosslink owing to their too low and high molecular weight, respectively, whereas Carbowax 20M represents a good compromise. Polyethylene glycol immobilized columns are widely available, e.g., from Supelco [41] (Supelcowax 10), Chrompack (CP Wax 57 CB or the more novel phase CP Wax 52 CB), J&W Scientific (Durawax 1-4), Hewlett Packard [482] (HP-20M). Phases that are crosslinked and bonded to the fused silica surface (Durabond Wax from J&W Scientific) have the additional advantage of a lower temperature limit of 0 to 20°C [483].

Immobilized polysiloxane capillary columns with superior column efficiencies are also marketed by renowned companies (Tables 19 and 20). In addition to these tables, Ultra 1 (crosslinked methylsilicone gum) and Ultra 2 (crosslinked 5% phenylmethylsilicone gum) are available from Hewlett-Packard. Scientific Glass Engineering (cf. [503]) also offers a wide range of bonded polysiloxane columns in both small- and wide-bore varieties (BP-1, dimethylsilicone; BP-5, 5% phenylmethylsilicone; BP-10, OV-1701; BP-15, OV-225; BP-75, OV-275). More recently, Supelco introduced bonded polysiloxane phases, arranged according to increasing polarity in Table 21. Throughout the manufacturers recommended their bonded capillary columns for applications that require a high degree of inertness, high column efficiency, and a sufficient retention index reproducibility from column to column.

**Table 19** Bonded Siloxane Columns Offered by J&W Scientific (Folsom, CA) and ICT (Frankfurt FRG)

| Bonded phase | Description | Substitute | Temperature range (°C) |
|---|---|---|---|
| DB-1[a] | Methylsilicone | OV-1, SE-30, CP SiL-5 | -60-320 |
| DB-5 | 5% Phenylmethyl | SE-54, CP SiL-8 | -60-320 |
| DB-1701 | 86% Dimethyl<br>14% Cyanopropylphenyl | No equivalent phase | -20-300 |
| DB-17 | 50% Methyl<br>50% Phenyl | OV-17, SP-2250, OV-11, OV-22 | 40-280 |
| DB-210 | 50% Trifluoropropyl | OV-210, SP-2401, QF-1 | 45-240 |
| DB-225 | 25% Cyanopropylphenyl | OV-225, SP-2300 | 40-220 |
| DB-608<br>DB-624[b]<br>DB-1301 | Cyanopropylphenyl dimethyl (devised for special applications, e.g., pesticides) | | Up to 280<br>Up to 280<br>-20-300 |

[a]DB = Durabond.
[b]As can be inferred from chromatograms depicted in ICT information bulletins, the phase DB-624 is presumably an oil. Therefore, it would be a promising approach to prepare an appropriate silicone gum with higher viscosity and better film-forming properties (see page 200).

Table 20 Bonded Siloxane Columns Offered by Chrompack (Middleburg, The Netherlands)

| Bonded phase | Description | Substitute | Maximum operating temperature, isothermal (°C) |
|---|---|---|---|
| CP Sil-5 CB | Polydimethylsiloxane | SE-30, OV-1, SP-2100, SF-96 | 300 |
| CP Sil-8 CB | 5% phenyl, 95% methylpolysiloxane | CP Sil-8, SE-52, OV-73, SE-54 | 300 |
| CP Sil-19 CB | Cyanopropylphenyl-methylpolysiloxane | OV-17, OV-1701 | 300 |
| CP Sil-43 CB | Phenylmethylcyano-propylpolysiloxane | OV-225, XE-60, AN-600 | 200 |

A very promising approach to preparing immobilized phases through condensation of hydroxyl-terminated phases has been reported by the Grobs [484], who based their development on the work of Verzele et al. [485] and Lipsky and McMurray [454]: The capillaries were statically coated with a hydroxyl-terminated viscous polysiloxane such as OV-61 OH on a leached, dehydrated, nonsilanized surface. Since the hydroxyl groups of the "diol" polysiloxane are readily condensed with the surface silanol groups at elevated temperatures, the in situ gummification is carried out by heat curing, resulting in an immobilized and solvent-resistant film. Thus, immobilization may be based primarily on surface bonding rather than on a large increase of chain length.

### Mixed Liquid Phases

Unique selectivities can be created by using mixed phases with only a few stationary phases needed. In this paragraph, "mixed" retention mechanisms caused by interactions between the (nondeactivated) support and solute molecules are excluded (cf. [486] in this instance). Mixed phases are preferentially applied to difficult separation problems (cf. references cited in [487]). There are several techniques:

1. Coupling usually two columns with different phase polarities in series [488]. Coupled column behavior is governed by the pressure gradient [489] because the first segment of the coupled column has a greater influence than the second. To determine the length of each segment the construction of window diagrams showing the relationship of relative retention of

**Table 21** McReynolds Values for Bonded Polysiloxane Phases (at 70°C) Offered by Supelco

| Bonded phase | Description | X | Y | Z | U | S | ΣΔI |
|---|---|---|---|---|---|---|---|
| SPB-1 | Bonded SE-30 | 4 | 58 | 43 | 56 | 38 | 199 |
| SPB-5 | Bonded SE-54 | 19 | 74 | 64 | 93 | 62 | 312 |
| SPB-20 | 20% diphenyl/ 80% dimethyl-polysiloxane with crosslinking moieties | 67 | 116 | 117 | 174 | 131 | 605 |
| SPB-35 | 35% diphenyl/ 65% dimethyl-polysiloxane with crosslinking moieties | 101 | 146 | 150 | 219 | 180 | 796 |

*Source*: From [369].

solutes of interest and stationary phase composition is recommended. If this coupling is used, care should be taken to ensure low dead volumes between the columns to be connected.

2. Mixing support particles coated with different liquids and subsequent filling of the column with the mixed particles ("mixed bed").
3. Mixing suitable liquids together prior to the coating procedure. As already mentioned in item 1, the optimum composition of the mixture can be taken from window diagrams.
4. Preparing a copolymer such as OV-330 (Carbowax/silicone polymer, Table 16). J&W and Quadrex offer mixed phases of polyethylene glycols with nonpolar silicones designated as DX-1 to DX-4 and Q-Wax-1 to Q-Wax-4, respectively.

The most common technique in current practice is mixing technique (3). Extensive studies have been reported on the theoretical and practical aspects of this method (cf. [492] and references cited therein). It should be noted that there is some confusion in the literature on the dependence of the retention data on the composition of the mixed phase. Purnell et al. (cf. [490]) predicted infinite dilution partition coefficients $K_L$ for a stationary phase composed of a binary (A + B) mixture:

$$K_L = \phi_A \cdot K^0_{L(A)} + \phi_B \cdot K^0_{L(B)} \qquad (16)$$

where $\phi$ = volume fraction and $K^0_L$ = partition coefficient in the pure liquid phase

Others [487] observed a nonlinear relationship instead of a linear one, the latter being indicated by Eq. (16). Clearly, the type of relationship depends on both stationary liquid-stationary liquid and solute-stationary liquid interactions. Investigating a number of binary stationary phases, Huber et al. [487] as well as Bogoslovsky et al. [491] observed similarities in retention between coupled columns with single phases and columns with composite liquids, provided there was a linear relationship between the capacity factor and the stationary phase composition.

With improved column technology it appears that the use of mixed phases may be a powerful tool in improving column efficiency, inertness, and thermal stability. For instance, to retain polar phases such as UCONs or Pluronics on the smooth silica or glass capillary wall, roughening is required, which leads to a loss in inertness and efficiency By mixing the gums OV-1 and Superox 20M (Tables 11 and 16) prior to coating procedure, columns with high efficiency, inertness, and thermal stability were obtained [343]. The resulting polarity of 50% Superox/50% OV-1 corresponds to that of UCON LB 550X, whereas by increasing the share of Superox to 75% a polarity similar to that of UCON 50 HB 280X results (Table 12).

In a similar manner, high-temperature silylated glass surfaces have been coated with several mixtures of OV-1/FFAP to obtain a sufficient retention of free fatty acids (pure OV-1 shows only a poor solving capacity for the polar free fatty acids) and a good deactivation [266]. OV-1/FFAP mixed-phase capillaries were employed in the temperature-programmed mode up to 260°C without any noticeable decrease in efficiency, allowing the symmetrical elution of a wide-boiling free fatty acid mixture in a single run (cf. Fig. 25). In this context it should be mentioned that the temperature stability of FFAP deposited on borosilicate glass surfaces extends only to 225°C. This stability is further decreased on soda "soft" glass.

The examples given clearly demonstrate the adjustment of the polarity by mixed phases. Recently, a successful attemps was made to adjust enantioselectivity by mixing an enantioselective phase with common achiral phases of different polarities [493].

An approach to creating changes in the overall selectivity by slightly altering the carrier gas flow rate (pressure) in one part of a capillary series with different stationary phase selectivity or film thickness was reported by Kaiser et al. [494,495]. The key feature of this so-called multichromatography is the change of selectivity by time as a

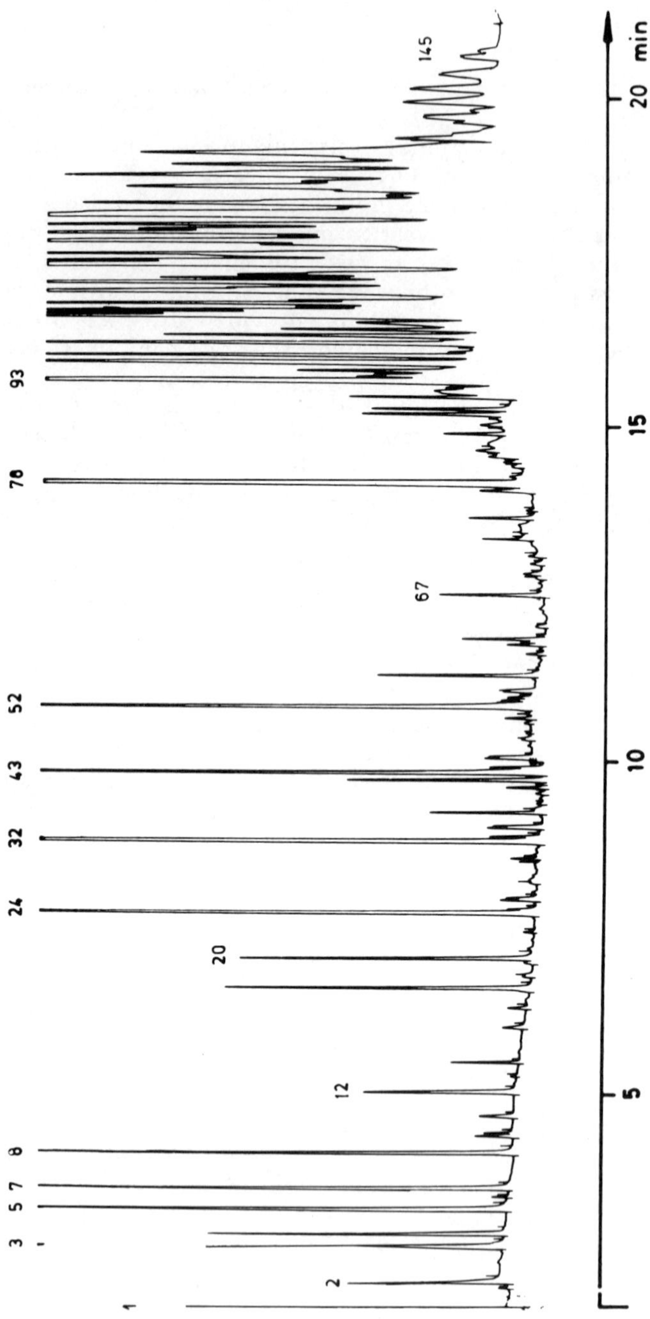


Fig. 25 Chromatogram of $C_2$-$C_{22}$ free fatty acids obtained from a sludge sample on a glass capillary column: 10 m × 0.25 mm i.d., statically coated with OV-1/FFAP = 2:1 (w/w) in methylene chloride; column temperature, hold 2 min at 70°C, then to 220°C at 7°/min. Peaks 2, 3, 5, 8, 12, 15, 20, 25, 32, 43, 52, 60, 67, 73, and 78 homologous, saturated, nonbranched $C_2$-$C_{16}$ free fatty acids. Peaks 93-145 $C_{18}$-$C_{22}$ free fatty acids. (From Ref. 266, with permission.)

physical measure rather than by chemical measures, i.e., the chromatographic selectivity is governed by both retention time (or, better, by residence time) and thermodynamic quantities (partition coefficient). The residence time is proportional to the dead time in the components of the tandem, whereas the system retention time is given by the sum of the residence times and can be calculated by adding the products of capacity factor and dead time in each part of the system. Varying the pressure at the coupling point of two capillaries one will measure different dead times for the single columns and therefore different system retention times.

For a better understanding of multichromatography, let us consider a series of coupled capillaries of differing polarity with the first capillary being nonpolar. On applying a slight pressure at the coupling point of the two capillaries, the first nonpolar capillary will be "slower," whereas the second polar capillary will be "faster." This will result in diminished overall polarity.

## Selection of Stationary Phases

The choice of a proper stationary phase is one of the most important decisions in GC. In chromatographic practice compromises must often be reached between several parameters such as selectivity, operating temperature range, detection conditions, and so on. Several compilations are available from suppliers and may be used as a guide when selecting the stationary phase.

There are over 700 stationary phases covering a wide polarity range. Renowned companies engaged in GC supplies offer about 150-200 phases. Historically, in the early days of GC, marked by the prevalence of the packed column type, the large number of stationary phases was primarily conceived to compensate for a lack of efficiency. With the advent of capillary GC a considerably smaller number of phases is necessary.

Obviously, the large number of liquids (around 700) catalogued in the "GC Data Compilation DS 25A" (ASTM) calls for standardization and restrictions. The problem consists of reducing the number of stationary phases, although the separation capacity should not be diminished. Since 1956, several attempts have been made to recommend "standard" or "preferred" phases for GC to cover the bulk of separations. The phases should meet the following criteria: well tested, readily available, stable over a wide temperature range, covering collectively an extended range of polarity, available as immobilized phases, and specifically manufactured for GC use. In addition, the standard phase should be defined by McReynolds constants. Some recommendations are arranged below chronologically:

The list of 12 according to Leary et al. [496]. The recommended phases SE-30, OV-3, OV-7, DC-710, OV-22, QF-1, XE-60, PEG 20M, DEGA, DEGS, and TCEP can supplant 226 liquid phases under study.

The extended list of 24 and 13 special stationary phases according to the Hawkes committee.

The list of six according to Bishara and Souter [498], which is similar to that of Hawkes. The compilation of Bishara and Souter is based on successful checking of the purity of drugs in a rapid screening test.

The list of seven according to Wayne [499] including six silicones (SP-2100, SP-2250, SP-2401, OV-225, SP-2300, SP-2340 and analogs, respectively), and Carbowax 20M. In comparison with the older suggestions just mentioned, monomeric phases were excluded. In addition, with the help of crosslinking procedures, these seven phases may also be used effectively in capillary GC.

The classification of stationary phases into seven types according to the order of increasing values of five Rohrschneider factors [500].

After thorough consideration of the recommendations, the basic role of a methylsilicone, a methylphenylsilicone, a nitrile silicone, and a high molecular weight polyethylene glycol becomes evident. Together they are capable of covering the majority of all analyses in GC.

In analytical practice, the selection of stationary phases is frequently carried out on a trial-and-error basis. However, qualitative knowledge of the molecular interactions combined with the theory of solutions (knowledge of activity coefficients for different systems) helps us to understand the workings of the stationary phase. In this respect, the stationary liquid classification system suggested by Rohrschneider and McReynolds is a powerful guide in selecting a convenient phase. For instance, the X value is of decisive importance in separating unsaturated fatty acid methyl esters from saturated ones. Consequently, owing to its high X values (504), DEGS is a more favorable phase than neopentyl glycol succinate (X: 272, cf. Table 13). This theoretical estimation is confirmed by chromatographic practice, which shows more efficient separations of the unsaturated solutes from the saturated ones on DEGS. A brief compilation of selected separation problems is given in Table 22.

A rough measure of average polarity is given by the sum of Rohrschneider or McReynolds constants. In Fig. 26 the sum of five McReynolds test solutes [$\Sigma(X + Y + Z + U + S)$] is plotted against an arbitrary number. As can be inferred from Fig. 26, the polysiloxanes commercially available (OV or SP series) taken together cover almost the entire polarity range. OV-275 is the most polar phase commonly used in GC. As depicted in Fig. 26, the number of stationary liquids at our disposal decreases with increasing polarity. Nonpolar phases, having a sum of five McReynolds values in the range 200-500, are numerous (SE-30, OV-1, OV-101, SE-54, OV-3, OV-105, Dexsil 300, etc.), whereas in the range 4000-5000 there are only a few phases such as

**Table 22** Selection of Liquid Stationary Phases for a Given Separation Problem

| Separation problem | Liquid stationary phase type |
|---|---|
| Aldehydes from alkanes | Nitriles and nitrile ethers |
| Aldehydes from ketones | Amines |
| Nitriles from nitro compounds | Fluorosilicones, cyanosilicones |
| Acetyls from alcohols | Polyesters |
| Aromatics from cycloalkanes and cycloalkenes | Nitrilesilicones, nitriles and nitrile ethers |
| Alcohols from alkanes | As above, nonpolar hydrocarbons |
| Esters from ketones | Amines |
| Ethers from alkanes | Nonpolar hydrocarbons, polyglycols |

diglycerol, *N,N*-bis(2-cyanoethyl)formamide, OV-275, and 1,2,3-tris-(2-cyanoethoxy)propane.

However, this graphic representation provides no information on the selectivity of stationary liquids. Stark et al. made a contribution to the subject of selective phases based on the ideas of McCloskey and Hawkes (cf. [364]). They plotted (cf. Fig. 27) the polarity of the phase under study (horizontal axis) vs. the dispersion forces for hydrocarbons such as *cis*-hydrindane (vertical axis top) and the acid-base (hydrogen-bonding) attraction for the alcohols (vertical dimension bottom). On this basis they underlined the fact that in selecting an appropriate stationary liquid an accordance must be reached between the functionality of both stationary liquid and sample. Some interpretations of Fig. 27 will be given in what follows: as can be inferred from the figure, the OV-25 phase (75% phenyl) has the greatest retention for *cis*-hydrindane, whereas diglycerol and THEED have the greatest retention for alcohols. The high retention of the alcohol-containing phases for alcohols is caused by hydrogen bond-donating properties. Phases containing ether and ester oxygens do not substantially retain alcohols (note the position of PEG and DEEEP). Cyanosilicones such as OV-275 (note the position at the top of the polar list), Silar 10C, SP-2340, XF-1150 are polar, but do not exhibit acid-base character. By introducing phenyl groups with strong dispersive effects into the cyanosilicone backbone, a phase results which combines both characteristics (note the position of OV-225 in comparison with that of XE-60, both having nearly the same polarity). Considering the position of OV-225, XF-1150, and PEG, it can be concluded that a sili-

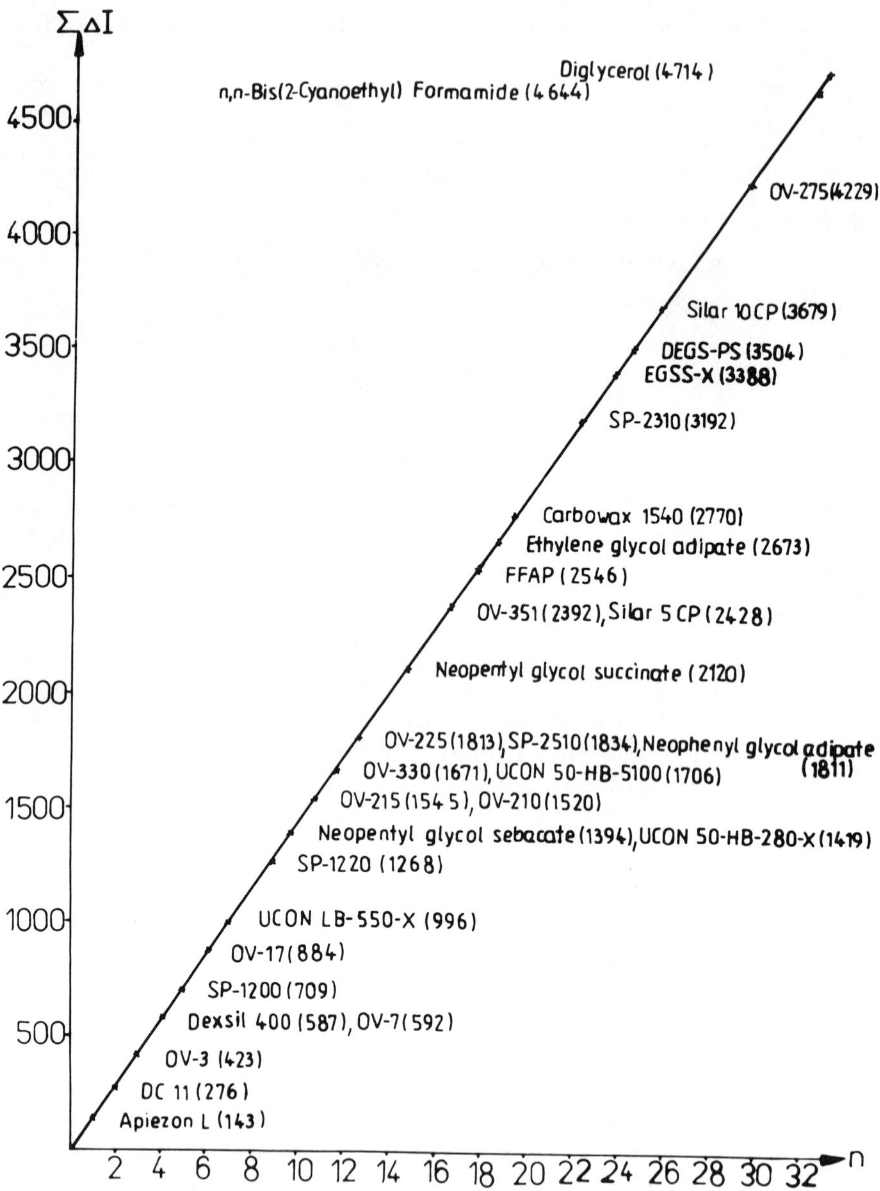


Fig. 26 Average polarity of stationary phases.

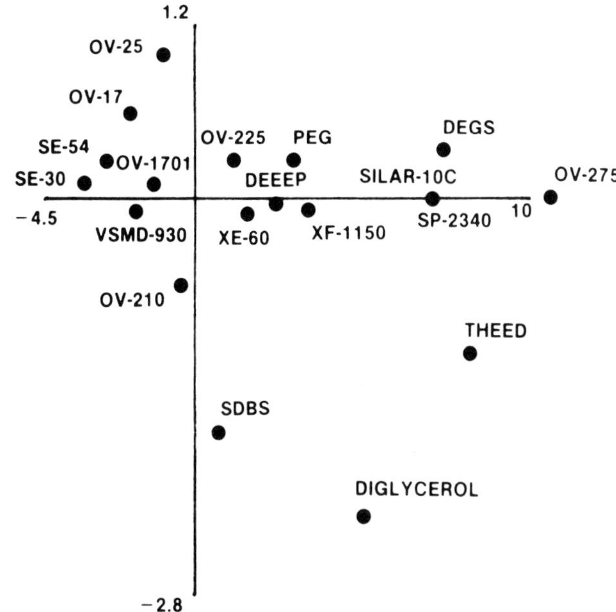


Fig. 27 Relationship of selected phases on the principal component plane. Key: SE-30 = 100% methylpolysiloxane; SE-54 = 5% phenyl, 1% vinylmethylpolysiloxane; OV-17 = 50% phenylmethylpolysiloxane; VSMD-930 = versamid 930; OV-25 = 75% phenylmethylpolysiloxane; OV-1701 = 7% phenyl, 7% cyanopropylmethylsiloxane; OV-210 = 50% trifluoropropylmethylsiloxane; SDBS = sodium dodecyl benzenesulfonate; OV-225 = 25% phenyl, 25% cyanopropylmethylpolysiloxane; XE-60 = 25% cyanoethylmethylpolysiloxane; DEEEP = di-ethoyxmethyl-phthalate; XF-1150 = 50% cyanoethyl methylpolysiloxane; SP-2340 = 100% cyanopropylpolysiloxane; SILAR-10C = 100% cyanopropylpolysiloxane; DEGS = diethylene glycol succinate; THEED = tetrahydroxyethylethylenediamine; OV-275 = 60% cyanopropyl, 40% cyanoethylpolysiloxane. (From Ref. 364, with permission.)

cone of 40-45% cyanopropyl and 20-25% phenyl substitution may give a selectivity close to that of Carbowax [364]. Nonpolar methyl- and phenylsilicones are all in the upper left quadrant with squalane being the farthest to the left. The position of OV-1701 on the polarity axis corresponds to about 40% phenyl, but OV-1701 is placed lower on the dispersion/alcohol axis than would be expected for a 40% phenylsilicone due to lower dispersion and increased alcohol retention. The positions of OV-25 and OV-210 indicate that the ability to be polarized by the solute is smallest for trifluoropropyl and greatest for phenyl.

In the authors' experience, most of today's GC analyses—as a rule of thumb nearly 75%—can be performed on nonpolar columns of high thermal stability which can separate compounds according to their boiling points. Such phases provide a first orientation on the composition of an unknown mixture but frequently fail in distinguishing finer structural features of solute molecules. Consequently, it would be valuable to have some polar phases at one's disposal which can undergo typical interactions (orientation, charge-transfer) with the solute molecules. According to the guiding comments of Lee [371], highly efficient and deactivated capillary columns coated with stable polar and medium polar phases offering new selectivities should soon become common for many applications in which peak identification must be confirmed by chromatography on a column of different polarity, and where critical solute pairs cannot be properly resolved on a nonpolar column. However, special separation problems call for specifically tailor-made phases, such as liquid crystals, enantioselective phases, self-made polarizable polysiloxanes (e.g., made by Lee et al. [371]), or adsorbents with specific adsorption features. Furthermore, the introduction of tailored phases devoted to specific applications (e.g., pesticides) are expected from renowned companies, such as J&W Scientific, Supelco, and others. In addition, by using mixed phases and by adopting the multicolumn techniques any desired selectivity may be obtained.

## REFERENCES

1. G. E. Baiulescu and V. A. Ilie, *Stationary Phases in Gas Chromatography*, Pergamon Press, Oxford, 1975.
2. J. A. Yancey, *J. Chromatogr. Sci.*, *23*:161 (1985).
3. L. S. Ettre, *J. Chromatogr.*, *165*:235 (1979).
4. N. N. Avgul, A. V. Kiselev, and D. P. Poshkus, *Adsorbcija gasov i parov na odnorodnich Poverchnostjach*, Khimija, Moskau, 1975.
5. D. Atkinson and G. Curthoys, *J. Chem. Educ.*, *55*:564 (1978).
6. L. S. Ettre, *Proceedings of the sixth international symposium on capillary chromatography*, Riva del Garda, 1985, Hüthig, 1985, p. 28.
7. E. V. Kalashnikova, A. V. Kiselev, and K. D. Shcherbakova, *Chromatographia*, *17*:521 (1983).
8. J. Rudolph and K. Bächmann, *J. Chromatogr.*, *187*:319 (1980).
9. J. J. Van Deemter, F. J. Zuyderweg, and A. Klinkenberg, *Chem. Eng. Sci.*, *5*:271 (1956).
10. L. S. Ettre, *Open Tubular Columns*, Perkin Elmer, 1973.
11. J. F. K. Huber, H. H. Lauer, and H. Poppe, *J. Chromatogr.*, *112*:377 (1975).
12. J. Jennings, Recent developments in high resolution gas chromatography, in *Flavour 81*, deGruyter, Berlin, p. 233.

13. L. S. Ettre, *Chromatographia, 17*:553 (1983); cf. *Chromatographia, 17*:560 (1983).
14. G. Goretti, A. Liberti, and A. Nota, in *Gas Chromatography 1968* (C. L. A. Harbourn and R. Stock, eds.), Elsevier, Amsterdam, 1969, p. 22.
15. K. Grob, *J. High Resol. Chromatogr. Chromatogr. Commun., 7*:252 (1984).
16. W. J. M. Houtermans, in: *Chrompack News Special* 85-01.
17. J. C. Giddings, *Dynamics of Chromatography*, Vol. 1, New York, p. 120 (1965).
18. G. Guiochon, in *Column Chromatography* (E.sz. Kovats, ed.). Sauerlaender AG, Lausanne, 1970, p. 250.
19. L. S. Ettre, *J. Chromatogr., 112*:1 (1975); *Anal. Chem., 57*: 1419A (1985).
20. E. Cremer, *J. High Resol. Chromatogr. Chromatogr. Commun., 2*:7 (1979).
21. E. V. Kalashnikova, A. V. Kiselev, A. M. Makogon, and K. D. Shcherbakova, *Chromatographia, 8*:399 (1975).
22. C. L. Guillemin, *J. Chromatogr., 158*:21 (1978).
23. I. Halasz and E. Heine, in *Advances in Chromatography*, Vol. 4 (J. C. Giddings and R. A. Keller, eds.), Marcel Dekker, New York, 1967, pp. 207-263.
24. I. Halasz and E. Heine, in *Progress in Gas Chromatography* (J. H. Purnell, ed.). Interscience, New York, 1968, p. 153.
25. Th. Welsch, W. Engewald, and J. Pörschmann, *J. Chromatogr., 148*:143 (1978).
26. C. A. Cramers and J. A. Rijks, in *Advances in Chromatography*, Vol. 17 (J. C. Giddings and R. A. Keller, eds.), Marcel Dekker, New York, 1979, pp. 102-161.
27. C. Reglero, M. Herraiz, M. D. Cabezudo, E. Fernandez-Sanchez, and J. A. Garcia-Dominguez, *J. Chromatogr., 348*:327 (1985).
28. Th. Welsch, W. Engewald, and J. Pörschmann, *J. Prakt. Chem., 320*:493 (1978).
29. A. Malik, V. G. Berezkin, and V. S. Gavrichev, *Chromatographia, 19*:327 (1984).
30. M. J. E. Golay, *J. Chromatogr., 348*:416 (1985).
31. W. Jennings, *Proceedings of the sixth international symposium on capillary chromatography*, Riva del Garda, 1985, Hüthig, 1985, p. 1.
32. C. P. M. Schutjes, E. A. Vermeer, and C. A. Cramers, *J. Chromatogr., 279*:49 (1983).
33. A. Liberti, G. Goretti, and M. V. Russo, *J. Chromatogr., 279*:1 (1983).
34. G. Goretti, F. Zoccolillo, F. Geraci, and S. Gravina, *Chromatographia, 15*:361 (1982).

35. C. Vidal-Madjar, S. Bekassy, M. F. Gonnord, P. Arpino, and G. Guiochon, *Anal. Chem.*, *49*:768 (1977).
36. J. V. Bullen, *Chromatogr. Int.*, *9*:5 (July 1985).
37. T. . Halasz and C. Horvath, *Nature*, *197*:71 (1963).
38. J. Chauhan and A. Darbre, *J. High Resol. Chromatogr. Chromatogr. Commun.*, *4*:11 (1981).
39. A. J. Nunez and K. Tesarik, *J. Chromatogr.*, *284*:235 (1984).
40. A. N. M. Dalemans and E. G. Boeron, *J. High Resol. Chromatogr. Chromatogr. Commun.*, *7*:703 (1984).
41. *Supelco Reporter*, Vol. 4, No. 6, December 1985.
42. V. G. Berezkin and A. A. Korolev, *Chromatographia*, *20*:482 (1985).
43. S. Bratoz, in *Advances in Quantum Chemistry*, Vol. 3 (P. O. Lawdin, ed.), Academic Press, New York, 1967.
44. S. C. Dhanesar, M. E. Coddens, and D. F. Poole, *J. Chromatog. Sci.*, *23*:320 (1985).
45. A. V. Kiselev, *J. Chromatogr.*, *49*:84 (1970).
46. K. Unger, *Angew. Chem.*, *84*:331 (1972).
47. A. V. Kiselev and Y. I. Yashin, *Gazo-adsorbcionnaja chromatografia*, Moskau, Nanka, 1967.
48. M. M. Dubinin, *J. Colloid Interf. Sci.*, *23*:487 (1967).
49. K. Unger, *Kontakte*, 2/79.
50. J. F. K. Huber and F. Eisenbeiss, *J. Chromatogr.*, *149*:127 (1978).
51. J. A. Perry, in *Introduction to Analytical Gas Chromatography*, Chromatographic Science Series, Vol. 14, Marcel Dekker, New York, 1981.
52. L. S. Ettre, *Chromatographia*, *6*:525 (1973); *7*:141, 407 (1974).
53. H. Lamparczyk, *Chromatographia*, *20*:283 (1985).
54. G. Schomburg and G. Dielmann, *J. Chromatog. Sci.*, *11*:151 (1973).
55. L. Rohrschneider, *J. Chromatogr.*, *17*:1 (1965); *22*:6 (1966); *39*:383 (1969).
56. M. B. Evans and M. J. Osborn, *Chromatographia*, *13*:177 (1980).
57. W. M. McReynolds, *J. Chromatog. Sci.*, *8*:685 (1970).
58. L. Kaplar, C. Szita, J. Takacs, and G. Tarjan, *J. Chromatogr.*, *65*:115 (1972).
59. V. G. Berezkin, *J. Chromatogr.*, *159*:359 (1978).
60. V. G. Berezkin, *Int. Lab.*, July/Aug. 1980, p. 19.
61. J. Krupcik, A. Matisova, J. Garaj, L. Sojak, and V. G. Berezkin, *Chromatographia*, *16*:166, 169 (1982).
62. R. V. Golovnya, *Chromatographia*, *12*:533 (1979).
63. R. G. Golovnya and T. A. Misharina, *J. High Resol. Chromatogr. Chromatogr. Commun.*, *3*:51 (1980).
64. R. V. Golovnya and T. A. Misharina, *J. Chromatogr.*, *190*:1 (1980).
65. R. V. Golovnya and T. A. Misharina, *J. High Resol. Chromatogr. Chromatogr. Commun.*, *3*:4 (1980).

66. R. V. Golovnya and D. N. Grigoryeva, *Chromatographia, 17*:613 (1983).
67. J. Ševčik and M. S. H. Löwentap, *J. Chromatogr., 217*:139 (1981).
68. I. Temmermann, P. Sandra, and M. Verzele, *Proceedings of the sixth international symposium on capillary chromatography*, Riva del Garda, 1985, Hüthig, 1985, p. 144; cf. also *Chromatographia, 16*:63 (1982).
69. M. Verzele, P. Musche, and P. Sandra, *J. Chromatogr., 190*:176 (1980).
70. K. Grob, *J. Chromatogr., 1980*:176 (1980).
71. N. N. Avgul and A. V. Kiselev, in *Chemistry and Physics of Carbon* (P. L. Walker, ed.), Vol. 6, Marcel Dekker, New York, 1970.
72. W. Engewald, J. Pörschmann, Th. Welsch, and K. D. Shcherbakova, *Z. Chem., 17*:375 (1977).
73. A. V. Kiselev and Y. I. Yashin, *Adsorbcionnaja gazovaja i chidkostnaja chromatografia*, Chimii, Moskau, 1975.
74. A. V. Kiselev, K. D. Shcherbakova, and Y. I. Yashin, *Zhurn. strukt. Khim., 10*:951 (1969).
75. A. Di Corcia, A. Liberti, and R. Samperi, *J. Chromatogr., 122*:459 (1976).
76. O. G. Eisen, A. V. Kiselev, A. E. Pilt, S. A. Rang, and K. D. Shcherbakova, *Chromatographia, 4*:448 (1971).
77. Z. Krawiec, M.-F. Gonnord, G. Guiochon, and J. R. Chretien, *Anal. Chem., 51*:1655 (1979).
78. S. A. Rang, O. G. Eisen, A. V. Kiselev, A. E. Meister, and K. D. Shcherbakova, *Chromatographia, 8*:327 (1975).
79. M. F. Gonnord, C. Vidal-Madjar, and G. Guiochon, *J. Chromatog. Sci., 12*:839 (1974).
80. W. Engewald, L. Wennrich, and J. Pörschmann, *Chromatographia, 11*:434 (1978).
81. E. V. Kalashnikova, A. V. Kiselev, and K. D. Shcherbakova, *Chromatographia, 7*:22 (1974).
82. A. V. Kiselev, N. S. Kulikov, and G. Curthoys, *Chromatographia, 18*:297 (1984).
83. J. Fryčka, *J. Chromatogr., 65*:341, 432 (1972).
84. E. V. Kalashnikova and K. D. Shcherbakova, *J. Chromatogr., 91*:695 (1974).
85. A. Di Corcia and F. Bruner, *Anal. Chem., 43*:1634 (1971).
86. A. Di Corcia, R. Samperi, and C. Severini, *J. Chromatogr., 198*:347 (1980).
87. A. G. Bezus, A. V. Kiselev, A. M. Makogon, E. I. Model, and K. D. Shcherbakova, *Chromatographia, 7*:246 (1974).
88. A. Di Corcia, R. Samperi, and C. Severini, *J. Chromatogr., 170*:325 (1979).
89. A. V. Kiselev, E. B. Polotnyuk, and K. D. Shcherbakova, *Chromatographia, 14*:478 (1981).

90. F. Bruner, A. Liberti, M. Possanzini, and I. Allegrini, *Anal. Chem.*, *44*:2070 (1972).
91. A. Di Corcia, R. Samperi, E. Sebastiani, and C. Severini, *Chromatographia*, *14*:86 (1981).
92. A. Di Corcia, A. Liberti, and G. Severini, *J. Chromatogr.*, *204*:255 (1981).
93. G. Cosmi, A. Di Corcia, R. Samperi, and G. Vinci, *Chromatographia*, *16*:323 (1982).
94. C. Vidal-Madjar and G. Guiochon, *Sep. Purif. Methods*, *2*:1 (1973).
95. F. Mangani and F. Bruner, *J. Chromatogr.*, *289*:85 (1984).
96. A. Di Corcia, R. Samperi, E. Sebastiani, and C. Severini, *Anal. Chem.*, *52*:1345 (1980).
97. N. Petsev, I. Topalova, S. Ivanov, C. Dimitrov, T. B. Gavrilova, and E. V. Vlasenko, *J. Chromatogr.*, *286*:57 (1984).
98. M. T. Gilbert, J. H. Knox, and B. Kaur, *Chromatographia*, *16*:138 (1982).
99. T. N. Grozdovich, A. V. Kiselev, and Y. I. Yashin, *Neftechimija*, *8*:476 (1968).
100. R. Kaiser, *Chromatographia*, *2*:453 (1969).
101. R. Kaiser, *Chromatographia*, *3*:38 (1970).
102. D. Li-Ru, *J. Chromatogr.*, *186*:317 (1979).
103. D. H. Bollmann and D. M. Mortimore, *J. Chromatog. Sci.*, *10*:523 (1972).
104. A. Zlatkis, H. R. Kaufman, and D. E. Durbin, *J. Chromatog. Sci.*, *8*:416 (1970).
105. D. R. Jenke and M. R. Hannifan, *Anal. Chem.*, *54*:843 (1982).
106. J. M. H. Daeman, W. Dankelman, and M. E. Hendrike, *J. Chromatog. Sci.*, *13*:79 (1975).
107. V. Patzelova, J. Jansta, and F. P. Dousek, *J. Chromatogr.*, *148*:53 (1978).
108. W. Asche, *J. High Resol. Chromatogr. Chromatogr. Commun.*, *7*:282 (1984).
109. O. L. Hollis, *Anal. Chem.*, *38*:309 (1966).
110. C. P. M. G. A'Campo, S. M. Lemkowitz, P. Verbrugge, and P. J. Van Den Berg, *J. Chromatogr.*, *203*:271 (1981).
111. M. G. Neumann, *Fres. Z. anal. Chem.*, *244*:302 (1969).
112. S. B. Dave, *J. Chromatog. Sci.*, *7*:389 (1969).
113. Chrompack News, *12*:55 (1985).
114. H. Hoffmann, M. Henke, K. Häupke, and G. Schwachula, *Plaste Kautschuk*, *25*:453 (1978).
115. K. I. Sakodynskii, V. L. Boeva, L. I. Panina, and L. D. Glazunova, *J. Chromatogr.*, *186*:227 (1979).
116. K. I. Sakodynskii, V. L. Boeva, L. I. Panina, and Yu. Godovskii, *J. Chromatogr.*, *206*:233 (1981).

117. J. Lukaš, F. Švec, E. Votavová, and J. Kálal, *J. Chromatogr.*, *153*:373 (1978).
118. J. Lukaš, *J. Chromatogr.*, *190*:13 (1980).
119. J. Lukaš, E. Votavová, F. Švec, and J. Kálal, *J. Chromatogr.*, *194*:297 (1980).
120. E. Gawdzik, Z. Zuchowski, T. Matynia, and J. Gawdzik, *J. Chromatogr.*, *286*:11 (1984).
121. R. van Wijk, *J. Chromatog. Sci.*, *8*:418 (1970).
122. K. I. Sakodynskii, L. Panina, and N. Klinskaya, *Chromatographia*, *7*:339 (1974).
123. L. H. Ponder, *J. Chromatogr.*, *97*:77 (1974).
124. N. C. Saha, S. K. Jain, and R. K. Dua, *Chromatographia*, *10*:368 (1977).
125. M. L. Knuth and M. D. Hoglund, *J. Chromatogr.*, *285*:153 (1984).
126. J. R. L. Smith, A. H. H. Tameesh, and D. J. Waddington, *J. Chromatogr.*, *148*:353, 365 (1978).
127. J. R. L. Smith, A. H. H. Tameesh, and D. J. Waddington, *J. Chromatogr.*, *151*:21 (1978).
128. K. I. Sakodynskii, L. I. Panina, and S. B. Makarova, Abstracts of V. Danube Symposium on Chromatography, Yalta, Nauka, 1985, p. 210.
129. T. N. Gvosdovich, A. V. Kiselev, and Y. I. Yashin, *Chromatographia*, *11*:596 (1972).
130. G. Castello and G. D'Amato, *J. Chromatogr.*, *269*:153 (1983).
131. G. Castello and G. D'Amato, *J. Chromatogr.*, *254*:69 (1983).
132. K. I. Sakodinsky, *Chromatographia*, *1*:483 (1968).
133. O. L. Hollis, *J. Chromatog. Sci.*, *11*:335 (1973).
134. G. Castello and G. D'Amato, *J. Chromatogr.*, *196*:245 (1980).
135. L. R. Snyder and H. Poppe, *J. Chromatogr.*, *184*:363 (1980).
136. S. K. Milonjič and M. M. Kopečni, *Chromatographia*, *19*:342 (1984).
137. E. Smolková, L. Feltl, and J. Zima, *Chromatographia*, *12*:463 (1979).
138. N. K. Bebris, A. V. Kiselev, B. Ya. Makeev, Yu. S. Nikitin, Ya. I. Yashin, and G. E. Zaizeva, *Chromatographia*, *4*:93 (1971).
139. C. L. Guillemin, M. Deleuil, S. Cirendini, and J. Vermont, *Anal. Chem.*, *43*:2015 (1971).
140. D. J. Brookman and D. T. Sawyer, *Anal. Chem.*, *40*:2013 (1968).
141. A. F. Isbell and D. T. Sawyer, *Anal. Chem.*, *41*:1381 (1969).
142. M. M. Kopečni, S. K. Milonjič, and R. J. Laub, *Anal. Chem.*, *52*:1032 (1980).
143. D. F. Cadogan and D. T. Sawyer, *Anal. Chem.*, *43*:941 (1971).
144. J. E. Picker and R. E. Sievers, *J. Chromatogr.*, *203*:29 (1981).
145. R. Leboda and A. Waksmundzki, *Chromatographia*, *12*:207 (1979).
146. R. Leboda, *Chromatographia*, *13*:549 (1980).
147. R. Leboda, *Chromatographia*, *14*:524 (1981).

148. E. Tracz and R. Leboda, *J. Chromatogr.*, *346*:346 (1985).
149. J. Rayss, *Chromatographia*, *15*:517 (1982).
150. I. Halasz and H. O. Gerlach, *Anal. Chem.*, *38*:281 (1966).
151. M. Mohnke and W. Saffert, *Kernenergie*, *5*:434 (1962).
152. H. Traitler and A. Prevot, *J. High Resol. Chromatogr. Chromatogr. Commun.*, *4*:433 (1981).
153. R. G. Mathews, J. Torres, and R. D. Schwartz, *J. Chromatogr.*, *186*:183 (1979).
154. R. G. Mathews, J. Torres, and R. D. Schwartz, *J. Chromatogr.*, *199*:97 (1980).
155. R. G. Mathews, J. Torres, and R. D. Schwartz, *J. Chromatog. Sci.*, *20*:160 (1982).
156. A. G. Ober, M. Cooke, and G. Nickless, *J. Chromatogr.*, *196*: 237 (1980).
157. E. D. John and G. Nickless, *J. Chromatogr.*, *138*:399 (1977).
158. F. Janowski and W. Heyer, *Z. Chem.*, *19*:1 (1979).
159. F. Janowski and W. Heyer, *Poröse Gläser*, VEB Deutscher Verlag für Grundstoffindustrie, Leipzig, 1982.
160. D. Hager, *J. Chromatogr.*, *187*:285 (1980).
161 T. Mizutani and A. Mizutani, *J. Chromatogr.*, *168*:143 (1979).
162. M. G. Neumann and W. Hertl, *J. Chromatogr.*, *65*:467 (1972).
163. C. G. Scott, *Gas Chromatographie 1962* (M. v. Swaay, ed.), Butterworth, London, 1962.
164. S. Moriguchi, K. Naito, and S. Takei, *J. Chromatogr.*, *131*:19 (1977).
165. S. Moriguchi and S. Takei, *J. Chromatogr.*, *295*:73 (1984).
166. K. Naito, M. Endo, S. Moriguchi, and S. Takei, *J. Chromatogr.*, *253*:205 (1982).
167. K. Naito, R. Kurita, S. Moriguchi, and S. Takei, *J. Chromatogr.*, *246*:199 (1982).
168. K. Naito, R. Kurita, M. Endo, S. Moriguchi, and S. Takei, *J. Chromatogr.*, *268*:359 (1983).
169. W. Schneider, J. C. Frohne, and H. Bruderreck, *J. Chromatogr.*, *155*:311 (1978).
170. R. C. M. De Nijs and J. De Zeeuw, *J. Chromatogr.*, *279*:41 (1983).
171. F. Schwochow and L. Puppe, *Angew. Chem.*, *87*:659 (1975).
172. T. G. Andronikashvili and L. G. Eprikashvili, *J. Chromatogr.*, *286*:3 (1984).
173. G. V. Tsitsishvili, S. D. Sabelashvili, and S. T. G. Andronikashvili, *Zavod. Lab.*, *41*:398 (1975).
174. W. Wardencki and R. Staszewski, *J. Chromatogr.*, *329*:128 (1985).
175. W. Wardencki and R. Staszewski, *J. Chromatogr.*, *325*:430 (1985).
176. V. G. Berezkin, *Fres. Z. anal. Chem.*, *296*:1 (1979).
177. M. Long, G. Raverdino, G. Di Tullio, and L. Tomarchio, *J. Chromatogr.*, *117*:305 (1976).

178. J. De Zeeuw and R. C. M. De Nijs, *Chrompack News, 12*:1 (1985).
179. T. G. Andronikashvili, V. G. Berezkin, L. Ya. Laperashvili, and N. A. Nadiradze, *J. Chromatogr., 331*:402 (1985).
180. R. Kaiser, *Ber. Bunsenges. phys. Chem., 69*:826 (1965).
181. J. C. Maillen, T. M. Reed, and J. A. Young, *Anal. Chem., 36*: 1883 (1964).
182. A. Bhattacharjee and A. N. Basu, *J. Chromatogr., 71*:534 (1972).
183. E. Smolková, H. Králová, S. Krýsl, and L. Feltl, *J. Chromatogr., 241*:3 (1982).
184. T. Koscielski, D. Sybilska, and J. Jurczak, *J. Chromatogr., 280*:131 (1983); *349*:3 (1985).
185. J. Mráz, L. Feltl, and E. Smolková-Keulemansová, *J. Chromatogr., 286*:17 (1984).
186. T. B. Gavrilova, A. V. Kiselev, and T. M. Roshchina, *J. Chromatogr., 192*:323 (1980).
187. T. B. Gavrilova, T. M. Roshchina, Ch. Dimitrov, I. Topalova, and N. Petsev, *J. Chromatogr., 286*:49 (1984).
188. L. D. Belyakova, A. V. Kiselev, and G. A. Soloyan, *Chromatographia, 3*:254 (1970).
189. R. L. Grob, G. W. Weinert, and J. W. Drelich, *J. Chromatogr., 30*:305 (1967).
190. B. T. Guran and L. B. Rogers, *Anal. Chem., 39*:632 (1967).
191. R. L. Grob and E. J. Mc. Gonigle, *J. Chromatogr., 59*:13 (1971).
192. A. G. Altenau and L. B. Rogers, *Anal. Chem., 36*:1726 (1964).
193. G. A. Eiceman, *J. Chromatogr., 242*:267 (1982).
194. N. M. Gogitidze, T. G. Andronikashvili, Z. Litvak, L. I. Panina, K. I. Sakodynskii, and S. B. Makarova, *J. Chromatogr., 206*: 239 (1981).
195. R. F. Hirsch, R. J. Gaydosh, J. R. Chretien, and J. E. Dubois, *Chromatographia, 16*:251 (1982).
196. A. Di Corcia, and A. Liberti, Gas-Liquid-Solid-Chromatography, in: *Advances in Chromatography*, Vol. 14 (J. C. Giddings, ed.), 1976, p. 306.
197. J. F. Parcher, *J. Chromatog. Sci., 21*:346 (1983).
198. J. F. Parcher, and K. Hyver-LoCoco, *J. Chromatog. Sci., 21*: 304 (1983).
199. T. Ramstad and L. W. Nicholson, *Anal. Chem., 54*:1191 (1982).
200. B. P. Semonian and L. B. Rogers, *J. Chromatog. Sci., 16*:49 (1978).
201. T. Tsuda, T. Ichiba, H. Muramatsu, and D. Ishii, *J. Chromatogr., 130*:87 (1977).
202. A. Betti, G. Lodi, C. Bighi, and F. Dondi, *J. Chromatogr., 106*:291 (1975).
203. C. L. Guillemin, F. Martinez, and S. Thiault, *J. Chromatog. Sci., 17*:677 (1979).
204. S. M. Rakhmankulov, Izv. Akad. Nank, SSSR, ser. Khim. p.

2785 (1975), p. 1530 (1976).
205. R. G. Ackman, *J. Chromatog. Sci.*, *10*:506 (1972).
206. C. L. Guillemin, J. L. Millet, and E. Hamon, *J. Chromatogr.*, *301*:11 (1984).
207. F. T. Eggertsen, H. S. Knight, and S. Groennings, *Anal. Chem.*, *28*:303 (1956).
208. H. Purnell, *Gas Chromatography*, John Wiley and Sons, New York, 1962.
209. J. F. Parcher and D. M. Johnson, *J. Chromatog. Sci.*, *23*:459 (1985).
210. A. Di Corcia, *Anal. Chem.*, *45*:492 (1973).
211. P. Ciccioli, J. M. Hayes, G. Rinaldi, K. B. Denson, and W. G. Meinschein, *Anal. Chem.*, *51*:400 (1979).
212. V. G. Berezkin and V. S. Gavrichev, *J. Chromatogr.*, *116*:9 (1976).
213. A. Aider, B. Devallez, and M.-H. Guermouche, *J. High Resol. Chromatogr. Chromatogr. Commun.*, *7*:222 (1984).
214. A. Di Corcia, L. Ripani, and R. Samperi, *J. Chromatogr. Biomed. Appl.*, *229*:365 (1982).
215. A. Di Corcia and R. Samperi, *Anal. Chem.*, *46*:997 (1974).
216. A. Di Corcia and R. Samperi, *Anal. Chem.*, *51*:776 (1979).
217. A. Di Corcia, *J. Chromatogr.*, *80*:69 (1973).
218. F. Bruner, P. Ciccioli, and F. Di Nardo, *Anal. Chem.*, *47*:141 (1975).
219. A. Di Corcia and R. Samperi, *Anal. Chem.*, *47*:1853 (1975).
220. G. Goretti, A. Liberti, and G. Pili, *J. High Resol. Chromatogr. Chromatogr. Commun.*, *1*:143 (1978).
221. C. Vidal-Madjar and G. Guiochon, *J. Chromatog. Sci.*, *9*:664 (1971).
222. J. Hille, M. Procházka, L. Feltl, E. Smolková-Keulemansová, A. V. Kiselev, N. V. Kovaleva, and E. V. Zagorevskaya, *J. Chromatogr.*, *283*:77 (1984).
223. M. Procházka and E. Smolková-Keulemansová, *J. Chromatogr.*, *189*:25 (1980).
224. C. L. Guillemin and F. Martinez, *J. Chromatogr.*, *139*:259 (1977).
225. I. Halasz and I. Sebestian, *Angew. Chem.*, *81*:464 (1969).
226. F. W. Karasek, *Res. Dev.*, *81*:464 (1969).
227. J. N. Little, W. A. Dark, Ph. W. Farlinger, and K. J. Bombough, *J. Chromatog. Sci.*, *8*:647 (1970).
228. E. W. Abel, F. H. Pollard, P. C. Uden, and G. Nickless, *J. Chromatogr.*, *22*:23 (1966).
229. W. A. Aue, S. Kapila, and K. O. Gerhardt, *J. Chromatogr.*, *78*:228 (1973).
230. V. Rehák and E. Smolková, *Chromatographia*, *9*:219 (1976).
231. D. C. Locke, J. T. Schmermund, and B. Banner, *Anal. Chem.*, *44*:90 (1972).

232. I. Sebestian and I. Halasz, *Chromatographia, 7*:371 (1974).
233. B. D. Škrbič and M. J. Zlatkovic, *Chromatographia, 17*:44 (1983).
234. C. J. Cowper and P. A. Wallis, *Chromatographia, 19*:85-94 (1984).
235. N. Sannolo, P. Vajro, G. Dioguardi, R. Mensitieri, and D. Longo, *J. Chromatogr. (Biomed. Appl.), 276*:257 (1983).
236. R. H. Liu and W. W. Ku, *J. Chromatogr., 271*:309 (1983).
237. H. J. Vreman, L. K. Kwong, and D. K. Tevenson, *Clin. Chem., 30*:382 (1984).
238. D. Payne-Bose, A. Tsegaya, R. D. Morrison, and G. R. Waller, *Anal. Biochem., 88*:659 (1978).
239. D. Krockenberger, H. Lorkowski, and L. Rohrschneider, *Chromatographia, 12*:787 (1979).
240. S. Moon, D. W. Boyd, A. Lichtman, and R. A. Porter, *Int. Lab., 15*:56 (1985).
241. P. D. Goldan, F. C. Fehsenfeld, W. C. Kuster, M. P. Phillips, and R. E. Sievers, *Anal. Chem., 52*:1751 (1980).
242. P. Fang, G. D. Mc. Ginnis, and W. W. Wilson, *Anal. Chem., 53*:2172 (1981).
243. Choosing Porapak Appl., Bull. Waters Assoc., *4* (1968).
244. R. T. Palo, J. D. Walters, E. W. March, and L. S. Ettre, *J. High Resol. Chromatogr. Chromatogr. Commun., 7*:358 (1984).
245. R. J. Leibrand, *J. Gas Chromatogr., 5*:518 (1967).
246. R. Mindrup, *J. Chromatog. Sci., 16*:380 (1978).
247. P. G. Jeffery and P. J. Kipping, *Gas Analysis by Gas Chromatrography*, Pergamon Press, Oxford, 1972.
248. C. F. Cowper and A. J. de Rose, *The Analysis of Gases by Chromatography*, Pergamon Series in Anal. Chem., Vol. 7, Pergamon Press, Oxford, 1983.
249. S. Nand and M. K. Sarkar, *J. Chromatogr., 89*:73 (1974).
250. D. R. Deans, M. T. Huckle, and R. M. Peterson, *Chromatographia, 4*:279 (1971).
251. F. F. Andrawes and E. K. Gibson, Jr., *Anal. Chem., 51*:462 (1979).
252. W. Gates, P. Zambri, and J. N. Armor, *J. Chromatog. Sci., 19*:183 (1981).
253. E. Schnelle and R. Hibber, *Microchim. Acta, 3-4*:263 (1982).
254. D. Squier and A. Hill, *J. Chromatog. Sci., 20*:429 (1982).
255. L. Huber and H. Obbens, *J. Chromatogr., 279*:167 (1983).
256. E. H. Osjord and D. Malthe-Sørenssen, *J. Chromatogr., 279*:219 (1983).
257. K. Frunzke and W. G. Zumft, *J. Chromatogr., 299*:477 (1984).
258. J. H. Purnell and P. S. Williams, *J. Chromatogr., 325*:1 (1985).
259. M. Riederer, *Labor Praxis*, July/August 1981, p. 578.
260. A. Di Corcia, R. Samperi, and G. Capponi, *Chromatographia, 10*:554 (1977).
261. T. L. C. de Souza, D. C. Lane, and S. P. Bhatia, *Anal. Chem., 47*:543 (1975).

262. D. Barnett and E. G. Davis, *J. Chromatog. Sci.*, *21*:205 (1983).
263. S. Latif, J. K. Haken, and M. S. Wainwright. *J. Chromatogr.*, *258*:233 (1983).
264. F. Raulin and G. Toupance, *J. Chromatogr.*, *90*:218 (1974).
265. J. L. Genna, W. D. Mc Aninch, and R. A. Reich, *J. Chromatogr.*, *238*:103 (1982).
266. J. Pörschmann, T. Welsch, W. Engewald, and Gy. Vigh, *J. High Resol. Chromatogr. Chromatogr. Commun.*, 7:509 (1984).
267. A. Kuksis, *Sep. Purif. Methods*, 5(2), 353 (1977).
268. J. C. Du Preez and P. M. Lategan, *J. Chromatogr.*, *124*:63 (1976).
269. J. Carlsson, *Appl. Microbiol.*, *25*:287 (1973).
270. O. Somorin, *Anal. Biochem.*, *88*:442 (1978).
271. K. Uobe, K. Nishida, H. Inoue, and M. Tsutsui, *J. Chromatogr.*, *193*:83 (1980).
272. M. H. Henderson and T. A. Steedman, *J. Chromatogr.*, *244*:337 (1982).
273. W. Böttcher, *Arch. Tierern.*, *32*:287 (1982).
274. A. Kettrup, J. Nolte, and W. Riepe, *Fres. Z. Anal. Chem.*, *307*:1 (1981).
275. J. Jurenitsch, G. Lumper, and H. Bachmann, *J. Chromatogr.*, *189*:43 (1980).
276. G. Goretti, A. Liberti, M. V. Russo, and J. L. Sanchez, *J. Chromatogr.*, *245*:109 (1982).
277. E. D. Lund, C. L. Kirkland, and P. E. Snaw, *J. Agric. Food Chem.*, *29*:361 (1981).
278. J. Snyder and B. Franco-Filipasic, *J. Am. Oil Chem. Soc.*, *60*: 1269 (1983).
279. I. A. Mc Donald, L. P. Hackett, and L. J. Dusci, *Clin. Chim. Acta*, *63*:235 (1975).
280. H. Tanner and H. Limacher, *Flüssiges Obst*, April 1984, p. 183.
281. S. Cirendini, J. Vermont, J. C. Gressin, and C. L. Guillemin, *J. Chromatogr.*, *84*:21 (1973).
282. I. M. Pirzada and J. H. Hills, *Analyst*, *108*:1096 (1983).
283. S. Lowis, M. A. Eastwood, and W. G. Brydon, *J. Chromatogr.* (Biomed. Appl.), *278*:139 (1983).
284. A. Tavakkol and D. B. Drucker, *J. Chromatogr.*, *274*:37 (1983).
285. J.-L. Pons, A. Rimbault, J. C. Darbord, and G. Leluan, *J. Chromatogr.*, *337*:213 (1985).
286. D. S. Treybig, *J. Chromatog. Sci.*, *21*:310 (1983).
287. J. R. Lindsay, J. R. L. Smith and D. J. Waddington, *Anal. Chem.*, *40*:522 (1968).
288. J. B. Soar and P. J. Buttery, *J. Sci. Food Agric.*, *30*:795 (1979).
289. P. C. Uden, *J. Chromatogr.*, *313*:3 (1984).
290. W. M. Moore, *Anal. Chem.*, *54*:602 (1982).
291. K. Funazo, T. Hirashima, H.-L. Wu, M. Tanaka, and T. H. Shono, *J. Chromatogr.*, *243*:85 (1982).

292. S. Tsalas and K. Bächmann, *Talanta, 27*:201 (1980).
293. G. Nota, C. Improta, and V. R. Miraglia, *J. Chromatogr., 242*: 359 (1982).
294. G. Nota, C. Improta, U. Papale, and C. Ferretti, *J. Chromatogr., 262*:360 (1983).
295. D. N. Sokolov and N. A. Vakin, *J. Chromatog. Sci., 10*:417 (1972).
296. M. H. Hahn, K. J. Mulligan, M. E. Jackson, and J. A. Caruso, *Anal. Chim. Acta, 118*:115 (1980).
297. G. Nickless, *J. Chromatogr., 313*:129 (1984).
298. A. I. Clark, A. E. McIntyre, J. N. Lester, and R. Perry, *J. Chromatogr., 252*:147 (1982).
299. G. Hunt and N. Pangaro, *Anal. Chem., 54*:369 (1982).
300. W. G. Jennings and A. Rapp, *Sample Preparation for GC Analysis*, Hüthig, 1983.
301. A. J. Nunez, L. F. Gonzalez, and J. Janak, *J. Chromatogr., 300*:127 (1984).
302. J. Schaefer, in *Flavour 81*, Walther de Gruyter, Berlin, 1981, p. 381.
303. G. Bertoni, F. Bruner, A. Liberti, and C. Perrino, *J. Chromatogr., 203*:263 (1981).
304. J. Namiesnik, L. Torres, E. Kozlowski, and J. Mathieu, *J. Chromatogr., 208*:239 (1981).
305. S. Coppi, A. Betti, G. Blo, and C. Bighi, *J. Chromatogr., 267*: 91 (1983).
306. G. A. Eiceman and F. W. Karasek, *J. Chromatogr., 200*:115-124 (1980).
307. E. R. Kennedy and R. H. Hill, Jr., *Anal. Chem., 54*:1739 (1982).
308. J. P. Guenier, P. Simon, J. Delcourt, M. F. Didierjean, C. Lefevre, and J. Muller, *Chromatographia, 18*:137 (1984).
309. C. D. Chriswell and J. S. Fritz, *J. Chromatogr., 136*:371 (1977).
310. Y. Hoshika, *Anal. Chem., 54*:2433 (1982).
311. Y. Hoshika, *Analyst, 106*:166 (1981).
312. A. Przyjazny, *J. Chromatogr., 333*:32 (1985); *346*:61 (1985).
313. F. Bruner, C. Crescentini, F. Mangani, and R. Petty, *Anal. Chem., 55*:793 (1983).
314. N. E. Spingarn, D. J. Northington, and T. Pressely, *J. Chromatog. Sci., 20*:286 (1982).
315. A. Zlatkis, et al., *Chromatographia, 20*:343 (1985); see also Chrompack News, *12*:4 (1985).
316 J. F. Pankow and L. M. Isabelle, *J. Chromatogr., 237*:25 (1982).
317. K. Altenburg, *J. Chromatogr., 45*:306 (1969).
318. S. Rang, K. Kuningas, A. Orav, and O. Eisen, *J. Chromatogr., 128*:59 (1976).
319. W. Ecknig, *Z. Chem., 25*:240 (1985).
320. A. V. Kiselev, V. I. Nazarova, K. D. Shcherbakova, E. Smolkova-Keulemansová, and L. Feltl, *Chromatographia, 17*:533 (1983).

321. D. P. Poshkus and A. J. Grumadas, *J. Chromatogr.*, *191*:169 (1980).
322. A. V. Kiselev and D. P. Poshkus, *Far. Trans. II*, *72*:950 (1976).
323. C. Vidal-Madjar, M.-F. Gonnord, and G. Guiochon, *Adv. Chromatogr.*, *13*:177 (1975).
324. A. V. Kiselev and D. L. Markosyan, *Chromatographia*, *17*:526 (1983).
325. H.-J. Hofmann, W. Engewald, D. Heidrich, J. Pörschmann, K. Thieroff, and P. Uhlmann, *J. Chromatogr.*, *115*:299 (1975).
326. A. V. Kiselev, *Chromatographia*, *11*:691 (1978).
327. E. V. Kalashnikova, A. V. Kiselev, K. D. Shcherbakova, and S. D. Vasileva, *Chromatographia*, *14*:510 (1981).
328. W. Engewald, K. Epsch, J. Graefe, and Th. Welsch, *J. Chromatogr.*, *91*:623 (1974).
329. W. Engewald, J. Graefe, A. V. Kiselev, K. D. Shcherbakova, and Th. Welsch, *Chromatographia*, *7*:229 (1974).
330. W. Engewald, E. V. Kalashnikova, A. V. Kiselev, R. S. Petrova, K. D. Shcherbakova, and A. L. Shilov, *J. Chromatogr.*, *152*:453 (1978).
331. F. Riedo, D. Fritz, G. Tarjan, and E. sz. Kovats, *J. Chromatogr.*, *126*:63 (1976).
332. F. I. Onuska, *J. Chromatogr.*, *186*:259 (1979).
333. F. Vernon and C. O. E. Ogundipe, *J. Chromatogr.*, *132*:181 (1977).
334. J. K. Haken, *J. Chromatogr.*, *300*:1 (1984); cf. also *J. Chromatogr.*, *141*:247 (1977); *J. Chromatogr.*, *73*:419 (1972).
335. J. K. Haken and F. Vernon, *J. Chromatogr.*, *186*:89 (1979).
336. J. R. Mann and S. T. Preston, *J. Chromatog. Sci.*, *11*:216 (1973).
337. G. Castello and G. D'Amato, *J. Chromatogr.*, *90*:291 (1974).
338. H. E. Persinger and J. T. Shank, *J. Chromatog. Sci.*, *11*:190 (1973).
339. L. Soják and J. Krupčik, *J. Chromatogr.*, *190*:283 (1980).
340. M. B. Evans and J. F. Smith, *J. Chromatogr.*, *36*:489 (1968).
341. R. Dandeneau, P. Bente, T. Rooney, and R. Hiskes, *Am. Lab.*, *11*:61 (1979).
342. M. Verzele and P. Sandra, *J. Chromatogr.*, *158*:111 (1978).
343. P. Sandra and M. van Roelenbosch, *Chromatographia*, *14*:345 (1981).
344. M. K. Withers, *J. Chromatogr.*, *66*:249 (1972).
345. K. Grob, Jr. and K. Grob, *J. Chromatogr.*, *140*:257 (1977).
346. Gy. Vigh, A. Bartha, and J. Hlavay, *J. High Resol. Chromatogr. Chromatogr. Commun.*, *4*:3 (1981).
347. Analabs Chromatography Chemicals and Accessories, Catal. No. 23, p. 104.
348. F. Rinderknecht, and B. Wenger, *J. High Resol. Chromatogr. Chromatogr. Commun.*, *6*:422 (1983).

349. R. F. Kruppa and R. S. Henly, *J. Chromatog. Sci.*, *12*:127 (1974).
350. M. G. Voronkov, V. P. Mileshkevich and Ya. Y. Yuzhelevskii, *The Siloxane Bond*, Consultants Bureau, New York, 1978.
351. L. C. Case, *J. Chromatogr.*, *6*:381 (1961).
352. R. G. Mathews, R. D. Schwartz, J. E. Stouffer, and B. C. Pettitt, *J. Chromatog. Sci.*, *8*:508 (1970).
353. H. M. Tenney, *Anal. Chem.*, *30*:2 (1958).
354. J. H. Beeson and R. E. Pecsar, *Anal. Chem.*, *41*:1678 (1969).
355. R. G. Mathews, R. D. Schwartz, C. P. Pfaffenberger, S.-N. Lin, and E. C. Horning, *J. Chromatogr.*, *99*:51 (1972).
356. H. Borwitzky and G. Schomburg, *J. Chromatogr.*, *170*:99 (1979).
357. F. Sellier, G. Tersac, and G. Guiochon, *J. Chromatogr.*, *219*:213 (1981).
358. C. F. Poole, H. Butler, S. A. Agnello, W.-F. Sye, A. Zlatkis, and G. Holzer, *J. Chromatogr.*, *217*:39 (1981).
359. W. C. Hammann, R. M. Schisla, J. S. Adams, and S. D. Koch, *J. Chem. Eng. Data*, *15*:351 (1970).
360. L. B. Itsikson, V. G. Berezkin, and J. K. Haken, *J. Chromatogr.*, *334*:1 (1985).
361. L. Butter and S. Hawkes, *J. Chromatog. Sci.*, *10*:512 (1972).
362. R. J. Leibrand, *J. Chromatog. Sci.*, *13*:556 (1975).
363. W. Millen and S. Hawkes, *J. Chromatog. Sci.*, *15*:148 (1977).
364. T. J. Stark, P. A. Larson, and R. D. Dandeneau, *J. Chromatogr.*, *279*:31 (1983).
365. A. Venema, L. G. J. v. d. Ven, and H. v. d. Steege, *J. High Resol. Chromatogr. Chromatogr. Commun.*, *2*:69 (1979).
366. K. Grob, *J. High Resol. Chromatogr. Chromatogr. Commun.*, *93*:221 (1978).
367. C. R. Trash, *J. Chromatog. Sci.*, *11*:196 (1973).
368. A. C. Moffat, A. H. Stead, and K. W. Smalldon, *J. Chromatogr.*, *90*:19 (1974).
369. *Supelco Reporter*, Vol. IV, No. 3, 1985.
370. J. E. Brady, D. Bjorkman, C. D. Herter, and P. W. Carr, *Anal. Chem.*, *56*:278 (1984).
371. M. L. Lee, J. C. Kuei, N. W. Adams, B. J. Tarbet, M. Nishioka, B. A. Jones, and J. S. Bradshaw, *J. Chromatogr.*, *302*:303 (1984).
372. J. M. Kong and S. J. Hawkes, *J. Chromatog. Sci.*, *14*:279 (1976).
373. M. B. Evans, *Chromatographia*, *15*:355 (1983).
374. A. E. Coleman, *J. Chromatog. Sci.*, *11*:198 (1973).
375. B. A. Colenutt and R. M. Clinch, *J. Chromatogr.*, *346*:25 (1985).
376. K. Grob and G. Grob, *J. High Resol. Chromatogr. Chromatogr. Commun.*, *4*:491 (1981); *5*:13 (1982).
377. J. Buijten, L. Blomberg, K. Markides, and T. Wännman, *J. Chromatogr.*, *237*:465 (1982).

378. E. F. Barry, P. Ferioli, and J. A. Hubball, *J. High Resol. Chromatogr. Chromatogr. Commun.*, *6*:172 (1983).
379. F. J. van Lenten, J. E. Conaway, and L. B. Rogers, *Sep. Sci.*, *12*:1 (1977); *10*:517 (1975).
380. J. Pörschmann, Th. Welsch, R. Herzschuh, W. Engewald, and K. Müller, *J. Chromatogr.*, *241*:73 (1982).
381. L. Blomberg, J. Buijten, K. Markides, and T. Wännman, *J. High Resol. Chromatogr. Chromatogr. Commun.*, *4*:578 (1981).
382. S. C. Dhanesar and C. F. Poole, *Anal. Chem.*, *55*:1462, 2148 (1983).
383. B. A. Jones, J. C. Kuei, J. S. Bradshaw, and W. L. Lee, *J. Chromatogr.*, *298*:389 (1984).
384. J. J. Myher, L. Maria, and A. Kuksis, *Anal. Biochem.*, *62*:188 (1974).
385. E. Bayer, K. P. Hope, H. Mack, in A. Zlatkis (ed.), *Proc. Int. Symp. Adv. Gas Chromatogr.*, Houston, 1963, Preston, Technical Abstr., 1963.
386. H. Jaeger, H.-U. Klör, G. Blos, and H. Ditschuneit, *Chromatographia*, *8*:507 (1975).
387. H. T. Slover and E. Lanza, *J. Am. Oil Chem. Soc.*, *56*:933 (1979).
388. W.-H. Kunau, *Angew. Chem.*, *88*:97 (1976).
389. H. Heckers, K. Dittmar, F. W. Melcher, and H. O. Kalinowski, *J. Chromatogr.*, *135*:93 (1977).
390. T. Kobayashi, *J. Chromatogr.*, *194*:404 (1980).
391. D. M. Ottenstein, D. A. Bartley, and W. R. Supina, *J. Chromatogr.*, *119*:401 (1976).
392. S. R. Lipsky and W. J. Mc Murray, *J. Chromatogr.*, *289*:129 (1984).
393. B. E. Richter, J. C. Kuei, J. I. Shelton, L. W. Castle, J. S. Bradshaw, and M. L. Lee, *J. Chromatogr.*, *279*:21 (1983).
394. K. Markides, L. Blomberg, J. Buijten, and T. Wännman, *J. Chromatogr.*, *267*:29 (1983); *254*:53 (1983).
395. K. Markides, L. Blomberg, S. Hoffmann, J. Buijten, and T. Wännmann, *J. Chromatogr.*, *302*:319 (1984).
396. J. A. Yancey and Th. R. Lynn, *Analabs Research Note*, *14*:1 (1974).
397. M. Novotny and A. Zlatkis, *J. Chromatogr.*, *56*:353 (1971).
398. M. Novotny, R. Segura, and A. Zlatkis, *Anal. Chem.*, *44*:9 (1972).
399. W. Szczepaniak, J. Nawrocki, and W. Wasiak, *Chromatographia*, *12*:484, 559 (1979).
400. J. Nawrocki and W. Szczepaniak, *J. Chromatogr.*, *330*:370 (1985).
401. G. P. Cartoni, R. S. Lowrie, C. S. G. Philips, and L. M. Venanzi, in *Gas Chromatography 1960* (R. P. W. Scott, ed.), Butterworths, London, 1960, p. 273.

402. V. G. Berezkin, V. R. Alishoyev, E. N. Victorova, V. S. Gavrichev, and V. M. Fateeva, *Chromatographia, 16*:126 (1982).
403. V. G. Berezkin, V. R. Alishoyev, E. N. Victorova, V. S. Garrichev, and V. M. Fateeva, *J. High Resol. Chromatogr. Chromatogr. Commun., 6*:42 (1983).
404. M. Riederer and W. Sawodny, *Angew. Chem. Int. Ed. Engl., 17*:610 (1978).
405. W. Wasiak and W. Szczepaniak, *J. Chromatogr., 257*:13 (1983).
406. J. N. Gillis, R. E. Sievers, and G. E. Pollock, *Anal. Chem., 57*:1572 (1985).
407. R. Liardon and R. F. Hurell, *J. Agric. Food. Chem., 31*:432 (1983).
408. N. J. Benevenga and R. D. Steele, *Ann. Rev. Nutr., 4*:157 (1984).
409. W. A. König, *J. High Resol. Chromatogr. Chromatogr. Commun., 5*:588 (1982).
410. V. Schurig, *Angew. Chem., 96*:733 (1984).
411. E. Bayer, *Z. Naturforsch., 38b*:1281 (1983).
412. C. H. Lochmüller, *Sep. Purif. Meth., 8*:21 (1979).
413. G. Palla, A. Dossena, and R. Marchelli, *J. Chromatogr., 349*:9 (1985).
414. B. Feibush, *Chem. Commun.*, 544 (1971).
415. I. Abe, S. Kuramoto, and S. Musha, *J. High Resol. Chromatogr. Chromatogr. Commun., 6*:366 (1983).
416. G. J. Nicholson, H. Frank, and E. Bayer, *J. High Resol. Chromatogr. Chromatogr. Commun., 2*:411 (1979).
417. H. Frank, G. J. Nicholson, and E. Bayer, *J. Chromatogr., 167*:187 (1978).
418. H. Frank, A. Rottenmeier, H. Weisker, G. J. Nicholson, and E. Bayer, *Clin. Chim. Acta, 105*:715 (1982).
419. G. Schomburg, I. Benecke, and G. Severin, *Proceedings of the sixth international symposium on capillary chromatography*, Riva del Garda, 1985, Hüthig, 1985, p. 104.
420. T. Saeed, P. Sandra, and M. Verzele, *J. Chromatogr., 186*:611 (1979).
421. W. A. König and I. Benecke, *J. Chromatogr., 209*:91 (1981).
422. E. Küsters, M. Allgaier, G. Jung, and E. Bayer, *Chromatographia, 18*:287 (1984).
423. C. Wang, H. Frank, E. Bayer, and P. Lu, *Chromatographia, 18*:387 (1984).
424. A.-M. Antonsson, O. Gyllenhaal, K. Kylberg-Hanssen, L. Johannsson, and J. Vessman, *J. Chromatogr., 308*:181 (1984).
425. P. J. Wedlund, B. J. Sweetman, C. B. Mc Allister, R. A. Branch, and G. R. Wilkinson, *J. Chromatogr., 307*:121 (1984).
426. R. Weber and V. Schurig, *Naturwissensch., 71*:408 (1984).
427. V. Schurig and R. Weber, *J. Chromatogr., 289*:321 (1984).

428. K. P. Naikwadi, S. Rokushika, H. Hatano, and M. Ohshima, *J. Chromatogr.*, *331*:69 (1985).
429. S. Rokushika, K. P. Naikwadi, A. L. Jadhav, and H. Hatano, *Proceedings of the sixth international symposium on capillary chromatography*, Riva del Garda, 1985, Hüthig, 1985, p. 849.
430. G. M. Janini, in *Advances in Chromatography* (J. C. Giddings, E. Grushka, J. Cazes, and P. R. Brown (eds.), Marcel Dekker, New York, 1979, Vol. 17, p. 231.
431. Z. Suprynowicz, W. M. Buda, M. Mardarowicz, and A. Patrykiejew, *J. Chromatogr.*, *333*:11 (1985).
432. Z. Witkiewicz, *J. Chromatogr.*, *251*:311 (1982).
433. G. Kraus and M. Schierhorn, *J. High Resol. Chromatogr. Chromatogr. Commun.*, *4*:123 (1981).
434. J. Rayss and A. Waksmundzki, *J. Chromatogr.*, *292*:207-216 (1984).
435. Z. Suprynowicz, W. M. Buda, M. Mardarowicz, and Z. Witkiewicz, *Chromatographia*, *19*:418 (1984).
436. G. Kraus, M. Schierhorn, T. Hong Van, and E. Tucek, *Fres. Z. anal. Chem.*, *312*:242 (1982).
437. L. Sojak, G. Kraus, P. Farkas, and I. Ostrovsky, *J. Chromatogr.*, *238*:51 (1982).
438. R. J. Laub, W. L. Roberts, and C. A. Smith, *J. High Resol. Chromatogr. Chromatogr. Commun.*, *3*:355 (1980).
439. F. Janssen and T. Kalidin, *J. Chromatogr.*, *235*:323 (1982).
440. Z. Suprynowicz and Z. Witkiewicz, Abstracts of V. Danube Symposium on Chromatography, Yalta 1985, Nauka, 1985, p. 176.
441. M. A. Apfel et al., *Anal. Chem.*, *57*:651 (1985).
442. K. E. Markides, H.-C. K. Chang, C. M. Schregenberger, M. Nishioka, B. J. Tarbet, J. S. Bradshaw, and M. L. Lee, *Proceedings of the sixth international symposium on capillary chromatography*, Riva del Garda, 1985, Hüthig, 1985, p. 137.
443. M. L. Lee and B. W. Wright, *J. Chromatogr.*, *184*:235 (1980).
444. W. G. Jennings, *Comparisons of Fused Silica and Other Glass Columns in GC*, Hüthig, 1981.
445. G. Alexander and G. A. F. M. Rutten, *J. Chromatogr.*, *99*:81 (1974).
446. M. Verzele and P. Sandra, *J. High Resol. Chromatogr. Chromatogr. Commun.*, *2*:303 (1979).
447. H. Traitler, *J. High Resol. Chromatogr. Chromatogr. Commun.*, *6*:60 (1983).
448. M. L. Lee, F. J. Yang, and K. D. Bartle, *Open Tubular Column Gas Chromatography*, Wiley, New York, 1984.
449. L. Blomberg, J. Buijten, K. Markides, and T. Wännman, *J. Chromatogr.*, *279*:9 (1983).
450. K. Grob, G. Grob, and K. Grob, Jr., *J. High Resol. Chromatogr. Chromatogr. Commun.*, *2*:31, 677 (1979).

451. L. Blomberg, *J. High Resol. Chromatogr. Chromatogr. Commun.*, *7*:232 (1984).
452. K. E. Markides, B. J. Tarbet, C. L. Wooley, C. M. Schregenberger, J. S. Bradshaw, K. D. Bartle, and K. L. Lee, *Proceedings of the sixth international symposium on capillary chromatography*, Riva del Garda, 1985, Hüthig, 1985, p. 7.
453. W. Blum, K. Grob, *J. Chromatogr.*, *346*:341 (1985).
454. S. R. Lipsky, W. J. McMurray, *J. Chromatogr.*, *279*:59 (1983).
455. J. Hrivnac, Abstracts of V. Danube Symposium on Chromatography, Yalta, Nauka, 1985, p. 175.
456. Th. Welsch, R. Müller, W. Engewald, and G. Werner, *J. Chromatog.*, *241*:41 (1982).
457. K. Grob and G. Grob, *J. High Resol. Chromatogr. Chromatogr. Commun.*, *3*:197 (1980).
458. P. Sandra, M. van Roelenbosch, I. Temmermann, and M. Verzele, *Chromatographia*, *16*:63 (1982).
459. G. Schomburg, H. Husmann, S. Ruthe, and M. Herraiz, *Chromatographia*, *15*:599 (1982).
460. T. Reiher, personal communication, 1985.
461. K. Grob, *Chromatographia*, *10*:625 (1977).
462. B. W. Wright, P. A. Peaden, and M. L. Lee, *J. High Resol. Chromatogr. Chromatogr. Commun.*, *5*:413 (1982).
463. R. Dandeneau and E. Zerenner, *J. High Resol. Chromatogr. Chromatogr. Commun.*, *2*:351 (1979).
464. J. Buijten, L. Blomberg, S. Hoffmann, K. Markides, and T. Wännman, *J. Chromatogr.*, *289*:143, 341 (1984).
465. B. W. Wright, P. A. Peaden, M. L. Lee, and G. M. Booth, *Chromatographia*, *15*:584 (1982).
466. B. W. Wright, P. A. Peaden, M. L. Lee, and Th. J. Stark, *J. Chromatogr.*, *248*:17 (1982).
467. C. L. Woley, R. C. Kong, B. E. Richter, and M. L. Lee, *J. High Resol. Chromatogr. Chromatogr. Commun.*, *7*:329 (1984).
468. M. L. Lee, R. C. Kong, C. L. Woolley, and J. S. Bradshaw, *J. Chromatog. Sci.*, *22*:136 (1984).
469. C. Madani and E. M. Chambaz, *Chromatographia*, *11*:725 (1978).
470. T. J. Stark and P. A. Larson, *J. Chromatog. Sci.*, *20*:341 (1982).
471. J. A. Hubball, P. R. Dimauro, S. R. Smith, and E. F. Barry, *J. Chromatogr.*, *302*:341 (1984).
472. O. Etler and Gy. Vigh, *J. High Resol. Chromatogr. Chromatogr. Commun.*, *7*:700 (1984).
473. K. Grob and G. Grob, *J. Chromatogr.*, *213*:211 (1981).
474. J. Buijten, L. Blomberg, K. Markides, and T. Wännman, *Chromatographia*, *16*:783 (1982).
475. B. E. Richter, J. C. Kuei, N. J. Park, S. J. Crowley, J. S. Bradshaw, and M. L. Lee, *J. High Resol. Chromatogr. Chromatogr. Commun.*, *6*:371 (1983).

476. B. E. Richter, J. C. Kuei, L. W. Castle, B. A. Jones, J. S. Bradshaw, and M. L. Lee, *Chromatographia, 17*:570 (1983).
477. A. Aerts, A. Bemgard, L. Blomberg, and J. Rijks, *Proceedings of the sixth international symposium on capillary chromatography*, Riva del Garda, 1985, Hüthig, 1985, p.
478. K. Grob and G. Grob, *J. High Resol. Chromatogr. Chromatogr. Commun., 6*:153 (1983).
479. R. C. M. de Nijs and J. de Zeeuw, *J. High Resol. Chromatogr. Chromatogr. Commun., 5*:501 (1982).
480. J. Buijten, L. Blomberg, K. Markides, and T. Wännman, *J. Chromatogr., 268*:387 (1983).
481. H. Traitler, L. Kolarovic, and A. Sorio, *J. Chromatogr., 279*:69 (1983).
482. Hewlett-Packard, Analytical Supplies, Order Guide, April 1985.
483. J. K. G. Kramer, R. C. Fouchard, and K. J. Jenkins, *J. Chromatog. Sci., 23*:54 (1985).
484. K. Grob and G. Grob, *J. Chromatogr., 347*:351 (1985).
485. M. Verzele, F. David, M. van Roelenbosch, G. Diricks, and P. Sandra, *J. Chromatogr., 279*:99 (1983).
486. J. A. Jönsson, L. Matthiasson, and Z. Suprynowicz, *J. Chromatogr., 207*:69 (1981).
487. J. F. K. Huber, E. Kenndler, and H. Markens, *J. Chromatogr., 167*:291 (1978).
488. R. J. Laub, *Trends Anal. Chem., 1*:74 (1981).
489. G. W. Pilgrim and R. A. Keller, *J. Chromatog. Sci., 11*:206 (1973).
490. J. H. Purnell, R. J. Laub, and P. S. Williams, *J. Chromatogr., 134*:249 (1977).
491. Yu. N. Bogoslovsky, V. M. Sakharov, and I. M. Shevchuk, *J. Chromatogr., 69*:17 (1972).
492. A. Gröbler and G. Balizs, *J. Chromatog. Sci., 17*:631, 671 (1979); *19*:46 (1981).
493. T. Hobo, Sh. Suzuki, K. Watabe, and E. Gil-Av, *Anal. Chem., 57*:362 (1985).
494. R. E. Kaiser, R. I. Rieder, L. Leming, L. Blomberg, and P. Kusz, *Proceedings of the sixth international symposium on capillary chromatography*, Riva del Garda, 1985, Hüthig, 1985, p. 638.
495. R. E. Kaiser and R. I. Rieder, *Labor Praxis, 9*:1130, 1312, 1465 (1985).
496. J. J. Leary, J. B. Justice, S. Tsuge, S. R. Lowry, and T. L. Isenhour, *J. Chromatog. Sci., 11*:201 (1973).
497. S. Hawkes (Chairman) et al., *J. Chromatog. Sci., 13*:115 (1975).
498. R. H. Bishara and R. W. Souter, *J. Chromatog. Sci., 13*:593 (1975).
499. R. S. Wayne (Chairman) et al., *J. AOAC, 59*:420 (1976).

500. L. V. Semenchenko and M. S. Vigdergauz, *J. Chromatogr.*, *245*:177 (1982).
501. R. W. Souter, *Chromatographic Separations of Stereoisomers*, CRC Press, Boca Raton, Florida, 1985.
502. D. Demus, *Z. Chem. (GDR)*, *26*:6 (1986).
503. A. Bacic, P. J. Harris, E. W. Hak, and A. E. Clarke, *J. Chromatogr.*, *315*:373 (1984).
504. H. Kelker and R. Hatz, *Handbook of Liquid Crystals*, Verlag Chemie, Weinheim, 1980, Chap. 9.
505. V. Martinez de la Gandara, J. Sanz, and I. Martinez-Castro, *J. High Resol. Chromatogr. Chromatogr. Commun.*, *7*:44 (1984).
506. J. C. MacDonald (ed.), *Inorganic Chromatographic Analysis*, John Wiley and Sons, New York, 1985, p. 229.
507. C. F. Poole, H. T. Butler, M. E. Coddens, S. C. Dhanesar, and F. Pacholec, *J. Chromatogr.*, *289*:299 (1984).

# 4

# Supports and Stationary Phases in Liquid-Liquid Partition Chromatography

Klaus K. Unger / *Institute of Inorganic Chemistry and Analytical Chemistry, Johannes Gutenberg University, Mainz, Federal Republic of Germany*

## BRIEF HISTORICAL REVIEW

Liquid-liquid partition chromatography (LLPC) was pioneered by Martin and Synge (1941) employing an organic-aqueous two-phase system and a silica support to separate amino acids [1]. Later on, organic-organic two-phase systems were applied to the resolution of water-insoluble sample mixtures and the range of supports was extended to cellulose and dextran-based supports. In the early days of high-performance liquid chromatography (HPLC), before microparticulate packings were developed, liquid-liquid partition chromatography was performed on pellicular-type packings with polar stationary and unpolar to moderately polar mobile phases [2]. Huber and coworkers [3] showed that remarkable selectivity differences were achieved by liquid-liquid phases generated from ternary solvent systems, e.g., 2,2,4-trimethylpentane, ethanol, and water. The distribution coefficients of solutes derived from chromatographic experiments coincided well with those obtained from static measurements that allowed the design of suitable liquid-liquid phase systems for a given sample mixture. The rapid development of bonded phases which avoided the constraints of liquid-liquid partitioning columns, e.g., bleeding of the stationary phase and maintenance of isothermal conditions, pushed this powerful technique somewhat into the background.

Recently, Huber et al. [4,5] demonstrated again the enormous capability of LLPC, overcoming the practical limitations by solvent-generated liquid-liquid phase systems on native and bonded silicas.

While these efforts were mainly centered on the separation of low molecular weight compounds, the adaptation of Albertsson's classical aqueous-aqueous two-phase system to LLPC with organic polymers or silica as supports had a substantial impact on the resolution and fractionation of nucleic acids and proteins [6,7].

The current LLPC systems share a number of common favorable features which makes them attractive for widespread application in HPLC:

- Simple preparation of liquid bulk phases
- High selectivity by employing ternary solvent systems and modifiers
- Less contamination and longer lifetime compared to adsorbent columns
- Higher reproducibility of retention and selectivity under isothermal conditions

## RETENTION MECHANISMS

In liquid-liquid partition systems the solute is distributed between the mobile phase m and the liquid stationary phase s, held by an inert porous support (Fig. 1). Three distribution processes are distinguished, contributing to the total retention of the solute:

- Partitioning between the two liquid bulk phases
- Adsorption of the solute at the surface of the support
- Adsorption of the solute at the liquid-liquid interface

The total capacity factor $k_i^{'(s/m)}$ is then written as

$$k_i^{'(s/m)} = k_i^{'(l/m)} + k_i^{'(\sigma/m)} + k_i^{'(\lambda/m)} \tag{1}$$

where

$k_i^{'(s/m)}$ = overall capacity factor for the distribution of the component i between the system of stationary phases s and the mobile phase m

$k_i^{'(l/m)}$ = capacity factor for the partition between the stationary liquid phase l and the mobile liquid phase m

$k_i^{'(\sigma/m)}$ = capacity factor for the adsorption on the solid surface $\sigma$ with respect to the mobile phase m

$k_i^{'(\lambda/m)}$ = capacity factor for the adsorption on the liquid-liquid interface $\lambda$ with respect to the mobile phase m

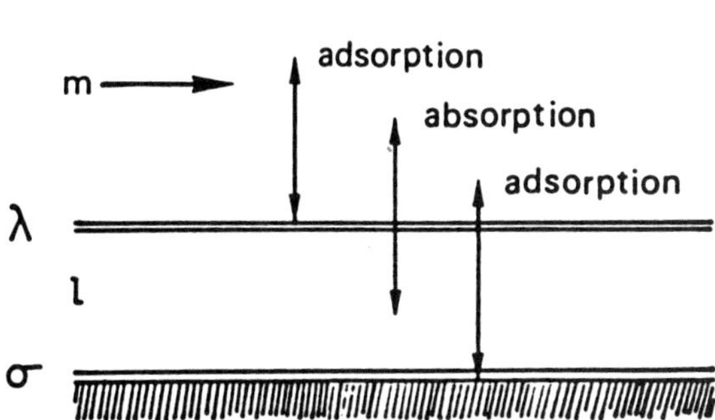


Fig. 1 Schematic representation of the phase system and distribution processes in liquid-liquid chromatography. (From Ref. 5, reprinted by permission of the publisher.)

Assuming an inert porous support, the term $k_i^{'(\sigma/m)}$ is equal to zero. Furthermore, when the adsorption of the solute at the liquid-liquid interface remains small, the term $k_i^{'(\lambda/m)}$ can be neglected. Under the conditions of a pure liquid-liquid partition mechanism, one obtains

$$k_i^{'(s/m)} = k_i^{'(l/m)} \tag{2}$$

Below follows a brief discussion that shows in which way adsorption interactions can be monitored in LLPC.

The retention volume of a solute in interaction chromatography is expressed by

$$V_R = V_m (1 + k_i') \tag{3}$$

where $V_m$ is the volume of the mobile phase in the column and $k_i'$ the solute capacity factor, the latter being defined by

$$k_i' = \frac{V_s}{V_m} \cdot {}^cK_{i0} \tag{4}$$

where $V_s$ and $V_m$ are the volume of stationary and mobile phase and ${}^cK_{i0}$ the partition coefficient of the solute at infinite dilution, being equal to the ratio of $C_i^{(s)}/C_i^{(m)}$. $C_i^{(s)}$ and $C_i^{(m)}$ are the concentration of solute in the stationary and mobile phase. At a pure liquid-liquid partitioning ${}^cK_{i0}$ should be equal to $K_{i0}^{(L/L)}$, this being the distribution coefficient of the compound derived from static batch experiments. $K_i^{(L/L)}$ is defined as

$$K_{i0}^{(L/L)} = \frac{C_i^{(\beta)}}{C_i^{(\alpha)}} = K_{i0}^{(\beta/\alpha)} \tag{5}$$

where $C_i^{(\beta)}$ and $C_i^{(\alpha)}$ are the concentration of the compound in the more polar ($\beta$) and less polar ($\alpha$) phase at infinite dilution. The plot of $V_R$ against $K_i^{(L/L)}$ then gives a straight line with $V_m$ as the intercept and $V_s$ as the slope at a given solid support and liquid-liquid phase system. Deviations from linearity indicate that additional retention mechanisms, e.g., adsorption, are involved. Huber et al. [5] compared the intercepts of the linear regression lines, $V_m^{calc}$, to the experimental value, $V_m^{exp}$, obtained with unretained marker compounds. They found that on large-pore-size native and *n*-alkyl silicas of a pore diameter of pd >> 10 nm, both values coincided well, while for smaller pore size silicas $V_m^{calc}$ was larger than $V_m^{exp}$ reflecting adsorption interactions of the marker. Allowing for adsorption and absorption of the solute, Eq. (4) extends to

$$k_i' = \left(K_{ad,i}\frac{a_s}{V_m}\right) + \left({}^cK_i\frac{V_s}{V_m}\right) \tag{6}$$

where $a_s$ is the specific surface area of the porous support per unit column volume and $K_{ad,i}$ the adsorption coefficient of solute i, in ml $m^{-2}$.

In order to minimize adsorption effects the first choice is to employ a low-surface-area and a large-pore-size support. With increasing load of the porous support with the liquid stationary phase, $V_m$ decreases concurrently up to about $0.5V_m$ at maximum load. Under these conditions the first term on the right-hand side of Eq. (6) doubles, provided $K_{ad,i}$ and $a_s$ remain unchanged at increasing load [8].

A further fundamentally important aspect in LLPC relates to the fact that straight phase and reversed phase liquid-liquid systems are easily accomplished. Provided adsorption effects are absent, the mutual dependence of the distribution coefficient in polar and unpolar LLPC systems is simply related as

$${}^cK_{i0}(I) = \frac{1}{{}^cK_{i0}(II)} \tag{7}$$

where I and II refer to the polar (I) and unpolar (II) stationary phase.

Conversion from a phase system I to II at a given liquid-liquid system composed of a less polar phase $\alpha$ and more polar phase $\beta$ is achieved as follows. The two-phase system is pumped through a column containing a hydrophilic support until the surface is loaded with β. Choosing a hydrophobic support α is preferentially adsorbed and phase system II is obtained.

It should be emphasized that the miscibility of a multisolvent system, e.g., binary, ternary, or quaternary, is strongly dependent on the temperature. Thus, an isothermal conditioning at the column is essentially required.

## SUPPORTS

In LLPC, the porous packing functions as a support for the liquid stationary phase and should satisfy the following two requirements:

1. The surface of the support should be wetted by the liquid, i.e., the contact angle $\theta$ should amount to $\theta < 90°$. This means that for a polar liquid stationary phase hydrophilic supports must be employed and vice versa.
2. The support surface should be homogeneously coated with the liquid stationary phase and must be sufficiently large to generate a maximum liquid-liquid interface per column volume for the distribution and retention of the solutes.

Given these considerations, macroporous supports of an average pore diameter of pd >50 nm are most suitable as packings. This corresponds to a specific surface area of <25 $m^2\ ml^{-1}$ of column volume at a packing density of 0.5 g $ml^{-1}$. Supports with higher specific surface areas and smaller pores of pd <50 nm might contribute to retention by adsorption of solutes as discussed in the previous section. Adsorption can be monitored by correlating the solute capacity factor (a) with the phase ratio of the liquid-liquid phase system [8-10] and (b) with the liquid-liquid partition coefficient from batch experiments [11-13]. The magnitude of adsorption effects is not only influenced by the specific surface area of the support but also depends on the load with the liquid stationary phase. Kraak and Crombeen [14] showed that the phase ratio $V_s/V_m$ was proportional to the specific surface area of native silica supports, expressed in $m^2\ ml^{-1}$ of column volume. However, no strong linearity between $V_s/V_m$ against $a_s$ resulted.

The load, expressed in g of liquid stationary phase per g of loaded support, can be varied between two limits. The lower limit corresponds to the load where the surface bears an adsorbed layer. The upper limit is obtained by filling the total volume of the porous support with stationary liquid. With highly porous silicas of specific pore volumes of up to ⩾ 2 ml/$g^{-1}$, maximum loads of 78% were achieved with di-2-cyanoethylether (ODPN) [15].

In principle both types of packings, i.e., organic and inorganic based, can be employed in LLPC provided they meet the necessary requirements. Soft packings such as cellulose, dextrans, and diatomaceous earth do not withstand high pressures and are preferably used as larger particles at low pressures. For the high-performance mode, microparticulate and bonded silicas are the supports of choice.

For the separation of large biopolymers, e.g., proteins and nucleic acids (DNA, RNA), specific supports have been developed, characterized by large pores to allow access of the solutes to the stationary phase and by a composite structure. Composite structure means that a hydrophilic polymer, e.g., Fractogel, or a bonded silica, e.g., LiChrospher Diol (both from E. Merck), were reacted with acrylamide to form polyacrylamide chains of defined lengths that facilitate the distribution of the dextran-rich phase (Fig. 2).

## LIQUID-LIQUID SYSTEMS

The decisive property of a liquid-liquid system employed in LLPC is the polarity difference between the two phases. The polarity of a liquid is expressed by the solubility parameter, characterizing its ability to participate in interactions of various types [16] or by the Snyder polarity index P' based on actual solubility data [17,18]. Both parameters are interrelated.

Applying the solubility parameter concept, retention is controlled by the polarity difference between the solute i and the mobile phase m and the solute i and the stationary phase s. The distribution coefficient is expressed by

$$\ln K_i = \frac{\bar{v}_i [(\delta_i - \delta_m)^2 - (\delta_i - \delta_s)^2]}{RT} \qquad (8)$$

where $\delta_i$, $\delta_m$, and $\delta_s$ are the solubility parameters of i, m, and s, $\bar{v}_i$ the molar volume of i, R the gas constant, and T the temperature. In practice most of the liquid-liquid systems show a polarity difference in the range of $4 \leqslant (\delta_s - \delta_m) \leqslant 17$ (16). Typical binary liquid-liquid systems together with their applications are listed in Table 1. Ternary liquid-liquid systems are composed of a polar solvent, e.g., water, an unpolar solvent, e.g., 2,2,4-trimethylpentane (2,2,4-TMP), and an intermediate polar solvent, e.g., acetonitrile (ACN). The phase diagram of ACN, EtOH, and 2,2,4-TMP is given in Fig. 3. While at the upper part of the triangle a complete miscibility, i.e., a homogeneous phase, exists, a solvent composition chosen within the lower domain will split into a two-phase system. One phase is less, the other more polar. The composition of the two phases follows the tie lines as indicated by the Roman number in Fig. 3. The corresponding equilibrium composition of the five-phase systems is listed in Table 2.

**1.) Chemical modification: grafting polymerisation of acrylamide[1] on premanufactured hydrophilized beads' (e. g. Fractogel, LiChrospher® Diol,)**

$CH_2OH$, $\overset{R}{|}CHOH$ $\xrightarrow[CH_2=CH-C(=O)NH_2]{Ce^{IV}}$ $(CH_2-CH(CONH_2))_n-CH_2-CH_2-C(=O)NH_2$ on $CHOH$; $R-CH$ with $(CH_2-CH(CONH_2))_n-CH_2CH_2-C(=O)NH_2$ (n≈15-25)

(1) Mino & Kaizerman.(1958)

**2. ) "Installation" of the aqueous PEG-Dx two-phase system**
**2.1) adsorption of the Dx-rich bottom phase**

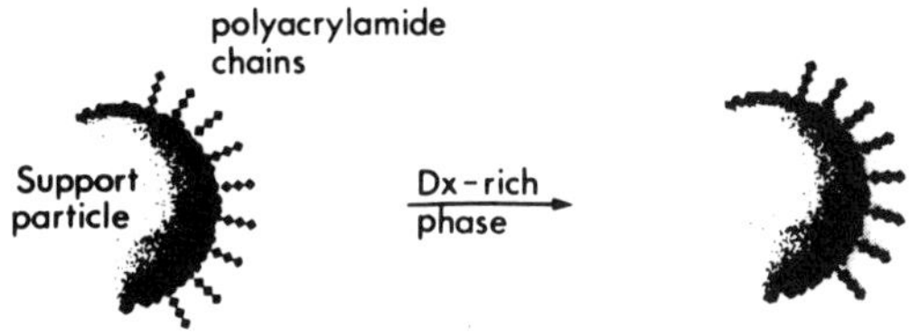


**2.2) removal of the excess of the Dx-rich phase by the PEG-rich top phase**

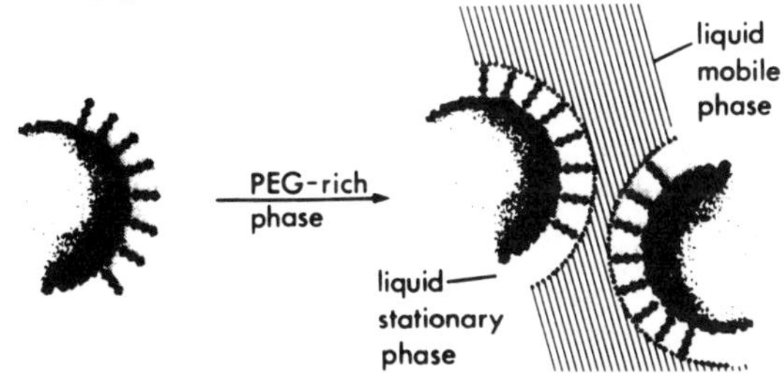


**packing of coated support particles (suspended in the mobile top phase) into thermostated glass colums**

Fig. 2 Preparation of supports for partition chromatography in the aqueous polyethylene glycol-dextran two-phase system. (From W. Müller, by courtesy of E. Merck, Darmstadt, FRG.)

Table 1 Liquid-Liquid Systems and Applied Separations

| Stationary phase | Mobile phase |
|---|---|
| **Binary systems** | |
| **Normal phase** | |
| Di-2-cyanoethylether (ODPN) | 2,2,4-Trimethylpentane, di-$n$-butylether, $n$-heptane, or $n$-hexane |
| Hydroxylated aromatics [24], polychlorinated dimethoxybenzenes and chlorinated aromatic amines [25], alkylated phenol derivatives [26], chlorinated pesticides [10,27,28], urea herbicides [29], 2,4-dinitrophenylhydrazone derivatives of 17-ketosteroids [30], phthalate esters [31], and polyaromatic hydrocarbons [28] | |
| 1,2,3-tris(2)-(cyanoethoxy)propane (Fraktonitril III) | $n$-Hexane |
| Polyaromatic hydrocarbons and $m$-oligophenylenes [32] and 2,4-dinitrophenylhydrazone derivatives of carbonyl compounds [33] | |
| Formamide, water (+buffer) | $n$-Hexane, dichloromethane, or ethyl acetate |
| Free steroids [34,9,35], methylated phenols [34,35], and urea herbicides [36,37] | |
| Dimethylsulfoxide acetonitrile or $\gamma$-butyrolactone | Cyclohexane |
| Polyaromatic hydrocarbons [38,12,39], benzoate esters [12], and mono- and polychlorinated nitrobenzenes [72] | |
| **Reversed phase** | |
| Tri-$n$-butylphosphate, tri-$n$-octylphosphine, or tri-$n$-octylamine | Water + mineral acids or buffer |
| Derivatives of phenyl carboxylic acids [40-42], phenol, benzene- and naphthalenesulfonic acids [41] | |
| Di-(2-ethylhexyl)phosphoric acid | Water + mineral acids |
| Radionuclides [43] | |
| $n$-Octanol or oleyl alcohol | Water + buffers |
| 1,4-Benzodiazepines and other N-containing compounds (44,45) | |

Table 1 (continued)

| Stationary phase | Mobile phase |
|---|---|
| Multicomponent systems | |
| Normal phase | |
| Dimethylsulfoxide of *N,N*-dimethylformamide + Carbowax or Carbowax 200-400 | *i*-Octane or tetra (or dichloromethane) + *i*-octane |
| Aromatic alcohols, free steroids and benzodiazepines [10], polycyclic aromatic hydrocarbons [46,47], polymers of alkylphenylglycol [48] | |
| Polar phase of | Apolar phase of |
| water (plus/or acetonitrile)/ethanol or *n*-pentanol/*i*-octane | |
| Metal chelates [3], free steroids [9,11], glycosides [50], polycyclic aromatic hydrocarbons [51], thioridazines and metabolites [49] | |
| Polar phase of | Apolar phase of |
| water/acetic acid (+ aceton)/di-*i*-propylether or *n*-heptane | |
| Dansyl amino acids [34] and urinary free porfyrines [52] | |
| Polar phase of | Apolar phase of |
| water (+ buffer)/methanol or 2-chloroethanol or propanol-2/dichloromethane or chloroform | |
| Dansyl amino acids [34], free steroids [53], urea herbicides [36], carboxylic and sulfonic acids, nucleobases and nucleosides [9], barbiturates and benzodiazepines [37,54] | |
| Polar phase of formamide/chloroform/*n*-hexane | Apolar phase of formamide/chloroform/*n*-hexane |
| Barbiturates [9,39] and chlorinated phenols [39] | |
| 0.1 or 0.2 *N* Sulfuric acid | *n*-pentanol or *t*-pentanol/chloroform/cyclohexane |
| (Hydroxy)-phenyl carboxylic acids, alifatic mono- and dicarboxylic acids [55,56] | |
| Reversed phase | |
| Apolar phase of ethanol/acetonitrile/*i*-octane | Polar phase of ethanol/acetonitrile/*i*-octane |
| Alkylbenzenes and polycyclic aromatic hydrocarbons [51] | |

*Source*: From [19], reprinted by permission of the publisher.

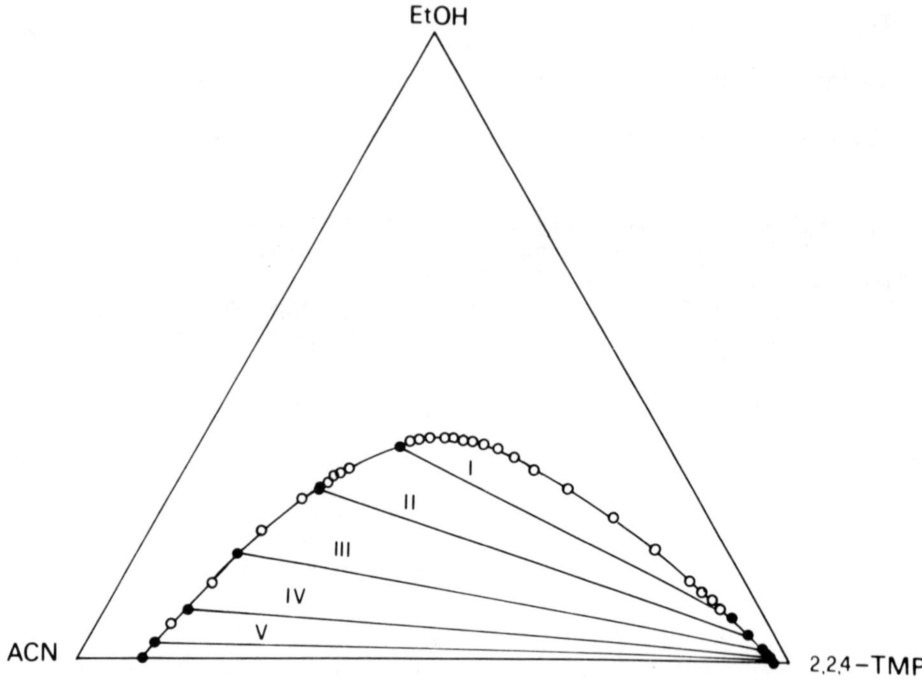


Fig. 3 Triangle phase diagram of the ternary system consisting of acetonitrile (ACN)-ethanol (EtOH)-2,2,4-trimethylpentane (2,2,4-TMP) at 25.0 ± 0.1°C. Open circles, data obtained by two-phase titration; full circles, data obtained by gas chromatography. (From Ref. 4, reprinted by permission of the publisher.)

With ternary solvent systems such as these a wide range of selectivity is matched in LLPC.

## COLUMN PREPARATION [20,21]

Several techniques are available to prepare an LLPC column. Their application depends on the selected phase system, the particle size of the support, and the load of the stationary phase.

### Solvent Evaporation Technique

This technique is common for preparing columns in gas-liquid chromatography and preferred in LLPC when columns with large particles of dp >20 μm are packed by the dry-packing method. A known amount

Table 2 Equilibrium Compositions of the Investigated Ternary and Binary Liquid-Liquid Systems at 25.0 ± 0.1°C (See Fig. 3).

| Liquid-liquid system | | Mass fraction (%) | | | | | |
|---|---|---|---|---|---|---|---|
| | | More polar phase (β) | | | Less polar phase (α) | | |
| | | ACN | EtOH | 2,2,4-TMP | ACN | EtOH | 2,2,4-TMP |
| Ternary | I | 37.52 | 33.66 | 28.82 | 4.43 | 6.85 | 88.72 |
| | II | 52.51 | 27.24 | 20.25 | 3.80 | 3.89 | 92.31 |
| | III | 71.63 | 14.18 | 14.19 | 3.11 | 1.33 | 95.56 |
| | IV | 80.76 | 6.90 | 12.34 | 3.03 | 0.63 | 96.34 |
| | V | 87.90 | 2.35 | 9.75 | 2.50 | 0.25 | 97.25 |
| Binary | | 90.63 | 0 | 9.37 | 2.17 | 0 | 97.83 |

*Source*: From [4], reprinted by permission of the publisher.

of a nonvolatile liquid to be chosen as stationary phase is dissolved in a volatile solvent. The solution is added to the dry support and mixed gently. Then the solvent is slowly removed by rotating the suspension in a rotary evaporator until the coated support remains as a dry and free-flowing powder. LLPC columns with microparticulate supports are coated in situ. The slurry-packing technique applied to pack such columns does not allow a preloading with a liquid stationary phase.

### Direct Coating

The pure liquid stationary phase is pumped through the column and then the excess is displaced by elution with the mobile phase saturated with the stationary phase. As a result a maximum load is attained. This technique is suited for stationary phases of low viscosity of <2 cP and does not permit a variation of the load.

### Precipitation Technique

The procedure consists of a precipitation of the stationary liquid within the pores of the support. The column is preequilibrated with a solvent (S1) and then flushed with the solution of the liquid stationary phase in S1. A second solvent (S2), which is completely miscible with S1 but not with the stationary liquid, is then pumped through the col umn. During this procedure the stationary liquid precipitates in the pores. Graduated loadings are achieved by varying the concentration of stationary liquid in S1. Finally the column is conditioned with the liquid mobile phase.

### Dynamic Coating Technique

The liquid mobile phase saturated with the liquid stationary phase is pumped through the prepacked column. The liquid stationary phase is adsorbed at the surface of the support, builds up a multilayer, and finally fills the whole specific pore volume. The formation of the liquid stationary phase is monitored via the change in the retention of test solutes.

With this technique the phase ratio cannot be varied as the maximum load is attained. As shown by Crombeen et al. [22], a small temperature difference applied between the stock reservoir and the column allows adjustment of $V_s/V_m$ within certain limits. A notable advantage of dynamically generated liquid-liquid systems is their excellent stability. The retention volumes of a test compound varied less than ±0.5 during 5 days of column operation [4]. Column-to-column reproducibility was found to vary between ±2 and ±5%, applying different batches of the support at a given liquid-liquid system.

## APPLICATIONS

In the beginning of LLPC, stationary phases with a viscosity of about 10 cP were preferably employed, e.g., di-2-cyanoethylether (ODPN), 1,2,3-tris(2'-cyanoethoxy)propane (Fraktonitril III), and poly(ethyleneglycol) (Carbowax 200-450) in combination with unpolar or moderately polar mobile phases. Today, solvent-generated liquid-liquid systems with a low viscosity of <2 cP are most common. The efficiency of such columns in terms of plate number is comparable to those observed in liquid adsorption chromatography.

LLPC is a preferred technique to resolve sample mixtures of a wide polarity, ranging from polycyclic aromatic hydrocarbons to nucleobases and nucleosides (see Table 1). The majority of applications was performed on straight phase LLPC rather than on reversed phase systems. Aqueous two-phase systems have been shown to be excellently suited for the fractionation of nucleic acids and proteins [6,7].

LLPC also serves as a means to assess liquid-liquid partition coefficients from retention data, provided the phase ratio is known and adsorption effects are negligible. The accuracy of coefficients determined by LLPC compared to those measured from static experiments was found to be within a few percentage points [23].

## CONCLUSION

Liquid-liquid partition chromatography with solvent-generated phase systems has developed into a powerful tool to resolve sample mixtures of almost every polarity. The systems are easy to handle, provide a high stability, a long lifetime, and a high column-to-column reproducibility.

Aqueous polyethylene glycol-dextran systems were developed that allow the fractionation of nucleic acids and proteins.

## SELECTED READINGS

L. R. Snyder and J. J. Kirkland, *Introduction to Modern Liquid Chromatography*, John Wiley and Sons, New York, 1979, pp. 323-348.

K. K. Unger, *Porous Silica*, J. Chromatogr. Libr. Vol. 16, Elsevier, Amsterdam, 1979, pp. 237-248.

J. J. Kraak and J. P. Crombeen, in *Practice of High Performance Liquid Chromatography* (H. Engelhardt, ed.), Springer-Verlag, Berlin, 1986, pp. 180-198.

## REFERENCES

1. A. I. P. Martin and R. L. M. Synge, *Biochem. J.*, *35*:1358-1368 (1941).
2. L. R. Snyder and J. J. Kirkland, *Introduction to Modern Liquid Chromatography*, John Wiley and Sons, New York, 1979, pp. 330-331.
3. J. F. K. Huber, J. C. Kraak, and H. Veening, *Analyt. Chem.*, *44*:1554-1559 (1972).
4. J. F. K. Huber, M. Pawlwoska, and P. Markl, *Chromatographia*, *17*:653-663 (1983).
5. J. F. K. Huber, M. Pawlwoska, and P. Markl, *Chromatographia*, *19*:19-28 (1984).
6. W. Müller, *Kontakte* (Merck) *3*:3-11 (1986).
7. W. Müller, *Eur. J. Biochem.*, *155*:213-222 (1986).
8. K. K. Unger, in *Porous Silica* (K. K. Unger, ed.), J. Chromatogr. Libr. Vol. 16, Elsevier, Amsterdam, 1979, pp. 244-245.
9. J. P. Crombeen, S. Heemstra, and J. C. Kraak, *J. Chromatogr.*, *282*:95-106 (1983).
10. M. Viricel and M. Lemar, *J. Chromatogr.*, *116*:343-352 (1976).
11. J. F. K. Huber, R. van der Linden, E. Ecker, and M. Oreans, *J. Chromatogr.*, *83*:267-277 (1973).
12. J. P. Crombeen, S. Heemstra, and J. C. Kraak, *Chromatographia*, *19*:219-224 (1984).

13. J. F. K. Huber, C. A. M. Meijers, and J. A. R. J. Hulsman, *Anal. Chem.*, *44*:111-116 (1972).
14. J. C. Kraak and J. P. Crombeen, in *Practice of High Performance Liquid Chromatography* (H. Engelhardt, ed.), Springer-Verlag, Berlin, 1986, p. 186.
15. K. K. Unger, in *Porous Silica* (K. K. Unger, ed.), J. Chromatogr. Libr. Vol. 16, Elsevier, Amsterdam, 1979, pp. 242-247.
16. C. Horvath, L. R. Snyder, and B. L. Karger (eds.), *Introduction to Separation Science*, Wiley-Interscience, New York, 1974, pp. 49-55, 268-276.
17. L. R. Snyder, *J. Chromatogr.*, *92*:223-230 (1974).
18. L. R. Snyder, *J. Chromatogr. Sci.*, *16*:223-234 (1978).
19. J. C. Kraak and J. P. Crombeen, in *Practice of High Performance Liquid Chromatography* (H. Engelhardt, ed.), Springer-Verlag, Berlin, 1986, pp. 189-190.
20. K. K. Unger, in *Porous Silica* (K. K. Unger, ed.), J. Chromatogr. Libr. Vol. 16, Elsevier, Amsterdam, 1979, pp. 241-243.
21. J. C. Kraak and J. P. Crombeen, in *Practice of High Performance Liquid Chromatography* (H. Engelhardt, ed.), Springer-Verlag, Berlin, 1986, pp. 182-184.
22. J. P. Crombeen, S. Heemstra, and J. C. Kraak, *J. Chromatogr.*, *282*:95-106 (1983).
23. J. C. Kraak and J. P. Crombeen, in *Practice of High Performance Liquid Chromatography* (H. Engelhardt, ed.), Springer-Verlag, Berlin, 1986, pp. 192-194.
24. J. J. Kirkland, *J. Chromatogr. Sci.*, *10*:593-599 (1972).
25. J. J. Kirkland and C. H. Jr. Dilks, *Anal. Chem.*, *45*:1778-1781 (1973).
26. B. L. Karger, K. Conroe, and H. Engelhardt, *J. Chromatogr. Sci.*, 242-250 (1970).
27. J. L. Waters, J. N. Little, and D. F. Horgan, *J. Chromatogr. Sci.*, *7*:292-296 (1969).
28. J. Vermont, M. Deleuil, A. J. De Vries, and C. L. Guillemin, *Anal. Chem.*, *46*:1329-1337 (1975).
29. J. J. Kirkland, *J. Chromatogr. Sci.*, *7*:7-12 (1969).
30. F. A. Fitzpatrick, S. Siggia, and J. Dingman, *Anal. Chem.*, *44*:2211-2216 (1972).
31. L. Fishbein and P. W. Albro, *J. Chromatogr.*, *70*:365-412 (1972).
32. D. Randau and W. Schnell, *J. Chromatogr.*, *57*:373-381 (1971).
33. L. J. Papa and L. P. Turner, *J. Chromatogr. Sci.*, *10*:747-750 (1972).
34. H. Engelhardt, J. Asshauer, U. Neue, and N. Weigand, *Anal. Chem.*, *46*:336-340 (1974).
35. H. Engelhardt and N. Weigand, *Anal. Chem.*, *45*:1149-1154 (1973).
36. C. Gonnet and J. L. Rocca, *J. Chromatogr.*, *109*:297-303 (1975).
37. Z. El Rassi, C. Gonnet, and J. L. Rocca, *J. Chromatogr.*, *125*:179-201 (1976).

38. J. P. Durand and N. Petroff, *J. Chromatogr.*, *190*: 85-95 (1980).
39. J. P. Crombeen, S. Heemstra, and J. C. Kraak, *J. Chromatogr.*, *286*: 119-129 (1984).
40. K.-G. Wahlund and B. Edlen, *J. Liq. Chromatogr.*, *4*: 309-323 (1981).
41. J. C. Kraak, Ph.D. thesis, Amsterdam.
42. H. W. Stuurman and K.-G. Wahlund, *J. Chromatogr.*, *218*: 455-463 (1981).
43. E. P. Horwitz, C. A. A. Bloomquist, and W. H. Delphin, *J. Chromatogr. Sci.*, *15*: 41-46 (1977).
44. M. S. Mirrlees, S. J. Moulton, C. T. Murphey, and P. J. Taylor, *J. Med. Chem.*, *19*: 615-619 (1976).
45. A. Hulshoff and J. H. Perrin, *J. Chromatogr.*, *129*: 263-276 (1976).
46. R. E. Jentoft and T. H. Gouw, *Anal. Chem.*, *40*: 923-927 (1968).
47. R. E. Jentoft and T. H. Gouw, *Anal. Chem.*, *40*: 1787-1790 (1968).
48. J. F. K. Huber, F. F. M. Kolder, and J. M. Miller, *Anal. Chem.*, *44*: 105-110 (1972).
49. R. G. Muusze and J. F. K. Huber, *J. Chromatogr.*, *83*: 405-420 (1973).
50. W. Lindner and R. W. Frei, *J. Chromatogr.*, *117*: 81-86 (1976).
51. J. F. K. Huber, M. Pawlovska, and P. Markl, *Chromatographia*, *17*: 653-663 (1983).
52. H. Norlov, P. M. Jordan, G. Burton, and A. J. Scott, *J. Chromatogr.*, *190*: 221-225 (1980).
53. N. A. Parris, *J. Chromatogr. Sci.*, *12*: 753-757 (1974).
54. C. Gonnet and J. L. Rocca, *J. Chromatogr.*, *120*: 419-433 (1976).
55. J. F. Morot-Gaudry, M. Z. Nicol, and E. Jolivet, *J. Chromatogr.*, *100*: 206-210 (1974).
56. J. F. Morot-Gaudry, V. Fiala, J. C. Huet, and E. Jolivet, *J. Chromatogr.*, *117*: 279-284 (1974).
57. J. F. K. Huber, *J. Chromatogr. Sci.*, *9*: 72-76 (1971).

# 5

# Sorbent Materials and Precoated Layers in Thin-Layer Chromatography

H. E. Hauck and Willi Jost / *E. Merck, Darmstadt, Federal Republic of Germany*

## INTRODUCTION

Thin-layer chromatography was one of the first chromatographic techniques to be applied in routine analysis and was primarily used in preparative chemical laboratories. Ismailov and Shraiber pioneered the TLC procedure in 1938-1939 for determining alkaloids [1,2]. Thin-layer chromatography only became a generally applicable method, however, through the work of Stahl [3,4] in cooperation with various equipment and sorbent manufacturers.

Important conditions prerequisite to the continued evolution of thin-layer chromatography were the ability to standardize the physical and chemical properties of sorbents used and the development of means to manufacture them in a consistent quality. From this time the way was set for reproducible results to be attained in thin-layer chromatography.

Hitherto users had prepared plates by hand and the coatings inevitably had many flaws until the introduction of factory-made plates in 1966 permitted the use of extremely high-quality coatings. A further qualitative improvement was made in 1975 in the form of HPTLC plates (HPTLC = high-performance thin-layer chromatography). Their uniform surface and consistent layer homogeneity make them particularly suitable for direct, in situ evaluation, quantitative as well as qualitative. Developments discussed so far had been centered around silica as sorbent and had been directed toward improving the separation efficiency of TLC plates.

The use of materials other than silica considerably broadened the range of selectivities attainable. For example, alumina, cellulose, polyamide, and kieselguhr were used first as additional sorbents. Also, various layers prepared from these sorbents can be impregnated with suitable liquid stationary phases for special types of separation. Such stationary phases are only physically attached to the support, however, and can be eluted to some extent by the mobile phase used. Reproducibility and substance detection suffer in consequence. Therefore, a logical step was the development of sorbent materials consisting of a support with a chemically bonded stationary phase of similar properties.

Separations performed using the described modified, nonmodified, and impregnated layers are due to a difference in chemical properties between the phases used and the sample substances to be separated, the actual retention mechanisms differing accordingly. We distinguish between the following types of chromatographic processes based on the type of determining interactions involved:

Straight phase chromatography
Partition chromatography
Reversed phase chromatography
Ion exchange chromatography

The sorbent materials and precoated layers for thin-layer chromatography discussed in this chapter are also classified under the same broad headings.

## SORBENT MATERIALS FOR STRAIGHT PHASE CHROMATOGRAPHY

In straight phase chromatography sample retention is determined on the one hand by its interactions with polar surface centers on the solid stationary phase and on the other by sample solubility in the nonpolar, liquid mobile phase.

The interactions affecting retention in straight phase chromatography are chiefly hydrogen-bonding and induced dipole-dipole interactions.

The retention of samply molecules can be deliberately influenced, e.g., by the selection of the appropriate solvent systems. Changes in the polarity and solvent properties of the mobile phase have a particularly marked influence on retention. The most frequently used organic solvents have been compiled in a special list known as the eluotropic series [5], according to their elution power based on silica as stationary phase (see Chap. 6).

To illustrate the situation, Fig. 1 presents a series of chromatographic separations of various polycyclic aromatic hydrocarbons using

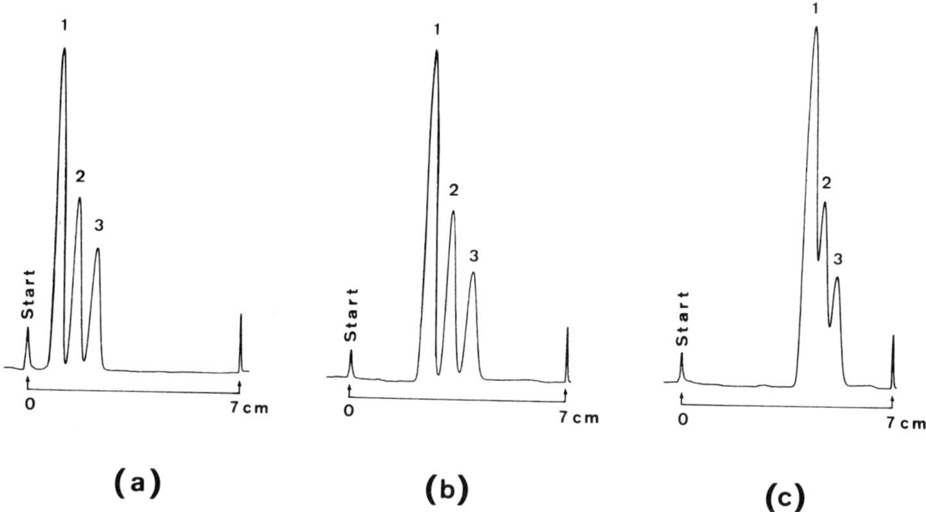


Fig. 1 Influence of the polarity of the mobile phase on the retention and resolution of some polycyclic aromatic hydrocarbons. Plate: HPTLC precoated plate silica gel 60 F 254 (E. Merck). Compounds: 1, benzanthracene; 2, pyrene; 3, acenaphthene. Eluent (a) cyclohexane; (b) cyclohexane/toluene 95:5 (v/v); (c) cyclohexane/toluene 80:20 (v/v). Detection: Camag TLC/HPTLC scanner, UV 265 nm.

solvents or solvent mixtures of increasing elution strength. Parts a-c of the figure show that the retention of the sample substances becomes continuously weaker as the polarity of the eluent increases. Also, the most polar eluent in this example, an 80:20 (v/v) mixture of cyclohexane and toluene, is least selective and hence produces the poorest resolution in this example.

## Special Features of Straight Phase Thin-Layer Chromatography

Since thin-layer chromatography (TLC) uses an open system (unlike column chromatography), due regard must be given to the effect of the gas phase on TLC results. Preconditioning is an important concept in this context. It means the adsorption of certain organic or inorganic gaseous substances to the sorbent layer prior to chromatographic development.

### *Effect of Preconditioning with Water*

An inherent characteristic of sorbents used in straight phase chromatography is their ability, owing to their surface activity, to capture water from the surrounding atmosphere by adsorption at their active adsorption centers. This results in a partial blockade, and hence deactivation, of these centers. In this deactivated state the stationary phase has a smaller number of active adsorption centers available and interactions with sample molecules are thereby diminished. Consequently, retention of the sample substances is also lessened and the selectivity of the chromatographic system altered.

Changes in the selectivity of the TLC plate by preconditioning produced by various relative humidities may lead to an improvement or deterioration of chromatographic resolution.

If the results of straight phase TLC are to be reproducible, it is essential that the moisture content of the separation layer be adjusted to a defined value [6].

Figure 2 illustrates the way in which the separation can be affected by varying the amount of water in the sorbent. Chromatograms of several *m*-oligophenylenes were developed at various relative humidities.

### *Effect of Preconditioning with Components of the Mobile Phase*

It is not only water, however, that can be adsorbed at the active centers. Indeed, other adsorbed molecules are capable of changing the chromatographic properties of the support. A feature of several types of chromatographic chambers is that they permit the solvent from the mobile phase to be taken up by the stationary phase via the gas phase. This effect can be achieved most noticeably in chambers specially designed to permit total saturation of the gas phase with a solvent (a process known as chamber saturation).

Figure 3 demonstrates the extent to which the separation of plant protectants can be influenced by varying the amount of solvent adsorbed at a precoated layer.

The chromatogram in Fig. 3a was prepared using a normal chamber without chamber saturation, i.e., a chamber in which the gas space is only partially saturated with a solvent. The chromatogram in Fig. 3b was obtained in a chamber with complete saturation of the gas phase with a solvent.

The fact that retention is increased in the chamber with chamber saturation is attributable primarily to the markedly lower relative humidity when the gas space is completely saturated with solvent. As a result, a certain amount of the water adsorbed onto the surface of the sorbent material is desorbed as the system equilibrates, thus elevating the sorbent material into a higher level of activation.

### *Effect of Preconditioning with Organic or Inorganic Components Not Present in the Mobile Phase*

A third way of modifying selectivity via the gas phase is to precondition the sorbent material with certain organic or inorganic components not present in the mobile phase. Examples of this method are given in Fig. 4a-e (separation of plant protectants). Ammonia, hydrochloric acid, methanol, and dichloromethane were used for the preconditioning.

Within this series retention of sample substances on all preconditioned plates is lower than on the equivalent nontreated layer (Fig. 4a). Also, selectivity decreases progressively with conditioning agent in the order ammonia, hydrochloric acid, methanol, dichloromethane.

The effect of the gas phase on TLC separations has been discussed in previous sections. In order to attain reproducible results, the preconditioning of the stationary phase with components via the gas phase must be kept under control prior to and during the chromatographic development. The most effective way to succeed at this is to utilize an appropriate developing chamber [4,6-8], if necessary in combination with a suitable pretreatment of the layer.

## Silicas

During the evolution of thin-layer chromatography it was initially found that the most suitable, universally applicable sorbents for straight phase chromatography were the porous, non-surface-modified silicas (see Chap. 6).

In straight phase chromatography it is the hydroxyl groups on the surface of the silica which are the polar, active centers where most of the interactions take place in the retention of the substances to be separated. The mean hydroxyl group density for all silica types amounts to about 8 $\mu mol/m^2$ [9].

The specific surface area $a_s$(BET) [10] of porous silicas is the parameter used to characterize and standardize their adsorptive capacity. Porous silicas commonly used in straight phase TLC have specific surface areas between 400 and 800 $m^2/g$. The interactive strength of substances under separation by straight phase chromatography and hence also their retention within the chromatographic system are directly proportional to the specific surface area of the stationary phase.

Another important parameter used to denote the chromatographic properties of porous sorbents is the specific pore volume vp [11]. For silicas used in straight phase thin-layer chromatography, vp is between approximately 0.5 and 1.2 ml/g.

A third parameter to characterize the structure of the porous adsorbents is the mean pore diameter, pd. Porous silicas used for chromatography, even if they belong to the same type, do not have a uniform pore size. What can be quoted, however, is the pore size distribution. Silica types are frequently referred to by their mean pore size (expressed in angstrom units). The mean pore sizes of silicas used in adsorption chromatography lie mostly within the range 4-12 nm (40-120 Å). It must be borne in mind that all the primary parameters used here to characterize the pore structure ($a_s$(BET), vp, pd) are mutually dependent.

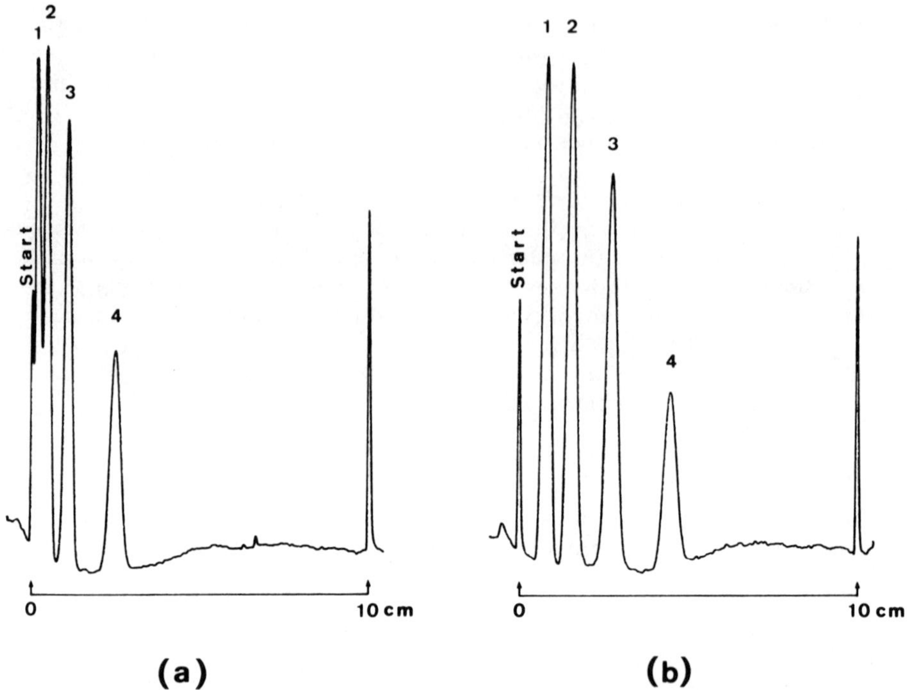


Fig. 2 Influence of relative humidity on the retention and resolution of some *m*-oligophenylenes. Plate: TLC precoated plate silica gel 60 F 254 (E. Merck). Compounds: 1, *m*-quinquephenyl; 2, *m*-quaterphenyl; 3, *m*-terphenyl; 4, biphenyl. Eluent: cyclohexane. Preconditioning (a) 20% relative humidity; (b) 50% relative humidity; (c) 80% relative humidity. Type of chamber: Camag Vario KS. Detection: Camag TLC/HPTLC scanner, UV 254 nm.

The following mathematical equation devised by Wheeler [12] indicates the interdependence of the parameters:

$$pd = \frac{4 \times 10^4 \cdot vp}{a_s(BET)} \tag{1}$$

where pd is the mean pore diameter (Å), vp the specific pore volume (ml/g), and $a_s$(BET) the specific surface area ($m^2/g$).

The chemical and physical primary parameters so far discussed determine the particular selectivity of a sorbent within the chromatographic system.

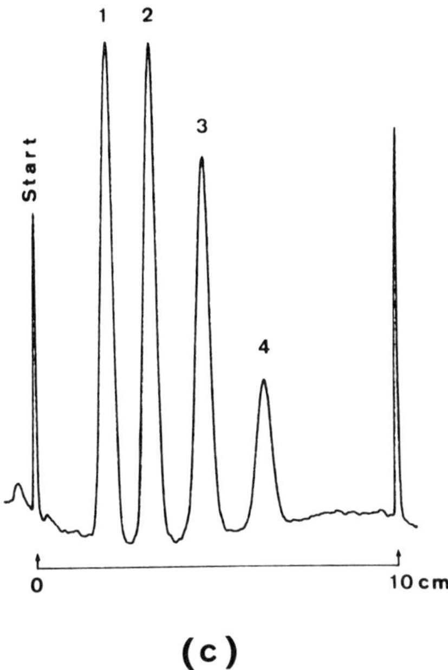


Fig. 2 (continued)

The chromatographic separation efficiency of a thin-layer plate or coated sheet is determined more, however, by two secondary parameters: the mean particle size dp, and particle size distribution.

Since the sorbents used in TLC are irregular particles of nonuniform size, any reference to size must include information about the particle size distribution. This can be measured by a number of methods including the following [11,13]:

Sieve analysis
Sedimentation
Microscopic enumeration
Measurement of conductivity

Results are expressed in terms of cumulative undersize distribution quoted not only for a mean particle size ($dp_{50}$) but also for a sorbent mass distribution at other defined values ($dp_5$, $dp_{10}$, $dp_{90}$, $dp_{95}$).

Layers made from sorbents having a relatively high proportion of fines (small $dp_5$ and $dp_{10}$ values) at a specified mean particle size are less permeable (because of the smaller interstices and consequent weakening of capillary forces) than layers comprising particles of the same

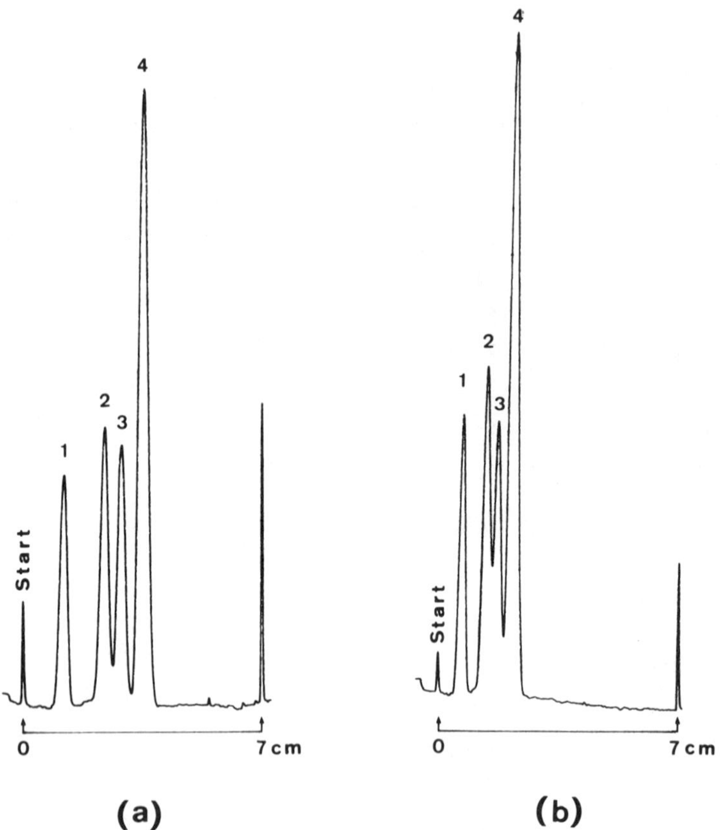


Fig. 3 Influence of conditioning with solvent molecules on the retention and resolution of some plant-protective agents. Plate: HPTLC precoated plate silica gel 60 F 254 (E. Merck). Compounds: 1, *p,p'*-DDD; 2, *p,p'*-DDT; 3, *o,p'*-DDT; 4, *p,p'*-DDE. Eluent: *n*-heptane. Type of chamber: (a) normal chamber without chamber saturation; (b) normal chamber with chamber saturation. Detection: Camag TLC/HPTLC scanner, UV 254 nm.

mean size but containing only a small amount of fine-grained particles (larger $dp_5$ and $dp_{10}$ values). On the other hand, large amounts of coarser particles (high $dp_{90}$ and $dp_{95}$ values) in the sorbent have a negative effect on separation efficiency in the respective chromatographic system. A sorbent having optimum chromatographic properties must consequently have a particle size distribution as narrow as possible for a given mean particle size (preferably with a $dp_{90}$ to $dp_{10}$

ratio between 1.5 and 2.0). An additional important factor affecting the performance of thin layers is the mean particle size of the sorbent. Sorbents with low $dp_{50}$ values (= mean particle size) offer more separation efficiency than those of a higher mean particle size, even though the particle size distribution may be the same. Once these facts had been taken into account, TLC precoated layers underwent further development aimed in particular at increasing their separation performance. These so-called HPTLC layers (high-performance thin-layer chromatography) have roughly half the mean particle size of conventional precoated TLC layers.

Since the HPTLC precoated layers have smaller mean particle sizes and since the interstices are much smaller, the HPTLC layers are less permeable than TLC layers. For this reason it is very important that sorbents used in HPTLC layers, in particular, have as narrow a particle size distribution as possible with very low fine-grained particle content.

Unlike HPLC, where solvent flow is generally linear, thin-layer chromatography involves a quadratic relationships between solvent migration distance ($z_f$) and the time (t) required for the solvent to migrate this distance [6]. The factor linking these two values is the flow constant ($\kappa$), which is a measure of the permeability of a thin layer. Equation (2) demonstrates the mathematical relationship between $z_f$, t, and $\kappa$.

$$\kappa = \frac{z_f^2}{t} \ (\mathrm{mm}^2/\mathrm{sec}) \tag{2}$$

For example, on a silica gel 60 HPTLC precoated plate with toluene as eluent the flow constant $\kappa$ has a value of 5.0 $mm^2$/sec, while on a silica gel 60 TLC precoated plate with the same solvent the value is 6.5 $mm^2$/sec.

One direct measure of the separating performance of a thin layer is the plate height H. In thin-layer chromatography the plate height is calculated on the basis of bandwidth b and migration distance $z_x$ of a substance spot. The bandwidth can be measured in terms of base width ($b_{base}$), half-width ($b_{0.5}$), or standard deviation $\sigma$. How these values are related is shown in Eq. (3) [6]:

$$H = \frac{\sigma^2}{z_x} = \frac{b_{base}^2}{16 \cdot z_x} = \frac{b_{0.5}^2}{5.54 \cdot z_x} \tag{3}$$

In Eq. (3) the migration distance $z_x$ of a substance spot used for the calculation is not only dependent on the respective substance retention in the TLC system but also on the total running distance $z_f$ of the

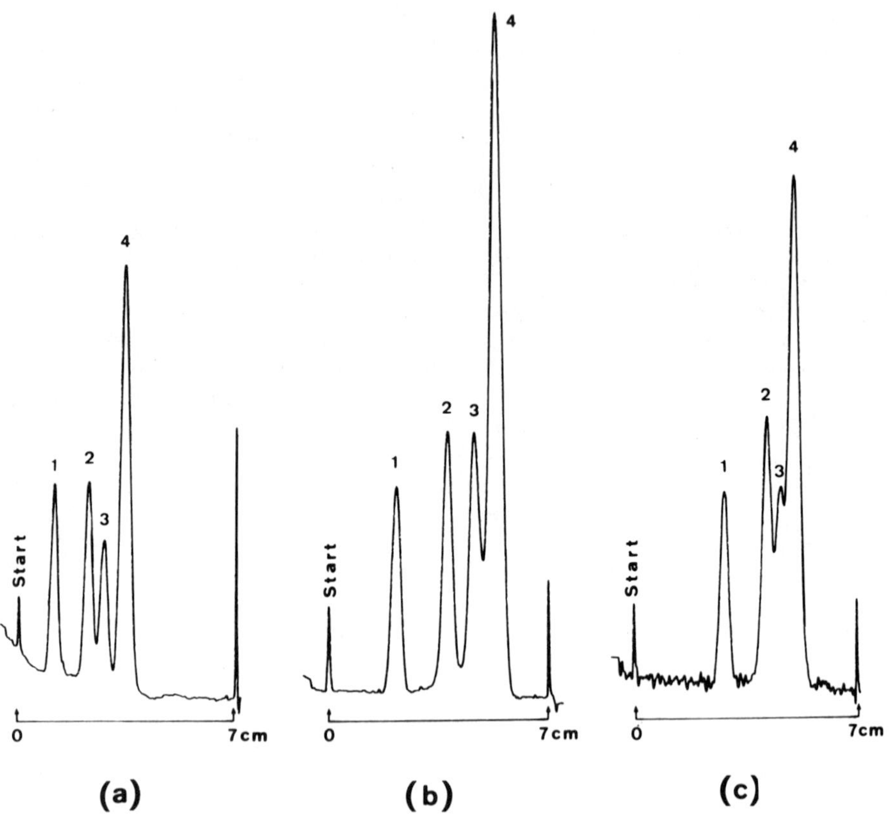


Fig. 4 Influence of preconditioning with organic or inorganic substances on the retention and resolution of some plant-protective agents. Plate: HPTLC precoated plate silica gel 60 F 254 (E. Merck). Compounds: 1, *p,p'*-DDD; 2, *p,p'*-DDT; 3, *o,p'*-DDT; 4, *p,p'*-DDE. Eluent: *n*-heptane. Preconditioning (a) without; (b) 25% ammonia; (c) 37% HCl; (d) methanol; (e) dichloromethane. Type of chamber: Camag Vario KS. Detection: Camag TLC/HPTLC scanner, UV 254 nm.

solvent in the thin layer. As a consequence there is an additional interdependence between plate height H and the total running distance $z_f$.

Figure 5 shows the H-$z_f$ curves for two precoated layers of identical silica type but differing in their mean particle size and particle size distribution (TLC silica gel 60 precoated plate compared with HPTLC silical gel 60 precoated plate). The sample substance used was a lipophilic dye, with toluene as mobile phase. The curves illustrate the fact that the plate heights for the HPTLC plate are smaller in the $z_f$ range in question and the separation efficiency of this layer is higher than that

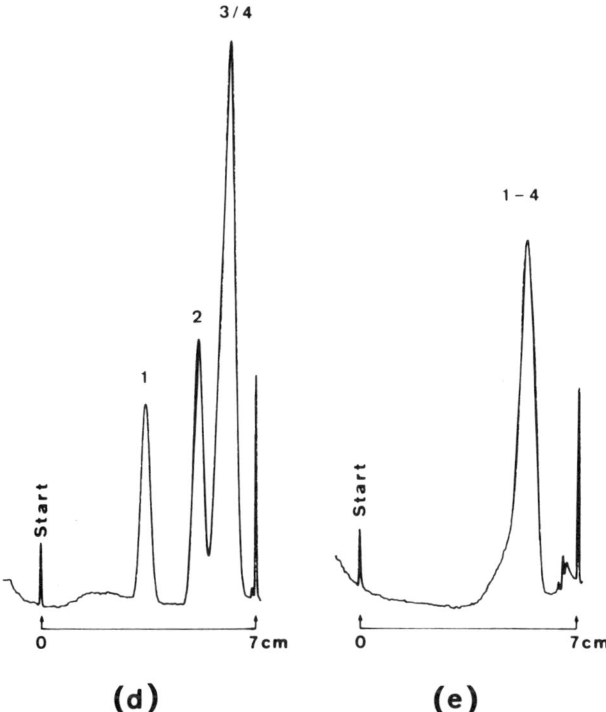


Fig. 4 (continued)

of the TLC precoated plate. This is because the standard deviation $\sigma$ [see Eq. (3)] used to calculate plate height H is dependent on the particle size of the sorbent material used.

Smaller sorbent particle sizes favor substance exchange proceedings between the stationary and mobile phase, and thus result in more compact spot development (smaller standard deviation $\sigma$) than sorbents with a larger particle size.

There is a minimum on both H-$z_f$ curves, meaning that for both TLC and HPTLC plates there are optimum migration distances for maximum separation efficiency to occur.

A further measure of plate performance also affecting the selectivity of the particular chromatographic system is resolution $R_S$ of two substance spots located adjacent to one another on the thin-layer chromatogram. $R_S$ is calculated as

$$R_S = \frac{z_{x2} - z_{x1}}{2 \cdot (\sigma_2 + \sigma_1)} \qquad (4)$$

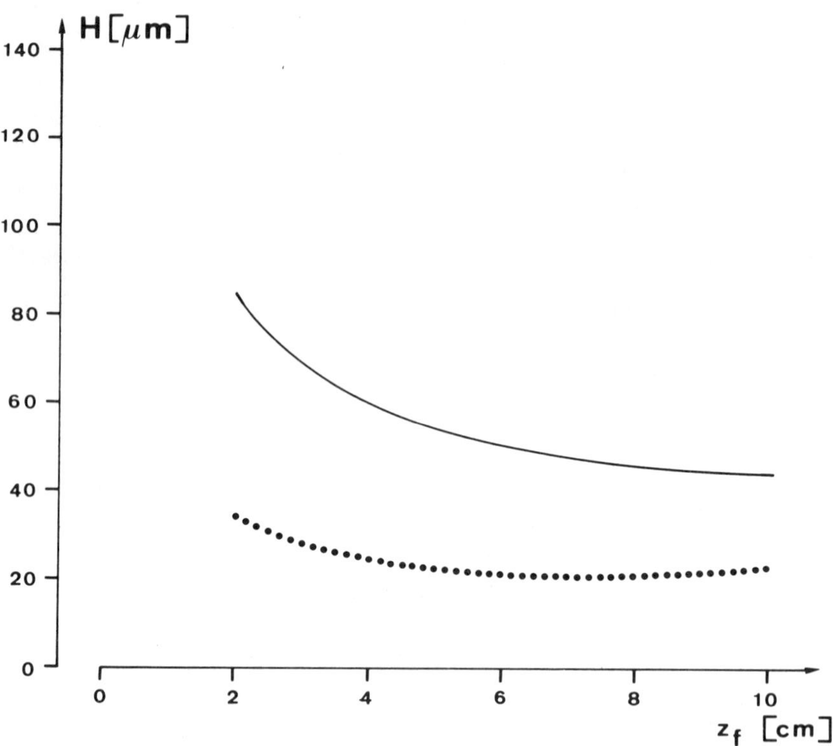


Fig. 5 H-$z_f$ curves of a lipophilic dye on TLC and HPTLC precoated plates silica gel 60. Plates: ——, TLC precoated plate silica gel 60 (E. Merck); . . . ., HPTLC precoated plate silica gel 60 (E. Merck). Compound: Ceres violet BRN. Eluent: toluene. Abscissa: plate height H. Ordinate: migration distance $z_f$.

where $z_{x1}$ ($z_{x2}$) is the migration distance substance 1(2) and $\sigma_1$ ($\sigma_2$) the standard deviation substance 1(2).

In the same way as for plate height H, resolution $R_S$ can be related to the total migration distance $z_f$ of a thin-layer chromatogram. Figure 6 is a graph of resolution $R_S$ for the dyes Nitro Fast Blue and Sico Fat Blue 5040 plotted as a function of total migration distance $z_f$ in an adsorption system. The two stationary phases compared were a TLC silica gel 60 precoated plate and an HPTLC silica gel 60 precoated plate, with toluene as solvent in both cases. The graph shows that, in the same way as for the plate heights, the resolution on the HPTLC precoated plate over the $z_f$ range of interest here is better than on the TLC plate because of the smaller sorbent particle size. On both

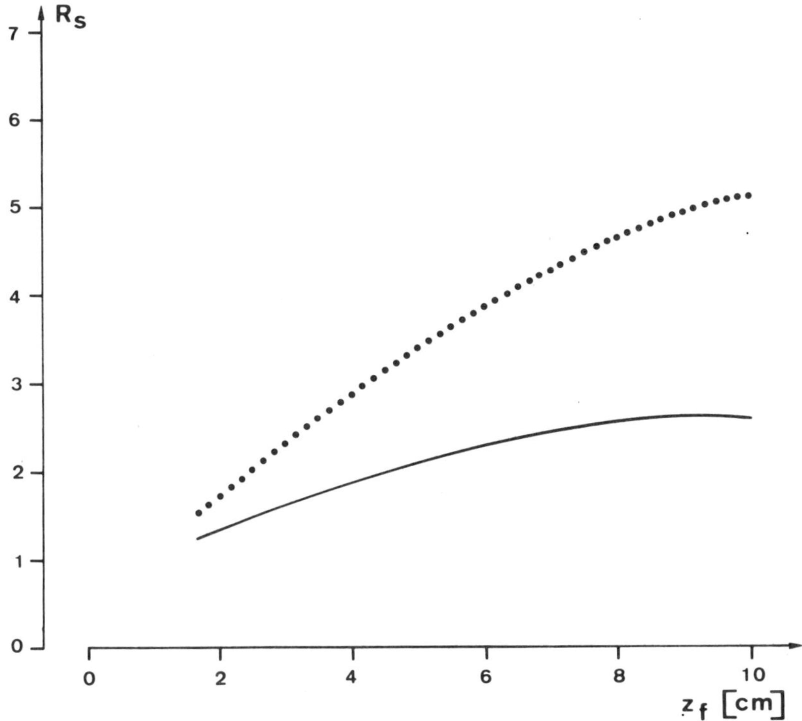


Fig. 6 $R_S$-$z_f$ curves of the resolution between two lipophilic dyes on TLC and HPTLC precoated plates silica gel 60. Plates: ———, TLC precoated plate silica gel 60 (E. Merck); . . . . , HPTLC precoated plate silica gel 60 (E. Merck). Compounds: Nitro fast blue, Sico fat blue. Eluent: toluene. Abscissa: resolution $R_S$. Ordinate: migration distance $z_f$.

the TLC plate and the HPTLC plate the resolution improves as the total running distance $z_f$ increases. This can be explained in Eq. (4): as the total running distance increases, the difference between the migration distances $z_x$ for the separated substances (in the numerator) becomes greater. Although there is a simultaneous spreading of peaks ($\sigma_1$ and $\sigma_2$ in the denominator of the equation), which in itself would lead to poorer results, it is more than compensated by the increase in the difference between the migration distances.

The physical and chemical parameters discussed above are decisive factors for the standardization and manufacturing of bulk sorbents and precoated layers for straight phase chromatography in a reproducible way. Since these parameters are in many cases interrelated, the materials sold for adsorption chromatography by various manufacturers are compromises in which the mutually dependent factors have been optimized. Table 1 lists commercially available bulk silicas for manual preparation by the user. The bulk silicas are made by various manufacturers and in some cases are offered with a variety of binding additives and fluorescent indicators. Data compiled in this table are based on product information issued by individual manufacturers.

With higher demands and rising expectations regarding quality and reproducibility of thin-layer materials, the use of precoated layers has become more and more widespread. For some years now most of the TLC separations have been performed on precoated layers. Within the scope of this development different manufacturers have started offering precoated layers for HPTLC and for preparative layer chromatography (PLC).

As a means of simplifying sample application and, more especially, of increasing separation efficiency when large volumes are applied, special precoated layers with so-called concentrating zones have also been developed. The materials used to manufacture the concentrating zones are kieselguhrs or inert silicas (see below). Figure 7 shows the sequence of development phases for an HPTLC silica gel 60 precoated plate with concentrating zone.

As the sample is applied as a spot onto the concentrating zone (phase a), a relatively large substance spot is produced. In subsequent development within the concentrating zone (phases b and c), the spots are transported to the phase interface by the solvent front and compressed into a narrow band. The extreme compactness of the band in the direction of development enables better results to be obtained than on thin-layer plates without a concentrating zone, especially when larger volumes are applied. The actual separation shown in phases d-f follows on directly from the concentrating process without requiring a change of solvent.

Another advantage of using layers with a concentrating zone is that applying the sample substances at various heights on the concentrating zone causes no shift in $R_f$ values when separation is finished. This is because separation does not begin at the point of application, as is the case on layers without a concentrating zone, but at the interface between the concentrating zone and the separating layer proper.

Tables 2-5 are lists of TLC, HPTLC, and PLC precoated silica layers for adsorption chromatography and precoated products with concentrating zones from various manufacturers on a variety of supports (glass plates, aluminum sheets, plastic sheets). The tables are partly incomplete due to missing information from the manufacturers.

Figure 8 shows the separation of various chlorophenols and serves to illustrate a typical use to which silica gel 60 precoated plates can be put in adsorption chromatography. The phase system consists of an HPTLC silica gel 60 $F_{254}$ precoated plate, with chloroform as solvent.

Table 1 Bulk Silicas

| Manufacturer | Product | Type of silica | Nominal pore size (nm) | Particle size ($\mu m$) | Indicator | Binder | Particularities |
|---|---|---|---|---|---|---|---|
| Analtech | Silica Gel G, High-Performance | 60 | 6 | 10 ± 4 | — | $CaSO_4$ | |
| | Silica Gel G, High-Performance | 60 | 6 | 10 ± 4 | F254 | $CaSO_4$ | |
| | Silica Gel, High-Performance | 60 | 6 | 10 ± 4 | — | — | |
| | Silica Gel, High-Performance | 60 | 6 | 10 ± 4 | F254 | — | |
| J. T. Baker | Silica Gel G/HR | 60 | 6 | 5-25 | — | $CaSO_4$ | |
| | Silica Gel PH 7 | 60 | 6 | 5-40 | — | — | |
| | Silica Gel PH 7 G | 60 | 6 | 5-40 | — | $CaSO_4$ | |
| | Silica Gel PH 7 GF | 60 | 6 | 5-40 | F254 | $CaSO_4$ | |
| Camag | Silica Gel D-0 | | | Fine | — | — | |
| | Silica Gel D-5 | | | Fine | — | $CaSO_4$ | |
| | Silica Gel DF-0 | | | Fine | F254 | — | |
| | Silica Gel DF-5 | | | Fine | F254 | $CaSO_4$ | |
| | Silica Gel DS-0 | | | Coarser | — | — | |
| | Silica Gel DS-5 | | | Coarser | — | $CaSO_4$ | |
| | Silica Gel DSF-0 | | | Coarser | F254 | — | |
| | Silica Gel DSF-5 | | | Coarser | F254 | $CaSO_4$ | |
| ICN | Silica TLC | 60 | 6 | 5-15 | — | — | |
| | Silica F-TLC | 60 | 6 | 5-15 | F254 | — | |
| | Silica G-TLC | 60 | 6 | 5-15 | — | $CaSO_4$ | |
| | Silica GF-TLC | 60 | 6 | 5-15 | F254 | $CaSO_4$ | |

Table 1 (continued)

| Manufacturer | Product | Type of silica | Nominal pore size (nm) | Particle size (μm) | Indicator | Binder | Particularities |
|---|---|---|---|---|---|---|---|
| Macherey-Nagel | Silica Gel G | 60 | 6 | 2-25 | — | $CaSO_4$ | |
| | Silica Gel G/UV 254 | 60 | 6 | 2-25 | F254 | $CaSO_4$ | |
| | Silica Gel N | 60 | 6 | 2-25 | — | — | |
| | Silica Gel N/UV 254 | 60 | 6 | 2-25 | F254 | — | |
| | Silica Gel N/UV 254 + 366 | 60 | 6 | 2-25 | F254+366 | — | |
| | Silica Gel N with starch | 60 | 6 | 2-25 | — | Starch | |
| | Silica Gel N/UV 254 with starch | 60 | 6 | 2-25 | F254 | Starch | |
| | Silica Gel G-HR | 60 | 6 | 5-25 | — | $CaSO_4$ | High purity |
| | Silica Gel G-HR/UV 254 | 60 | 6 | 5-25 | F254 | $CaSO_4$ | High purity |
| | Silica Gel N-HR | 60 | 6 | 5-25 | — | — | High purity |
| | Silica Gel N-HR/UV 254 | 60 | 6 | 5-25 | F254 | — | High purity |
| | Silica Gel P/UV 254 | 60 | 6 | | F254 | Organ. | For prep. layers |
| | Silica Gel P/UV 254 + 366 | 60 | 6 | | F254+366 | Organ. | For prep. layers |
| | Silica Gel P/UV 254 with gypsum | 60 | 6 | | F254 | $CaSO_4$ | For prep. layers |
| E. Merck | Silica Gel 60 G | 60 | 6 | 5-40 | — | $CaSO_4$ | |
| | Silica Gel 60 G, 15 μm | 60 | 6 | 15 | — | $CaSO_4$ | |
| | Silica Gel 60 GF 254 | 60 | 6 | 5-40 | F254 | $CaSO_4$ | |
| | Silica Gel 60 GF 254, 15 μm | 60 | 6 | 15 | — | $CaSO_4$ | |
| | Silica Gel 60 H | 60 | 6 | 5-40 | — | Without alien binder | |
| | Silica Gel 60 H, 15 μm | 60 | 6 | 15 | — | Without alien binder | |
| | Silica Gel 60 HF 254 | 60 | 6 | 5-40 | F254 | Without alien binder | |

| | | | | | | | |
|---|---|---|---|---|---|---|---|
| | Silica Gel 60 HF 254, 15 μm | 60 | 6 | 15 | F254 | Without alien binder | |
| | Silica Gel 60 HF 254 + 366 | 60 | 6 | 5-40 | F254+366 | Without alien binder | |
| | Silica Gel 60 HR | 60 | 6 | 5-40 | — | Without alien binder | High purity |
| | Silica Gel 60 PF 254 | 60 | 6 | | F254 | — | For prep. layers |
| | Silica Gel 60 PF 254 + 366 | 60 | 6 | | F254+366 | — | For prep. layers |
| | Silica Gel 60 PF 254 with gypsum | 60 | 6 | | F254 | $CaSO_4$ | For prep. layers |
| Riedel-de Haën | Silica Gel D | | | 30 | — | — | |
| | Silica Gel DF | | | 30 | F254 | — | |
| | Silica Gel DG | | | 30 | — | $CaSO_4$ | |
| | Silica Gel DGF | | | 30 | F254 | $CaSO_4$ | |
| Schleicher + Schüll | 150 | | | 40 | — | — | |
| | 150/LS 254 | | | 40 | F254 | — | |
| | 150 G | | | 40 | — | $CaSO_4$ | |
| | 150 G/LS 254 | | | 40 | F254 | $CaSO_4$ | |
| | 150 S | | | 40 | — | Starch | |
| | 150 S/LS 254 | | | 40 | F254 | Starch | |

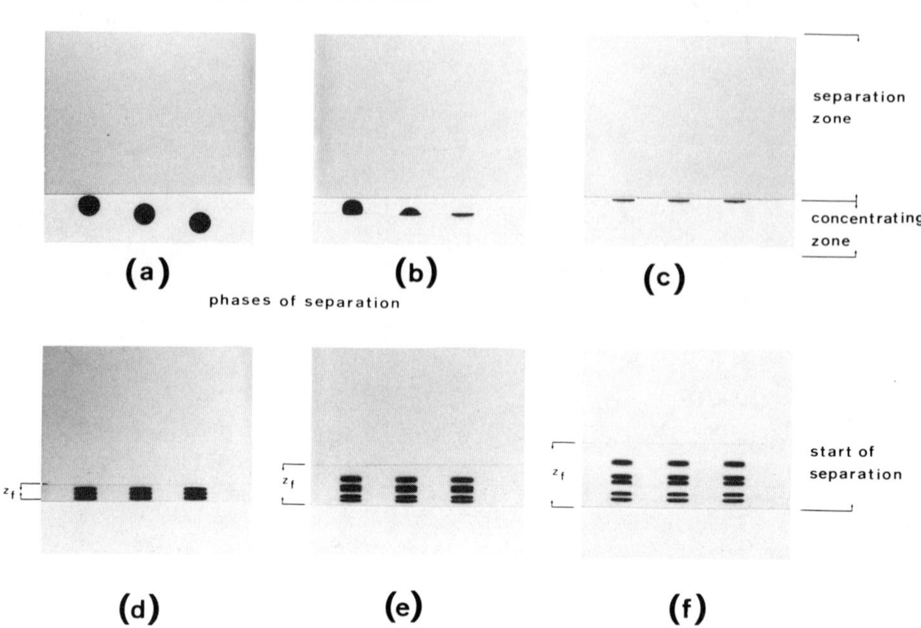


Fig. 7 Different phases of concentration and separation on a precoated layer with concentrating zone. Plate: HPTLC precoated plate silica gel 60 with concentrating zone (E. Merck). Compounds: Violet I (from Ceres violet BRN); Violet II (from Ceres black G); Nitro fast blue 2B; Nitro fast violet FBL; Ceres red G. Eluent: toluene. Phases of development: a-c concentration; d-f separation.

Figure 9 shows the separation of a polystyrene standard, underlining the fact that precoated layers with concentrating zones can be put to particularly effective use in adsorption chromatography. The eluent in this case was carbon tetrachloride. The separating layer was a TLC silica gel 60 $F_{254}$ precoated plate with concentrating zone. In spite of the very large applied sample quantity of 5 μl of a 0.1% solution and pronounced spreading of the applied spot, the use of a precoated plate with concentrating zone virtually maximizes utilization of available separating efficiency.

Table 6 shows the solvent composition and literature references for separations of various important classes of substance by straight phase chromatography on silica thin layers.

Table 2 Silica Precoated TLC Layers

| Manufacturer | Product | Type of silica | Nominal pore size (nm) | Particle size (μm) | Indicator | Binder | Support | Layer thickness (mm) |
|---|---|---|---|---|---|---|---|---|
| Analtech | Silica Gel G Uniplates | 60 | 6 | | — | $CaSO_4$ | Glass | 0.25 |
| | Silica Gel GF Uniplates | 60 | 6 | | F254 | $CaSO_4$ | Glass | 0.25 |
| | Silica Gel GHR Uniplates | 60 | 6 | | — | $CaSO_4$ | Glass | 0.25 |
| | Silica Gel GHRF Uniplates | 60 | 6 | | F254 | $CaSO_4$ | Glass | 0.25 |
| | Silica Gel H Uniplates | 60 | 6 | | — | $SiO_2$ | Glass | 0.25 |
| | Silica Gel HF Uniplates | 60 | 6 | | F254 | $SiO_2$ | Glass | 0.25 |
| | Hard Layer Uniplates HL | 60 | 6 | | — | Organ. | Glass | 0.25 |
| | Hard Layer Uniplates HLF | 60 | 6 | | F254 | Organ. | Glass | 0.25 |
| | Hard Layer Uniplates GHL | 60 | 6 | | — | Inorg. | Glass | 0.25 |
| | Hard Layer Uniplates GHLF | 60 | 6 | | F254 | Inorg. | Glass | 0.25 |
| J. T. Baker | Si 250 - Silica Gel | 80 | 8 | | — | Organ. | Glass | 0.25 |
| | Si 250 F - Silica Gel | 80 | 8 | | F254 | Organ. | Glass | 0.25 |
| | Baker-flex Silica Gel IB2 | | | | — | Organ. | Plastic | 0.20 |
| | Baker-flex Silica Gel IB2-F | | | | F254 | Organ. | Plastic | 0.20 |
| | Baker-flex Silica Gel IB | | | | — | Organ. | Plastic | 0.25 |
| | Baker-flex Silica Gel IB-F | | | | F254 | Organ. | Plastic | 0.25 |
| Eastman-Kodak | Chromagram Silica Gel F | | | | F254 | Organ. | Plastic | 0.10 |
| | Chromagram Silica Gel | | | | — | Organ. | Plastic | 0.10 |
| ICN | Silica-Plates | 60 | 6 | | — | Organ. | Glass | 0.25 |
| | Silica-Plates F 254 | 60 | 6 | | F254 | Organ. | Glass | 0.25 |
| | Silica-Plates F 254/366 | 60 | 6 | | F254+366 | Organ. | Glass | 0.25 |
| | Silica-Rapid-Plates F 254 | 60 | 6 | | F254 | Organ. | Glass | 0.25 |
| | Silica-Sheets | 60 | 6 | | — | Organ. | Alumin. | 0.20 |
| | Silica-Sheets F254/366 | 60 | 6 | | F254+366 | Organ. | Alumin. | 0.20 |

Table 2 (continued)

| Manufacturer | Product | Type of silica | Nominal pore size (nm) | Particle size (μm) | Indicator | Binder | Support | Layer thickness (mm) |
|---|---|---|---|---|---|---|---|---|
| Macherey-Nagel | SIL G-25 | 60 | 6 | 5-25 | — | Organ. | Glass | 0.25 |
| | SIL G-25 UV 254 | 60 | 6 | 5-25 | F254 | Organ. | Glass | 0.25 |
| | SIL G-25 UV 254 + 366 | 60 | 6 | 5-25 | F254+366 | Organ. | Glass | 0.25 |
| | SIL G-25 HR | 60 | 6 | 5-25 | — | $CaSO_4$ | Glass | 0.25 |
| | SIL G-25 HR UV 254 | 60 | 6 | 5-25 | F254 | $CaSO_4$ | Glass | 0.25 |
| | Polygram SIL G | 60 | 6 | 5-25 | — | Organ. | Plastic | 0.25 |
| | Polygram SIL G/UV 254 | 60 | 6 | 5-25 | F254 | Organ. | Plastic | 0.25 |
| | Polygram SIL N-HR | 60 | 6 | 5-25 | — | | Plastic | 0.20 |
| | Polygram SIL N-HR/UV 254 | 60 | 6 | 5-25 | F254 | | Plastic | 0.20 |
| | Alugram SIL G | 60 | 6 | 5-25 | — | Organ. | Alumin. | 0.20 |
| | Alugram SIL G/UV 254 | 60 | 6 | 5-25 | F254 | Organ. | Alumin. | 0.20 |
| E. Merck | TLC Plates Silica Gel 40 F 254 | 40 | 4 | | F254 | Organ. | Glass | 0.25 |
| | TLC Plates Silica Gel 60 | 60 | 6 | | — | Organ. | Glass | 0.25 |
| | TLC Plates Silica Gel 60 F 254 | 60 | 6 | | F254 | Organ. | Glass | 0.25 |
| | TLC Alusheets Silica Gel 60 | 60 | 6 | | — | Organ. | Alumin. | 0.20 |
| | TLC Alusheets Silica Gel 60 F 254 | 60 | 6 | | F254 | Organ. | Alumin. | 0.20 |
| | TLC Plastic Sheets Silica Gel 60 | 60 | 6 | | — | Organ. | Plastic | 0.20 |
| | TLC Plastic Sheets Silica Gel 60F254 | 60 | 6 | | F254 | Organ. | Plastic | 0.20 |
| | TLC Plates Silica Gel 100 F 254 | 100 | 10 | | F254 | Organ. | Glass | 0.25 |
| Riedel-de Haën | Silica Gel - | | | | — | Organ. | Glass | 0.25 |
| | Silica Gel + | | | | F254 | Organ. | Glass | 0.25 |
| | SI Silica Gel | | | | — | Organ. | Alumin. | 0.20 |
| | SIF Silica Gel | | | | F254 | Organ. | Alumin. | 0.20 |

| | | | | | | | |
|---|---|---|---|---|---|---|---|
| Schleicher + Schüll | G 1500 | | | — | Organ. | Glass | 0.25 |
| | G 1500/LS 254 | | | F254 | Organ. | Glass | 0.25 |
| | F 1500 | | | — | Organ. | Plastic | 0.20 |
| | F 1500/LS 254 | | | F254 | Organ. | Plastic | 0.20 |
| Whatman | MK 6 F | 60 | 6 | F254 | Organ. | Glass | 0.25 |
| | K 6 | 60 | 6 | — | Organ. | Glass | 0.25 |
| | K 6 F | 60 | 6 | F254 | Organ. | Glass | 0.25 |
| | K 5 | 150 | 15 | — | Organ. | Glass | 0.25 |
| | K 5 F | 150 | 15 | F254 | Organ. | Glass | 0.25 |

Table 3 Silica Precoated HPTLC Layers

| Manufacturer | Product | Type of silica | Nominal pore size (nm) | Particle size (μm) | Indicator | Binder | Support | Layer thickness (mm) |
|---|---|---|---|---|---|---|---|---|
| Analtech | High Efficiency Uniplates HL | 60 | 6 | 10 | — | Organ. | Glass | 0.15 |
| | High Efficiency Uniplates HLF | 60 | 6 | 10 | F254 | Organ. | Glass | 0.15 |
| | High Efficiency Uniplates GHL | 60 | 6 | 10 | — | Inorg. | Glass | 0.15 |
| | High Efficiency Uniplates GHLF | 60 | 6 | 10 | F254 | Inorg. | Glass | 0.15 |
| J. T. Baker | Si HPF - Silica Gel | 80 | 8 | 5 | F254 | Organ. | Glass | 0.20 |
| Macherey-Nagel | Nano-SIL-20 | 60 | 6 | 5-10 | — | Organ. | Glass | 0.20 |
| | Nano-SIL-20 UV 254 | 60 | 6 | 5-10 | F254 | Organ. | Glass | 0.20 |
| E. Merck | HPTLC Plates Silica Gel 40 F 254 | 40 | 4 | | F254 | Organ. | Glass | 0.20 |
| | HPTLC Plates Silica Gel 60 | 60 | 6 | | — | Organ. | Glass | 0.20 |
| | HPTLC Plates Silica Gel 60 F 254 | 60 | 6 | | F254 | Organ. | Glass | 0.20 |
| | HPTLC Alusheets Silica Gel 60 | 60 | 6 | | — | Organ. | Alumin. | 0.20 |
| | HPTLC Alusheets Silica Gel 60 F 254 | 60 | 6 | | F254 | Organ. | Alumin. | 0.20 |
| | HPTLC Plates Silica Gel 100 F 254 | 100 | 10 | | F254 | Organ. | Alumin. | 0.20 |
| Schleicher + Schüll | G 1570 | | | | — | | Glass | |
| | G 1570/LS 254 | | | | F254 | | Glass | |
| Whatman | HP - K | | | 4.5 | — | Organ. | Glass | 0.20 |
| | HP - KF | | | 4.5 | F254 | Organ. | Glass | 0.20 |

Table 4 Silica Precoated Preparative Layers

| Manufacturer | Product | Type of silica | Nominal pore size (nm) | Particle size (μm) | Indicator | Binder | Support | Layer thickness (mm) |
|---|---|---|---|---|---|---|---|---|
| Analtech | Preparative Uniplates Silica Gel G | 60 | 6 | | — | | Glass | 0.5 |
| | Preparative Uniplates Silica Gel G | 60 | 6 | | — | | Glass | 1.0 |
| | Preparative Uniplates Silica Gel G | 60 | 6 | | — | | Glass | 1.5 |
| | Preparative Uniplates Silica Gel G | 60 | 6 | | — | | Glass | 2.0 |
| | Preparative Uniplates Silica Gel GF | 60 | 6 | | F254 | | Glass | 0.5 |
| | Preparative Uniplates Silica Gel GF | 60 | 6 | | F254 | | Glass | 1.0 |
| | Preparative Uniplates Silica Gel GF | 60 | 6 | | F254 | | Glass | 1.5 |
| | Preparative Uniplates Silica Gel GF | 60 | 6 | | F254 | | Glass | 2.0 |
| J. T. Baker | Si HPF - Silica Gel | 80 | 8 | | F254 | Organ. | Glass | 0.5 |
| ICN | Silica Plates F 254 | 60 | 6 | | F254 | Organ. | Glass | 0.5 |
| Macherey-Nagel | SIL G-50 | 60 | 6 | | — | Organ. | Glass | 0.5 |
| | SIL G-50 UV 254 | 60 | 6 | | F254 | Organ. | Glass | 0.5 |
| | SIL G-100 | 60 | 6 | | — | Organ. | Glass | 1.0 |
| | SIL G-100 UV 254 | 60 | 6 | | F254 | Organ. | Glass | 1.0 |
| | SIL G-200 | 60 | 6 | | — | Organ. | Glass | 2.0 |
| | SIL G-200 UV 254 | 60 | 6 | | F254 | Organ. | Glass | 2.0 |
| E. Merck | PLC Plates Silica Gel 60, 0.5 | 60 | 6 | | — | Organ. | Glass | 0.5 |
| | PLC Plates Silica Gel 60 F 254, 0.5 | 60 | 6 | | F254 | Organ. | Glass | 0.5 |
| | PLC Plates Silica Gel 60 F 254, 1.0 | 60 | 6 | | F254 | Organ. | Glass | 1.0 |
| | PLC Plates Silica Gel 60, 2.0 | 60 | 6 | | — | Organ. | Glass | 2.0 |
| | PLC Plates Silica Gel 60 F254, 2.0 | 60 | 6 | | F254 | Organ. | Glass | 2.0 |
| | PLC Plates Silica Gel 60 F 254 + 366 | 60 | 6 | | F254+366 | Organ. | Glass | 2.0 |

Table 4 (continued)

| Manufacturer | Product | Type of silica | Nominal pore size (nm) | Particle size (μm) | Indicator | Binder | Support | Layer thickness (mm) |
|---|---|---|---|---|---|---|---|---|
| Schleicher + Schüll | G 1505/LS 254 | | | | F254 | Organ. | Glass | 0.5 |
| | G 1510/LS 254 | | | | F254 | Organ. | Glass | 1.0 |
| Whatman | PK 6 F | 60 | 6 | | F254 | Organ. | Glass | 0.5 |
| | PK 6 F | 60 | 6 | | F254 | Organ. | Glass | 1.0 |
| | PK 5 F | 150 | 15 | | F254 | Organ. | Glass | 0.5 |
| | PK 5 | 150 | 15 | | — | Organ. | Glass | 1.0 |
| | PK 5 F | 150 | 15 | | F254 | Organ. | Glass | 1.0 |

Table 5 Silica Precoated Layers with Concentrating Zones

| Manufacturer | Product | Types of silica | Indicator | Support | Layer thickness (mm) | Application |
|---|---|---|---|---|---|---|
| Analtech | Uniplates G with Preadsorbent Zone | 60 /inert mat. | — | Glass | 0.25 | TLC |
| | Uniplates GF with Preads. Zone | 60 /inert mat. | F254 | Glass | 0.25 | TLC |
| | Hard Layer GHL with Preads. Zone | 60 /inert mat. | — | Glass | 0.25 | TLC |
| | Hard Layer GHLF with Preads. Zone | 60 /inter mat. | F254 | Glass | 0.25 | TLC |
| | Hard Layer HL with Preads. Zone | 60 /inert mat. | — | Glass | 0.25 | TLC |
| | Hard Layer HLF with Preads. Zone | 60 /inert mat. | F254 | Glass | 0.25 | TLC |
| | High Efficiency HL with Preads. Zone | 60 /inert mat. | — | Glass | 0.15 | HPTLC |
| | High Efficiency HLF with Preads. Zone | 60 /inert mat. | F254 | Glass | 0.15 | HPTLC |
| | Preparative G with Preads. Zone | 60 /inert mat. | — | Glass | 0.5 | PLC |
| | Preparative G with Preads. Zone | 60 /inert mat. | — | Glass | 1.0 | PLC |
| | Preparative GF with Preads. Zone | 60 /inert mat. | F254 | Glass | 0.5 | PLC |
| | Preparative GF with Preads. Zone | 60 /inert mat. | F254 | Glass | 1.0 | PLC |
| | Taper Plate | 60 /inert mat. | — | Glass | 0.3-1.7 | PLC |
| | Taper Plate F | 60 /inert mat. | F254 | Glass | 0.3-1.7 | PLC |
| J. T. Baker | Si 250-PA Silica Gel | 80 /inert mat. | — | Glass | 0.25 | TLC |
| | Si 250 F-PA Silica Gel | 80 /inert mat. | F254 | Glass | 0.25 | TLC |
| Macherey-Nagel | SILGUR-25 | 60 /kieselguhr | — | Glass | 0.25 | TLC |
| | SILGUR-25 UV 254 | 60 /kieselguhr | F254 | Glass | 0.25 | TLC |
| | Nano-SILGUR-20 | 60 /kieselguhr | — | Glass | 0.20 | HPTLC |
| | Nano-SILGUR-20 UV 254 | 60 /kieselguhr | F254 | Glass | 0.20 | HPTLC |
| E. Merck | TLC Plates Silica Gel 60 with Concentrating Zone | 60 /Si 50,000 | — | Glass | 0.25 | TLC |

Table 5 (continued)

| Manufacturer | Product | Types of silica | Indicator | Support | Layer thickness (mm) | Application |
|---|---|---|---|---|---|---|
| E. Merck | TLC Plates S.G.[a]60 F254 with Concentrating Zone | 60 /Si 50,000 | F254 | Glass | 0.25 | TLC |
| | TLC Alusheets S.G.[a]60 with Concentrating Zone | 60 /Si 50,000 | — | Alumin. | 0.20 | TLC |
| | TLC Alusheets S.G.[a]F254 with Concentrating Zone | 60 /Si 50,000 | F254 | Alumin. | 0.20 | TLC |
| | HPTLC Plates S.G.[a]60 with Conc. Zone | 60 /Si 50,000 | — | Glass | 0.20 | HPTLC |
| | HPTLC Plates F254 with Conc. Zone | 60 /Si 50,000 | F254 | Glass | 0.20 | HPTLC |
| | PLC Plates Silica Gel 60 F254s with Concentrating Zone | 60 /Si 50,000 | F254s | Glass | 0.5 | PLC |
| | PLC Plates F254s with Conc. Zone | 60 /Si 50,000 | F254s | Glass | 1.0 | PLC |
| | PLC Plates F254s with Conc. Zone | 60 /Si 50,000 | F254s | Glass | 2.0 | PLC |
| Schleicher + Schüll | G 1520 | — /inert sil. | — | Glass | 0.25 | TLC |
| | G 1520/LS 254 | — /inert sil. | F254 | Glass | 0.25 | TLC |
| Whatman | LK 6 | 60 /inert mat. | — | Glass | 0.25 | TLC |
| | LK 6F | 60 /inert mat. | F254 | Glass | 0.25 | TLC |
| | LK 5 | 150 /inert mat. | — | Glass | 0.25 | TLC |
| | LK 5F | 150 /inert mat. | F254 | Glass | 0.25 | TLC |
| | LHP-K | — /inert mat. | — | Glass | 0.20 | HPTLC |
| | LHP-KF | — /inert mat. | F254 | Glass | 0.20 | HPTLC |
| | PLK 5 | 150 /inert mat. | — | Glass | 1.0 | PLC |
| | PLK 5F | 150 /inert mat. | F254 | Glass | 1.0 | PLC |

[a]S.G. = silica gel

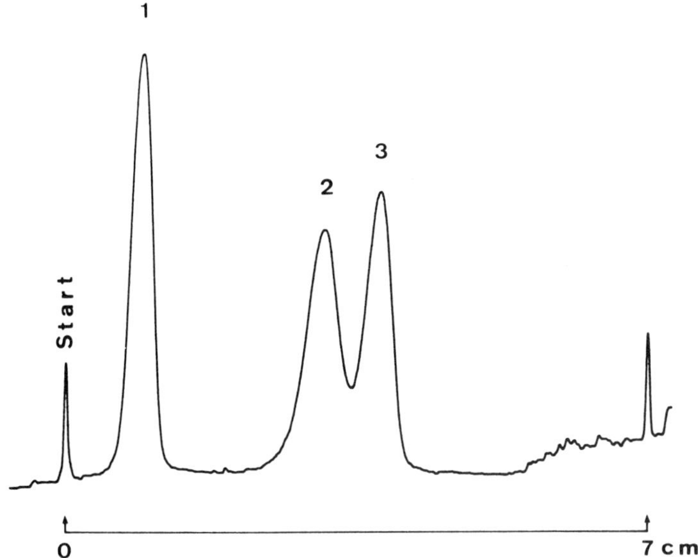


Fig. 8 Straight phase chromatographic separation of some chlorophenols on a silica gel precoated plate. Plate: HPTLC precoated plate silica gel 60 F 254 (E. Merck). Compounds: 1, 2,4,6-trichlorophenole; 2, 2,4-dichlorophenole; 3, 4-chlorophenole. Eluent: chloroform. Detection: Camag TLC/HPTLC scanner, UV 254 nm.

## Polar-Bonded Silicas

Apart from the polar hydroxyl groups located on the surface of native silica, there are other polar, functional groups attached to the silica skeleton capable of entering into adsorptive interactions with suitable sample molecules.

Silicas with hydrophilic polar ligands were also developed for thin-layer chromatography along the lines of the stationary phases already well proven in HPLC. The functional groups used are first of all amino, cyano, and diol functions attached by alkyl chains to the silica skeleton (silica gel 60). Presently, only amino- and cyano-modified layers are available for thin-layer chromatography, these being available in an HPTLC version.

The interactions responsible for substance separation in an adsorption chromatographic separation mode are principally the same as on the surface of a nonmodified silica gel (hydrogen bonding, dipole-dipole interactions, van der Waals forces).

In the case of amino-bonded phases the intensity of the adsorptive interactions is almost comparable to that in the nonmodified silica, while

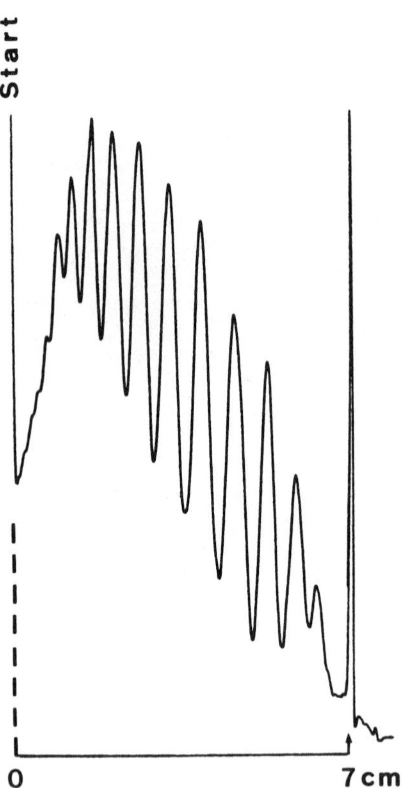


Fig. 9 Straight phase chromatographic separation of a polystyrene standard on a silica gel precoated plate with concentrating zone. Plate: TLC precoated plate silica gel 60 F 254 with concentrating zone (E. Merck). Compound: polystyrene standard 800. Eluent: carbon tetrachloride. Detection: Camag TLC/HPTLC scanner, UV 254 nm.

in the cyano-modified layer it is significantly lower. This can be seen in Fig. 10, in which the separation of three digitalis compounds is shown on silicagel 60, $NH_2$, and CN precoated plates for comparison purposes, the solvent being the same in each case.

Unlike the nonmodified silicas, the sorbents with hydrophilic polar ligands are not available as bulk sorbents for manual spreading but only as HPTLC precoated plates. Table 7 lists the precoated plates with hydrophilic modified silicas currently available.

The separation of various polyvalent phenols in Fig. 11 serves to illustrate the use of an HPTLC $NH_2$ $F_{254}s$ precoated plate in a straight phase chromatographic mode. As one would expect when regarding

Table 6 Applications on Silica Layers in Straight Phase Chromatography

| Substance classes | Mobile phase system | Ref. |
|---|---|---|
| Aflatoxins | Chloroform/acetone | 14 |
| Alkaloids | Chloroform/ethanol | 15 |
| Carbamate pesticides | Toluene/ethyl acetate | 16 |
| Cortisol | Benzene/dioxane/methanol | 17 |
| Coumarins | Hexane/ethyl acetate | 18 |
| Diethylstilbestrol | Benzene/ethyl acetate | 19 |
| Ergot alkaloids | Chloroform/methanol/diethylamine | 20 |
| Estrogens | Ethanol/cyclohexane | 21 |
| Ethoxylated nonionic surfactants | Toluene/hexane/acetone | 22 |
| Fatty acids | Benzene/hexane | 15 |
| Griseofulvin | Butyl acetate/acetone | 23 |
| Hydrazines | Benzene/hexane | 15 |
| Lipids | Petr. ether/ethyl ether/glac. acetic acid | 24 |
| Mycotoxins | Toluene/ethyl acetate/acetone | 25 |
| Neomycins | Benzene/ethanol | 26 |
| Nitrogen bases | Benzene/methanol | 15 |
| Nitrosamines | Hexane/ethyl ether/methylene chloride | 27 |
| Organochlorine and organophosphorus pesticides | *n*-Heptane/acetone | 28 |
| Penicillins | Acetone/methanol | 29 |
| Polycyclic aromatic hydrocarbons | *n*-Pentane/ethyl ether | 15 |
| Spermines | Chloroform/toluene/triethylamine | 30 |
| Steroids | Chloroform/ethyl acetate | 31 |
| Streptomycins | Benzene/ethanol | 32 |
| Sulfonamides | Chloroform/tert.-butanol | 33 |
| Terpenes | Benzene/ethyl acetate | 34 |
| Theophylline | Chloroform/isooctane/methanol | 35 |
| Tocopherols | Hexane/isopropyl ether | 36 |
| Tryptamides | Benzene/ethyl acetate | 37 |
| Vitamin D | Benzene/ethanol | 38 |

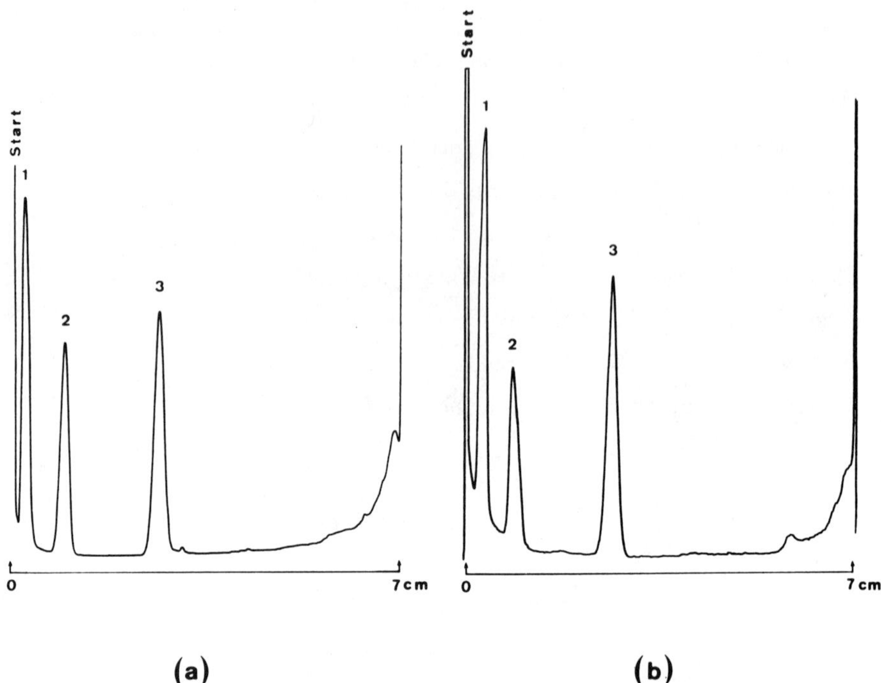


Fig. 10 Influence of the chemical structure of the stationary phase on the retention and resolution of some digitalis compounds in adsorption chromatography. Plates (a) HPTLC precoated plate silica gel 60 (E. Merck); (b) HPTLC precoated plate $NH_2$ F 254s (E. Merck); (c) HPTLC precoated plate CN F 254s (E. Merck). Compounds: 1, digitoxigenin; 2, digoxigenin; 3, digoxin. Eluent: petroleum benzine/ acetone 65:35 (v/v). Detection: Camag TLC/HPTLC scanner, UV 366 nm.

retention mechanism involved here, the two bivalent phenols are more polar and hence more strongly retained than the two monovalent phenols.

Figure 12 is intended to serve as an example of the use of an HPTLC CN $F_{254}$s precoated plate in straight phase chromatography. The chromatogram shows the separation of three male sex hormone derivatives.

The solvent systems used for the separation by straight phase chromatography of further substance types of interest are listed in Table 8 together with the relevant literature references.

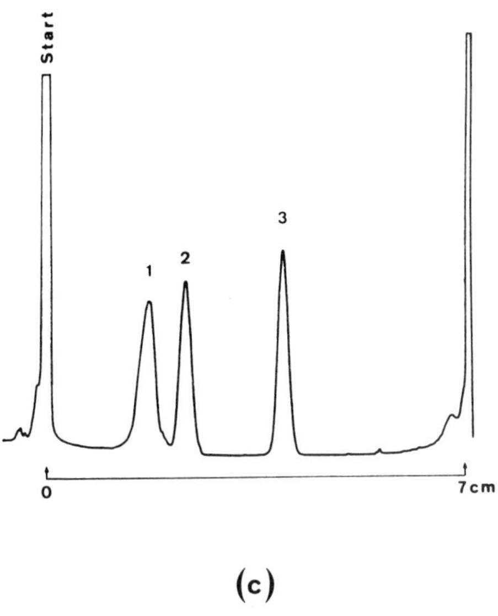


Fig. 10 (continued)

## Aluminas

Aluminas have always been fairly important in straight phase thin-layer chromatography. A number of separations are still performed on various alumina types. Owing to their chemical structure, all alumina types have surface hydroxyl groups which cause sample substances to be retained, as in the case of silica in straight phase chromatography. The mean hydroxyl group density at the surface of aluminas is approximately 13 $\mu$mol/m$^2$ [45].

In aluminas too it is the specific surface area of the material which largely determines its chromatographic properties. The specific surface areas of commercially available bulk and precoated aluminas for thin-layer chromatography range from about 50 to 350 m$^2$/g. Their specific pore volumes are between 0.1 and 0.4 ml/g. From the physical parameters listed we can calculate, as for the silicas, that the mean pore diameter ranges from about 2 to 35 nm.

Unlike the silicas, aluminas are available both as bulk sorbents and as precoated layers with variously adjusted pH values for thin-layer chromatography. "Neutral" signifies a pH of about 7.0-8.0. "Basic" signifies aluminas with a pH of about 9.0-10.0. "Acidic" aluminas are adjusted to a pH within the range of approximately 4.0-4.5.

There is another parameter apart from specific surface and pH which plays a decisive role in determining the chromatographic behavior

Table 7 Precoated Layers with Hydrophilic Modified Silicas

| Manufacturer | Product | Type of modification | Indicator | Support | Layer thickness (mm) | Application |
|---|---|---|---|---|---|---|
| E. Merck | HPTLC Plates $NH_2$ F254s | γ-Aminopropyl | F254s | Glass | 0.20 | HPTLC |
| | HPTLC Plates CN F254s | γ-Cyanopropyl | F254s | Glass | 0.20 | HPTLC |

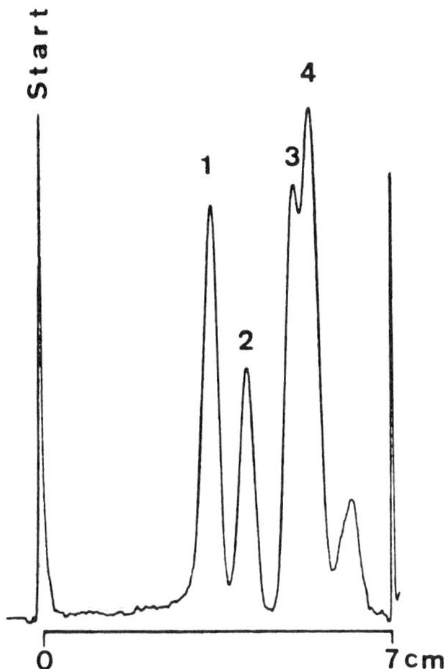


Fig. 11 Straight phase chromatographic separation of some polyvalent phenols on a precoated plate with amino modification. Plate: HPTLC precoated plate $NH_2$ F 254s (E. Merck). Compounds: 1, resorcinol; 2, hydroquinone; 3, 1-naphthol; 4, *m*-cresol. Eluent: chloroform-methanol 50:50 (v/v). Detection: Camag TLC/HPTLC scanner, UV 254 nm.

of aluminas. This is moisture content. Because this is the case, some commercially available aluminas are adjusted to defined water contents. These predefined levels are referred to as activity stages and are classified according to a method devised by Brockmann and Schodder [46].

Table 9 contains a list of bulk aluminas commercially available for preparation of home-made thin layers. As with silicas, however, the majority of aluminas used are in the form of precoated layers. Table 10 lists the alumina layers available for TLC and PLC.

Figure 13 shows the chromatogram obtained by developing various *m*-oligophenylenes with *n*-heptane, and serves as a typical separation by straight phase chromatography on a precoated plate $Al_2O_3$ type 60. Table 11 lists a number of interesting fields of applications for aluminas used as stationary phases in straight phase TLC.

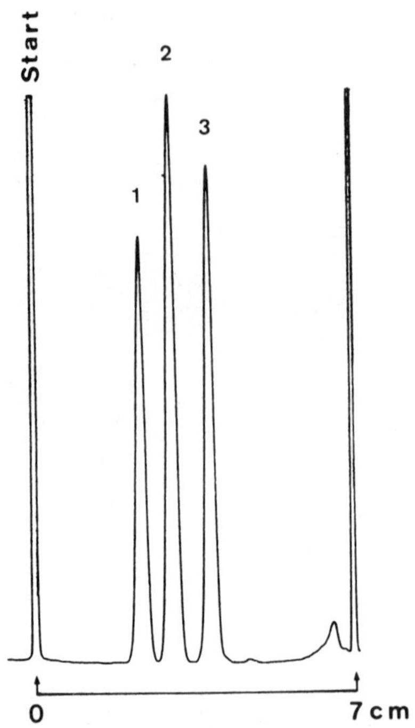


Fig. 12 Straight phase chromatographic separation of sex hormone derivatives on a precoated plate with cyano modification. Plate: HPTLC precoated plate CN F 254s (E. Merck). Compounds: 1, 4-androstene-3,17-dione; 2, progesterone; 3, pregnenolone. Eluent: petroleum benzine-ethanol 80:20 (v/v). Detection: spray reagent: 5% methanolic perchloric acid, heating up to 110°C for 5 min, Camag TLC/HPTLC scanner, UV 366 nm.

Table 8 Applications on Hydrophilic Modified Silica Layers in Straight Phase Chromatography

| Substance class | Stationary phase | Mobile phase system | Ref. |
|---|---|---|---|
| Methylated phenols | $NH_2$ | Chloroform/methanol | 39 |
| Nitroanilines | $NH_2$ | Hexane/ethanol | 40 |
| Phenols | $NH_2$ | Acetone/chloroform | 41 |
| Xanthine deriv. | $NH_2$ | Methanol | 42 |
| Estrogens | CN | Petroleum benzine/acetone | 43 |
| Phenols | CN | Ethyl acetate/toluene | 43 |
| PTH amino acids | CN | Chloroform/methanol | 44 |

Table 9 Bulk Aluminas

| Manufacturer | Product | Type of $Al_2O_3$ | Nominal pore size (nm) | pH | Particle size (μm) | Indicator | Binder | Application |
|---|---|---|---|---|---|---|---|---|
| J. T. Baker | Aluminum Oxide pH 9F | 60 | 6 | 9.0 | 5-40 | F254 | — | TLC |
| Camag | Alumina DS-0 | | | | | — | — | TLC |
| | Alumina DS-5 | | | | | — | $CaSO_4$ | TLC |
| | Alumina DSF-0 | | | | | F254 | — | TLC |
| | Alumina DSF-5 | | | | | F254 | $CaSO_4$ | TLC |
| ICN | Alumina A-DC | 60 | 6 | 4.5 | 9-22 | — | — | TLC |
| | Alumina B-DC | 60 | 6 | 9.2 | 9-22 | — | — | TLC |
| | Alumina N-DC | 60 | 6 | 7.6 | 9-22 | — | — | TLC |
| | Alumina G-DC | 60 | 6 | 7.5 | 9-22 | — | $CaSO_4$ | TLC |
| Macherey-Nagel | Alumina G | | | 7.5-8.0 | <60 | — | $CaSO_4$ | TLC |
| | Alumina G/UV 254 | | | 7.5-8.0 | <60 | F254 | $CaSO_4$ | TLC |
| | Alumina N | | | 9.0 | <60 | — | — | TLC |
| | Alumina N/UV 254 | | | 9.0 | <60 | F254 | — | TLC |
| E. Merck | Alumina 60G neutral | 60 | 6 | 7.5 | 5-40 | — | $CaSO_4$ | TLC |
| | Alumina 60GF 254 neutral | 60 | 6 | 7.5 | 5-40 | F254 | $CaSO_4$ | TLC |
| | Alumina 60H basic | 60 | 6 | 9.0 | 5-40 | — | | TLC |
| | Alumina 60HF 254 basic | 60 | 6 | 9.0 | 5-40 | F254 | | TLC |
| | Alumina 60PF 254 | 60 | 6 | 9.0 | 5-40 | F254 | — | PLC |
| | Alumina 60PF 254+366 | 60 | 6 | 9.0 | 5-40 | F254+366 | — | PLC |
| | Alumina 150 basic | 150 | 15 | 9.0 | 5-40 | — | — | TLC |
| | Alumina 150 neutral | 150 | 15 | 7.5 | 5-40 | — | — | TLC |
| | Alumina 150 acid | 150 | 15 | 4.0 | 5-40 | — | — | TLC |
| | Alumina 150 PF 254 | 150 | 15 | 9.0 | 5-40 | F254 | — | PLC |
| | Alumina 150 PF 254+366 | 150 | 15 | 9.0 | 5-40 | F254+366 | — | PLC |

Table 9 (continued)

| Manufacturer | Product | Type of $Al_2O_3$ | Nominal pore size (nm) | pH | Particle size (μm) | Indicator | Binder | Application |
|---|---|---|---|---|---|---|---|---|
| Riedel-de Haën | Alumina D basic | | | 10.0 | <30 | — | — | TLC |
| | Alumina D acid | | | 4.5 | <30 | — | — | TLC |
| | Alumina D neutral | | | 7.5 | <30 | — | — | TLC |
| | Alumina DG neutral | | | 7.5 | <30 | — | $CaSO_4$ | TLC |
| | Alumina DGF neutral | | | 7.5 | <30 | F254 | $CaSO_4$ | TLC |
| | Alumina DF basic | | | 10.0 | <30 | F254 | — | TLC |

Table 10 Alumina Precoated Layers

| Manufacturer | Product | Type of $Al_2O_3$ | Nominal pore size (nm) | pH | Indicator | Binder | Support | Layer thickness (mm) | Application |
|---|---|---|---|---|---|---|---|---|---|
| Analtech | Alumina G Uniplates | | | | — | $CaSO_4$ | Glass | 0.25 | TLC |
| | Alumina G Uniplates with F. | | | | F254 | $CaSO_4$ | Glass | 0.25 | TLC |
| | Woelm Alumina Uniplates | | | Neutral | — | — | Glass | 0.25 | TLC |
| | Woelm Alumina Uniplates | | | Neutral | F254 | — | Glass | 0.25 | TLC |
| | Woelm Alumina Uniplates | | | Basic | — | — | Glass | 0.25 | TLC |
| | Woelm Alumina Uniplates | | | Basic | F254 | — | Glass | 0.25 | TLC |
| | Woelm Alumina Uniplates | | | Acid | — | — | Glass | 0.25 | TLC |
| | Woelm Alumina Uniplates | | | Acid | F254 | — | Glass | 0.25 | TLC |
| J. T. Baker | Baker-flex Aluminum Oxide IB | | | Basic | — | | Plastic | 0.20 | TLC |
| | Baker-flex Aluminum Oxide IBF | | | Basic | F254 | | Plastic | 0.20 | TLC |
| Eastman-Kodak | Chromagram Alumina | | | Basic | — | Organ. | Plastic | 0.10 | TLC |
| | Chromagram Alumina | | | Basic | F254 | Organ. | Plastic | 0.10 | TLC |
| Macherey-Nagel | Alox 25 | 60 | 6 | 9.0 | — | Organ. | Glass | 0.25 | TLC |
| | Alox 25 UV 254 | 60 | 6 | 9.0 | F254 | Organ. | Glass | 0.25 | TLC |
| | Alox 100 UV 254 | 60 | 6 | 9.0 | F254 | Organ. | Glass | 1.0 | PLC |
| | Alox N Polygram | 60 | 6 | 9.0 | — | Organ. | Plastic | 0.20 | TLC |
| | Alox N/UV 254 Polygram | 60 | 6 | 9.0 | F254 | Organ. | Plastic | 0.20 | TLC |
| E. Merck | TLC Plates Alox 60 F 254 | 60 | 6 | Basic | F254 | Organ. | Glass | 0.25 | TLC |
| | PLC Plates Alox 60 F 254 | 60 | 6 | Basic | F254 | Organ. | Glass | 1.5 | PLC |
| | TLC Alusheets Alox 60 F 254 | 60 | 6 | Neutral | F254 | Organ. | Alumin. | 0.20 | TLC |

Table 10 (continued)

| Manufacturer | Product | Type of $Al_2O_3$ | Nominal pore size (nm) | pH | Indicator | Binder | Support | Layer thickness (mm) | Application |
|---|---|---|---|---|---|---|---|---|---|
| E. Merck | TLC Plastic Sheets Alox 60 F254 | 60 | 6 | Neutral | F254 | Organ. | Plastic | 0.20 | TLC |
| | TLC Plates Alox 150 F 254 | 150 | 15 | Basic | F254 | Organ. | Glass | 0.25 | TLC |
| | PLC Plates Alox 150 F 254 | 150 | 15 | Basic | F254 | Organ. | Glass | 1.5 | PLC |
| | TLC Alusheets 150 F 254 | 150 | 15 | Neutral | F254 | Organ. | Alumin. | 0.20 | TLC |
| Riedel-de Haën | TLC Plates AL | | | Basic | — | Organ. | Glass | 0.25 | TLC |
| | TLC Plates ALF | | | Basic | F254 | Organ. | Glass | 0.25 | TLC |
| | TLC Card ALF | | | | F254 | Organ. | Alumin. | 0.20 | TLC |

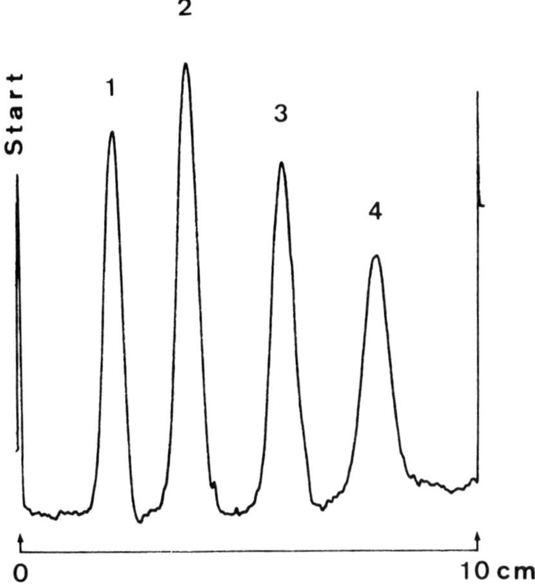


Fig. 13 Straight phase chromatographic separation of some *m*-oligophenylenes on an alumina precoated plate. Plate: TLC precoated plate $Al_2O_3$ 60 F 254 (E. Merck). Compounds: 1, *m*-quinquephenyl; 2, *m*-quaterphenyl; 3, *m*-terphenyl; 4, biphenyl. Eluent: *n*-heptane. Detection: Camag TLC/HPTLC scanner, UV 254 nm.

Table 11 Applications on Alumina Layers in Straight Phase Chromatography

| Substance class | Mobile phase system | Ref. |
|---|---|---|
| Aflatoxins | Chloroform/methanol | 47 |
| Chlorinated insecticides | Heptane | 48 |
| DNPH derivatives of carbonyl compounds | Benzene/hexane | 49 |
| Ergot alkaloids | Chloroform/methanol | 50 |
| Hydrazines | Benzene/hexane | 51 |
| Nitrosamines | Hexane/ether/methylene chloride | 52 |
| Sulfonamides | Chloroform/methanol | 53 |
| Vitamins | Benzene | 54 |

## SORBENT MATERIALS AND PRECOATED LAYERS FOR PARTITION CHROMATOGRAPHY

In partition chromatography retention of the sample molecules results primarily from the differences in partition coefficients between a liquid stationary and a mobile phase. Under ideal conditions the sorbent of the thin layer in partition chromatography acts exclusively as a support for the liquid stationary phase and does not interact with the sample molecules.

Selectivity and degree of retention in a system designed for partition chromatography are determined predominantly by the composition of the liquid stationary and mobile phases used as well as by the relative proportions of one to the other (phase ratio).

### Special Features of Partition Thin-Layer Chromatography

In thin-layer chromatography there are basically two ways of applying liquid stationary phase to the sorbent:

1. Impregnation of the sorbent layer before chromatographic development. This method involves immersing the TLC plate or sheet in a solution of stationary phase in a suitable solvent to give an even distribution of liquid stationary phase, followed by removal of the solvent by evaporation.
2. In situ application to the sorbent layer during chromatographic development. Here a continuous gradient of liquid stationary phase is set up on the sorbent in the direction of solvent flow by developing the TLC plate or sheet with a suitable multicomponent solvent system. This causes a second (composition) gradient to occur in the liquid stationary phase. This gradient represents the affinity of the solvent components for the surface of the sorbent such that the component with the highest affinity in the region nearest to the surface is the one most strongly concentrated.

Unlike adsorption chromatography, effects caused by the gas phases (see above) play only a minor role in partition thin-layer chromatography. This is because the active sorbent surfaces responsible for gas phase effects in the case of partition chromatography are covered with liquid stationary phase.

### Silicas

The bulk silica sorbents used for partition TLC are the same types as used in straight phase chromatography (see text above and Table 1).

Also, the precoated silica layers used for partition chromatography are virtually the same products as those used in straight phase chromatography (see Tables 2-4). There is a tendency, however, to use the slightly larger pore size silicas (pore size above 6 nm) with a correspondingly lower specific surface area.

From a variety of possible uses we selected the separation of non-derivatized amino acids on a TLC silica gel 100 precoated plate with an eluent suitable for partition chromatography (*n*-propanol/water/ammonia solution) (Fig. 14).

Table 12 lists solvent components and references, thus focusing the typical applications of silica layers in partition chromatography.

### Inert Silicas (Kieselguhrs, Silica 50,000)

Sorbents with a very low specific surface area are eminently suited not only for use in so-called concentrating zones but also as supports for the liquid stationary phase in partition chromatography.

As already mentioned, the hydroxyl group density for all amorphous silicas is about 8 $\mu mol/m^2$. In silicas with low specific surface areas this means that there is only a small number of hydroxyl groups available for retention interactions. Therefore, one can expect that when support materials of this type are used there will be a uniform partition retention mechanism without further types of interaction occurring.

#### *Kieselguhrs*

Various kieselguhrs have been used as inert silica supports for thin-layer chromatography. These are natural products composed of the skeleton substance of siliceous algae. According to their deposits kieselguhrs differ in chemical composition and amount of their accompanying and, in some cases, embedded impurities. Kieselguhrs contain about 90% $SiO_2$ as well as significant amounts of $Al_2O_3$, $Fe_2O_3$, $TiO_2$, $CaO$, $MgO$, $Na_2O$, $K_2O$, and carbonates.

Not all of the impurities can be removed. Some can be eliminated by fairly laborious purification but the remainder stay within the bulk phase, possibly influencing the chromatographic behavior of the layer.

Table 13 shows the ranges of physical parameters relevant to the chromatographic behavior of kieselguhrs.

#### *Silica 50,000*

To circumvent the impairments of the chromatographic behavior caused by impurities in kieselguhrs, a synthetic silica (silica 50,000) was developed. Silica 50,000 is a porous, extremely inactive silica with chromatographic properties similar to those of kieselguhr. The procedure used to manufacture silica 50,000 ensures not only high reproducibility

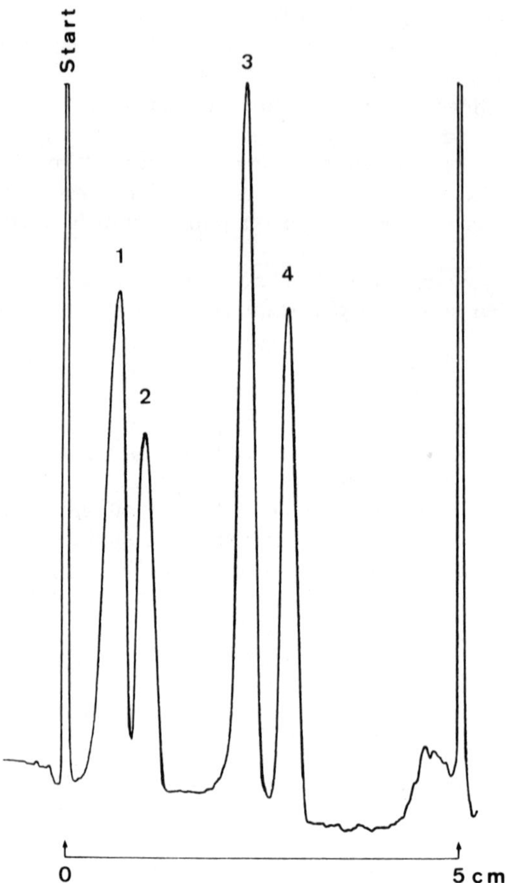


Fig. 14 Partition chromatographic separation of some amino acids on a silica gel precoated plate. Plate: TLC precoated plate silica gel 100 F 254 (E. Merck). Compounds: 1, histidine; 2, asparagine; 3, isoleucine; 4, phenylalanine. Eluent: *n*-propanol-water-10% $NH_4OH$ 40:10:1 (v/v). Detection: spray reagent: ninhydrin, heating up to 100°C for 10 min, Camag TLC/HPTLC scanner, 510 nm.

of physical parameters but also high chemical purity (100% $SiO_2$) of the material, so that chromatographic effects such as those caused by impurities in kieselguhr are eliminated. Silica 50,000 has a mean pore diameter of 5000 nm, a specific pore volume of about 0.6 ml/g, and a specific surface area of about 0.5 $m^2/g$.

As an example of a partition chromatographic separation on an inert silica-based support (in this case an HPTLC silica 50,000 pre-

Table 12 Applications on Silica Layers in Partition Chromatography

| Substance class | Mobile phase system | Ref. |
|---|---|---|
| Aflatoxins | Benzene/ethanol/water | 55 |
| Aflatoxins | Ether/methanol/water | 56 |
| Alkaloids | Methylethyl ketone/methanol/water/ammonia | 57 |
| Analgesics | Ethanol/pyridine/dioxane/water | 58 |
| Antibiotics with peptide structure | Ethanol/water | 59 |
| $B_1$-vitamins | Pyridine/glacial acetic acid/water | 60 |
| Carbohydrates | Toluene/glacial acetic acid/water | 61 |
| Lipids | Chloroform/methanol/water | 62 |
| Nitrosamines | Butanol/ammonia/water | 63 |
| Peptides | Pyridine/acetic acid/water | 64 |
| Pesticides | Methylene chloride/methanol/ammonia | 65 |
| Phenothiazine | Toluene/ethyl acetate/ammonia | 66 |
| Steroids | Chloroform/methanol/water | 67 |
| Sulfonamides | Chloroform/methanol/water | 68 |

coated plate) Fig. 15 shows the chromatogram of *p*-hydroxybenzoic acid and several of its esters.

### *Mixed Layers*

In addition to the inert supports for partition chromatography which have already been discussed, use has been made of precoated layers consisting of a mixture of kieselguhr and surface-active silica. These mixed layers can be used for both adsorption chromatography and partition thin-layer chromatography. Because of the properties of the two sorbent components the mixed layers are of medium surface activity when used in straight phase chromatography. Also, the mixed layers are somewhat faster running than the sorbent layers consisting solely of silica or kieselguhr. Tables 14 and 15 list common, inert $SiO_2$-based bulk sorbents and precoated layers. Table 16 specifies typical applications for inert supports in partition thin-layer chromatography with separate mention of the respective solvent components and references to the literature.

Table 13 Physical Parameters of Kieselguhrs for Thin-Layer Chromatography

| Specific surface area ($m^2/g$) | Specific pore volume (ml/g) | Mean pore diameter (nm) |
|---|---|---|
| 1-5 | 1-3 | 1000-10,000 |

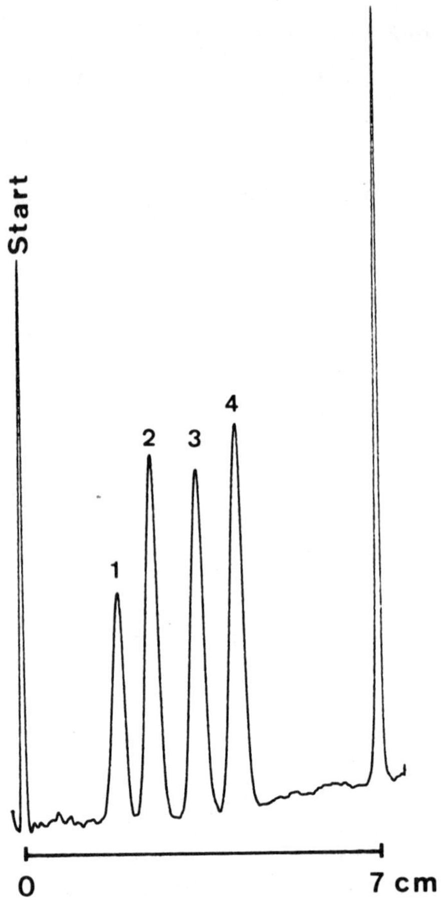


Fig. 15 Partition chromatographic separation of some preservatives on a precoated plate with an inert silica. Plate: HPTLC precoated plate Si 50000 F 254s (E. Merck). Compounds: 1, PHB (4-hydroxybenzoic acid); 2, PHB-methyl ester; 3, PHB-ethyl ester; 4, PHB-propyl ester. Eluent: *n*-pentane-glacial acetic acid 88:12 (v/v). Detection: Camag TLC/HPTLC scanner, UV 254 nm.

Table 14 Bulk Kieselguhrs

| Manufacturer | Product | Particle size (μm) | Indicator | Binder | Application |
|---|---|---|---|---|---|
| J. T. Baker | Kieselguhr | <40 | — | — | TLC |
| Macherey-Nagel | Kieselguhr G | | — | $CaSO_4$ | TLC |
| | Kieselguhr G UV 254 | | F254 | $CaSO_4$ | TLC |
| | Kieselguhr N | | — | — | TLC |
| | Kieselguhr N UV 254 | | F254 | — | TLC |
| E. Merck | Kieselguhr G | 5-40 | — | $CaSO_4$ | TLC |
| Riedel-de Haën | Kieselguhr DG | <32 | — | $CaSO_4$ | TLC |

Table 15 Precoated Layers with Inert $SiO_2$ Sorbents

| Manufacturer | Product | $SiO_2$ type | Indicator | Binder | Support | Layer thickness (mm) | Application |
|---|---|---|---|---|---|---|---|
| Macherey-Nagel | TLC Plate GUR N-25 | Kieselguhr | — | Organ. | Glass | 0.25 | TLC |
| | TLC Plate GUR N-25 UV 254 | Kieselguhr | F254 | Organ. | Glass | 0.25 | TLC |
| E. Merck | TLC Plate Kieselguhr F254 | Kieselguhr | F254 | Organ. | Glass | 0.25 | TLC |
| | TLC Alusheet K.guhr F254 | Kieselguhr | F254 | Organ. | Alumin. | 0.20 | TLC |
| | TLC Plate Si60/K.guhr F254 | Si60/K.guhr 1/1 | F254 | Organ. | Glass | 0.25 | TLC |
| | TLC Alusheet Si60/K.guhr F | Si60/K.guhr 1/1 | F254 | Organ. | Alumin. | 0.20 | TLC |
| | HPTLC Plate Si 50,000 | Silica 50,000 | — | Organ. | Glass | 0.20 | HPTLC |
| | HPTLC Plate Si 50,000 F254s | Silica 50,000 | F254s | Organ. | Glass | 0.20 | HPTLC |

Table 16 Applications on Inert $SiO_2$ Layers in Partition Chromatography

| Substance class | Inert sorbent | Mobile phase system | Ref. |
|---|---|---|---|
| Aflatoxins | Kieselguhr | Benzene/formamide/water | 69 |
| $B_{12}$ vitamins | Kieselguhr | Chloroform/methanol/water | 70 |
| Carbohydrates | Silica 50,000 | *n*-Propanol/water/ammonia | 71 |
| Herbicides | Kieselguhr | Ether/methanol/water | 72 |
| Organic acids | Si60/K.guhr | Benzene/ethanol/ammonia | 73 |
| PCBs | Kieselguhr/ paraffin | Acetonitrile/acetone/methanol/water | 74 |
| Tetracyclines | Kieselguhr | Ethyl acetate/acetone/ethylene glycol/water | 75 |

## Aluminas

In the same way as silicas, aluminas are available both as bulk sorbents (see text above and Table 9) and as precoated layers (see text above and Table 10), with the same materials being used for partition chromatography as for adsorption chromatography. Figure 16 serves as an example in which an alumina precoated layer (TLC $Al_2O_3F_{254}$, type 150 precoated plate) is applied in a separation of various water-soluble vitamins by partition chromatography. Further uses for alumina precoated layers in partition chromatography are given in Table 17.

The use of alumina layers in partition TLC is of only minor importance compared to that of surface-active and inert silica sorbents.

## Celluloses

Apart from the inorganic sorbents already listed, a number of organic sorbent materials have also proven useful in partition TLC, with celluloses at the top of the list. These are naturally occurring polysaccharides having the general chemical formula $(C_6H_{10}O_5)_x$. Before being used in TLC, these "native" celluloses must first be purified and comminuted. In many cases the native celluloses also undergo recrystallization. This process yields the "microcrystalline" celluloses, which can also be used in thin-layer chromatography and offer a number of advantages with respect to running times and separating performance. Unlike native cellulose, which has a fibrous structure, microcrystalline cellulose is in the form of rods. Since the celluloses only have specific surface areas of up to approximately 2 $m^2/g$, they are most suitably used in partition chromatography.

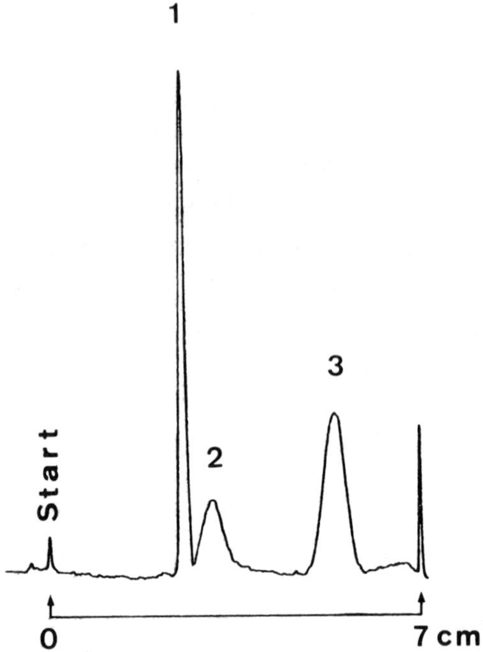


Fig. 16 Partition chromatographic separation of some water-soluble vitamins on an alumina precoated plate. Plate: TLC precoated plate $Al_2O_3$ 150 F 254 (E. Merck). Compounds: 1, vitamin $B_1$; 2, vitamin $B_{12}$; 3, nicotinic acid amide. Eluent: chloroform-methanol 50:50 (v/v) 70% saturated with 1 N aqueous acetic acid. Detection: Camag TLC/HPTLC scanner, UV 254 nm.

Table 17 Applications on Alumina Layers in Partition Chromatography

| Substance class | Mobile phase system | Ref. |
|---|---|---|
| $B_{12}$ vitamins | Chloroform/methanol/water | 76 |
| Chlorinated catechols | Benzene/acetone/acetic acid | 77 |
| Diterpenes | Chloroform/*i*-propanol/ether/water | 78 |
| Rotenones | Benzene/ethanol/water | 79 |
| Water-soluble vitamins | Glacial acetic acid/acetone/methanol/benzene | 80 |

The manufacturers, types, and properties of celluloses available for hand-preparing thin layers are compiled in Table 18.

Precoated cellulose layers are also widely used for thin-layer chromatography on account of the superior layer qualities and higher reproducibility they afford. Table 19 provides a summary of the precoated cellulose layers available commercially. Cellulose layers are used predominantly for the partition separation of relatively polar substances. For example, Fig. 17 shows the chromatogram of various inorganic phosphates on an HPTLC cellulose precoated plate. Table 20 lists typical applications for cellulose layers.

### Polyamides

Apart from the celluloses, a number of polyamide-type sorbents are used in partition thin-layer chromatography. For this purpose a polycaprolactam (PA 6) and a polyundecanamide (PA 11) are used. Polar substances in particular, whose structure permits them to interact with the amide groups of the polyamides by hydrogen bonding, can be separated by TLC on polyamide layers.

Manufacturers of bulk polyamides and precoated polyamide layers for TLC together with the relevant parameters are listed in Tables 21 and 22. Figure 18 shows the chromatogram of a few dansyl amino acids, a typical separation on polyamide layers. Table 23 lists various classes of substance which can be separated either on precoated or on home-made polyamide layers.

## PRECOATED LAYERS FOR REVERSED PHASE TLC

The differences in the selectivities and chromatographic properties of the TLC stationary phases previously discussed are attributable to differences in their respective overall chemical structure or to the adsorption of suitable liquids on surfaces. A further way of deliberately modifying the selectivity and retention characteristics of sorbents is to perform suitable chemical modifications of the surface. Depending on the type of modification undertaken, a distinction is made between predominantly hydrophobic and predominantly hydrophilic surface properties.

Except in the rather special case of modifications performed for ion exchange chromatography, all of the surface modifications for thin-layer chromatography up to now have been based on a silica matrix.

### Hydrophobic Modified Silica Precoated Layers

In thin-layer chromatography and column chromatography alike, reversed phase materials, i.e., supports with hydrophobic ligands in

Table 18 Bulk Celluloses

| Manufacturer | Product | Type of cellulose | Particle size (μm) | Indicator | Application |
|---|---|---|---|---|---|
| Analtech | Cellulose Avicel | Microcrystalline | | — | TLC |
| | Cellulose Avicel | Microcrystalline | | F254 | TLC |
| Camag | Cellulose DS-0 | Microcrystalline | | — | TLC |
| | Cellulose DSF-0 | Microcrystalline | | F254 | TLC |
| | Cellulose D-0 | Native | | — | TLC |
| | Cellulose DF-0 | Native | | F254 | TLC |
| Macherey-Nagel | Cellulose MN 300 | Native | 2-20 | — | TLC |
| | Cell. MN 300 UV 254 | Native | 2-20 | F254 | TLC |
| | Cellulose MN 300 HR[a] | Native | 2-20 | — | TLC |
| | Avicel for TLC | Microcrystalline | | — | TLC |
| E. Merck | Cellulose | Native | <20 | — | TLC |
| | Cellulose Avicel | Microcrystalline | <20 | — | TLC |
| Riedel-de Haën | Cellulose D | Native | 2-20 | — | TLC |
| | Cellulose DF | Native | 2-20 | F254 | TLC |
| | Cell. D extra pure[a] | Native | 2-20 | — | TLC |
| | Cellulose D Avicel | Microcrystalline | | — | TLC |
| Schleicher + Schüll | Cellulose 180 | Native | 2-25 | — | TLC |
| | Cellulose 180-LS 254 | Native | 2-25 | F254 | TLC |
| | Cellulose 180a[a] | Native | 2-25 | — | TLC |
| | Cellulose 142 dg[a] | | 10-50 | — | TLC |
| | Cellulose 144 | Microcrystalline | 20 | — | TLC |
| | Cellulose 144-LS254 | Microcrystalline | 20 | F254 | TLC |
| Whatman | TLC Cellulose CC 41 | Microgranular | | — | TLC |

[a]Purified.

Table 19 Cellulose Precoated Layers

| Manufacturer | Product | Type of cellulose | Indicator | Support | Layer thickness (mm) | Application |
|---|---|---|---|---|---|---|
| Analtech | Avicel Cellulose Uniplates | Microcrystalline | — | Glass | 0.25 | TLC |
| | Avicel Cellulose Uniplates | Microcrystalline | F254 | Glass | 0.25 | TLC |
| J. T. Baker | Baker-flex Cellulose | Native | — | Plastic | | TLC |
| | Baker-flex Cellulose F | Native | F254 | Plastic | | TLC |
| Macherey-Nagel | CEL 300-10 | Native | — | Glass | 0.10 | TLC |
| | CEL 300-10 UV 254 | Native | F254 | Glass | 0.10 | TLC |
| | CEL 300-25 | Native | — | Glass | 0.25 | TLC |
| | CEL 300-25 UV 254 | Native | F254 | Glass | 0.25 | TLC |
| | CEL 300-50 | Native | — | Glass | 0.50 | PLC |
| | CEL 300-50 UV 254 | Native | F254 | Glass | 0.50 | PLC |
| | CEL 400-10 | Microcrystalline | — | Glass | 0.10 | TLC |
| | CEL 400-10 UV 254 | Microcrystalline | F254 | Glass | 0.10 | TLC |
| | Polygram CEL 300 | Native | — | Plastic | 0.10 | TLC |
| | Polygram CEL 300 UV 254 | Native | F254 | Plastic | 0.10 | TLC |
| | Polygram CEL 400 | Microcrystalline | — | Plastic | 0.10 | TLC |
| | Polygram CEL 400 UV 254 | Microcrystalline | F254 | Plastic | 0.10 | TLC |
| E. Merck | TLC Plate Cellulose | Microcrystalline | — | Glass | 0.10 | TLC |
| | TLC Plate Cellulose F | Microcrystalline | F254+366 | Glass | 0.10 | TLC |
| | HPTLC Plate Cellulose | Microcrystalline | — | Glass | 0.10 | HPTLC |
| | HPTLC Plate Cellulose F254s | Microcrystalline | F254s | Glass | 0.10 | HPTLC |
| | PLC Plate Cellulose | Microcrystalline | — | Glass | 0.50 | PLC |
| | TLC Alusheet Cellulose | Microcrystalline | — | Alumin. | 0.10 | TLC |

Table 19 (continued)

| Manufacturer | Produce | Type of cellulose | Indicator | Support | Layer thickness (mm) | Application |
|---|---|---|---|---|---|---|
| | TLC Alusheet Cellulose F | Microcrystalline | F254+366 | Alumin. | 0.10 | TLC |
| | TLC Plastic Sheet Cellulose | Microcrystalline | — | Plastic | 0.10 | TLC |
| | TLC Plastic Sheet Cellulose F | Microcrystalline | F254+366 | Plastic | 0.10 | TLC |
| | HPTLC Alusheet Cellulose | Microcrystalline | — | Alumin. | 0.10 | HPTLC |
| Riedel-de Haën | TLC Plate CE | | — | Glass | 0.10 | TLC |
| | TLC Plate CE F | | F254 | Glass | 0.10 | TLC |
| | TLC Card CE | | — | Alumin. | 0.10 | TLC |
| | TLC Card CE F | | F254 | Alumin. | 0.10 | TLC |
| Schleicher + Schüll | G 1440 | Microcrystalline | — | Glass | 0.10 | TLC |
| | G 1440-LS 254 | Microcrystalline | F254 | Glass | 0.10 | TLC |
| Whatman | K2 | Microcrystalline | — | Glass | 0.25 | TLC |
| | K2 F | Microcrystalline | F254 | Glass | 0.25 | TLC |
| | PK2 F | Microcrystalline | F254 | Glass | 0.50 | PLC |
| | PK2 F | Microcrystalline | F254 | Glass | 1.00 | PLC |
| | LK2[a] | Microcrystalline | — | Glass | 0.25 | TLC |
| | LK 2 F[a] | Microcrystalline | F254 | Glass | 0.25 | TLC |

[a]With preadsorbent area.

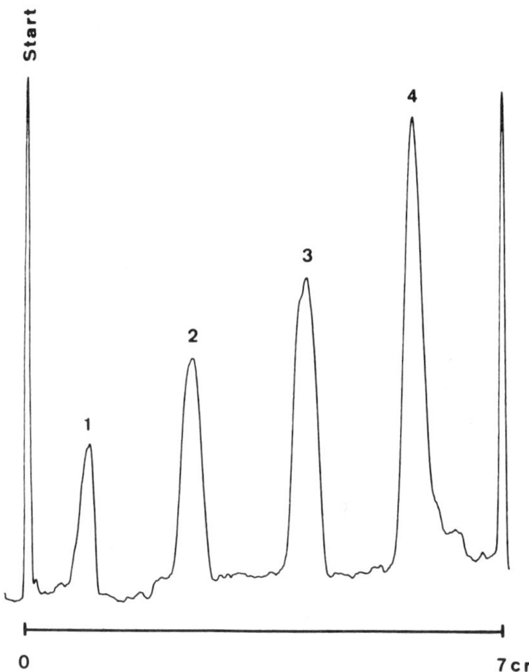


Fig. 17 Separation of some phosphates on a cellulose precoated plate. Plate: HPTLC precoated plate cellulose (E. Merck). Compounds: 1, $(NaPO_3)_3$; 2, $Na_5P_3O_{10}$; 3, $Na_4P_2O_7$; 4, $Na_2HPO_4$. Eluent: dioxane-solution of 160 g trichloroacetic acid and 8 ml ammonia 25% liter water 70:30 (v/v). Detection: spray reagents (2 × a) then (1 × b): (a)

(a) $Na_2MoO_4 \cdot 2H_2O$ 18 g, $NH_4NO_3$ 22.5 g } in 460 ml water
plus 40 ml conc. $HNO_3$.

(b) $Na_2S_2O_5$ 150 g, $Na_2SO_3$ 5 g } in 500 ml water

Camag TLC/HPTLC scanner, 586 nm.

the form of alkyl groups, are the most important. Surface modification in this case is such that all or, alternatively, only a portion of the available hydroxyl groups on the silica are reacted with suitable alkyl- or arylsilanes. Eliminating these hydroxyl groups causes new siloxane groups to be formed with direct silicon-carbon bonds. The relationship between the available reacted and available nonreacted hydroxyl groups is a measure of the particular degree of modification. The chain length of alkyl ligands commonly used in thin-layer chromatography ranges from 1 to 18 carbon atoms.

**Table 20** Applications on Cellulose Layers in Partition Chromatography

| Substance class | Mobile phase system | Ref. |
|---|---|---|
| Amino acids | Butanol/glacial acetic acid/water | 81 |
| Amino acids | Pyridine/water | 82 |
| Amino acids | *i*-Propanol/water | 83 |
| Carbohydrates | Formic acid/butanone/*t*-butanol/water | 84 |
| Chlorophylles | Heptane/pyridine | 85 |
| Oligosaccharides | Ethyl acetate/pyridine/water/glacial acetic acid/propionic acid | 86 |
| Peptides | Butanol/acetic acid/water | 87 |
| Synthetic sweeteners | Pyridine/ethanol/water | 88 |
| Tetracyclines | Impr. w. $Na_2HPO_4$, citric acid, and ethylene glycol; ethyl acetate saturated w. water | 89 |
| Thiols | *n*-Butanol/acetic acid/water | 90 |

**Table 21** Bulk Polyamides

| Manufacturer | Product | Type of PA | Particle size (μm) | Indicator | Application |
|---|---|---|---|---|---|
| J. T. Baker | Polyamide 6 | PA 6 | 5-30 | — | TLC |
| Macherey-Nagel | Polyamide-TLC 6 | PA 6 | | — | TLC |
| | Polyamide-TLC 6 UV 254 | PA 6 | | F254 | TLC |
| | Polyamide-TLC 6 AC[a] | PA 6 | | — | TLC |
| Riedel-de Haën | Polyamide 6 D | PA 6 | <40 | — | TLC |
| | Polyamide 6 DF | PA 6 | <40 | F254 | TLC |

[a]Acetylated.

Table 22 Polyamide Precoated Layers

| Manufacturer | Product | Type of PA | Indicator | Binder | Support | Layer thickness (mm) | Application |
|---|---|---|---|---|---|---|---|
| J. T. Baker | Baker-flex polyamide 6 | PA 6 | — | | Plastic | | TLC |
| | Baker-flex Polyamide 6 F | PA 6 | F254 | | Plastic | | TLC |
| Macherey-Nagel | Polygram Polyamide 6 | PA 6 | — | | Plastic | 0.10 | TLC |
| | Polygram Polyamide 6 UV 254 | PA 6 | F254 | | Plastic | 0.10 | TLC |
| | Polygram Polyamide Wang[a] | PA 6 | — | | Plastic | 2 × 0.05 | TLC |
| E. Merck | TLC Plate Polyamide 11 F254 | PA 11 | F254 | | Glass | 0.15 | TLC |
| | TLC Alusheet Polyamide 11 F254 | PA 11 | F254 | | Alumin. | 0.15 | TLC |
| Schleicher + Schüll | Plate G 1600 Polyamide | | — | Starch | Glass | 0.12 | TLC |
| | Plate G 1600 LS 254 Polyamide | | F254 | Starch | Glass | 0.12 | TLC |
| | Sheet F 1700 Micropolyamide[a] | | — | Without | Plastic | 2 × 0.025 | TLC |
| | Sheet A 1700 Micropolyamide[a] | | — | Without | Alumin. | 2 × 0.025 | TLC |

[a]Coated on both sides.

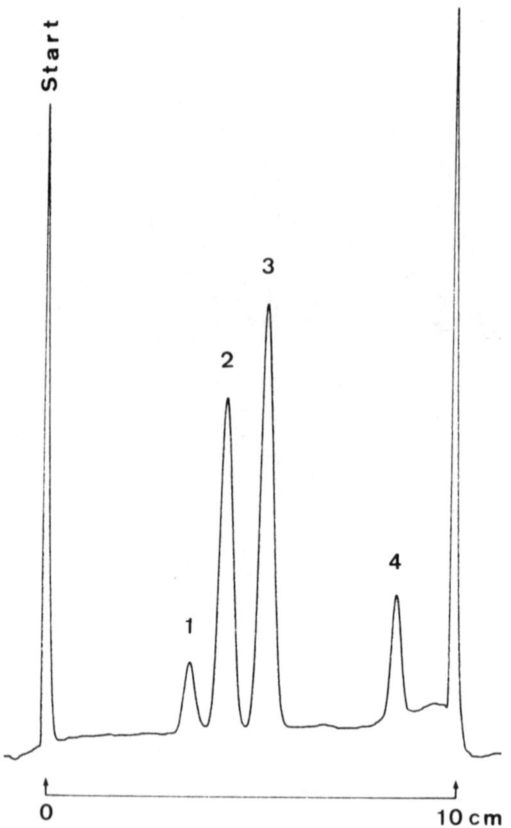


Fig. 18 Separation of some dansyl amino acids on a polyamide precoated plate. Plate: TLC precoated plate polyamide 11 F 254 (E. Merck). Compounds: 1, dansyl aspartic acid; 2, dansyl glycine; 3, dansyl alanine; 4, dansyl arginine. Eluent: methanol-acetic acid-water 8:1:1 (v/v). Detection: Camag TLC/HPTLC scanner, UV 366 nm.

The hydrophobic character of a reversed phase silica is increased both as the alkyl chain becomes longer and as the degree of modification becomes higher.

Unlike liquid column chromatography, certain limitations are imposed on the stationary phase in thin-layer chromatography as the hydrophobic character becomes more pronounced. In TLC solvent flow is induced exclusively by capillary forces. When highly aqueous solvents such as those commonly used in reversed phase chromatography are applied, high-carbon-loaded reversed phase sorbents give rise to hydrophobic forces tending to repel the solvent. Consequently, the stationary

Table 23 Applications on Polyamide Layers

| Substance class | Mobile phase system | Ref. |
|---|---|---|
| Amino acids | Benzene/glacial acetic acid | 91 |
| Chinones | Methanol/water | 92 |
| Corticosteroids | Methylene chloride/methanol | 93 |
| DNS Amino acids | Benzene/acetic acid | 94 |
| Flavones | Cyclohexane/ethanol | 95 |
| Nitrophenols | Cyclohexane/glacial acetic acid | 96 |
| Phenols | Cyclohexane/methylene chloride | 97 |
| Preservatives | Acetone/water | 98 |
| PTH amino acids | Toluene/heptane/acetic acid | 99 |
| Sex hormones | Hexane/acetone | 100 |
| Sweeteners | Formic acid/benzene | 101 |

phase is no longer wetted by the mobile phase, thus rendering chromatographic development impossible. There are two ways of overcoming this limitation on the potential applications of reversed phase TLC:

1. Reduction or restriction of the water content in the mobile phase when used with high-carbon-loaded reversed phase precoated layers.
2. Defined lower degree of modification than the maximum possible, to maintain a certain residual hydrophilic character of the sorbent; hence no limitation on the water content of the eluent.

Apart from differentiations according to modification rests and degrees of modification, a distinction is also made in reversed phase thin-layer chromatography between layers for HPTLC, TLC, and PLC.

At the moment there are only few bulk reversed phase materials available for the user to prepare home-made thin layers, but a large number of manufacturers offer reversed phase precoated plates (see Table 24). Based on the hydrophobic interactions in reversed phase chromatography, particularly substances with nonpolar groups of molecules can be separated. A typical separation of such substances is shown in Fig. 19, the chromatogram of cholesterol and several of its bile acid metabolites on an HPTLC RP-2 precoated plate. Adding acetic acid to the solvent in this case leads to a reduction in the dissociation of carboxyl groups in the free bile acids.

Table 24 Precoated Plates for Reversed Phase Chromatography

| Manufacturer | Product | Type of modification | Indicator | Layer thickness (mm) | Appl. | Annotation |
|---|---|---|---|---|---|---|
| Analtech | Reversed phase Uniplates | C-18 | — | | TLC | |
| | Reversed phase Uniplates | C-18 | F254 | | TLC | |
| Antec | Opti-UP C 12 | C-12 | F254 | 0.30 | TLC | Wettable by water |
| J. T. Baker | Si-C 18 | C-18 | — | 0.20 | TLC | |
| | Si-C 18 F | C-18 | F254 | 0.20 | TLC | |
| Macherey-Nagel | Nano-Sil C 18-100 | C-18 | — | 0.20 | HPTLC | 100% modification |
| | Nano-Sil C 18-100 UV 254 | C-18 | F254 | 0.20 | HPTLC | 100% modification |
| | Nano-Sil C 18-50 | C-18 | — | 0.20 | HPTLC | 50% modification |
| | Nano-Sil C 18-50 UV 254 | C-18 | F254 | 0.20 | HPTLC | 50% modification |
| E. Merck | TLC Silica Gel 60 silanized | C-2 | — | 0.25 | TLC | 100% modification |
| | TLC Silica Gel 60 F254 silanized | C-2 | F254 | 0.25 | TLC | 100% modification |
| | TLC RP-8 F 254s | C-8 | F254s | 0.25 | TLC | Wettable by water |
| | TLC RP-18 F 254s | C-18 | F254s | 0.25 | TLC | Wettable by water |
| | HPTLC RP-2 F 254s | C-2 | F254s | 0.20 | HPTLC | 100% modification |
| | HPTLC RP-8 F 254s | C-8 | F254s | 0.20 | HPTLC | 100% modification |
| | HPTLC RP-18 F 254s | C-18 | F254s | 0.20 | HPTLC | 100% modification |
| | HPTLC RP-18 w. concentrating zone | C-18 | — | 0.20 | HPTLC | 100% modification |
| | HPTLC RP-18 F254s w. conc. zone | C-18 | F254s | 0.20 | HPTLC | 100% modification |
| | PLC RP-18 F 254s | C-18 | F254s | 2.00 | PLC | Wettable by water |

| | | | | | | |
|---|---|---|---|---|---|---|
| Whatman | KC 2 | C-2 | — | 0.20 | TLC | |
| | KC 2F | C-2 | F254 | 0.20 | TLC | |
| | KC 8 | C-8 | — | 0.20 | TLC | |
| | KC 8F | C-8 | F254 | 0.20 | TLC | |
| | KC 18 | C-18 | — | 0.20 | TLC | |
| | KC 18F | C-18 | F254 | 0.20 | TLC | |
| | LKC 18 | C-18 | — | 0.20 | TLC | w. preads. area |
| | LKC 18F | C-18 | F254 | 0.20 | TLC | w. preads. area |
| | CS 5[a] | C-18 | — | 0.25 | TLC | |
| | SC 5[b] | C-18 | — | 0.25 | TLC | |
| | PLKC 18F | C-18 | F254 | 1.00 | PLC | w. preads. area |
| | Diphenyl | Diphenyl | — | 0.25 | TLC | |
| | Diphenyl | Diphenyl | F254 | 0.25 | TLC | |

[a] 3 cm KC 18 and 17 cm Silica K5.
[b] 17 cm KC 18 and 3 cm Silica K5.

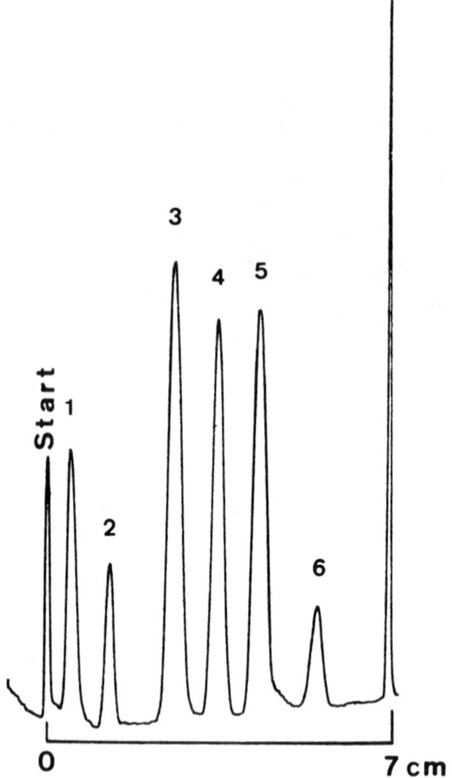


Fig. 19 Reversed phase chromatographic separation of cholesterol and its bile acid metabolites on a RP-2 precoated plate. Plate: HPTLC precoated plate RP-2 F 254s (E. Merck). Compounds: 1, cholesterol; 2, 7-hydroxycholesterol; 3, lithocholic acid; 4, cholic acid methyl ester; 5, cholic acid; 6, dehydrocholic acid. Eluent: methanol-1 N acetic acid 80:20 (v/v). Detection: spray reagent: $MnCl_2$-sulfuric acid, heating up to 120°C for 5 min, Camag TLC/HPTLC scanner, UV 366 nm.

Polar, charged substances can be converted to nonpolar compounds by adding suitable ion pair reagents to the solvent. In this form they are able to interact with the hydrophobic reversed phase support. Figure 20 demonstrates the use of ion pair reagent for separating substituted benzoic acids with the addition of tetrapropylammonium bromide on TLC RP-8 precoated plates. Further applications for reversed phase precoated plates are compiled in Table 25.

As in the case of the silica precoated plates, reversed phase precoated plates are also available with a concentrating zone (see also

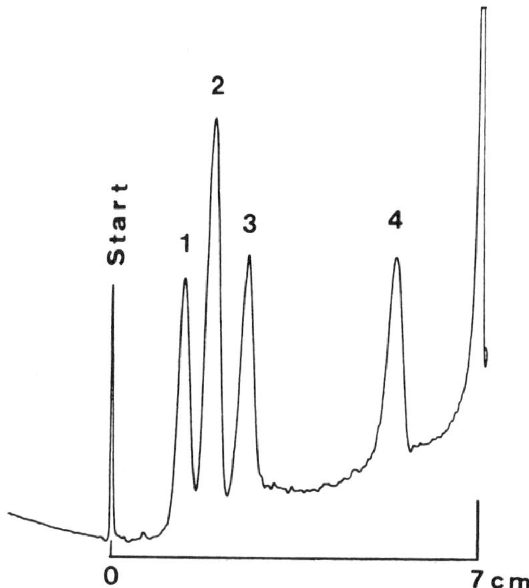


Fig. 20 Ion pair chromatographic separation of some benzoic acids on RP-8 precoated plate. Plate: TLC precoated plate RP-8 F 254s (E. Merck). Compounds: 1, 2-methyl benzoic acid; 2, benzoic acid; 3, acetylsalicylic acid; 4, phthalic acid. Eluent: acetone-water 40:60 (v/v) with addition of 0.005 mol/liter tetrapropylammonium bromide. Detection: Camag TLC/HPTLC scanner, UV 254 nm.

Table 24). These layers consist of a hydrophobic modified separating layer and an inert region, the so-called concentrating zone. The way in which these reversed phase precoated plates with concentrating zone work is similar to that of the silica precoated layers with concentrating zone (see Fig. 7).

Figure 21 gives an example of a separation performed on an HPTLC RP-18 $F_{254}$s precoated plate with concentrating zone. The figure shows two chromatograms of 10 polycyclic aromatic hydrocarbons. It is interesting to note that the separation was carried out at a temperature of -18°C and that the same chromatogram was evaluated at two different wavelengths.

## Hydrophilic Modified Silica Precoated Layers

The two hydrophilic modified silica precoated layers introduced up to now in thin-layer chromatography (amino and cyano modifications, see text above and Table 7) have terminal polar functional groups as

Table 25 Applications on Reversed Phase Precoated Layers

| Substance class | Mobile phase system | Ref. |
|---|---|---|
| Alkaloids | Methanol/water/acetic acid/heptane-sulfonic acid | 102 |
| Amines | Water/methanol/acetic acid | 103 |
| Barbiturates | Acetonitrile/water | 104 |
| Black inks | Phosphate buffer/ethanol | 105 |
| Cannabinoides | Acetonitrile/water | 106 |
| Carboxylic and sulfonic acids | Methanol/water/tetraalkylammonium salt | 107 |
| Catecholamines | Chloroform/propanol/formic acid | 108 |
| DNS amino acids | Methanol/acetic acid | 109 |
| Nitrogen bases | Acetone/water/alkylsulfonates | 110 |
| Polycyclic aromatic hydrocarbons | Acetonitrile/methylene chloride/water | 111 |
| Rare earths | Diisopropylether/THF/nitric acid | 112 |
| Steroids | Ethanol/water | 113 |
| Sulfonamides | Methanol/water | 114 |

well as a hydrophobic spacer group in the form of a short-chain alkyl group. The presence of this hydrophobic spacer group enables reversed phase chromatography to be conducted even with a hydrophilic modified silica layer.

The use of an HPTLC $NH_2F_{254}s$ precoated plate for reversed phase chromatography is illustrated in Fig. 22, which shows the separation of various nucleoside monophosphates. All of the sample substances used here are negatively charged, but they have the same charge number. Therefore, in this case separation is based solely on different hydrophobic interactions between the sample molecules and the stationary phase. Figure 23 shows a typical RP separating mechanism on an HPTLC CN $F_{254}s$ precoated plate. Here a number of narcotic substances have been separated by adding tetraethylammonium bromide, showing that reversed phase ion pair chromatography can also be successfully applied using a cyano-modified layer.

## SORBENTS AND PRECOATED LAYERS FOR ION EXCHANGE CHROMATOGRAPHY

In thin-layer chromatography supports for ion exchange chromatography play only a minor role compared to silica-based sorbents for straight or reversed phase chromatography.

### Amino-Modified Silica Precoated Layers

The only support presently available for ion exchange thin-layer chromatography based on a modified silica is the HPTLC $NH_2$ $F_{254}s$ precoated plate. This stationary phase has already been described in the sections dealing with straight phase chromatography and reversed phase chromatography, but first of all according to its type of modification the HPTLC $NH_2$ $F_{254}s$ precoated plate is able to act as a weak-base ion exchanger [41]. Figure 24 shows the separation of several oligonucleotides on an HPTLC $NH_2$ $F_{254}s$ precoated plate. In the aqueous solvent system used here, separation of the oligoadenylic acids occurs with the aid of an ion exchange mechanism according to their differing formal charge number.

### Modified Celluloses

A number of chemically modified or impregnated native or microcrystalline celluloses have been developed to serve as ion exchange sorbents or precoated layers in thin-layer chromatography. Surface modifications are of the following types:

AE (aminoethyl)
CM (carboxymethyl)
DEAE (diethylaminoethyl)
ECTEOLA (reaction product of epichlorohydrin, triethanolamine, and alkali cellulose)
P (phosphorylated)

The impregnated celluloses serving as ion exchangers are:

PEI (polyethyleneimine)
Poly-P (polyphosphate)

The AE, DEAE, ECTEOLA, and PEI celluloses are basic (anion) exchangers and the CM, P, and Poly-P celluloses are acidic (cation) exchangers. Figure 25 shows the separation of uridine nucleotides, this being based on an ion exchange mechanism. The stationary phase used here was a TLC PEI cellulose precoated plate and the mobile phase a mixture of acetic acid and ethanol.

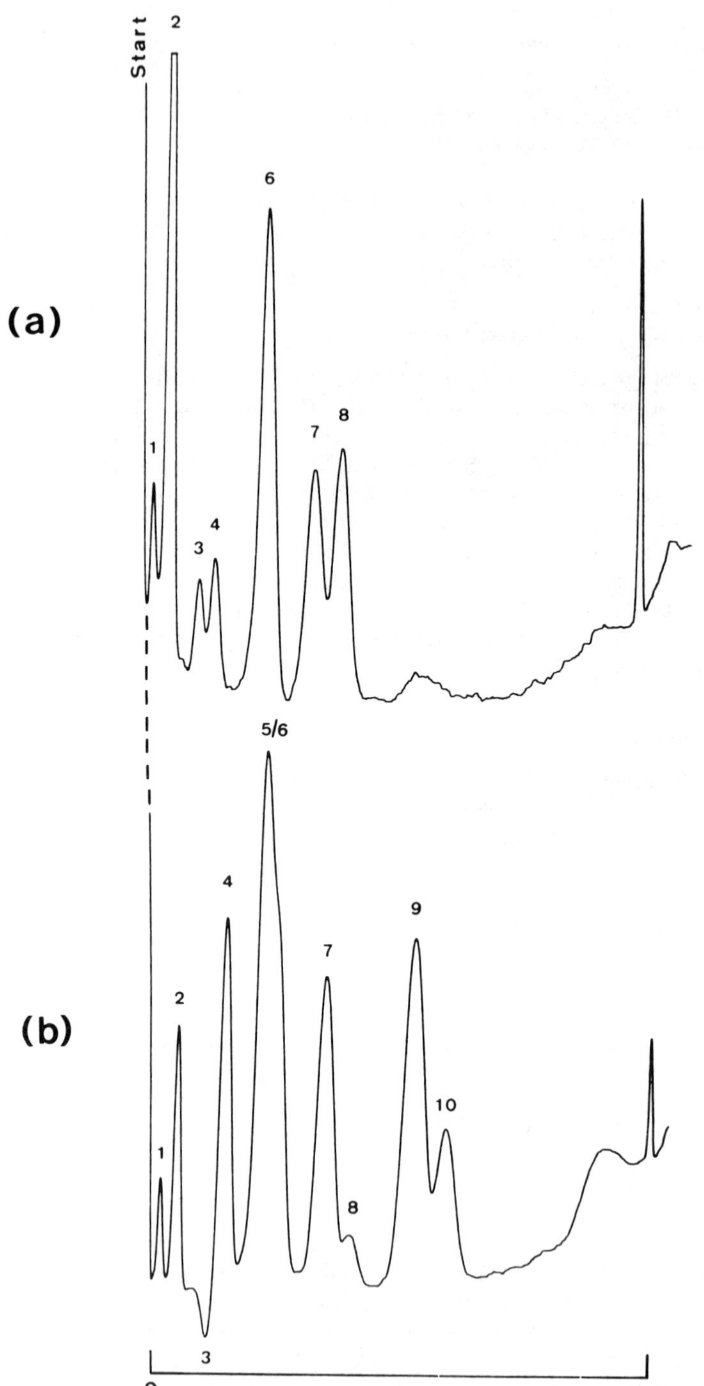
Start
2
6
(a)
1
7
8
3
4
5/6
4
9
7
(b)
2
10
1
8
3
0
6.5 cm

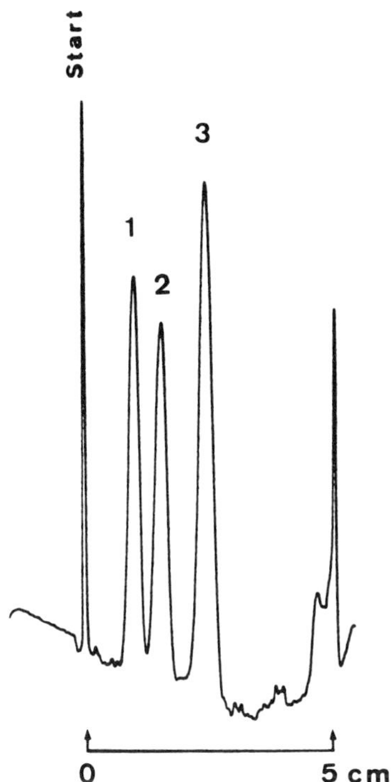


Fig. 22 Reversed phase chromatographic separation of some nucleoside monophosphates on a $NH_2$ precoated plate. Plate: HPTLC precoated plate $NH_2$ $F_{254}s$ (E. Merck). Compounds: 1, guanosine monophosphate (GMP); 2, uridine monophosphate (UMP); 3, adenosine monophosphate (AMP). Eluent: ethanol-water 30:70 (v/v) with addition of 0.18 mol/liter NaCl. Detection: Camag TLC/HPTLC scanner, UV 254 nm.

---

Fig. 21 Reversed phase chromatographic separation of some polycyclic aromatic hydrocarbons on a RP-18 precoated plate with concentrating zone. Plate: HPTLC precoated plate RP-18 F 254s with concentrating zone (E. Merck). Compounds: 1, coronene; 2, indeno(1,2,3-c,d)pyrene; 3, benzo(k)fluoranthene; 4, benzo(e)pyrene; 5, chrysene; 6, benzo(a)anthracene; 7, pyrene; 8, fluoranthene; 9, phenanthrene; 10, fluorene. Eluent: acetonitrile-methanol 50:50 (v/v); temperature of development: -18°C. Detection: Camag TLC/HPTLC scanner (a) UV 366 nm; (b) UV 254 nm.

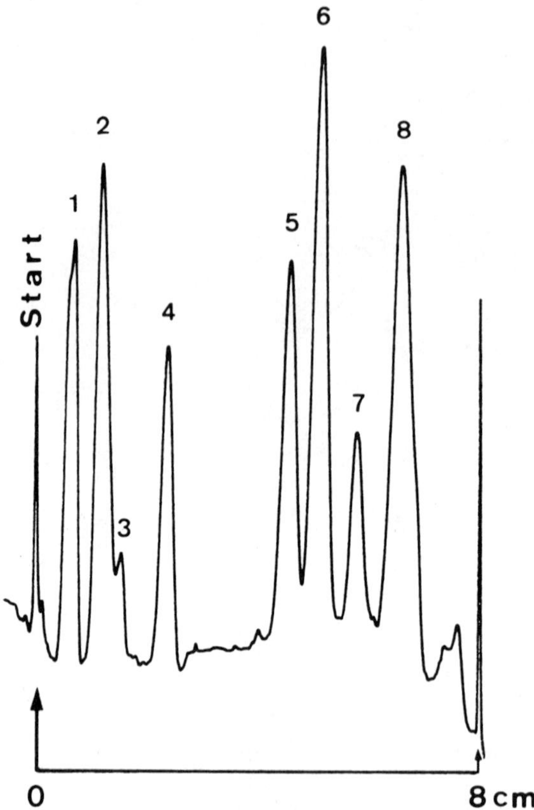


Fig. 23 Reversed phase chromatographic separation of some drugs on a CN precoated plate. Plate: HPTLC precoated plate CN F 254s (E. Merck). Compounds: 1, thiogenal; 2, methaqualone; 3, diphenhydramine hydrochloride; 4, luminal; 5, ethylmorphine hydrochloride; 6, codeine; 7, morphine hydrochloride; 8, barbituric acid. Eluent: methanol-water 40:60 (v/v) with addition of 0.1 mol/liter tetraethylammonium bromide. Detection: Camag TLC/HPTLC scanner, UV 254 nm.

### Polymer-Based Ion Exchangers

In addition to the silica- and cellulose-based ion exchangers mentioned so far as being suitable for thin-layer chromatography, polymer resins are also employed as supports for ion exchange chromatography in TLC. The polymer resins are frequently used in the form of mixtures with silica to afford better layer stability. Tables 26 and 27 list all

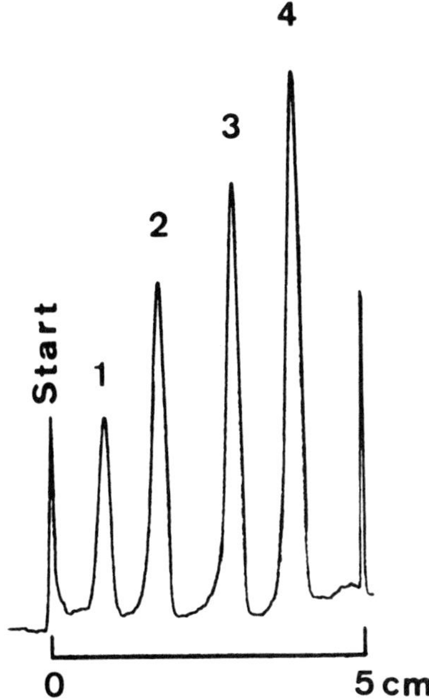


Fig. 24 Ion exchange chromatographic separation of some oligoadenylic acids on a $NH_2$ precoated plate. Plate: HPTLC precoated plate $NH_2$ F 254s (E. Merck). Compounds: 1, A2'p5'A2'p5'A2'p5'A; 2, A2'p5'-A2'p5'A; 3, A2'p5'A; 4, adenosine. Eluent: methanol-water 90:10 (v/v) with addition of 0.2 mol/liter lithium chloride. Detection: Camag TLC/HPTLC scanner, UV 254 nm.

bulk sorbents and precoated layers for the separation by ion exchange discussed earlier. Table 28 gives an insight into the possible uses of ion exchangers in thin-layer chromatography.

## THIN LAYERS FOR SPECIAL APPLICATIONS

All of the stationary phases for thin-layer chromatography discussed so far have a more or less broad range of applications. There are, however, a number of other sorbents and precoated layers offering a very narrow spectrum of applications at the moment. These are acetylated cellulose in pure form and as a mixture with other sorbents, a specially impregnated RP precoated layer for separating racemates, and florisil (magnesium silicate). Table 29 lists the manufacturers of the respective bulk sorbents and Table 30 states the manufacturers

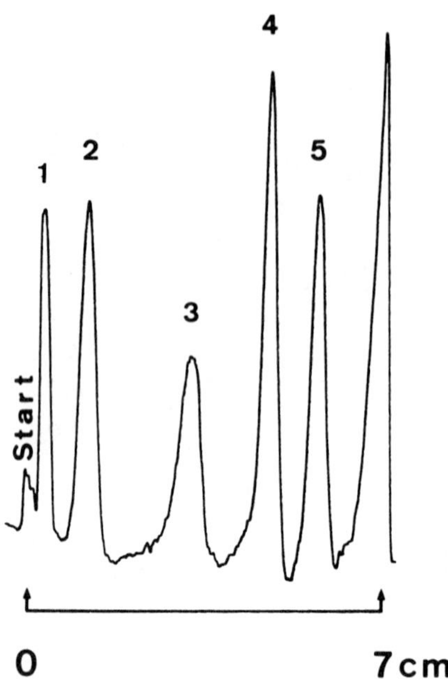


Fig. 25 Ion exchange chromatographic separation of some uridine nucleotides on a PEI cellulose precoated plate. Plate: TLC precoated plate PEI cellulose F (E. Merck). Compounds: 1, UTP; 2, UDP; 3, UDP-glucose; 4, UMP; 5, uridine. Eluent: 1 N acetic acid-ethanol 80:20 (v/v) with addition of 0.5 mol/liter lithium chloride. Detection: Camag TLC/HPTLC scanner, UV 270 nm.

Table 26 Bulk Sorbents for Ion Exchange Thin-Layer Chromatography

| Manufacturer | Product | Type of ion exchanger | Particle size (μm) |
|---|---|---|---|
| J. T. Baker | ECTEOLA | ECTEOLA | 2-20 |
| Macherey-Nagel | Cellulose MN 300 DEAE | DEAE | |
| | Cellulose MN 300 ECTEOLA | ECTEOLA | |
| | Cellulose MN 300 PEI | PEI | |
| Riedel-de Haën | Cellulose DEAE | DEAE | |
| | Cellulose ECTEOLA | ECTEOLA | |

Table 27 Precoated Layers for Ion Exchange Thin-Layer Chromatography

| Manufacturer | Product | Type of modification | Indicator | Support | Layer thickness (mm) | Application |
|---|---|---|---|---|---|---|
| Analtech | DEAE Cellulose Uniplates | DEAE | — | Glass | | Anion exch. |
| | DEAE Cellulose Uniplates 9:1 | DEAE | — | Glass | | ", 90% cell |
| | DEAE Cellulose Uniplates 7.5:1 | DEAE | — | Glass | | ", 88% cell |
| | DEAE HR Cell. Uniplates 15:2 | DEAE | — | Glass | | ", 88% cell |
| J. T. Baker | Baker-flex Cellulose PEI | PEI | — | Plastic | | Anion exch. |
| | Baker-flex Cellulose PEI F | PEI | F254 | Plastic | | Anion exch. |
| | Baker-flex Cellulose DEAE | DEAE | — | Plastic | | Anion exch. |
| | Baker-flex Cellulose ECTEOLA | ECTEOLA | — | Plastic | | Anion exch. |
| | Baker-flex Cellulose CM | CM | — | Plastic | | Cationexch. |
| Chromatronix | Fixion 50x8 | Exchanger resin | — | Plastic | | Cationexch. |
| | Fixion 2x8 | Exchanger resin | — | Plastic | | Anion exch. |
| Macherey-Nagel | CEL DEAE/HR-MIX-20 | DEAE | — | Glass | 0.20 | Anion exch. |
| | Polygram CEL 300 DEAE | DEAE | — | Plastic | 0.10 | Anion exch. |
| | Polygram CEL 300 DEAE/HR-2/15 | DEAE | — | Plastic | 0.10 | Anion exch. |
| | Polygram CEL 300 PEI | PEI | — | Plastic | 0.10 | Anion exch. |
| | Polygram CEL 300 PEI UV 254 | PEI | F254 | Plastic | 0.10 | Anion exch. |
| | Polygram Ionex-25 SA-Na | Exchanger resin | — | Plastic | 0.25 | Cationexch. |
| | Polygram Ionex-25 SB-AC | Exchanger resin | — | Plastic | 0.25 | Anion exch. |
| | Polygram Ionex-25 SB-AC/UV254 | Exchanger resin | F254 | Plastic | 0.25 | Anion exch. |
| E. Merck | HPTLC Plate $NH_2$ F 254s | Amino | F254s | Glass | 0.20 | Anion exch. |

Table 27 (continued)

| Manufacturer | Product | Type of modification | Indicator | Support | Layer thickness (mm) | Application |
|---|---|---|---|---|---|---|
| E. Merck | TLC Plate PEI-Cellulose | PEI | — | Glass | 0.10 | Anion exch. |
| | TLC Plate PEI-Cellulose F | PEI | F254+366 | Glass | 0.10 | Anion exch. |
| | Plastic Sheet PEI-Cellulose F | PEI | F254+366 | Plastic | 0.10 | Anion exch. |
| Schleicher + Schüll | G 1440/A5 | 5% Dowex 1-X8 | — | Glass | 0.10 | Anion exch. |
| | G 1440/A10 | 10% Dowex 1-X8 | — | Glass | 0.10 | Anion exch. |
| | G 1440/K5 | 5% Dowex 50W-X8 | — | Glass | 0.10 | Cationexch. |
| | G 1440/K10 | 10% Dowex 50W-X8 | — | Glass | 0.10 | Cationexch. |
| | G 1440 PEI | PEI-Cellulose | — | Glass | 0.10 | Anion exch. |
| | G 1440 PEI/LS 254 | PEI-Cellulose | F254 | Glass | 0.10 | Anion exch. |
| | F 1440 PEI | PEI-Cellulose | — | Plastic | 0.10 | Anion exch. |
| | F 1440 PEI/LS 254 | PEI-Cellulose | F254 | Plastic | 0.10 | Anion exch. |
| Whatman | CM Cellulose | CM | — | Plastic | 0.10 | Cationexch. |
| | DEAE Cellulose | DEAE | — | Plastic | 0.10 | Anion exch. |

Table 28 Applications on Ion Exchange Precoated Layers

| Substance class | Type of ion exchanger | Mobile phase system | Ref. |
|---|---|---|---|
| Amino acids | Fixion 50X8 | Citrate buffer | 115 |
| Carboxylic and sulfonic acids | $NH_2$-Sil. gel | Ethanol/water with add. of NaCl | 116 |
| Inorganic ions | Ionex 25SA-Na | Water with addition of $NaNO_2$ | 117 |
| Inorganic ions | P-Cellulose | Methanol/sulfuric acid | 118 |
| Nucleotides | $NH_2$-Sil. gel | Alcohol/water with add. of NaCl | 119 |
| Nucleotides | DEAE-Cell. | 0.1 N HCl | 120 |
| Oligo thymidinic acids | PEI-Cell. | Water with addition of NaCl | 121 |
| RNA, DNA fragments | ECTEOLA-Cell. | Water/$NH_4OH$/NaCl/phosphate buffer | 122 |
| Steroids | DEAE-Cell. | *i*-Propanol/water/formic acid | 123 |

Table 29 Bulk Sorbents for Special Applications in Thin-Layer Chromatography

| Manufacturer | Product | Type | Particle size (μm) | Annotation |
|---|---|---|---|---|
| J. T. Baker | AC 10 | Acetylated cell. | 2-20 | 10% acetylation |
| Macherey-Nagel | Cellulose MN 300AC-10 | Acetylated cell. | | 10% acetylation |
| | Cellulose MN 300AC-20 | Acetylated cell. | | 20% acetylation |
| | Cellulose MN 300AC-30 | Acetylated cell. | | 30% acetylation |
| | Cellulose MN 300AC-40 | Acetylated cell. | | 40% acetylation |
| E. Merck | Florisil | Mg-silicate | 10-63 | |
| Riedel-de Haën | Florisil | Mg-silicate | | |
| | DAC 20 acetylated | Acetylated cell. | | 20% acetylation |
| | DAC 40 acetylated | Acetylated cell. | | 40% acetylation |
| Schleicher + Schüll | 144/21 ac | Acetylated cell. | 20 | 21% acetylation |
| | 144/45 ac | Acetylated cell. | 20 | 45% acetylation |

Table 30 Precoated Layers for Special Applications

| Manufacturer | Product | Type | Support | Layer thickness (mm) | Annotation |
|---|---|---|---|---|---|
| J. T. Baker | Baker-flex AC 10 | Acetylated cell. | Plastic | | 10% acetylation |
| Macherey-Nagel | CEL 300-10/AC 10% | Acetylated cell. | Glass | 0.10 | 10% acetylation |
| | CEL 300-10/AC 20% | Acetylated cell. | Glass | 0.10 | 20% acetylation |
| | CEL 300-10/AC 30% | Acetylated cell. | Glass | 0.10 | 30% acetylation |
| | CEL 300-10/AC 40% | Acetylated cell. | Glass | 0.10 | 40% acetylation |
| | ALOX/CEL-AC-MIX-25 | $Al_2O_3$ + acet. cell. | Glass | 0.25 | For sep. of PAH |
| | Chiralplate | RP with impregnation | Glass | 0.25 | For racemates |
| Schleicher + Schüll | G 1800-21 ac | Acetylated cell. | Glass | 0.10 | 21% acetylation |
| | G 1800-45 ac | Acetylated cell. | Glass | 0.10 | 45% acetylation |
| | F 1800-21 ac | Acetylated cell. | Plastic | 0.10 | 21% acetylation |
| | F 1800-45 ac | Acetylated cell. | Plastic | 0.10 | 45% acetylation |

Table 31 Applications on Special Sorbents and Precoated Layers

| Substance class | Stationary phase | Mobile phase system | Ref. |
|---|---|---|---|
| Anthrachinone dyes | Acetylated cell. | Ethyl acetate/THF/water | 124 |
| Enantiomeric resolution of D,L-amino acids | Chiralplate | Methanol/water/acetonitrile | 125 |
| Ketocarboxylic acids | Acetylated cell. | Propanol/butanol/ammonium carbonate | 126 |
| Polycyclic aromatic hydrocarbons | Acetylated cell. | Methanol/ether/water | 127 |
| Polycyclic aromatic hydrocarbons | $Al_2O_3$ + acet. cell. | Hexane/benzene or methanol/ether/water | 128 |
| Sweeteners | Ac. cell. + polyamide | Propanol/glacial acetic acid/formic acid | 129 |
| Tocopheroles | Florisil | Benzene | 130 |

of precoated layers. Table 31 suggests some typical applications for the special layers presented in this section.

## SUMMARY AND PROSPECTS

Thin-layer chromatography started off by being a purely qualitative analytical technique, being useful additionally in the preparation of relatively small quantities of substance. Just a few bulk sorbents such as silicas, aluminas, and celluloses, were available at that time. The user was obliged to prepare his own thin layers from these materials.

Once thin-layer chromatography had been shown to be powerful and versatile in a wide variety of applications, this method of separation needed to become more reproducible and standardizable. In meeting these demands various manufacturers introduced precoated layers consisting of sorbents whose physical and chemical parameters are deliberately fixed at the manufacturing stage. In this way the TLC requirements with regard to reproducibility are accomplished.

Based on experience already gathered in liquid column chromatography which shows that increased performance can be attained by using sorbents with narrower gradings and smaller mean particle sizes, manufacturers have also developed high-performance plates for use in thin-layer chromatography. The advent of these high-performance thin layers opened the way for TLC to be used in quantitative analysis. These plates, used in conjunction with various devices now available for sample application, development, and, more especially, evaluation, permit routine quantitative HPTLC of substances right down into the picogram range with a degree of precision and reproducibility comparable with other chromatographic methods.

Besides the improvement of the performance and quality of the plates by various new developments, the introduction of thin layers with new stationary phases (e.g., polyamides, derivatized celluloses, various silica types, and aluminas) broadened the selectivity spectrum. Similarly, various impregnations with liquid stationary phases also opened the way to new types of separation. Subsequently, various sorbents with a chemically modified surface were introduced in thin-layer chromatography to achieve comparable selectivities while at the same time avoiding the disadvantages encountered when using the impregnations (e.g., their tendency to elute out).

A successful way of simplifying sample application and of utilizing the full separation efficiency of the plate when fairly large sample volumes are applied has been to use precoated layers with so-called concentrating zones.

However, even with these spectacular developments the full potential of thin-layer chromatography has not yet been exhausted. Moreover, new possibilities have opened in a number of directions:

Increased use of high-performance thin-layer chromatography in quantitative analysis with still further improvements of the detection limits

"Automated" thin-layer chromatography (e.g., using microprocessors and computers for controlling sample application, in situ evaluation, data processing; use of laboratory robots)

Further improvements in layer quality

Further modifications with a view to new selectivities (e.g., for separating racemates and for biochemical applications)

A further important aspect of thin-layer chromatography will be the potential it affords in serving as a "pilot system" for working out suitable phase systems for HPLC [131].

In view of the ongoing developing work and the high degree of flexibility associated with thin-layer chromatography, it will continue to maintain and increase its importance as a qualitative and quantitative method of analysis.

*Acknowledgment.* The authors thank Mrs. Heidrun Herbert and Mrs. Antonina Castillo for preparing the chromatograms appearing in this chapter.

## REFERENCES


1. N. A. Izmailov and M. S. Schraiber, *Farm. Farmakol.*, *4*:8 (1938).
2. N. A. Izmailov and M. S. Shraiber, *Farmatsiya*, *6*:1 (1939).
3. E. Stahl, *Pharmazie*, *11*:633 (1956).
4. E. Stahl, *Dünnschicht-Chromatographie*, Springer-Verlag, Berlin, 1962.
5. H. Halpaap and J. Ripphahn, *Kontakte (Merck)*, *3*:16 (1976).
6. F. Geiss, *Die Parameter der Dünnschicht-Chromatographie*, Friedr. Vieweg and Sohn, Braunschweig, 1972.
7. J. C. Touchstone and M. F. Dobbins, *Practice of Thin-Layer Chromatography*, John Wiley and Sons, New York, 1978.
8. J. C. Touchstone and D. Rogers, *Thin Layer Chromatography*, John Wiley and Sons, New York, 1980.
9. H. P. Boehm, *Angew. Chem.*, *78*:617 (1966).
10. S. Brunauer, P. H. Emmett, and E. Teller, *J. Am. Chem. Soc.*, *60*:309 (1938).
11. K. K. Unger, *Porous Silica*, Elsevier, New York, 1979.
12. A. Wheeler, *Catalysis*, Vol. 2, Reinhold, New York, 1955.
13. W. Reich, *Kontakte (Merck)*, *3*:26 (1977).
14. Official Methods of Analysis of the Association of Official Analytical Chemists, Washington, D.C., 1980.

15. I. Schmeltz, A. Wenger, and D. Hoffman, in *Thin-Layer Chromatography* (J. C. Touchstone and D. Rogers, eds.), John Wiley and Sons, New York, 1980.
16. W. Hyde, H. M. Stahr, R. Moore, M. Domoto, and R. Pfeiffer, in *Advances in Thin Layer Chromatography* (J. C. Touchstone, ed.), John Wiley and Sons, New York, 1982.
17. W. Funk, R. Kerler, E. Boll, and V. Dammann, *J. Chromatogr.*, *217*:349 (1981).
18. G. Calabro and P. Curre, *Essenza, 46*:215 (1976).
19. H. Stan, *Z. Lebensm. Unters. Forsch.*, *166*:287 (1978)p
20. M. Sarsunowa, W. Swarcz, and C. Michalec, *Thin-Lafer Chromatrography in Biology, Pharmacy, and Clinical Biochemistry*, Mir, Moscow, 1980.
21. B. Wortberg, *J. Chromatogr.*, *156*:205 (1978).
22. G. E. Stolzenberg, in *Advances in Thin Layer Chromatography* (J. C. Touchstone, ed.), John Wiley and Sons, New York, 1982.
23. F. Kreuzig, *J. Chromatogr.*, *163*:332 (1979).
24. H. R. Sloan and C. S. Seckel, in *Advances in Thin Layer Chromatography* (J. C. Touchstone, ed.), John Wiley and Sons, New York, 1982.
25. H. M. Stahr and M. Domoto, in *Advances in Thin Layer Chromatography* (J. C. Touchstone, ed.), John Wiley and Sons, New York, 1982.
26. J. Lawrence and R. Frei, *Chemical Derivatization in Liquid Chromatography*, Elsevier, Amsterdam, 1976.
27. J. C. Young, *J. Chromatogr.*, *124*:115 (1976).
28. R. Pfeiffer and H. M. Stahr, in *Advances in Thin Layer Chromatography* (J. C. Touchstone, ed.), John Wiley and Sons, New York, 1982.
29. J. G. Kirchner, *Thin-Layer Chromatography*, John Wiley and Sons, New York, 1978.
30. C. Beyer and A. Van den Enden, *Clin. Chem. Acta, 192*:211 (1982).
31. J. C. Touchstone, G. J. Hansen, C. M. Zelop, and J. Sherma, in *Advances in Thin Layer Chromatography* (J. C. Touchstone, ed.), John Wiley and Sons, New York, 1982.
32. J. Lawrence and R. Frei, *Chemical Derivatization in Liquid Chromatography*, Elsevier, Amsterdam, 1976.
33. M. H. Thomas and K. E. Soroka, in *Advances in Thin Layer Chromatography* (J. C. Touchstone, ed.), John Wiley and Sons, New York, 1982.
34. P. Ghosh and S. Thakur, *Fresenius Z. Anal. Chem.*, *313*:144 (1982).
35. E. Heilweid and J. C. Touchstone, *J. Chromatogr.*, *Sci.*, *19*: 2193 (1981).
36. T. R. Watkins, B. A. Ruggeri, R. J. H. Gray, and R. C. Tomlins, in *Techniques and Applications of Thin Layer Chroma-*

*tography* (J. C. Touchstone and J. Sherma, eds.), John Wiley and Sons, New York, 1985.
37. A. Studer and H. Traitler, *J. High Resol. Chromatogr. Chromatogr. Commun.*, *5*:581 (1982).
38. P. Czuczy and E. Morava, *Anal. Chem. Symp. Ser.*, 1982, 483.
39. H. E. Hauck and W. Jost, *GIT Fachz. Lab. Supplement Chromatographie*, *3*:3 (1983).
40. M. Okamoto, F. Yamada, and T. Omori, *J. High Resol. Chromatogr. Chromatogr. Commun.*, *5*:163 (1982).
41. W. Jost and H. E. Hauck, *J. Chromatogr.*, *261*:235 (1983).
42. W. Jost and H. E. Hauck, *Anal. Biochem.*, *135*:120 (1983).
43. W. Jost and H. E. Hauck, *Proceedings of the Third International Symposium on Instrumental High Performance Thin-Layer Chromatrography*, Würzburg, 1985.
44. T. Omori, M. Okamoto, and F. Yamada, *J. High Resol. Chromatogr. Chromatogr. Commun.*, *6*:47 (1983).
45. C. Combellas and B. Drochon, *Anal. Lett.*, *16* (A20):1647 (1983).
46. H. Brockmann and H. Schodder, *Ber.*, *74*:73 (1941).
47. W. T. Trager, L. Stoloff, and A. D. Campbell, *J. Assoc. Off. Anal. Chem.*, *47*:993 (1964).
48. M. F. Kovac, *J. Assoc. Off. Agr. Chem.*, *46*:884 (1963).
49. J. Rosmus and Z. Deyl, *J. Chromatogr.*, *6*:187 (1961).
50. P. M. Scott, in *Thin-Layer Chromatography* (J. C. Touchstone and D. Rogers, eds.), John Wiley and Sons, New York, 1980.
51. I. Schmeltz, A. Wenger, and D. Hoffman, in *Thin-Layer Chromatography* (J. C. Touchstone and D. Rogers, eds.), John Wiley and Sons, New York, 1980.
52. J. C. Young, *J. Chromatogr.*, *124*:115 (1976).
53. L. R. Alexander and E. R. Stanley, *J. Assoc. Off. Agr. Chem.*, *48*:278 (1965).
54. A. Seher, *Microchim. Acta*, *1961*:308.
55. B. Stefaniak, *J. Chromatogr.*, *44*:403 (1969).
56. H. Müller and V. Siepe, *Deutsch. Lebensm. Rundsch.*, *74*:133 (1978).
57. N. Oswald and H. Flück, *Pharm. Acta Helv.*, *39*:293 (1964).
58. J. Cochia and J. W. Daly, *Experientia* *18*:294 (1962).
59. T. Ikekawa, *J. Antibiotics (Jpn) Ser. A*, *16*:56 (1963).
60. H. Henning, *Vitaminbestimmung*, Verlag Chemie, Weinheim, 1963.
61. O. O. Blumenfeld, M. A. Paz, P. M. Gallup, and S. Seifter, *J. Biol. Chem.*, *238*:3835 (1963).
62. H. Wagner, L. Hörhammer, and P. Wolff, *Biochem. Z.*, *334*:175 (1961).
63. B. Liberek, J. Augustyniak, J. Ciarkowski, K. Plucinska, and K. Stachowiak, *J. Chromatogr.*, *95*:223 (1974).
64. D. E. Nitecki, in *Thin-Layer Chromatography* (J. C. Touchstone and D. Rogers, eds.), John Wiley and Sons, New York, 1980.

65. R. Fischer and W. Klingelhöller, *Arch. Toxikol.*, *19*:119 (1961).
66. D. C. Fenimore and C. M. Davis, in *Thin-Layer Chromatography* (J. C. Touchstone and D. Rogers, eds.), John Wiley and Sons, New York, 1980.
67. R. D. Bennett and E. Heftmann, *J. Chromatogr.*, *9*:348 (1962).
68. A. Wehrli, *Canad. Pharm. J. Sci. Sect.*, *97*:208 (1964).
69. A. Adye and R. I. Mateles, *Biochim. Biophys. Acta*, *86*:418 (1964).
70. H. R. Bolliger and A. König, in *Dünnschicht-Chromatographie* (E. Stahl, ed.), Springer-Verlag, Berlin, 1967.
71. H. E. Hauck and H. Halpaap, *Chromatographia*, *13*(9):538 (1980).
72. K. Polzhofer, *Z. Lebensm.-Unters. Forsch.*, *167*(3):162 (1978).
73. E. Bancher, H. Scherz, and V. Prey, *Mikrochim. Acta*, *1963*:712.
74. H. Thielemann, *Z. Anal. Chem.*, *282*:144 (1976).
75. I. R. Murray, *Aust. J. Pharm. Sci.*, *10*(3):82 (1981).
76. M. Hayashi and T. Kamikubo, *J. Vitaminol.*, *11*:286 (1963).
77. M. A. Sattar, J. Paasivirta, R. Vesterinen, and J. Knuutinen, *J. Chromatogr.*, *135*:395 (1977).
78. A. D. Kinghorn, F. H. Jawad, and N. J. Doorenbos, *J. Chromatogr.*, *147*:299 (1978).
79. D. Yoshtake, *Kagaku Keisatsu Kenkyusho Hokoku*, *16*:51 (1963).
80. S. Ishikawa and G. Katsui, *Vitamins*, *29*:203 (1964).
81. D. Myhill and D. S. Jackson, *Anal. Biochem.*, *6*:193 (1963).
82. I. Sjöholm, *Acta Chem. Scand.*, *18*:889 (1964).
83. J. Dittmann, *Z. Klin. Chem.*, *1*:190 (1963).
84. D. W. Vomhof and T. C. Tucker, *J. Chromatogr.*, *17*:300 (1965).
85. P. Hynninen, *J. Chromatogr.*, *134*:359 (1977).
86. J. W. Walkley and J. Tillman, *J. Chromatogr.*, *132*:172 (1977).
87. D. E. Nitecki, in *Thin-Layer Chromatography* (J. C. Touchstone and D. Rogers, eds.), John Wiley and Sons, New York, 1980.
88. K. Nagasawa, H. Yoshidome, and K. Anryu, *J. Chromatogr.*, *52*:173 (1970).
89. A. Szabo, M. Kovacs Nagy, and E. Tömörkeny, *J. Chromatogr.*, *151*:256 (1978).
90. S. Ito and K. Fujita, *J. Chromatogr.*, *187*:418 (1980).
91. M. Weise and D. E. Okan, *J. Chromatogr.*, *152*:175 (1978).
92. K.-T. Wang, P. H. Wu, and T.-B. Shih, *J. Chromatogr.*, *44*:635 (1969).
93. M. S. Scandrett and E. J. Ross, *Clin. Chim. Acta*, *72*:165 (1976).
94. N. N. Osborne, W. L. Stahl, and V. Neuhoff, *J. Chromatogr.*, *123*:212 (1976).
95. H. Szumilo and E. Soczewinski, *J. Chromatogr.*, *124*:394 (1976).
96. R. J. T. Graham, *J. Chromatogr.*, *33*:118 (1968).
97. E. Soczewinski and H. Szumilo, *J. Chromatogr.*, *81*:99 (1973).
98. K. Nagasawa, H. Yoshidome, and R. Takeshita, *J. Chromatogr.*, *43*:473 (1969).
99. H. Nakamura, J. J. Pisano, and Z. Tamura, *J. Chromatogr.*, *175*:153 (1979).

100. B. Wortmann, W. Wortmann, and J. C. Touchstone, *J. Chromatogr.*, *70*:199 (1972).
101. R. Takeshita, *J. Chromatogr.*, *66*:283 (1972).
102. D. Volkmann, *J. High Resol. Chromatogr. Chromatogr. Commun.*, *4*:350 (1981).
103. L. Lepri, P. G. Desideri, and D. Heimler, *J. Chromatogr.*, *173*: 119 (1979).
104. G. Grassini-Strazza, and M. Cristalli, *J. Chromatogr.*, *214*:209 (1981).
105. A. Siouffi and G. Guiochon, *J. Chromatogr.*, *209*:441 (1981).
106. H. Neuninger, *Arch. Krim.*, *167*:99 (1981).
107. H. M. Ruijten, P. H. van Amsterdam, and H. de Bree, *J. Chromatogr.*, *252*:193 (1982).
108. H. Jork and M. Weiss, *GIT Fachz. Lab.*, *23*:385 (1979).
109. K. Macek, Z. Deyl, and M. Smrz, *J. Chromatogr.*, *193*:421 (1980).
110. W. Jost and H. E. Hauck, *J. Chromatogr.*, *264*:91 (1983).
111. R. Tomingas and Y. P. Grover, *Fresenius Z. Anal. Chem.*, *315*: 515 (1983).
112. K. Jung and H. Specker, *Fresenius Z. Anal. Chem.*, *288*:28 (1977).
113. I. D. Wilson, S. Scalia, and E. D. Morgan, *J. Chromatogr.*, *212*:211 (1981).
114. L. Lepri, P. G. Desideri, and D. Heimler, *J. Chromatogr.*, *169*: 271 (1979).
115. A. Varadi and S. Pongor, *J. Chromatogr.*, *173*:419 (1979).
116. W. Jost and H. E. Hauck, *J. Chromatogr.*, *261*:235 (1983).
117. V. Cardaci, M. Lederer, and L. Ossicini, *J. Chromatogr.*, *101*: 411 (1974).
118. T. Shimizu, T. Kurosaki, and H. Suwa, *Chromatographia, 20*: 364 (1985).
119. W. Jost and H. E. Hauck, *Anal. Biochem.*, *135*:120 (1983).
120. K. Randerath, *Nature 194*:768 (1962).
121. G. Weimann and K. Randerath, *Experientia, 19*:49 (1963).
122. R. D. Bauer and K. D. Martin, *J. Chromatogr.*, *16*:519 (1964).
123. G. W. Oertel, M. C. Tornero, and K. Groot, *J. Chromatogr.*, *14*:509 (1964).
124. P. Wollenweber, *J. Chromatogr.*, *7*:557 (1962).
125. K. Günther, J. Martens, and M. Schickedanz, *Naturwissensch.*, *72*:149 (1985).
126. M. Rink and S. Herrmann, *J. Chromatogr.*, *14*:523 (1964).
127. D. T. Kaschani, *GIT Labor-Med.*, *2*:128 (1979).
128. H. Hellmann, *Fresenius Z. Anal. Chem.*, *295*:24 (1979).
129. T. Salo, E. Airo, and K. Salminen, *Z. Lebensm.-Unters. Forsch.*, *125*:20 (1965).
130. P. G. Roughan, *J. Chromatogr.*, *29*:293 (1967).
131. W. Jost, H. E. Hauck, and F. Eisenbeiss, *Kontakte (Merck)*, *3*:45 (1984).

# 6

# Adsorbents in Column Liquid Chromatography

Klaus K. Unger / *Institute of Inorganic Chemistry and Analytical Chemistry, Johannes Gutenberg University, Mainz, Federal Republic of Germany*

## I. INTRODUCTION

Throughout this chapter the term *column liquid adsorption chromatography* is used for all techniques which employ columns with porous packings operated with solvents in the elution, displacement and frontal analysis mode to enrich, separate, and isolate substances. A packing is termed an adsorbent when its particles are porous exhibiting a large surface area to volume ratio of >10 $m^2$/ml of packing. The surface area is exclusively internal, even for microparticulate packings of 3-5 μm. Adsorption is understood as an enrichment of solute at the adsorbent surface or interface relative to its concentration in the liquid bulk phase. The enrichment results from weak attractive molecular interactions between the solute and the surface of the adsorbent. The strength and specificity of these interactions are largely controlled and adjusted by the type and composition of the solvent and by changes of the solvent composition. In order to achieve a discrimination among the solutes in column liquid adsorption chromatography, it is essential that the adsorbent surface resembles or imitates the chemical structure of the analytes. In other words, the surface of the adsorbent is considered to function as a chemical sensor for analyte recognition.

Classical column liquid adsorption chromotography has been performed on hydrophilic adsorbents such as silicas and aluminas with nonpolar to moderately polar solvents (see Chap. 1) [1,2]. Typically, glass columns filled with coarse particles of silica and alumina permitted the resolution of substances within hours. Around 1970, modern column liquid chromatography (high-performance liquid chromatography

= HPLC) began with the application of microparticulate silicas and aluminas of defined particle shape and size, surface area, porosity, and pore size [3,7,4]. This led to an enormous improvement in chromatographic resolution and analysis speed. The most remarkable progress, however, was brought about by the development and manufacture of chemically bonded silicas, in particular those with hydrophobic groups, called reversed phase (RP) silicas [3,5]. Since reversed phase chromatography (RPC) is performed with organic solvents, hydroorganic solvent mixtures, and buffers, it allows the resolution of unpolar, polar nonionic, and ionic analytes. Thus RPC has extended the application of HPLC into pharmaceutical and toxicological analyses and other areas.

While the major application of RPC lay in the separation of low molecular weight analytes, it also developed to a powerful separation technique for biopolymers, e.g., peptides and proteins [6]. The limited pH stability of RP silicas in buffered eluents then initiated the development of polymer-based adsorbents and polymer-coated silicas as alternative packings [7]. Today, column liquid adsorption chromatography on silica- and polymer-based packings is one of the most widely applied techniques in HPLC.

This chapter focuses on the role of packings in column liquid adsorption chromatography and reviews the current developments in this important field. Definitions and general features of liquid-solid adsorption systems are briefly discussed in Sec. II. A survey of the types of adsorbents and stationary phases follows in Sec. III. The decisive stationary phase properties controlling retention and selectivity of analytes are discussed in Sec. IV. The chapter ends with appendices on commercial packings and silanes applied in bonding chemistry of silica.

## II. ADSORPTION IN LIQUID-SOLID SYSTEMS

It is essential for the understanding of the basic retention phenomena in column liquid adsorption chromatography to deal briefly with the adsorption in liquid-solid systems [8-10]. When an adsorbent is immersed in a pure liquid which wets the adsorbent surface, the liquid is adsorbed. The adsorbed phase is considered as a monolayer of densely packed solvent molecules. By adding a second component to the system, either a solid or a liquid, the solute is adsorbed when the attraction forces between the solute and the surface are stronger than those between the solvent and the solute, and the solvent and the surface. Attraction forces are van der Waals, hydrogen bonding, hydrophobic association, acid-base, and complexation interactions associated with free energy changes of $\Delta G^\circ$ of about 10-80 kJ/mol.

The nature of these interactions is dependent on the chemical structure of the surface, on the type of solvent and solute, and on the temperature. When dealing with liquid-solid adsorption under the aspect of interactions, the solutes are conveniently grouped into nonelectrolytes and electrolytes. Further subdivision is made into low molecular weight compounds and ions, nonionic and ionic surfactants, and polymers and polyelectrolytes [10].

Adsorption data are usually reported in the form of the surface excess isotherms where $n^0 \Delta x_i/m$ is plotted against $x_i$, the latter being the liquid mole fraction of solute i. $\Delta x_i$ is the decrease in mole fraction of solute i, when $n^0$ moles of the original solution are equilibrated with m grams of the adsorbent. The resulting isotherms are termed as composite isotherms from which the individual isotherms of the single components can be computed. Figure 1 shows the types of composite isotherms according to the classification of Schay and Nagy [11]. A negative value of $n^0 \Delta x_i/m$ means that the solvent is preferentially

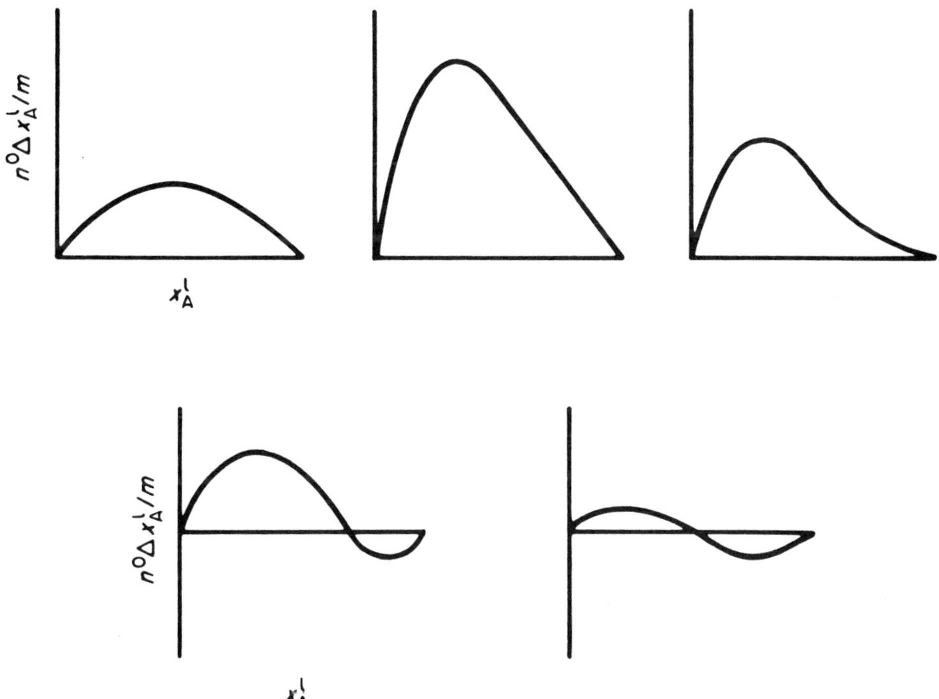


Fig. 1 Composite isotherms according to the classification of Schay and Nagy [11].

adsorbed. Surface excess isotherms of binary solvents on adsorbents have been measured by chromatographic techniques [12-18]. From the limiting linear regime of the isotherm the capacity factor $k_i'$ is derived:

$$k_i' = \frac{n_i^\sigma}{n_i} \tag{1}$$

where $n_i^\sigma$ is the surface excess amount of solute i, and $n_i$ is the total amount of solute i in the column [19].

Since $n_i^\sigma$ can be positive or negative depending on whether the solute is preferentially adsorbed or not, positive or negative capacity factors can result. This is different from the general definition of the solute capacity factor in HPLC: k' is always positive, being defined as the ratio of the amount of solute in the stationary to the amount of solute in the mobile phase.

In recent years a large number of adsorption studies with liquid-solid systems have been performed, taking into account the nonideality of solvent-solvent and solvent-solute interactions, the differences in the size and shape of solvent and solute molecules, and the heterogeneity of the surface of the adsorbent [20-23]. From these examinations, equations have been derived in which the chromatographic retention of solutes is expressed in terms of surface excess quantities. They also allow estimation of the effect of the solvent, the solute, and the surface properties on retention. However, since most of these studies are based on binary solvent systems, their application to practical HPLC systems, where often ternary and quaternary solvent mixtures and modifiers are employed, is rather limited.

## III. ADSORBENTS AND STATIONARY PHASES

The most common adsorbents are treated groupwise according to their chemical bulk composition as listed in Table 1. The specific features of each group are surveyed briefly, followed by a discussion of the major aspects of the synthesis, properties, and characterization of individual adsorbents of chromatographic grade.

### A. Oxides

Solid oxides represent the traditional type of hydrophilic adsorbent. Oxides are classified into basic (MgO), amphoteric ($Al_2O_3$), and acidic ($TiO_2$) according to their chemical reactivity. The surface metal atoms of oxides are coordinated to hydroxyl groups, which are identified by a variety of physiochemical methods, e.g., by infrared spectroscopy in the range 2000-4000 $cm^{-1}$ [24]. The wave number shifts ($\Delta\nu_{OH}$) of hydroxyl groups in the infrared spectrum, produced by the adsorption of benzene as a hydrogen bond-donating probe, were found to correlate to the $pk_a$ value of the oxide. In this way the following $pk_a$ sequence was obtained [25]: MgO 18.5, $ZrO_2$ 10.5, $\gamma$-$Al_2O_3$ 8.5, $SiO_2$ 7.0, $TiO_2$ (anatas) 0.5-2.0.

Several types of hydroxyl groups may occur at a given oxide, depending on its phase composition and surface chemical structure. As a typical case, Fig. 2 (top) shows the hydroxyl groups at the surface of $\gamma$-alumina. Five different configurations of hydroxyl groups are discriminated, depending on whether they are linked to tetrahedrally and/or octahedrally coordinated aluminum atoms. As a consequence, five distinct absorption bands assigned to these groups appear in the infrared spectrum between 3000 and 4000 $cm^{-1}$ [24,26]. Amorphous silica, with only tetrahedrally coordinated silicon atoms, has three types of hydroxyl groups: single or isolated, geminal and vicinal [24,75] (see Fig. 2, bottom).

Thermal treatment of oxides first eliminates any adsorbed water. At temperatures above 400 K, surface hydroxyl groups start to dehydroxylate, forming Lewis acid and Lewis basic sites. For instance, annealing of silica yields reactive so-called strained siloxane groups as surface intermediates, converting into stable siloxane groups at temperatures above 900 K:

$$\equiv Si{-}OH + HO{-}Si\equiv \xrightarrow[-H_2O]{\Delta} \equiv Si^{\oplus} + {}^{\ominus}O{-}Si\equiv \xrightarrow{\Delta} \equiv Si{-}O{-}Si\equiv$$

A similar pattern has been discussed for the dehydroxylation of alumina surfaces [24]. Broensted and Lewis acid sites are distinguishable by the adsorption of pyridine as monitored by infrared spectroscopy [28]. Fully hydroxylated surfaces of oxides, entirely free of physisorbed water, possess the highest Broensted acidity and surface activity. On increasing uptake of water or polar solvents, the surface activity successively diminishes, which is attributable to the blocking of active surface sites.

Impurities of the adsorbent often produce noticeable changes of their acidity. Metals other than the parent, either present as oxides and hydrous oxides or linked to the parent metal atom via oxygen bonds, decrease or increase the acidity via inductive and other phenomena [24]. Furthermore, basic or acidic impurities may remain from the manufacturing process, resulting in a surface acidity different from that of the pure adsorbent. Thus, before a reliable estimate of the surface acidity is made, the purity of the material under investigation

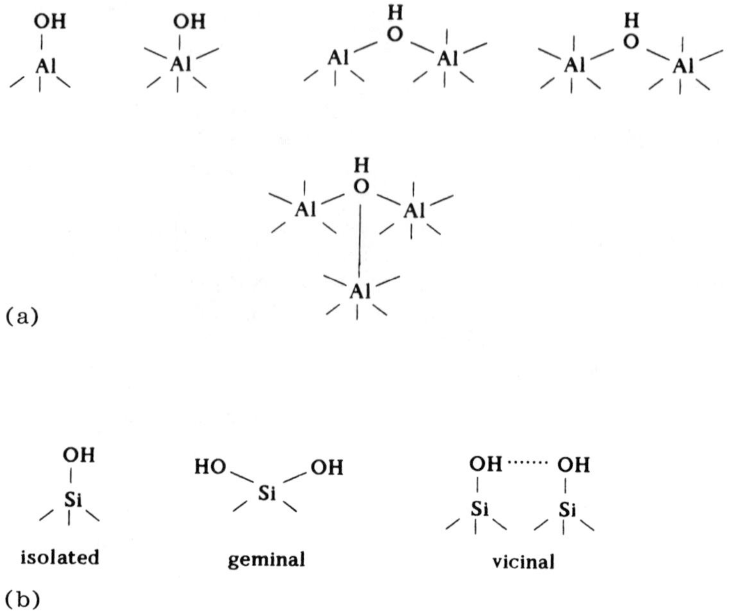

(a)

isolated geminal vicinal

(b)

Fig. 2 Types of Broensted acid sites at the surface of γ-alumina (a) and silica (b).

must be carefully checked. On the other hand, a controlled adsorption of basic or acidic substances permits adjustment of the surface acidity of a given adsorbent, provided that the surface composition remains unchanged in chromatographic use.

### 1. *Silica*

Silica as an adsorbent is an amorphous product of stoichiometric composition $(SiO_2 \cdot H_2O)_n$. Lacking an x-ray crystalline structure, silicas are grouped according to their morphology and aggregation characteristics into pyrogenic silica, precipitated silica, silica hydrogel, and xerogel [3,29,30]. As the name implies, pyrogenic silicas are produced by high-temperature hydrolysis of silicon tetrachloride and marketed as Aerosil and Cab-O-Sil. Aerosil is a low-density powder composed of agglomerates of loosely connected nonporous colloidal silica particles of spherical shape and hence cannot be employed as a chromatographic packing. Precipitated silica is obtained by an appropriate reaction of sodium silicate and an acid, maintaining a low silica concentration and a high pH. It has a somewhat higher bulk density than Aerosil, and the particles exhibit a permanent porosity. Silica hydrogel represents a coherent system of three-dimensionally linked colloidal silica particles in a dispersion medium such as water, an electrolyte, or an alcohol. Removal of the dispersing liquid by drying or other processes yields a xerogel, with a pore space built up by the interstitial voids between the assembled colloidal particles.

There are several ways to form a hydrogel. One route starts by mixing water glass solution with an acid, whereby polysilicic acids are formed via hydrolysis and condensation. Nuclei are formed which grow to larger colloidal particles. Another alternative starts from stabilized colloidal dispersions and destabilizes the silica sol, which then solidifies to a silica hydrogel. Usually, the obtained hydrogel has a tendency to shrink due to syneresis, whereby the immobilized liquid is squeezed out. After aging, the hydrogel is washed and dehydrated to the xerogel. For efficient dehydration the gelatinous mass is broken into pieces, so that irregularly shaped particles are obtained. These are milled and sized to the desired particle size distribution.

The sol-gel transition, converting a quasi-liquid to a quasi-solid, provides a unique means to prepare beads. Figure 3 illustrates the principal ways of achieving spherical particles in the course of the manufacturing process. The most common process is the gelling of liquid droplets of a colloidal silica dispersion in a two-phase system under agitation or other manipulation. Hereby, the gelling time is the critical operational parameter to be adjusted. After sedimentation or separation, the hydrogel beads are subjected to dehydration at about 423 K. Another process is based on agglutination. An organic monomer is added to a colloidal silica and polymerization is started by raising

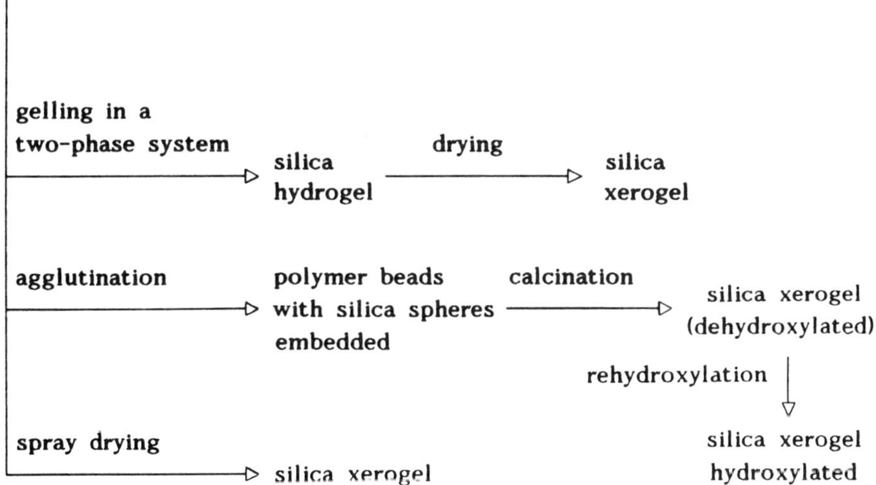


Fig. 3 Schemes of processes to form spherical silicas.

the pH. Polymer beads are obtained in which the silica particles are embedded [31]. Burning-out of the polymer and further annealing cements the colloidal particles together to form stable beads. In order to achieve a fully hydroxylated surface, the beads are then subjected to rehydroxylation. The most common and economic method of beading is spray-drying of a silica sol, a process which combines shaping and drying in one step. The sol is sprayed through a nozzle at higher temperatures. The liquid evaporates and spherical particles are obtained.

A common feature of porous oxides is their corpuscular structure. In the case of silica, nonporous spherical colloidal particles of a given size and size distribution are assembled to aggregates which ultimately form the porous particles. The pore structure parameters are thus directly related to the size of the primary particles and their coordination number, i.e., their packing density [32-35]. The particle porosity P and the specific pore volume vp is proportional to the packing density of the spheres; the specific surface area $a_s$ is inversionally proportional to the size of the primary particles. The pore diameter pd is a function of both vp and $a_s$ [33,34].

$$pd = \frac{2.8\ vp}{a_s} \qquad (2)$$

Porous silicas differ widely in vp, $a_s$, and pd in accordance with the size of the primary colloidal particles and their packing density. The ranges are:

| | | |
|---|---|---|
| $a_s$: | 1-1000 (10-500) | $m^2/g$ |
| vp: | 0.1-3.0 (0.5-1.5) | ml/g |
| pd: | 1-1000 (6-100) | nm |

The values in parentheses are those for chromatographic silica packings. As shown by Eq. (2), the smaller the pore diameter, the higher the specific surface area. This relation is illustrated by the following set of $a_s$ and pd values:

| $a_s/(m^2/g)$ | pd/nm |
|---|---|
| 500 | 6 |
| 100 | 30 |
| 50 | 50 |
| 20 | 100 |

The specific pore volume vp directly controls the packing density $\rho_p$, i.e., the mass of silica per column volume. $\rho_p$ was found to be inversely proportional to vp [36]:

| $v_p$/(ml/g) | $\rho_p$/(g/ml) |
|---|---|
| 1.5 | 0.2 |
| 0.5 | 0.8 |
| 0* | 1.6 |

*Valid for nonporous silica beads.

The surface chemistry of silica has been the subject of thorough investigations in the past and was reinvestigated by the methods of modern surface analysis [3]. Several independent models based on crystalline silica modifications have been developed to predict the actual surface. The silica surfaces obtained by these models differ in the types of hydroxyl groups attached to the surface silicon atoms and in their relative concentration according to thermal pretreatment. None of the models (e.g., de Boer, Kiselev, Peri, etc.), however, provides a consistent pattern to confirm the experimental findings (cf. Ref. 37). Since an extensive treatment of the subject is beyond the scope of this chapter, we will focus on those aspects which are most relevant to the chromatographic application of silica.

Due to its composition, the total water content of a porous silica originates from hydroxyl groups at the surface, physisorbed water at the surface, and hydroxyl groups or water in the bulk of the solid. The total water content is usually measured by the weight loss on annealing at 1473 K and varies from 2 to 10% (w/w). The mean concentration of surface hydroxyl groups of a fully hydroxylated silica corresponds to about 8-9 μmol/m$^2$ [3]. For a silica product of $a_s$ = 300 m$^2$/g, the hydroxyl group content amounts to about 4% (w/w).

Both the hydroxyl group concentration and the content of physisorbed water control the retention chromatographic behavior on silica in straight-phase chromatography. Employing unpolar solvents, retention is highest on silicas which are free of physisorbed water and exhibit the maximum concentration of hydroxyl groups. With an increasing amount of physisorbed water under otherwise constant conditions, retention declines. The physisorbed water forms a multilayer and water successively fills the pores. On silicas with a multilayer of adsorbed water, solute adsorption is highly improbable and partitioning becomes the dominant retention principle. For the application of silica in straight-phase HPLC, two consequences result from these considerations. First, the water content of the silica, being in equilibrium with the nonpolar eluent, must be adjusted to a certain level. Second, this level must be held constant; otherwise reproducible analyses cannot be carried out. It should be added that an equilibration of silica with water requires an extended period of time compared to the column equilibration in RPC. The uptake of water of silica can be accurately assessed by measuring water adsorption isotherms in batch experiments. However, such a system is not comparable to a silica column with an unpolar mobile phase. In this situation, the effective water content of the silica is largely dependent on the water content of the mobile phase. On applying unpolar solvents, the relative water content is matched by blending the "water-free" solvent with the water-saturated solvent and excluding atmospheric moisture. A closed system was introduced by Engelhardt and Elgass [4], comprising a thermostatted funnel containing a coarse grain silica with a defined water content between 2 and 20% (w/w) through which the eluent is recycled. Water can be replaced by other polar solvents, e.g., alcohols [4]. The strong dependence of retention on the water content in straight-phase HPLC is much reduced when substituting silicas with hydrophilic bonded silicas, e.g., diol silicas [38]. This behavior is the result of modification wherby the most active hydroxyl groups are removed. When moderately polar solvents, e.g., dioxan, are added in excess of 5% (v/v) to the unpolar mobile phase, e.g., n-heptane, control and adjustment of the water content are no longer necessary.

Little is known on the batch-to-batch reproducibility and the conditions for storing silica packings. The reproducibility of a silica with regard to $a_S$ and vp was shown to be ±5% of the mean for about 36 batches over a period of 1 year [39]. Storage of silica at normal humidity in closed bottles over 2 years did not change $a_S$ and vp noticeably, i.e., variations were within reproducibility of the measurement of $a_S$ and vp [40].

Amorphous silica is soluble in water [41]. The concentration at saturation amounts to 100 ppm at room temperature and neutral pH. Solubility increases drastically above pH 8. The speed of dissolution is known to be a function of the specific surface area and the particle size. In order to suppress the dissolution in HPLC a saturator column is recommended, filled with a coarse silica and inserted before the analytical column. Attempts have been made to reduce the solubility of silica by doping the surface with alumina [42], and by zirconia cladding [43]. The pretreatment of silica with a buffer solution, whereby the solvent is removed and the salt remains [44-46], or the conditioning of the silica column with a buffer solution in situ [47], is an effective means for standardizing silica in straight-phase HPLC. Through such depositions, the apparent "surface acidity" can be adjusted and the surface heterogeneity of the native silica is reduced, i.e., the peaks are less tailed. However, the use of such columns is limited to less water-rich mobile phases to avoid bleeding.

### 2. *Alumina*

Aluminum oxide hydrates compose a class of solid compounds of the composition $Al_2O_3 \cdot nH_2O$, where n ranges from 0 to 3 [48-50]. The limiting members are aluminum hydroxide, $Al(OH)_3$, i.e., gibbsite (hydrargillite), bayerite, and nordstrandite, and $\alpha$-$Al_2O_3$, corundum. All are nonporous products. Porous active aluminas employed as adsorbents and catalysts contain $\gamma$-alumina as the main component apart from other crystalline aluminum oxides such as $\chi$, $\eta$, $\kappa$, $\delta$, collectively named as transition aluminas. $\gamma$-Alumina is produced by controlled thermal decomposition of gibbsite or boehmite, AlO(OH) at temperatures between 700 and 1100 K, illustrated in Fig. 4. It possesses a lattice structure close to the spinel type, $MgAl_2O_4$, with tetragonal deformation [48,49]. Although as polar as silica, its surface chemistry is somewhat different due to the low concentration of Broensted acid sites and the presence of Lewis acid and Lewis basic sites, which is reflected by its chromatographic retention behavior.

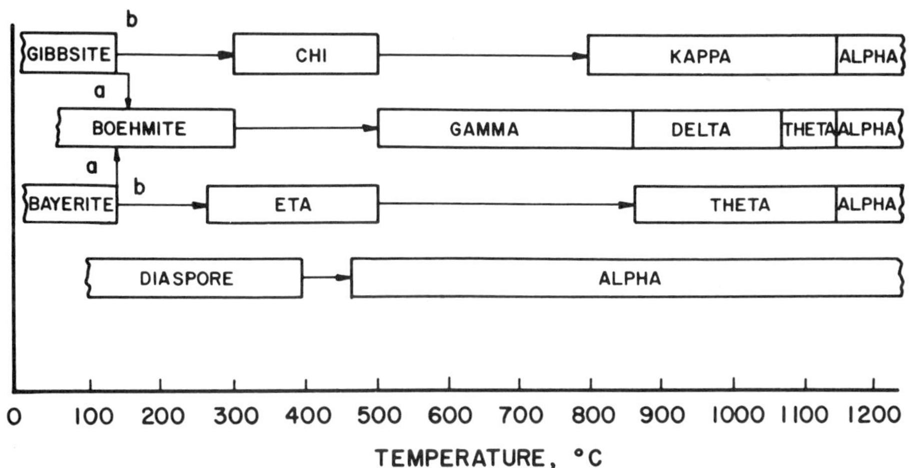


Fig. 4 Decomposition sequence of alumina hydroxides. (From Ref. 48 by permission of the publisher.)

| Conditions | Conditions favoring transformations | |
|---|---|---|
| | Path a | Path b |
| Pressure | >1 atm | 1 atm |
| Atmosphere | moist air | dry air |
| Heating rate | >1°C/min | < 1°C/min |
| Particle size | >100 μm | < 10 μm |

*Surface Structure* Models of the surface of γ-alumina have been developed by Peri [53] and Knözinger and Ratnasamy [26]. Five distinct bands in the infrared spectrum between 3000 and 4000 $cm^{-1}$ were identified and assigned to the stretching vibration bands of five types of hydroxyl groups, as indicated in Fig. 2a. On raising the calcination temperature, the Broensted sites convert into aprotic Lewis acid and Lewis basic sites ($Al^{3+}$ atoms as Lewis acids and oxygen ions as Lewis bases). Lewis acid sites were monitored by interaction of electron pair donor molecules such as ammonia and pyridine [24]. The total hydroxyl group content of γ-aluminas, measured by exchange with tritiated water and by n-butylamine adsorption, amounted to about 3 $\mu mol/m^2$ [54]. As already mentioned, an aqueous suspension of γ-alumina has a pH ∿ 9, attributed to the presence of $Na_2O$ as impurity. As a result, γ-alumina possesses cation exchange properties [55]. The basic γ-alumina can be converted to a "neutral alumina" by a neutralization process or further to "acidic alumina." Acid-set γ-aluminas act as anion exchangers [55]. It is worth noting that ion exchange reactions can lead to a dissolution of the alumina, attributed to its amphoteric character [56].

For use as adsorbent in straight-phase HPLC with unpolar mobile phases, the surface activity must be adjusted by controlling the water content of both alumina and eluent, analogous to silica as discussed above. According to Brockmann [57], five distinct activity states were distinguished, depending on the water content of the aluminas.

*Specific Surface Area and Pore Size* γ-Aluminas develop specific surface areas between 50 and 200 $m^2/g$, with specific pore volumes, vp, up to 0.6 ml/g. The mean pore diameter is an inverse function of $a_s$ and achieves maximum values of about 10 nm. γ-Aluminas often contain a large number of micropores of pd < 2 nm. Large-pore aluminas, prepared by employing a swing from acidic to basic pH during gelation [58], have not yet found use as adsorbents in chromatography.

*Chromatographic Retention* [*59, 60*] γ-Alumina preferentially retains organic compounds with carbon-carbon double bonds, particularly aromatic compounds, probably through Lewis acid-base interactions, and resolves aromatic isomers. Untreated γ-aluminas preferentially retain weak acids, while strong acids are irreversibly bound. The retention of acidic and basic solutes can be manipulated by employing pH-adjusted γ-aluminas as neutral, acidic, or basic material.

### 3. *Magnesia, Ceria, Zirconia, and Titania*

Although these oxides find widespread use as industrial adsorbents, carriers and pigments, there is little application in HPLC. Magnesia, having predominantly basic properties, is manufactured by thermal decomposition of magnesium salts, preferably carbonate, yielding a porous product [61,62]. As discussed for γ-alumina, the texture of porous magnesia is strongly dependent on the chemical composition and structure of the starting material, the heating temperature, heating period, and activation atmosphere. The few applications in TLC and GLC are surveyed in Refs. 62 and 63.

The oxides $CeO_2$, $ZrO_2$, and $TiO_2$ exhibit properties different from those of the foregoing oxides. They precipitate as oxide hydrates from salt solutions of a low pH and hence behave more acidically and more stably toward alkaline solutions than silica. Furthermore, they have strong tendency to form stable octahedral complexes with a variety of ligands.

Ceria beads with an extremely low specific surface area have been used as support for ion pair partition chromatography [64]. Zirconia, made by hydrolysis of zirconyl chloride or zirconium alkoxides, or by thermal decomposition of $ZrOCl_2 \cdot 8H_2O$, yields porous products with a texture very sensitive to thermal and hydrothermal treatment [65,66]. No applications of these oxides in chromatography have yet been reported.

Titania, manufactured as anatase and rutile on a large scale, is mainly employed as white pigment [67]. By means of the sol-gel process, carried out on a two-phase system, porous amorphous titania beads have been synthesized but are just starting to undergo tests in column liquid chromatography [68].

## B. Silicates

Silicates [69] with a layer structure, e.g., clays, and with a framework structure, e.g., zeolites, have gained a considerable amount of attention as adsorbents and catalysts. The clay minerals are composed of charged layers interleaved with hydrated ions [70]. They exhibit cation exchange properties with simple ions and charged organic species, and form intercalation complexes by increasing the distance between adjacent layers. On account of their structure, they are soft materials and cannot withstand high pressure. The clay minerals include the kaolins and montmorillonites (the latter were formerly called bentonites). Bentonites were employed as fillers due to their swelling behavior. By ion exchange, long-chain quaternary ammonium ions have been introduced. Such organoclays have repeatedly been applied in gas and low-pressure column liquid chromatography [71,72] to separate isomers.

While clay minerals have the drawback of a soft structure, framework silicates such as zeolites are based on an infinite three-dimensional silicon-oxygen framework with positions occupied by aluminum atoms. The negative charge of the $SiO_2 \cdot Al_2O_3$ framework is compensated by cations.

By three-dimensional coordination of the $SiO_4$ and $AlO_4$ tetrahedra, polyhedra are formed, exhibiting a distinct crystalline structure and offering a pore system with defined openings and channels [73]. The pore openings are 0.4-1.0 nm wide and are accessible after the removal of adsorbed water. Restricted by the pore size, zeolites adsorb polar substances, but also unpolar compounds, particularly when the $SiO_2$/$Al_2O_3$ ratio of the bulk structure is increased. Due to the presence of micropores, their use in liquid chromatography is limited.

Florisil, a magnesium silicate, is applied as adsorbent for the separation of natural products [74]. It combines some of the acidic properties known from silica-aluminas.

### C. Porous Glass

Porous glass is a nearly amorphous silica with traces of $B_2O_3$ and $Na_2O$. It is produced by leaching of certain borosilicate glass compositions [75]. Careful control of the heat treatment and leaching conditions produces porous granules of narrow and graduated pore diameters between 5 and 500 nm. In its surface chemistry it behaves similarly to porous silica, with small differences associated with surface boron atoms [76]. A detailed survey of its chromatographic application is given in Ref. 77.

### D. Porous Carbon

Porous carbon is used here as a synonym for a variety of solid carbonaceous materials with a pore structure. Active carbon adsorbents, manufactured from carbonaceous precursors by heat treatment, play an important role in industrial purification and cleaning processes [78] and were also employed as adsorbent in the early phases of column liquid chromatography [79,80]. However, active carbons suffer from the fact that they possess a large hydrophobic surface area which is accessible by a microporous system, leading to slow kinetics and undesired peak shapes when applied in column liquid chromatography.

#### *1. Manufacture [81]*

Guiochon and coworkers [82,83] synthesized porous carbon packings through pyrolysis of aggregated particles composed of graphitized thermal carbon black. In addition, thermally pretreated silicas were employed for carbon decomposition to yield composite deposits [84,85]. The reduction of polytetrafluoroethylene (PTFE) at room temperature, followed by a heat treatment, produced porous carbon packings [86, 87]. An analogous procedure is based on the reduction of fluorocarbon Kel-F 300 LD [88]. Ciccioli et al. [89] examined commercial graphitized carbon black, ground to small particle sizes. Unger et al.[90,91] employed cokes and active carbons as precursors which were subjected to extraction and calcination treatments. The most successful approach has been made by Knox and coresearchers [92,93] using porous silica as a template loaded with a carbonizable resin. After carbonization, the silica was removed by dissolution and the remaining porous carbon subjected to a heat treatment at about 2500 K. The material obtained was termed porous graphitized carbon (PGC), which is now commercially available.

### 2. *Surface Chemistry* [94,95]

When porous carbons are calcined at sufficiently high temperatures (above 2200 K), they exhibit a pronounced hydrophobic and aromatic surface character due to graphitization. Apart from this, acidic as well as basic surface functional groups are present, formed through chemisorption of oxygen. These groups can be eliminated by hydrogenation at 1300 K, but might be reconstituted when exposed to the atmosphere at room temperature. The lack of reproducibility of the surface chemistry of porous carbon packings puts the most serious limitation on the widespread use of carbon packings in LC at present.

### 3. *Specific Surface Area and Pore Size*

Another serious drawback is the elimination of the micropores of the packing in order to achieve a reasonable column dispersion. It appears that in the case of porous glassy carbons, subjected to a high-temperature treatment, these micropores were totally removed. In the case of the template synthesis employing silica, the pore size of the template is carried over to the carbon packings. Thus, in order to obtain sufficiently porous carbons, silicas with high pore volumes must be employed as templates.

### 4. *Chromatographic Retention*

As a result of the pronounced hydrophobic character, porous carbons strongly retain hydrocarbons, in particular aromatic hydrocarbons, compared to reversed phase silica packings [96,97]. Isomeric compounds are separated on carbons due to sterical and nonlocalized solute-surface interactions. Selectivity is also found for polar compounds, with a somewhat different pattern than for reversed phase silicas [81].

## E. Crosslinked Organic Polymers

### 1. *General Aspects*

Synthetic crosslinked organic polymers were introduced as packings in column liquid chromatography a decade or so later than oxides. The first organic-based packings were synthetic ion exchangers made by condensation polymerization of phenol and formaldehyde [98]. In the 1960s, procedures were developed by Moore [99] to synthesize crosslinked polystyrenes with graduated pore sizes for size exclusion chromatography. The synthesis of crosslinked dextran [100,101] and agarose [102] were milestones in the manufacture of the polysaccharide type of packing. At the same time, polyacrylamide packings were synthesized from acrylamide and N,N'-bismethylene acrylamide [103,104]. All the above products except the first one served as packings in size exclusion chromatography. The major breakthrough in the synthesis of crosslinked organic polymers with tailor-made properties for column liquid chromatography occurred in the decade between 1960 and 1970 (for comparison, see Refs. 105-107). Since then, crosslinked organic polymers have maintained a leading position as packings in ion exchange and size exclusion chromatography while their use in column liquid adsorption chromatography to resolve unpolar and polar low molecular weight compounds was rather limited and bare silica dominated the market. Currently, the situation seems to be changing slightly, but distinctly. It is the author's belief that organic-based packings will gain greater importance in the future.

In textbooks one often finds that the structure of organic- and oxide-based packings are treated according to different aspects [108]. On viewing a particle of an organic polymer and an oxide, its structure is best described by a coherent system either of three-dimensionally crosslinked chains or of a three-dimensional array of packed colloidal particles as limiting cases. Thus, previous classifications into xerogel, xerogel-aerogel hybrid, and aerogel [108] appear to be inadequate form this point of view. In order to provide a sufficient rigidity of particles, the chains or colloidal particles should be linked by chemical bonds rather than by physical attraction forces.

Some examples serve to illustrate the structure of organic-based packings. Figure 5 shows a scheme of the structure of a crosslinked polyacrylamide gel and a crosslinked agarose gel of an equivalent polymer concentration. Both polymers possess a random coil structure. In the agarose, the double-helix-shaped chains are collected in bundles which generate a quite open structure [109]. The structure is stabilized by hydrogen bonds between the chains. When agarose is subjected to crosslinking, links are formed among the chains in these bundles [111]. Another type of structure is met in macroporous, marcoreticular, or isoporous polymer packings (see Fig. 6). As the name implies, these polymers contain so-called macropores of larger than 10 nm pore size and micropores of less than 1 nm pore sizes, the latter being inaccessible to large solutes. In other words, the macroporous polymer particles constitute an agglomerate made of secondary particles which themselves represent agglomerates of microspheres. This structure resembles that of porous silica particles, which are composed of agglomerates of spherical colloidal silica particles. With respect to the mechanical rigidity of the polymeric packings, crosslinking becomes an essential means in the synthesis. Other requirements which must

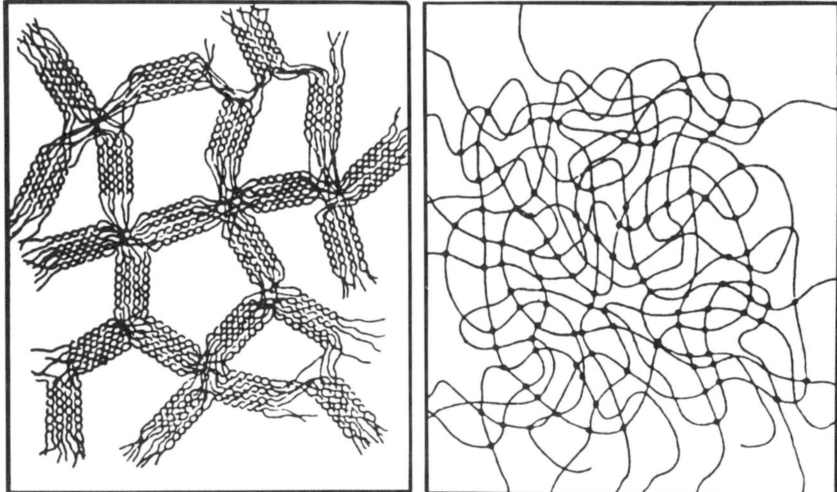

Fig. 5 Schematic representation of the agarose gel network (right) as compared to a network formed by random chains of Sephadex or Bio-Gel P at similar polymer concentrations. Note that the aggregates in agarose gels may actually contain 10-$10^4$ helices rather than the smaller numbers shown here. (From Ref. 110 by permission of the author and the publisher.)

be met are insolubility, resistance to oxidation and reduction, and a defined, controllable, and reproducible pore structure.

Polymerization is performed either by condensation or addition polymerization, depending on the type of starting monomer. For crosslinking, comonomers such as divinylbenzene (styrene), bisethylene glycol methacrylate (ethylene glycol methacrylate), epichlorhydrin, 2,3-dibromopropanol, and divinylsulfone (saccharides) are added [106, 117-119]. The crosslinking reagent can amount to as much as 70% (w/w) (see Ref. 107). Macroporous copolymers are synthesized in the presence of an inert solvent which functions as a volume modifier. Both the crosslinker and the inert solvent have a substantial impact on the kinetics of the polymerization and the resulting properties of the copolymer. The decisive parameters relevant for the synthesis of macroporous copolymers are reviewed in Refs. 105 and 112.

As in the synthesis of silica packings, specific processes must be chosen in polymerization to manufacture polymeric packings with beads of controlled size and size distribution [113]. Emulsion polymerization starts with a solution of a detergent to which the monomers are added. As a result, micelles swollen with the monomer are formed. After a water soluble initiator is added (for styrene as a monomer),

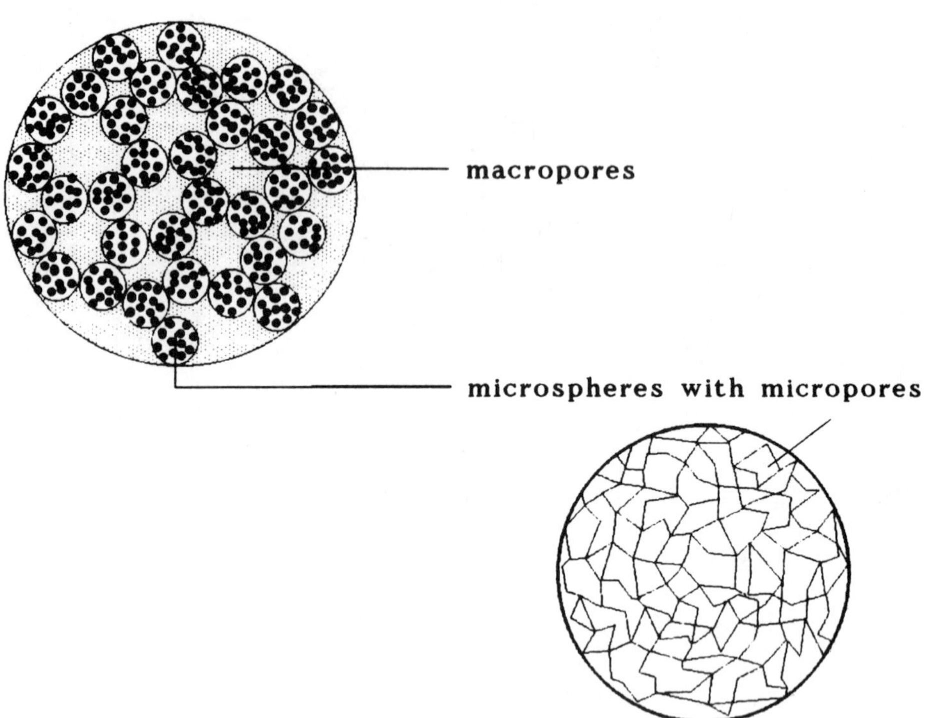


Fig. 6 Structure of a macroporous polymer.

polymerization leads to particles of exactly the same size as the swollen micelles. Emulsion polymerization processes generate particles of up to 0.5 μm in one step.

Suspension polymerization is usually designed to prepare larger beads of >5 μm mean particle diameter. The monomer or comonomer solution is vigorously agitated in water in the presence of a colloidal suspending agent. The colloidal agent coats the hydrophobic monomer droplets (in the case of, say, styrene or divinylbenzene). Coalescence of the droplets is prevented by the surface charge of the droplets. Adding a lipophilic catalyst or initiator starts the polymerization in the droplets and this continues until the beads are solidified in bulk. The size of the beads is thus controlled by the size of the droplets via the stirring speed.

A third variant in polymerization technology is the swollen emulsion polymerization pioneered by Ugelstad [114,115]. The procedure is performed in two steps. First the polymerization is started by adding a swelling agent, which causes the submicrometer polymer particles to swell by large volumes of the monomer. The increase in volume can reach a factor of 1000. Second, in a consecutive step the monomer-swollen beads of defined size are polymerized.

Having briefly examined the structure of organic packings and the various routes in their manufacture, the most important features may be summarized as follows:

1. Hydrophilic as well as lipophilic organic packings are synthesized with a controlled pore and surface structure depending on the type of monomer/comonomer and the polymerization reaction. The surface structure can be altered by controlled consecutive surface reactions.
2. In accordance with the bulk composition, polymer packings are stable across almost the entire pH range, particularly under strong alkaline conditions.
3. The chemical stability is affected by oxidizing and reducing solutions.
4. Although crosslinking reacitons have been optimized inasmuch as rigid pressure-stable particles can be manufactured, some remaining swelling property is often noted when changing the solvent composition in HPLC.
5. As in the manufacture of silica, porosity, pore size, and surface area of polymer packings can be adjusted over a wide range, and micro-, meso-, and macro- as well as nonporous beads are synthesized reproducibly.

### 2. *Hydrophilic Polymer Packings: Polysaccharides*

This family comprises crosslinked agarose and dextran, which were originally designed for the separation of biopolymers by size exclusion chromatography [100-102]. By means of appropriate chemical derivatization, packings for ion exchange, hydrophobic interaction, and affinity chromatography have been obtained [116]. Novel packings based on dextran and agarose have since been developed, possessing a small bead diameter and high rigidity.

Agarose is made form agar. The byproduct agaropectin is removed by washing agar with a buffer. Beading is achieved by an emulsion-gelation procedure [102,110] as follows:

> A warm organic solvent immiscible with water is poured into a round-bottom flask containing a warm agarose solution. A "water in oil" emulsion is created by stirring. The presence of a stabilizer prevents the agarose-containing water beads from coalescing. When the temperature is lowered the water solution in the beads forms a

gel. The organic phase is then removed by suction filtration. By elutriation a more narrow size distribution of the gel beads can be obtained.

For crosslinking of the beads, epichlorohydrin, 2,3-dibromopropanol, and divinylsulfone have been employed [111,117-119]. Rigid agarose packings contain 8-20% agarose. The bead size can be controlled in the range 2-100 μm.

Commercial products are Sepharose 2B, 4B, and 6B of Pharmacia LKB Biotechnology—2B, 4B, and 6B designating the corresponding agarose concentration in the beads (2, 4, and 6%). Sepharose is stable in water and salt solutions over the pH range 4-9 in the absence of oxidizing reagents.

Sepharose CL is made from Sepharose by reaction with 2,3-dibromopropanol under strongly alkaline conditions. The crosslinked derivative has the same porosity as the parent substance but offers a substantially higher thermal and chemical stability. After crosslinking, the material is desulfated by alkaline hydrolysis under reducing conditions, yielding a product with an extremely low content of ionizable groups (see Fig. 7). Phenyl- and Octyl-Sepharose CL-4B are derivatives of Sepharose CL designed for hydrophobic interaction chromatography (HIC) of proteins. They are prepared by coupling crosslinked Sepharose CL-4B with the glycid ether of the phenyl and octyl moieties.

Sephadex (*Sep*aration, *Pha*rmacia, *Dex*tran) is a dextran crosslinked with epichlorhydrin pioneered by Porath and Flodin [100] as a size exclusion packing. Its partial structure is shown in Fig. 8. Sephadex products differ in the degree of crosslinking and hence in their swelling properties. The high content of hydroxyl groups renders the packing hydrophilic. It is stable in water, salt solutions, organic solvents, alkaline, and weakly acidic solutions. In strong acids the glycosidic linkages of the matrix are hydrolyzed. Prolonged exposure to oxidizing reagents also degrade the product. Sephadex LH 20 and LH 60 as well as Sephasorb HP Ultrafine are prepared by hydroxypropylation of crosslinked dextran. Sephadex LH 20 has been successfully applied in preparative chromatography for the isolation of polar low molecular weight compounds. Dichloromethane was used as solvent. Mass loadability is reported to amount to about 300 mg/g of dry packing [122].

Sephasorb HP Ultrafine resolves aldehydes, phenols, ketones, nucleotides and aromatic compounds.

### 3. Lipophilic Polymer Packings

*Poly(styrene-divinylbenzene)* The synthesis of crosslinked copolymers of styrene and divinylbenzene has been studied intensively and is well documented [105-107,114,115]. The starting monomer is sty-

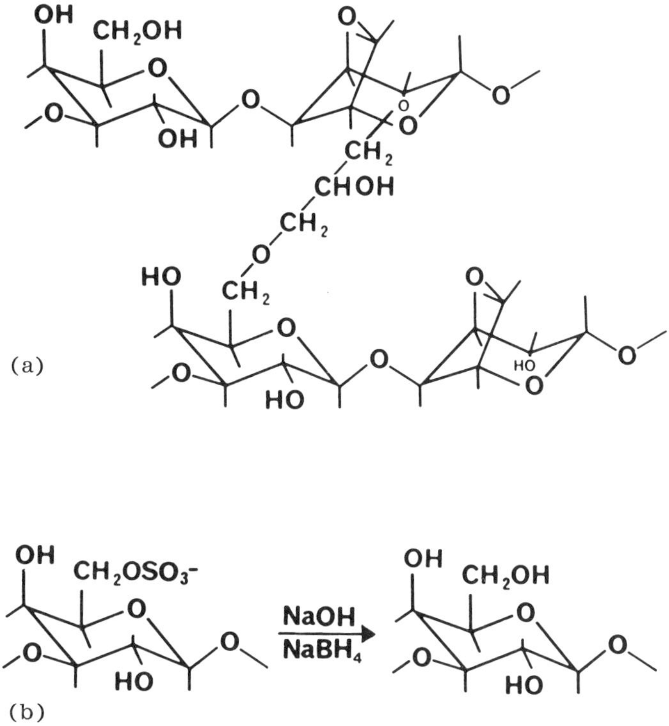

Fig. 7 (a) Structure of crosslinks in Sepharose CL. (b) Loss of sulfate groups by alkaline hydrolysis under reducing conditions. (From Ref. 120 by permission of Pharmacia LKB Biotechnology.)

rene. Divinylbenzene (DVB) is used as crosslinker. The amount of DVB can reach about 55% (w/w). At 55% DVB, the copolymer shows practically no swelling and possesses a permanent porosity [107]. The network structure of poly(styrene-divinylbenzene) (St-DVB) is illustrated in Fig. 9. Commercial products differ in bead size and pore size (see Appendix 7). There are even nonporous products on the market, designed for the rapid separation of peptides and proteins by reversed phase HPLC [123]. St-DVB copolymers are stable in the pH range 0-14. They find increasing application in the separation of low molecular weight compounds, peptides, and proteins by means of reversed phase chromatography [124,125] and as parent materials for the synthesis of derivatized packings in interactive chromatography of biopolymers [116,126,127].

A large number of other organophilic polymer packings have been synthesized for size exclusion and interactive HPLC after suitable de-

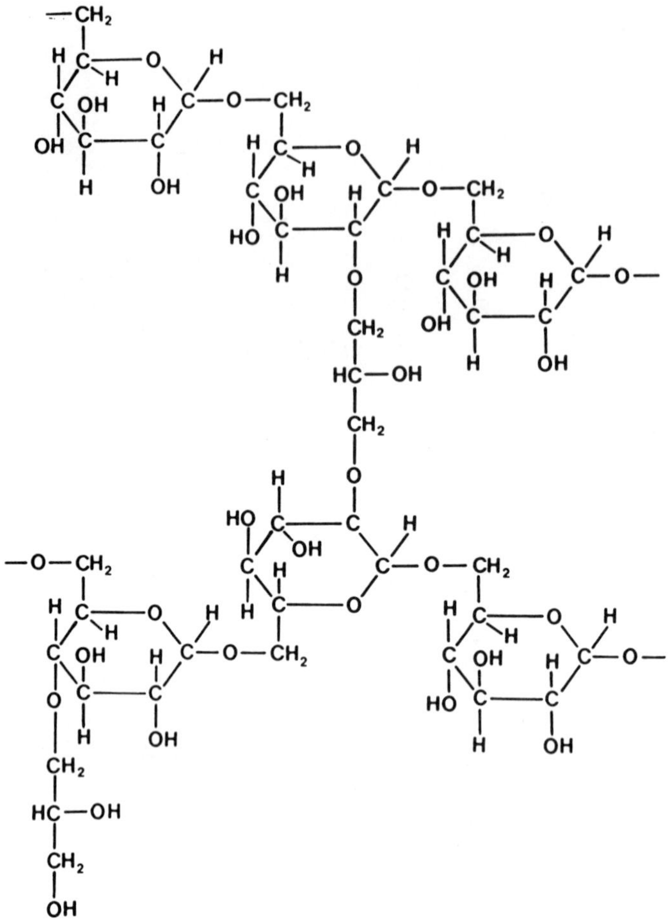

Fig. 8 Partial structure of Sephadex. (From Ref. 121 by permission of Pharmacia LKB Biotechnology.)

rivatizations. They are reviewed in depth in Ref. 116. Some of these packings are mentioned briefly below.

*Polyacrylamide* Acrylamide as monomer and N,N'-methylenebis-acrylamide as crosslinker [103,104].

*Polyvinylacetate* Vinylacetate as monomer and divinyl adipate as crosslinker [128].

*Polymethyl Methacrylate* Methylmethacrylate as monomer and glycol dimethacrylate as crosslinker [129,130].

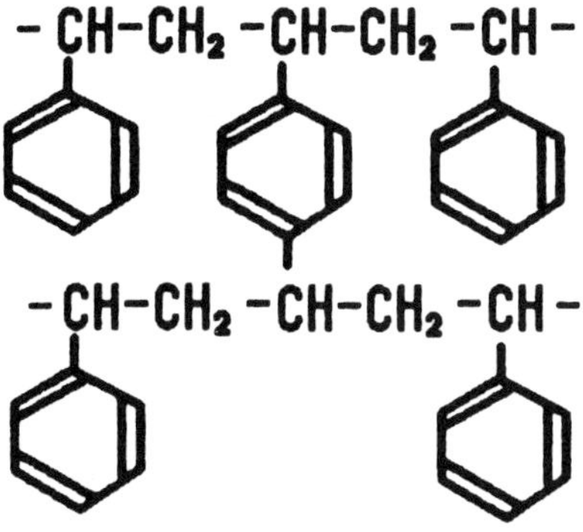

$-CH-CH_2-CH-CH_2-CH-$

$-CH-CH_2-CH-CH_2-CH-$

Fig. 9 Structure of poly(styrene-divinylbenzene).

*Polyethylene Glycol Methacrylate* Ethylene glycol methacrylate as monomer and bisethylene glycol methacrylate as crosslinker [131].

A list of commercial polymer packings can be found in Appendix VII.

## F. Bonded Packings

### *1. General Aspects*

In contrast to gas chromatography, liquid chromatography provides the unique advantage of adjusting and controlling retention and selectivity by both the mobile phase and the stationary phase composition. At the onset of HPLC the variation of the stationary phase was simply achieved by choosing an appropriate liquid stationary phase (see Chap. 4). The classical liquid-liquid partitioning systems, however, had the drawback of limited column stabilities caused by bleeding. Therefore, attempts were made at the end of the 1960s to develop bonded stationary phases for HPLC with silica as parent material [132,133]. The chemical modification of silica adsorbents, being either porous or finely dispersed, was thoroughly examined almost two decades ago (for comparison, see Ref. 134). The broad knowledge accumulated in the modification of adsorbents, pigments, and fillers was then readily adapted to modify chromatographic packings and greatly advanced the synthesis of tailor-made bonded phases. There were other favorable circumstances which furthered the rapid development of bonded packings. First, advances came from various areas of material science where novel high-tech products were synthesized with the aid of modern surface analysis methods. Second, a direct impact came from physiochemical studies of solute-surface interactions which allowed a prediction of design elements for the desired stationary phase and facilitated the synthesis of tailor-made products. Third, modeling of molecular structures initiated the modeling of stationary phases, particularly in the fields of affinity chromatography and chromatography of enantiomers.

The modification of a support in HPLC serves a number of purposes, such as

Attachment of desired and tailored functional groups at the surface, mimicking the structure of solutes and hence greatly enhancing selectivity
Deactivation of the parent surface to minimize or avoid undesired matrix effects and too strong adsorption of basic compounds
Enhancement of the chemical stability of the parent support

From a purely geometric point of view, bonded phases can be divided into several types such as monolayer, multi- and polymer layer, coating, sandwich structure, structure with diffusion barrier, and mixed-mode structure (see Fig. 10). This grouping conveniently follows the route by which bonded phases are synthesized.

*Monolayer Type* It is common in chemisorption that the uptake of an adsorptive which is chemically bound to the surface of an adsorbent is confined to a monolayer as described by a Langmuir type of isotherm. In surface modification of silica the activated support or its chlorinated derivative is reacted with an appropriate silane or a Grignard reagent, whereby the reactant molecule is coupled by one or two covalent links to the surface. Such anchoring links are $\equiv$Si-O-C$\equiv$, $\equiv$Si-C$\equiv$, $\equiv$Si-N=C=, or $\equiv$Si-O-Si-C$\equiv$ bonds. The latter is the most common. Bonding occurs through the reaction of the surface hydroxyl groups of the silica with a silane of type $X_nSiR_{4-n}$ (n = 1-3) under anhydrous conditions, where X is a chloro, alkoxy, or dimethylamino group and R an organic radical. For a monofunctional silane of type $XSiR_3$, the reaction can be written as

$$\equiv\text{Si-OH} + \text{X-SiR}_3 \rightleftarrows \equiv\text{Si-O-SiR}_3 + \text{HX}$$

When the parent silica is fully hydroxylated, the concentration of surface hydroxyl groups amounts to 8-9 μmol $m^2$, being equivalent to 4.8-6.3 OH groups $nm^2$. The maximum surface concentration of silanes of type $XSiR_3$ was found to be 4.75 μmol $m^2$ for trimethylsilyl groups and 4.1 μmol $m^2$ for n-alkyldimethylsilyl groups with longer n-alkyl chains [135,136].

Thus, about 50% of the original hydroxyl group population remains unreacted, being accessible to small solutes in liquid-solid adsorption equilibria. Kovats and coworkers [137,138] claimed that shielding

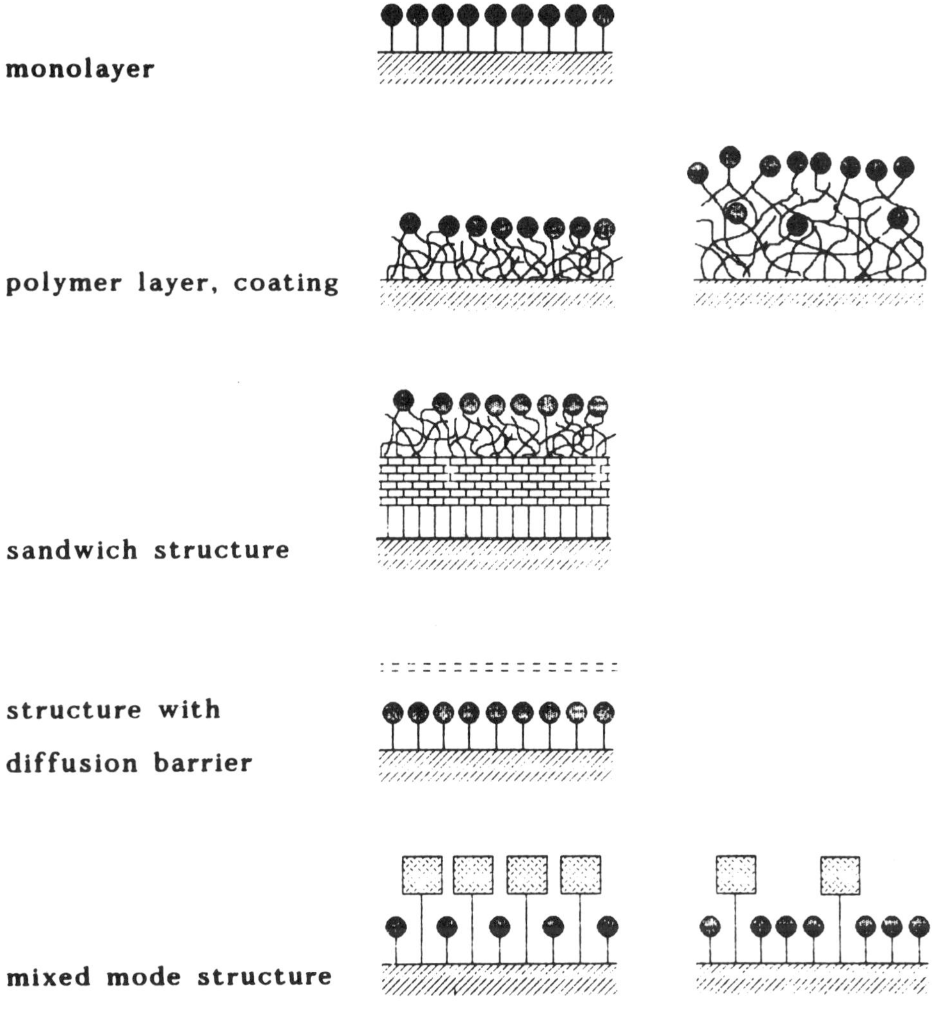

**monolayer**

**polymer layer, coating**

**sandwich structure**

**structure with diffusion barrier**

**mixed mode structure**

Fig. 10 Structure of bonded phases

of the remaining hydroxyl groups is most efficient by chemical bonding of the following silanes: N-[(3,3-dimethylbutyl)dimethylsilyl]-N,N-dimethylamine, N-[(5-cyano-3,3-dimethylpentyl)dimethylsilyl]-N,N-dimethylamine, and N-[tetradecyldimethylsilyl]-N,N-dimethylamine (see Fig. 11).

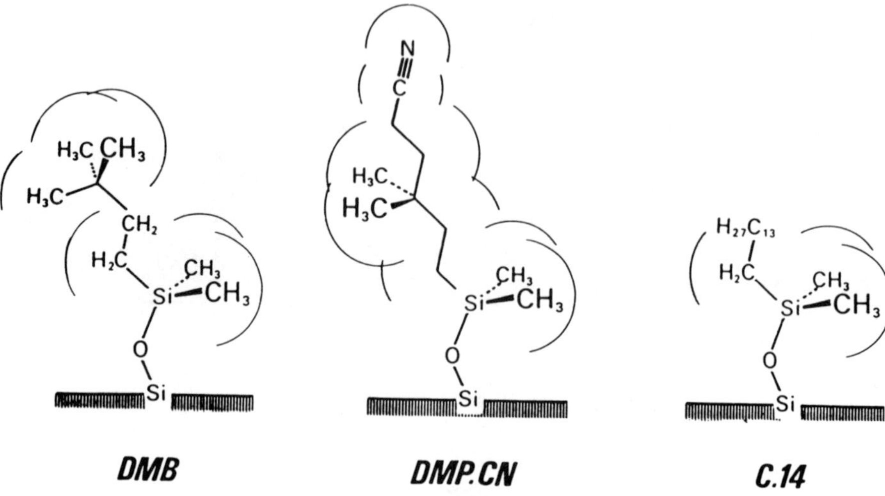


Fig. 11 Organosiloxy groups which effectively shield the silica surface: DMB substituent = (3,3-dimethylbutyl)dimethylsiloxy, DMP CN substituent = (5-cyano-3,3-dimethylpentyl)dimethylsiloxy, and $C_{14}$ substituent = tetradecyldimethylsiloxy. (From Ref. 137 by the permission of the publisher.)

Bifunctional and trifunctional silanes of type $X_2SiR_2$ and $X_3SiR$ are also capable of forming monolayers, provided water is excluded. Hereby two surface hydroxyl groups react with one silane molecule at maximum. Unreacted Si-X groups at the bonded organosilyl groups are subsequently hydrolyzed to "new" hydroxyl groups due to the water at the washing procedure after reaction. It is noteworthy that these products then possess the same concentration of hydroxyl groups as organo groups, plus the additional hydroxyl groups left unreacted. In order to remove the remaining surface hydroxyl groups, a second silanization called "end-capping" is carried out using small-size reactive silanes. These procedures have been adapted from hydroxyl group protection chemistry and from techniques aiming at the deactivation of supports in gas chromatography. Typical silanes are hexamethyldisilazane, trimethylsilylimidazole, trimethylsilyldimethylamine, trimethylchlorosilane, dimethylchlorosilane, N-trimethylsilyl-N,N'-diphenylurea, bis(trimethylsilyl)urea, monotrimethylsilylacetamide, N,O-bis(trimethylsilyl)acetamide, N,O-bis(trimethylsilyl)trifluoroacetamide, N-methyl-O-trimethylsilyltrifluoroacetamide, N,O-bis(trimethylsilyl)-carbamate, N,O-bis(trimethylsilyl)sulfamate, trimethylsilyltrifluoromethanesulfonate, and triisopropylsilyltrifluoroacetate.

*Polymer Layer and Coating Type* The surface reactions between silica and trifunctional silanes in the presence of water yield polymer layers through condensation. Suitable silanes with double bonds in the organic radical have been chemically bonded to silica and then subjected to polymerization. Chemical attachment was not found to be needed to sufficiently hold the polymeric layer at a support. Numerous examinations have been reported where monomers, oligomers, and polymers were adsorbed at the surface of inorganic supports such as porous glasses, silicas, and aluminas [139-156]. Polymerization of adsorbed monomers, oligomers, or other precursors is accomplished by thermal treatment, by adding an initiator, and by $\gamma$ irradiation. Polymers that have been deposited on inorganic supports were polyacrylamide, polymethylsiloxane, polybutadiene, polyvinylpyrrolidone, poly-2-hydroxy-3-N-ethylenediaminobutadiene, etc. A review on the chemical modification by polymer coating was recently given by Schomburg [7]. Another principle to hold the polymeric layer on the support is based on electrostatic attraction between the charged surface and the charged polymer, provided they carry opposite surface charges. Typical instances are polyethyleneimine [147] and dextran or agarose with dimethylaminoethyl groups [157], which both carry positive charges that have been attached to the negatively charged silica surface. When depositing polymer layers or coatings on porous supports, it is essential to adjust and to control the layer thickness and homogeneity of the deposit. The aim is to avoid pore blocking and matrix effects, and ensure rapid mass transfer in chromatographic separation.

*Sandwich Structure Type* This principle was mainly advocated by El Rassi and Horvath [158] to prepare bonded phases for the separation of nucleic acids. An n-propyl group is first bonded to the silica as a sublayer and this is linked to a polar top layer composed of large, highly polar compounds such as polyethyleneimine, polyols, etc. Such sandwich structures provide a high flexibility in the design of bonded phases and permit the binding of functional groups at the top layer with the desired spacing.

*Structures with Diffusion Barriers* In the analysis of drugs in biological fluids it is necessary to remove the proteins associated with the drug [159]. As a first attempt, protein-coated reversed phase packings were applied which did not adsorb plasma proteins but retained the drug by hydrophobic interaction [160]. Pinkerton introduced the principle of internal reversed phase supports [161]: a 7-nm pore size silica was modified in such a way that the external surface of the beads was hydrophilic and the internal surface of the pores hydrophobic. Proteins elute quickly on these columns because they are not adsorbed and cannot penetrate the pores, while the drug enters the internal pore volume and is retained.

Another concept recently introduced by Desilets and Regnier [162] is based on the use of polyoxyethylene sorbitan monoalkylate surfactants which are adsorbed at the surface of a conventional reversed phase packing and form a semipermeable hydrophilic layer over the silanized surface. The layer sterically hinders protein access to the hydrophobic stationary phase.

*Mixed-Mode Structure* Usually, the chemical heterogeneity of the surface of an adsorbent is undesirable in chromatographic separation because more than one retention mechanism is then operating in solute retention, making it difficult to develop a separation strategy. On the other hand, it is known that solutes with multifunctional groups such as polar, ionogenic, and nonpolar call for multifunctional types of stationary phases to be resolved.

Thus the concept of controlled heterogeneity was established and mixed-mode bonded phases have been synthesized [163-166]. Such phases mostly consist of two types of functional groups such as hydrophobic/ionic and hydrophilic/hydrophobic. In order to make one retention principle dominant, the mobile phase composition must be properly adjsuted. The mixed-mode bonded phases possess a great similarity to the simple approach whereby different bonded phases are blended and applied as mixed-bed columns [167].

It should be emphasized that all bonded phases in HPLC have a bifunctional character, either intrinsic or introduced by the bonding chemistry. The bifunctional character of a bonded phase is reflected in the so-called bimodal retention plot. When measuring the solute capacity factor k' on these phases as a function of the mobile phase composition in the isocratic or gradient elution mode, the k' value is seen to decrease first, passing through a minimum and then increasing. Figure 12 illustrates three typical cases in a more schematic way. Numerous examples are found in the literature. Figure 12a represents a typical situation met in reversed phase gradient elution of peptides and proteins [168]. The logarithm of k' is plotted against the volume fraction $\Phi$ of the organic modifier of the mobile phase. The initial descending part of the curve is dominated by hydrophobic interactions. The latter decrease when $\Phi$ increases. At the minimum, the solutes are unretained. In organic-rich mobile phases k' increases, associated with silanophilic interactions between the surface and the solute. Figure 12b shows a similar curve which is typical for ion exchange chromatography of proteins [165]. On the left-hand side at a low salt concentration, solute retention is dominated by electrostatic interactions, which decrease with increasing salt concentration. After k' has passed through a minimum, the ascending part of the curve is caused by hydrophobic solute-surface interactions.

The third case, typified by Fig. 12c, holds for polar-bonded phases such as diol, cyano, etc. Here, k' is plotted against the mobile phase composition when a cyano-bonded phase is used [169]. Retention is highest when starting with an unpolar mobile phase, e.g., n-heptane. By adding increasing amounts of dioxan to the mobile phase the retention declines. The column is operated in the straight-phase mode. Retention is at a minimum in pure dioxan. Adding increasing amounts of a buffer to the mobile phase (pure dioxan), k' is seen to increase again. The column is now run under reversed phase conditions. The hydrophobic interactions arise from the propyl spacer of the functional groups and the hydrophobic residues of the solute. Given these considerations, it becomes obvious that a strict discrimination of bonded phases into unpolar, polar, and ionogenic is not realistic.

A typical instance is a reversed phase silica packing which carries hydrophobic as well as hydrophilic groups, the latter capable of forming ionogenic groups at pH > 4 of the mobile phase. Thus, the powerful tool in our hands is the mobile phase composition which enables us to play with the specificity and strength of solute-surface interactions while aiming at a high resolution.

## 2. *Reversed Phase Packings*

*Evolution of Reversed Phase Packings* The term *reversed phase* has been introduced to collectively name those lipophilic stationary phases in HPLC which are employed with polar mobile phases (see Sec. VI). Pioneering work was done by Howard and Martin [170], Boldingh [171], Tiselius and coworkers [172-174], and Horvath and Lipsky [175]. They used silanized kieselguhr, rubber, and charcoal as supports; these were equilibrated with lipophilic stationary phases and then employed with polar mobile phases in partition chromatography [176].

The idea of utilizing silanized silicas as lipophilic bonded phases in column liquid chromatography traces back to the work of Stewart and Perry [177], which is based on experiments by Abel et al. [178] employing n-hexadecyl-bonded silicas in gas chromatography.

The breakthrough of reversed phase packings began when Kirkland and De Stefano [179] synthesized a bonded phase by coupling a poly-n-octadecylsiloxane to the surface of a pellicular support. The product was marketed under the trade name of Permaphase ODS. By reaction of microparticulate silicas with organotrichlorosilanes, Majors [180] prepared a series of RP packings where the organofunctional group was anchored via a ≡Si-O-Si(-)(-)-C≡ bond to the silica surface. Bonded packings of the ester type with a ≡Si-O-C≡ bond were popularized in gas chromatography by Rossi et al. [181] and Halasz and Sebestian [182,183]. Attempts to use these bonded phases in HPLC proved unsuccessful due to their limited pH stability. Although the

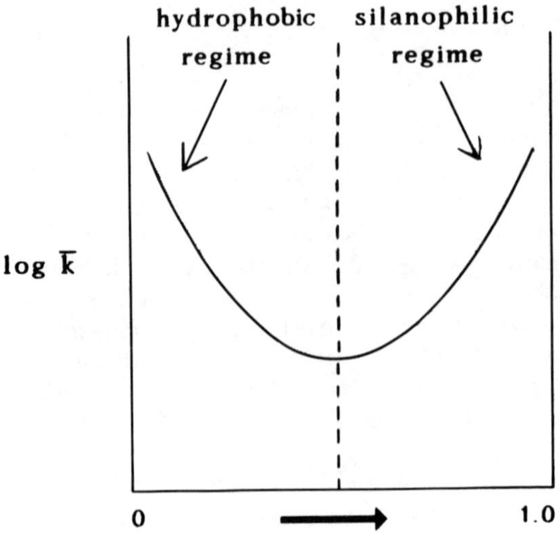


(a)

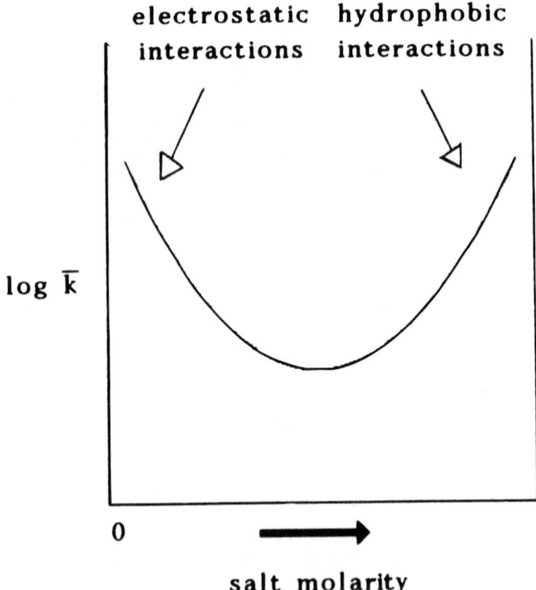


(b)

Fig. 12 Bimodal retention plots in reversed phase chromatography (a), ion exchange chromatography (b), and chromatography of polar bonded phases (c).

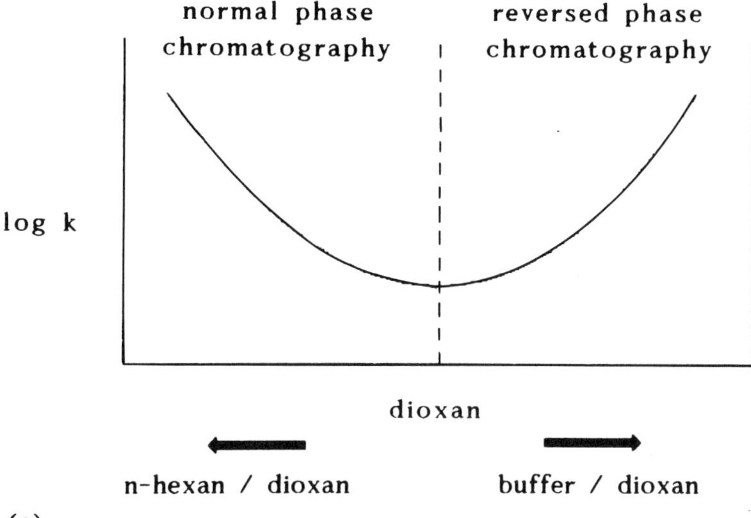


(c)

Fig. 12 (continued)

vast majority of packings are based on microparticulate silicas as support, other support materials such as aluminas, crosslinked agarose, and poly(styrene-divinylbenzene) are also attracting a great deal of attention.

*Stationary Phase Requirements* The rapid expansion of RPC as the dominating mode in HPLC was accompanied by a step-by-step improvement of reversed phase packings. In retrospect, the last two decades can be clearly divided into two periods. The first period covers the decade 1970-1980 when the major focus was put on the development of reversed phase packings for the analysis of low molecular weight compounds. Typically, reversed phase packings were n-octadecyl- and n-octyl-bonded phases on 6- to 15-nm pore size silicas. The second period started around 1980. As a consequence of the rapid development in biotechnology, biochemistry, and molecular biology, fast and high-resolution methods were needed for biopolymers. Thus, research efforts were centered on tailor-made reversed packings based on silicas of a pore size larger than 15 nm and with a more sophisticated surface chemistry in order to adapt ot the requirements of biopolymer separations. This phase is still in progress.

Covering all aspects of the application areas of RP packings, a number of specifications and requirements that must be met by RP packings can be defined as follows.

(1) In order to adapt to the variety of compounds to be resolved in HPLC, the lipophilic character of the stationary phase should be reproducibly adjusted and controlled by the bonding chemistry. This can be accomplished by the choice of the lipophilic group, the density, and the carbon load. For example, in the separation of biopolymers soft hydrophobic solute-surface interactions are often required to prevent unfolding and denaturation of the biomacromolecules. This has lead to a family of specific lipophilic bonded phases with low hydrophobic character termed hydrophobic interaction packings [126,127].

(2) In order to match the molecular size of solutes to be resolved, the parent support must provide sufficiently large pores and an accessible pore system throughout the particles. The focus is on achieving a fast mass transfer of solutes in the stagnant mobile phase and fast adsorption-desorption kinetics. For HPLC of low molecular weight compounds of <3 kD, bonded phases with a pore size between 6 and 15 nm are sufficient. Biopolymers of larger than 20 kD require pore sizes of 30-100 nm [127]. Furthermore, it is important in this context that the pore structure of the support constitutes an open system of regular channels providing a high connectivity.

(3) The modification of the support should be performed in such a way that a uniform and homogeneous stationary phase is obtained. Homogeneity here has a twofold meaning. First, it refers to a constant thickness of the bonded layer or coating on the parent surface. Second, it implies a homogeneous distribution of surface sites along the surface, avoiding highly as well as less active groups and yielding uniform and distinctively selective solute-surface interactions. The quality of a bonded phase in terms of these parameters not only improves the chromatographic separation process itself but also affects to a large extent equilibration and regeneration of the column, as well as the stability and the lifetime of a given bonded phase. A bonded packing should withstand a variety of elution conditions, i.e., changes of the pH, the ionic strength, and the temperature.

(4) When bonded phases are operated in the elution and displacement mode of HPLC, it is important that the solutes be quantitatively eluted, that the chemical structure remain unchanged, and that the biological activity be preserved.

(5) Bonded packings should be synthesized with a high batch-to-batch consistency in order to attain the highest possible reproducibility of chromatographic parameters. This demand poses severe problems for producers, particularly when 10- to 100-kg batches are manufactured. A feature not yet fully investigated is the possible deterioration in the properties of bonded packings when stored in bulk or in columns over an extended period of time.

Given these considerations it is apparent that all requirements cannot be satisfied by currently available packings and compromises must be made. However, on comparing the properties of RP packings made in the mid 1970s to those of current products, a substantial improvement cannot be denied.

*Preparation of Reversed Phase Packings* The support and its activation. In order to prepare reversed phase packings, various supports such as silica [139], alumina [156], agarose [184,185], poly-(styrene-divinylbenzene) [186], and hydrophilic polymers [124] are employed as parent materials. The pretreatment of the parent material prior to modification differs, depending on the surface composition of the support and on whether bonding or simple coating is intended. Pretreatment of silica before modification serves both to clean and to activate the surface. Several procedures have been recommended:

1. Washing the product with water to remove salts, basic or acidic constituents of the surface
2. Treatment with complexing agents to extract metal impurities
3. Treatment with acids in order to extract inorganic impurities and to rehydroxylate the surface
4. Hydrothermal treatment to fully rehydroxylate the surface [187]
5. Calcination followed by rehydroxylation to stabilize the structure and to remove volatile impurities [187]
6. Rehydroxylation of silica by hydrofluoric acid [187]

The treatment applied is specific to any silica product, i.e., it depends on its origin and the resulting surface properties. The removal of inorganic impurities, e.g., metals, is monitored by measuring the content before and after the treatment by inductively coupled plasma-atomic emission spectroscopy (= ICP-AAS). Several methods have been developed to assess the concentration of surface hydroxyl groups [188,189]. The most reliable are based on isotopic exchange followed by spectroscopic analysis, e.g., exchange with deuterated water and mass spectrometric analysis [190], and exchange with deuterated trifluoroacetic acid and $^1$H nuclear magnetic resonance (NMR) spectroscopy [191]. The latter method permits an estimation of the concentration of surface hydroxyl groups and the amount of physisorbed water simultaneously. Spectroscopic methods such as transmission infrared and solid-state NMR have the disadvantage that they also detect bulk hydroxyl groups and bulk water in addition to surface species.

Reagents for hydrophobization of packings. Reagents fall into two main groups: reactive organosilanes and hydrocarbons with functional groups suitable for polymerization. Monomeric silanes (see Figure 13) are commonly of the type:

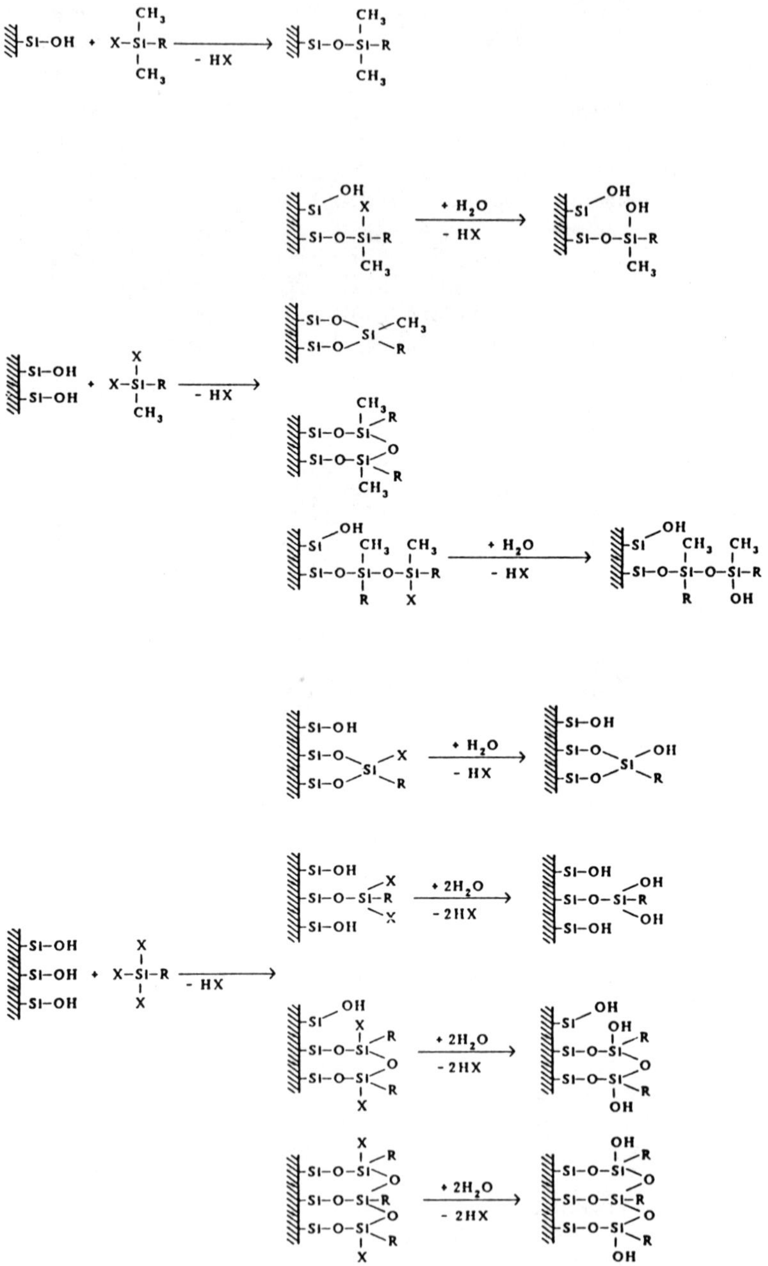

Fig. 13 Pathways in the reaction of silica and monofunctional (a), bifunctional (b), and trifunctional silanes (c). (From Ref. 206 by permission of the author.)

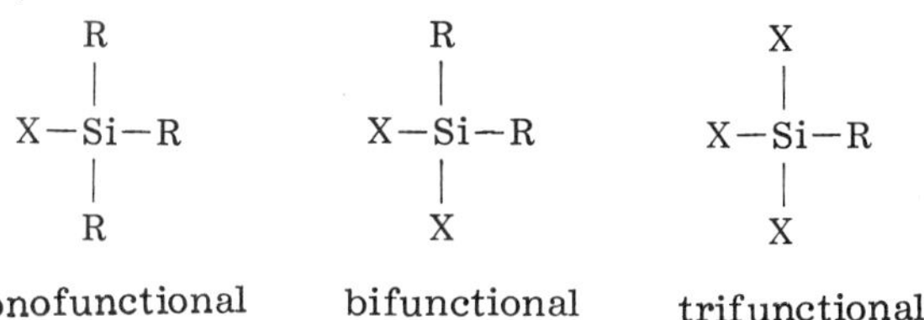


where X is chloro or alkoxy.

Other rarely employed silanes are:

$RR'_2Si\text{-}OCOCF_3$ organosilyltriflates [192]

$RR'_2Si\text{-}N(CH_3)_2$ N-(organosilyl)-N,N-dimethylamine [193]

$RR'_2Si\text{-}O\text{—}O$ (enolate structure) organosilylenolates [194]

Appendix V contains a list of some, but not all, silanes employed for the silanization of silicas together with their properties. Silanes were first produced by the Müller-Rochow synthesis [195] and by the hydrosilylation reaction [196]. Since that time, the number of silanes of technical importance has increased substantially. For silanization of chromatographic packings, silanes must be purified by distillation and by other means. Also, their purity must be assessed by spectroscopic means before use and after an extended period of storage.

Other reactive silanes are those with double bonds, to be subjected to polymerization with suitable monomers after bonding. Wheals [197] studied the reaction of vinyltrichlorosilane with silica and the subsequent polymerization with monomers such as acrylonitrile, acrylic acid, butyl methacrylate, etc. The other groups comprise hydrocarbons with double bonds, e.g., styrene, butadiene, etc.

Reaction routes and conditions. Organosilanes differ in their reactivity, which is attributed to the type of reactive group, Si-X, of the silane. The most reactive silanes are triorganyldimethylaminosilanes constituting a $\equiv$Si-N= bond [193]. The reaction mechanism between the silica surface and the silane is proposed as a nucleophilic attack of the oxygen atom of the hydroxyl group on the silica atom of the silane, forming a reactive intermediate with a pentavalent silicon atom:

$$\equiv Si{-}O{-}H + CH_3 \cdots Si(CH_3)(R){-}N(CH_3)_2 \longrightarrow \equiv Si{-}O(H) \cdots Si(CH_3)_2(R){-}N(CH_3)_2 \rightleftharpoons \equiv Si{-}O{-}Si(CH_3)_2(R){-}\overset{H}{N}(CH_3)_2$$

$$\longrightarrow \equiv Si{-}O{-}Si(CH_3)_2{-}R + HN(CH_3)_2$$

In the homologous series of organochlorosilanes the reactivity decreases in the sequence $RSiCl_3 > R_2SiCl_2 > R_3SiCl$. A comparative study on the reactivity silanes of type $n\text{-}C_8H_{17}(CH_3)_2$ Si-X was performed under standardized conditions by Lork et al. [198]. Taking the maximum ligand density as the decisive criterion, the following reactivity sequence was found:

$$C_8\text{-}N(CH_3)_2 > C_8\text{-}OCOCF_3 > C_8\text{-}Cl >> C_8\text{-}OCH_3 \sim C_8\text{-}OC_2H_5 >$$

$$C_8\text{-}OH >> C_8\text{-}O\text{-}C_8$$

where $C_8$ is $n\text{-}C_8H_{17}(CH_3)_2Si$.

The reactivity of monochlorosilanes is much enhanced by adding a base as a nucleophil. The added base serves two purposes. First, it works as an acid scavenger yielding a hydrochloride. Second, if forms a reactive intermediate with an $\equiv$Si-N= bond which reacts as described above for the aminosilane. With imidazole as base, the following reaction scheme was proposed [199]:

$$RR'_2Si\text{-}Cl + 2\ HN\langle imidazole \rangle N \rightleftharpoons RR'_2Si\text{-}N\langle imidazole \rangle N + HN\langle imidazole \rangle N\cdot HCl$$

$$\equiv Si\text{-}O\text{-}H$$

$$+ RR'_2\text{-}Si\text{-}N\langle imidazole \rangle N \longrightarrow \equiv Si\text{-}O\text{-}SiR'_2R + HN\langle imidazole \rangle N$$

Figure 14 shows the kinetics of a silanization reaction between a 15-nm pore size silica and n-octyldimethylchlorosilane with 2,6-lutidine as a

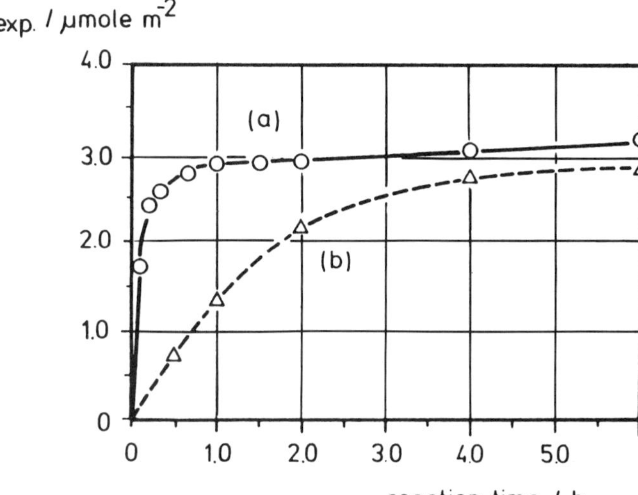


Fig. 14 Kinetics of (a) catalyzed and (b) noncatalyzed silanization reaction. n-Octyldimethylchlorosilane, silica of 15 nm pore size (a) solvent dichloromethane, 2,6-lutidine, reaction temperature 323 K, (b) silane only, reaction temperature 433 K. (From Ref. 199 by the permission of the publisher.)

catalyst dissolved in dichloromethane at 323 K (a), and without the catalyst and a solvent at 433 K (b). Under the given conditions, the density of the ligand approaches a saturation value within 1-2 hr in the catalyzed reaction. Testing different bases, the reactivity increased in the order chinuclidine < 1,2,4-triazol < pyridine < triethylamine < 2,4-lutidine < 2,6-lutidine < imidazole [199].

Buszewski et al. [200] examined the catalytic effect of the bases quinoline, pyridine, morpholine, N-methylpyrrolidone, piperidine, and 2-pyrrolidone on the reaction of n-octadecyldimethylchlorosilane and silica. They concluded that the reactivity is influenced by the $pk_a$ of the base and that a certain molar ratio of base to silane is required for achieving the maximum density.

Traces of water in the reaciton of monochlorosilane and silica in excess of 0.5 mol of water/mol of n-octyldimethylchlorosilane drastically reduced ligand density [198]. This is due to the fact that water hydrolyzes a part of the chlorosilane forming hydroxysilane which exhibits a much lower reactivity.

For n-alkyldimethylchlorosilane with a given base, the solvent type was found to exert a notable influence on the kinetics of the silanization reaction, which can be interpreted in terms of three cooperative effects [199]. The reactivity of solvents increased in the sequence tetrahydrofuran < benzene < diethyl ether < acetonitrile < dichloromethane < N,N-dimethylformamide.

The silanization of silica to a bonded phase in HPLC must be viewed under the following practical aspects: economic consumption of the silane, adequate reaction temperature, high reproducibility, and adjustment of graduated values of the density or carbon load.

When the silane can be vaporized, the silanization is carried out as a gas-solid phase reaction between the silica and the pure silane vapor. The vapor is purged through a fixed-bed reactor which contains the silica particles. A more efficient way in terms of heat and mass transfer is to use a fluidized bed reactor. This method was advocated by Khong and Simpson [201] and can be beneficial in the silanization of larger amounts of silicas. The excellent heat and mass transfer conditions also permit a high homogeneity of the product.

The most commonly employed system is the slurry reactor, where the silica is suspended in a liquid and predetermined amounts of the silane and a catalyst are added. The slurry is agitated by bubbling dry gas through the suspension or by mechanical stirring. The reaction in suspension is accomplished under anhydrous or hydrous conditions. Anhydrous conditions are chosen when a monolayer type of bonded phase is intended. When using organotrialkoxysilanes, traces of water equivalent to a monolayer of water are needed as activator [202]. The adsorbed water layer initiates fast hydrolysis of the silane and inhibits polycondensation. Engelhardt and Orth [202] found that the addition of an acidic catalyst such as p-toluene sulfonic acid substantially enhanced the reactivity of the alkoxysilane. In this way, a maximum ligand density of 4.2 $\mu$mol $m^2$ was achieved with n-octyltrialkoxysilane on LiChrosorb Si 100. The preparation of silanized silicas of the n-octyl and n-octadecyl type is exemplified by the following recipes:

Silanization with a monochlorosilane under anhydrous conditions [118]. "The silica (20 g) was activated at 473 K at a reduced pressure of $<10^{-3}$ mbar for 12 hours. A solution of 0.047 mole of 2.6-lutidine in 100 ml of dry dichloromethane was added to the activated silica at room temperature. The suspension was then heated under reflux for 24 hours while stirring under a dry nitrogen atmosphere. The silica was then filtered and thoroughly washed with dichloromethane, methanol, methanol-water (50/50 V/V), methanol and diethylether."

Silanization with a trichlorosilane under hydrous conditions [203]. "A 10 ml aliquot of the n-octyltrichlorosilane reagent was added to 100 ml of carbontetrachloride in a 500 ml vessel. The mixture was heated to boiling and 3.0 g of dry silica was added. After adding 0.5 ml of water the slurry was refluxed for 4 hours. At the conclusion of the reaction of the silica was filtered, washed and dried in the usual manner."

Silanization with a triethoxysilane with traces of water [202]. "The starting silica was first subjected to a 'cleaning process' by heating 10 g of silica in 100 ml of a 9:1 mixture of concentrated sulfuric acid and nitric acid until the development of nitrosic gases ceased. The material was then washed with water to neutrality. Water was removed with methanol and after washing with dichloromethane the silica was dried at 120°C for 24 hours. The silica was then suspended in toluene containing 750 ppm of water. An excess of n-octyltriethoxysilane (5 mole per g of silica) was added, together with p-toluene sulfonic acid as a catalyst (4% relative to the amount of silane)."

Silanization with a dimethylaminodimethylalkylsilane [193]. "The silica was first treated hydrothermally in boiling water for 70 hours to fully hydroxylate the surface. In a dry-box, 100 mg of the silica was weighed into an ampoule. The (dimethylamino) dimethylalkyl silane with an n-alkyl chain length of $n > 6$ was dissolved in 200 mg of isopentane. This solution was added to the ampoule outside the dry-box. The isopentane was evaporated in an argon stream at 60°C and dried at $10^{-2}$ torr and 40°C for 20 min. Lower boiling (dimethylamino)-silanes were added without a solvent. The ampoule was cooled in liquid nitrogen, evacuated to $10^{-2}$ torr and sealed. The ampoule was heated in an oven at the desired temperature for the desired period, cooled, and opened. The product was washed 5 times by suspension in 10 ml of diethylether, sedimented by centrifugation (20 000 g) for 10 min, and the supernatant decanted. The powder was then kept at 110°C and $10^{-2}$ torr for 12 hours."

Under defined conditions, the reaction between the surface hydroxyl groups of the silica and the silane follows a distinct stoichiometry, simply expressed by the number of surface hydroxyl groups that react with the organosilane molecule. For monofunctional silanes this ratio is unity. For bifunctional and trifunctional reagents it varies between 1 and 2. This was examined in detail for the homologous series of n-alkyl- and phenylchlorosilanes [204]. The phenylchlorosilanes phenyltrichloro, diphenyldichloro, and triphenylchloro also served to establish the effect of the bulkiness of the silane on the maximum density in silica modification. The maximum density of phenylhydroxysilyl, diphenylhydroxysilyl, and triphenylsilyl groups followed the sequence 3.6, 2.6, and 1.9 $\mu mol/m^2$. From these values, the corresponding molecular cross-sectional areas per organosilyl group were calculated to be 0.16, 0.61, and 0.87 $nm^2$/group [204]. The effect of the bulkiness of silane also becomes apparent on comparison of the maximum densities achieved for trimethylsilyl and triphenylsilyl groups, these being 4.5 (4.75) to 4.9 $\mu mol/m^2$ [204,205].

The smallest silane, trimethylchlorosilane, reacts with about 50% of the total amount of surface hydroxyl groups at the silica surface. The concentration of the remaining hydroxyl groups can be detected by isotopic exchange methods [190,191]. Thus, in the case of monochlorosilanes of type $R_3SiCl$, the concentration of organosilyl groups plus the concentration of remaining surface hydroxyl groups is equal to the original concentration of surface hydroxyl groups of the silica (see Table 1).

While the bulkiness effect is clearly evident in the case of phenylchlorosilanes as reagents, a different situation is met for n-alkyldimethylchlorosilanes or N-(n-alkyldimethylsilyl)-N,N'-dimethylamines,

**Table 1** Surface Concentration of Organosilyl Groups, $\alpha_{org}$, of Type $R_n(CH_3)_2$ Si (1 < n < 20) and Surface Hydroxyl Group Concentration Before ($\alpha_{OH(a)}$) and After Silanization ($\alpha_{OH(b)}$)

| Organo group R | $\alpha_{org}$ (μmol/m²) | $\alpha_{OH(b)}$ (μmol/m²) | $\alpha_{OH(b)} + \alpha_{org}$ (μmol/m²) |
|---|---|---|---|
| – | 0 | 8.39[a] | 8.39[a] |
| $CH_3$ | 4.09 | 4.38 | 8.47 |
| $C_2H_5$ | 3.58 | 4.72 | 8.30 |
| $C_3H_7$ | 3.47 | 4.95 | 8.42 |
| $C_4H_9$ | 3.61 | 4.68 | 8.29 |
| $C_5H_{11}$ | 3.34 | 4.95 | 8.29 |
| $C_6H_{13}$ | 3.54 | 4.69 | 8.33 |
| $C_7H_{15}$ | 3.59 | 4.66 | 8.25 |
| $C_8H_{17}$ | 3.42 | 4.80 | 8.22 |
| $C_9H_{19}$ | 3.48 | 4.62 | 8.10 |
| $C_{10}H_{21}$ | 3.64 | 4.74 | 8.38 |
| $C_{12}H_{25}$ | 3.60 | 4.59 | 8.19 |
| $C_{14}H_{29}$ | 3.54 | 4.80 | 8.34 |
| $C_{16}H_{33}$ | 3.50 | 4.67 | 8.17 |
| $C_{18}H_{37}$ | 3.26 | 5.04 | 8.30 |
| $C_{20}H_{41}$ | 3.25 | 5.10 | 8.35 |

[a] $\alpha_{OH(a)}$

*Note*: $\alpha_{OH}$ was measured by the method of Holik and Matejkova [191].
*Source*: Ref. 206.

where n varies between 4 and 20. The maximum attained surface concentration amounts to about 4.1 ± 0.2 $\mu mol/m^2$ [135]. This is an indication that the surface concentration is fairly independent of the length of n-alkyl chain. Thus, a dense parallel stacking of bonded chains can be assumed. In comparison, the concentration of the compressed monolayers of alkyl derivatives with small polar groups on the surface of water is of the order of 7.6-8.3 $\mu mol/m^2$ [207,208]. The structure of bonded phases will be treated in more detail in a later section.

The coating of silica and alumina by polymers entails two main steps [156,209,210]. First, a predetermined amount of silica is added to a solution of the precursor (monomer, comonomer, oligomer) in a suitable solvent. After homogenization, the solvent is evaporated. Second, the coated silica is subjected to polymerization, and crosslinked by thermal treatment, $\gamma$-irradiation, and by adding initiators. After the treatment the silica is washed with a series of solvents to remove soluble monomers and oligomers.

The precursors and final polymers can be conveniently divided into two groups. The first group comprises polymers with siloxane bonds. Examples are polymethyloctylsiloxane [209], polymethyloctadecylsiloxane [209], and silicone monomers [210]. The second type is represented by purely organic monomers and oligomers [209].

Bien-Vogelsang et al. [209] employed five different methods to immobilize polymethyl-n-alkylsiloxanes and polybutadiene, such as thermal radical generation by azo-t-butane, thermal radical generation by dicumylperoxide, radical generation by $\gamma$-irradiation, stabilization of radicals by allylmethacrylate, copolymerization without a special radical generator, thermal polymerization of polymethyloctadecylsiloxane. The silicas subjected to polymer immobilization were either native or presilanized with short-chain silanes. Another route was attempted by Ohtsu et al. [211] using silicone monomers. The surface was first coated with the monomer and subjected to polymerization. In a second step the silicone layer was derivatized with n-octyl and n-octadecyl groups, respectively. The coated supports are termed "capsule-type" supports.

The carbon load of polymer-coated silicas and aluminas achieves values between 8 and 20% [156,211]. The film thickness of the polymer is evaluated from the amount of immobilized polymer and the surface area of the support. The thickness can also be calculated indirectly by assessing the pore size distribution of the support before and after the coating [211]. Film thicknesses between 0.7 and 1.4 nm have been reported [156,211]. No reliable methods are available to assess the homogeneity of the immobilized layers.

There are several ways of preparing polymer-based hydrophobic bonded phases. Poly(styrene-divinylbenzene) copolymers are per se reversed phase packings. They are available as microparticulate packings with graduated pore sizes (for comparison, see Ref. 212). Such polymers have also been derivatized with n-octadecyl groups (see ACT-1 phase of Interaction Chemicals). Polysaccharide type polymers as well as polyvinylalcohols and polyhydroxyalkylesters have been employed to introduce hydrophobic functionalities.

A typical recipe is as follows [213]. "To a suspension of 3 g of a G 4000 PW gel of TOSOH corp. in 50 ml of N,N-dimethylformamide, 1.4 g of sodiumhydride were added while stirring under a stream of nitrogen at room temperature. The mixture was stirred until foaming ceased, then 7.8 g of n-octylmethanesulphonate were added at 80°C. Stirring was continued for 5 h. The particles were filtered and washed with water, methanol and chloroform."

A great deal of work has been invested in the preparation of bonded phases with a weakly hydrophobic character, applied in the hydrophobic interaction chromatography of biopolymers [127]. The phases consist of short bonded n-alkyl or phenyl groups at low surface concentrations. They are prepared either by reaction of silane to the silica surface followed by alkylation or acylation of the polymer.

Weakly hydrophobic agarose packings were first studied by Hjerten [214] and Shaltiel [215]. Engelhardt and Mathes [216] synthesized a variety of n-acyl bonded phases originally designed as size exclusion packings for biopolymers. An ether phase with methyl, ethyl, and n-butyl ligands of type $Si(CH_2)_3$-O-$(CH_2$-$CH_2O)_n$-R was synthesized by Miller et al. [217] in a two-step procedure. First, the ether alcohol was converted to the allyl ether followed by hydrosilylation of the corresponding triethoxysilane. Second, the silane was reacted with the silica using traces of water as catalyst.

Polyethyleneimine [147] and polysucchinimide [153] coated silicas have been employed as precursors to synthesize hydrophobic bonded phases for HIC.

### 3. *Polar Bonded Packings*

*Introduction* Parallel to the evolution of reversed phase types of packings ran extensive trials to synthesize polar bonded phases to enhance the selectivity options in normal or straight-phase chromatography [218]. Attempts were made also to replace and mimic the polar liquid stationary phases employed in liquid-liquid partition chromatography. Moreover, polar bonded silica packings offer unique opportunities to overcome some of the undesirable intrinsic properties of silica in straight-phase chromatography:

1. Control of the water content of the phase system to enable a reproducible analysis with unpolar mobile phases [4]
2. A variation of the apparent acidity of commercial silicas which causes a product-specific retention behavior toward acidic and basic analytes

3. The irreversible adsorption of strongly polar analytes which is attributed to the heterogeneity of the silica surface
4. The slow reequilibration when changing the solvent composition

Polar bonded phases carry functional groups with terminating polar nonionogenic substituents such as alcoholic hydroxyl, nitro, cyano, and amino. The latter phase already exhibits a weak anion exchange character when employed in acidic mobile phases. The polar substituents are linked through a short hydrocarbon spacer to the surface of the support. Most common is propyl. Polar bonded phases resemble those which have been developed for size exclusion chromatography of biopolymers [219].

Bonded phases applied in normal phase chromatography are most useful for the analysis of organic soluble solutes and in particular for isomers. The mobile phases employed range from unpolar to polar solvent mixtures [4]. The polar character of the bonded phases dominates when mobile phases of low to moderate polarity are used, in accordance with the eluotropic series of solvents and solvent mixtures introduced by Snyder [2]. When changing to water/organic mobile phases, the hydrophobic spacer of the polar bonded phase generates a reversed phase behavior. Compared to normal phase conditions, the elution sequence of analytes reverses [169]. The polar bonded phases occupy an intermediate position with respect to polarity between silica, alumina, etc., and reversed phase packings.

Chromatographic studies on a polyol type of bonded phase indicated that its polarity was similar to or even higher than that of the parent silica [220,221].

In this context it is worth mentioning that the discussed shortcomings of silica applied in normal phase chromatography can be circumvented by the following simple means. In order to handle the varying acidic characters of silica from product to product, it is recommended that the silica be coated with a buffer salt or a buffer solution under static or dynamic conditions [44-47]. Following the work of Schwarzenbach, the bulk silica is first loaded with a buffer solution, followed by evaporation of the water [44-46]. Hansen et al. [47] suggested a dynamic procedure whereby the buffer solution is purged through the silica column. The excess of buffer is displaced by rinsing the column with methanol.

In order to circumvent the problems encountered with the elution of polar ionic solutes on silica column, e.g., strong acids, buffers were added to the mobile phase to shift the pH to a range where the ionization of solutes is suppressed [222,223].

*Preparation of Polar Bonded Packings* The introduction of polar functional groups to the silica surface by silanization can be accomplished in two ways. First, the ligand coupling is performed in one step by reacting the silica with an appropriate silane (method A). Second, the polar bonded phase is prepared in two consecutive steps. An intermediate is formed by bonding a silane to the surface. The polar functional groups are then introduced by the well-established substitution reactions (method B).

The direct way according to method A is the simpler procedure. With method B, the yield of the intermediate is often much less than 100% because of extensive cleaving reactions. The most commonly used silanes for the synthesis of polar bonded packings are trialkoxysilanes (methoxy, ethoxy) [202]. Compared to trichlorosilanes they are less reactive. The bonding chemistry of trialkoxyorganosilanes to inorganic fillers has been reviewed by Pluedemann [224]. The reaction of silica with the silane of type $RSi(OR)_3$ is carried out either with an organic solvent or with an aqueous solution of adjusted pH. Accordingly, monomeric or polymeric bonded layers are formed. The reaction of the alkoxysilane and silica in unpolar solvents such as toluene, xylene, etc., requires an activator and a catalyst. Water, adsorbed at the silica or added in controlled amounts to the solvent, functions as activator. Usually a monolayer of adsorbed water is sufficient. The monolayer capacity $x_m$, of silica is given by Eq. (3):

$$x_m = \frac{a_s}{a_m N_L} \tag{3}$$

where $a_s$ is the specific surface area of silica, $a_m$ the molecular cross-sectional area of an adsorbed water molecule (being 0.2 $nm^2$/molecule), and $N_L$ the Avogadro number. For silica with $a_s$ = 300 $m^2/g$, $x_m$ = 40 mg/g or 4% (w/w). Adding to that the amount of water originating from the hydroxyl groups, the total water content of the silica becomes 80 mg/g or 8% (w/w).

The reaction between the surface hydroxyl groups of the silica and the alkoxy groups of the silane is catalyzed by acids and bases. Thus, aminofunctional alkoxysilanes are self-catalytic in the reaction. Among several reagents tested, such as pyridine, triethylamine, ammonia, formamide, benzoylperoxide, and p-toluenesulfonic acid, the latter was found to be most suitable to achieve the highest surface concentration of functional groups [202]. Leyden and coworkers [225-227] postulated a reaciton mechanism whereby one trialkoxysilane molecule is bonded to three surface hydroxyl groups, which has been proven by NMR spectroscopic measurements.

In aqueous solutions, trialkoxysilanes hydrolyze stepwise to the corresponding silanols which then condense to siloxanes. The hydrolysis is fast while the condensation proceeds more slowly and is influenced by the pH of the solution. Organosilanols of type $RSi(OH)_3$

are stable at a pH between 3 and 5 but condense rapidly at pH 7-9 [224]. Herman et al. [228] examined in detail the reaction of 3-glycidoxytrialkoxysilane in an aqueous solution of pH 5-6 and established a scheme of the reaction kinetics on the basis of inter- and intramolecular condensation reactions. For sufficient hydrolytic stability, the bonded phases obtained by reaction of trialkoxysilanes and silica must be crosslinked [224]. The crosslinking of aminopropyl-bonded phase with epoxides was studied by Wonnacott and Patton [229]. The epoxide were 1,4-butanedioldiglycidether and allylglycidyl ether. The latter did not crosslink under the reaction conditions employed and merely served as a model for comparison to the reaction chemistry of the epoxide. The diepoxide-treated bonded phase showed an exceptionally high stability compared to the monoepoxide-treated and nontreated.

Cyclodextrin-bonded silicas have developed to an extremely versatile and flexible stationary phase, applied in reversed and normal phase modes. Cyclodextrins are natural macrocyclic oligomers of glucose that consist of 6-12 glucose units bonded through $\alpha$ (1,4) linkages. The oligomers are chiral and possess toroidal-shaped cavities. Three types are commercially available as $\alpha$-, $\beta$-, $\gamma$-cyclodextrin (= cd) with 6, 7, and 8 glucose units and internal cavity diameters of $\sim$0.45-0.6, $\sim$0.6-0.85, and $\sim$0.8-1.05 nm [230]. The inside of the cyclodextrin cavities is relatively hydrophobic. The mouth of the cavities is lined with secondary 2- and 3-hydroxyl groups and thus behaves hydrophilically. Primary 6-hydroxyls, as sketched in Fig. 15 for $\beta$-cyclodextrin, are at the opposite site of the oligomers. They serve as active groups bonding the cyclodextrin to the silica via 6-10 atom spacer arms [231-242]. It is essential that a hydrolytically stable bond is achieved.

The characteristic feature of the cyclodextrin-bonded phases is that they form inclusion complexes or interact via hydrogen bonds with the analytes. As mentioned, cyclodextrin-bonded phases are operated in the normal phase mode using n-hexane/isopropanol as mobile phase.

The majority of separations, however, are carried out in the reversed phase mode using water/organic solvent mixtures as mobile phase.

Polar polymeric coatings according to the methods described on page 357 have been achieved on silicas and aluminas. The type of polymers were polyacrylamide [243], polyvinylpyrrolidone [244,245], polar silicons[246], and polar organosiloxanes [247].

## C. Characterization of Bonded Packings

### 1. *Introduction*

The characterization of the structural and chromatographic properties of bonded packings has attracted a considerable amount of systematic investigation. Detailed characterization is aimed at the following:

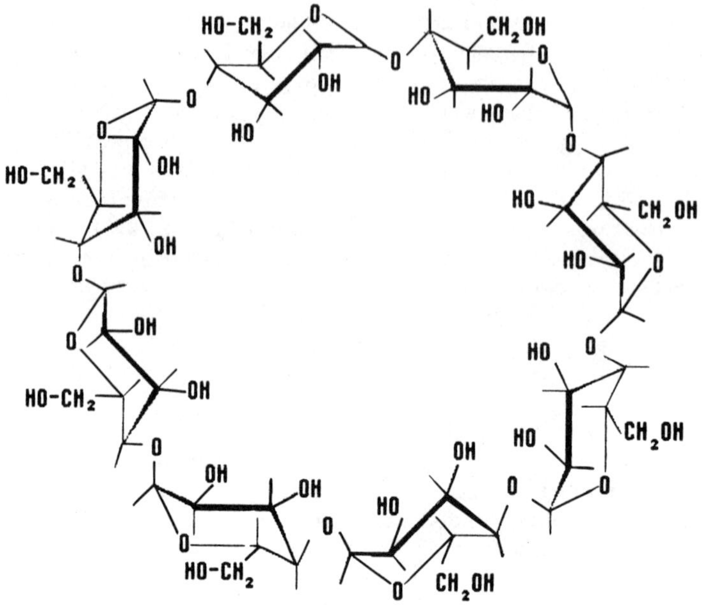

Fig. 15 Schematic of γ-cyclodextrin showing the chair conformation of the glucose units, the α-(1,40-glucosidic linkages, and the opposite orientation of the 2- vs. 3-hydroxyls at the mouth of the cavity. (From Ref. 230 by the permission of the publisher.)

- Establishing the physical and chemical structure of bonded phases relative to the parent materials
- Estimating the chromatographic properties in terms of retention, selectivity, column efficiency, column stability, and column lifetime
- Improving the properties of bonded phases with respect to their chromatographic use
- Assessing and controlling reproducibility and repeatability of the properties of bonded phases and given manufacturing processes
- Designing novel bonded phases

Characterization is performed either on the bulk powder or using the packed columns. It is debatable to which extent the results of powder characterization can be adapted to interpret the chromatographic behavior of bonded phases. Clearly, the physical and chemical methods applied to the bulk bonded phase provide information on the following structural properties: particle size, particle morphology, pore size, porosity, surface area, chemical and phase composition, concentration, topography and mobility of functional groups, and mechanical and chemical stability of the product. However, it should be emphasized that these properties are often assessed under conditions which deviate substantially from the conditions under which the bonded phase is applied in the chromatographic column.

Tests on columns allow a direct derivation of those chromatographic parameters of the bonded phase which refer to the resolution and speed of separation. In order to reliably evaluate the implication of the bulk bonded phase and stationary phase properties on the chromatographic performance, one must consider the relationships describing the kinetics and thermodynamics of the chromatographic separation process. These relationships are discussed in depth in Chap. 2.

The conclusions drawn from such correlations and comparisons are not very detailed due to the complexity of the systems studied. The ideal solution to the problem implies the direct application of highly sophisticated and diagnostic physicochemical methods to the column during chromatographic use. However, these in situ measurements are still in the initial phase as exemplified by NMR spectroscopic imaging [248].

### *2. Characterization of Bulk Powders*

We will briefly discuss the most relevant methods applied to the characterization of bonded packings and examine the assessment of corresponding properties.

### *Elemental and Chemical Analysis*

Elemental analysis data on bonded phase packings are obtained from microanalyzers, yielding the weight percent of carbon, hydrogen, nitrogen, etc., with a precision of about ±0.2% and a detection limit of about 0.25% (w/w). The carbon content ranges from a few percent up to 20% and higher depending on the ligand, the ligand concentration, and the extent of polymer coating. Elemental analysis of C and N on bonded phases permits an estimation of the stoichiometric composition of the stationary phase [202].

On the basis of the carbon (nitrogen) content and the specific surface area $a_s$ of the native silica, the surface concentration, $\alpha_{exp}$, or ligand density of the functional group is estimated, provided that the silane for bonding and the stoichiometry of the bonding reaction are known [249]. The ligand density of a silica reacted with a monofunctional silane is calculated as follows:

$$\alpha_{exp} = \frac{P_C}{M_C \cdot a_s^x} \quad /\mu mol/m^2 \tag{4}$$

where $P_C$ is the carbon content (w/w), $M_C$ the molecular weight of the bonded carbon residue, and $a_s^x$ the specific surface area according to BET of the native silica, corrected by the weight increase due to bonding. The weight increase $\Delta G$ is given by

$$\Delta G = \frac{P_C (M_{mod} - nM_H)}{M_C} \quad (5)$$

where $M_{mod}$ is the molecular weight of the bonded group and $nM_H$ the number of hydrogen atoms of the hydroxyl groups replaced by the bonded ligand, multiplied by the molecular weight of hydrogen. The corrected specific surface area $a_s$, then, is

$$a_s^x = a_s (1 - \Delta G) \quad (6)$$

For bonded phases prepared by the reaction of monofunctional silanes, Eqs. 4 to 6 can be combined:

$$\alpha_{exp} = \left[ \frac{M a_s}{P_C} \cdot (M_{mod} - 1) a_s \right]^{-1} \quad (7)$$

Another method to calculate the expected surface concentration of a ligand is to use the molecular cross-sectional area $a_m$ of the bonded ligand or polymer unit. The reciprocal of $a_m$ multiplied by Avogadro's number gives the ligand concentration in $mol/m^2$. For polymer coatings, the thickness $d_f$ is calculated from the amount of immobilized polymer and the surface area of the support [156]. Most of the commercial bonded packings are merely characterized by the carbon content in percent (w/w). In some instances the silane employed for bonding is given (see Appendix III).

Elemental analysis is insensitive to the identity of more than one bonded ligand and to the determination of their surface concentration. Thus, methods are needed which permit an assessment of both the ligand identity and quantity. A suitable approach is to subject the bonded phase to alkaline or acidic hydrolysis, whereby the anchoring siloxane bonds are cleaved and silanes or siloxanes formed. The silanes are extracted and analyzed by GC and GC/MS. Alkaline hydrolysis of bonded phases was applied by Verzele et al. [250] and Genieser et al. [251]. Acid hydrolysis using concentrated hydrochloric acid was developed as an approach to the analysis of reversed phase packings modified with monofunctional silanes [252]. Fazio et al. used hydrogen fluoride in a methanol/water mixture [253]. With the methanolic hydrofluoric acid solution as digestion/derivatization reagent, volatile fluorosilanes are formed which are monitored and quantitated by capillary GC after extraction with n-hexane. This method is applicable to reversed phase packings prepared by bonding of mono-, di-, and trifunctional silanes. Also, endcapped products were analyzed. On comparing the method with elemental analysis, the accuracy for reversed phase silicas obtained from monofunctional silanes was found to be ±2-5%. For silicas made by reaction of trifunctional silanes the proposed method yielded only semiquantitative results. Analysis of commercial products indicated that mono-, di-, and trifunctional silanes were applied for silanization.

The method developed by Erard and Kovats [254] consists of the following steps: dissolution of the bonded silica in a solution of hydrogen fluoride in ether, elimination of the main part of $SiF_4$, displacement of $SiF_4$ with nitrogen, and analysis of the fluorosilanes formed quantitatively from the ligands by GC. The precision of the method is claimed to be about ±0.2% (95% confidence interval).

Trojer and Hanson [255] and Mussche and Verzele [256] used pyrolysis GC and GC/MS in the identification of bonded ligands on silica.

There is an ongoing debate regarding to which extent inorganic impurities affect the chromatographic properties of bonded silicas. Inductively coupled plasma-atomic emission spectroscopy (ICP-AES) is the method of choice to measure the content of sodium, potassium, aluminum, titanium, iron, calcium, barium, and other elements at the ppm level.

Typical values for a number of commercial silicas are given in Table 2 [257]. Additional data are reported in Refs. 189 and 258. The values constitute the bulk and surface content and might serve as reference values to monitor changes in the trace metal content as a result of an aftertreatment of the silica.

The surface concentration alone must be measured by electron spectrometry (SIMS). The results of these methods provide information on the surface species, their coordination, and their relative concentration (for comparison, see Ref. 259).

### *Spectroscopic analysis*

Spectroscopic methods have emerged as the most powerful tools in the investigation of the structure and dynamics of molecular systems. With the development of high-resolution techniques these methods have become extremely useful for probing the surface of solids, yielding information on surface species and their structural parameters such as bond lengths and bond angles. Moreover, the use of appropriate probe molecules for photophysical studies permits a detailed insight into the subtle organization of derivatized surfaces. Of these spectroscopic methods, Fourier transform infrared (FTIR) spectroscopy, solid-state

Table 2 Content of Sodium, Aluminum, and Iron in Some Commercial HPLC Silica

| | Na (ppm) | Al (ppm) | Fe (ppm) |
|---|---|---|---|
| Kromasil 10 μm | 10 | 20 | 40 |
| Vydac TPB-2030 | 30 | 10 | 45 |
| Nucleosil 100-30 | 250 | 30 | 110 |
| Nucleosil 100-10 | 50 | <10 | 50 |
| Hypersil, 5 μm | 3360 | 300 | 260 |
| Matrex, 5 μm | 500 | 350 | 110 |
| Partisil, 5 μm | 15 | 60 | 75 |
| LiChrospher Si 100 | 130 | 300 | 420 |
| Spherisorb, 5 μm | 5600 | 300 | 420 |
| Zorbax BP-SIL | 20 | 60 | 80 |

*Source*: From Ref. 257 with permission of the publisher.

nuclear magnetic resonance spectroscopy (NMR), and fluorescence spectroscopy have gained most attention. They are applicable to bonded packings and their parent materials and permit identification of surface species, elucidation of their short-range chemical structure, and their semiquantitative determination.

Between 1960 and 1970, IR spectroscopy was applied extensively in surface studies of silica and its chemically modified derivatives and is reviewed in several monographs [260-263]. Compared to dispersive IR spectrometers, modern FTIR systems provide noticeably higher signal-to-noise ratios [264,265]. Novel sampling techniques such as DRIFT have been evolved which were shown to obtain qualitative and semiquantitative structural information on bonded packings [266]. Figure 16 shows DRIFT spectra of trimethylsilyl derivatized silica and its parent material [189]. Since DRIFT is a bulk analytical technique, detailed investigations are required for quantitation [267-269]. Measurements at various sample temperatures and atmospheres can be carried out by using suitable DRIFT accessories. On studying the molecular structure of n-alkyl-bonded silicas by FTIR, Sander et al. [270] found that a significant fraction of carbon-carbon bonds were in the "gauche" conformation, and the degree of conformational disorders of RP phases was comparable to that observed in the corresponding n-alkane liquids at room temperature and above.

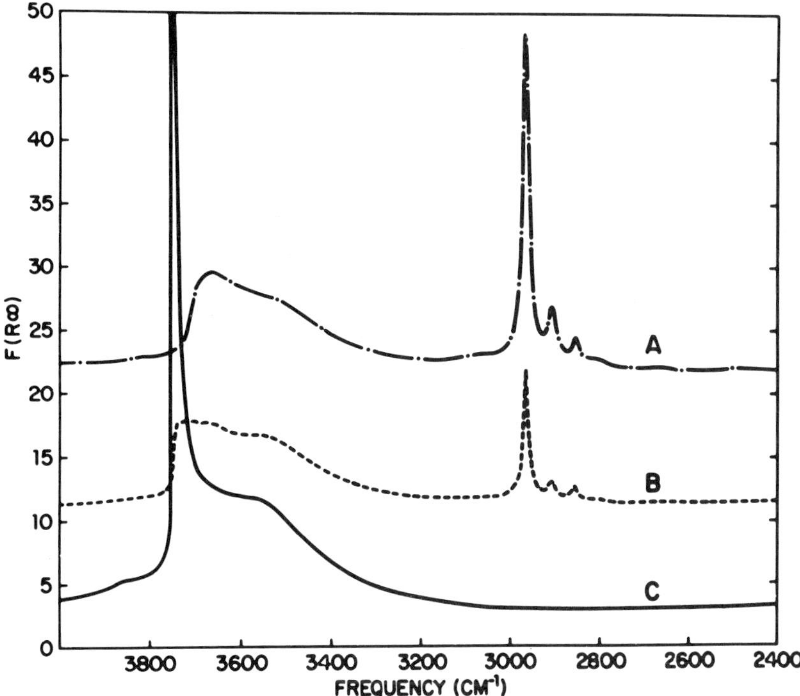


Fig. 16 Drift spectra of silicas. (A) Fully TMS-modified Zorbax PSM-60. (B) Partially TMS-modified Zorbax PSM-60. (C) Zorbax PSM-60. (From Ref. 189 by the permission of the publisher.)

The introduction of magic angle spinning (MAS) and cross-polarization (CP) techniques into solid-state NMR spectroscopy has rendered this method comparable with respect to spectral resolution to NMR spectroscopy in solution [271,272]. Among the nuclei which provide most structural information on bonded phases are $^{13}C$, $^{29}Si$, and $^{1}H$. A solid-state spectrum is characterized by three main parameters: the chemical shift (line position), the intensity (integrated peak area), and the linewidth (width at half-height). Other parameters are the relaxation times and spin-spin coupling constants, which provide more specific information. Extensive solid-state NMR examination on bonded packings and their parent materials have been conducted by Maciel and coworkers at Colorado State University, USA, Bayer and coworkers at Universität Tübingen, FRG, and Cramers and coworkers at Eindhoven University of Technology, The Netherlands. Since it is beyond the scope of the chapter to provide an in-depth treatment of all aspects, the most important results will be summarized briefly.

Maciel and Sindorf [273,274] found three peaks in the $^{29}Si$/CP MAS NMR spectra of native silica at -90.6, -99.8, and -109.3 ppm which were assigned to the following groups:

```
      H              H
      O              O              O
  O  Si  O H     O  Si  O       O  Si  O
      O              O              O
```

They also performed $^{29}Si$ and $^{13}C$ CP MAS NMR studies on silicas treated with hexamethyldisilazane [274] and trimethylchlorosilane [275, 276]. Bayer et al. [277] investigated a series of home-made and commercial reversed phase silicas by $^{29}Si$ and $^{13}C$ CP MAS NMR spectroscopy. They were able to identify the structures listed in Fig. 17 and could also discriminate between reversed phase packings of different origin. By measuring the relaxation times $T_{1\rho}H$ on silicas reacted with n-alkyldimethylchlorosilanes by $^{13}C$ CP MAS NMR spectroscopy, the mobility of n-alkyl groups as a function of chain length was assessed [278]. Mobility was highest at a chain length between 6 and 8. Particularly useful are examinations on RP packings in the solvated state carried out by soild-state $^{13}C$ CP MAS and solution $^{13}C$ NMR spectroscopy [279]. Using a silica reacted with n-octadecyldimethylethoxysilane, the mobility was found to depend on ligand density and was reduced when the water content of the acetonitrile/water mixture was increased. Figure 18a shows the $^{13}C$ solid-state CP NMR spectrum and 18b the $^{13}C$ NMR spectrum of a suspension of the $C_{18}$ bonded silica in acetonitrile.

A correlation between the mobility of n-alkyl chains of reversed phase silica and chromatographic retention data of test solutes was also established [280,281].

A great deal of effort has been concentrated on understanding the molecular structure of reversed phase packings, their solvation, and their interaction with probe molecules. Photo-induced diffusion-controlled processes, using fluorescence, luminescence, and pulsed laser photolysis, have been employed to investigate a number of surface properties, such as polarity and heterogeneity of the surface, effective viscosity of the interface, mobility and orientation of functional groups, intercalation of solvents and solutes, etc. The generation of excited dimers (eximers) has frequently been applied to the study of the properties of reversed phase packings. Probe molecules were pyrene [282-298], 1,3-di(1-pyrenyl)propane [299-300], and perylene [301] which were adsorbed at the surface. Studies have been performed also with 1-alkylpyrenylsilanes such as [3-(1-pyrenyl)-propyl]dimethylmonochlorosilane and [10-(1-pyrenyl)decyl]dimethylmonochlorosilane which were bound to the silica surface [302-304].

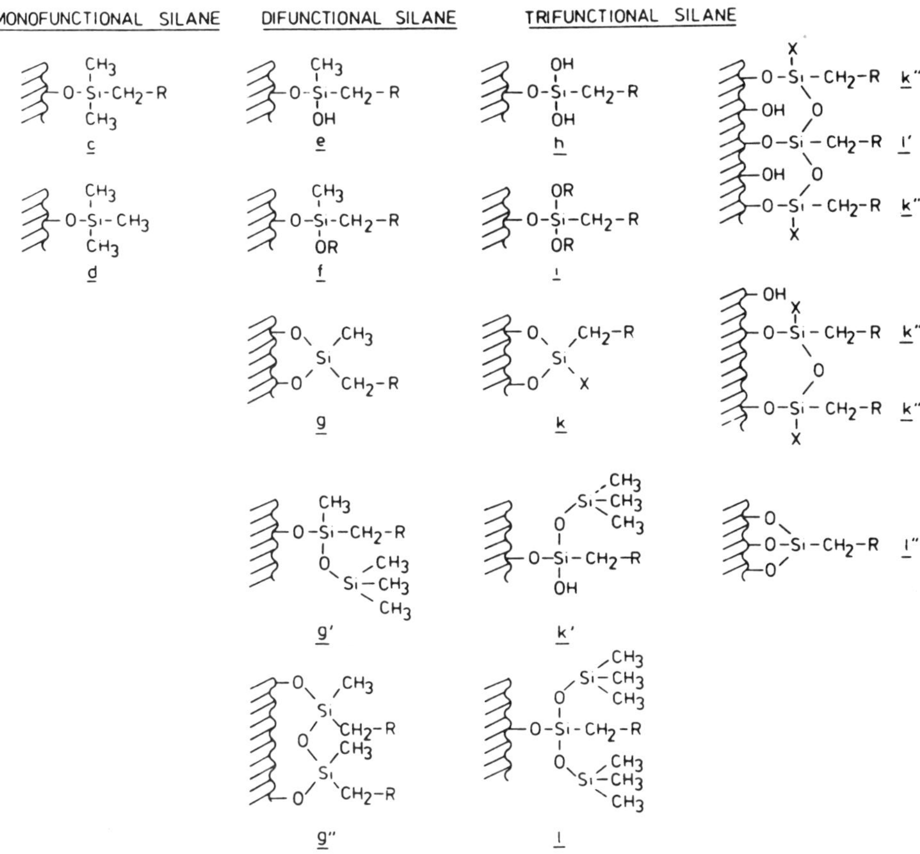


Fig. 17 Organofunctional species at the surface of silanized silicas identified by $^{13}C$ CP MAS NMR spectroscopy. (From Ref. 277 by permission of the publisher.)

The rate constant of the eximer formation obtained from both static and dynamic measurements was used to estimate diffusion coefficients and derive parameters on the microviscosity of the stationary phase. By use of static and dynamic fluorescence polarization methodologies with perylene as probe molecule, it was shown that the rotational motion of n-alkyl chains is highly constrained and anisotropic [301].

In order to gain fundamental information on the interface of reversed phase packings, Gilpin et al. [305] developed an approach based on the immobilization of surface isotopic, functional, and spin probe centers and the NMR and ESR investigation of orientational and

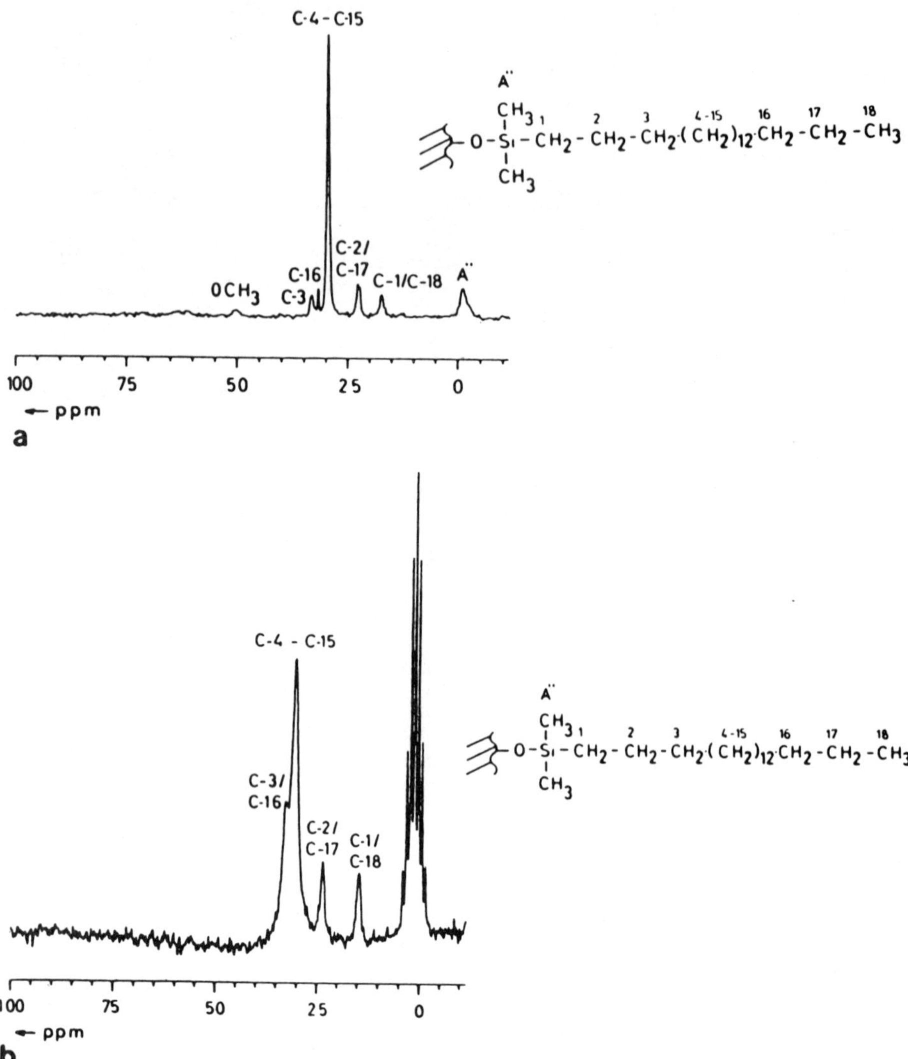


Fig. 18 (a) $^{13}C$ CP MA spectrum (75.46 MHz) of the reaction product of n-octadecyldimethylmethoxysilane with silica gel Nucleosil 100 7 μm (phase LAB I); (b) $^{13}C$ NMR spectrum (22.63 MHz) of the suspension of phase LAB I in acetonitrile-$d_3$. (From Ref. 279 by the permission of the publisher.)

motional changes which occur in or at the interfacial region under various perturbation conditions.

### *Microscopy*

Microscopy methods are among the well-established diagnostic techniques for characterizing the particle and pore texture of porous solids [306], and thus have been extended to the analysis of the physical structure of chromatographic packings. Microscopy benefits from providing a direct image of the object. A major shortcoming, however, relates to the fact that any image covers only a small portion or cross-section of the total material. In order to draw generally valid conclusions a sufficient number of representative micrographs must be taken.

Microscopy is applied as light or electron microscopy. The latter falls into the transmission and scanning technique. Light microscopy is the method of choice in the assessment of particle size, size distribution, and particle morphology [307]. Particles of a diameter $dp < 1$ μm are difficult to measure due to their Brownian motion. Scanning electron microscopy (SEM), with a resolution of about 50 nm, serves to estimate the surface roughness of particles, to examine the structure of an assemblage of packed particles [308], to monitor the damage of particle as a result of ultrasonic treatment and too harsh packing conditions, and to provide information on the pore structure of macroporous packings with a pore diameter of $pd > 50$ nm [309].

Recently, a thorough examination of the internal structure of some commercial HPLC packings by transmission electron microscopy (TEM) was performed by Tanaka et al. [310]. Both polymer- and silica-based packings were investigated. Figure 19a-d shows as part of the results the cross-section of a particle of three commercial silicas, revealing the typical globular structure. In contrast, macroporous silicas of $pd = 50\text{-}500$ nm often show the sponge-type structure as typified by LiChrospher Si 1000 (Fig. 19d). The images also provide clear evidence for a bimodal pore size distribution and a notable gradient of the pore size across the particle for some commercial products.

ETM has been applied successfully in the manufacture of novel packings. A typical example is the work of Knox et al. [97] on the synthesis of porous graphitized carbons. High-resolution ETM together with electron diffraction was a powerful means to monitor structural changes as a result of the high-temperature treatment conditions.

### *Adsorption and Wetting*

Much valuable information on the physical and chemical structure is available from sorption and wetting experiments. Sorption studies on bonded phases are conveniently divided into gas-solid and liquid-solid. In gas-solid adsorption, nitrogen and argon isotherms are measured from which the specific surface area $a_S$, the specific pore volume

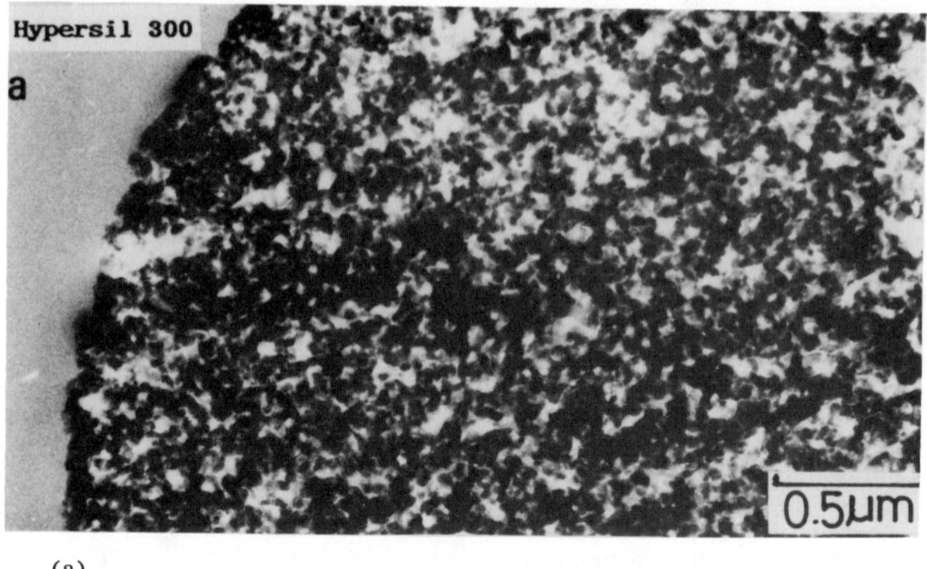


(a)

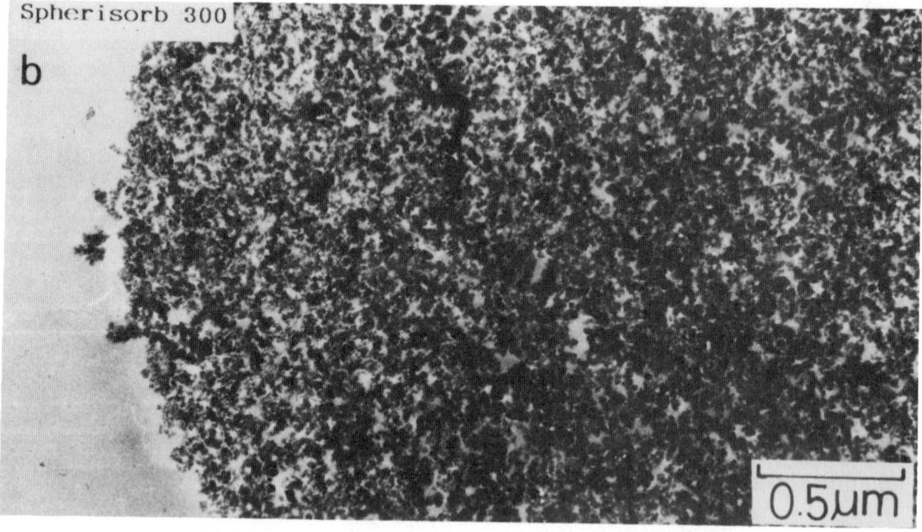


(b)

Fig. 19 Transmission electron micrographs of Hypersil 300 (a), Spherisorb 300 (b), and Vydac TP (c). (d) Transmission electron micrographs of LiChrospher Si 1000. (From Ref. 310 by the permission of the publisher.)

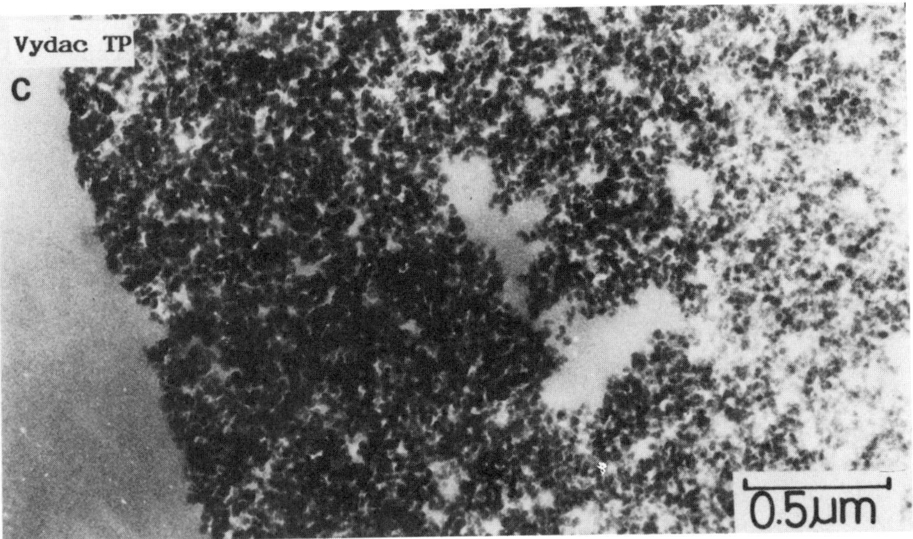


(c)

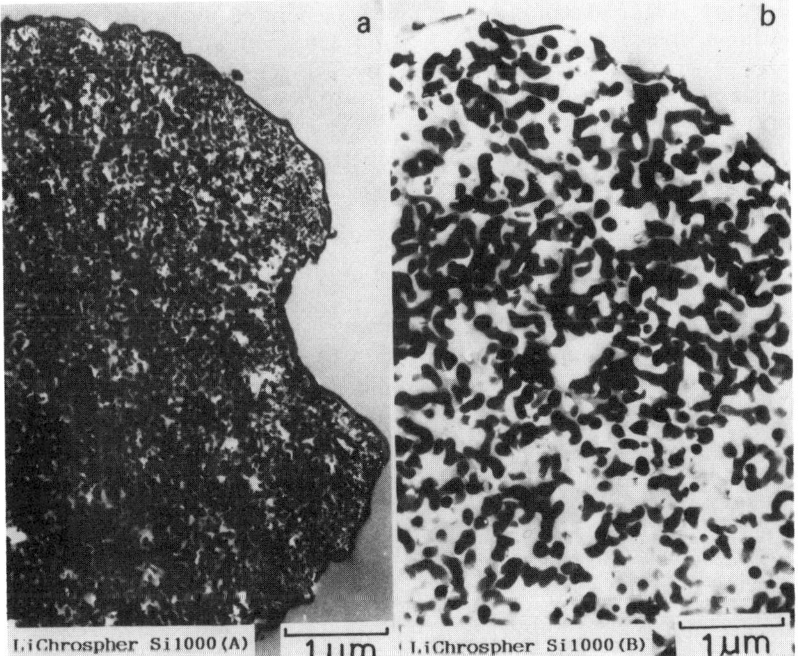


(d)

Fig. 19 (continued)

vp, and the pore size distribution of the bonded packings are derived. In contrast to oxidic surfaces, where a dense monolayer of nitrogen is formed with a molecular cross-sectional area of nitrogen of $a_m = 0.162$ $nm^2$/molecule, loose monolayers are obtained with $a_m \sim 0.2$ $nm^2$/molecule on nonpolar surfaces with exposed methyl, phenyl, methoxy, dimethylamino, and cyano groups [311]. Contact angle measurements on chemically modified glass surfaces revealed that they behave as low-energy surfaces compared to the parent material [312]. Glass surfaces with bonded short n-alkyl chains (methyl, ethyl, propyl) are termed rough, whereas those with bonded long chains (tetradecyl, octadecyl, docosyl) are considered swollen [208].

*Thermal Analysis*

Thermal analysis comprises three techniques: thermogravimetry (TG), differential thermal analysis (DTA), and differential scanning calorimetry (DSC) [313]. The first two monitor the weight loss during ignition, the latter the exothermic and endothermic changes during ignition. DSC has frequently been used to elucidate the structure of n-alkyl-bonded phases in context with the nonlinear change of solute retention as a function of the column temperature. This nonlinear behavior of n-alkyl-bonded phases was found in GC and LC experiments. Morel and Serpinet [314] examining $C_{18}$- and $C_{20}$-bonded phases, observed a "melting transmission" between 280 and 300 K ($C_{18}$) and 313 and 321 K ($C_{22}$) in GLC which was altered by using linear and branched n-alkanes as stationary liquids [315]. Similar results were observed in HPLC experiments [316].

Specific heat measurements by adiabatic calorimetry on n-octyl- and n-octadecyl-bonded phases revealed second-order transitions between 70 and 310 K and 150 and 305 K. This transition somewhat resembles a melting process, although the transition enthalpy and entropy values of both n-alkyl phases are only 25% of the melting enthalpies and entropies of the corresponding n-alkanes [317].

Schunk and Burke [318] identified a transition temperature of an n-octadecyl-bonded phase of 298-308 K, and showed that it was affected by both the mobile phase composition and the solute type.

In general, phase transition on n-alkyl-bonded phases occur when the chain length of the ligand exceeds 16, the ligand density is larger than 3 $\mu mol/m^2$, and the mean pore diameter of the parent silica is much larger than 6 nm [319].

*3. Chromotographic Characterization*

Chromatographic examination of columns with bonded packings addresses two issues. First, one objective is to obtain information on the bonded phase itself. For example, frontal analysis of deuterated water and solvents on bonded phase columns permits the measurement of excess isotherms, from which the accessibility of hydroxyl groups of the parent silica can be evaluated [320]. Second, the major goal is to characterize the bonded phase column with respect to its kinetic and thermodynamic performance. Columns can be tested under isocratic and gradient elution conditions. Table 3 lists the most common parameters determined from such an analysis.

Among the quality control data, the selectivity of a bonded phase column is of prime interest. Attempts have been made to establish generally acknowledged test procedures using test solutes and standardized conditions. The wide application of reversed phase columns necessitates several tests to judge the selectivity of a column [321]. Table 4 collects tests and conditions for reversed phase columns.

Chromatographic tests are also extremely useful for ascertaining the lifetime of a column. The procedure consists of flushing the column with appropriate solvent mixtures, preferably at higher temperatures, and measuring the retention, plate number, and peak shape of test solutes over an extended period of time. Trimethylsilyl-modified packings were subjected to chromatographic tests in water/methanol of 50:50 (v/v) after purging the column with water at 323 K over a prolonged period [187]. Solutes were 1-phenylhexane, 1-phenylheptane, 5-phenylpentanol, 2,6-di-tert-butylpyridine, and N,N-diethylaniline. It was found that the hydrolytic stability of TMS-bonded phase was highly dependent on the correct pretreatment of the parent silica.

In situ stability studies on reversed phase columns were performed by Sagliano et al. [334] to monitor the hydrolysis as a function of silica type, bonded group type, ligand density, and silane reactivity. The mobile phase consisted of water/acetonitrile of 68:32 (v/v), with the aqueous component being 10 mM $H_3PO_4$ and trifluoroacetic acid at a concentration of 0.5% (v/v). The silicas had a pore diameter of about 20 nm and were subjected to reaction with n-butyldimethylchlorosilane, n-butylmethyldichlorosilane, and n-butyltriethoxysilane as silanes. The results were as follows:

The loss of bonded alkylsilane (n-butyldimethylsilyl groups) was less at the high ligand density due to the better shielding effect.

The hydrolysis resistance at comparable ligand density followed the sequence n-butylsilyl $\sim$ n-butylmethylsilyl > n-butyldimethylsilyl.

Column degradation of reversed phase packings in low ionic acidic mobile phases using peptides and proteins as test solutes and gradient elution was examined by Glajch et al. [335]. Nearly all the tested columns lost substantial amounts of the bonded phase. However, the separation characteristics of some proteins remained unchanged.

Table 3 Methods and Characteristic Properties of Bonded Packings and Parent Materials

| Method applied to powders, suspensions, etc. | Property |
|---|---|
| *Elemental Analysis* | |
| (C, H, N, Al, Ti, Ba, Sr, K, Na, etc.) | Carbon load, ligand density, thickness of coating inorganic, impurities |
| Chemical treatment with HF/methanol and HF/ether GC of fluorosilanes | Ligand identity, ligand density |
| *Spectroscopy* | |
| Inductively coupled plasma-atomic emission spectroscopy (ICP-AES) | Content of elements in the bulk phase and surface |
| Electron spectroscopy for chemical analysis (ESCA) | Identification of surface species |
| Infrared spectroscopy Diffuse reflection infrared Fourier transform spectroscopy (DRIFT) | Types of hydroxyl groups, types of bonded groups, interactions with probe molecules |
| Nuclear magnetic resonance spectroscopy (NMR) Solution NMR Suspension NMR Solid-state NMR (cross-polarization magic angle spinning (CP MAS)) $^{1}H$, $^{13}C$, $^{29}Si$ | Types of hydroxyl groups, and bonded groups, ligand density, monomer-polymer grafting, conformation and mobility of groups |
| Deuterium exchange using $CF_3COOD$ followed by $^{1}H$ NMR spectroscopy | Content of hydroxyl groups and physisorbed water |
| Deuterium exchange using $D_2O$ followed by mass spectrometric analysis | Content of hydroxyl groups |
| Fluorescence and luminescence spectroscopy | Viscosity of the interface, mobility and orientation of functional groups, intercalation of solvents and solutes |

Table 3 (continued)

| Methods applied to powders, suspensions, etc. | Property |
|---|---|
| *Microscopy* | |
| Light microscopy | Particle size, size distribution, particle shape |
| Electron microscopy (transmission, scanning) (TEM, SEM) | Pore structure, connectivity and size of pores, distribution of pores across the particles |
| *Adsorption and wetting* | |
| Gas-solid adsorption (nitrogen, 77 K) | Specific surface area, specific pore size distribution |
| Wetting experiments | Surface energy |
| *Thermal analysis* | |
| Thermogravimetry (TG) | Weight loss during ignition |
| Differential thermal Analysis (DTA) | |
| Differential scanning Calorimetry (DSC) | Exothermic and endothermic changes during ignition, phase transitions |

| Methods applied to columns | Property |
|---|---|
| Isocratic elution | Solute capacity factor k'<br>Selectivity coefficient $\alpha$<br>Standard deviation of solute peak in time units $\sigma_t$<br>Column plate height H<br>Column plate number N<br>H as a function of the linear velocity of eluent u<br>Column pressure drop $\Delta p$ as a function of the flow rate of eluent F<br>Column resistance factor $\Phi$<br>Chromatographic resolution $R_S$<br>Lifetime of column = dependence of k', $\alpha$, $R_S$, and $\Delta p$ as a function of column usage |
| Gradient elution | Solute capacity factor $\bar{k}$<br>Standard deviation of solute peak in time units $\sigma_t$<br>Chromatographic resolution $R_S$<br>Peak capacity PC |

Table 4 Retention and Selectivity Tests for Reversed Phase Columns

| Name | Probes | Eluent and elution conditions | Detection |
|---|---|---|---|
| Tests to compare columns for reversed phase liquid chromatography [322] | terphenyl<br>biphenyl | methanol/ water 90:10 | UV, 254 nm |
| | dimethylphthalate<br>-------------------<br>diethylphthalene | methanol/water 65:35 | UV, 254 nm |
| | toluene<br>-------------------<br>naphthalene<br>-------------------<br>anthracene | methanol/water 85:15 | UV, 254 nm |
| | theophylline<br>-------------------<br>caffeine | acetonitrile/ 10 mM sodium acetate buffer pH 4.5 20:80 | UV, 254 nm |
| | benzoic acid<br>-------------------<br>toluic acid | acetonitrile/ 10 mM sodium acetate buffer pH 4.5 20:80 | UV, 254 nm |
| | benzene<br>-------------------<br>toluene | methanol/water 70:30 | UV, 254 nm and 280 nm |
| | theophylline<br>-------------------<br>toluene | methanol/water 70:30 | UV, 254 nm and 280 nm |
| | theophylline<br>-------------------<br>caffeine | methanol/water 40:60 | UV, 254 nm and 280 nm |
| Tests to compare commercial octadecyl-bonded phases in reversed phase liquid chromatography [323] | Acidic compounds:<br>3-methyluric acid<br>-------------------<br>1-methyluric acid<br>-------------------<br>3-methylxanthine<br>-------------------<br>1-methylxanthine<br>-------------------<br>1,3-dimethyluric acid | methanol/ 10 g/liter sodium dihydrogenphosphat 9:91 | UV, 254 nm and 280 nm |

Table 4 (continued)

| Name | Probes | Eluent and elution conditions | Detection |
|---|---|---|---|
| Tests to characterize $C_8/C_{18}$ bonded phases [324] | diphenhydramine<br>5-(p-methylphenyl)-5-phenylhydantoin<br>diazepam | acetonitrile/potassium-dihydrogen-phosphate and phosphoric acid buffer pH 2.3 156:340 (w/w) | UV, 220 nm |
| Homologues retention index standards [325] | alkan-2-ones ($C_3$-$C_{23}$) | methanol/water from 10:90 to 90:10 (%/%) | UV, 260 nm |
| | alkylarylketones ($C_8$-$C_{14}$) | | UV, 242 and 279 nm |
| | 2,4-dinitrophenylhydrazones from alkan-2-ones ($C_9$-$C_{11}$) | | UV, 363 nm |
| | toluene<br>nitrobenzene<br>2-phenylethanol<br>p-cresol | (A) methanol/water 70:30 (%/%)<br>(B) acetonitril/water 50:50 (%/%)<br>(C) tetrahydro furan/water 40:60 (%/%) | —<br>—<br>— |
| Solvophobic effects on silica-based reversed phase columns [326] | benzyl alcohol<br>butyl paraben<br>chlorpropham CIPC<br>corticosterone<br>cortisone<br>dexamethasone<br>dioctylphthalate | linear gradients over 15 min from 5% to 100% of methanol and/or acetonitrile in water | UV, 254 nm |

Table 4 (continued)

| Name | Probes | Eluent and elution conditions | Detection |
|---|---|---|---|
| | dibutylphthalate | | |
| | dimethylphthalate | | |
| | ethyl paraben | | |
| | fluorobenzene | | |
| | hexafluorobenzene | | |
| | methylparaben | | |
| | 1-methylnapthalene | | |
| | 1-methyl-phenanthrene | | |
| | o-nitrophenol | | |
| | propachlor, Ramrod | | |
| | propyl paraben | | |
| | toluene | | |
| | tri-p-tolyl phosphate | | |
| Retention mechanisms on $C_{18}$-silica packings [327] | 47 different chalcomes ($X-C_6H_4-CH=CH-CO-C_6H_4-Y$) diversely substituted | methanol/water 7:3 | UV, 300 nm |
| | nitrobenzene | methanol/water 7:3 | UV, 254 nm |
| | naphthalene | | |
| | phenanthrene | | |

Table 4 (continued)

| Name | Probes | Eluent and elution conditions | Detection |
|---|---|---|---|
| | methylbenzoate<br>biphenyl<br>diethylphthalate<br>anthracene<br>p-cresol<br>2-phenylethanol<br>benzophenone<br>benzyl alcohol<br>4-phenylbutanol<br>6-phenylhexanol<br>9-phenylnonanol<br>2-nitronaphthalene | | |
| Selectivity on alkyl- or phenyl-bonded stationary phases [328] | 53 different chalcones in their E-s-cis and 7-s-cis isomer forms | methanol/water 7:3 | UV, 280 nm |
| Influence of silica gel pretreatment and bonding technique on PAH selectivity of octadecyl-bonded phases [329] | naphthalene<br>acenaphthalene<br>fluorene<br>phenanthrene<br>anthracenc | acetonitrile/water 78:22 (%/%) | UV, 254 nm |

Table 4 (continued)

| Name | Probes | Eluent and elution conditions | Detection |
|---|---|---|---|
| | fluoranthene | | |
| | pyrene | | |
| | benz(a)anthracene | | |
| | chrysene | | |
| | benzo(b)fluoranthene | | |
| | benzo(k)fluoranthene | | |
| | benzo(a)pyrene | | |
| | dibenzo(a,h)-perylene | | |
| | indeno(1,2,3-cd)-pyrene | | |
| | pyridine | acetonitrile/water 70:39 (%/%) | UV, 254 nm |
| | toluene | | |
| | N,N-diethylaniline | | |
| Retention indices on octadecyl silica phases [33] | acetophenone | methanol/phosphate pH 8.5 40:60 | UV, 240 nm |
| | propiophenone | | |
| | butyrophenone | | |
| | barbital | | |
| | phenobarbital | | |
| | cyclobarbital | | |

Table 4 (continued)

| Name | Probes | Eluent and elution conditions | Detection |
|---|---|---|---|
| | butobarbital | | |
| | falbutal | | |
| | amobarbital | | |
| | pentobarbital | | |
| | secobarbital | | |
| | methohexital | | |
| Comparison of monomeric vs. polymeric $C_{18}$ columns [331] | 12 phenols of benzo(a)pyrene | methanol/water 4:1 (v/v) | UV, 254 or 280 nm |
| | 10 phenols of benz(a)anthracene | methanol/water 3:1 (v/v) | |
| | 5 phenols of chrysene | methanol/water 4:1 (v/v) | |
| Penicillins on bonded, polymeric and silica columns [333] | penicillin G | acetonitrile/0.01 M phosphoric and pH 1.6 from 0:100 to 90:10 (%/%) and acetonitrile/0.002 M potassium dihydrogen phosphate pH 4.6 from 0:100 to 90:10 (%/%) | UV, 220 nm |
| | ampicillin | | |
| | amoxicillin | | |
| Nucleotide test[a] [232] | NAD | acetonitrile/50 mM phosphate buffer pH 6.5 90:10 (v/v) + 1 mM $TBAHSO_4$ | UV, 254 nm |
| | adenosine | | |
| | NADH | | |
| | NADPH | | |

Table 4 (continued)

| Name | Probes | Eluent and elution conditions | Detection |
|---|---|---|---|
| ASTM test[a] [232] | benzyl alcohol<br>benzaldehyde<br>acetophenone<br>benzene<br>benzoic acid-methyl ester<br>dimethylterephthalate | methanol/water 50:50 (v/v) | UV, 254 nm |
| Aromatic hydrocarbon test[a-c] [232] | benzene<br>naphthalene<br>fluorene<br>anthracene<br>benzanthracene | acetonitrile/water 75:25 | UV, 254 nm |
| p-Hydroxybenzoic acid test[a] [232] | p-benzoic acid (PBA)<br>PBA methyl ester<br>PBA ethyl ester<br>PBA propyl ester | acetonitrile/50 mM phosphate buffer pH 3.5 35:65 (v/v) | UV, 254 nm |
| Digitalis test[a] [232] | digoxigenine<br>lanatoside C<br>digoxine<br>gitoxigenine<br>digitoxigenine | acetonitrile/50 mM phosphate buffer pH 3.5 32:68 (v/v) | UV, 254 nm |

Table 4 (continued)

| Name | Probes | Eluent and elution conditions | Detection |
|---|---|---|---|
| Protein test[a] [232] | ribonuclease<br>insulin<br>lysozyme<br>trypsin inhibitor<br>myoglobin | (A) acetonitrile/water/trifluoro acetic acid 0.5:99:0.5 (v/v/v)<br>(B) acetonitrile/water/trifluoro acetic acid 75:24.5:0.5 (v/v/v)<br>gradient:<br>0-20 min 10% B to 70% B<br>20-25 min 70% B to 90% B | Fluorescence ex: 275 nm |
| Nucleoside test[a] [232] | cytosine<br>uracil<br>cytidine<br>uridine<br>thymine<br>guanosine<br>theophilline<br>caffeine | (A) 20 mM phosphate buffer pH 6.5 + 1% methanol<br>(B) methanol<br>gradient: 0-2.5 min A 2.5-4.5 min 100% A to 10% B | UV, 254 nm |
| Vitamin test[b] [232] | ascorbic acid<br>nicotinic acid<br>pyridoxol hydrochloride<br>nicotinamide<br>thiamine hydrochloride | methanol/10 mM phosphate buffer pH 3.0 2:98 (v/v) + 2 $mgl^{-1}$ of triethylamine | UV, 270 nm |

Table 4 (continued)

| Name | Probes | Eluent and elution conditions | Detection |
|---|---|---|---|
| Pesticide test[b] [232] | 1,1-DDOH<br>α-enodosulfane<br>4,4-DDM<br>heptachlor<br>p,p-DDT<br>o,p-DDT<br>aldrin | acetonitrile/water 70:30 (v/v) | UV, 220 nm |
| Gallic acid test[b] [232] | gallic acid (GA)<br>GA methyl ester<br>GA propyl ester | acetonitrile/50 mM phosphate buffer pH 3.0 30:70 (v/v) | UV, 254 nm |
| Catecholamine test[b] [232] | 1,4-hydroxy-3-methoxymandelic acid<br>2-hydroxyacet-anilide<br>DL-threodihydroxy-phenyl serine<br>L-noradrenaline-1-tartrate<br>adrenaline<br>DL-octopamine<br>L-dopa<br>hydroxytyraminium chloride | acetonitrile/50 mM phosphate buffer pH 2.0 10:90 (v/v) + 30 mM sodium n-octylsulfonic acid | UV, 220 nm |

Table 4 (continued)

| Name | Probes | Eluent and elution conditions | Detection |
|---|---|---|---|
| Engelhardt test[c] [232] | aniline<br>phenol<br>o-toluidine<br>m-toluidine<br>p-toluidine<br>N,N-dimethylaniline<br>toluene<br>benzoic acid methyl ester<br>ethylbenzene | methanol/50 mM phosphate buffer pH 4.8 50:50 (v/v) | UV, 254 nm |
| Nicotinic acid test[c] [232] | isonicotinic acid<br>nicotinic acid<br>isonicotinamide<br>nicotinamide | methanol/0.7 M acetic acid 2:100 (v/v) | UV, 260 nm |
| Drug test[c] [232] | barbituric acid<br>codeine phosphate<br>atropinium sulfate<br>papverine<br>noscapinium chloride<br>diphenylhydramine | acetonitrile/50 mM phosphate buffer pH 2.3 20:80 (v/v) | UV, 220 nm |

Table 4 (continued)

| Name | Probes | Eluent and elution conditions | Detection |
|---|---|---|---|
| Beta blocker test[c] [232] | atenolol<br>metaprolol<br>bisoprolol<br>propanolol | acetonitrile/20 mM phosphate buffer pH 4.7 25:75 (v/v) | UV, 220 nm |
| DMD test[c] [232] | diphenhydramine<br>MPPH<br>diazepam | acetonitrile/50 mM phosphate buffer pH 2.3 20:80 (v/v) | UV, 220 nm |

[a]Test applied to LiChrospher RP-8
[b]Test applied to LiChrospher RP-18
[c]Test applied to LiChrospher 60 RP-select B

## IV. IMPACT OF THE STATIONARY PHASE PROPERTIES OF PACKINGS ON RETENTION AND SELECTIVITY

### A. Brief Survey of the Retention Mechanisms

The aim of any separation is to gain in resolution, which is governed by differences in solute retention. Hence, the accurate prediction and control of retention is the major subject of the theory of chromatography. A strictly thermodynamic or molecular statistical treatment of chromatographic interactions—to estimate changes in the free enthalpy of the solute ($\Delta G^0$) and to assess the related solute retention—is limited by the complexity of the systems. As an alternative, retention mechanisms to describe solute-surface interactions have been suggested. Equations based on these approaches have been derived, correlating retention with properties of the solute, the adsorbent, and the eluent. In column liquid chromatography, several retention mechanisms exist, each applicable to specific elution cases, i.e., dependent on the type of phase system.

The most straightforward and comprehensive retention model, applicable to the classic polar adsorbents, was developed by Snyder [2] in the 1960s. The model was improved and extended to chemically bonded polar packings and is now known as the Snyder-Soczewinski model [336]. Another approach, the so-called solvent interaction model, was proposed by Scott and coworkers [337]. Both models describe the retention of nonpolar and polar nonionic solutes.

The introduction of reversed phase chromatography expanded the application to polar and ionic solutes. An extended retention model to account for elution on hydrophobic surfaces was introduced by Horvath and coworkers [338-340]. Later this approach was refined by Horvath [5,341] and Hearn [6,342] for the retention of amphophilic solutes, such as amino acids, peptides, and proteins.

Hydrophobic interactions are also known to be powerful tools in the separation of proteins on agarose and on polar chemically bonded silicas (for examples, please see Ref. 343-345).

Snyder's model has also been adapted to describe solute retention on a type of hydrophobic packing based on pyrocarbon-coated silicas [346] and carbon [81].

### *1. The Snyder-Soczewinski Model*

This model is applicable to solute retention on oxidic adsorbents, including their polar chemically bonded derivatives, and nonpolar, moderately polar, to nonamphoteric polar solvents as eluents [336]. (In this context, the term amphoteric denotes solvents capable of self-hydrogen-bonding.) The following assumptions apply:

1. The adsorbent surface is covered by an adsorbed monolayer composed of solvent molecules (M).
2. The volume of adsorbed monolayer in $ml/m^2$ of adsorbent is fairly constant. Minor variations may occur, depending on the type of solvent and solute and their orientation in the monolayer.
3. The adsorbent surface is homogeneous, i.e., the adsorbent surface activity ($\alpha$) does not vary with experimental conditions.

The model differentiates between several solute-surface interaction cases [336,347].

*Displacement Without Localization* Retention of the solute molecule (X) takes place by displacement of an equivalent volume of solvent molecules (M) from the monolayer. If n is the number of adsorbed solvent molecules being displaced, the equilibrium is described as

$$X_n + nM_a \rightleftarrows X_a + nM_n \qquad (8)$$

where the subscripts n and a refer to molecules in the nonadsorbed and adsorbed phases, respectively. n is the ratio of the molecular cross-sectional area of solute ($A_s$) and solvent ($A_1$) at the adsorbent surface. The dimensionless standard-state energy, $\Delta E$ of X, characterizing the equilibrium, then becomes

$$\Delta E = E_X + nE_M - E_X - nE_M \qquad (9)$$

$\Delta E$ corresponds to the sum of energies of educts on the right-hand side of Eq. (8) minus the sum of energies of the products on the left-hand side. It is further assumed that the various interactions between solvent and solute molecules in solution are canceled by the corresponding interactions in the adsorbed phase. As a first approximation, Eq. (9) then becomes

$$\Delta E \sim E_X - nE_M$$

$E_X$ is the interaction energy of solute X with the adsorbent surface, $E_M$ the corresponding energy for the solvent. Related to the solute capacity factor k', the basic equation for solute retention is given by

$$\log k' = \log \frac{V_a \cdot W}{V_m} + \alpha'(S^0 - \varepsilon^0 A_s) + \Delta_{eas} \qquad (10)$$

$V_a$ is the volume of the adsorbed monolayer, W the mass of adsorbent, $V_m$ the column dead volume, $\alpha'$ the activity parameter of adsorbent, $S^0$ a relative energy parameter being equal to $E_X$, $\varepsilon^0$ the solvent strength parameter associated with the interaction of solvent molecules with the adsorbent surface being equal to $E_M$, $A_s$ the molecular cross-sectional area of adsorbed solute, and $\Delta_{eas}$ a second-order term for corrections in case the system deviates from the regular behavior predicted according to Eq. (10).

By extention of solvents A and B to a binary solution as eluent, the solvent strength $\varepsilon^0$ becomes

$$\varepsilon^0 = \varepsilon_A + \frac{\log(N_B \cdot K + 1 - N_B)}{\alpha' \cdot n_B} \qquad (11)$$

where B is the more polar, A the less polar solvent. K is the equilibrium constant for displacement of adsorbed solvent molecules A by adsorbing molecules B according to Eq. (8); $N_B$ is the mole fraction of solvent B in the solution; $n_B$ is equal to $A_m$, the molecular cross-sectional area of B; and $\alpha'$ is the adsorbent activity parameter specific for a given grade of adsorbent.

The displacement model was found to be in good agreement with the experimental data, e.g., for silica, alumina, and amine-alkyl-bonded silica [318]. Linear plots of log k' against $\varepsilon^0$ were obtained for less polar solutes.

*Restricted Access Delocalization* Fore more polar solvents—classified by Snyder and Poppe [348] as P solvents—and polar solute molecules, a localization of M or X at the adsorbed phase is assumed to occur. Localization is associated with a strong short-range interaction of the adsorbate molecule and the active surface sites. For binary solutions of A + B as eluents, where B is the polar localizing solvent, localization takes place at a low concentration of B. This results in a high effective solvent strength ($\varepsilon_B'$). An enrichment of B molecules in the adsorbed phase leads to a point where localized adsorption no longer takes place; this is attributed to the mutual interference of the adsorbed solvent B molecules. At surface coverages of $\Theta_B > 0.75$, adsorption then takes place without localization and the effective solvent strength of B; for larger values of $\Theta_B$, $\varepsilon_B''$ becomes smaller than at $\Theta_B << 0.75$.

The effective solvent strength of B, $\varepsilon_B$, corresponds to

$$\varepsilon_B = \varepsilon_B'' + 0/\text{olc}\ (\varepsilon_B' - \varepsilon_B'') \tag{12}$$

where 0/olc is a measure of the relative overall localization of B on the surface.

In the case of less polar nonlocalizing solvents B in binary eluents, $\varepsilon_B$ is essentially constant as a function of $\Theta_B$ over the whole range, while with increasing polarity of B, $\varepsilon_B$ drops sharply at $\Theta_B \sim 0.7$, as exemplified for silica and alumina as adsorbents.

*Site Competition Delocalization* Site competition is observed when solvent molecules M or solute molecules X are adsorbed adjacent to localized adsorbed molecules X or M, which results in a decrease of $E_X$, or $E_B$ is proportional to the solvent strength of M, $\varepsilon_M$. When this so-called site competition delocalization is in operation, the dimensionless interaction energy of solute X, $E_X$, becomes

$$E_X = E_X^0 - f_1(X)\varepsilon_M \tag{13}$$

$E_X^0$ is equal to the value of $E_X$ when $\varepsilon_M = 0$, and $f_1(X)$ is a localization function. $f_1(X)$ increases with increasing localization of X. Hammers and coworkers examined the function $f_1(X)$ for different adsorbents [349,350]. Site competition delocalization also occurs for binary eluents A + B, where B is polar and capable of localization. Equation (7) then becomes

$$E_B = E_B^0 - f_1(B)\varepsilon_A \tag{14}$$

*Intramolecular Delocalization* The interaction energy of a solute X of constituent groups i in the molecule gives

$$E_X = \sum^{m} E_i \tag{15}$$

It is conceivable that there are cases where a solute molecule X is localized in the adsorbed state by a group k, while remaining groups of X are not capable of achieving an optimum configuration. This intramolecular localization effect is expressed by

$$E_X = \sum E_i - f(Q_k^0) \sum^{m \neq k} E_i \tag{16}$$

where $f(Q_k^0)$ is a localization function which increases with the adsorption energy $Q_k^0$ of group k of the solute molecule. The effect of intramolecular delocalization reduces $E_X$; it has been established for solutes on different adsorbents [349,350].

### 2. The Scott-Kucera Model

The Scott-Kucera model [336,337,351,352] applies to the retention of solutes on silica adsorbents in binary solvent systems. It is assumed that the silica adsorbent, activated by heating at 423 K or by treatment with a dry solvent, carries a monolayer of adsorbed water at the surface which is hydrogen-bonded to the underlying hydroxyl groups. When such a silica is equilibrated with a unpolar solvent A, sorption of A takes place on the water monolayer by dispersion forces. On applying a polar solvent B instead of A, B is hydrogen-bonded to the water monolayer. The surface hydroxyl groups on the silica are believed to be largely deactivated by the hydrogen-bound water. Interaction of a solute X with such a stationary phase occurs in three different ways [337]. In the case of a weakly held solvent M, the solute X displaces solvent molecules and interacts directly with the "hydrated hydroxyl groups." For polar strongly interacting solutes, X associates directly with the solvent molecules, i.e., M is not displaced but functions as a stationary phase. A third case is distinguished where polar solutes X are stronger than the polar solvent; when the solvent layer is close to completion, both association and displacement take place.

The capacity factor k' of the solute is related to the concentration of the strong solvent B, $C_P$, by

$$\frac{1}{k'} = A' + B' \cdot C_P \tag{17}$$

where A' and B' are constants for a given solute X and solvent B.

Equation (17) states that changes in the retention of X are solely due to solvent-solute interactions; the application of the model is restricted to class P solvents, i.e., nonamphoteric polar solvents such as ethyl acetate, acetonitrile, tetrahydrofuran, and methyl ethyl ketone.

In its basic assumptions the model includes several inconsistencies with regard to the surface chemistry of silica, only some of which are discussed in the following. A silica annealed at 423 K under vacuum is largely free of physically adsorbed water (cf. Ref. 353) and has attached isolated and hydrogen-bonded hydroxyl groups. The average concentration of these hydroxyl groups—after annealing between 373 and 473 K—is about 8-9 $\mu mol/m^2$ (354) and not 2.5 $\mu mol/m^2$ as reported by Scott. Also, simple thermogravimetry cannot be used to identify surface hydroxyl groups, since water bound within the silica is also expelled during annealing [354]. An integration of the absorption bands of silica between 3400 and 3750 $cm^{-1}$ cannot be used to assess quantitatively the amount of water adsorbed, as the absorption coefficients of isolated and hydrogen-bonded hydroxyl groups are not known.

Snyder and Poppe [336], having made a detailed analysis of the Snyder-Soczewinski and Scott-Kucera models, came to the following conclusions:

> The solvent interaction model of Scott and Kucera does not provide an adequate description of the effects of solvent-solute interactions on solution activity coefficients or solute retention in LSC or ion-exchange systems. The postulate that solute-solvent polarizibility quantitatively predicts polar interactions in solution is questionable. The inference that dispersion interactions between the silica surface and adsorbing molecules are negligible or altogether absent is misleading.

### *3. The Jaronieč-Jaronieč Model*

Jaronieč and coworkers [355] (and further references therein) elaborated a general theory of liquid-solid adsorption processes from multicomponent liquid mixtures onto solid surfaces. The approach takes into account the energetic heterogeneity of the solid surface and the nonideal behavior of the phases while recognizing differences in the molecular size of solutes and solvents, and comprises several solute-solvent and solvent-solvent interactions. Whereas previous papers were mainly related to the theory of the adsorption process itself, recent studies [356] deal explicitly with the correlation between adsorption and chromatographic parameters.

The model assumes the following: that nonspecific interactions between solute molecules and solvent molecules are neglected; that the surface phase is of the monolayer type and behaves ideally; that the molecules of the sth solute form complexes with molecules of the first solvent of the eluent; that the total number of moles of all solvents in the surface phase is constant and independent of the presence of solute molecules; that the molecules of solvent and solute are spherically shaped; that the adsorbent surface may be energetically homogeneous or heterogeneous.

Equations for the retention of solutes in binary eluents have been derived, relating the capacity factor of the sth solute $k_s'$ to the mole fraction of the solute in the mobile phase $x_1$. On plotting $k_s'$ against $x_1$ in a log-log scale, straight lines are obtained (depending on several characteristic parameters).

### 4. *The Solvophobic Theory*

Several models have been proposed to describe the retention of solutes on reversed phase silicas in polar eluents [340], whereby the principal dispute concerns the question of whether the retention of solutes on hydrophobic surfaces is governed by bulk or interfacial phenomena. Detailed studies by Horvath and associates [338,339] have shown that a partitioning mechanism is highly improbable and that retention is controlled by solute-solvent and solute-stationary phase interactions. The solvophobic theory developed by Sinanoglu [357] has been adapted for the chromatographic retention of polar and amphiphilic solutes on reversed phase systems.

The process assumes a reversible association of solute X and the hydrocarbonaceous ligand L:

$$X + L \leftrightharpoons XL \qquad (18)$$

characterized by an equilibrium constant K. K is related to the change in the standard free energy $\Delta G^0$:

$$\Delta G^0 = -\ln \frac{K}{RT} \qquad (19)$$

where R is the gas constant and T the absolute temperature. $\Delta G^0$ is related to the solute capacity factor k' by

$$\ln k' = \ln \frac{V_s}{V_m} - \frac{\Delta G^0}{RT} \qquad (20)$$

where $V_s/V_m$ is the phase ratio of the system. $\Delta G^0$ consists of two terms. The first term reflects the free energy change occurring on association of the solute and the ligand L in the gas phase; the second term arises from the transfer of the solute, the ligand L, and the solute-ligand complex from the gas phase into the mobile phase. Figure 20

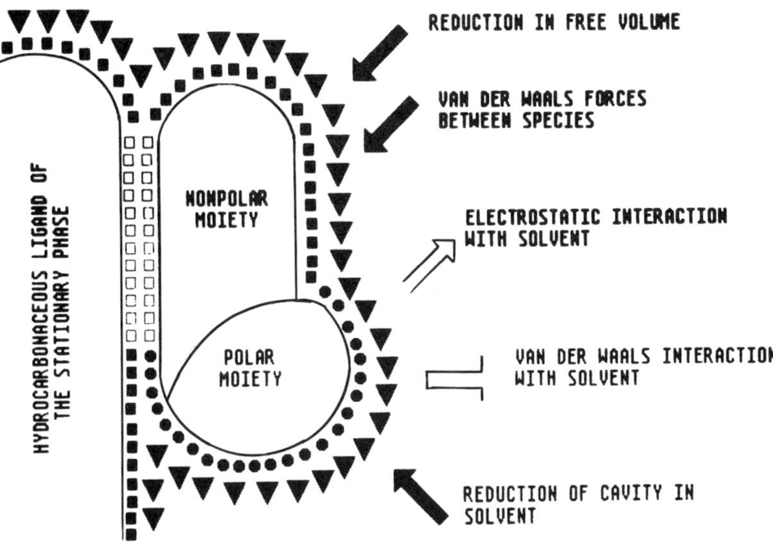


Fig. 20 Schematic illustration of the association between an amphilic solute and the hydrocarbonaceous ligand of the stationary phase in reversed phase chromatography. Water represented by open squares covers the molecular surface area by which the total cavity area is reduced upon contact of the two species. (From Ref. 340 by the permission of the publisher.)

illustrates the various interactions contributing to the reversible association of the solute in the stationary phase. The strength of the resulting net interaction is determined by the balance of opposing forces, which are attributed to the following phenomena: the reduction in free volume of the solute, the reduction of cavity in the solvent, and the van der Waals forces between the species vs. the van der Waals and electrostatic interaction with the solvent. An explicit expression is derived, relating the retention of the solute, measured as k', to the properties of solute, solvent, and ligand, expressed as polarizibility, molar volume, contact surface area, dipole moment of the solute, surface tension and dielectric constant of the solvent, and contact surface area of the ligand L. The equation permits a prediction of the retention of solutes based on their molecular properties. The logarithm of the solute capacity factor was found to be a linear function of the hydrocarbonaceous surface area of the solute for homologue series under otherwise constant conditions. Furthermore, log k' follows a linear dependence on the solvent composition in a certain range; for neutral solutes, log k' increases linearly with the rising salt concentration of the eluent.

A change in the pH of the eluent has drastic effects on the retention of ionizable compounds, attributed to electrostatic interactions. With increasing pH of the eluent the capacity factor of acids decreases, the extent depending on their $pk_a$ value. The stationary phase properties controlling retention and selectivity comprise the type of bonded hydrophobic ligand L, the ligand density $\alpha$, and the density and reactivity of hydroxyl groups. The mobile phase properties comprise the surface tension $\gamma$, the viscosity $\eta$, and the solvent strength.

The wide application potential of reversed phase chromatography is due to the fact that not only polar but also ionized and amphiphilic solutes can be resolved. Moreover, separation has expanded to peptides and proteins.

In reversed phase chromatography, applied under isocratic and gradient elution conditions, retention of amino acids, peptides, and proteins is governed by the hydrophobic character of the solute. More precisely, the retention order follows the difference between the molecular area of the solute-ligand complex and that of the solute and ligand L [342]. Several approaches have been developed to predict the retention of peptides based on their hydrophobic fragments. As stated earlier, the organic solvent affects retention through the bulk surface tension $\gamma_m$; the dielectric constant and k' values follow the sequence water > methanol > n-propanol > isopropanol > acetonitrile > tetrahydrofuran.

Elution is carried out under isocratic, stepwise, or gradient elution conditions, with organic solvent concentrations of up to 50% and sometimes higher.

Since the solutes are ionized, the capacity factor is dependent on the pH of the eluent under the given conditions. In accordance with the pk and pI values of the solute, pH-dependent optima of k' exist, i.e., pH change is a powerful tool for manipulating retention and selectivity. The pH range is usually limited to pH 1-8 for reversed phase silica packings. A pH of 2-3 is favored for amino acids, peptides, and proteins in the majority of separations.

The solute capacity factor of a peptide is the sum of three contributions:

$$k' = \rho_{so} k'_{so} + \rho_{si} k'_{si} + \rho_{ex} k'_{ex} \tag{21}$$

where $k'_{so}$, $k'_{si}$, and $k'_{ex}$ are the capacity factors of the solvophobic, silanophilic, and size exclusion mode, and $\rho_{so}$, $\rho_{si}$, and $\rho_{ex}$ are the weighted mole fractions of the solute in each retention mode. Retention of peptides and proteins in reversed phase gradient elution follows the empirical relation:

$$\log k' = k'_w - S\Phi \tag{22}$$

where $k'_w$ is the extrapolated value of k' for neat water, S the slope of the log k' vs. $\Phi$ function, and $\Phi$ the volume fraction of the organic modifier. S relates to the molecular weight M of the peptide as follows:

$$S = 0.48M^{0.44} \tag{23}$$

for acetonitrile as solvent.

*5. Adsorption vs. Partition*

For the sake of simplicity we have classified chromatographic separations on columns with porous packings of a surface area of >10 $m^2$/ml of packing as adsorption chromatography. As a matter of fact, the packings need not be porous. Nonporous microparticulate packings of a diameter of dp <2 μm also generate sufficiently high (external) surface areas. Studies on 1- to 2-μm nonporous reversed phase silicas with external surface areas of 2 to 1 $m^2$/ml of packing have shown that even these low surface area supports are capable of retaining low molecular weight compounds [308].

Under certain conditions in normal phase chromatography on silica and in particular in reversed phase chromatography, it is debatable as to whether solute retention is due to adsorption or partition effects. In an adsorption process, solute retention is related to the total interfacial surface area and expressed by

$$k' = \frac{V_s}{V_m} \cdot K_{ads} \tag{24}$$

where $K_{ads}$ is the adsorption coefficient in moles adsorbed per unit volume of packing to the moles of solute per unit volume of eluent and $V_s$ ($V_m$) the volume of stationary (mobile) phase. $V_s$ is then given by

$$V_s = \delta_m \cdot A_s \tag{25}$$

where $\delta_m$ is the thickness of a monolayer of adsorbed solvent molecules and $A_s$ the accessible surface area of the column packing in $m^2$.

In a partition process, solute retention is related to the stationary phase volume via the partition coefficient P. The solute capacity factor is

$$k' = \frac{V_s}{V_m} \cdot P \tag{26}$$

where P is the partition coefficient of the solute in moles per unit volume of stationary phase to the moles per unit volume of mobile phase and $V_s(V_m)$ the volume of stationary (mobile) phase. The fundamental assumption for both coefficients is that they are dependent on the solute concentration, i.e., they are associated with the linear part of the isotherm.

There is a simple procedure to prove whether retention is based on adsorption of partition: the solute partition coefficient for a given stationary and mobile phase is determined in a batch experiment and compared to the coefficient calculated from chromatographic measurements, where the phase ratio $V_s/V_m$ is known (see Chap. 4 and also compare Ref. 358). As shown by Huber et al. [359], a transition between adsorption and partition of solutes using ternary mobile phases is indicated by distinct changes of k' when k' is plotted against the mobile phase composition.

The partitioning process requires a stationary phase with properties comparable to those of a bulk liquid. For silica and alumina packings, being hydrophilic, the quantity of water adsorbed should exceed the monolayer capacity, which amounts to about 8 $\mu mol/m^2$. By appropriate procedures, silica can be loaded with water, thus gradually filling the pore volume [360]. The loading is best monitored by the consecutive decrease of the total porosity of the column. The retention and selectivity of such a liquid-liquid partition column differs substantially from a column operated in the adsorption mode.

In reversed phase chromatography the situation is more complex. While the solvophobic theory postulated by Horvath et al. [350-352] assumes an adsorption mechanism, Dill [361] recently provided convincing arguments for a partition mechanism. Solute capacity factors in RPC have often been compared with the solute partition coefficient P in n-octanol-water as reference system. The results obtained are controversial. Sometimes a good correlation is obtained, but other instances show a large deviations [362,363]. Reversed phases, when equilibrated with organic solvent water mixtures, preferentially adsorb the organic solvent component and this enrichment increases as the chromatographic solvent strength increases [318,364]. Thus, the stationary phase in RPC is viewed as a dynamic multicomponent system composed of an arrangement of hydrocarbon chains solvated by an organic-enriched solvent mixture. At this point it is useful to give thought to the conformational behavior and mobility of solvated n-alkyl chains. Melander and Horvath [365] summarized various structures under which the stationary phase in RPC has been viewed. The picket fence configuration is not very likely because of the low ligand density of n-alkyl chains compared to the compressed monolayers of compounds with n-alkyl groups. The fur configuration is more open than the picket fence configuration and permits solute molecules to associate with the chains laterally. In the stack conformation the n-alkyl chains are in close contact with each other, forming clusters at the surface. The latter structure resembles that proposed by Lochmüller and Wilder [366], termed droplet-like. In this context, attention is drawn to some results obtained from physicochemical and chromatographic measurements. Amati and Kovats [367] classified the silanized surfaces into low surface energy systems, as opposed to the surface of the parent oxides. Wetting experiments on these surfaces revealed a sharp change in the wetting angle as a function of temperature and chain length [208]. Gilpin [368,369] further showed that n-octyl- and n-decyl-bonded phases with pure water mobile phases undergo an irreversible transition as a function of the column temperature. These findings are consistent with those by Morel and Serpinet [314-316] who also found a nonlinear change of solution retention as a function of the column temperature in GC and HPLC experiments. Carbon-13 NMR studies of reversed phases clearly evidenced the mobility of bonded n-alkyl chains, this being a function of the solvation and the chain length [370,371]. Thus, two kinds of conformations can be discriminated, one with bent chains and a high mobility, the other with stiffened chains and lower mobility [318].

Schunk and Burke [318] concluded from their studies on an n-octadecyl-bonded silica that with decreasing column temperature the chains become elongated and stiffened. This conformational change comes on at about 300 K and is accompanied by an alteration in the composition of the solvated layer. As a consequence, the intercalation/association and partition of the solutes change also. In conclusion, stationary phases in RPC exhibit both characters: they function as interfaces for solute adsorption and as liquid-like layers for solute partition, depending on their structure and on chromatographic conditions. Bonded packings with linear polyacrylamide chains used in the liquid-liquid partition chromatography of proteins and nucleic acids might serve as an additional example that bonded chains can hold a liquid stationary phase [372].

### 6. *Stationary Phase Properties Controlling Retention and Selectivity in HPLC on Silica, Alumina, and Polar Bonded Phases*

Following the Snyder equation (see p. 404) stationary phase properties decisive for the retention of a given solute and mobile phase can be predicted. Clearly, these are the $V_a$ and $\alpha'$. $V_a$ is equal to the average thickness of the adsorbed monolayer of the eluent multiplied by the specific surface area $a_s$ of the adsorbent. $\alpha'$ is the surface activity parameter, which depends on the reactivity, strength, and distribution of surface sites of the adsorbent, which in turn are influenced by the chemical surface structure and the pore structure.

At a constant $\alpha'$, $V_m$, and W, the solute capacity factor will increase proportionally with increasing $a_s$ (see Figure 21). A linear dependence

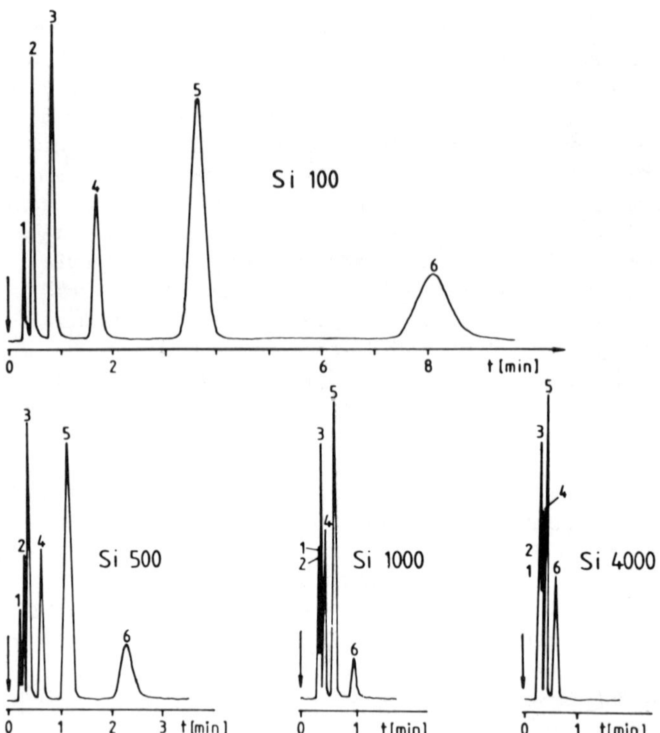


Fig. 21 Effect of specific surface area of silica on the retention of solutes in straight-phase chromatography. Columns: 200 × 3 mm packed with LiChrospher Si 100, 500, 1000, and 4000, dp 10 μm, eluent: n-heptane (adjusted to 20% relative humidity), flow rate: 5 ml/min, inlet pressure: 125 bar, solutes: benzene, diphenyl, m-terphenyl, m-quaterphenyl, m-quinquephenyl, m-sexiphenyl. (Courtesy of E. Merck, Darmstadt, FRG.)

of k' on $a_s$ was found for a series of superficially porous packings prepared with differing thicknesses of the silica layer on the impermeable glass beads, and thus varying the surface area between 5 and 35 $m^2/g$ (see Fig. 22). Such a clear relationship is only found when the adsorbent has been prepared in the same way. Usually, the specific surface areas of silicas vary from 5 to 500 $m^2/g$ depending on the average pore diameter (see p. 339). When plotting the solute capacity factor obtained on columns with silicas of various $a_s$ (normalized to unit column volume) at constant mobile phase, deviations from linearity arise associated with the different procedures in manufacturing and thus with the variation in the parameter $\alpha'$ (compare Ref. 374).

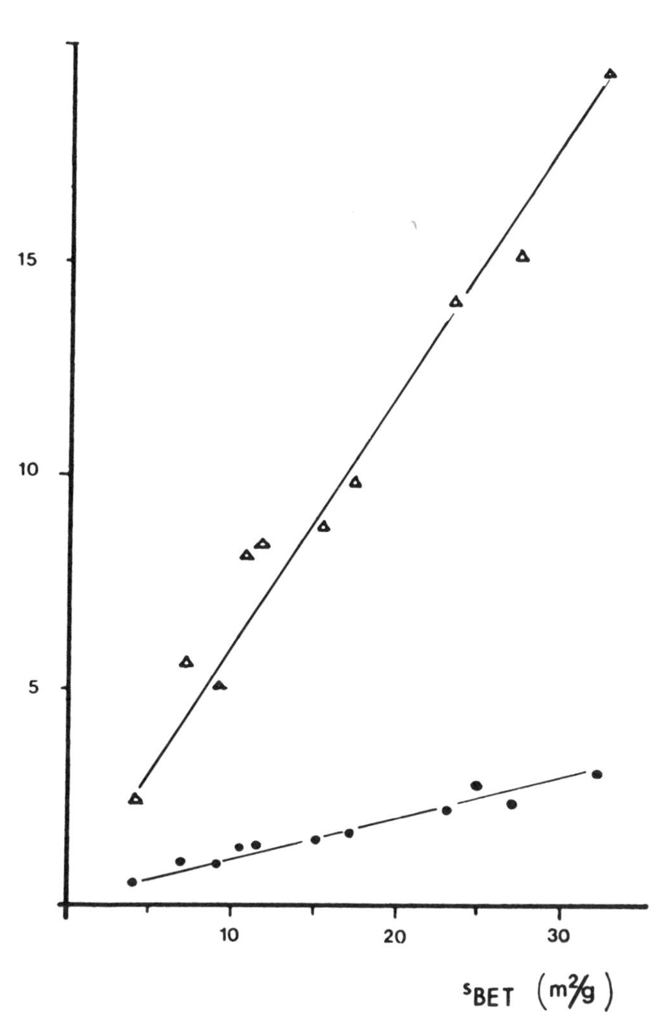


Fig. 22 Dependence of capacity factor on specific surface area on pellicular-type materials of varying $S_{BET}$ and constant average pore diameter. Columns: length 500 mm, I.D. 4.0 mm. Packings: porous layer beads (home-made preparation); $d_p$ = 31.5 μm, $d_s$ = 0.1-1.0 μm; $S_{BET}$ = 4-34 $m^2/g$; D = 6 nm. Eluent: n-heptane (30% relative water content). Detector: UV (254 nm). Samples: (●) m-terphenyl; (Δ) m-quinquephenyl. (From Ref. 373 by the permission of the publisher.)

As mentioned before, the surface activity parameter reflects the type and distribution of surface sites of the adsorbent. The surface sites are specific for any type of hydrophilic adsorbent. At the silica surface the active sites are hydroxyl groups of the Broensted type (isolated, geminal, and vicinal). This pattern is valid for pure silica. It has been shown, however, that impurities from the manufacturing process often remain, causing a drastic change of the surface acidity as measured by the pH of the silica suspension [375]. Even at pure silica the hydroxyl groups are heterogeneously distributed across the surface. Various after-treatment procedures have been elaborated by the manufactures to yield a silica with a more homogeneous surface. These procedures are mainly proprietary. To standardize the surface of silicas made by different manufacturers, the method suggested by Hansen et al. [47] can be employed: an acidic or buffer solution is purged through a given silica column. The excess is displaced with methanol. In this way a distinct apparent surface pH is achieved. Broensted acid, Lewis acid, and Lewis base sites have been discriminated at the surface of alumina. Again, aluminas manufactured for chromatographic purposes are not pure but contain sodium and other impurities. Aqueous suspension of alumina show a pH value above 9. Such a base alumina has cation exchange properties. By careful "neutralization" a "neutral" alumina is obtained. Treatment of basic alumina with an excess of acid yields "acidic" alumina exhibiting anion exchange properties.

Silica and alumina have the highest surface activity when the adsorbents are free from physisorbed water. Solute retention in "water-free" hydrocarbon mobile phases is then at its maximum. The addition of water blocks the most active surface sites since water as a polar adsorptive is preferentially adsorbed. Other polar compounds such as alcohols also adsorb irreversibly at the surface. Consecutive adsorption of water deactivates the surface and retention decreases concurrently. When plotting the logarithm of the solute capacity factor against the relative water content of the mobile phase, a linear decrease is observed with increasing water content [376,377]. The water content only plays this role with unpolar mobile phases, e.g., when using n-hexane, n-heptane, etc. At a composition of n-hexane/dioxan of 90:10 (v/v) and increasing amounts of moderately polar eluent, the dependency disappears. The role of water and other polar modulators on solute retention and selectivity in straight-phase chromatography is discussed in depth by Snyder [2] and Engelhardt and Elgass [4]. It should be emphasized that the polar modifiers in hydrocarbon mobile phases not only affect retention but also the peak shape. The higher the surface activity of silica and alumina, the more heterogeneous will be the surface. As a result the solutes are strongly retained and show tailed peaks.

Another means to deactive the silica surface is chemical bonding. Polar bonded silica packings do not show such a strong dependence when plotting the log k' against the water content of the hydrocarbon mobile phase.

The solute retention in straight phase chromatography is due to specific solute-surface site interactions. The strength of the interaction is reflected by $S^0$ in the Snyder equation. $S^0$ depends on the type, number, and position of functional groups in the solute molecule at a given adsorbent and mobile phase. This is the basis for the discrimination of solutes with different types of functional groups by straight-phase chromatography.

An optimum interaction is dependent on the position of functional groups in the solute molecule and their steric orientation to the surface sites of the adsorbent. In this way the separation of geometric isomers is obtained. The influence of the solvent composition on the solute capacity factor is reflected by the solvent strength parameter $\varepsilon^0$. The following relationships have been found:

$$\varepsilon^0_{\text{silica}} \sim 0.7\,\varepsilon^0_{\text{alumina}} \text{ [2] and} \tag{27}$$

$$\varepsilon^0_{\text{diol silica}} \sim 0.3\,\varepsilon^0_{\text{silica}} \text{ [378]}$$

Although straight-phase chromatography is known to resolve weak and medium polar solutes, polar compounds can also be separated on bare silica and alumina. Brugman and Kraak [222] employed polar mobile phases which contain a buffer to suppress the ionization of the compounds. Another means is to add ion-pairing reagents, which results in the formation of neutral solute complexes [223]. Hansen and Helboe [379,380] used cetyltrimethylammonium bromide to modify the silica surface to separate nonionic and anionic solutes.

The ion exchange properties of silica and alumina have been utilized to resolve strong basic compounds in aqueous mobile phases buffered to a high pH [381-383]. Lingeman et al. [384] investigated the influence of the pH, the competing ions, counterions, and cosolvent on the retention of basic drugs (amines) on bare silica and alumina columns. Similar studies have been performed by Schmid and Wolf [385] on silicas. The phase systems provided an excellent stability and a high reproducibility.

Attempts have been made to compare the retention and selectivity of polar bonded phases with that of bare silica. Hara et al. [386-392] studied cyano- and aminopropyl-bonded silicas using steroids and protected oligopeptides as test substances. Verzele et al. [393] compared the retention of selected test solutes on amino-, cyano-, diol-, polyol-, polyphenol-, nitro-, sulfonic acid-, and polyoxiran-coated silicas with the parent silica in n-hexane/dioxan and n-hexane/tetrahydrofuran mobile phases. Silica showed a medium retention strength compared to the polar bonded phases.

### 7. *Stationary Phase Properties Controlling Retention and Selectivity on Reversed Phase Packings*

Currently, there are about 300 different brands of reversed phase packings available as packed columns. Commercial silica-based reversed phase packings are characterized by the type of bonded hydrocarbon ($C_{18}$, $C_8$, $C_4$, phenyl, etc.), whether they are of monomeric or polymeric type, and whether endcapping has been performed. The carbon load is given for some packings (see Appendix III). On poly-(styrene-divinylbenzene)-based packings the nominal pore diameter is indicated. The most frequently used reversed phase packings are n-octadecyl-bonded phases followed by n-octyl [394].

Despite the diversity of brands, the choice of an appropriate reversed phase column is greatly eased when the basic relationships between the stationary phase and the chromatographic properties are known. The former comprise the particle size, size distribution and particle shape, the pore size, porosity, the type of stationary phase, the carbon load, etc., the latter are the column efficiency, retention and selectivity. Particle size, size distribution, and particle shape primarily dictate the column performance in terms of column plate number, peak shape, column pressure drop, and the column efficiency-flow rate dependency. Pore size comes into play when the separation of larger molecules is intended. Low molecular weight compounds are best resolved on 6- to 15-nm pore size packings. When larger pore size packings are applied, the corresponding loss of surface area of the stationary phase must be compensated by a decrease in the solvent strength of the mobile phase to gain comparable retention. For the separation of peptides and proteins up to 20 kD by reversed phase gradient elution, a pore diameter of 15 nm is sufficient [395]. Larger proteins require pore sizes in excess of 30 nm [126,127,396]. By bonding a long hydrocarbon chain or by depositing a polymer layer, the pore structure of the parent silica can be changed in such a way that narrow constrictions of the pores occur leading to a reduction of the intraparticle diffusion of the solute molecules. The result will be an impaired column performance, particularly at higher flow rates of the mobile phase. Thus, to maintain fast kinetics in the chromatographic process the pores carrying the stationary phase must be freely connected and accessible. The decrease in pore size as a result of bonding the stationary phase is monitored by measuring the pore size distribution of the parent silica before and after bonding, e.g., by nitrogen adsorption. Depending on the length of the ligand or the thickness of the coating, the pore size distribution curve shifts to smaller pore diameters. The lengths of n-alkyl ligands has been calculated to 0.3 nm (trimethylsilyl), 0.7 nm (n-butyldimethylsilyl), 1.2 nm (n-octyldimethylsilyl), 1.7 nm (n-dodecyldimethylsilyl), and 2.3 nm (n-octadecyldimethylsilyl) [397].

Polymer-based reversed phase packings in general exhibit a different retention pattern than silica-based packings for both low molecular weight and high molecular weight solutes at a constant mobile phase composition [213]. This is attributable to a different retention mechanism.

A considerable amount of work has been done to explore the effect of the stationary phase properties of n-alkyl-bonded silicas on the retention and selectivity of nonpolar, polar, basic, and peptidic solutes. The simplest approach consists of plotting the logarithm of the solute capacity factor against the carbon load of the reversed phase packing at a constant mobile phase composition. The resulting dependencies either show a linear relationship (for comparison, see Ref. 398) or approach a plateau at high carbon loads [399]. Two variables contribute to the carbon load: the n-alkyl chain length of the ligand and the ligand density. Berendsen and Galan [400-402] studied the dependence of k' of solutes on the chain length n of n-alkyldimethylsilyl-bonded silicas at constant ligand densities and mobile phase composition. The graphs log k' against n showed a linear increase up to a plateau. From the bending points a critical chain length was derived, which varied between 6 and 14. These findings were borne out by Lork [206] for low molecular weight solutes (see Fig. 23a). Surprisingly, the same dependency measured for a family of paracelsin peptides on n-alkyldimethylsilyl-bonded silicas resulted in a completely different pattern: the relative retention as a function of n went through two maxima at n = 2 and 4 and then decreased at increasing n (see Fig. 23b). Furthermore, careful inspection of the graphs showed that retention was always higher on n-alkyl ligands with an even rather than an odd number of carbon atoms. This was also observed by Simpson [403].

A hyperbolic relationship was obtained when plotting the logarithm of the capacity factor against the ligand density $\alpha$ at constant n-alkyl chain length of the ligand and mobile phase composition [206,404]. A critical ligand density was derived which varied between 2 and 3 $\mu mol/m^2$, above which the solute capacity factor remained constant (Fig. 24).

The interpretation of the hyperbolic relationships of log k' against n and $\alpha$ is straightforward and can be explained by the model of solvophobic theory. According to the model the logarithm of k' is linearly dependent on the molecular contact area $\Delta A$, between the solute and the hydrocarbonaceous ligand, until $\Delta A$ reaches a maximum and the curve log k' vs. n or $\alpha$ plateaus. This implies that short-chain solutes reach the maximum $\Delta A$ at short n-alkyl chains, while long-chain solutes

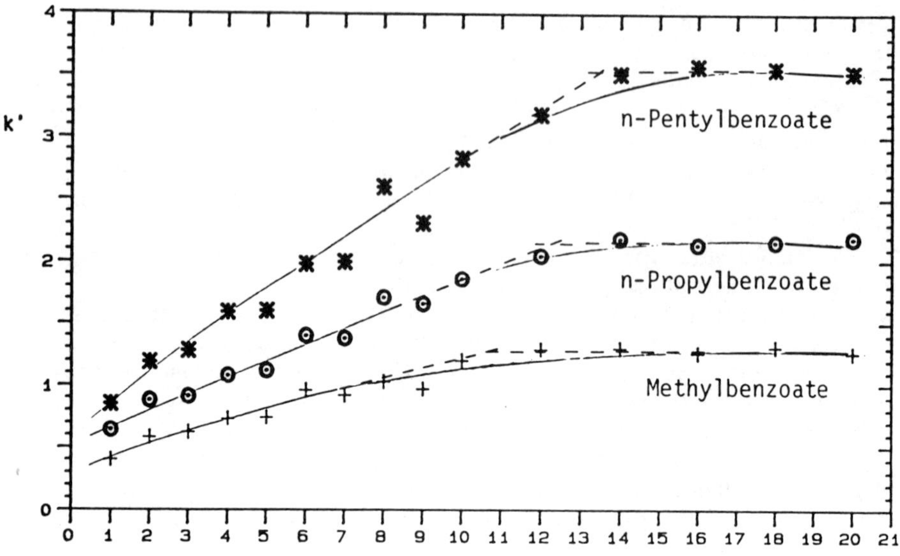


(a)

Fig. 23 Dependence of the capacity factor of n-alkylbenzoates (a) and paracelsin peptides A, B, C, and D (b) on the n-alkyl chain length of n-alkyldimethylsilyl bonded silicas at constant ligand density of $\alpha = 3.5 \pm 0.2\ \mu mol/m^2$.

| | |
|---|---|
| column | 205 × 4.6 mm |
| eluent | water/methanol/acetonitrile, 22:39:39 (v/v/v) |
| detector | UV, 254 nm |

require longer n-alkyl chains. This hypothesis is evidenced by the results of Karch et al. [405] and Lork [206]. Similar explanations can be given for the observation of the critical ligand density: with increasing density the accessibility of the chains reduces and a maximum $\Delta A$ is approached. The latter depends on the n-alkyl chain length, the type of solute, and the mobile phase composition.

As discussed earlier, n-octadecyl-bonded silicas undergo phase transitions and/or conformational changes when increasing the column temperature. This results in an enhanced solvation of n-alkyl chains and in an improved access of solutes associated with net increase in retention. The plot of logarithm of k' vs. the reciprocal of the solumn temperature consists of two straight lines of different slopes which

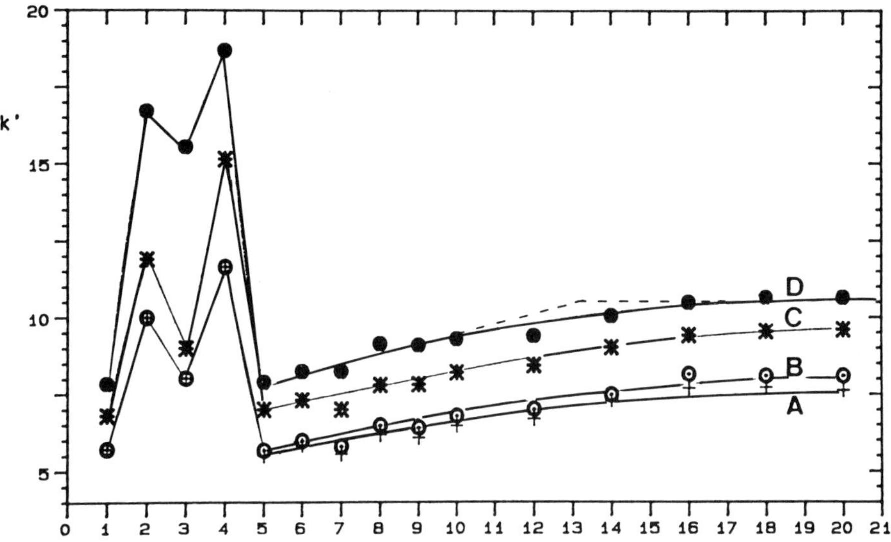


(b)

Fig. 23 (continued)

intersect at a distinct temperature. This deviation temperature spans a range from 283 to 306 K and is influenced by the mobile phase composition and the polarity of the solute. Putting all these considerations together, it is highly recommended that an n-octadecyl-bonded silica column be thermostatted.

Spacek et al. [406] prepared mixed n-octadecylsilyl-, trimethylsilyl-bonded phases and compared the chromatographic data to those obtained on bonded phases with single n-alkyl chain ligands. The capacity factors and selectivity coefficients were always higher on the mixed bonded phases than on the single type at the same carbon content. Polymeric phases with various n-alkyl chain lengths have been probed as attractive alternatives to monomeric phases in the separation of polynuclear aromatic hydrocarbons (PAH) [407-409]. Studies on the effect of the n-alkyl chain length of monomeric and polymeric phases on the selectivity for PAHs showed that the selectivity of long-chain monomeric phases approaches that of long-chain polymeric phases [410] (see Fig. 25).

By reducing the n-alkyl chain length from n-octadecyl- to n-octyl-bonded silicas, the hydrophobic character of the stationary phase diminishes and the selectivity toward polar solutes increases. A test

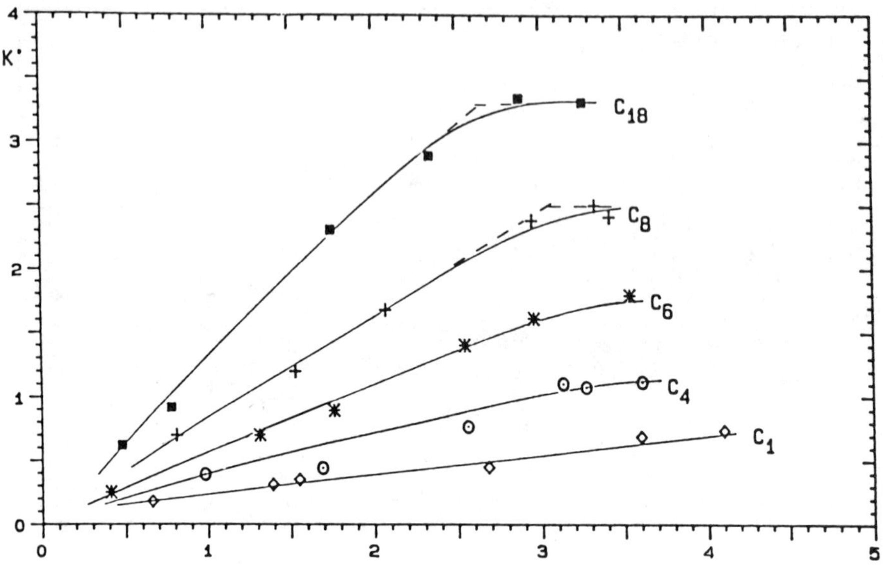


Fig. 24 Dependence of the capacity factor of n-butylbenzoate on the ligand density of $C_1$-, $C_4$-, $C_6$-, $C_8$-, and $C_{18}$- (dimethylsilyl) bonded silicase.

| | |
|---|---|
| column | 250 × 4.6 mm |
| eluent | water/methanol/acetonitrile, 22:39:39 (v/v/v) |
| detector | UV, 254 nm |

to judge the polarity of a reversed phase and to discriminate between an n-octadecyl and an n-octyl bonded phase was developed by Engelhardt [411]. The test mixture consisted of the following solutes: benzene, toluene, ethylbenzene, benzamide, and ethylbenzoate. The mobile phase applied is water/methanol of 50:50 (v/v). The ratio of the capacity factors $k'_{toluene}/k'_{benzene}$ reflects the hydrophobic selectivity of the stationary phase. The capacity ratio of toluene to ethylbenzoate serves as an indicator to discriminate between an n-octyl- and an n-octadecyl-bonded silica: toluene elutes ahead of ethylbenzoate on an n-octyl column while on n-octadecyl columns the ester elutes ahead of toluene.

A major problem in the RPC of silica-based packings is related to the elution of basic substances. Highly active and acidic residual hydroxyl groups at the silica surface result in a pronounced tailing and an enhanced retention of basic solutes, e.g., aniline and pyridine.

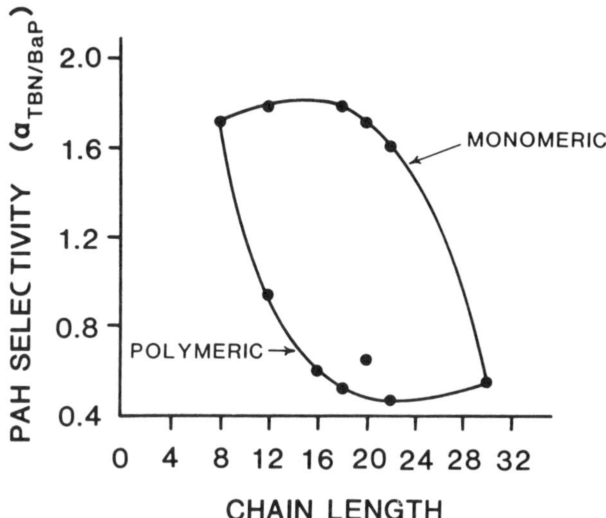


Fig. 25 Plot of column selectivity $\alpha_{TBN/BaP}$ vs. chain length of monomeric and polymeric phases. TBN = tetrabenzonaphthalene (dibenzo-(g,p)chrysene). BaP = benzo(a)pyrene. Eluent: water/acetonitrile 15:85 (v/v). (From Ref. 410 by the permission of the publisher.)

Thus, the retention and peak shape of aniline and benzamide is used in the Engelhardt test to monitor the residual activity of silica. On a "good" column, aniline should elute with a symmetric peak in pure water as mobile phase. Pretreatment of the silica with acids before silanization considerably improves the chromatographic performance of reversed phase silica columns toward basic solutes [208,412-414]. The behavior of reversed phase columns toward basic and acidic solutes is best checked by running the column with a gradient of the following mobile phases [415]:

Neutral compounds and extremely weak acids and bases.
  eluent A: water/acetonitrile 90:10 (v/v)
  eluent B: acetonitrile
  gradient from (A) to water/acetonitrile of 10:90 (v/v)

Weak to medium-strong acids of $pk_a > 4$ such as carboxylic acids, phenols, acidic amides, sulfonamides, acidic enoles, where ionization is suppressed and lipophilic bases (amines, anilines) in the protonated form.
  eluent A: 10 mM phosphoric acid, pH 2.3
    700 μl phosphoric acid [85% (w/w) per 1 water] and acetonitrile 10:90 (v/v)

eluent B: acetonitrile
gradient from (A) to phosphoric acid solution/acetonitrile of 10:90 (v/v)

Weak to medium-strong bases of pk < 6 (anilines, pyridine, heterocyclic bases) where ionization is suppressed, and lipophilic acids in the deprotonated form.
eluent A: 5 mM phosphate buffer, pH 7.5 (0.68 g of $KH_2PO_4$ + 40 ml of 0.1 M KOH per liter of water)
eluent B: acetonitrile
gradient from (A) to phosphate buffer/acetonitrile of 10:90 (v/v)

An initial partial deactivation of the silica with trimethylchlorosilane and hexamethyldisilazane followed by exhaustive octadecylation was found to notably improve the column efficiency compared to the non-predeactivated support [416]. The authors claim that a much more homogeneous lipophilic stationary phase is generated than at direct silanization.

Aryl-bonded silicas are less retentive than n-octadecyl-bonded silicas. They exert a pronounced selectivity toward solutes with aromatic ring systems, attributed to $\pi$-$\pi$ interactions.

Perfluorocarbon-bonded phases with ligands of the type heptadecafluorodecyl(dimethyl)silyl, $CF_3(CF_2)_7(CH_2)_2Si(CH_3)_2$, synthesized by Berendsen et al. [417], showed an enhanced retention for fluorine-containing compounds through specific fluorine-fluorine interactions. Similar results on phases of the type heptafluoroisopropoxypropyl(dimethyl)silyl, $F(CF_3)_2C$-$O$-$(CH_2)_3Si(CH_3)_2$, and pentafluorophenyl(dimethyl)silyl, $C_6F_5Si(CH_3)_2$ were obtained by Sadek and Carr [418].

## REFERENCES


1. M. S. Tswett, *Ber. Deut. Botan. Ges.*, *24*:322 (1906).
2. L. R. Snyder, *Principles of Adsorption Chromatography*, Marcel Dekker, New York, 1968.
3. K. K. Unger, *Porous Silica*, Journal fo Chromatography Library, Vol. 16, Elsevier, Amsterdam, 1979.
4. H. Engelhardt and H. Elgass, *HPLC: Advances and Perspectives*, Vol. 2 (Cs. Horvath, ed.), Academic Press, London, 1980, pp. 57-108.
5. W. Melander and C. Horvath, *HPLC: Advances and Perspectives*, Vol. 2 (Cs. Horvath, ed.), Academic Press, London, 1980, pp. 114-303.
6. M. T. W. Hearn, *Advances in Chromatography*, Vol. 20 (J. C. Giddings, E. Grushka, J. Cazes, and Ph. Brown, eds.), Marcel Dekker, New York, 1982, pp. 1-64.

7. G. Schomburg, *LC/GC*, *6*(1):36 (1988).
8. J. J. Kipling, *Adsorption from Solution of Non-electrolytes*, Academic Press, London, 1965.
9. J. Oscik, *Adsorption*, Ellis Horwood Ltd., Chichester, UK, 1982.
10. G. D. Parfitt and C. H. Rochester, *Adsorption from Solution at the Solid/Liquid Interface*, Academic Press, London, 1983.
11. G. Schay, L. Nagy, and T. Szekremesy, *Polytechnica*, *4*:95 (1960).
12. J. F. Huber and R. G. Gerritse, *J. Chromatogr.*, *58*:137 (1976).
13. H. L. Wang, J. L. Duda, and C. J. Radke, *J. Colloid Interface Sci.*, *66*:1277 (1978).
14. F. Riedo and E. Kovats, *J. Chromatogr.*, *239*:1 (1982).
15. N. Le Ha, J. Ungváral, and E. Kovats, *Anal. Chem.*, *54*:2410 (1982).
16. F. Köster and G. H. Findenegg, *Chromatographia*, *15*:743 (1982).
17. W. Markowski, K. Czapinska, and H. Poppe, *Chromatographia*, *17*:221 (1983).
18. J. Jacobson, J. Frenz, and Cs. Horvath. *J. Chromatogr.*, *316*:53 (1984).
19. G. H. Findenegg and F. Köster, *J. Chem. Soc. Faraday Trans.*, *82*:2691 (1986).
20. R. E. Boehm and D. E. Martire, *J. Phys. Chem.*, *84*:3620 (1980).
21. M. Borowko, *J. Colloid Interface Sci.*, *102*:519 (1984).
22. W. Rudzinski, J. Narkiewicz-Michalek, Z. Suprynowicz, and K. Pilorz, *J. Chem. Soc. Faraday Trans.*, *181*:533 (1985).
23. M. Jaronieč, D. E. Martire, and M. Borowko, *Adv. Colloid Interface Sci.*, *22*:177 (1985).
24. H. P. Boehm and H. Knöezinger, *Catalysis*, Vol. 4 (J. R. Anderson and M. Boudart, eds.), Springer-Verlag, Berlin, 1983, pp. 40-189.
25. N. E. Tretyakov and V. N. Filimonov, *Kinetics and Catalysis* (Russ.), *13*:815 (1972).
26. H. Knöezinger and P. Ratnasamy, *Catal. Rev. Sci. Eng.*, *17*:31 (1978).
27. K. K. Unger, *Porous Silica*, J. Chromatogr. Libr. Vol. 16, Elsevier, Amsterdam, 1979, pp. 1-4.
28. H. Karge, *Z. Phys. Chem. Neue Folge*, *76*:133 (1971).
29. D. Barby, *Characterization of Powder Surfaces* (G. D. Parfitt and K. S. W. Sing, eds.), Academic Press, London, 1976, pp. 353-419.
30. H. Ferch and A. Kreher, *Chemische Technologie*, *Bd. 3, Anorgan. Technologie* 2 (H. Harnisch, R. Steiner, and K. Winnacker, eds.), C. Hanser Verlag, München, 1983, pp. 75-90.
31. R. K. Iler and H. J. Mc Queston, U.S. Patent 3,855,172 (1974).
32. A. V. Kiselev, *The Structure and Properties of Porous Materials* (D. H. Everett and F. S. Stone, eds.), Butterworth, London, 1958, pp. 195-226.

33. A. P. Karnaukhov, *Kinetics and Catalysis* (Russ.), *12*:1025 (1971).
34. A. P. Karnaukhov, *Kinetics and Catalysis* (Russ.), *12*:1235 (1971).
35. K. K. Unger, J. N. Kinkel, B. Anspach, and H. Giesche, *J. Chromatogr.*, *296*:3 (1984).
36. K. K. Unger, B. Anspach, and H. Giesche, *J. Pharm. Biom. Analysis*, *2*:139 (1984).
37. D. W. Sindorf and G. E. Maciel, *J. Am. Chem. Soc.*, *105*:1487 (1983).
38. N. Becker and K. K. Unger, *Chromatographia*,
39. Grace Chromatographie Gel, Basisgel-Daten, Grace GmbH, Worms, FRG, 1983.
40. R. Arras, Ph.D. thesis, Johannes-Gutenberg-Universität, Mainz, FRG.
41. R. K. Iler, *The Chemistry of Silica*, Wiley-Interscience, New York, 1979.
42. Yu. S. Nikitin and T. D. Khoklova, Vth Danube Symposium on Chromatogrpahy, Yalta, paper 074 (1985).
43. R. W. Stout and J. J. DeStefano, *J. Chromatogr.*, *326*:63 (1985).
44. R. Schwarzenbach, *J. Liq. Chromatogr.*, *2*:205 (1979).
45. R. Schwarzenbach, *J. Liq. Chromatogr.*, *202*:379 (1980).
46 R. Schwarzenbach, *J. Liq. Chromatogr.*, *334*:35 (1985).
47. S. H. Hansen, P. Helboe, and M. Thomsen, *J. Chromatogr.*, *368*:39 (1986).
48. K. Wefers and G. M. Bell, Oxides and Hydroxides of Aluminum, Alcoa Research Laboratories, Technical Paper No. 19 (1972).
49. B. C. Lippens, *Physical and Chemical Aspects of Adsorbents and Catalysts* (B. G. Linsen, ed.), Academic Press, London, 1979, pp. 171-211.
50. K. Bielefeldt and G. Winkhaus, *Chemische Technologie, Bd. 3 Anorganische Technologie* (K. Winnacker and L. Kuechler, eds.), C. Hanser Verlag, München, 1983, pp. 2-39.
51. Condea Chemie GmbH, Brunsbüttel, FRG.
52. German Patent DOS 2,647,701 (1976).
53. J. B. Peri, *J. Phys. Chem.*, *69*:220 (1965).
54. W. K. Kreis and K. K. Unger, *Adsorption at the Gas-Solid and Liquid Solid Interface*, Elsevier, Amsterdam, 1982, p. 373.
55. C. Laurent, H. A. H. Billet, and L. de Galaan, *Chromatographia*, *17*:253 (1983).
56. R. Fricke, W. Neugebauer, and H. Schäfer, *Z. Anorg. Allg. Chem.*, *273*:215 (1953).
57. F. Geiss, *Die Parameter der Dünnschichtchromatographie*, Vieweg Verlag, Braunschweig, 1972, pp. 122-133.
58. T. Ono, Y. Ohguchi, and O. Togari, *Preparation of Catalysts 3*, Elsevier, Amsterdam, 1983, p. 631.
59. L. R. Snyder, *Principles of Adsorption Chromatography*, Marcel Dekker, New York, 1968, pp. 163-168.

60. Activated and Catalytic Aluminas, Alcoa product data chemicals, Alcoa, Pittsburgh, 1969.
61. W. F. N. M. de Vleesschauwer, *Physical and Chemical Aspects of Adsorbents and Catalysts* (B. G. Linsen, ed.), Academic Press, London, 1970, pp. 265-314.
62. J. Masewska and G. Bazylok, *Chromatographia, 17*:191 (1983).
63. L. R. Snyder, *J. Chromatogr., 28*:300 (1967).
64. M. T. Gilbert and R. A. Wall, *J. Chromatogr., 149*:341 (1978).
65. H. Th. Rijnten, *Physical and Chemical Aspects of Adsorbents and Catalysts* (B. G. Linsen, ed.), Academic Press, London, 1970, pp. 315-372.
66. J. Egly, Ph.D. thesis, Technische Hochschule, Darmstadt, FRG (1977).
67. T. J. Wiseman, *Characterization of Powder Surfaces* (G. D. Parfitt and K. S. W. Sing, eds.), Academic Press, London, 1976, pp. 159-201.
68. U. Trüdinger and K. K. Unger, to be published.
69. A. F. Wells, *Structural Inorganic Chemistry*, Clarendon Press, Oxford, 1962, pp. 787-816.
70. H. van Olphen, *Characterization of Powder Surfaces* (G. D. Parfitt and K. S. W. Sing, eds.), Academic Press, London, 1976, pp. 428-457.
71. M. Lafosse, G. Keravis, J. P. Coic, and E. Stain-Ruf, *Chromatographia, 11*:141 (1978).
72. G. Goretti, F. Geraci, and M. V. Russo, *Chromatographia, 14*: 285 (1981).
73. D. W. Breck, *Zeolite Molecular Sieves*, Wiley-Interscience, New York, 1974.
74. L. R. Snyder, *Principles of Adsorption Chromatography*, Marcel Dekker, New York, 1968, pp. 170-171.
75. F. Janowski and W. Heyer, *Poröse Gläser*, VEB Deutscher Verlag für Grundstoffindustrie, Leipzig, 1982.
76. A. M. Filbert, in *Immobilized Enzymes for Industrial Reactors* (R. A. Messing, ed.), Academic Press, New York, 1975, pp. 39-61.
77. CPG, *Bibliography on Controlled Pore Glass Chromatography and Related Subjects*, Electro-Nucleonics, Inc., Fairfield, NJ.
78. P. N. Cheremisinoff and F. Ellenbusch, *Carbon Adsorption Handbook*, Ann Arbor Science, Ann Arbor, Michigan, 1978.
79. A. Tiselius and S. Claesson, *Arkiv Kemi, Mineral. Geol. B., 15*(18) (1942).
80. A. Tiselius, *Kolloid-Z., 105*:101 (1943).
81. J. H. Knox, K. K. Unger, and H. Müller, *J. Liq. Chromatogr., 6*:1 (1983).
82. H. Colin, C. Eon, and G. Guiochon, *J. Chromatogr., 119*:41 (1976).

83. H. Colin, C. Eon, and G. Guiochon, *J. Chromatogr.*, *122*:223 (1976).
84. H. Colin and G. Guiochon, *J. Chromatogr.*, *126*:43 (1976).
85. G. Guiochon and H. Colin, *Chromatogr. Rev. Spectra Physics*, *4*(2):2 (1978).
86. Z. Plzák, F. P. Dousek, and J. Jansta, *J. Chromatogr.*, *147*:137 (1978).
87. V. Patzelova, J. Jansta, and F. P. Doušek, *J. Chromatogr.*, *148*:53 (1978).
88. T. H. Zwier and M. F. Burke, *Anal. Chem.*, *53*:812 (1981).
89. P. Ciccioli, R. Tappa, A. di Corcia, and A. Liberti, *J. Chromatogr.*, *206*:35 (1981).
90. K. Unger, P. Roumeliotis, H. Müller, and H. Goetz, *J. Chromatogr.*, *202*:3 (1980).
91. K. Unger and H. Goetz, German Patent 2,802,846 (1979).
92. J. H. Knox and M. T. Gilbert, German Patent 2,946,688 (Cl Col B31/08) June 12, 1980, Brit. Appl. 78/45397, Nov. 21, 1978, U.S. Patent.
93. M. T. Gilbert, J. H. Knox, and B. Kaur, *Chromatographia, 16*: 138 (1982).
94. B. R. Puri, in *Chemistry and Physics of Carbon*, Vol. 6 (Ph. L. Walker, ed.), Marcel Dekker, New York, 1971, pp. 191-275.
95. H. P. Boehm, E. Diehl and W. Heck, *Industr. Carbons Graphites*, *2*:369 (1966).
96. K. K. Unger, *Anal. Chem.*, *55*:361 A (1983).
97. J. H. Knox, B. Kaur, and G. R. Millward, *J. Chromatogr.*, *352*: 3-25 (1986).
98. B. A. Adams and E. L. Holmers, *J. Soc. Chem. Ind. (London)*, *54*:1T (1935).
99. J. C. Moore, *J. Polym. Sci. Part A*, *2*:835 (1964).
100. J. Porath and P. Flodin, *Nature, 183*:1657 (1959).
101. J. C. Janson, *Chromatographia, 23*:361 (1937).
102. S. Hjerten, *Biochim. Biophys. Acta, 79*:393-398 (1964).
103. S. Hjerten and R. Mosbach, *Anal. Biochem.*, *3*:109-118 (1962).
104. D. J. Lea and A. H. Sehon, *Can. J. Chem.*, *40*:159-160 (1962).
105. J. Seidl, J. Malinsky, D. Dušek, and W. Heitz, *Adv. Polymer Sci.*, *5*:113-213 (1967).
106. W. Heitz, *Angew. Chem. Int. Ed.*, *9*:689 (1970).
107. W. Heitz, *J. Chromatogr.*, *53*:37-49 (1970).
108. R. Epton, *Chromatography of Synthetic and Biological Polymers*, Vol. 2, *Hydrophobic, Ion Exchange and Affinity Methods*, E. Horwood, Ltd., Chichester, 1978, pp. 1-9.
109. S. A. Arnott, A. Fulmer, W. E. Scott, I. C. M. Dea, R. Moorehouse, and D. A. Rees, *J. Mol. Biol.*, *90*:269-284 (1974).
110. S. Hjertén, in *Protides of Biological Fluids*, Vol. 30 (H. Peeters, ed.), Pergamon Press, Oxford, 1983, pp. 9-17.

111. J. Porath, T. Laas, and J. C. Janson, *J. Chromatogr.*, *103*:49-62 (1975).
112. O. Mikeš, P. Štrop, J. Zbrožek, and J. Čoupek, *J. Chromatogr.*, *119*:339-354 (1976).
113. L. B. Bangs, Uniform Latex Particles, Seradyn, Inc., P. O. Box 1210, Indianapolis, IN, 46206.
114. J. Ugelstad, P. C. Mørk, K. H. Kaggernd, T. Ellingsen, and A. Berge, *Adv. Colloid Interface Sci.*, *13*:101-140 (1980).
115. J. Ugelstad, P. C. Mørk, A. Berge, T. Tllingsen, and A. A. Khan, in *Emulsion Polymerization* (I. Piirma, ed.), Academic Press, New York, 1982, pp. 383-390.
116. O. Mikeš, *HPLC of Biopolymers and Biooligomers*, J. Chromatogr. Library, Vol. 41A, Elsevier, Amsterdam, 1988, pp. A142-A146.
117. V. Gethie and H. D. Schell, *Rev. roum. Biochim.*, *4*:179-184 (1967).
118. J. Porath, J. C. Janson, and T. Laas, *J. Chromatogr.*, *60*:167-177 (1971).
119. T. Laas, *J. Chromatogr.*, *111*:373-387 (1975).
120. *Gelfiltration: Theory and Practice*, Pharmacia LKB, Biotechnology, Uppsala, Sweden, p. 20.
121. *Gelfiltration: Theory and Practice*, Pharmacia LKB, Biotechnology, Uppsala, Sweden, p. 8.
122. H. Henke, Akzo AG Research Laboratories, Obernburg, FRG.
123. Yih-Fen Maa and Cs. Horvath, *J. Chromatogr.*, *445*:71-86 (1988).
124. N. Tanaka, K. Hashizume, and M. Araki, *J. Chromatogr.*, *400*:38-45 (1987).
125. K. A. Tweeten and T. N. Tweeten, *J. Chromatogr.*, *359*:111 (1986).
126. K. K. Unger, R. Janzen, and G. Jilge, *Chromatographia*, *24*:144-154 (1988).
127. F. E. Regnier, *Chromatographia*, *24*:241-251 (1988).
128. W. Heitz and K. L. Platt, *Makromolekulare Chem.*, *127*:113 (1969).
129. H. Determann, G. Lüben, and Th. Wieland, *Makromolekulare Chem.*, *73*:168 (1964).
130. W. Heitz and H. Winau, *Makromolekulare Chem.*, *131*:75 (1979).
131. J. Čoupek, M. Křivaková, and S. Pokorný, *J. Polym. Sci., Polym. Sym.*, *42*:185-190 (1973).
132. H. N. M. Stewart and S. G. Perry, *J. Chromatogr.*, *37*:97-98 (1968).
133. I. Halasz and I. Sebestian, *Angew. Chem., Int. Ed. Engl.*, *8*:453-464 (1969).
134. K. K. Unger, *Porous Silica*, J. Chromatogr. Library, Vol. 16, Elsevier, Amsterdam, 1979, pp. 57-141.
135. J. Gobet and E. Kovats, *Ads. Sci. Technol.*, *1*:77 (1984).
136. J. I. Erard, L. Nagy, and E. Kovats, *Colloid Surfaces*, *9*:109 (1984).

137. G. Foti, C. Martinez, and E. Kovats, *J. Chromatogr.*, in print.
138. G. Foti and E. Kovats, Langmuir, in print.
139. N. K. Boardman, *Biochem. Biophys. Acta, 18*:290 (1955).
140. N. K. Boardman, *Biochem. Biophys. Acta, 18*:290 (1955).
141. J. J. Kirkland, *J. Chromatogr. Sci., 8*:72 (1970).
142. C. W. Hiatt, A. Shelokov, E. J. Rosenthal, and J. M. Galimore, *J. Chromatogr., 56*:362 (1971).
143. G. L. Hawk, J. A. Cameron, and L. B. Dufault, *Prep. Biochem., 2*:193 (1972).
144. I. Schechter, *Anal. Biochem., 58*:30 (1974).
145. T. Darling, J. Albert, P. Russel, D. Albert, and T. W. Reid, *J. Chromatogr., 131*:383 (1977).
146. J. L. Tayot, M. Tardy, and M. C. Mynard, in *Affinity Chromatography* (O. Hoffman-Ostenhof, et al., eds.), Pergamon Press, Oxford, 1978.
147. A. J. Alpert and F. E. Regnier, *J. Chromatogr., 185*:375 (1979).
148. B. S. Welinder, *J. Liq. Chromatogr., 3*:1399 (1980).
149. T. Mizutami, *J. Chromatogr., 196*:485 (1980).
150. D. L. Gooding, M. N. Schmuck, and K. H. Gooding, *J. Chromatogr., 296*:107 (1984).
151. L. Letot, J. Lesec, and C. Quivoron, *J. Liq. Chromatogr., 4*:1311 (1981).
152. E. Pfefferkorn, Q. K. Tran, and R. Varoqui, *J. Chim. Phys., 78*:549 (1981).
153. A. J. Alpert, *J. Chromatogr., 359*:85 (1986).
154. R. M. Chicz, Z. Shi, and F. E. Regnier, *J. Chromatogr., 359*:121 (1986).
155. E. Pfefferkorn, A. Carroy, and R. Varoqui, *Macromolecules, 18*:2252 (1985).
156. H. Figge, A. Deege, J. Köhler, and G. Schomburg, *J. Chromatogr., 351*:393 (1986).
157. X. Santavelli, D. Muller, and J. Jozetouviez, *J. Chromatogr., 443*:55 (1988).
158. Z. El Rassi and Cs. Horvath, *Chromatographia, 19*:9-18 (1984).
159. D. Westerlund, *Chromatographia, 24*:155-169 (1987).
160. H. Toshida, I. Movita, G. Tamai, T. Masujima, T. Tsuru, N. Tokai, and H. Imai, *Chromatographia, 19*:466-472 (1984).
161. I. H. Hagestam and T. C. Pinkerton, *Anal. Chem., 57*:1757 (1985).
162. C. Desilets and F. E. Regnier, 12th International Symposium on Column Liquid Chromatography, June 19-24, 1988, Washington, D.C., paper W-P 326.
163. J. B. Crowther and R. A. Hartwick, *Chromatographia, 16*:349 (1982).
164. R. Bischoff and L. W. Mc Laughlin, *J. Chromatogr., 270*:117 (1983).

165. L. A. Kennedy, W. A. Kopaciewicz, and F. E. Regnier, *J. Chromatogr.*, *359*:73 (1986).
166. Th. R. Floyd and R. A. Hartwick, in *HPLC: Advances and Perspectives*, Vol. 4 (Cs. Horvath, ed.), Academic Press, London, 1986, pp. 45-90.
167. R. Hirz, Ph.D. thesis, Universität, Wien, 1982.
168. M. T. W. Hearn and B. Grego, *J. Chromatogr.*, *218*:497 (1981).
169. W. Jost, G. Schwinn, and M. Tüylu, *Labor Praxis*, *11*:43-45 (1987).
170. G. H. Howard and A. J. P. Martin, *Biochem. J.*, *46*:532 (1950).
171. J. Boldingh, *Experientia*, *4*:270 (1948).
172. A. Tiselius, *The Svedberg*, Almquist & Wiksells, Uppsala, 1944, p. 370.
173. A. Tiselius and Hagdahl, *Acta Chem. Scand.*, *4*:374 (1950).
174. L. Hagdahl, R. J. P. Williams, and A. Tiselius, *Arkiv Kemi*, *4*:193 (1952).
175. Cs. Horvath and S. R. Lipsky, *Nature*, *211*:748 (1966).
176. W. R. Melander and Cs. Horvath, in *HPLC: Advances and Perspectives*, Vol. 2, Academic Press, London, 1980, pp. 117-120.
177. H. N. M. Stewart and S. G. Perry, *J. Chromatogr.*, *37*:97 (1968).
178. E. W. Abel, F. H. Pollard, P. C. Uden, and G. Nickless, *J. Chromatogr.*, *22*:23 (1966).
179. J. J. Kirkland and J. J. De Stefano, *J. Chromatogr. Sci.*, *8*:309 (1970).
180. R. E. Majors, *Anal. Chem.*, *44*:1722 (1971).
181. C. Rossi, S. Munari, C. Cengari, and G. F. Tealdo, *Chim. Ind. (Milano)*, *42*:724 (1960).
182. I. Halasz and I. Sebestian, *Angew. Chem., Int. Ed.*, *8*:453 (1969).
183. J. Halasz and J. Sebestian, *J. Chromatogr. Sci.*, *12*:161 (1974).
184. B. Hofstee, *Anal. Biochem.*, *52*:430 (1973).
185. S. Shaltiel and Z. Er-El, *Proc. Natl. Acad. Sci. U.S.A.*, *70*:778 (1973).
186. ACT-1™ phase, Interaction Chemicals, Mountain View, CA.
187. J. Köhler and J. J. Kirkland, *J. Chromatogr.*, *385*:125-150 (1987).
188. K. Unger, *Porous Silica*, J. Chromatogr. Library, Vol. 16, Elsevier, Amsterdam, 1979, pp. 58-76.
189. J. Köhler, D. B. Chase, R. D. Farlee, J. Vega, and J. J. Kirkland, *J. Chromatogr.*, *352*:275 (1986).
190. L. T. Zhuravlev, *Langmuir*, *3*:316-318 (1987).
191. H. Holik and B. Matejkova, *J. Chromatogr.*, *213*:33 (1981).
192. E. J. Corey, H. Cho, Ch. Rücker, and D. H. Hua, *Tetrahedron Lett.*, *22*:3455-3458 (1981).
193. K. Szabo, N. Le Ha, Ph. Schneider, P. Aeltner, and E. Kovats, *Helv. Chim. Acta*, *67*:2128-2142 (1984).

194. G. Schomburg, A. Deege, J. Köhler, and U. Bien-Vogelsang, *J. Chromatogr.*, *282*:27 (1983).
195. V. Bazant, J. Joklik, and J. Rathousky, *Angew. Chem. Int. Ed. Engl.*, *7*:112 (1968).
196. E. Lukevics, Z. U. Belyakova, M. G. Pomerantseva, and M. G. Noronkov, *J. Organomet. Chem. Libr.*, *5*:1 (1977).
197. B. B. Wheats, *J. Chromatogr.*, *107*:402 (1975).
198. K. D. Lork, K. K. Unger, and J. N. Kinkel, *J. Chromatogr.*, *352*:199 (1986).
199. J. N. Kinkel and K. K. Unger, *J. Chromatogr.*, *316*:193-200 (1984).
200. B. Buszewski, A. Jurasek, J. Garaj, L. Nondek, J. Novak, and D. Berek, *J. Liq. Chromatogr.*, *10*:2325 (1987).
201. T. M. Khong and C. F. Simpson, *Chromatographia, 24*:385-394 (1987).
202. H. Engelhardt and P. Orth, *J. Liq. Chromatogr.*, *10*:1999-2022 (1987).
203. L. C. Sander and S. A. Wise, *J. Chromatogr.*, *316*:163 (1986).
204. K. K. Unger, *Porous Silica*, J. Chromatogr. Library Vol. 16, Elsevier, Amsterdam, 1979, pp. 99-104.
205. J. Gobet and E. Kovats, *Adsorption Sci. Technol.*, *1*:77 (1984).
206. K. D. Lork, Ph.D. thesis, Johannes Gutenberg-Universität, Mainz, FRG, 1988.
207 I. Langmuir, *J. Am. Chem. Soc.*, *39*:1848 (1917).
208. F. Riedo, M. Czencz, O. Liardon, and E. Kovats, *Hel. Chim. Acta, 61*:1912-1941 (1978).
209. U. Bien-Vogelsang, A. Deege, H. Figge, J. Köhler, and G. Schomburg, *Chromatographia, 19*:170-179 (1984).
210. G. Schomburg, J. Köhler, H. Figge, A. Deege, and U. Bien-Vogelsang, *Chromatographia, 18*:265 (1984).
211. Y. Ohtsu, H. Fukui, T. Kanda, K. Nakamura, M. Nakano, O. Nakata, and Y. Fujiyama, *Chromatographia, 24*:380-384 (1987).
212. D. P. Lee, *J. Chromatogr.*, *443*:143-153 (1988).
213. N. Tanaka, K. Hashizuma, and M. Araki, *J. Chromatogr.*, *400*:33-45 (1987).
214. S. Hjertén, *J. Chromatogr.*, *87*:325 (1973).
215. S. Shaltiel, in *Methods in Enzymology*, Vol. 104, *Enzyme Purification and Related Techniques* (W. B. Jakoby, ed.), Academic Press, London, 1984, pp. 69-96.
216. H. Engelhardt and D. Mathes, *J. Chromatogr.*, *142*:311 (1977).
217. N. T. Miller, B. Feibush, and B. L. Karger, *J. Chromatogr.*, *316*:519 (1985).
218. R. E. Majors, in *HPLC: Advances and Perspectives*, Vol. 1 (Cs. Horvath, ed.), Academic Press, London, 1980, pp. 86-94.
219. K. K. Unger and J. N. Kinkel, in *Aqueous Size Exclusion Chromatography*, Vol. 40 (P. L. Dubin, ed.), J. Chromatogr. Library, Elsevier, Amsterdam, 1988, pp. 193-234.

220. M. Verzele and F. van Damme, *J. Chromatogr.*, *362*:23-31 (1986).
221. M. Verzele, F. van Damme, C. Dewaele, and M. Ghijs, *Chromatographia*, *24*:302-317 (1987).
222. W. T. Brugman and J. C. Kraak, *J. Chromatogr.*, *205*:170 (1981).
223. W. T. Brugman and J. C. Kraak, *Chromatographia*, *15*:282 (1982).
224. E. P. Pluedemann, in *Silylated Surfaces*, (D. E. Leyden, ed.), Gordon and Breach, London, 1980, pp. 31-53.
225. T. G. Waddell, D. E. Leyden, and M. T. DeBello, *J. Am. Chem. Soc.*, *103*:5303 (1981).
226. T. G. Waddell, D. E. Leyden, and D. M. Hercules, in *Silylated Surfaces*, (D. E. Leyden, ed.), Gordon and Breach, London, 1980, pp. 55-72.
227. G. S. Carjaval, D. E. Leyden, and G. F. Maciel, in *Silanes, Surfaces, and Interfaces*, , (D. E. Leyden, ed.), Gordon and Breach, London, 1986.
228. D. P. Herman, L. R. Field, and S. Abbott, *J. Chromatogr. Sci.*, *19*:470 (1981).
229. D. M. Wonnacott and E. V. Patton, *J. Chromatogr.*, *389*:103-113 (1987).
230. D. W. Armstrong and W. Li, *Chromatographia*, *2*:43-48 (1987).
231. D. W. Armstrong and W. De Mond, *J. Chromatogr. Sci.*, *22*:411 (1984).
232. *Merck Spectrum*, *3*:34-38 (1987), E. Merck, Darmstadt, FRG.
233. D. W. Armstrong, A. Alak, K. Bui, W. DeMond, T. Ward, T. E. Riehl, and W. L. Hinze, *J. Inclus. Phenom.*, *2*:533 (1984).
234. D. W. Armstrong, *J. Liq. Chromatogr.*, *7*(S-2):353 (1984).
235. D. W. Armstrong, W. De Mond, and B. P. Czech, *Anal. Chem.*, *57*:481 (1985).
236. D. W. Armstrong, W. DeMond, A. Alak, W. L. Hinze, T. E. Riehl, and K. H. Bui, *Anal. Chem.*, *57*:234 (1985).
237. W. L. Hinze, T. E. Riehl, D. W. Armstrong, W. DeMond, A. Alak, and T. Ward, *Anal. Chem.*, *57*:237 (1985).
238. D. W. Armstrong, T. J. Ward, A. Czech, B. P. Czech, and R. A. Bartsch, *J. Org. Chem.*, *50*:5556 (1985).
239. D. W. Armstrong, T. J. Ward, R. D. Armstrong, and T. E. Beesley, *Science*, *232*:1132 (1986).
240. J. Kirschbaum and L. Kerr, *LC Magazine*, *4*:30 (1986).
241. D. W. Armstrong, A. Alak, W. DeMond, W. L. Hinze, and T. E. Riehl, *J. Liq. Chromatogr.*, *8*:261 (1985).
242. D. W. Armstrong, *Anal. Chem.*, A-pages Report, Jan. 15 (1987).
243. J. C. Voegel, E. Pfefferkorn, and A. Schmitt, *J. Chromatogr.*, *428*:17 (1988).
244. L. Letot, I. Lesec, and C. Quivoron, *J. Liq. Chromatogr.*, *4*:1311-1322 (1981).
245. J. Köhler, *Chromatographia*, *21*:573 (1986).
246. M. Novotny, S. L. Bektesch, K. B. Denson, K. Grohmann, and W. Parr, *Anal. Chem.*, *45*:971 (1973).

247. J. J. Kirkland and P. C. Yates, US Patent 3,795,313 (1974).
248. K. Albert, and B. Pfleiderer, 12th International Symposium on Column Liquid Chromatography, June 19-24, Washington, D.C., USA, paper M-L-3 (1988).
249. K. K. Unger, *Porous Silica*, J. Chromatogr. Library Vol. 16, Elsevier, Amsterdam, 1979, p. 102.
250. M. Verzele, P. Mussche, and P. Sandra, *J. Chromatogr.*, *190*:331-337 (1980).
251. H. G. Genieser, D. Gabel, and B. Jastorff, *J. Chromatogr.*, *269*:127-152 (1983).
252. J. B. Crowther, S. D. Fazio, R. Schicksuis, S. Marcus, and R. A. Hartwick, *J. Chromatogr.*, *289*:367 (1983).
253. S. Fazio, S. Tomellini, H. Shih-Hsieu, J. Crowther, T. Raglione, T. Floyd, and R. A. Hartwick, *Anal. Chem.*, *57*:1559 (1985).
254. J. F. Erard and E. Kovats, *Anal. Chem.*, *54*:193 (1982).
255. L. Trojer and L. Hansson, *J. Chromatogr.*, *262*:183 (1983).
256. P. Mussche and M. Verzele, *J. Anal. Appl. Pyrolysis*, *4*:273 (1983).
257. M. Nyström, W. Herrmann, D. Sanchez, and P. Möller, Kromasil, a new high performance silica structure of liquid chromatography Eka Nobel, S-44501 Surte, Sweden.
258. M. Verzele, M. De Potter, and J. Ghyseis, *J. High Resol. Chromatogr. Chromatogr. Commun.*, *152*:10076 (1979).
259. M. R. Ross and J. F. Evans, in *Silylated Surfaces*, (D. E. Leyden, ed.), Gordon and Breach, London, 1980, pp. 99-123.
260. A. V. Kiselev and V. I. Lygin, *Infrared Spectra of Surface Compunds*, Wiley-Interscience, New York, 1975.
261 M. L. Hair, *Infrared Spectroscopy of Surface Chemistry*, Marcel Dekker, New York, 1967.
262. L. H. Little, *Infrared Spectra of Adsorbed Species*, Academic Press, London, 1966.
263. H. P. Boehm and H. Knöezinger, in *Catalysis*, Vol. 4, (J. R. Anderson and M. Boudart, eds.), Springer-Verlag, Heidelberg, 1983, pp. 40-189.
264. P. R. Griffiths and J.A. de Haseth, *Fourier Transform Infrared Spectrometry*, Wiley-Interscience, New York, 1986.
265. J. R. Ferraro and L. J. Basile (eds.), *Fourier Trnasform Infrared Spectroscopy*, Vols. 1-4, Academic Press, New York, 1978, 1979, 1982, 1985.
266. D. E. Leyden and R. S. S. Murthy, *Trends Analyt. Che.*, *7*:164-169 (1988).
267. J. P. Blitz, R. S. S. Murthy, and D. E. Leyden, *Appl. Spectrosc.*, *40*:829-831 (1986).
268. R. S. S. Murthy and D. E. Leyden, *Anal. Chem.*, *58*:1228-1233 (1986).
269. J. Gronholz and W. Heres, *Instrum. Comp.*, 3:10-19, 45-55 (1985).

270. L. C. Sander, J. B. Callis, and L. R. Field, *Anal. Chem.*, *55*: 1068-1075 (1983).
271. C. A. Fyfe, *Solid-State NMR for Chemists*, CFC Press, Guelph, Canada, 1983.
272. G. Engelhardt and D. Michel, *High-Resolution Solid-State NMR of Silicates and Zeolites*, John Wiley and Sons, Chichester, 1987.
273. G. E. Maciel and D. W. Sindorf, *J. Am. Chem. Soc.*, *102*:7607-7608 (1980).
274. D. W. Sindorf and G. E. Maciel, *J. Phys. Chem.*, *87*:5516-5521 (1983).
275. G. E. Maciel, D. W. Sindorf, and V. J. Bartuska, *J. Chromatogr.*, *205*:438-443 (1981).
276. D. W. Sindorf and G. E. Maciel, *J. Am. Chem. Soc.*, *103*:4263-4265 (1981).
277. E. Bayer, K. Albert, J. Reiners, M. Nieder, and D. Müller, *J. Chromatogr.*, *264*:197-213 (1983).
278. B. Pfleiderer, K. Albert, K. D. Lork, K. K. Unger, H. Brückner, and E. Bayer, *Angew. Chem. Int. Ed. Engl.* *28*:327 (1989).
279. K. Albert, B. Evens, and E. Bayer, *J. Magnet. Resonance*, *62*:428-436 (1985).
280. H. A. Claessons, L. J. M. van de Ven, J. W. de Haan, and C. A. Cramers, *J. High Resol. Chromatogr. Chromatog. Commun.*, *6*:433-435 (1983).
281. E. Bayer, A. Paulus, B. Peters, G. Laupp, J. Reiners, and K. Albert, *J. Chromatogr.*, *364*:25-37 (1987).
282. D. Oelkrug, M. Radjaipour, and H. Erbse, *Z. Phys. Chem. (Wiesbaden)*, *88*:23-36 (1974).
283. R. G. Bogar, J. C. Thomas, and J. B. Callis, *Anal. Chem.*, *56*:1080-1084 (1984).
284. L. D. Weis, T. R. Evans, and P. A. Leermakers, *J. Am. Chem. Soc.*, *90*:6109-6118 (1968).
285. C. H. Nicholls and P. A. Leermakers, *Adv. Photochem.*, *8*:315-336 (1971).
286. K. Hara, P. de Mayo, W. R. Ware, A. C. Weedon, G. C. K. Wong, and K. C. Wu, *Chem. Phys. Lett.*, *69*:105-108 (1980).
287. R. K. Bauer, R. Borenstein, P. de Mayo, K. Okada, M. Rafalska, W. R. Ware, and K. C. Wu, *J. Am. Chem. Soc.*, *104*:4635-4644 (1982).
288. R. K. Bauer, P. de Mayo, W. R. Ware, and K. C. Wu, *J. Phys. Chem.*, *86*:3781-3789 (1982).
289. R. K. Bauer, P. de Mayo, K. Okada, W. R. Ware, and K. C. Wu, *J. Phys. Chem.*, *87*:460-466 (1983).
290. C. Francis, J. Lin, and L. A. Singer, *Chem. Phys. Lett.*, *94*: 162-167 (1983).
291. J. M. Drake and J. Klafter, *J. Lumin.*, *31*:642-644 (1984).
292. P. Levitz, H. Van Damme, and D. Keravis, *J. Phys. Chem.*, *88*:2228-2235 (1984).

293. R. K. Bauer, P. de Mayo, L. V. Natarajan, and W. R. Ware, *Can. J. Chem.*, *62*:1279-1286 (1984).
294. P. de Mayo, L. V. Natarajan, and W. R. Ware, *Chem. Phys. Lett.*, *107*:187-192 (1984).
295. D. Avnir, R. Busse, M. Ottolenghi, E. Wellner, and K. A. Zachariasse, *J. Phys. Chem.*, *89*:3521-3526 (1985).
296. P. de Mayo, L. V. Natarajan, and W. R. Ware, *J. Phys. Chem.*, *89*:3526-3530 (1985).
297. E. Wellner, M. Ottolenghi, D. Avnir, and D. Huppert, *Langmuir*, *2*:612-619 (1986).
298. K. A. Zachariasse, in *Photochemistry on Solid Surfaces*, (T. Matsura and M. Anpo, eds.), Elsevier, Amsterdam, in print.
299. R. K. Bauer, P. de Mayo, K. Okada, W. R. Ware, and K. C. Wu, *J. Phys. Chem.*, *87*:460-466 (1983).
300. D. Avnir, R. Busse, M. Ottolenghi, E. Wellner, and K. A. Zachariasse, *J. Phys. Chem.*, *89*:3521-3526 (1985).
301. J. B. Callis and C. Webb, *10th International Symposium on Column Liquid Chromatography*, San Francisco, May 1983, Abstracts, p. 1703 (1986).
302. C. H. Lochmüller, A. S. Colborn, M. L. Hunnicutt, and J. M. Harris, *Anal. Chem.*, *55*:1344-1348 (1983).
303. C. H. Lochmüller, A. S. Colborn, M. L. Hunnicutt, and J. M. Harris, *J. Am. Chem. Soc.*, *106*:4077-4082 (1984).
304. C. H. Lochmüller and M. L. Hunnicutt, *J. Phys. Chem.*, *90*:4318-4322 (1986).
305. R. K. Gilpin, A. Kasturi, and S. S. Yang, *10th International Symposium on Column Liquid Chromatography*, San Francisco, May 1983, Abstracts p. 1701 (1986).
306. K. K. Unger, J. Rouquerol, K. S. W. Sing, and H. Kral (eds.), *Characterization of Porous Solids*, Studies in Surface Science and Catalysis, Vol. 39, Elsevier, Amsterdam, 1988.
307. T. Allen, *Particle Size Measurement*, Chapman and Hall, London, 1981.
308. H. Giesche, K. K. Unger, B. Eray, U. Trüdinger, and J. N. Kinkel, *J. Chromatogr.*, *465*:39 (1989).
309. K. K. Unger and M. Gimpel, *J. Chromatogr.*, *180*:93-102 (1979).
310. N. Tanaka, K. Hashidzume, M. Araki, H. Tsuchiya, A. Okuno, K. Iwaguchi, S. Ohniski, and N. Takai, *J. Chromatogr.*, *389*:115-125 (1987).
311. D. Amati and E. Kovats, *Langmuir*, *3*:687-695 (1987).
312. G. Körösi and E. Kovats, *Colloids Surfaces*, *2*:315-355 (1981).
313. R. C. Mackenzie, *Differential Thermal Analysis*, Vols. 1 and 2, Academic Press, London, 1972.
314. D. Morel and J. Serpinet, *J. Chromatogr.*, *200*:95-104 (1980).
315. D. Morel and J. Serpinet, *J. Chromatogr.*, *214*:202 (1981).
316. D. Morel and J. Serpinet, *J. Chromatogr.*, *248*:231-240 (1982).
317. J. C. Van Miltenburg and W. E. Hammers, *J. Chromatogr.*,

*268*:147-155 (1983).
318. T. C. Schunk and M. F. Burke, *Int. J. Environ. Anal. Chem. 25*:81 (1986).
319. J. Serpinet, personal communication.
320. G. Foti, C. Martinez, and E. Kovats, *J. Chromatogr.*, in print.
321. K. K. Unger and K. D. Lork, *Eur. Chrom. News*, *2*:14 (1988).
322. A. P. Goldberg, *Anal. Chem.*, *54*:342-345 (1982).
323. C. Gonnet, C. Bory, and G. Lachatre, *Chromatographia, 16*: 242-246 (1982).
324. T. Daldrup and B. Kardel, *Chromatographia, 18*:81-83 (1984).
325. R. M. Smith, *TRAC, 3*:186-190 (1984).
326. P. E. Antle, A. P. Goldberg, and L. R. Snyder, *J. Chromatogr.*, *321*:1-32 (1985).
327. J. R. Chretien, B. Walczak, L. Morin-Allory, M. Dreux, and M. La Fosse, *J. Chromatogr.*, *371*:253-267 (1986).
328. B. Walczak, M. Dreux, J. R. Chretien, K. Szymoniak, M. La Fosse, *J. Chromatogr.*, *371*:252-267 (1986).
328. B. Walczak, M. Dreux, J. R. Chretien, K. Szymoniak, M. La Fosse, L. Morin-Allory, and J. P. Doucet, *J. Chromatogr.*, *353*:109-121 (1986).
329. P. J. van den Driest and H. J. Ritchie, *Chromatographia, 24*: 324-328 (1987).
330. M. Bogusz, *J. Chromatogr.*, *387*:404-409 (1987).
331. M. Mushtag, Z. Bao, and Sh. Yang, *J. Chromatogr.*, *385*:293-298 (1987).
332. W. M. Moats and L. Leskinen, *J. Chromatogr.*, *386*:79-86 (1987).
333. B. Walczak, L. Morin-Allory, M. La Fosse, M. Dreux, and J. R. Chretien, *J. Chromatogr.*, *395*:183-202 (1987).
334. N. Sagliano, R. A. Hartwick, J. M. Di Bussolo, and N. T. Miller, 7th International Symposium on HPLC of Proteins, Peptides and Polynucleotides, Nov. 2-4, Washington, D.C., poster 130 (1987).
335. J. L. Glajch, J. J. Kirkland, and J. Köhler, *J. Chromatogr.*, *384*:81 (1987).
336. L. R. Snyder and H. Poppe, *J. Chromatogr.*, *184*:363 (1980).
337. R. P. W. Scott, *J. Chromatogr. Sci.*, *18*:297 (1980).
338. Cs. Horvath, W. Melander, and I. Molnar, *J. Chromatogr.*, *15*:129 (1976).
339. Cs. Horvath and W. Melander, *J. Chromatogr. Sci.*, *15*:339 (1977).
340. Cs. Horvath and W. Melander, *Am. Lab.*, *10*(10):17 (1978).
341. K. E. Bij, Cs. Horvath, W. R. Melander, and A. Nahum, *J. Chromatogr.*, *203*:65 (1981).
342. M. T. W. Hearn, *HPLC: Advances and Perspectives*, Vol. 3 (Cs. Horvath, ed.), Academic Press, London, 1983, pp. 87-155.
343. W. R. Melander, D. Corradini, and Cs. Horvath, *J. Chroma-*

*togr.*, *317*:67 (1984).
344. J. L. Fausnaugh, L. A. Kennedy, and F. E. Regnier, *J. Chromatogr.*, *317*:141 (1984).
345. N. T. Miller and B. L. Karger, *J. Chromatogr.*, *326*:45 (1985).
346. H. Colin, G. Guiochon and P. Jandera, *Chromatographia, 15*: 133 (1982).
347. L. R. Snyder, *HPLC: Advances and Perspectives*, Vol. 3 (Cs. Horvath, ed.), Academic Press, New York, 1983, pp. 157-222.
348. L. R. Snyder and T. C. Schunk, *Anal. Chem.*, *54*:1764 (1982).
349. W. E. Hammers, M. C. Spanjer, and C. L. Liguy, *J. Chromatogr.*, *174*:219 (1979).
350. W. E. Hammers, R. H. A. M. Janssen, A. G. Baars, and C. L. de Ligny, *J. Chromatogr.*, *167*:273 (1978).
351. R. P. W. Scott and P. Kucera, *J. Chromatogr.*, *149*:93 (1978).
352. R. P. W. Scott and P. Kucera, *J. Chromatogr.*, *171*:37 (1979).
353. K. K. Unger, K. D. Lork, and J. N. Kinkel, *J. Chromatogr.*, in print.
354. K. K. Unger, *Porous Silica*, Elsevier, Amsterdam, 1979, pp. 57-76.
355. M. Jaronieč and J. A. Jaronieč, *J. Liq. Chromatogr.*, *4*:2121 (1976).
356. M. Jaronieč and J. Oscik, *J. High Resol. Chromatogr. Chromatog. Commun.*, *5*:3 (1982).
357. O. Sinanoglu, *Molecular Associations in Biology* (B. Pullman, ed.), Academic Press, 1968, p. 429.
358. J. C. Kraak and J. P. Crombeen, in *Practice of High Performance Liquid Chromatography*, (H. Engelhardt, ed.), Springer-Verlag, Heidelberg, 1986, pp. 192-194.
359. J. F. K. Huber, M. Pawowska, and P. Markl, *Chromatographia*, *19*:19-28 (1984).
360. H. Engelhardt and H. Elgass, in *HPLC: Advances and Perspectives*, Vol. 2, (Cs. Horvath, ed.), Academic Press, London, 1980, pp. 85-87.
361. K. D. Dill, *J. Phys. Chem.*, *91*:1980 (1987).
362. T. Braumann, *J. Chromatogr.*, *373*:191-225 (1986).
363. R. Kaliszan, in *High Performance Liquid Chromatography* (Ph. R. Brown and R. A. Hartwick, eds.), John Wiley & Sons, New York, 1989, pp. 563-599.
364. E. H. Slaats, W. Markowski, J. Fekete, and H. Poppe, *J. Chromatogr.*, *207*:299-323 (1981).
365. W. R. Melander and Cs. Horvath, in *HPLC: Advances and Perspectives*, Vol. 2 (Cs. Horvath, ed.), Academic Press, London, 1980, pp. 156-159.
366. C. H. Lochmüller and D. R. Wilder, *J. Chromatogr. Sci.*, *17*: 574-579 (1979).
367. D. Amati and E. Kovats, *Langmuir*, *3*:687-695 (1987).
368. R. K. Gilpin and J. A. Squires, *J. Chromatogr. Sci.*, *19*:195-199 (1981).
369. R. K. Gilpin, M. E. Gangoda, and A. E. Krishen, *J. Chroma-*

*togr. Sci.*, *20*:345-348 (1982).
370. R. K. Gilpin and M. E. Gangoda, *J. Chromatogr. Sci.*, *21*:352-361 (1983).
371. T. A. Zwier, Ph.D. thesis, University of Arizona, Tuscon, 1982.
372. W. Müller, GIT-Verlag, Darmstadt, FRG, 1988.
373. K. K. Unger, *Porous Silica*, J. Chromatogr. Library Vol. 16, Elsevier, Amsterdam, 1979, pp. 198-199.
374. J.F.K. Huber and F. Eisenbeiss, *J. Chromatogr.*, *149*:127-141 (1978).
375. H. Engelhardt and D. Müller, *J. Chromatogr.*, *218*:395-407 (1981).
376. K. K. Unger, *Porous Silica*, J. Chromatogr. Library Vol. 16, Elsevier, Amsterdam, 1979, pp. 198-203.
377. H. Engelhardt and H. Elgass, in *HPLC: Advances and Perspectives* (Cs. Horvath, ed.), Academic Press, London, 1980, pp. 80-85.
378. N. Becker and K. K. Unger, *Fresenius Z. Anal. Chem.*, *304*:374-381 (1980).
379. S. H. Hansen, *J. Chromatogr.*, *209*:203 (1981).
380. S. H. Hansen and P. Helboe, *J. Chromatogr.*, *285*:53 (1984).
381. C. J. C. M. Laurent, H. A. H. Billiet, and L. de Galan, *Chromatographia*, *17*:253 (1983).
382. C. J. C. M. Laurent, H. A. H. Billiet, and L. de Galan, *Chromatographia*, *17*:394 (1983).
383. C. J. C. M. Laurent, H. A. H. Billiet, and L. de Galan, *J. Chromatogr.*, *285*:161 (1983).
384. H. Lingemann, H. V. A. Munster, J. H. Beynen, W. J. M. Underberg, and A. Hulshoff, *J. Chromatogr.*, *352*:261-274 (1986).
385. R. W. Schmid and Ch. Wolf, *Chromatographia*, *24*:713-719 (1987).
386. S. Hara and S. Ohnishi, *J. Liq. Chromatogr.*, *7*(1):59 (1984).
387. S. Hara and S. Ohnishi, *J. Liq. Chromatogr.*, *7*(1):69 (1984).
388. S. Hara and S. Ohnishi, *J. Liq. Chromatogr.*, *200*:85 (1980).
389. S. Hara and S. Miyamoto, *Anal. Chem.*, *53*:1365 (1981).
390. S. Hara, N. Yamaguchi, C. Nakae, and S. Sakai, *Anal. Chem.*, *52*:33 (1980).
391. K. Oka and S. Hara, *J. Chromatogr.*, *202*:187 (1980).
392. S. Hara, K. Kunihiro, and Soczewinski, *J. Chromatogr.*, *239*:687 (1982).
393. M. Verzele, F. Van Damme, C. Dewaele, and M. Ghijs, *Chromatographia*, *24*:302-308.
394. R. E. Majors, *LC Magazine*, *3*:774 (1985).
395. M. A. Stadalius, H. S. Gold, and L. R. Snyder, *J. Chromatogr.*, *327*:27-45 (1985).
396. K. K .Unger, B. Anspach, R. Janzen, G. Jilge, and K. D. Lork, in *HPLC: Advances and Perspectives*, Vol. 5 (Cs. Hor-

vath, ed.), Academic Press, London, 1989, pp. 2-87.
397. K. K. Unger, *Porous Silica*, J. Chromatogr. Library Vol. 16, Elsevier, Amsterdam, 1979, pp. 106-108.
398. H. Hemetsberger, M. Kellermann, and H. Ricken, *Chromatographia, 10*:726 (1977).
399. M. C. Hennion, C. Picard, and M. Caude, *J. Chromatogr., 166*:21 (1978).
400. G. E. Berendsen and L. de Galan, *J. Liq. Chromatogr., 1*:561-586 (1978).
401. G. E. Berendsen, K. A. Pikaart, and L. de Galan, *J. Liq. Chromatogr., 3*:1437-1464 (1980).
402. G. E. Berendsen and L. de Galan, *J. Chromatogr., 196*:21-37 (1980).
403. C. F. Simpson, to be published.
404. J. G. Dorsey and K. B. Sentell, 12th International Symposium on Column, Liquid Chromatography, June 19-24, Washington, D.C. (1988).
405. K. Karch, I. Sebestian, and I. Halasz, *J. Chromatogr., 122*:33 (1976).
406. P. Spaček, M. Kubin, S. Vozka, and B. Porsch, *J. Liq. Chromatogr., 3*:1465-1480 (1980).
407. L. C. Sander and S. A. Wise, *Advances in Chromatography*, Vol. 25, Marcel Dekker, New York, 1986, pp. 139-218.
408. S. A. Wise and L. C. Sander, *HRC CC, J. High Resol. Chromatogr. Chromatog. Commun., 8*:248-255 (1985).
409. S. A. Wise, W. J. Bonnett, F. R. Guenther, and W. E. May, *J. Chromatogr. Sci., 19*:457-465 (1981).
410. L. C. Sander and S. A. Wise, *Anal. Chem., 59*:2304-2313 (1987).
411. H. Engelhardt, to be published.
412. M. Verzele and C. Dewaele, *J. Chromatogr., 217*:399-404 (1981).
413. I. Eisenbeiss, K.-F. Krebs, and J. F. K. Huber, 15th International Symposium on Chromatography, Nürnberg, October (1984).
414. P. S. Sadek, C. J. Kaester, and L. D. Bowers, *J. Chromatogr. Sci., 25*:489-493 (1987).
415. E. Weber, in K. K. Unger (eds.), *Handbuch der HPLC Teil I: Leitfaden für Anfänger und Praktiker*, GIT Verlag, Darmstadt, 1989.
416. D. B. Marshall, K. A. Stutler, and C. H. Lochmüller, *J. Chromatogr. Sci., 22*:217-220 (1984).
417. G. E. Berendsen, K. A. Pikaart, and L. de Galan, *Anal. Chem., 52*:190-193 (1980).
418. P. C. Sadek and P. W. Carr, *J. Chromatogr., 288*:25-41 (1984).

## APPENDIX I: SURVEY OF COMMERCIAL SILICAS

| Supplier | Trade name | dp (μm) | $a_s$ ($m^2/g$) | pd (nm) |
|---|---|---|---|---|
| Alltech | Adsorbosphere Silica | 3,5,10 | 220 | 10 |
| | Econosphere Silica | 3,5 | 220 | 10 |
| | Versapack Silica | 10 | 300 | 15 |
| Amicon/ Grace | Matrex Silica Media | 5,10,15,20,30, 50,105 | 540 | 6 |
| | | | 320 | 10 |
| | | | 320 | 25 |
| | XWP 500 A | 30-100 | 60 | 50 |
| | XWP 1000 A | 30-100 | 40 | 100 |
| | XWP 1500 A | 30-100 | 28 | 150 |
| Baker | Silica Gel | 3,5 | 170 | 12 |
| | Silica Gel | 10 | 300 | 15 |
| | Silica Gel | 40 | 480 | 6 |
| | Flash Silica Gel | 100-425 | — | 6 |
| | | 63-200 | — | 6 |
| Beckmann | Ultrasil-Si | 10 | — | — |
| | Ultrasphere-Si | 5 | 180 | 8 |
| Bio-Rad | Bio-Sil HA | <100 | — | — |
| | Bio-Sil A | 20-44, 80-150 | — | — |
| Chrompack | ChromSpher Si | 5 | 160 | 12 |
| | CP-Spher Si | 8 | — | 8 |
| Du Pont | Zorbax SIL | 3,5-6 | 350 | 7-8 |
| ICN | ICN Silica | 3-6,7-12,10-18, 18-32,32-63,63-100,63-200,100-200,200-600 | 550 | 6 |
| Kali Chemie | KC-Mikroperl M | 30-60 | 210 | 10 |
| | | | 130 | 18 |
| | | | 76 | 27 |
| | | | 65 | 34 |
| | | | 55 | 43 |
| | | | 47 | 54 |
| | | | 38 | 68 |
| | KC-Mikroperl L | 100-200 | | |
| Macherey Nagel | Nucleosil 50 | 5,7,10 | 450 | 5 |
| | Nucleosil 100 | 5,7,10,15-25, 25-40,40-63 | 350 | 10 |
| | Nucleosil 120 | 3,5,7,10 | 200 | 12 |

APPENDIX I (continued)

| Supplier | Trade name | dp (μm) | $a_s$ ($m^2$/g) | pd (nm) |
|---|---|---|---|---|
| | Nucleosil 300 | 5,7,10,15-25, 25-40,40-63 | 100 | 30 |
| | Nucleosil 500 | 5,7,10,15-25, 25-40,40-63 | 35 | 50 |
| | Nucleosil 1000 | 5,7,10,15-25, 25-40,40-63 | 25 | 100 |
| | Nucleosil 4000 | 5,7,10,15-25 25-40 | 10 | 400 |
| | Polygosil 60 | 5,7,10,15-25, 25-40,40-63, 63-100 | 450 | 6 |
| | Polygosil 100 | 5,7,10,25-40, 40-63 | 300 | 10 |
| | Polygosil 300 | 7,15-25,25-40, 40-63 | 35 | 30 |
| | Polygosil 500 | 7,15-25,25-40, 40-63 | 35 | 50 |
| Merck | LiChrosorb Si 60 | 5,7,10 | 500 | 6 |
| | LiChrosorb Si 100 | 5,7,10 | 300 | 10 |
| | Lichrospher Si 100 | 5,10,20 | 250 | 10 |
| | LiChrospher Si 300 | 10 | 250 | 25 |
| | LiChrospher Si 500 | 10 | 60 | 50 |
| | LiChrospher Si 1000 | 10 | 30 | 100 |
| | LiChrospher Si 4000 | 10 | 10 | 400 |
| | Fractosil 200 | 40-63,63-125 | 150 | 18 |
| | Fractosil 500 | 40-63,63-125 | 50 | 50 |
| | Fractosil 1000 | 40-63,63-125 | 20 | 100 |
| | Fractosil 2500 | 63-125 | 8 | 250 |
| | Fractosil 5000 | 63-125 | 3 | 500 |
| | Fractosil 10,000 | 63-125 | 1.5 | 1000 |
| | Fractosil 25,000 | 63-125 | 0.6 | 2500 |
| | Superspher Si 60 | 4 | 550 | 6 |
| | Superspher Si 100 | 4 | 350 | 10 |
| | Kieselgel 40 | 63-200,200-500 | >500 | 4 |
| | Kieselgel 60 | <63,63-200, 200-500 | 500 | 6 |
| | Kieselgel 100 | 63-200,200-500 | 250 | 10 |
| | Kieselgel 60 | 15-40,40-63, 63-100 | 500 | 6 |
| | Keiselgel 60 reinst | 63-200 | 500 | 6 |

APPENDIX I (continued)

| Supplier | Trade name | dp (μm) | $a_s$ ($m^2/g$) | pd (nm) |
|---|---|---|---|---|
| | LiChroprep Si 40 | 15-25,40-63 | — | — |
| | LiChroprep Si 60 | 15-25,40-63 | — | — |
| | LiChroprep Si 100 | 15-25,40-63 | — | — |
| Millipore/ Waters | μ -Porasil | 10 | 300 | 6 |
| | Resolve Silica | 5,10 | — | 9 |
| | Porasil A | 37-75,75-125 | 300-500 | |
| | Porasil B | 37-75,75-125 | 140-230 | |
| | Prep-PAK 500 Silica | 55-105 | — | — |
| Orpegen | HD-Sil-60 | 10,20,30,60 | — | 6 |
| | HD-Sil-100 | 10,20,30,60 | — | 10 |
| | HD-Sil-SS-80 | 5 | — | 8 |
| | HD-Sil-SS-100 | 5 | — | 10 |
| | HD-Gel-WP | 10,30,60 | — | 25 |
| | HD-Gel-300 | 7 | — | 30 |
| Phase Separations | Spherisorb SW | 3,5,10 | 220 | 8 |
| | Spherisorb SX | 5,10 | 190 | 30 |
| | Spherisorb VLS | 20 | 190 | 30 |
| Reichelt | Thomasorb Si 60 | 5,7.5,10 | 500 | 6 |
| | Thomaspher Si 50 | 5,7.5,10 | 500 | 5 |
| | Thomaspher Si 500 | 5,7.5,10 | 300 | 10 |
| Separations Group | Vydac TP Silica | 5,10,20-30 | 80 | 30 |
| | Vydac HS Silica | 5,10,20-30 | 500 | 8 |
| Serva | Daltosil 75 | 40-100,100-200 | — | 7.5 |
| | Daltosil 100 | 40-100,100-200 | | 10 |
| | Daltosil 150 | 40-100,100-150, 150-200,100-200 | — | 15 |
| | Daltosil 300 | 40-100,100-150, 150-200,100-200 | — | 30 |
| | Daltosil 500 | 40-100,100-200 | — | 50 |
| | Daltosil 1200 | 40-100,100-200 | — | 120 |
| | Daltosil 3000 | 40-100,100-200 | — | 300 |
| | Si 60 | 3,5,10,15,20-40, 50-100,125-150 | — | 6 |
| | Si 100 | 3,5,10 | — | 10 |
| | Si 300 | 3,5,10,20-40 | — | 30 |
| Shandon | Hypersil | 3,5,10 | 170 | 12 |
| | Hypersil WP-300 | 5,10 | 60 | 30 |
| Showa | Silicapak E-411 | 5 | — | — |

APPENDIX I (continued)

| Supplier | Trade name | dp (μm) | $a_s$ ($m^2/g$) | pd (nm) |
|---|---|---|---|---|
| Supelco | LC-Si | 3,5 | 170 | 10 |
| | LC-3-Si | 5 | — | 30 |
| | LC-5-Si | 5 | — | 50 |
| | PLC-Si | 15 | 170 | 10 |
| TOSOH | TSK-Gel LS 310 SIL | 5,10,15 | — | — |
| Whatman | Partisil | 5,10 | 350 | 9 |
| | LPS-1 | 13-24 | 250 | — |
| | LPS-2 | 37-53 | 450 | — |
| YMC | 100 A Silica | 3,5,15,25-44 | — | — |

## APPENDIX II: SURVEY OF COMMERCIAL ALUMINAS

| Supplier | Trade name | dp (μm) | $a_s$ ($m^2/g$) | pd (nm) | pH of suspension |
|---|---|---|---|---|---|
| Baker | Aluminum oxide, | | | | |
| | acid-washed | 50-200 | — | 6 | pH 4.5 |
| | neutral | 50-200 | — | 6 | pH 7 |
| | acid, activity 1 | 50-200 | — | 6 | pH 4.5 |
| | basic, activity 1 | 50-200 | — | 6 | pH 10.0 |
| | neutral, activity 1 | 50-200 | — | 6 | pH 7.5 |
| Bio-Rad | Neutral alumina AG7 | — | — | — | pH 6.9-7.1 |
| | Basic alumina AG10 | — | — | — | pH 10.0-10.5 |
| | Acid alumina | — | — | — | pH 3.5-45 |
| ICN | ICN alumina N | 18-32, 32-63 | 200 | 6 | pH 7.5 |
| | ICN alumina A | 18-32 | 200 | 6 | pH 4.5 |
| | ICN alumina B | 18-32 | 200 | 6 | pH 10 |
| Macherey Nagel | Alox 60-5, 10 | 5, 10 | 155 | 6 | Basic alumina pH 9.5 |
| Merck | Aluminum oxide 60, active basic | 63-200 | — | 6 | pH 9.0 ± 0.5 |
| | Aluminum oxide 90, active basic | 63-200 | — | 9 | pH 9.0 ± 0.5 |
| | Aluminum oxide 90, active neutral | 63-200 | — | 9 | pH 9.0 ± 0.5 |
| | Aluminum oxide 90, active acid | 63-200 | — | 9 | pH 4.0 ± 0.5 |
| | Aluminum oxide 150, basic | 63-200 | — | 15 | pH 9.0 ± 0.5 |
| | LiChroprep Alox T | 25-40 | — | — | Basic alumina |
| | LiChrosorb Alox T | 5, 10 | — | 15 | Basic Alumina |
| Phase Separations | Spherisorb AY | 5, 10 | 90 | 13 | — |

# APPENDIX III: SURVEY OF COMMERCIAL REVERSED PHASE PACKINGS

| Supplier | Trade name | dp (μm) | dp (nm) | Comments |
|---|---|---|---|---|
| Alltech | Adsorbosphere $C_8$ | 3,5,10 | 8 | Octylsilyl |
| | Adsorbosphere $C_{18}$ | 3,5,10 | 8 | Octadecylsilyl-endcapped |
| | Adsorbospher Phenyl | 3,5,10 | 8 | Phenylsilyl-endcapped |
| | Adsorbospher TMS | 3,5,10 | 8 | Trimethylsilyl |
| | Econosphere $C_8$ | 5 | 10 | |
| | Econosphere $C_{18}$ | 5 | 10 | |
| | Ro Sil $C_3$ | 3,5,8 | 8 | |
| | Ro Sil $C_8$ | 3,5,8 | 8 | |
| | Ro Sil $C_{18}$ | 3,5,8 | 8 | |
| | R Sil $C_3$ | 5,10 | 6 | |
| | R Sil $C_8$ | 5,10 | 6 | |
| | R Sil $C_{18}$ HL | 5,10 | 6 | 16% carbon |
| | R Sil $C_{18}$ LL | 5,10 | 6 | 10% carbon |
| | R Sil Phenyl | 5,10 | 6 | |
| | Versapack $C_{18}$ | 10 | 6 | Endcapped, demineralized |
| Analytichem | Sepralyte $C_1$ Methyl | 5,10,40 | — | Endcapped |
| | Sepralyte $C_2$ Ethyl | 5,10,40 | — | Endcapped |
| | Sepralyte $C_4$ Butyl | 5,10,40 | — | Endcapped |
| | Sepralyte $C_6$ Hexyl | 5,10,40 | — | Endcapped |
| | Sepralyte $C_8$ Octyl | 3,5,10,40 | — | Endcapped |
| | Sepralyte $C_{18}$ Octacecyl | 5,10,40 | — | Endcapped |
| | Sepralyte CH Cyclohexyl | 5,10,40 | — | Endcapped |
| | Sepralyte PH Phenyl | 5,10,40 | — | |
| | Sepralyte 2PH Diphenyl | 5,10,40 | — | |

| | | | | |
|---|---|---|---|---|
| Amicon | Amicon $C_8$ | 5,10,15, 20,20-45, 35-70, 90-130 | 6,10,25 | Octylsilyl |
| | Amicon $C_{18}$ | 5,10,15,20, 20-45,35-70, 90-130 | 6 | Octadecylsilyl |
| Baker | Bakerbond Methyl | 10,40 | 6 | 4.6% carbon load |
| | Bakerbond Ethyl | 10,40 | 6 | 6.0% carbon load |
| | Bakerbond Butyl | 5 | 33 | 3.0% carbon load |
| | Bakerbond Butyl | 10,40 | 6 | 7.9% carbon load |
| | Bakerbond Hexyl | 10,40 | 6 | 10% carbon load |
| | Bakerbond Octyl | 5 | 5 | 10.7% carbon load |
| | Bakerbond Octyl | 5 | 33 | 3.4% carbon load |
| | Bakerbond Octyl | 10,40 | 6 | 12.0% carbon load |
| | Bakerbond Octadecyl | 5 | 5 | 17.2% carbon load |
| | Bakerbond Octadecyl | 5 | 33 | 7.3% carbon load |
| | Bakerbond Octadecyl | 10,40 | 6 | 18.0% carbon load |
| | Bakerbond Phenylethyl | 5 | 5 | 11.4% carbon load |
| | Bakerbond Phenylethyl | 10,40 | 6 | 10.7% carbon load |
| | Bakerbond Diphenyl | 5 | 33 | 3.8% carbon load |
| | Bakerbond Diphenyl | 10 | 6 | 11.5% carbon load |
| | WP-Butyl | 5,15,40 | 27.5,30 | |
| | WP-Octyl | 5,15,40 | 27.5,30 | |
| | WP-Octadecyl | 5,15,40 | 27.5,30 | |
| | WP-Cyanopropyl | 5,15,40 | 27.5,30 | |
| | WP-Diphenyl | 5,15,40 | 27.5,30 | |
| Beckman | Ultrasphere $C_3$ | 5 | — | |
| | Ultrasphere $C_8$ | 5 | — | |
| | Ultrasphere $C_{18}$ | 5 | — | |

APPENDIX III (continued)

| Supplier | Trade name | dp (μm) | dp (nm) | Comments |
|---|---|---|---|---|
| Bio-Rad | Hi-Pore RP-304 | 5 | 33 | Butyl, 3% carbon |
| | Hi-Pore RP-318 | 5 | 33 | Octadecyl, 8% carbon |
| Brownlee | Aquapore Butyl BU 300 | 10 | 30 | Butylsilyl |
| | Aquapore Octyl RP-300 | 10 | 30 | Octysilyl |
| | Aquapore RPC 18, RP-300 | 10 | 30 | Octadecylsilyl |
| | Aquapore Phenyl PH-300 | 10 | 30 | Phenylsilyl |
| | Spheri RP8 | 5,10 | 10 | |
| | Spheri RP18 | 5,10 | 10 | |
| Chrompack | Chrom Spher $C_8$ | 5 | 12 | |
| | Chrom Spher $C_{18}$ | 5 | 12 | |
| | CP-Microspher $C_{18}$ | 3 | — | |
| | CP-Spher $C_8$ | 10 | — | |
| | CP-Spher $C_{18}$ | 10 | — | |
| | Prep ODS | 10,20,32-74 | 6 | |
| Du Pont | Permaphase ODS | 30 | 200 | Solid core (glass) with a porous layer of silica |
| | Zorbax ODS | 5,7 | 7-8 | 20% carbon loading |
| | Zorbax Phenyl | 3,5,7 | 7-8 | |
| | Zorbax TMS | 3,5,7 | 7-8 | |
| | Zorbax $C_8$ | 5 | 7-8 | |
| | Zorbax Bio Series | | | |
| | PEP RP-1 | 10 | 15 | $C_8$, endcapped |
| | Protein PLUS | — | 30 | |

| | | | | |
|---|---|---|---|---|
| ES Industries | Chromegabond $C_1$ | 3,5,10 | 6,10 | Methyl |
| | Chromegabond $C_2$ | 3,5,10 | 6,10 | Ethyl |
| | Chromegabond $C_2$ | 3,5,10 | 6,10 | Dimethyl |
| | Chromegabond TMS | 3,5,10 | 6,10,30 | Trimethyl |
| | Chromegabond $C_3$ | 3,5,10 | 6,10 | n-Propyl |
| | Chromegabond $C_4$ | 3,5,10 | 6,10 | n-Butyl |
| | Chromegabond $C_6$ | 3,5,10 | 6,10 | n-Hexyl |
| | Chromegabond C-$C_6$ | 3,5,10 | 6,10 | Cyclohexyl |
| | Chromegabond $C_8$ | 5,10 | 10 | n-Octyl (polymeric) |
| | Chromegabond M-$C_8$ | 3,5,10 | 6,10,30 | n-Octyl (monolayer) |
| | Chromegabond $C_{10}$ | 5,10 | 6 | n-Decyl |
| | Chromegabond $C_{12}$ | 5,10 | 6 | n-Dodecyl |
| | Chromegabond $C_{18}$ | 6,10 | 10 | n-Octadecyl (polymeric) |
| | Chromegabond M-$C_{18}$ | 3,5,10 | 6,10,30 | n-Octadecyl (monolayer) |
| | Chromegabond $C_{22}$ | 5,10 | 10 | Docosyl |
| | Chromegabond AP | 3,5,10 | 6,10,30 | Alkylphenyl |
| | Chromegabond DP | 10 | 6,10,30 | Diphenyl |
| HP Chemicals | HP-1205 | 5 | 6 | Available as propylbutyl, octyl, oxtadecyl, and phenyl phases, endcapped |
| | HP-1210 | 10 | 6 | |
| | HP-1215 | 15 | 6 | |
| | HP-1220 | 20 | 6 | |
| | HP-2205 | 5 | 15 | |
| | HP-2210 | 10 | 15 | |
| | HP-2215 | 15 | 15 | |
| | HP-2220 | 20 | 15 | |
| | HP-3205 | 5 | 25 | |
| | HP-3210 | 10 | 25 | |
| | HP-3215 | 15 | 25 | |
| | HP-3220 | 20 | 25 | |

APPENDIX III (continued)

| Supplier | Trade name | dp (μm) | dp (nm) | Comments |
|---|---|---|---|---|
| ICN | ICN Silica $RP_8$ | 7-12, 18-32, 32-63 | 6, 10 | Endcapped, 14.5/16.5% carbon |
| | ICN Silica $RP_{18}$ | 7-12, 18-32, 32-63 | 6, 10 | Endcapped, 14.5/16.5% carbon |
| Isco | Isco MP $C_4$ | 5 | 30 | |
| | Isco MP $C_{18}$ | 5 | 30 | |
| | Isco $C_1$ | 5 | 8 | |
| | Isco $C_8$ | 5 | 8 | |
| | Isco $C_{18}$ | 5 | 8 | |
| Manville | Chromosorb LC-4 | 37-44 | — | Octadecylsilyl |
| | Chromosorb LC-5 | 37-44 | — | Phenylsilyl |
| | Chromosorb LC-7 | 5, 10 | — | Octadecylsilyl, 15% carbon |
| | Chromosorb LC-10 | 5, 10 | — | Octylsilyl |
| Macherey Nagel | Nucleosil 30 $C_{18}$ | 30 ± 10 | — | |
| | Nucleosil 300 $C_{18}$ | 5, 7, 10 | 30 | |
| | Nucleosil 300 $C_6H_5$ | 7 | 30 | |
| | Nucleosil 500 $C_4$ | 7 | 50 | |
| | Nucleosil 500 $C_8$ | 7 | 50 | |
| | Nucleosil 500 $C_{18}$ | 7 | 50 | |
| | Nucleosil 500 $C_6H_5$ | 7 | 50 | |
| | Nucleosil 1000 $C_4$ | 7 | 100 | |
| | Nucleosil 1000 $C_{18}$ | 7 | 100 | |
| | Nucleosil 1000 $C_6H_5$ | 7 | 100 | |
| | Nucleosil 4000 $C_4$ | 7 | 400 | |

| | | | | |
|---|---|---|---|---|
| | Nucleosil 4000 $C_{18}$ | 7 | 400 | |
| | Nucleosil 4000 $C_6H_5$ | 7 | 400 | |
| | Nucleosil $C_8$ | 5,7,10 | 10 | |
| | Nucleosil $C_{18}$ | 3,5,7,10,30 | 10 | |
| | Nucleosil $C_6H_5$ | 7 | 10 | |
| | Nucleosil 120 $C_8$ | 3,5,7,10 | 12 | |
| | Nucleosil 120 $C_{18}$ | 3,5,7,10 | 12 | |
| | Nucleosil 102 $C_6H_5$ | 7 | 12 | |
| | Nucleosil 300 $C_4$ | 5,7,10 | 30 | |
| | Nucleosil 300 $C_8$ | 5,7,10 | 30 | |
| | Polygosil 60 $C_8$ | 5,7.5,10,5-20, 25-40,40-63 | 6 | |
| | Polygosil 60 $C_{18}$ | 5,7.5,10,5-20, 25-40,40-63 | 6 | |
| | Polygosil 100 $C_{18}$ | 5,7,10 | 10 | |
| Merck | Kieselel 60 silanized | 63-200 | 6 | Dimethylsilyl |
| | LiChroprep RP2 | 25-40 | 6 | Modified with dimethyldichloro-silane |
| | LiChroprep RP8 | 5-20,15-25, 25-40 | 6 | Octyldimethylsilyl |
| | LiChroprep RP18 | 5-20,15-25, 25-40,40-63 | 6 | Octadecyldimethylsilyl |
| | LiChrosorb RP2 | 5,10 | 6 | Modified with dimethyldichloro-silane |
| | LiChrosorb RP8 | 5,7,10 | 6 | Octyldimethylsilyl |
| | LiChrosorb RP18 | 5,7,10 | 6 | Octadecyldimethylsilyl |
| | LiChrospher RP 8 | 5,10 | 10 | Octyldimethylsilyl |
| | LiChrospher 100 RP18 | 5,10 | 10 | Octadecyldimethylsilyl |
| | LiChrospher 500 RP8 | 10 | 50 | Octyldimethylsilyl |

APPENDIX III (continued)

| Supplier | Trade name | dp (μm) | dp (nm) | Comments |
|---|---|---|---|---|
| | LiChrospher 1000 RP8 | 10 | 100 | Octyldimethylsilyl |
| | LiChrospher 4000 RP8 | 10 | 400 | Octyldimethylsilyl |
| | Perisorb RP2 | 30-40 | — | Solid core (glass) with a porous layer of silica |
| | Perisorb RP8 | 30-40 | — | |
| | Perisorb RP18 | 30-40 | — | |
| Millipore/ Waters | Bondapak $C_{18}$/Porasil B | 37-75 | — | n-Octadecyl |
| | μ-Bondapak $C_{18}$ | 10 | 12.5 | Endcapped |
| | μ-Bondapak Phenyl | 10 | 12.5 | Endcapped |
| | Nova-Pak $C_{18}$ | 4 | 6 | Endcapped |
| | Nova-Pak Phenyl | 4 | 6 | Endcapped |
| | Prep-PAK Vydac $C_4$ | 30 | — | n-Butyl |
| | Prep-PAK Vydac $C_{18}$ | 30 | — | n-Octadecyl |
| | Prep PAK 500 $C_{18}$ | 55-105 | — | n-Octadecyl |
| | Resolve $C_8$ | 5,10 | 9 | |
| | Resolve $C_{18}$ | 10 | 9 | |
| Orpegen | HD-Sil-18-55-80 | 5 | 8 | Octadecyl |
| | HD-Sil-18-55-100 | 5 | 10 | Octadecyl |
| | HD-Sil-18-10-60 | 10 | 6 | Octadecyl |
| | HD-Sil-18-10-100 | 10 | 10 | Octadecyl |
| | HD-Sil-18-20-60 | 15-25 | 10 | Octadecyl |
| | HD-Sil-18-20-60 | 20-45 | 6 | Octadecyl |
| | HD-Sil-18-60-60 | 35-70 | 6 | Octadecyl |
| | HD-Sil-18-20-100 | 15-25 | 10 | Octadecyl |
| | HD-Sil-18-30-100 | 20-45 | 10 | Octadecyl |
| | HD-Sil-18-60-100 | 35-70 | 10 | Octadecyl |

| | | | | |
|---|---|---|---|---|
| Pharmacia/ LKB | Pep RPC HR5 | 5 | 10 | Mixed $C_2/C_{18}$ phase, 12% carbon |
| | ProRPC HR 5/2 | 5 | 30 | 2% carbon |
| | ProRPC HR 5/10 | 5 | 30 | 2% carbon |
| Phase Separations | Spherisorb SC1 | 3,510 | 8 | 2% carbon loading |
| | Spherisorb SC6 | 3,510 | 8 | 6% carbon loading |
| | Spherisorb SC8 | 3,5,10 | 8 | 6% carbon loading |
| | Spherisorb S ODS 1 | 3,5,10 | 8 | 7% carbon loading |
| | Spherisorb S ODS 2 | 3,5,10 | 8 | 12% carbon loading, endcapped |
| | Spherisorb S Phenyl | 3,5,10 | 8 | 3% carbon loading |
| | S5XC1 | 5 | 30 | $C_1$-bonded silica |
| | S5XC3 | 5 | 30 | $C_3$-bonded silica |
| | S5XC6 | 5 | 30 | $C_6$-bonded silica |
| | S5XC18 | 5 | 30 | $C_{18}$-bonded silica |
| | S10XC1 | 10 | 30 | $C_1$-bonded silica |
| | S10XC3 | 10 | 30 | $C_3$-bonded silica |
| | S10XC6 | 10 | 30 | $C_6$-bonded silica |
| | S10XC18 | | 30 | $C_{18}$-bonded silica |
| Reichelt | Thomasorb Si60 $C_8$ | 10 | 6 | |
| | Thomasorb Si60 $C_{18}$ | 10 | 6 | |
| Scientific Systems | Soft Seal Alkyl 8 | 3,5 | 10 | 12.5% carbon loading |
| | Soft Seal Alkyl 18 | 3,5 | 10 | 12.5% carbon loading |
| Separations Group | Vydac 201 HSB | 5,10,15-20, 20-30 | 8 | Octadecylsilyl, endcapped |
| | Vydac 201 TPB | 5,10,15-20, 20-30 | 30 | Octadecylsilyl, medium level phase loading |
| | Vydac 214 TPB | 5,10,15-20, 20-30 | 30 | Butylsilyl, endcapped |
| | Vydac 218 TPB | 5,10,15-20, 20-30 | 30 | Octadecylsilyl, endcapped |

APPENDIX III (continued)

| Supplier | Trade name | dp (μm) | dp (nm) | Comments |
|---|---|---|---|---|
| | Vydac 219 TPB | 5,10,15-20, 20-30 | 30 | Diphenylsilyl, endcapped |
| | Vydac 201 SC | 30-40 | | Octadecylsilyl, solid core (glass) with a porous layer of silica |
| Serva | Serva Butyl=Si 300 Polyol | 5,10 | 30 | Complete chemical derivatization of the silica matrix by polyhydroxylation |
| | Serva Octyl=Si 500 Polyol | 10,30 | 50 | |
| | Serva Octyl=Si 100 Polyol | 3,5,10,30 | 30 | |
| | Serva Octyl=Si 300 Polyol | 3,5,10 | 30 | |
| | Serva Octyl=Si 500 Polyol | 10 | 50 | |
| | Serva Octadecyl=Si 60 Polyol | 3,5,10 | 6 | |
| | Serva Octadecyl=Si 100 Polyol | 3,5,10,30 | 10 | |
| | Serva Octadecyl=Si 300 Polyol | 3,5,10 | 30 | |
| | Serva Octadecyl=Si 500 Polyol | 10 | 50 | |
| | Serva Phenyl=Si 100 Polyl | 3,5,10 | 10 | |
| | Serva Diphenyl=Si 100 Polyl | 3,5,10,30 | 10 | |
| | Serva Diphenyl=Si 300 Polyol | 3,5,10 | 30 | |
| | Serva Octyl=SP 500 | 04-100,100-200 | 50 | |
| | Serva Octadecyl=SP 500 | 40-100,100-200 | 50 | |
| | Serva Phenyl=SP 500 | 40-100,100-200 | 50 | |

| | | | | |
|---|---|---|---|---|
| Shandon | Hypersil-MOS | 3,5,10 | 12 | Octylsilyl, monolayer, 7.0% carbon |
| | Hypersil-ODS | 3,5,10 | 12 | Octadecylsilyl, monolayer, 10% carbon |
| | Hypersil-SAS | 3,5,10 | 12 | Trimethylsilyl, monolayer, 2.6% carbon |
| | Hypersil-Phenyl | 3,5,10 | 12 | Phenylsilyl, monolayer 5.0% carbon |
| | WP-300 | 5,10 | 30 | 2.0% carbon loading |
| | WP-300 Octyl | 5,10 | 30 | 2.7% carbon loading |
| Supelco | Supelcosil LC-1 | 5 | 10 | |
| | Supelcosil LC-8 | 3,5 | 10 | |
| | Supelcosil LC-8 DB | 3,5 | 10 | Specially deactivated silica for use with basic compounds |
| | Supelcosil LC-18 | 3,5 | 10 | |
| | Supelcosil LC-18 DB | 3,5 | — | Specially deactivated silica for use with basic compounds |
| | Supelcosil LC-DP | 5 | 10 | Diphenylmethylsilyl |
| | Supelcosil LC-PAH | 5 | 10 | Octadecyldimethylsilyl, polymeric |
| | Supelcosil LC-304 | 5 | 30 | Butyldimethylsilyl |
| | Supelcosil LC-308 | 5 | 30 | Octyldimethylsilyl |
| | Supelcosil PLC-318 | 5 | 30 | Octadecyldimethylsilyl |
| | Supelcosil PLC-8 | 15,40 | 50-200 | Octyldimethylsilyl |
| | Supelcosil PLC-18 | 15,40 | 50-200 | Octadecyldimethylsilyl |
| Synchrom | SynChropak RP-PC1 | 6.5 | 10,30,100 | |
| | SynChropak RP-PC4 | 6.5 | 10,30,100 | |
| | SynChropak RP-PC8 | 6.5 | 10,30,100 | |
| | SynChropak RP-PC18 | 6.5 | 10,30,100 | |
| | SynChroprep RP-PC1 | 30 | 30 | |
| | SynChroprep RP-PC4 | 30 | 30 | |

APPENDIX III (continued)

| Supplier | Trade name | dp (μm) | dp (nm) | Comments |
|---|---|---|---|---|
| TOSOH | TSKgel ODS-80T | 5 | 8 | Octadecyl, endcapped |
| | TSKgel ODS-120T | 5 | 12 | Octadecyl, endcapped |
| | TSKgel ODS-120A | 5 | 12 | Octadecyl |
| | TSKgel TMS-250 | 5 | 25 | Trimethylsilyl |
| Varian | Micro Pak CH | 10 | — | Octadecylsilyl, polymeric layer |
| | Micro Pak MCH | 5,10 | — | Octadecylsilyl, monomeric layer |
| | Micro Pak MCH-N-Cap | 5 | — | Octadecylsilyl, monomeric layer, endcapped |
| | Micro Pak $C_{18}$ | 4 | — | Octadecylsilyl |
| | Micro Pak SP $C_{18}$ | 3,5 | — | Octadecylsilyl |
| Whatman | LRP-1 | 13-24 | — | n-Octadecyl |
| | LRP-2 | 37-53 | — | n-Octadecyl |
| | Partisil 10 ODS | 5,10 | 8,5 | Octadecylsilyl, polymeric, 5% carbon |
| | Partisil 10 ODS 3 | 5,10 | 8,5 | Octadecylsilyl, endcapped, 10% carbon |
| | Partisil 10 CCS/$C_8$ | 5,10 | 8,5 | $C_8$-chain bonded on a CCS group endcapped, 9% carbon |
| | Protesil 300 Octyl | 10 | 30 | Endcapped, 7.5% carbon |
| | Protesil 300 Diphenyl | 10 | 30 | Endcapped, 8% carbon |
| YMC | 300 A ODS | 5 | 30 | $C_{18}$ bonded silica |
| | 300 A Butyl | 5,15 | 30 | $C_4$ bonded silica |

## APPENDIX IV: COMMERCIALLY AVAILABLE ORGANIC PACKINGS INCLUDING THEIR DERIVATIVES

| Supplier | Trade name | Bulk composition | dp (μm) | Modification | pH range of application | Mode of application |
|---|---|---|---|---|---|---|
| Asahi | Asahipak | | | | | |
| | ODP-50 | VA | 5 | $C_{18}$ | 2-13 | RPC |
| | ODP-90 | VA | 9 | $C_{18}$ | 2-13 | RPC |
| | GS-220,310 | VA | 9 | — | 2-9 | SEC, RPC |
| | 320,520 | VA | 9 | — | 2-12 | SEC, RPC |
| Dyno | Dynospheres P | StDVB | 5,10,20 | — | 1-11 | RPC |
| Hamilton | PRP-1 | StDVB | 5,10 | — | 1-13 | RPC |
| Hitachi | Hitachi Gel | | | | | |
| | 3011-0 (3013-0) | StDVB | 10, (5) | $CH_2OH$ | 2-12 | RPC |
| | 3011 (3013) | StDVB | 10, (5) | — | 2-12 | RPC |
| | 3019 | StDVB | 50 | — | 2-12 | RPC |
| Interaction | ACT-1 | StDVB | 10 | $C_{18}$ | 0-14 | RPC |
| | ACT-2 | VP | 11 | pyridyl | 0-14 | RPC |
| Jasco | Finepak Gel 110 | StDVB | 10 | — | 1-14 | RPC |
| Mitsubishi | MCI-Gel | | | | | |
| | CHP-3C, 5C | StDVB | 10 | — | 0-14 | RPC |
| | CHP-214G | MA | 10 | — | 1-14 | RPC |
| | CQP-06,10,30 | MA | 10 | — | 1-14 | RPC, SEC |
| | CQH-3E5, 3BS, 3PS | MA | 10 | PEG, butyl-phenyl | 1-14 | HIC |
| Pharmacia/LKB | Sephadex LH20 | D | 30-100 | Hydroxy-propyl | 5-14 | SEC, adsorption |
| | Sephadex LH60 | D | 40-100 | Hydroxy-propyl | 5-14 | SEC, adsorption |

APPENDIX IV (continued)

| Supplier | Trade name | Bulk composition | dp (μm) | Modification | pH range of application | Mode of application |
|---|---|---|---|---|---|---|
| | Sephasorb HP ultrafine | D | 10-23 | Hydroxy-propyl | 5-14 | NPC, RPC |
| | Octyl-Sepharose CL-4B | A | 45-165 | Octyl | 3-14 | HIC |
| | Phenyl-Sepharose CL-4B | A | 45-165 | Phenyl | 3-14 | HIC |
| Polymer Labs | PLRP-S 100, | StDVB | 5,8,10 | — | 1-13 | RPC |
| | 300, 1000 | — | — | — | — | — |
| Shimadzu | SPC-RP1 | Not given | 13 | — | — | RPC |
| Shodex | RSpak DS-613 | StDVB | 6 | — | 2-12 | RPC |
| | RSpak DE-613 | MA | 6 | — | 2-12 | RPC |
| | RSpak DM-614 | HMA | 10 | — | 2-12 | RPC |
| | PH-814 | Not given | — | — | 2-12 | HIC |
| TOSOH | Styrene-250,60 | StDVB | 5,10 | — | — | RPC |
| | Ether-250 | PEG | 5,10 | — | — | NPC |
| | Acetate-60 | VAc | 5,10 | — | — | NPC |
| | Phenyl-5PW | Not given | 10 | phenyl | 2-12 | HIC |
| | Ether-5PW | Not given | 10 | PEG | 2-12 | HIC |
| | Octadecyl-4PW | Not given | 7 | octadecyl | 2-12 | RPC |
| | Octadecyl-NPR | Not given | 2.5 | Octadecyl | 2-12 | RPC |
| | Phenyl-5PWRP | Not given | 10 | phenyl | 2-12 | RPC |

Abbreviations: A = agarose, D = dextran, HMA = hydroxyalkyl methacrylate, MA = methacrylate, PEG = polyethylene glycol, StDVB = stryene-divinylbenzene, VA = vinylalcohol, VAc = vinylacetate, VP = vinylpyridine.

## APPENDIX V: COLUMN AND PACKING SUPPLIERS

| | |
|---|---|
| Alltech | Alltech Associates Inc., Applied Science Labs, 2051 Waukegan Road, Deerfield, IL 60015, USA<br>Alltech Europe, Begoniastraat, B-9731 Eke, Belgium |
| Amicon | Amicon Corporation, 17 Cherry Hill Drive, Danvers, MA 01923, USA<br>Grace AG, Amicon-Schweiz, Av. Montchoisi 35, CH-1001 Lausanne, Switzerland |
| Analytichem | Analytichem International, Inc., 24201 Frampton Ave., Harbor City, CA 90710, USA |
| Asahi | Asahi Kasei Industries Co., Ltd. 1-3-2 Yakao, Kawasaki-ku Kawasaki, Kanagawa 210, Japan |
| Baker | J. T. Baker, Research Products, 222 Red School Lane, Phillipsburgh, NJ 08865, USA<br>Baker Chemikalien, Postfach 1661, D-6080 Gross-Gerau, FRG |
| Beckman | Beckman Instruments Inc., 2350 Camino Ramon, P.O. Box 5101, San Ramon, CA, USA<br>Beckman Instruments GMBH, Frankfurter Ring 115, D-8000 München 40, FRG |
| Bio-Rad | Bio-Rad Laboratories, 2200 Wright Avenue, Richmond, CA 94804, USA<br>Bio-Rad Laboratories GMbH, Dachauer-Strasse 364, D-8000 München 50, FRG |
| Brownlee | Brownlee Labs Inc., 2045 Martin Ave., Santa Clara, CA 95050, USA |
| Chrompack | Chrompack, P.O. Box 3, Middleburg, 4330AA, The Netherlands |
| Du Pont | Du Pont Company, Analytical Instruments Division, Mc Kean Bldg.-Concord Plaza, Wilmington, DE 19801, USA<br>Du Pont GmbH, Postfach 1509, Dieselstrasse 18, D-6350 Bad Nauheim 1, FRG |
| Dyno | Dyno Particles AS, P.O. Box 160, N-2001 Lillestrøm, Norway |
| ES Industries | ES Industries, 8 South Maple Ave., Marlton, NJ 08053, USA |

APPENDIX V (continued)

| | |
|---|---|
| Hamilton | Hamilton Co., P.O. Box 10030, Reno, NV 89520-0012, USA |
| Hitachi | Hitachi, 1-5-1 Marunouchi, Chiayoda-ku, Tokyo 100, Japan |
| HP Chemicals | HP Chemicals Inc., 4221 Forest Park Blvd., St. Louis, Missouri 63108, USA |
| ICN Biomedicals | ICN Biomedicals GmbH, Postfach 369, D-3440 Eschwege, FRG |
| Interaction | Interaction Chemicals, 1615 Plymouth Street, Mountain View, CA 94043, USA |
| ISCO | ISCO Inc., P.O. Box 5347, Lincoln, Nebrasaka 68505, USA |
| Jasco | Jasco Inc., 218 Bay St., Easton, MD 21601, USA |
| Kali Chemie | Kali Chemie AG, Hans-Böckler Allee 20, D-3000 Hannover, FRG |
| Macherey-Nagel | Macherey-Nagel, Neumann-Neander-Strasse 6-8, Postfach 307, D-5160 Düren, FRG |
| Manville | Manville Filtration and Minerals, P.O. Box 5108, Denver, Colorado 80217, USA |
| Merck | E. Merck, Frankfurter Strasse 250, D-6100 Darmstadt, FRG<br>E. M. Science, 111 Woodcrest Road, P.O. Box 5018, Cherry Hill, NJ 08034-0395, USA |
| Millipore-Waters | Millipore, Waters Associated, Milford, MA 01757, USA<br>Millipore, Waters Chromatographie, Hauptstrasse 71-79, D-6236 Eschborn, FRG |
| Mitsubishi | Mitsubishi Kasei Industries, Ltd., 2-5-2 Marunouchi, Chiyoda-ku, Tokyo 101, Japan |
| Orpegen | Orpegen, Czernyring 22, D-6900 Heidelberg, FRG |
| Pharmacia LKB | Pharmacia LKB Biotechnology AB, Box 175, S-75104, Uppsala 1, Sweden<br>Deutsche Pharmacia LKB Biotechnologie GmbH, Munzinger Strasse 9, D-7800 Freiburg, FRG |
| Phase Separations | Phase Separations Ltd., Deeside Industrial Estate, Queensferry, CLWYD CH5 2LR, GB, UK |

APPENDIX V (continued)

| | |
|---|---|
| Polymer Labs | Polymer Laboratories, Essex Road, Church Stretton, Shropshire, SY6 6AX, UK |
| Reichelt | Reichelt Chemie Technik, Englerstrasse 18, D-6900 Heidelberg 1, FRG |
| Scientific Systems | Scientific Systems, Inc., 1120 West College Ave., State College, PA 16801, USA |
| Separations Group (Vydac) | Separations Group, 17434 Majove Street, P.O. Box 867, Hesperia, CA 92345, USA |
| Serva | Serva, Carl-Benz-Strasse 7, Postfach 105260, D-6900 Heidelberg, FRG |
| Shandon | Shandon Southern Instruments Inc., 515 Broad Street, Drawer 43, Sewickley, PA 15143-0043, USA<br>Shandon Labortechnik GmbH, Karl von Prais Str. 18, Postfach 501029, D-6000 Frankfurt 50, FRG |
| Shimadzu | Shimadzu (Europe) GmbH, Ackerstr. 11, D-4000 Düsseldorf, FRG |
| Showa Denko | Showa Denko K. K., 13-9 Shiba Daimon 1-chome, Minato-ku, Tokyo 105, Japan<br>Showa Denko (Europe) GmbH, Niederkasseler Lohrweg 8, D-4000 Düsseldorf 11, FRG |
| Supelco | Supelco Inc., Supelco Park, Bellafonte, PA 16823-0048, USA<br>Supelchem, Am Laubach 3, Postfach 1127, D-6231 Sulzbach/Taunus, FRG |
| SynChrom | SynChrom Inc., P.O. Box 110, Linden, IN 47955, USA |
| TOSOH | TOSOH Manufacturing Co. Ltd., 1-7- Akasaka, Minato-ku, Tokyo 107, Hapan |
| Varian | Varian, 220 Humboldt Ct., Sunnyvale, CA 94089, USA<br>Varian GmbH, Alsfelder Strasse 6, D-6100 Darmstadt 11, FRG |
| Whatman | Whatman Inc., 9 Bridgewell Place, Clifton, NJ 07014, USA |
| YMC | YMC Inc., P.O. Box 492, Mt. Freedom, NJ 07970, USA |

## APPENDIX VI: SILANES EMPLOYED FOR THE SYNTHESIS OF BONDED PHASES

| Silane | Formula | MW | b.p. (°C/Torr) |
|---|---|---|---|
| *n-Alkyldimethylchlorosilanes* [1] (DMCS = dimethylchlorosilane) | | | |
| Trimethylchlorosilane | $(CH_3)_3SiCl$ | 108.7 | 57/760 |
| Ethyl DMCS | $C_2H_5(CH_3)_2SiCl$ | 122.7 | 91/760 |
| Propyl DMCS | $C_3H_7(CH_3)_2SiCl$ | 136.7 | 114/760 |
| Butyl DMCS | $C_4H_9(CH_3)_2SiCl$ | 150.7 | 140/760 |
| Pentyl DMCS | $C_5H_{11}(CH_3)_2SiCl$ | 164.7 | 165/760 |
| Hexyl DMCS | $C_6H_{13}(CH_3)_2SiCl$ | 192.7 | 95/18 |
| Octyl DMCS | $C_8H_{17}(CH_3)_2SiCl$ | 206.7 | 105/15 |
| Nonyl DMCS | $C_9H_{19}n(CH_3)_2SiCl$ | 220.7 | 113/15 |
| Decyl DMCS | $C_{10}H_{21}(CH_3)_2SiCl$ | 234.7 | 128/15 |
| Dodecyl DMCS | $C_{12}H_{25}(CH_3)_2SiCl$ | 262.7 | 89/0.1 |
| Tetradecyl DMCS | $C_{14}H_{29}(CH_3)_2SiCl$ | 290.7 | 107/0.1 |
| Hexadecyl DMCS | $C_{16}H_{33}(CH_3)_2SiCl$ | 318.7 | 12/0.1 |
| Octadecyl DMCS | $C_{18}H_{37}(CH_3)_2SiCl$ | 346.7 | 140-145/0.1 |
| Eicosyl DMCS | $C_{20}H_{41}(CH_3)_2SiCl$ | 374.7 | 163/0.1 |
| *Aryldimethylchlorosilanes* [2] (DMCS = dimethylchlorosilane) | | | |
| Phenyl DMCS | $C_6H_5(CH_3)_2SiCl$ | 170.5 | 87/18 |
| Benzyl DMCS | $C_6H_5CH_2(CH_3)_2SiCl$ | 184.7 | 80-84/14 |
| Phenetyl DMCS | $C_6H_5(CH_3)_2(CH_3)_2SiCl$ | 198.1 | 56/0.2 |
| Naphthyl DMCS | $C_{10}H_7(CH_3)_2SiCl$ | 221.5 | 145-148/14 |
| *Dialkylchlorosilanes* [3] | | | |
| Dimethyldichlorosilane | $(CH_3)_2SiCl_2$ | 129.1 | 70 |
| n-Hexylmethyldichlorosilane | $C_6H_{13}(CH_3)SiCl_1$ | 199.2 | 204-206 |
| n-Octylmethyldichlorosilane | $C_8H_{17}(CH_3)SiCl_2$ | 227.3 | 94/6 |

APPENDIX VI (continued)

| Silane | Formula | MW | b.p. (°C/Torr) |
|---|---|---|---|
| n-Octadecylmethyl-dichlorosilane | $C_{18}H_{37}(CH_3)SiCl_2$ | 367.5 | 185/2.5 |
| Dibutyldichlorosilane | $(C_4H_9)_2SiCl_2$ | 213.2 | 212 |
| 2-Adamantylethylmethyl-dichlorosilane | $(C_{10}H_{15})(C_2H_4)$-$(CH_3)SiCl_2$ | 277.4 | 125-126/1.5 |
| Dihexyldichlorosilane | $(C_6H_{13})_2SiCl_2$ | 269.3 | 111-113/6 |
| Phenylmethyldi-chlorosilane | $C_6H_5(CH_3)SiCl_2$ | 191.1 | 205 |
| Diphenyldichlorosilane | $(C_6H_5)_2SiCl_2$ | 253.2 | 309-310 |
| Methyl(2-bicyclohep-tyl)-dichlorosilane | $C_7H_{11}(CH_3)SiCl_2$ | 209.2 | 115-118/3 |
| 2-(4-Cyclohexenyl) ethylmethyldichloro-silane | $(C_6H_9C_2H_4)(CH_3)SiCl_2$ | 223.2 | 110/10 |
| *Alkyltrichlorosilanes* [3] | | | |
| Methyltrichlorosilane | $CH_3SiCl_3$ | 149.0 | 66 |
| n-Butyltrichlorosilane | $C_4H_9SiCl_3$ | 191.6 | 142-143 |
| n-Hexyltrichlorosilane | $C_6H_{13}SiCl_3$ | 219.6 | 191-192 |
| n-Octyltrichlorosilane | $C_8H_{17}SiCl_3$ | 247.7 | 224-226/730 |
| n-Dodecyltrichlorosilane | $C_{12}H_{25}SiCl_3$ | 303.8 | 155/10 |
| n-Octadecyltrichloro-silane | $C_{18}H_{37}SiCl_3$ | 388.0 | 160-162/3 |
| Phenyltrichlorosilane | $C_6H_5SiCl_3$ | 211.6 | 201 |
| Cylcohexyltrichloro-silane | $C_6H_{11}SiCl_3$ | 231.6 | 90-91/10 |
| β-Phenethyltrichloro-silane | $C_8H_9SiCl_3$ | 239.6 | 93-96/3 |
| 2-Bicycloheptyltri-chlorosilane | $C_7H_{11}SiCl_3$ | 229.6 | 63-64/9.5 |
| 2-(4 Cyclohexenyl)-ethyltrichlorosilane | $C_6H_9(CH_2)_2SiCl_3$ | 243.6 | 74-75/0.7 |

APPENDIX VI (continued)

| Silane | Formula | MW | b.p. (°C/Torr) |
|---|---|---|---|
| Adamantylethyltrichlorosilane | $C_{10}H_{15}(CH_2)_2SiCl_3$ | 297.7 | 135-136/3 |
| Eicosyltrichlorosilane | $C_{20}H_{41}SiCl_3$ | 416.0 | 225-227/3 |
| *Arylalkyltrichlorosilanes* [4] | | | |
| 2-Phenylethyltrichlorosilane | $C_6H_5(CH_2)_2SiCl_3$ | 238.5 | — |
| 4-Phenylbutyltrichlorosilane | $C_6H_5(CH_2)_4SiCl_3$ | 286.5 | — |
| 6-Phenylhexyltrichlorosilane | $C_6H_5(CH_2)_6SiCl_3$ | 294.5 | — |
| *n-Alkyldimethyltriflates* [2] (n-alkyldimethyltrifluoroacetoxysilanes) | | | |
| Trimethyltriflate | $(CH_3)_3SiOCOCF_3$ | 186.1 | 88-89/760 |
| Ethyldimethyltriflate | $C_2H_5(CH_3)_2SiOCOCF_3$ | 200.1 | 103/760 |
| n-Butyldimethyltriflate | $C_4H_9(CH_3)_2SiOCOCF_3$ | 228.1 | 145/760 |
| n-Pentyldimethyltriflate | $C_5H_{11}(CH_3)_2SiOCOCF_3$ | 242.1 | 56-58/14 |
| n-Hexyldimethyltriflate | $C_6H_{13}(CH_3)_2SiOCOCF_3$ | 256.1 | 72/13 |
| n-Heptyldimethyltriflate | $C_7H_{15}(CH_3)_2SiOCOCF_3$ | 270.1 | 90/15 |
| n-Octyldimethyltriflate | $C_8H_{17}(CH_3)_2SiOCOCF_3$ | 284.1 | 107-110/15 |
| n-Octadecyldimethyltriflate | $C_{18}H_{37}(CH_3)_2SiOCOCF_3$ | 424.1 | — |

*Organo(dimethylamino)silanes of type $R_1R_2R_3SiN(CH_3)_2$* [5,6]

| $R_1$ | $R_2$ | $R_3$ | Formula | MW | b.p. (°C/Torr) |
|---|---|---|---|---|---|
| H | $CH_3$ | $CH_3$ | $C_4H_{13}NSi$ | 103.23 | 65-67/720 |
| $CH_3$ | $CH_3$ | $CH_3$ | $C_5H_{15}NSi$ | 117.26 | 77-79/720 |
| Phenyl | $CH_3$ | $CH_3$ | $C_{10}H_{17}NSi$ | 179.33 | 82-83/15 |
| 3,3-Dimethylbutyl | $CH_3$ | $CH_3$ | $C_{10}H_{25}NSi$ | 187.40 | 156-160/720 |

APPENDIX VI (continued)

| $R_1$ | $R_2$ | $R_3$ | Formula | MW | b.p. (°C/Torr) |
|---|---|---|---|---|---|
| 5-Methoxy-3,3-dimethylpentyl | $CH_3$ | $CH_3$ | $C_{12}H_{29}NOSi$ | 231.45 | 93-95/7 |
| 5-Cyano-3,3-dimethylpentyl | $CH_3$ | $CH_3$ | $C_{12}H_{26}N_2Si$ | 226.43 | 89-91/8 |
| 5-(Dimethylamino)-3,3-dimethylpentyl | $CH_3$ | $CH_3$ | $C_{13}H_{32}N_2Si$ | 244.49 | 58-60/0.03 |
| Trifluoropropyl | $C_3F_3H_4$ | $CH_3$ | $C_9F_6H_{17}NSi$ | 281.32 | 70-72/12 |

| $R_1R_2R_3$ | Formula | MW | b.p.(°C/Torr) |
|---|---|---|---|
| Trimethyl | $C_5H_{17}NSi$ | 117.26 | 77-79/720 |
| Propyldimethyl | $C_7H_{19}NSi$ | 145.23 | 123-127/720 |
| Decyldimethyl | $C_{14}H_{33}NSi$ | 243.51 | $103\text{-}105/5 \times 10^{-2}$ |
| Docosyldimethyl | $C_{26}H_{57}NSi$ | 411.83 | $185\text{-}186/5 \times 10^{-3}$ |
| (3,3-dimethylbutyl)-dimethyl | $C_{10}H_{25}NSi$ | 187.40 | 156-160/720 |
| [2(Adamantyl-1)-ethyl]-dimethyl | $C_{16}H_{31}NSi$ | 265.50 | $95/10^{-3}$ |
| Phenyldimethyl | $C_{10}H_{21}NSi$ | 179.33 | 82/15 |
| Triphenyl | $C_{20}H_{21}NSi$ | 303.47 | (mp: 78) |
| (4-Diphenylphosphinophenyl)-dimethyl | $C_{22}H_{26}NPSi$ | 363.51 | $168\text{-}170/10^{-2}$ |
| (3-Diphenylphosphinopropyl)-dimethyl | $C_{19}H_{28}NPSi$ | 329.49 | $175\text{-}177/4 \times 10^{-2}$ |
| [3-(4'-Diphenylphosphinophenyl)propyl]-dimethyl | $C_{25}H_{32}NPSi$ | 405.59 | $211\text{-}215/5 \times 10^{-2}$ |
| *Alkylenolates* [7] | | | |
| Trimethylsilylenolate | $(CH_3)_3SiOC(CH_3)=CHCOCH_3$ | 172 | — |
| Allyldimethylsilylenolate | $(CH_2CH=CH_2)SiOC(CH_3)=CHCOCH_3$ | 168 | — |

APPENDIX VI (continued)

| $R_1R_2R_3$ | Formula | MW | b.p. (°C/Torr) |
|---|---|---|---|
| n-Octadecyldimethyl-silylenolate | $(C_8H_{17})(CH_3)_2SiOC-(CH_3)=CHCOCH_3$ | 270 | — |
| Substituted 1-silacyclopentanes and -hexanes [2] | | | |
| Si $R_1R_2$ H<br>1-silacyclopentane derivatives | Si $R_1R_2$ H<br>1-silacyclohexane derivatives | | |
| 1,1-Dichloro-1-sila-cyclopentane | $C_4H_8SiCl_2$ | 155.1 | 140/145/760 |
| 1-Methyl-1-chloro-1-silacyclopentane | $C_5H_{10}SiCl$ | 134.1 | 132/760 |
| 1-Phenyl-1-chloro-1-silacyclopentane | $C_{10}H_{13}SiCl$ | 196.7 | 130/13 |
| 1,1-Dichloro-1-sila-cyclohexane | $C_5H_{10}SiCl_2$ | 169.1 | 169-170/760 |
| 1-Methyl-1-chloro-1-silacyclohexane | $C_6H_{13}SiCl$ | 148.7 | 167/760 |
| 1-Phenyl-1-chloro-1-silacyclohexane | $C_{11}H_{15}SiCl$ | 210.7 | 158/15 |
| *n-Alkylhydrochlorosilanes* [8] | | | |
| Ethyldihydrochlorosilane | $C_2H_5(H)_2SiCl$ | 94.5 | 43/760 |
| n-Octyldihydrochlorosil-ane | $C_8H_{17}(H)_2SiCl$ | 178.5 | 60.5/4 |
| n-Octadecyldihydro-chlorosilane | $C_8H_{35}(H)_2SiCl$ | 316.5 | 142/0.3 |
| *Organoalkoxysilanes* [9] | | | |
| N-(3-tiethoxysilyl-propyl)acetamide | $CH_3CONH(CH_2)_3-Si(OC_2H_5)_3$ | 263.4 | 156/4 |
| N-(3-triethoxysilyl-propyl)trifluoroamide | $CF_3CONH(CH_2)_3-Si(OC_2H_5)_3$ | 317.4 | 135/16 |

APPENDIX VI (continued)

| | Formula | MW | b.p. (°C/Torr) |
|---|---|---|---|
| N-(3-triethoxysilyl-propyl)sulfonamide | $CH_3SO_2NH(CH_2)_3$-$Si(OC_2H_5)_3$ | 271.4 | — |
| N-(3-triethoxysilyl-propyl)glycinamide | $CH_3CONHCH_2CONHSi$-$(OC_2H_5)_3$ | 230.3 | — |
| 3-aminopropyltri-ethoxysilane | $NH_2(CH_2)_3Si$-$(OC_2H_5)_3$ | 221.4 | 122-123/30 |
| *Epoxyalkylalkoxysilanes* [2] | | | |
| 3,4-Epoxybutyltri-ethoxysilane | $CH_2$-$CH$-$(CH_2)_2Si$- (epoxide ring via O) $(OC_2H_5)_3$ | 234 | 115-121/14 |
| 3-Glycidoxypropyltri-methoxysilane | $CH_2CH$-$CH_2O(CH_2)_3$- (epoxide ring via O) $Si(OC_2H_5)_3$ | 263.3 | 260-226/760 |
| *Isocyanato- and isothiocyanatoalkyltrialkoxysilanes* [2] | | | |
| Isocyanatopropyl-trimethoxysilane | $(OCN)(CH_2)_3Si(OCH_3)_3$ | 205.4 | 70-75/5 |
| Isocyanatopropyl-triethoxysilane | $(OCN)(CH_2)_3Si$-$(OC_2H_5)_3$ | 247.4 | $85\text{-}88/10^{-2}$ |
| Isothiocyanatopropyl-triethoxysilane | $(SCN)(CH_2)_3Si$-$(OC_2H_5)_3$ | 263.4 | $72\text{-}73/10^{-1}$ |
| *Cyanoalkoxysilanes* [2] | | | |
| Cyanoethyldimethyl-chlorosilane | $(CNC_2H_4)(CH_3)_2SiCl$ | 146.6 | 83/18 |
| Cyanopropyldimethyl-chlorosilane | $(CNC_3H_6)(CH_3)_2SiCl$ | 160.6 | 95/18 |

APPENDIX VI (continued)

| | Formula | MW | b.p. (°C/Torr) |
|---|---|---|---|
| *Aminoalkylalkoxysilanes* [2] | | | |
| Diethylaminopentyltriethoxysilane | $(C_2H_5)_2N(C_5H_{10}Si(OC_2H_5)_3$ | 302.5 | $92\text{-}99/10^{-2}$ |
| Diethylaminohexyltriethoxysilane | $(C_2H_5)_2N(C_6H_{12})Si(OC_2H_5)_3$ | 316.6 | $108\text{-}112/10^{-2}$ |
| Diethylaminoheptyltriethoxysilane | $(C_2H_5)_2N(C_7H_{14})Si(OC_2H_5)_3$ | 330.6 | $110\text{-}117/10^{-3}$ |
| Diethylaminononyltriethoxysilane | $(C_2H_5)N(C_9H_{18})Si(OC_2H_5)_3$ | 358.6 | $135\text{-}145/10^{-3}$ |
| *Acetoxyalkylsilanes* [2] | | | |
| 3-Acetoxypropyltriethoxysilane | $CH_3COO(CH_3)_3Si(OC_2H_5)_3$ | 264 | 192-132/13 |
| 5-Acetoxypentyltriethoxysilane | $CH_3COO(CH_2)_5Si(OC_2H_5)_3$ | 292 | 150-153/13 |
| 3-Acetoxypropyldimethylethoxysilane | $CH_3COO(CH_2)_3(CH_2)_2\text{-}Si(OC_2H_5)$ | 204 | 97-99/13 |
| 5-Acetoxypentyldimethylethoxysilane | $CH_3COO(CH_2)_5(CH_2)_2\text{-}Si(OC_2H_5)$ | 232 | 122-124/13 |
| *Commercial organotrialkoxysilanes* [10] | | | |
| Vinyl | $CH_2{=}CHSi(OCH_2CH_2OCH_3)_3$ | | |
| Methacrylate | $CH_2{=}C(CH_3)COO(CH_2)_3Si(OCH_3)_3$ | | |
| Epoxy | $CH_2(O)CH\text{-}CH_2O(CH_2)_3Si(OCH_3)_3$ | | |
| Mercaptan | $HS\ CH_2CH_2CH_3Si(OCH_3)_3$ | | |
| Chloropropyl | $ClCH_2CH_2CH_2Si(OCH_3)_3$ | | |
| Aminopropyl | $H_2NCH_2CH_2CH_2Si(OC_2H_5)_3$ | | |

APPENDIX VI (continued)

| | Formula | MW | b.p. (°C/Torr) |
|---|---|---|---|
| Fluorinated silanes (commercially Avaliable only as fluorinated bonded phases) [11] | | | |
| Heptafluoroisopropoxy-propyldimethylsilane | $F(CH_3)_2CO(CH_2)_3Si(CH_3)_2X$ | | |
| Heptadecafluorodecyldi-methylsilane | $CF_3(CH_2)_7(CH_2)_2Si(CH_3)_2X$ | | |
| Pentafluorophenyldi-methylsilane | $C_6F_5Si(CH_3)_2SiX$ | | |

Reagents used for hydroxyl group protection and endcapping [2]

| Reagent | Formula |
|---|---|
| Hexamethyldisilazane (HMDS) | $(Me_3Si)_2NH$ |
| Trimethylsilylimidazole | $Me_3Si\ C_3H_3H_2$ |
| Trimethylsilyldimethylamine | $Me_3SiNMe_2$ |
| Trimethylchlorosilane + HMDS | $Me_3SiCl$ + HMDS |
| Trimethylchlorosilane + amine | $Me_3SiCl$ + amine |
| Dimethylchlorosilane + base | $Me_2SiCl_2$ + base |
| N-trimethylsilyl-N,N'-diphenylurea | $(Me_3Si)(Ph)NOCHNPh$ |
| Bis(trimethylsilyl)urea | $(Me_3SiNH)_2CO$ |
| Monotrimethylsilylacetamide | $Me_3SiNHCOMe$ |
| N,O-bis(trimethylsilyl)acetamide | $Me_3SiN{=}C(Me)OSiMe_3$ |
| N,O-bis(trimethylsilyl)trifluoroacetamide | $Me_3SiN{=}C(CF_3)OSiMe_3$ |
| N-methyl-O-trimethylsilyltrifluoroacetamide | $MeN{=}C(CH_3)OSiMe_3$ |
| N,O-bis(trimethylsilyl)carbamate | $Me_3SiNHCO_2SiMe_3$ |
| N,O-bis(trimethylsilyl)sulfamate | $Me_3SiNHSO_3SiMe_3$ |
| Trimethylsilyltrifluoromethanesulfonate + triethylamine | $Me_3SiOSO_2CF_3 + Et_3N$ |
| Triisopropylsilyltrifluoroacetate + 2,6-dimethyl-pyridine | $[(CH_3)_2CH]_3SiOOCCF_3 + C_7H_9N$ |

## BIBLIOGRAPHY


1. K. D. Lork, Ph.D. thesis, Johannes Gutenberg Universität, FRG, 1988.
2. J. N. Kinkel, Ph.D. thesis, Johannes Gutenberg Universität, FRG, 1984.
3. W. R. Melander and Cs. Horvath, in *HPLC: Advances and Perspectives*, Vol. 2 (Cs. Horvath, ed.), Academic Press, London, 1980, p. 135.
4. H. Hemetsberger, W. Maasfeld, and H. Ricken, *Chromatographia*, *9*:303 (1976).
5. J. F. Erard, L. Nagy, and E. Kovats, *Colloids Surfaces*, *9*: 109-132 (1984).
6. D. Amati and E. Kovats, *Langmuir*, *3*:687-695 (1987).
7. G. Schomburg, A. Deege, J. Köhler, and U. Bien-Vogelsang, *J. Chromatogr.*, *282*:27-39 (1983).
8. R. D. Golding, A. J. Barry, and M. F. Burke, *J. Chromatogr.*, *384*:105-116 (1987).
9. H. Engelhardt and D. Mathes, *J. Chromatogr.*, *142*:311 (1977).
10. E. P. Plueddemann, in *Silylated Surfaces*, (D. E. Leyden and W. Collins, eds.), Gordon and Breach, London, 1980, p. 50.
11. P. C. Sadek and P. W. Carr, *J. Chromatogr.*, *288*:25-41 (1984).

# 7

# Packings in Size Exclusion Chromatography

J. V. Dawkins / *Loughborough University of Technology, Loughborough, Leicestershire, England*

## INTRODUCTION

Although chromatographic experiments reported before 1959 had indicated that some polymer separations could be determined in part by an exclusion mechanism, the first effective demonstration that polymers may be separated by the size dependence of the degree of solute penetration into a porous packing was reported by Porath and Flodin [1]. The dextran gels pioneered by Porath and Flodin for *gel filtration chromatography* (GFC) are lightly crosslinked polymer networks (i.e., a very small fraction of crosslinking agent) which are highly swollen by water. These soft gels have low mechanical stability which deteriorates as pore size increases, so that GFC packings in relatively large columns can only be used at low pressures with slow eluent flow rates (<1 $cm^3$ $min^{-1}$). Consequently, the separations generally require several hours. GFC packings based on crosslinked dextran, agarose, and polyacrylamide have been widely used for analytical and preparative fractionations of biopolymers [2].

Gels with extensive crosslinking which undergo limited or no swelling with the solvent have good mechanical stability and may be used at high pressures and fast flow rates. Moore [3] introduced rigid crosslinked polystyrene gels and defined the term *gel permeation chromatography* (GPC). These gels are available in a wide range of pore sizes and are suitable for separations of synthetic polymers in organic media. GPC packings based on crosslinked polystyrene gels have been widely used for routine polymer characterization and quality control, in particular in the determination of molar mass distributions and for characterizing low polymers and small molecules, e.g., for prepolymers in resins and for polymer additives [4,5]. The excellent mechanical stability of inorganic gels led to the development of macroporous silicas and porous glasses for polymer separations with both aqueous and organic eluents. De Vries and coworkers [6-8] described the use of porous spherical silica beads in GPC and Haller [9-11] developed a controlled method of preparing porous glass with uniform size. However, experimental studies indicated that interactions could occur between the polymer and surface sites on the inorganic packing [12]. The dominant separation mechanism in GFC and GPC is assumed to be *size exclusion* (or *steric exclusion*), which requires minimum interactions between solute and gel. Recent work on fractionations of water-soluble polymers has therefore involved inorganic packings deactivated with bonded phases [13] in order to minimize *secondary retention mechanisms* arising from interactions. Because the crosslinked polymeric packings employed in GFC are generally less susceptible to secondary separation mechanisms than inorganic packings, efforts have been directed to the development of rigid macroporous packings constituted from hydrophilic polymers [14-16].

Considerable advances in the theoretical understanding and in the experimental practice of liquid chromatography have occurred in the past 15 years [17]. High-performance separations in liquid chromatography (HPLC) have been achieved with microparticulate packings, in particular with silica packings containing bonded phases. Chromatographic theory predicts that column efficiency and resolution are markedly raised by lowering the particle diameter of the column packing. Consequently, microspherical packings in short columns will give high-resolution separations, as long as the particles are sufficiently rigid to withstand the high column pressures. In the past decade there has been considerable interest in microparticulate rigid gel packings for *high-performance size exclusion chromatography* (HPSEC). Here emphasis will be placed on microparticulate SEC packings because of their widespread use in the characterization of synthetic polymers and because many analytical GFC separations with aqueous eluents may be accomplished with rigid gels and instrumentation in HPLC.

## RETENTION MECHANISMS

### Size Exclusion

The separation of a solute of given size in solution is determined by a distribution coefficient $K_{SEC}$ which governs the volume of stationary zone solvent that is accessible within the porous gel particles to this solute. The elution (or retention) volume $V_R$ for this solute calculated from the point of sample application to the column to the appearance of the peak height maximum of the chromatogram is given by

$$V_R = V_0 + K_{SEC} V_i \qquad (1)$$

where $V_0$ is the total volume of the mobile zone, i.e., the interstitial or void volume, which is accessible to all molecules and $V_i$ is the total volume of the stationary zone. For very large molecules, the value of $K_{SEC}$ will be zero because the sizes of these molecules prohibit solute diffusion into the gel pores. Very small molecules, on the other hand, have free access to both stationary and mobile zones, i.e., $K_{SEC}$ is unity. As the chromatographic column is washed with solvent, the large molecules are eluted first, followed by solutes of decreasing size which penetrate an increasing fraction of the solvent within gel particles. The dependence of $V_R$ on solute size is shown in Fig. 1 for solutes with a range of sizes, e.g., for calibration standards. The shape of the curve in the plateau region is related to the pore size distribution within the porous packing. Equation (1) leads to a solute capacity factor (or column capacity ratio) k" defined by

$$V_R = V_0(1 + k'') \qquad (2)$$

which is different from k' in HPLC, as shown in Fig. 2. Consequently, only one column volume of solvent is required to elute solutes in size exclusion separations.

Many theoretical models have been proposed for polymer separations by SEC. While flow mechanisms could be important in some experiments, most SEC work is performed at eluent flow rates around 1 $cm^3 \ min^{-1}$ when $V_R$ is independent of flow rate so that an equilibrium model may be presumed to represent separation behavior. For a separation operating at equilibrium conditions, the standard free-energy change $\Delta G°$ for the transfer of solute molecules from the mobile zone to the stationary zone at constant temperature T is related to $K_{SEC}$ by

$$\Delta G° = \Delta H° - T\Delta S° = - RT \ln K_{SEC} \qquad (3)$$

where R is the gas constant and $\Delta H°$ and $\Delta S°$ are the standard enthalpy and entropy differences between the zones, respectively. Thermodynamic theories for SEC [18,19] calculate the pore volume accessible to a solute of given size in solution in terms of pore size for various models of pore shape. A simplistic representation of this approach is shown in Fig. 3 for a solid sphere confined in a two-dimensional circular pore. The sphere size permits the transfer of the polymer molecule from the mobile zone to a pore within the packing, but the center of gravity of the sphere cannot approach the pore surface and so is excluded from a volume of solvent around the pore surface. This excluded volume is clearly dependent on solute size. A similar excluded

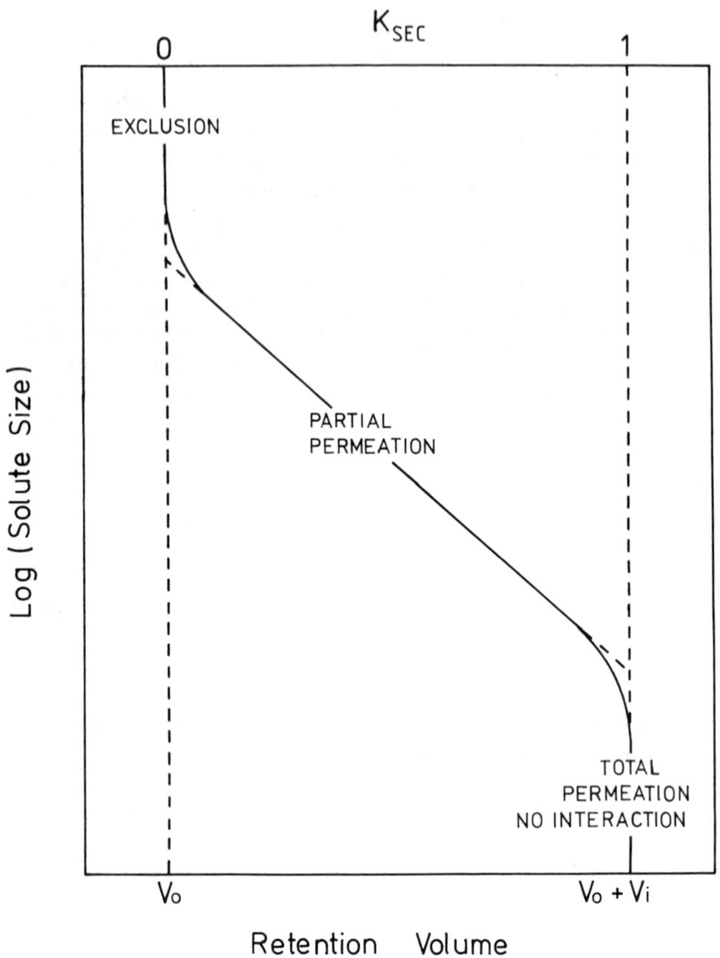


**Fig. 1** Calibration curve for a size exclusion mechanism.

volume model may be assumed for rigid rod and random coil polymers [18,19]. The distribution coefficient $K_{SEC}$ may be defined as the ratio of accessible solute arrangements within the porous packing to those in the mobile zone and is given by the relation

$$K_{SEC} \cong \exp \frac{\Delta S^\circ}{R} \tag{4}$$

In this size exclusion mechanism, it is assumed that $\Delta H^\circ \cong 0$, i.e., there is no enthalpic contribution when the solute transfers from one

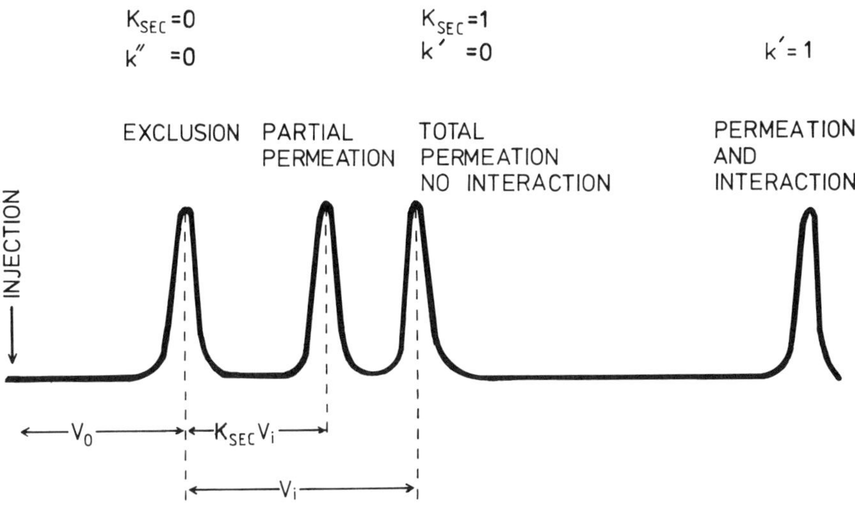


Fig. 2 Elution of peaks in gel permeation chromatography.

zone to the other and there is no solute interaction with the surface of the porous packing. For various pore and solute geometries [19-21], it can be shown that $K_{SEC}$ for polymer separations may be generalized in terms of the relation

$$K_{SEC} = \exp - \frac{s\bar{L}}{2} \tag{5}$$

where $\bar{L}$ is the mean external length or molecular projection, e.g., $\bar{L}$ is equal to the diameter D of a sphere, and s is the surface area per unit pore volume.

Equations (1) and (5) predict a universal calibration relation between solute size and $V_R$, and experimental results are typically presented as a plot of log (solute size) against $V_R$ (see Fig. 1). Benoit and coworkers [22] proposed that hydrodynamic volume $V_h$, which is proportional to the product of $[\eta]$ and M where $[\eta]$ is the intrinsic viscosity of the polymer in the SEC eluent and M is the molecular weight (or molar mass) of the polymer, may be used as the universal calibration parameter. For solid spheres and random coils represented by solid spheres, the product of $[\eta]$ and M is related to $V_h$ of the molecule by the Einstein equation

$$[\eta]M = 2.5N_0V_h = 2.5\pi N_0\frac{D^3}{6} \tag{6}$$

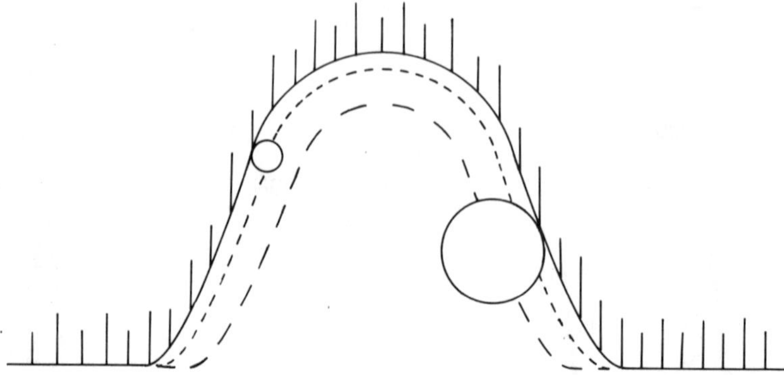

Fig. 3 Size exclusion in a circular pore in which the center of a hard sphere cannot approach the pore surface for distances below the sphere radius.

where $N_0$ is the Avogadro constant. If hydrodynamic volume, or a size parameter related to this volume such as the radius or diameter of the hydrodynamic sphere, controls the SEC separation, a plot of log $[\eta]M$ versus $V_R$ will be the same for all polymers. The experimental results in Fig. 4 for homopolymers and copolymers separating with crosslinked polystyrene gels and with tetrahydrofuran as eluent confirm the universal calibration method. Studies of universal calibration for a wide range of polymer-solvent-packing combinations have been performed [23]. Deviations from universal behavior may depend on polymer geometry, polymer concentration, solvent type, packing type, or ionic effects in aqueous separations. Procedures to minimize these deviations are well documented [23,24], and in the calibration plot shown in Fig. 5 for sodium polystyrene sulfonate and dextran separating on porous glass universal calibration behavior has been facilitated by the addition of sodium sulfate to the aqueous eluent [25]. This universal plot has been confirmed for porous silica by Rinaudo and coworkers [26,27] who recommended that the ionic strength of the eluent should exceed 0.05 M.

Equations (4) and (5) suggest that $K_{SEC}$ is independent of temperature, a characteristic of a mechanism controlled by entropy changes. The dependence of the relation between solute size (corrected for expansion or contraction) and $V_R$ on temperature may be studied with inorganic packings whose pore dimensions remain constant. Cooper and Bruzzone [28] eluted polystyrene and polyisobutene in trichlorobenzene from porous glass at 25 and 150°C, concluding that universal calibration behavior was observed. Therefore, little or no change in $V_R$ for a solute is expected as the temperature of a rigid packing

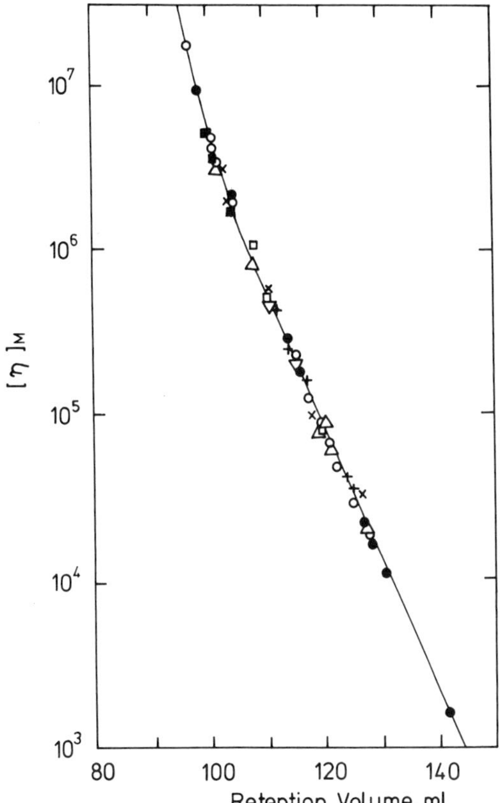


Fig. 4 Hydrodynamic volume universal calibration curve for cross-linked polystyrene gels with tetrahydrofuran. (●) Linear polystyrene; (○) branched polystyrene (comb type); (+) branched polystyrene (star type); (△) branched block copolymer of styrene-methyl methacrylate; (X) polymethyl methacrylate; (○) polyvinylchloride; (▽) graft copolymer of styrene-methyl methacrylate; (□) polybutadiene. (Reproduced from Ref. 22, © John Wiley and Sons, Inc.)

is changed. For soft gel packings with eluents which do not readily swell the network, substantial changes in pore volume and pore size may result on raising the temperature.

## Secondary Mechanisms

Although size exclusion dominates many polymer separations, many experiments indicate that interactions involving the pore surface con-

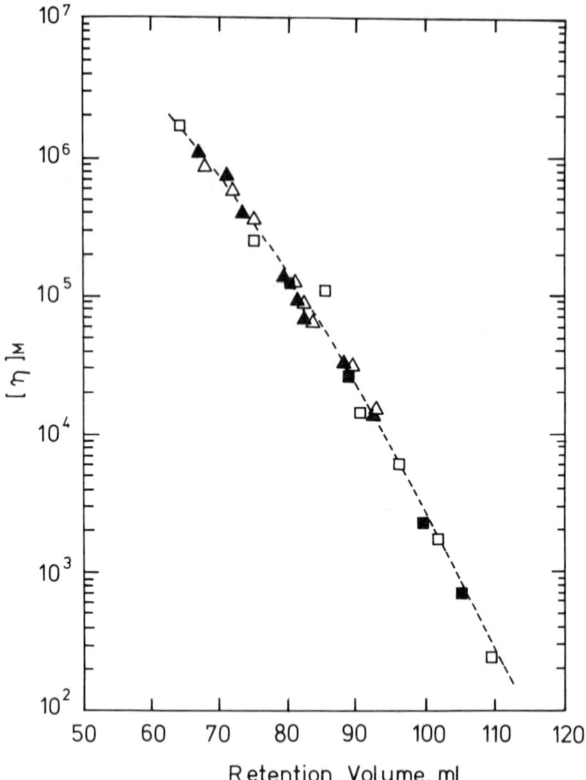


Fig. 5 Universal calibration for porous glass packing. (□ ■) Sodium polystyrene sulfonate; (△▲) dextran; open symbols, 0.2 M aqueous eluent; closed symbols, 0.8 M aqueous eluent. (Reproduced from Ref. 25, © John Wiley and Sons, Inc.)

tribute to retention. Separations with dimethylformamide (DMF), which is widely used as an eluent for polar synthetic polymers, are influenced by interactions between solute and the stationary zone [29]; see, for example, the retention data in Figs. 6 and 7 [30,31]. Studies of polymer separations with inorganic packings have indicated that the plot of log (solute size) vs. $V_R$ may be influenced by the eluent polarity with deviations from the universal calibration plot of hydrodynamic volume often being observed [12].

Separations in which $V_R$ is higher than expected from a size exclusion mechanism may arise from secondary partition (liquid-liquid) and adsorption (liquid-solid) mechanisms. Such interactions between polymer and stationary zone must be weak and reversible so that the polymer

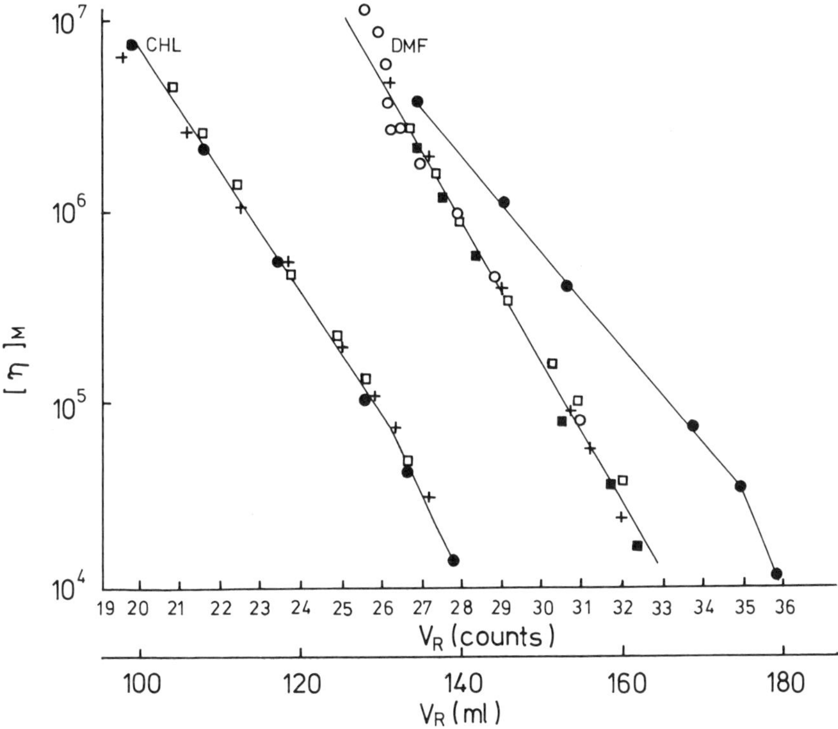


Fig. 6 Universal calibration plots for crosslinked polystyrene gels with chloroform and *N,N*-dimethylformamide at 20°C. (●) Polystyrene; (□ + ■) copolymers of styrene and acrylonitrile; (○) polyacrylonitrile. (Reproduced from Ref. 30.)

is not completely retained in the stationary zone. For some polymer-solvent-gel systems, $V_R$ is greater than $V_0 + V_i$, which is inconsistent with a size exclusion mechanism for which $K_{SEC}$ must lie between zero and unity. A typical example of polystyrene separating by a mixed mechanism consisting of size exclusion with a secondary adsorption or partition mechanism is shown in Fig. 6, in which the hydrodynamic volume curve for polystyrene in DMF is displaced with respect to a curve for the other polymers and copolymers in DMF separating solely by size exclusion. For water-soluble polymers, separation behavior may be dependent not only on adsorption but also on ionic effects. In addition, the size of a polyelectrolyte will be highly dependent on ionic strength. The pH of the eluent will influence the degree of ionization of the solute and groups on the column packing (e.g., hydroxyl groups on silica).

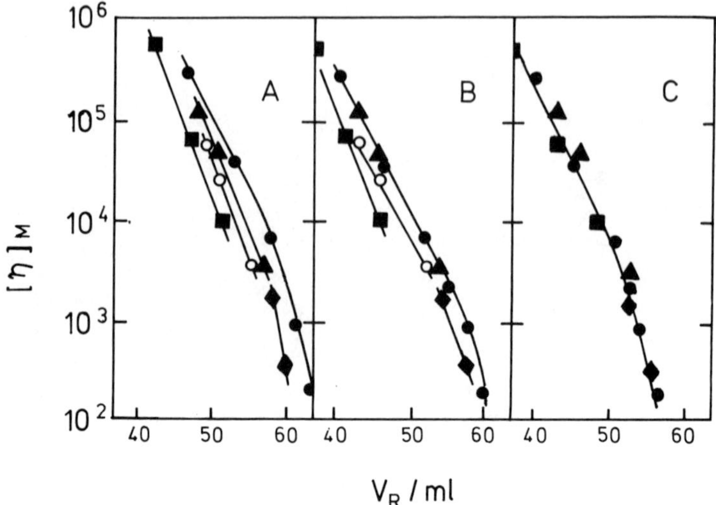


Fig. 7 Universal calibration plots for crosslinked polystyrene gels (A), silanized porous glass (B), and porous glass (C), with *N,N*-dimethylformamide. (●) Polystyrene; (▲) polymethylacrylate; (○) polyvinylpyrrolidone; (■) poly-*p*-nitrostyrene; (◆) polyethylene oxide. (Reproduced from Ref. 31, © John Wiley and Sons, Inc.)

A mixed mechanism consisting of size exclusion and a secondary mechanism may be considered as a network-limited separation [29] from which the following retention equation may be derived:

$$V_R = V_0 + K_p K_{SEC} V_i \tag{7}$$

where $K_p$ is the distribution coefficient for polymer-gel interaction. Therefore, Eqs. (1) and (7) are identical with $K_p = 1.0$ for a size exclusion mechanism with no interactions with the stationary zone. Equation (7) may be given thermodynamic interpretation [32]. Comparison of Eqs. (1), (3), (4), and (7) suggests that $K_p$ is given by

$$K_p \cong \exp \frac{-\Delta H°}{RT} \tag{8}$$

i.e., solute-gel interaction effects are determined predominantly by the standard enthalpy change $\Delta H°$ on solute transfer to the pore in the gel (as in adsorption mode of HPLC). It follows from Eqs. (7) and (8) that $K_p > 1.0$ corresponds to retardation of polymer in the stationary zone because of attractive interactions and that $K_p < 1.0$

corresponds to early elution of polymer because of repulsive interactions. Equation (7) has been shown to be a reasonable representation of a mixed mechanism for polymers in organic media [29]. Equation (8) clearly suggests that secondary mechanisms should be temperature-dependent, and data exhibiting a decrease in $K_p$ as T is raised are discussed elsewhere [32,33].

In the past decade considerable interest has been shown in separations of water-soluble polymers with microparticulate inorganic packings having a bonded phase [34,35], e.g., resulting from reaction with γ-glycidoxytrimethoxysilane (abbreviated here to γ-G). Columns containing γ-G silicas have been employed for separations of proteins in aqueous media at neutral pH [36-40]. Typical results [41] are shown in Fig. 8. Ovalbumin, albumin, and myoglobin are assumed to separate by size exclusion alone. Catalase and pepsin are assumed to separate by a mixed mechanism, and their retention behavior may be considered in terms of Eqs. (7) and (8). The γ-G silica will have negative charges on the pore surface because of unmodified hydroxyl groups which dissociate at pH ∿ 7. At this pH pepsin is negatively charged and will be

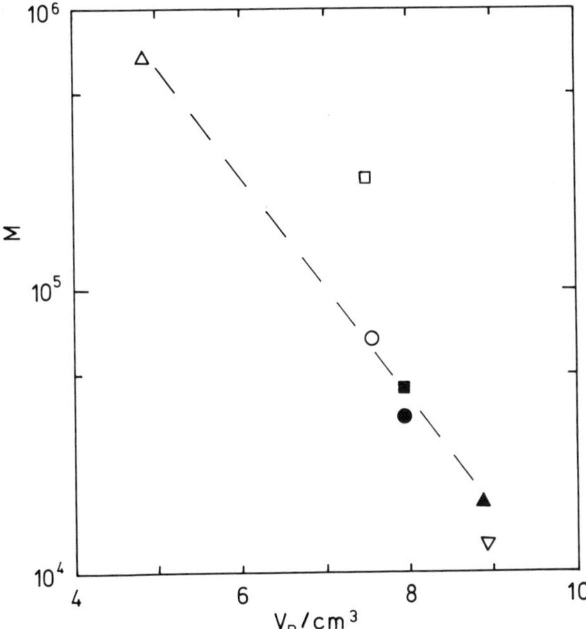


Fig. 8 Molar mass calibration for aqueous separations of proteins with γ-G silica. (△) Thyroglobulin; (□) catalase; (○) albumin; (●) pepsin; (▲) myoglobin; (▽) cytochrome c; (■) ovalbumin. (Reproduced from Ref. 41.)

subjected to ion exclusion by repulsive ionic interactions ($K_p < 1.0$). Catalase is close to a neutral protein at pH 6.3 and the retardation of this protein ($K_p > 1.0$) appears to arise because of hydrophobic interactions with the bonded phase. The pattern of protein retention behavior is significantly changed by varying the pH and ionic strength of the eluent which determine the magnitude of secondary interactions [38,39]. For polyelectrolytes interactions between charges on the surface of the column packing and ionic groups on the polymer may generate secondary mechanisms arising from ion exchange, ion exclusion, and ion inclusion [34,35]. The magnitude of these mechanisms is extremely dependent on the composition of the eluent. In the ion exclusion mechanism the position of the calibration curve for polymers small enough in size to enter pores can be shown to depend on the ionic strength of the eluent, a typical example [42] being shown in Fig. 9. Here sodium polystyrene sulfonate in water is prevented from entering pores because of electrostatic repulsion, whereas there is no contribution from ion exclusion when the electrolyte concentration is sufficiently high. Ion inclusion results because a Donnan equilibrium must be achieved even when an excluded high polymer containing charges cannot permeate pores. Consequently, low polymer is forced into the pores to equilibrate with counterions and is retarded in the column. This retardation may be reduced by adding electrolyte, but the electrolyte then becomes included generating "spurious" peaks at the total permeation volume of the column.

## PACKINGS

Essential conditions for the effective fractionation of polymers by SEC are that the pore sizes in the column packing be comparable to polymer sizes in solution and that the packings have substantial pore volume, typically $0.5 < V_i/V_0 < 1.65$ for macroporous packings. The pore size distribution in a given sample of a porous packing only gives a useful separation range, e.g., the plateau region in Fig. 1, of about one or two decades in molecular size, as illustrated by the calibration curves for rigid crosslinked polystyrene gels displayed in Fig. 10 [43]. Consequently, to separate samples over a wide size range, e.g., in Fig. 4, it is necessary either to use a single column containing several gels having various pore size distributions (see, for example, the mixed-gel calibration in Fig. 10), or to have a series arrangement of columns with each column covering a different molecular size range. Typical column dimensions for HPSEC are column length = 25-30 cm and internal diameter = 0.7-0.8 cm, and separations are often performed with three or four columns connected in series with short lengths of low-volume, narrow-bore tubing. It is advantageous for the pore size distribution to generate a linear relation between log M and $V_R$ in order to facilitate

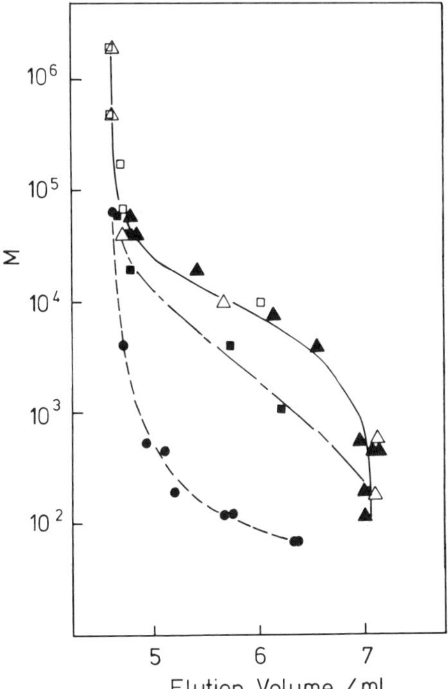


Fig. 9 Molar mass calibrations for aqueous separations of sodium polystyrene sulfonate and anionic standards (closed symbols) and dextran and nonionic standards (open symbols). (●○) 0.01 M aqueous eluent; (■□) 0.1 M aqueous eluent; (▲△) 1.0 M aqueous eluent. (Reproduced from Ref. 42, © John Wiley and Sons, Inc.)

molar mass analysis. This may be achieved in the bimodal pore size distribution approach [44] which involves coupling columns containing only two discrete pore size distributions and approximately equal pore volumes for each pore size distribution. Wide-range linear calibrations are possible for bimodal column sets with pore sizes differing by about one decade.

Rigid packings are generally employed in HPSEC. Microparticulate packings with particle diameters ∿10 μm are typically used for high polymers, with high resolution separations of low polymers, prepolymers, and small molecules being performed with particles having diameters of 4-6 μm. Separations of long-chain polymers with microparticulate packings must be performed carefully in order to avoid shear degradation during macromolecular diffusion through the column [45]. Guiochon and Martin [46] indicated how experimental conditions may be optimized to produce

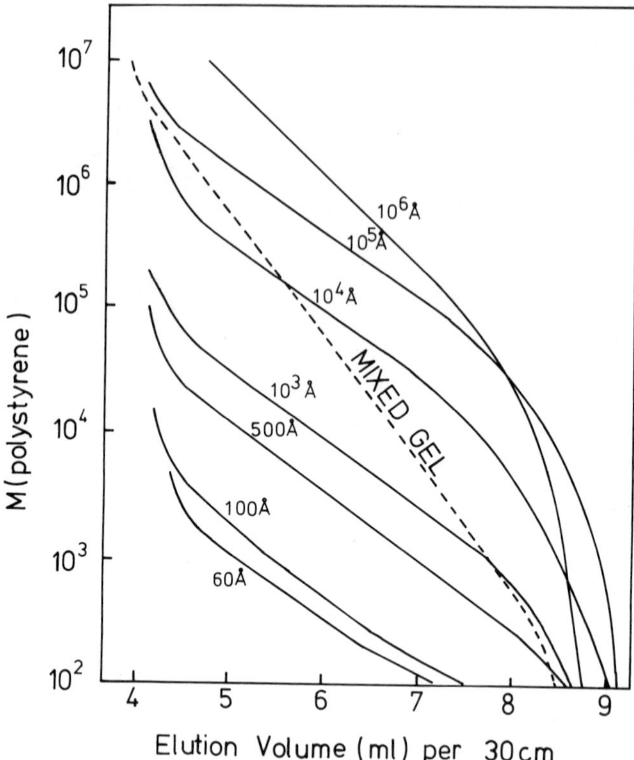


Fig. 10 Molar mass calibrations with polystyrene standards for cross-linked polystyrene-divinylbenzene packings (PLgel). (Reproduced from Ref. 43.)

very high-resolution separations of proteins with columns packed with 1- to 2-μm particles. Full details of rigid packings have been reviewed [4,13,40,43,47-49]. Slurry-packing methods are then preferred for column preparation; experimental packing methods are given elsewhere [4,49]. Many rigid packings are available in prepacked columns. The best SEC performance is obtained with columns containing regular microspheres having a narrow particle size distribution. The permeability of a homogeneous bed formed from such microspheres is optimized and the pressure drop is much lower than for a column packed with irregular particles having a wide distribution. Typically, the limiting conditions for columns containing rigid packings are pressure drop < 2000 psi and linear flow velocity <600 cm $h^{-1}$.

The choice of packing for the separation of a given polymer will have to consider the solvent for the polymer in order to assess whether polymer-eluent and eluent-packing interactions may influence secondary mechanisms. Ideally, the eluent should be a good solvent for the polymer, should permit high detector response from the polymer, and should wet the packing surface. The most common eluents in SEC are tetrahydrofuran for polymers which dissolve at room temperature, *o*-dichlorobenzene and trichlorobenzene at 130-150°C for crystalline polyolefins, and *m*-cresol and *o*-chlorophenol at 90°C for crystalline condensation polymers such as polyamides and polyesters. For more polar polymers, dimethylformamide and aqueous eluents may be employed, but care is required in avoiding solute-gel interaction effects. Secondary mechanisms are always likely to occur when polymer-solvent interactions are not favorable, when polar polymers are separated with less polar eluents, and when packings have active surface sites.

### Inorganic Packings

Rigid inorganic packings have numerous advantages in SEC separations. The packings do not swell and have excellent mechanical and thermal stability. Inorganic packings are much easier to handle than cross-linked organic gels and can be regenerated by heating or by treating with acids. The pore size distribution of rigid particles can be determined, e.g., by mercury porosimetry, so that theoretical studies of the separation mechanism can be compared with experimental SEC results. Furthermore, the volumes $V_i$ and $V_0$ are not dependent on eluent and temperature, so that the results from a variety of experiments can be compared.

Examples of inorganic packings are given in Table 1. Calibration curves reported in Ref. 50 for six Porasil porous silicas, with pore diameters in parentheses, are shown in Fig. 11. Columns of inorganic packings can be used at high flow rates and high pressures. Inorganic

**Table 1** Underivatized Inorganic Packings

| Packing | Fractionation range | Trade name |
|---|---|---|
| Porous glass | $100\text{-}4 \times 10^6$ (polystyrenes)<br>$1000\text{-}1.5 \times 10^6$ (polysaccharides)<br>$3000\text{-}8 \times 10^8$ (proteins) | CPG |
| Porous silica | Mean pore sizes from 6 to 2500 μm | Porasil<br>Fractosil<br>LiChrospher<br>SE Series<br>Spherosil<br>Zorbax |

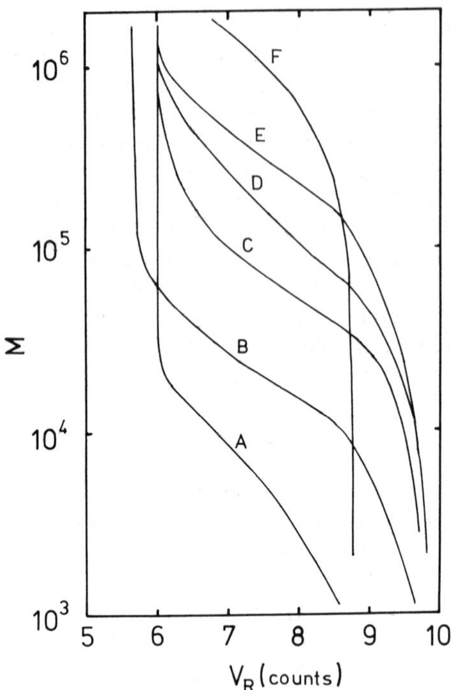


Fig. 11 Molar mass calibrations with polystyrene standards for Porasil silicas A (<100 Å), B (100-200 Å), C (200-400 Å), D (400-800 Å), E (800-1500 Å), F (>1500 Å). (Reproduced from Ref. 50.)

packings are suitable for both aqueous and organic eluents, and are particularly suitable for separations at high temperature, e.g., polyolefins. The main deficiency of inorganic packings is the presence of surface sites which may facilitate the interaction of some polymers with the packing. If the eluent has considerable affinity for the surface, then no polymer is adsorbed. This has been demonstrated with binary solvent mixtures in which one component is much more polar than the polymer separating on porous silica. The results of Berek and coworkers [51] for polystyrene in organic media displayed in Fig. 12 may be interpreted in terms of Eq. (7), since they observed separations by both size exclusion and adsorption when the binary solvent mixture was replaced by a single solvent having a polarity similar to that of the polymer. Also, adsorption is more prevalent with poor solvents, so that good solvents must be used when preferential eluent-adsorbent interactions are absent. The choice of eluent may be restricted because of polymer solubility considerations, when small quan-

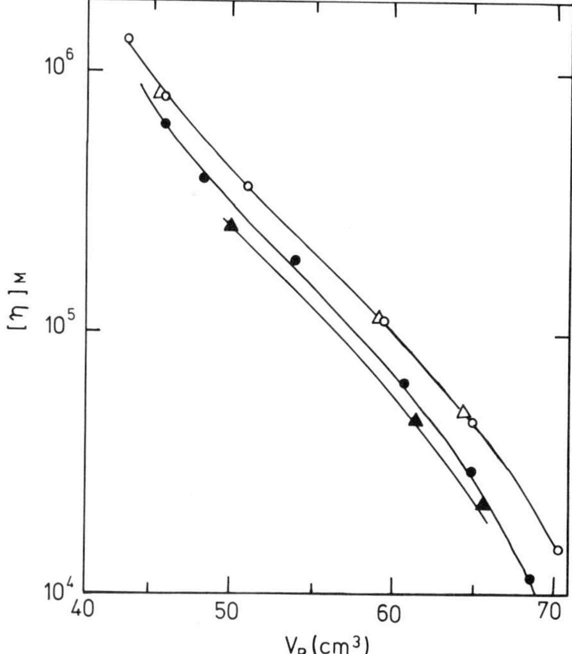


Fig. 12 Hydrodynamic volume universal calibrations with polystyrene standards for porous silica. (○) Benzene; (△) chloroform; (●) theta mixture benzene-methanol; (▲) theta mixture chloroform-methanol. (Reproduced from Ref. 51, © Huthig and Wepf Verlag.)

tities of an adsorption-active substance may be added to the eluent in order to suppress adsorption.

For polar polymers in both organic and aqueous phases, it is advantageous to coat inorganic packings with a surface bonded phase in order to minimize solute-gel interaction effects. Packings produced by reaction of porous silica or porous glass with γ-glycidoxypropyltrimethoxysilane have given size exclusion separations for a range of water-soluble polymers [34,35,40,47]. Inorganic packings with various bonded phases are given in Table 2. Calibration curves for aqueous separations of dextrans and polyethylene glycols with TSK gel type SW [52] are displayed in Fig. 13. Packings such as those in Table 2 have been advocated for SEC separations of biological macromolecules in aqueous media. However, three problems then have to be considered. First, not all hydroxyl groups are reacted, so ionic interactions may persist; second, interaction of the polymer with the bonded phase may occur at high ionic strength of the eluent via a

**Table 2** Inorganic Packings with Bonded Phases

| Packing | Bonded phase | Fractionation range | Trade name |
|---|---|---|---|
| Porous silica | $-CH_2CH(OH)CH_2O-$ | $100-2 \times 10^5$ (polyethylene glycols)<br>$10^3-1 \times 10^6$ (proteins) | TSK Gel SW |
| Porous silica | $(CH_2)_3OCH_2CH(OH)CH_2OH$ | $10^3-10^7$ (polysaccharides)<br>5000-500,000 (proteins) | SynChropak GPC<br>Aquapore<br>LiChrosorb Diol<br>LiChrospher Diol<br>Bio-Sil GFC<br>Aquachrom |
| Porous silica | Polyether | $500-10^6$ (polysaccharides) | μBondagel |
| Porous silica | Undisclosed | $600-5 \times 10^5$ (proteins) | Protein Column |
| Porous silica | Polymerized amine | (Similar SynChropak GPC) | SynChropak CATSEC |
| Porous glass | $(CH_2)_3OCH_2CH(OH)CH_2OH$ | (Similar SynChropak) | Glycophase CPG<br>Glyceryl CPG |

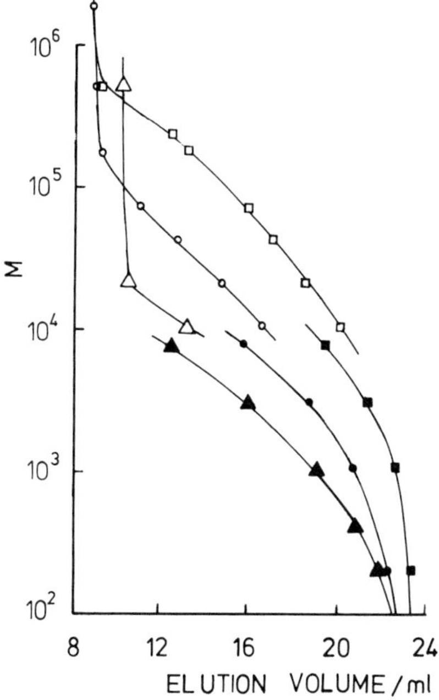


Fig. 13 Molar mass calibrations with dextran standards (open symbols) and polyethylene glycol standards (closed symbols) for TSK-Gel SW-type columns. (▲△) Gel G-2000SW; (○●) gel G-3000SW; (□■) gel G-4000SW. (Reproduced form Ref. 52.)

hydrophobic interaction; and third, dissolution of the silica surface, and therefore removal of the bonded phase, can occur for eluents with a pH above 8. For porous silica with the diol bonded phase, hydrophobic interactions are reduced by organic cosolvents [34,35]. Therefore, in order to achieve an aqueous SEC separation, in particular for polyelectrolytes, it is necessary to carefully select the eluent composition, e.g., by the addition of electrolyte or organic components, after assessing the nature of possible polymer-packing interactions.

## Polymer Packings: Macroporous Gels

Macroporous gel packings are highly crosslinked polymer particles which exhibit a predominantly permanent porosity with little or no swelling porosity (as defined in Chap. 2). These gels therefore have high rigidity and retain porosity in the dry state, permitting their use in HPSEC for a wide range of synthetic polymers. Some limited swelling may occur with selected eluents.

The crosslinked polystyrene gels in Fig. 10 and Table 3 are compatible with a wide range of organic eluents. It is preferable to use an eluent with a similar polarity to that of polystyrene, i.e., similar solubility parameter, when adsorption and partition effects are generally absent. Possible disadvantages of rigid organic gels are susceptibility to thermal degradation and a decrease in mechanical stability at the elevated temperatures which are required in separations of polyolefins and some condensation polymers.

Polymer packings which are hydrophilic and nonionic should provide a more versatile column packing for aqueous separations than inorganic packings. A range of polymer and protein fractionations [14,53,54] has been accomplished with TSK gel type PW (see Table 4 and Fig. 14). A comparison of the TSK gel types SW (see Table 2) and PW has been reported [55]. Both these column packings provided acceptable separations for dextran, polyvinyl alcohol, and polyvinyl pyrrolidone. For polyacrylamide, polyacrylic acid, and polyethyleneimine it was necessary to add salt to the eluent, but these polymers still displayed adsorption effects with the SW gel whereas satisfactory separation behavior was observed with the PW gel.

### Polymer Packings: Soft Gels

Soft gels have a high swelling porosity with little or no permanent porosity (as defined in Chap. 2). The control of pore size is the most important factor in the preparation of gel particles for GFC. To a first approximation, the porosity of a soft gel is related to its ability to swell. If a gel imbibes a large volume of solvent, large molecules will be able to penetrate a higher fraction of the pore volume. This degree of swelling is controlled by the amount of crosslinking agent in the gel preparation. The high degree of swelling generates much higher pore volumes for soft gels, with $V_i/V_0$ typically above 2, than for rigid macroporous gels. Consequently, a given gel is marketed in

Table 3 Rigid Polystyrene Packings

| Fractionation range | Trade name | Supplier |
|---|---|---|
| $100\text{-}10^7$ (polystyrenes) | PLaquagel | Polymer Laboratories |
| | μStyragel | Waters Associates |
| | Shodex A | Showa Denko |
| | HSG | Shimadzu |
| | TSK Type H | Toyo Soda |

Table 4 Hydrophilic Polymer Packings: Macroporous Gels

| Gel | Fractionation range | Trade name |
|---|---|---|
| Hydroxylated polyether copolymer | $1000-7 \times 10^6$ (polysaccharides)<br>$100-8 \times 10^6$ (polyethylene glycols) | TSK Gel PW<br>Bio-Gel TSK<br>Micropak TSK PW<br>Spherogel PW |
| Polyacrylamide | $100-10^5$ (polyethylene glycols) | PLaquagel |
| Poly-2-hydroxyethyl methacrylate | $20,000-5 \times 10^6$ (polysaccharides) | Spheron |
| Methacrylate glycerol copolymer | $<400,000$ (polysaccharides) | Shodex-OH pak |
| Sulfonated polystyrene | $<5 \times 10^6$ (polysaccharides) | Shodex Ion-pak |
| Polyvinylalcohol | $<5 \times 10^5$ | Asahipak GS |

various types with different swelling characteristics and different fractionation ranges [2]. The degree of swelling is also determined by the solvent, either aqueous eluent or a binary mixture of aqueous and organic solvents.

A listing of gels currently used in GFC is given in Table 5. These gels have higher particle diameters (∿50 μm for dry particles) than rigid packings. Typically, the limiting conditions for columns containing soft gel packings are pressure drop <20 psi and linear flow velocity <25 cm $h^{-1}$. These conditions result from the low mechanical stability of swollen soft gels, and consequently the packings in Table 5 are not suitable for HPSEC with short, low-capacity columns operated at fast eluent flow rates. Attempts have been made to improve the rigidity of soft gels, e.g., the Sephacryl and Ultrogel packings in Table 5 have increased stability because of additional crosslinking by polyacrylamide. A major advantage of soft gels is that the gel network appears to be more inert than the pore surfaces of rigid macroporous gels. Consequently, soft gels have been widely used for low-pressure SEC separations of biological macromolecules which require gel stability over a wide range of pH; typical values are 2-10 (Sephadex), 3-11 (Sephacryl), 1-10 (Bio-Gel P), 4-13 (Bio-Gel A), 4-9 (Sepharose), 3-14 (Sepharose CL), and 1-14 (Fractogel TSK). These gels also have to operate with eluents having high salt concentrations and in the presence of denaturing agents, such as urea and guanidine hydrochloride, and detergents, e.g., sodium dodecyl sulfate.

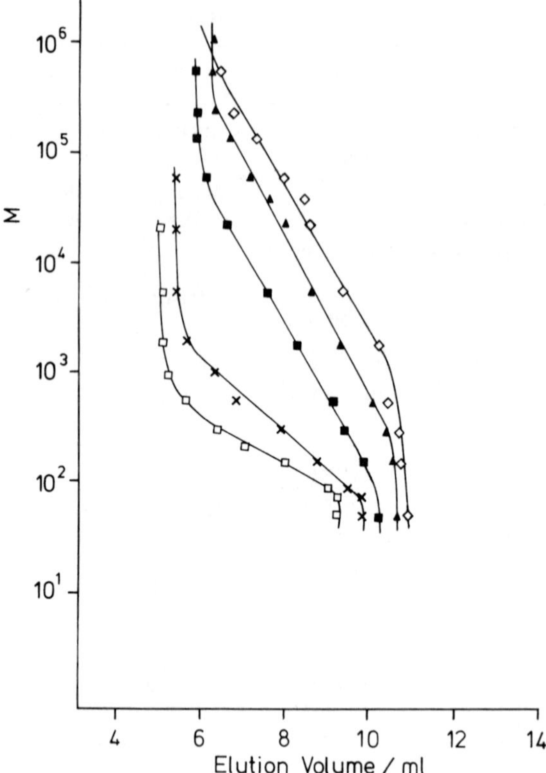


Fig. 14 Molar mass calibrations with polyethylene glycol standards for TSK-Gel PW-type columns. (□) gel 1000PW; (×) gel 2000 PW; (■) gel 3000PW; (▲) gel 4000PW; (◇) gel 5000PW. (Reproduced from Ref. 55.)

## COLUMN EFFICIENCY

### Plate Height and Number

A measure of the efficiency of a chromatography column is the height equivalent to a theoretical plate or plate height H [56]. The plate height for an experimental chromatogram is calculated from the expression:

$$H = \frac{L}{N} \tag{9}$$

where L is the column length and N is the plate number, as defined in Chap. 2. If the chromatogram is symmetrical, corresponding to a normal error (or Gaussian) function, then N may be determined from

Table 5 Polymer Packings: Soft Gels

| Gel | Fractionation range | Trade name |
|---|---|---|
| Dextran | 100-600,000 (proteins)<br>100-200,000 (polysaccharides) | Sephadex |
| Dextran-acrylamide | $5000-1.5 \times 10^6$ (proteins) | Sephacryl |
| Polyacrylamide | 100-400,000 (proteins) | Bio-Gel P |
| Agarose | $10^4-40 \times 10^6$ (proteins)<br>$10^4-20 \times 10^6$ (polysaccharides) | Bio-Gel A<br>Sepharose |
| Agarose, crosslinked | $10^4-40 \times 10^6$ (proteins)<br>$10^4-20 \times 10^6$ (polysaccharides) | Sepharose CL |
| Agarose-polyacrylamide | $6000-10^6$ (proteins)<br>$20,000-2 \times 10^6$ (polysaccharides) | Ultrogel |
| Polyacrylomorpholine | 100-20,000 (polysaccharides and polyethylene glycols) | EnzacrylGel |
| Hydrophilic vinyl polymer | $100-50 \times 10^6$ (proteins)<br>$100-10^6$ (polysaccharides) | Fractogel TSK<br>Toyopearl |

$$N = 5.54 \left[\frac{V_R}{w_{0.5}}\right]^2 \tag{10}$$

where $w_{0.5}$ is the width of the chromatogram at half its height. A typical microparticulate packing with particle diameter $\sim 10$ μm will generate a HPSEC column with $N > 20,000$ plates $m^{-1}$ for a solute eluting at $V_R = V_0 + V_i$, whereas a GFC column containing a soft gel packing with particle diameter $\sim 50$ μm will have $N \sim 1500$ plates $m^{-1}$.

## Resolution

If a polymer sample contains several species of very different sizes, then peaks for each monodisperse species will be obtained when w is minimized. For the case of two monodisperse solutes 1 and 2 having different sizes, as shown in Fig. 15, column resolution R is given by

$$R = \frac{2(V_{R2} - V_{R1})}{w_1 + w_2} \tag{11}$$

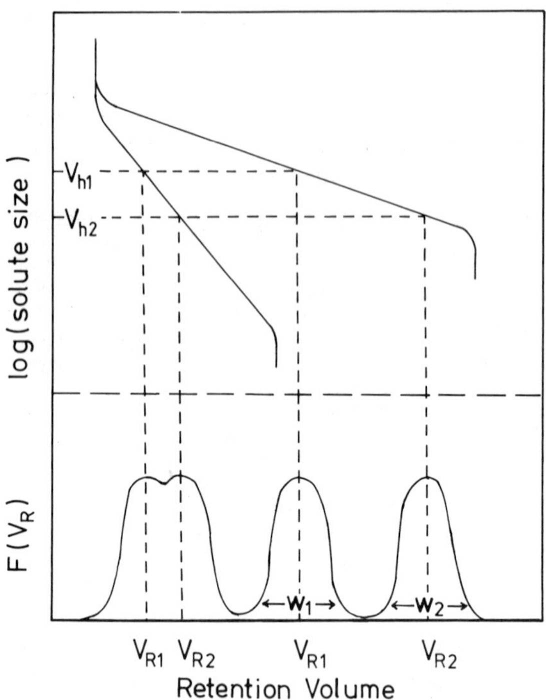


Fig. 15 Separation and resolution in size exclusion chromatography.

The numerator in Eq. (11) will depend on separation power, which is inversely proportional to the slope of the plot of log solute size vs. $V_R$ for permeating solutes in Fig. 1, and the denominator will depend on column efficiency according to Eq. (10). Therefore, an increase in separation power, or distance between the peak maxima in Fig. 15, is dependent on the pore size distribution of the packing, on gel capacity, i.e., $V_i/V_0$ which is larger for soft gels than for rigid gels, and on the column length L. Variables influencing the denominator in Eq. (11) will be discussed in the section on dispersion mechanisms. To achieve good resolution, R in Eq. (11) must be greater than unity, i.e., $(V_{R2} - V_{R1}) > w$. The definition of R in Eq. (11) can be extended by considering molar mass differences, according to the factor $M_2/M_1$ from Fig. 15, together with the slope $D_2$ of the linear semilog calibration. It follows, as shown by Yau et al. [57], that a general measure of SEC resolution may be developed, and they proposed the specific resolution $R_{SP}$ given by

$$R_{SP} = \frac{0.576}{D_2 \sigma} \tag{12}$$

where $\sigma$ is the standard deviation of a Gaussian peak with $4\sigma = w_1 = w_2$.

The value of $K_{SEC}$ is always between zero and unity (see Figs. 1 and 2) unless solute-gel interactions occur. This limited range of distribution coefficients is unique to SEC, since values exceeding unity are common in other forms of LC. Because of this restriction, the number n of components in a sample which can be resolved is related to column efficiency as defined by plate count. Giddings [58] suggested the following relation:

$$n \cong 1 + 0.2N^{0.5} \tag{13}$$

so that $n \sim 21$ for a column with 10,000 plates. The data in Table 6 clearly demonstrate that N must be maximized for high-resolution separations. Guiochon and Martin [46] indicated the possible three- to fourfold increase in peak capacity by the use of very small particles in the 1- to 2-μm range.

## Dispersion Mechanisms

Theoretical interpretations of column efficiency consider the influence of the mechanisms of solute dispersion in the mobile and stationary zones on the plate height H [56]. The basic concepts for general chromatographic separations can be applied to GFC and GPC, as proposed by Giddings and Mallik [59] who examined the dependence of H on the linear flow velocity u of the eluent. For separations of monodisperse high polymers [41,60], the dependence of H on u is given by Eq. (14).

$$H = 2\lambda d_p + \frac{R(1 - R)ud_p^2}{30D_s} \tag{14}$$

where $\lambda$ is a constant close to unity which depends on the packing, $d_p$ is the particle diameter, R is the retention ratio defined by $V_0/V_R$, and $D_s$ is the diffusion coefficient of the solute in the stationary zone. The first term in Eq. (14) results from the solute dispersion due to an eddy diffusion mechanism in the mobile zone. In this mechanism, some solute molecules are in mobile zone streamlines which move for some distance directly between particles, whereas other molecules are in streamlines whose path is obstructed by the particles and so the streamlines must go round the particles. The second term in Eq. (14) represents solute dispersion due to mass transfer in and out of the stationary zone. In this mechanism, a fraction of the molecules at any

**Table 6** Maximum Number of Peaks in Chromatography

| N | Peak capacity | | |
|---|---|---|---|
| | GPC | Gas | Liquid |
| 400 | 5 | 21 | 13 |
| 2,500 | 11 | 51 | 31 |
| 10,000 | 21 | 101 | 61 |

instant will be in the stationary zone and are left behind by the remaining fraction in the mobile zone.

The validity of Eq. (14) for H for high polymers may be discussed with reference to the experimental results [61] presented in Fig. 16. The polystyrene standard PS-1987000 may be regarded as a nonpermeating solute. The plate height data for this excluded solute shown in Fig. 16 suggest that for high polymers chromatogram broadening due to solute dispersion in the mobile zone exhibits little or no change with the eluent flow rate u. Consequently, it is concluded that the eddy diffusion term dominates mobile zone dispersion and that the mobile zone mechanisms of molecular diffusion and mass transfer [41], which both depend on u, may be neglected for macromolecular solutes. For the permeating polystyrenes in Fig. 16 mass transfer dispersion from the mobile zone to the stationary zone and vice versa increases considerably as u rises, and furthermore the slope of the plot of H against u increases as mass transfer increases with larger solutes having lower $D_S$. If Eq. (14) is related to Eqs. (9) and (10), it is evident that the denominator in Eq. (11) for column resolution will depend on gel particle size, eluent flow rate, solute size, and the porosity characteristics of the packing, which may restrict solute diffusion. Equation (14) clearly confirms the preference for microparticulat packings for high-performance separations because mass transfer dispersion is minimized.

Equation (14) presumes a monodisperse polymer. For a permeating polydisperse polymer, the width of an experimental chromatogram will depend on several variables, including the molar mass distribution of the polymer and chromatogram broadening arising in the column (or columns) from solute dispersion mechanisms. Consequently, it is necessary to include a polydispersity contribution to the experimental value of H. If it is assumed that the molar mass distribution of the polymer is represented by a logarithmic normal function [61], then H for a permeating polydisperse high polymer is given by

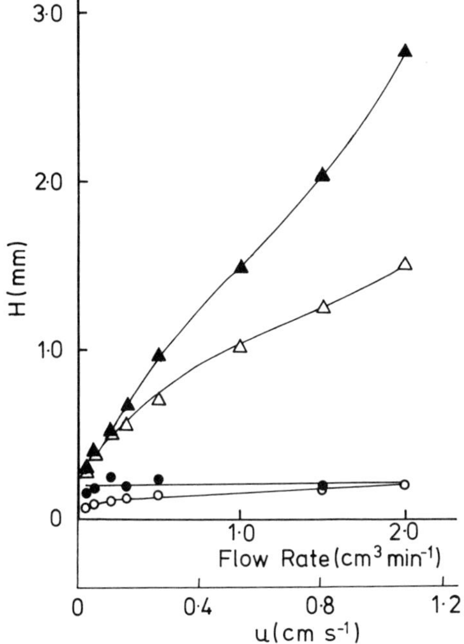


Fig. 16 Dependence of plate height on eluent flow rate for polystyrene standards with porous silica. (△) Polystyrene standard (M = 9800); (▲) polystyrene standard (M = 35,000); (●) polystyrene standard (M = 1,987,000); (o) toluene. (Reproduced from Ref. 61.)

$$H = 2\lambda d_p + \frac{R(1 - R)\, ud_p^2}{30D_s} + \frac{L \ln [\bar{M}_w/\bar{M}_n]_T}{D_2^2 V_R^2} \qquad (15)$$

where the true polydispersity $[\bar{M}_w/\bar{M}_n]_T$ is defined as the ratio of the weight average molar mass $\bar{M}_w$ and the number average molar mass $\bar{M}_n$, and $D_2$ is the slope of the SEC calibration relation between ln M and $V_R$. Procedures for estimating the eddy diffusion term in Eq. (15) have been discussed [41,60], so that it is possible to calculate $[\bar{M}_w/\bar{M}_n]_T$ from the experimental dependence of H on u with Eq. (15). This method for determining $[\bar{M}_w/\bar{M}_n]_T$ has demonstrated that reasonable values of polydispersity for polystyrene standards are obtained and that selected proteins may be regarded as monodisperse [41,60,61].

Equation (14) indicates that the chromatogram for a monodisperse solute is not a rectangle but a bell-shaped peak, so that a polydisperse polymer having a range of solute sizes generates a chromatogram which is a collection of a large number of overlapping peaks. This chromatogram broadening may require data correction methods when molar mass distributions are being determined by SEC [62,63]. Because of dispersion mechanisms, the tails of the chromatogram result from broadening alone and the solute concentration at a given retention volume depends on the component eluting at $V_R$ and on the broadening contributions from neighboring components. If the experimental chromatogram is represented by $F(V_R)$ and if $w(Y)$ represents the ideal chromatogram in the absence of broadening, i.e., as N tends to infinity, then these two functions are related by the equation proposed by Tung [64]:

$$F(V_R) = \int w(Y)\, G(V_R, Y)\, dY \tag{16}$$

where $G(V_R, Y)$ is a function describing the broadening contribution for an individual solute component having Y as its retention volume. The practical application of Eq. (16) involves choosing an appropriate function for $G(V_R, Y)$ and a numerical technique to solve the equation, but it is not necessarily easy to achieve reliable $w(Y)$ chromatograms with these correction methods. Equation (15) clearly suggests that reducing dispersion mechanisms will minimize chromatogram broadening, e.g., by using microparticulate packings (low $d_p$) and low eluent flow velocities (low u). Therefore, a chromatogram having a very low broadening contribution may be obtained by selecting the optimum chromatographic conditions, so that the calculations involved in Eq. (16) can be omitted.

## SEPARATIONS

In this section several selected examples will be presented to illustrate the scope of SEC. For polymers in organic media, many separations may be performed with crosslinked polystyrene gels, and the choice of eluent in order to minimize secondary retention mechanisms and to provide adequate detector response by the solute is generally straightforward. Consequently, examples of retention behavior will be restricted to aqueous SEC. Historically, the initial application of GPC was to the determination of molar mass distributions of high polymers, and so the use of HPSEC will be discussed. Considerations of selectivity arise in the SEC of small molecules, prepolymers, and low polymers when high-resolution and/or high-speed separations are required.

### Water-Soluble Polymers

Standards of dextran are widely used to characterize columns for aqueous SEC, and satisfactory elutions are obtained with most column packings. Calibration data for dextrans in Fig. 5 are in good agreement with the universal calibration principle, with electrolyte added to the aqueous eluent for separations with CPG porous glass. It is expected that nonionic hydrophilic polymers should be less influenced by eluent ionic strength than polyelectrolytes. However, some workers have indicated that low concentrations of salt should be present in the aqueous eluent because secondary electrostatic effects arising from possible negative charges on the dextran chain may occur with porous silica [65] and porous glass [66]. Satisfactory elutions of dextrans that have inorganic packings with bonded phases were obtained with eluents containing salts [67,68]. These observations for dextran indicate the separation conditions required to separate oligosaccharides, polysaccharides, and cellulosic polymers.

Standards of polyethylene glycol are available to calibrate columns for aqueous SEC. Satisfactory separation behavior is obtained for the hydrophilic polymer packings TSK gel type PW [14] and for PLaquagel [15]. These polymers adsorb on underivatized inorganic packings, but acceptable separation behavior has been reported for silicas with bonded phases [52,55,69,70]. Extreme care is required when selecting a separation system for N-containing polymers. While satisfactory separations of polyacrylamides with porous glass with an aqueous eluent containing added salt have been reported [71-74], some important samples of polyacrylamides for industrial applications will contain carboxylate groups which may generate interaction effects with underivatized porous silica even with electrolyte added to the eluent. Separations of polyacrylamide with inorganic packings with bonded phases [55,75,76] and with the TSK gel type PW appear to operate satisfactorily [14]. Polyvinyl pyrrolidone tends to adsorb strongly on underivatized inorganic packings. Even for some silicas having a bonded phase with an eluent containing an electrolyte, retardation of polyvinyl pyrrolidone may occur which may be minimized by adding an organic cosolvent, suggesting that retardation arose from hydrophobic interactions [77]. Reasonable separation behavior has been observed with the TSK gel types SW and PW [55].

Sodium polystyrene sulfonate is a typical example of an anionic polyelectrolyte. As long as electrolyte is added to the aqueous eluent, universal calibration behavior is observed for underivatized inorganic packings (see Fig. 5). The results in Fig. 9 demonstrate that the electrolyte concentration should not be too low, otherwise ion exclusion of sodium polystyrene sulfonate is observed [78]. Adsorption of sodium polystyrene sulfonate may occur with silica having a bonded phase [52], so electrolyte should always be added to the eluent [79]. Because of the negative charges generated by the dissociation of hydroxyl groups, porous silica is unsuitable for the separation of cationic polyelectrolytes. Attractive interactions also occur for cationic polyelectrolytes with porous glass. Attempts to use inorganic packings with bonded phases such as $-CH_2CH(OH)CH_2OH$ have not always been successful, e.g., one polyelectrolyte may be eluted when another is retained. Consequently, separations have been performed with porous silica having a repulsive surface charge, e.g., by incorporating a bonded phase containing quaternary ammonium groups [80-82]. While separations for individual cationic polyelectrolytes may be achieved, it is not always easy to minimize secondary mechanisms. An alternative procedure is to use a hydrophilic polymer packing which is uncharged. The TSK gel type PW would appear to be a reasonable packing for separating cationic polyelectrolytes [83,84].

### Molar Mass Distributions of High Polymers

Scientists working on the characterization of synthetic polymers will require precise information on molar mass distribution, and average molar masses, calculated from the experimental chromatogram. Consequently, the determination of the chromatogram will involve high-efficiency columns in order to minimize peak broadening, and an experimental calibration curve of log molar mass against $V_R$ which may be found accurately by appropriate methods [23]. It follows from Eq. (15) that the most accurate molar mass data for high polymers will be obtained at low eluent velocities with well-packed columns containing the smallest particles.

A typical molar mass distribution computed from a chromatogram obtained at low eluent flow rate is displayed in Fig. 17, where this experimental molar mass distribution is broader than a theoretical distribution predicted from the polymerization mechanism [85]. Even when mass transfer dispersion is minimized by performing separations at low u, chromatogram broadening will still be present because of the eddy diffusion term in Eq. (15). To demonstrate how the eluent velocity, and therefore mass transfer dispersion, influences the precision of average molar mass data, Dawkins and Yeadon [85] determined the molar mass distribution of a broad-distribution polystyrene as a function of u and computed values of $\overline{M}_w$ and $\overline{M}_n$ from the distribution obtained at each flow rate. The fall in $\overline{M}_w/\overline{M}_n$ as u decreases shown in Fig. 18 confirms that molar mass data for high polymers will not be accurate when fast eluent velocities are employed.

### High-Resolution Separations

It is clear from Eqs. (14) and (15) that the eluent flow velocity and the solute diffusion coefficient influence peak broadening, so that the chromatographic conditions will have to be chosen by considering whether high-speed or high-resolution separations are required. For small molecules the flow rate should not be extremely slow because of the dispersion mechanism arising from longitudinal molecular diffu-

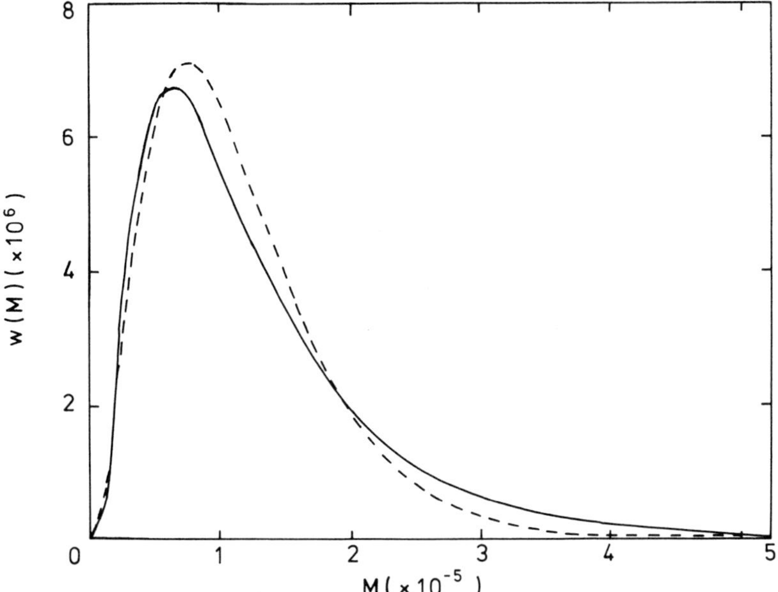


Fig. 17 Molar mass distribution for polydisperse polystyrene. (——) From experimental chromatogram determined by HPSEC at an eluent flow rate 0.1 $cm^3$ $min^{-1}$. (---) Theoretical distribution predicted from the polymerization mechanism. (Reproduced from Ref. 85.)

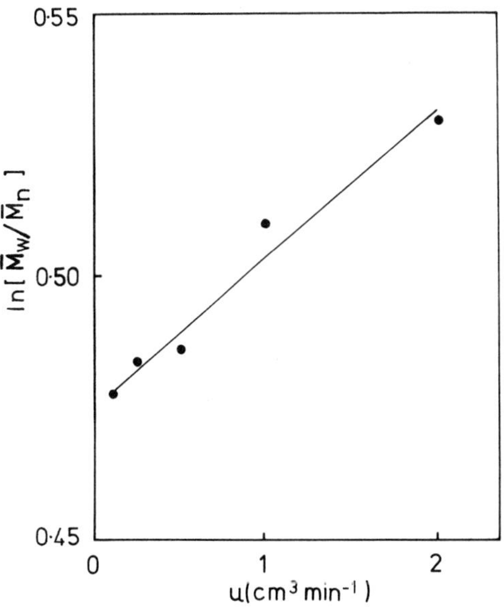


Fig. 18 Dependence of polydispersity of polystyrene on eluent flow rate. (Reproduced from Ref. 85.)

sion [56], which becomes less important for larger solutes having lower diffusion coefficients. For low polymers with high $D_S$, the mass transfer term in Eqs. (14) and (15) will not be too significant, and so high-speed separations may be performed with little loss in efficiency.

A typical example of a low-polymer separation is shown in Fig. 19 where well-resolved peaks corresponding to the individual components in an epoxy resin are produced [43]. The improved resolution which may be gained by lowering the particle diameter $d_p$ of the support in order to decrease both mobile zone and mass transfer dispersion [see Eq. (14)] is illustrated in Fig. 20. For complete characterization a chromatogram with nonoverlapping peaks is preferred, and this will require optimizing the range of u to lower H in order to find the

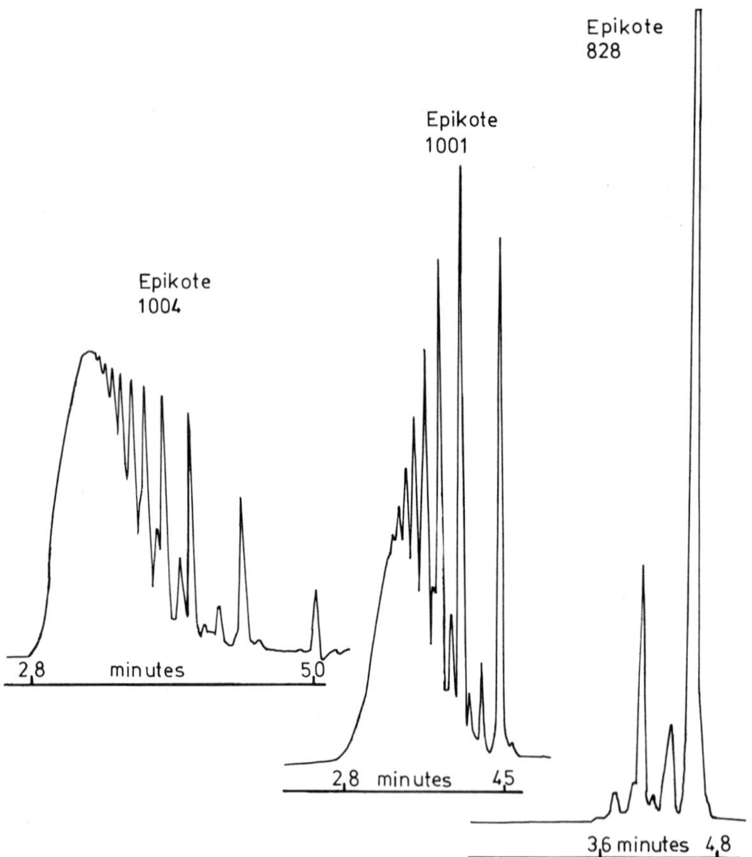


Fig. 19 Separation of epoxy resin prepolymers with PLgel packings (5-μm particle diameter). (Reproduced from Ref. 43.)

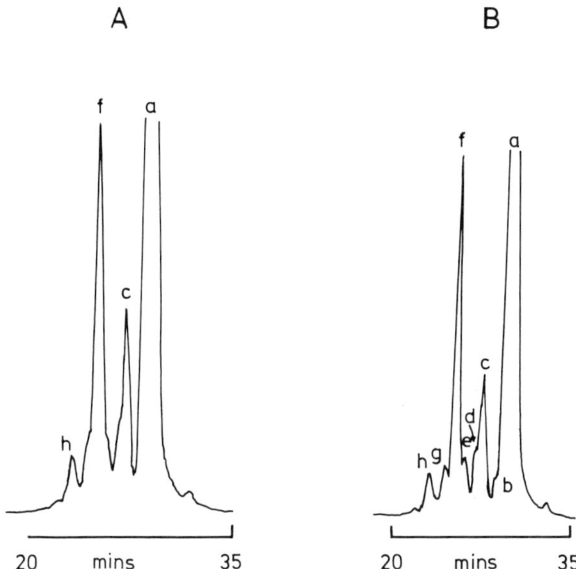


Fig. 20 Separation of epoxy resin prepolymer with PLgel packings. (A) 10-μm particle diameter; (B) 5-μm particle diameter. (Reproduced from Ref. 43.)

best combination of speed, efficiency, and column pressure drop. Further improvement in resolution involves raising separation power by increasing L to increase $V_2 - V_1$ in Eq. (11); this may be accomplished by adding further columns or by recycling techniques [4], but separation time will be increased.

An example of well-resolved peaks corresponding to polystyrene oligomers is shown in Fig. 21, demonstrating that fast, high-resolution separations of small molecules may be performed by HPSEC. Although SEC has a lower peak capacity than HPLC (see Table 6), Figs. 20 and 21 demonstrate that excellent separations of small molecules may be achieved by HPSEC as long as the number of components is not excessive. The separation of small molecules may be influenced by secondary mechanisms, but for some mixtures these mechanisms may provide greater peak resolution.

For permeating high polymers, it follows from Eqs. (14) and (15) that the most efficient separations should be performed at low u because of the pronounced increase in the mass transfer term at fast flow rates. Therefore, the smallest column packings in high-efficiency columns should be preferred for short separation times. This is illustrated for packings having $d_p$ = 5 μm in Fig. 22 by the resolution of

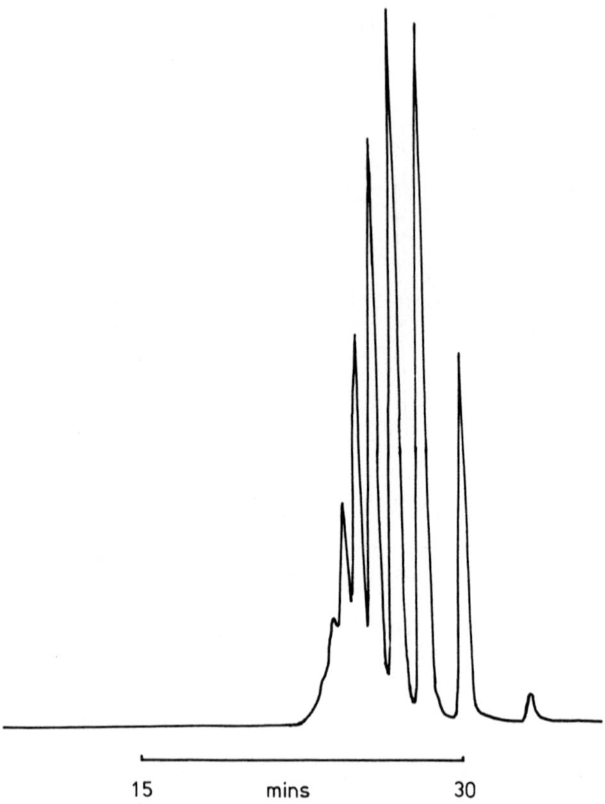


Fig. 21 Separation of oligomers in polystyrene 500 standard with PLgel packing (5-μm particle diameter). (Reproduced from Ref. 43.)

a mixture of polystyrene standards [43]. The eluent flow rate in this separation is not too high, not only to avoid extensive peak broadening for permeating high polymers at high u but also to minimize possible shear degradation of high polymers which may occur at fast flow rates [86]. High-resolution separations of monodisperse biopolymers may be performed with high-efficiency columns, and the chromatogram in Fig. 23 illustrates the fine performance which may be obtained with rigid packings for separations with aqueous eluents [87]. There is considerable interest in new microparticulate rigid packings for aqueous HPSEC because much of the analytical GFC work with soft gels may be accomplished by HPSEC with HPLC instrumentation. Fast routine analytical separations of many biological macromolecules are therefore possible.

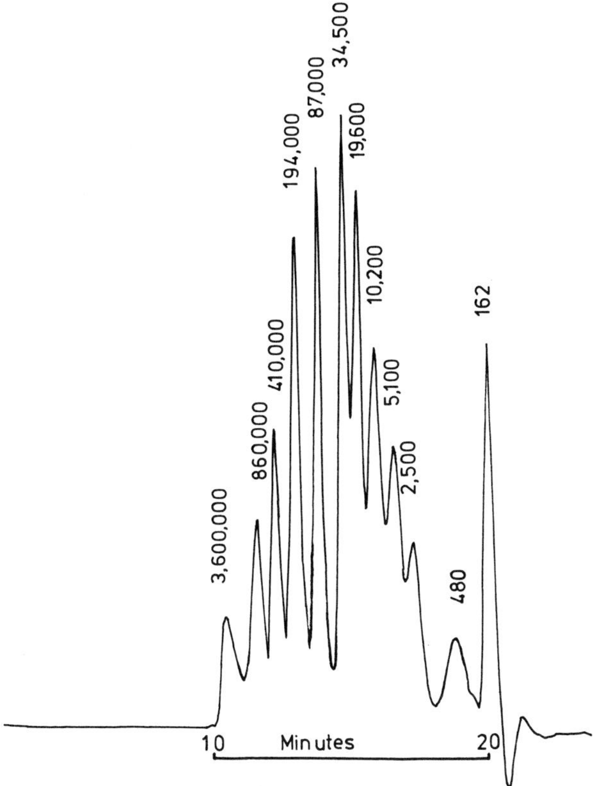


Fig. 22 Separation of polystyrene standards with PL mixed-gel packing (5-μm particle diameter) at an eluent flow rate 0.5 $cm^3$ $min^{-1}$. (Reproduced from Ref. 43.)

## CONCLUSIONS

SEC has advanced a long way since the GPC work reported by Moore [3]. The move to microparticulate packings has meant that SEC may be regarded as one of the main HPLC separation techniques. While packings for separations of synthetic polymers in organic media are now well established, it is evident that further research is required to develop improved packings for aqueous separations of biological macromolecules. It is expected, therefore, that further advances in the HPSEC technique will continue to attract interest.

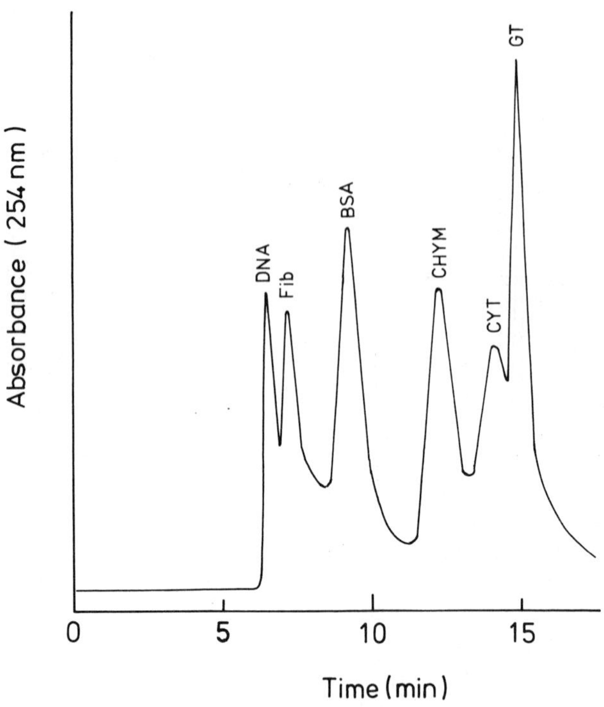


Fig. 23 Separation of biopolymers by aqueous HPSEC. DNA, deoxyribonucleic acid; Fib, fibrinogen; BSA, bovine serum albumin; CHYM, chymotrypsinogen; CYT, cytochrome c; GT, glycyltyrosine. (Reproduced from Ref. 87.)

## ACKNOWLEDGMENTS

The author wishes to thank Dr. Frank P. Warner, Polymer Laboratories Limited, Essex Road, Church Stretton, Shropshire, SY6 6AX for extremely helpful discussions and for kindly providing the results shown in Figs. 10, 19, 20, 21, and 22.

## REFERENCES

1. J. Porath and P. Flodin, *Nature, 183*:1657 (1959).
2. T. Kremmer and L. Boross, *Gel Chromatography: Theory, Methodology, Applications*, John Wiley and Sons, New York, 1979.
3. J. C. Moore, *J. Polym. Sci., Part A, 2*:835 (1964).
4. W. W. Yau, J. J. Kirkland, and D. D. Bly, *Modern Size-Exclusion Liquid Chromatography. Practice of Gel Permeation and Gel Filtration Chromatography*, John Wiley and Sons, New York, 1979.

5. J. Janca, *Steric Exclusion Liquid Chromatography of Polymers*, Marcel Dekker, New York, 1984.
6. A. J. de Vries, M. Le Page, R. Beau, and C. L. Guillemin, *Anal. Chem.*, *39*:935 (1967).
7. M. Le Page, R. Beau, and A. J. de Vries, *J. Polym. Sci.*, *Part C*, *21*:119 (1968).
8. R. Beau, M. le Page, and A. J. de Vries, *Appl. Polym. Symposia*, *8*:137 (1969).
9. W. Haller, *Nature*, *206*:693 (1965).
10. W. Haller, *J. Chem. Phys.*, *42*:686 (1965).
11. W. Haller, *J. Chromatogr.*, *32*:676 (1968).
12. J. V. Dawkins, *Chromatography of Synthetic and Biological Polymers Vol. 1* (R. Epton, ed.), Ellis Horwood, Chichester, England, 1978, p. 30.
13. R. E. Majors, *J. Chromatogr. Sci.*, *15*:334 (1977); *18*:488 (1980).
14. T. Hashimoto, M. Sasaki, M. Aiura, and Y. Kato, *J. Polym. Sci., Polym. Phys. Ed.*, *16*:1789 (1978).
15. J. V. Dawkins and N. P. Gabbott, *Polymer*, *22*:291 (1981).
16. H. Hatano, *J. Chromatogr.*, *332*:227 (1985).
17. C. F. Simpson, *Techniques in Liquid Chromatography*, Wiley Heyden, Chichester, England, 1982.
18. E. F. Casassa, *J. Phys. Chem.*, *75*:3929 (1971).
19. J. C. Giddings, E. Kucera, C. P. Russell, and M. N. Myers, *J. Phys. Chem.*, *72*:4397 (1968).
20. M. E. Van Krevald and N. Van Der Hoed, *J. Chromatogr.*, *83*:111 (1973).
21. E. F. Casassa, *Macromolecules*, *9*:182 (1976).
22. Z. Grubisic, P. Rempp, and H. Benoit, *J. Polym. Sci.*, *Part B*, *5*:753 (1967).
23. J. V. Dawkins, *Steric Exclusion Liquid Chromatography of Polymers* (J. Janca, ed.), Marcel Dekker, New York, 1984, p. 53.
24. S. Mori, *Steric Exclusion Liquid Chromatography of Polymers* (J. Janca, ed.), Marcel Dekker, New York, 1984, p. 161.
25. A. L. Spatorico and G. L. Beyer, *J. Appl. Polym. Sci.*, *19*:2933 (1975).
26. C. Rochas, A. Domard, and M. Rinaudo, *Eur. Polym. J.*, *16*: 135 (1980).
27. M. Rinaudo, J. Desbrieres, and C. Rochas, *J. Liq. Chromatogr.*, *4*:1297 (1981).
28. A. R. Cooper and A. R. Bruzzone, *J. Polym. Sci., Polym. Phys. Ed.*, *11*:1423 (1973).
29. J. V. Dawkins and M. Hemming, *Makromol. Chem.*, *176*:1795 (1975).
30. D. Kranz, U. Pohl, and H. Baumann, *Angew, Makromol. Chem.*, *26*:67 (1972).

31. P. L. Dubin, S. Koontz, and K. L. Wright, *J. Polym. Sci., Polym. Chem. Ed., 15*:2047 (1977).
32. J. V. Dawkins, *J. Polym. Sci., Polym. Chem. Ed., 14*:569 (1976).
33. J. V. Dawkins and M. Hemming, *Makromol. Chem., 176*:1815 (1975).
34. H. G. Barth, *J. Chromatogr. Sci., 18*:409 (1980).
35. P. L. Dubin, *Sep. Purif. Meth., 10*:287 (1981).
36. F. E. Regnier and R. Noel, *J. Chromatogr. Sci., 14*:316 (1976).
37. N. Becker and K. K. Unger, *Chromatographia, 12*:539 (1979).
38. H. Engelhardt, G. Ahr, and M. T. W. Hearn, *J. Liq. Chromatogr., 4*:1361 (1981).
39. D. E. Schmidt, R. W. Giese, D. Conron, and B. L. Karger, *Anal. Chem., 52*:177 (1980).
40. K. K. Unger and J. N. Kinkel, *Aqueous Size Exclusion Chromatography, J. Chromatogr. Library* (P. L. Dubin, ed.), Elsevier, Amsterdam, in press.
41. J. V. Dawkins and G. Yeadon, *Faraday Symp., 15*:127 (1980).
42. A. R. Cooper and D. P. Matzinger, *J. Appl. Polym. Sci., 23*:419 (1979).
43. F. P. Warner, Polymer Laboratories Limited, personal communication.
44. W. W. Yau, C. R. Ginnard, and J. J. Kirkland, *J. Chromatogr., 149*:465 (1978).
45. J. G. Rooney and G. Ver Strate, *Liquid Chromatography of Polymers and Related Materials*, 3rd ed. (Chromatographic Science Series, Vol. 19, J. Cazes, ed.) Marcel Dekker, New York, 1981, p. 207.
46. G. Guiochon and M. Martin, *J. Chromatogr., 326*:3 (1985).
47. K. K. Unger, B. Anspach, and H. Giesche, *J. Pharm. Biomed. Anal., 2*:139 (1984).
48. K. K. Unger, *Methods in Enzymology*, Vol. 104 (Enzyme Purification and Related Techniques, Part C, W. B. Jakoby, ed.), Academic Press, New York, 1984, p. 154.
49. B. G. Belenkii and L. Z. Vilenchik, *Modern Liquid Chromatography of Macromolecules* J. Chromatogr. Library Vol. 25), Elsevier, Amsterdam, 1983.
50. G. D. Wignall, G. W. Longman, M. Hemming, and J. V. Dawkins, *Colloid Polym. Sci., 252*:298 (1974).
51. D. Berek, D. Bakos, T. Bleha, and L. Soltes, *Makromol. Chem., 176*:391 (1975).
52. K. Fukano, K. Komiya, M. Sasaki, and T. Hashimoto, *J. Chromatogr., 166*:47 (1978).
53. Y. Kato, H. Sasaki, M. Aiura, and T. Hashimoto, *J. Chromatogr., 153*:546 (1978).
54. Y. Kato, K. Komiya, M. Sasaki, and T. Hashimoto, *J. Chromatogr. 193*:311 (1980).

55. T. V. Alfredson, C. T. Wehr, and L. Tallman, *Polymeric Separation Media* (A. R. Cooper, ed.), Plenum Press, New York, 1982, p. 123.
56. J. C. Giddings, *Dynamics of Chromatography, Part 1, Principles and Theory,* Marcel Dekker, New York, 1965.
57. W. W. Yau, J. J. Kirkland, D. D. Bly, and H. J. Stoklosa, *J. Chromatogr., 125*:219 (1976).
58. J. C. Giddings, *Anal. Chem., 39*:1027 (1967).
59. J. C. Giddings and K. L. Mallik, *Anal. Chem., 38*:997 (1966).
60. J. V. Dawkins and G. Yeadon, *J. Chromatogr., 206*:215 (1981).
61. J. V. Dawkins and G. Yeadon, *J. Chromatogr., 188*:333 (1980).
62. N. Friis and A. Hamielec, *Adv. Chromatogr., 13*:41 (1975).
63. A. E. Hamielec, *Steric Exclusion Liquid Chromatography of Polymers* (J. Janca, ed.), Marcel Dekker, New York, 1984, p. 117.
64. L. H. Tung, *J. Appl. Polym. Sci., 10*:375 (1966).
65. R. A. Buytenhuys and F. P. B. Van der Maeden, *J. Chromatogr., 149*:489 (1978).
66. S. N. E. Omorodion, A. E. Hamielec, and J. L. Brash, *J. Liq. Chromatogr., 4*:41 (1981).
67. H. D. Crone and R. M. Dawson, *J. Chromatogr., 129*:91 (1976).
68. H. G. Barth and F. E. Regnier, *J. Chromatogr., 192*:275 (1980).
69. H. Engelhardt and D. Mathes, *J. Chromatogr., 142*:311 (1977).
70. Y. Kato, K. Komiya, H. Sasaki, and T. Hashimoto, *J. Chromatogr., 190*:297 (1980).
71. S. N. E. Omorodion, A. E. Hamielec, and J. L. Brash, *Am. Chem. Soc. Symp. Series, 138*:267 (1980).
72. C. J. Kim, A. E. Hamielec, and A. Benedek, *J. Liq. Chromatogr.,* 5:1277 (1982).
73. N. Onda, K. Furusawa, N. Yamaguchi, M. Tokiwa, and Y. Hirai., *J. Appl. Polym. Sci., 25*:2363 (1980).
74. J. Klein and A. Westerkamp, *J. Polym. Sci., Polym. Chem. Ed., 19*:707 (1981).
75. R. Biran and J. V. Dawkins, *Eur. Polym. J., 20*:192 (1984).
76. C. D. Chow and G. L. Jewett, *J. Liq. Chromatogr., 3*:419 (1980).
77. D. P. Herman, L. R. Field, and S. Abbott, *J. Chromatogr. Sci., 19*:470 (1981).
78. A. R. Cooper and D. S. Van Derveer, *J. Liq. Chromatogr., 1*:693 (1978).
79. R. V. Vivilecchia, B. G. Lightbody, N. Z. Thimot, and H. M. Quinn, *J. Chromatogr. Sci., 15*:424 (1977).
80. C. P. Talley and L. M. Bowman, *Anal Chem., 51*:2239 (1979).
81. H. Stickler and F. Eisenbeiss, *Eur. Polym. J., 20*:849 (1984).
82. A. Domard and M. Rinaudo, *Polym. Commun., 25*:55 (1984).
83. P. L. Dubin and I. J. Levy, *J. Chromatogr., 235*:377 (1982).

84. P. L. Dubin, I. J. Levy, and R. Oteri, *J. Chromatogr. Sci.*, *22*:432 (1984).
85. J. V. Dawkins and G. Yeadon, *Polymer*, *20*:981 (1979).
86. H. G. Barth and F. J. Carlin, *J. Liq. Chromatogr.*, *7*:1717 (1984).
87. F. E. Regnier and K. M. Godding, *Anal. Biochem.*, *103*:1 (1980).

# 8

# Packings in Donor–Acceptor Complex Chromatography

**Helfried Hemetsberger** / *Institute of Organic Chemistry II, Ruhr University, Bochum, Federal Republic of Germany*

## INTRODUCTION

The formation of molecular complexes in solution by the association of certain molecules is a well-documented phenomenon that is still the topic of intense study. A large number of books [1] and reviews [2] have been published on this subject. Molecular complex formation has been widely applied in analytical chemistry, e.g., for the preparation of derivatives of certain organic compounds, spot tests [3], determination of molecular masses [4], etc. By early 1949 Godlewicz published the first report on a liquid chromatographic separation using molecular complex formation. In this work the fractionation of a hydrocarbon mixture on silica, which was impregnated with *s*-trinitrobenzene, was studied [5].

Molecular complexes are formed by weak interactions between the species as there are dipole-dipole, dipole-induced dipole (polarization), and dispersion forces. Many of these molecular complexes show an additional absorption in the UV-VIS spectra which are not present in the spectra of the pure, constituent components. These additional absorptions can be found at longer wavelengths in the spectra and are assumed to originate from an electron transfer from one component, which is considered to be the donor (D), to the other component, which is the acceptor (A) within the molecular complex by the impact of a light quantum.

The components D and A form a donor-acceptor complex (DAC) reversibly in the electronic ground state, according to Eq. (1).

$$A + D \rightleftharpoons AD \quad (1)$$

The forward and backward reactions proceed very fast at normal conditions, i.e., 300 K, as was shown by NMR experiments [6]. The position of the equilibrium is determined by the association constant $K^{AD}$ which depends on the temperature and, particularly, on the polarity of the solvents. The values of $K^{AD}$ are in the range of 0-100 liters $mol^{-1}$—larger values are observed occasionally—and are high only in solvents with low dielectric constants.

Since formation of donor-acceptor complexes has been successfully used in liquid chromatography, it seems reasonable to outline the factors which influence the stability of the donor-acceptor complexes in detail and to evaluate the types of complexes which are of concern.

Molecular orbital theories have been successfully applied in predicting the most stable geometry of donor-acceptor complexes of intermediate strength [7]. Weak interactions can be treated by the perturbation theory [8]. The energy level diagram of the molecular orbitals of the donor (D) and the acceptor (A) is presented in Fig. 1. As a result of the theoretical treatment of the interactions of the orbitals between the donor and the acceptor molecule, it was found that the interactions are strongest when the MO energy differences between the highest doubly occupied orbital (HOMO) of the donor and the lowest unoccupied orbital (LUMO) of the acceptor are small.

Because of the particular order of the energies of the MOs of the donor and the acceptor as shown in Fig. 1, the energy difference between the HOMO of D and the LUMO of A is smaller than any other energy differences between occupied and unoccupied MOs. Absorption of a light quantum will transfer an electron form the HOMO of the donor to the LUMO of the acceptor within the complex leading to the charge transfer absorption at long wavelength.

Since we are dealing in chromatography with complexes in the electronic ground state, we are not interested in these charge-transfer absorptions in particular, but rather in knowing the factors which influence the stability of donor-acceptor complexes. Morukuma investigated the formation of DAC using *ab initio* SCF molecular orbital calculations [9]. Using this treatment it could be shown that the total interaction energy can be separated into five ocmponents: electrostatic, polarization, exchange repulsion, charge-transfer, and coupling. It was found that the electrostatic interaction is the most important contribution to the energy stabilization of the complex in most cases. These electrostatic interactions lead to smaller distances between the donor and the acceptor, which in turn causes an increase in polarization and orbital interactions.

Hydrogen bonding is a very important interaction in many chromatographic processes. Applying the same procedure as above to hydrogen bonding reveals that at equilibrium geometries the electrostatic

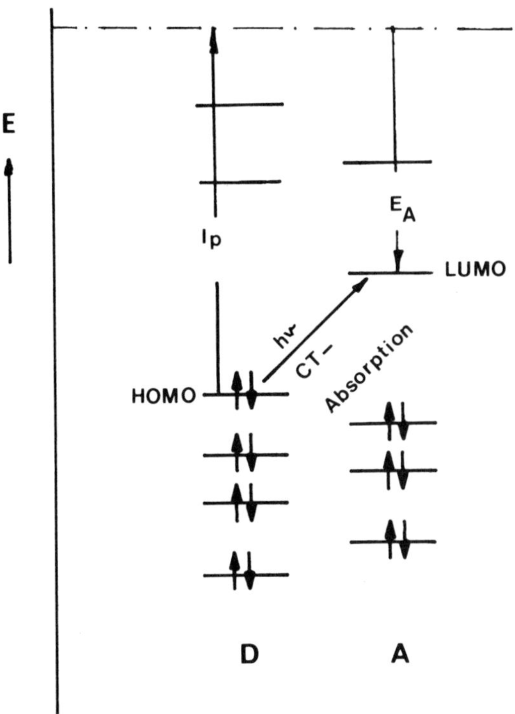


Fig. 1 MO interaction diagram of a donor and an acceptor.

interactions are close to the total interaction energies and control the most favorable approach of the components.

For the solution of a chromatographic problem using DAC formation, it is not necessary to go into further theoretical details. But it is helpful to realize that the energy of the HOMO of the donor corresponds to the negative value of the ionization potential $I_p$ and the energy of the LUMO of the acceptor to the electron affinity $E_A$. A chromatographic stationary phase with acceptor properties should be suitable for the separation of donor solutes. It can be expected that such a phase will retard solutes the strongest, which have low ionization potentials, since these solutes will form the strongest molecular complexes. Contrary, if a stationary phase is used for the separation of acceptor solutes, it can be expected that the solutes with the lowest LUMO energy—i.e., highest $E_A$—will be eluted latest. This is of interest to the analysts who apply DAC chromatography using a stationary phase possessing either donor or acceptor properties to predict the retention order, if no other factors, e.g., sterical factors, take influence on the complex stability.

Donor-acceptor complexes are classified by spectroscopists by the type of the observed charge-transfer absorption in the spectra as $\pi - \pi^*$, $n - \pi^*$, $n - \sigma^*$, $d - \pi^*$, where the first Greek letter signifies the type of the HOMO of the donor and the second the type of the LUMO of the acceptor. For chromatographic purposes, $\pi - \pi^*$ and $n - \pi^*$ interactions are most important, since we are interested in chromatographic separations of compounds possessing extended $\pi$ systems and/or lone-pair electrons in most cases.

Many different terms have been proposed for this type of adsorption chromatography. Klemm and Reed [10] used "molecular complexation" chromatography whereas Porath et al. [11-13], Mikes et al. [14-17], and Nondek [18] preferred charge-transfer chromatography. Hemetsberger and Holstein adopted the term "donor-acceptor" chromatography, since the contribution of charge transfer to complex formation and stabilization is of minor importance in the electronic ground state of the complexes [19].

## CHROMATOGRAPHIC CONSIDERATIONS

The application of the concept of DAC formation in chromatographic separations can be realized in different ways.

(1) Acceptor compounds can be layered or chemically bonded to the surface of an insoluble matrix. This yields a stationary phase appropriate to the separation of solutes possessing donor properties.

(2) Donor compounds are immobilized on the surface of an adsorbent either by impregnation or by covalent chemical bonding yielding a stationary phase for separation of acceptor solutes.

(3) A stationary phase without donor or acceptor properties, as C18 phase under reversed phase conditions, may be used and acceptor compounds may be added to the mobile phase.

If donor compounds are to be separated by chromatography, complexes of different stoichiometric compositions will be formed with the acceptor compounds in the mobile phase leading to a marked change in the retention behavior and order of elution. This concept was applied by Schomburg et al. [20], who added $Ag^+$ ions to the polar mobile phase and separated different alkenes successfully. Many examples of ligand exchange chromatography can be included here too.

(4) A stationary phase without donor or acceptor properties, as C18 phase under reversed phase conditions, may be used and donor compounds may be added to the polar mobile phase.

Since donor compounds which are apt to form DACs are poorly soluble in polar solvents, this chromatographic technique is not very promising, and only few reports are dealing with this mode of DAC chromatography [21-23].

If donor or acceptor compounds are only layered to an adsorbent and are not chemically bonded, a portion of these compounds will be dissolved in the mobile phase too. This leads to the well-known phenomenon of "bleeding" of the columns. Most of the acceptor compounds which are apt to form DACs are only sparingly soluble in polar and nonpolar solvents, so "stripping off" of the stationary phase is not a real problem in these cases. To the contrary, donor compounds are well soluble in nonpolar solvents in most cases. Consequently, the layered stationary phase will bleed off, and this mode of DAC chromatography is not very useful.

From the considerations outlined above on solubility of donor and acceptor compounds, the limitations of the different procedures become evident, and full advantage of the specific interactions can only be taken if chemically bonded stationary phases are used, since the full span of polarities of the mobile phases can be utilized.

In the following only examples belonging to paragraph 1 and 2 will be discussed, and chemically bonded stationary phases will be stressed over layered ones. The range of applications of DAC chromatography can be further increased if different matrices are used, i.e., organic polymers [11-13] or rigid gels as silica with different pore sizes, and if mobile phase compositions are varied.

In principal, DAC chromatography may be viewed as a special kind of adsorption chromatography. In Fig. 2 the mobile phase carries the donor solutes through the packing of the chromatographic column, and the donor molecules will be retained by complex formation, comparable to adsorption chromatography.

The capacity factors k' can be defined in DAC chromatography as $k' = n_{AD}/c_D V_m$, where $n_{AD}$ is the number of moles of the donor adsorbed by complex formation, $c_D$ is the molar concentration of the donor in the mobile phase, $V_m$ is the volume of the mobile phase, and

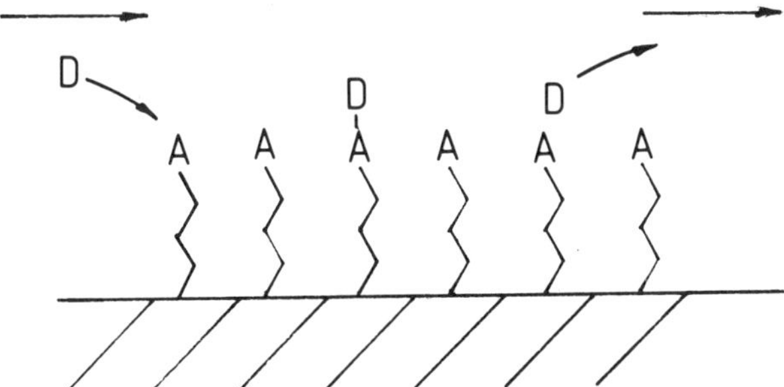


Fig. 2 Simplified picture of the surface of an adsorbent with chemically bonded acceptor molecules and donor solutes in the mobile phase.

the product $c_D V_m$ equals the number of moles of the donor in the mobile phase. The ratio of $n_{AD}/c_D$ is roughly proportional to $K^{AD} n_A$, the product of the association constant measured in solution and $n_A$ the number of moles of acceptor which is layered or chemically bonded to the surface of the matrix. As the first approximation, we expect that D will be retained more, if $K^{AD}$ is large and the number of moles of A high. This assumption could be verified qualitatively by Hemetsberger et al. [19], who measured the complex association constants of some aromatic compounds with *O*-allyl-2,4,5,7-tetranitrofluorenone-oxime in solution and chemically bonded. In a different approach, Lochmüller et al. [24] came to the same result. They found that the capacity factors of some aromatic donor compounds increased if the number of nitro groups on the acceptor molecules at constant stationary phase mass was increased. The increase of the number of nitro groups is accompanied by an increase of the association constant. Awareness of this relationship is helpful in the selection of suitable acceptors, donors, and solvents for a particular separation problem.

In recent years particular attention has been paid to packing materials which can be used in high-performance liquid chromatography (HPLC). Packing materials which are appropriate for use in HPLC should have a rigid matrix to avoid reduction of the permeability with the increase of pressure, and the stationary phase should be chemically bonded to the matrix to prevent stripping off. In almost all cases the donor and acceptor phases were bonded to silica by covalent bonds. In general, the bonding of the phase to the surface of a silica matrix is performed by reaciton of the surface silanol groups with some reactive groups of an organosilane. Because of steric hindrance it is very difficult and almost impossible to attain all silanol groups. Consequently, one is confronted with the problem of these unreacted silanol groups being active sites and contributing to the overall retention, especially if nonpolar solvent mixtures are used in the chromatographic process. Some investigations were performed to evaluate the contribution of the unprotected silanol groups to the retention mechanism. Matlin et al. [25] studied a picroamidopropyl silica phase and found that the retention of polycyclic hydrocarbons was increased over uncovered silica and defined a "binding constant" as a measure for the contribution fo DAC interaction on the retention similar to an approach of Harvey and Halonen [26]. These results can be taken as evidence that DAC complex formation contributed to the retention. Unfortunately, the amount of the contribution of the silanol groups to the overall retention is difficult to estimate. But one should be aware that the effect of the silanol groups will not only be important with solutes having polar functional groups but will show up with compounds which are easily polarizable as polycyclic hydrocarbons. The association constant of DACs correlates well with the ionization potential (vide supra). Therefore it is of interest when the capacity factors correlate with the ionization potentials as well. Nondek performed this investigation using a 3-(2,4-dinitroanilino)propyl silica phase and found an excellent correlation between $(\log k' + 1.28)^{-1}$ with the ionization potentials for eight polycyclic hydrocarbons [27]. With some larger polycyclic hydrocarbons the correlation failed, and larger capacity factors were observed, possibly as the result of multiple-site interactions.

In adsorption chromatography the selection of the composition of the solvent mixture used as mobile phase is of the utmost importance to gain the desired resolution of the individual components, whereas in reversed phase chromatography the separations are governed mostly by the polarity of the mobile phase. The effect of the polarity of the solvents on the association constants of donor-acceptor complexes was studied by Fialkov and Borovikov [28] and by Hemetsberger et al. by chromatographic methods [19]. Since the capacity factors became very small with DE values greater than 6, the chromatographic resolution became small as well, and an increase of the mobile phase polarity did not appear promising at first glance. Consequently, it is very impressive to view the results of chromatographic separations of a reaction mixture which contains compounds possessing donor properties on a chemically bonded 2,4,5,7-tetranitrofluorene acceptor phase starting at the nonpolar end of the eluotropic series of solvents and ending at the polar side as demonstrated in Fig. 3.

It is remarkable that the ordering of the elution which is reversed commonly in going from normal phase to reversed phase conditions is not changed in these series of chromatographic experiments. Under reversed phase conditions hydrophobic interaction will dominate the retention behavior. But the elution order of the compounds implies that donor-acceptor complex formation occurs even in polar solvents [29]. DAC formation in polar solvents could be demonstrated as well by the appearance of CT bands in the absorption spectra [30]. The altered and most often opposite elution order compared to the order in RP chromatography was observed by a number of research groups [25,32,33,41,69,114].

These results, although not understood entirely, signify that chemically bonded phases either with acceptor or donor properties are real alternatives for the separation of solutes for which the formation of DACs is possible, since the retention order does not correlate with the chemically bonded hydrocarbon phases as C8 or C18 which are mostly used.

DAC chromatography with stationary phases using polynitroaromatic compounds is primarily suited for the separation of unsaturated cyclic hydrocarbons as PAHS or azapolycyclic hydrocarbons possessing no functional groups to interact with the stationary phase. Solutes, which show no donor properties, can often be easily converted to appropriate derivatives exhibiting the desired characteristics. Primary or secondary amines can be chromatographed as α- or β-napthamides where the naphthalene moiety behaves as donor group [31-33]. In-

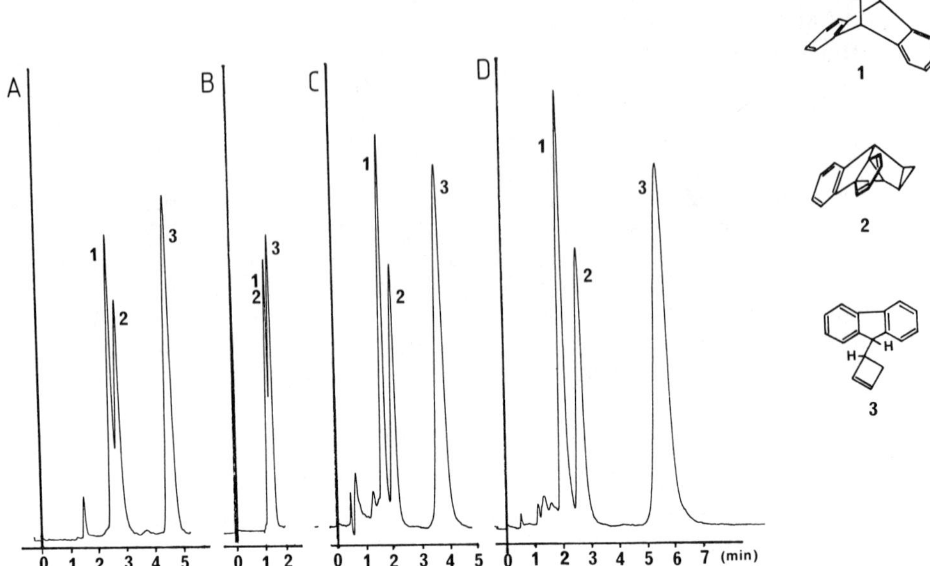


Fig. 3 Elution order and effect of solvent polarity on the separation of the compounds 1-3 shown in the figure. Column 250 mm × 4.1 mm i.d.; flow rate 3 ml $min^{-1}$. (A) *n*-hexane/methylene chloride (50:50); (C) MeOH; (D) MeOH-$H_2O$ (95:5).

versely, if a stationary donor phase is to be used, acceptor properties can be introduced into the solutes by conversion into dinitrobenzoic acid derivatives [34].

An important development in chromatography represents the ability of direct resolution of enantiomers without taking circuitous routes with diastereomeric derivatives. The chiral acceptor (-)-A, layered on the surface of the sorbent or chemically bonded to the support, interacts with the racemic donor solute, (+/-)-D, forming diastereo-

(−)–A -- (+)–D

(−)– A + (±)–D ⇌ [K(+)] / ⇌ [K(−)]

(−)–A --- (−)–D

meric DACs of different stability. Different association constants $K^{AD}$ caused by different geometries of the complexes (-)-A/(+)-D and (-)-A/(-)-D will result in different capacity ratios in the chromatographic process.

Newman and coworkers were the first to use DAC formation for the separation of enantiomers. They separated racemic mixtures via their diastereomeric complexes with (+)- or (-)-2,4,5,7-tetranitro-9-fluorenylideneaminooxypropionic acid (TAPA) (1) as acceptor by fractional crystallization according to the different stabilities and solubilities of the complexes [35-37]. Later attempts by Newman to separate racemic mixtures by chromatography applying optically active TAPA chemically bonded to an ester polymer were unsuccessful [38]. The first example of a successful separation of enantiomers by donor-acceptor complex formation was achieved by Klemm and Reed in 1957 [39]. They could present a separation of naphthyl-2-butylether and 3,4,5,6-dibenzo-6,10-dihydrophenanthrene on silica which was impregnated with optically active TAPA.

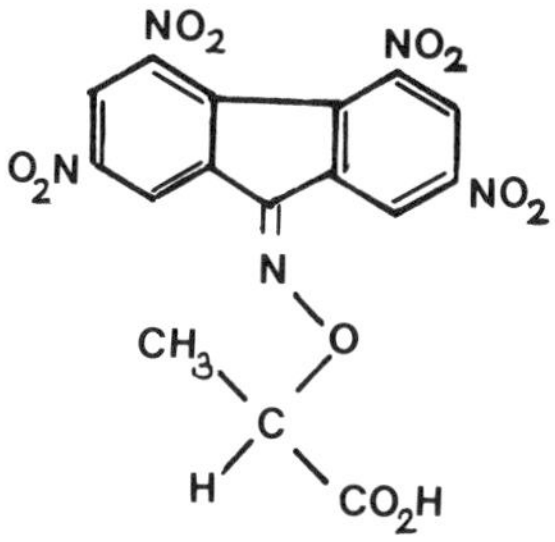

Later the three-point interaction model (three points are the minimum) originally stated by Dalgliesh in 1952 was used as the basis for the rational design of a fair number of chiral stationary phases [40]. Pirkle et al. [41-43] synthesized chemically bonded stationary phases with anthryl and naphthyl groups as donor and dinitrobenzamide groups as acceptors. The necessary three-point contacts with the chemically bonded stationary phase were believed to be the result of two hydrogen bonds and a $\pi$-$\pi^*$ donor-acceptor interaction [41-43].

In subsequent work the model was challenged, and it was proposed that dipole and Van der Waals interactions must contribute too [44]. In an investigation of the separation of alkylaryl carbinols and their acetates on a dinitrobenzamidoglycine phase it was found that the enantiomers retained more strongly in each series differ in absolute configuration, implying that the molecular model for chiral recognition was not reliable in this case [45].

A theoretical and practical evaluation of donor-acceptor chromatography can be found in review articles [46,47]. Special reviews cover the direct chromatographic separation of enantiomers and are treating the donor-acceptor complex chromatography in part [48,49].

## ELECTRON ACCEPTOR-COATED SUPPORTS AS STATIONARY PHASES IN HPLC

The first report using donor-acceptor chromatography in column chromatography was published in the early days of 1949 by Godlewics. He impregnated silica with trinitrobenzene and used this packing for the separation of lubricating oils in four fractions, containing hydrocarbons, partially hydrogenated aromatic compounds, aromatics, and resinous compounds. The color change caused by charge-transfer absorption was used to monitor the separation [50].

Sondheimer and Pollak [51] used 7-(2,3-dihydroxy)theophilline as coating agent. This compound is similar in structure to caffeine, which forms complexes with a variety of aromatic compounds but is less soluble in strong mobile phases when used in preparative chromatography. In the separation of cinnamic acid and methoxylated cinnamic acids using cyclohexane/chloroform = 3:1 as mobile phase, less than 10 mg/liter of the acceptor was soluble. In that case it did not pose serious problems in the preparative separations [51].

Karger and coworkers were the first to use this chromatographic technique in HPLC. They used a column packed with Corasil I impregnated with 2,4,7-trinitrofluorenone (2) and investigated the effect of water content in the mobile phase, the degree of coating on retention, and the influence of temperature on peak shape. The increase in temperature was accompanied by a decrease in capacity factors and selectivity values. The increase in the amount of coating agent from 0.05% to 1% led to a severe reduction in efficiency, but the selectivity values were found to be larger. The capacity factors obtained for water saturated heptane as mobile phase were markedly lower than for dry solvents [52].

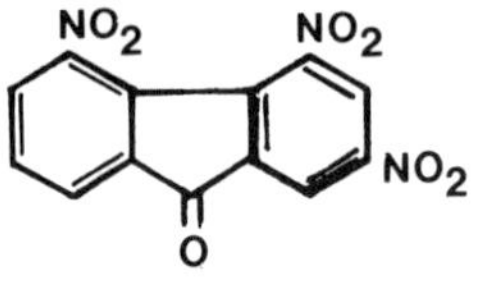

Other $\pi$-electron acceptors besides polynitroaromatic compounds were used as well. Burger and Tomlinson coated silica particles with tetracyanoethylene, a strong π-electron acceptor, and studied the changes produced in the retention times of electron donor compounds in HPLC and related these results to donor-acceptor complex association constants [53].

Klemm and coworkers applied DAC chromatography using silica impregnated with chiral (+)- or (-)-TAPA for the optical resolution of ethers and esters containing one center of chirality on a carbon atom for each compound [54]. In a later report, Klemm et al. determined the absolute configuration of (+)-TAPA in a well-defined sequence of chemical transformations and designed a molecular model for the geometries of the diastereomeric donor-acceptor complexes with (R)-(-)- and (S)-(+)-9,10-dihydrobenzo[c,g]phenanthrene of known absolute configuration [55]. Based on this molecular model they determined the relative stability of the diastereomeric complexes and explained the elution order of the enantiomers of helicenes on a silica column coated with (S)-(+)-TAPA. This model was used and discussed by several authors to evaluate the elution order on chiral TAPA and related columns [10,56-58].

In later work, Mikes and Gil-Av used an in situ technique to impregnate silica with 10-25% of TAPA and the homologic (tetranitro-9-fluorenylideneaminooxy)butyric acid (TABA), -isovaleric acid (TAIVA), and -hexanoic acid (TAHA) as complex-forming stationary phases. They used these columns in HPLC to separate racemic helicenes ([5] to [14]) and two double helicenes with nonpolar mobile phases [14,58]. The bulkiness of the group at the chiral center of the acceptor was found to be critical for the success of the resolution of the donor compounds. The racemic [6] to [14] helicenes and the double helicenes could be resolved completely in the enantiomers using TAPA; on the other hand, the enantiomers of the [5] helicenes could only be separated on layered TABA homologue phases with several recycling steps. The resolution factors $k'_2/k'_1$ were reported for all compounds, and it was found that in general they increase with increasing number of rings in the helicenes. A minimum was found with [10] and [11] helicenes on all stationary phases. The temperature effect was also studied. The resolution factors of the racemic [5] and [6] helicenes were found to be maximal around 0°C, lower factors were found with increasing and decreasing temperatures.

Packings consisting of sorbents coated with optically active TAPA were used by a fair number of groups to achieve chromatographic separations of racemic mixtures thereon. Wynberg and coworkers were able to resolve racemic heterohelicenes [59]. Haenel and Staab separated racemic 2.2 (2,6)-naphthalinophanes and determined the absolute configuration [60]. Rebafka and Staab reported the successful separation of enantiomers of intramolecular quinhydrones [61]. Krebs and coworkers were able to resolve the optical isomers of 3,4,7,8-dibenzocyclooctyne with *n*-pentane as eluent [62].

## CHEMICALLY BONDED ELECTRON ACCEPTOR PHASES

Coating of sorbents with π-electron acceptor compounds has many disadvantages. The main one is the solubility of the electron acceptors in polar mobile phases. This problem can be best handled by chemical bonding of the π-electron acceptor moieties. Ayres and Mann solved this problem of chemical bonding in 1964 by reacting polystyrene with a mixture of fuming nitric and sulfuric acid into a polynitrostyrene resin of unknown degree of nitration for open tubular chromatography. They could demonstrate the separation of antracene and pyrene but observed extensive peak broadening [63].

A weak π-electron acceptor consists of a mononitrophenyl group. A commercially available nitrophenyl silica phase was investigated by Blümer and Zander in HPLC for the separation of a large number of biaryls, condensed aromatics, and aza aromatics. For the three classes of compounds three different correlations between the capacity factors and the number of C atoms could be found. Since the elution order of the polynuclear aromatic hydrocarbons (PAH) and the aza aromatics is not dependent on the number of C atoms, a specific separation of the classes of compounds became possible [64]. Nitrated aromatics were grafted to the surface of silica by Pecka and coworkers. These sorbents were efficient in reversed phase and normal phase chromatography, respectively. The packings had reproducible properties, and the amount of the chemically bonded stationary phase can be gradually increased. The authors were successful in the analytical and preparative separations of model mixtures of aromatic compounds [65].

Using silica appropriate for HPLC, Nondek and Malek reported the preparation of the phase modified by chemical bonding of 3-(2,4-dinitroanilino)propyl (DNAP) groups (3) [18]. Using this chemically

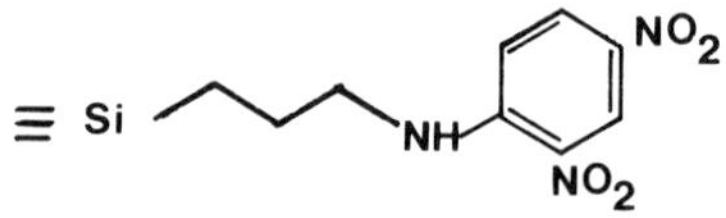

$\equiv$ Si NH $NO_2$ $NO_2$

3

bonded stationary phase, Nondek and coworkers examined the chromatographic behavior of a series of PAHs, alkyl aromatic hydrocarbons, and polyarylalkanes in a subsequent work. Interestingly, they found a linear correlation between the term 1/(log k' + 1.28) and the vertical ionization potential $I_P$ of medium-sized PAHs, providing evidence for donor-acceptor interactions of the solutes with DNAP groups [27]. Thomson and coworkers used the DNAP phase for compound class HPLC separation of aromatic hydrocarbons with 86 model compounds either expected or known to be present in petroleums, coal liquids, or shale oils. The importance of substituent and structure effects on separations was determined by varying the degree of alkyl and naphthenic substitution on the basic ring structure, and it was found in addition that the retention characteristic is strongly dependent on steric effects [66]. In a subsequent paper a comparison with a (2,4,6-trinitroanilino)-propyl (TNAP) phase (4) (vide infra) and a dinitroanilino group chem-

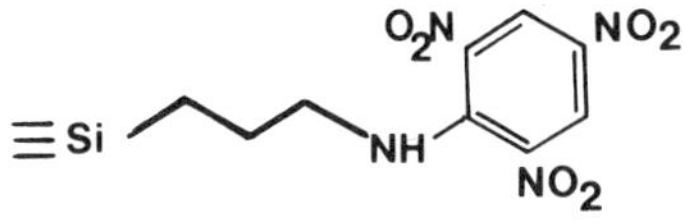

$O_2N$ $NO_2$
≡Si NH
$NO_2$

4

ically bonded with a $C_8$ spacer to silica was performed. The findings confirm the results of the earlier work, and the effect of the spacer length was only minor [67]. Two chiral chemically bonded phases were prepared in which dinitroaniline groups were attached near the chirality center. Lochmüller and Ryall [68] used L-alanine (5a), whereas Matlin and coworkers [69] introduced L-phenylalanine (5b). Both phases were useful for the resolution fo enantiomeric mixtures of helicenes; 5a could separate azahelicenes, whereas the separation of dicyanohexahelicenes was successful on 5b.

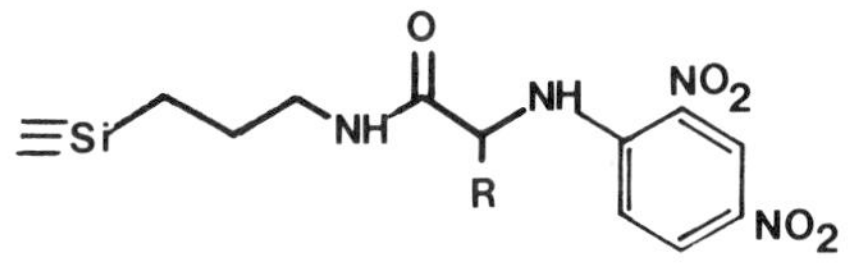

O
≡Si NH NH $NO_2$
R
$NO_2$

5a: R = Me; 5b: R = Ph

According to the same principles, Matlin and coworkers prepared phase (4) by reaction of aminopropyl silica with picrylchloride. Because of the larger number of nitro groups picrylamide is a stronger $\pi$-electron acceptor than dinitroaniline. As expected, aromatic compounds were strongly retarded on TNAP-silica compared with aminopropyl silica. The k' values were found to increase with the number of aromatic rings of solutes except for species which were hindered to attain planarity of steric reasons as 9,9'-bifluorene and 9,10-dibenzylanthracene. These authors also established that the elution order of the solutes was roughly the same in nonpolar (cyclohexane/ethylacetate) and polar ($H_2O$/MeOH) solvents [25,69]. This phase was evaluated as well in other laboratories for the ability of separation of complicated mixtures of polycyclic hydrocarbons [70,71].

A different approach, which turned out to be extremely successful, was found by Pirkle and coworkers. They used 3,5-dinitrobenzamide as π-electron acceptor and linked this group near the chirality center of phenylglycine or leucine. This unit was fixed to aminopropyl silica either covalently (Pirkle, phase 1) [41] (6a, 6b) or only ionically (Pirkle, phase 2) (7a, 7b) [42]. The Pirkle phases were used for the resolution of racemic mixtures of a wide variety of π-donor compounds.

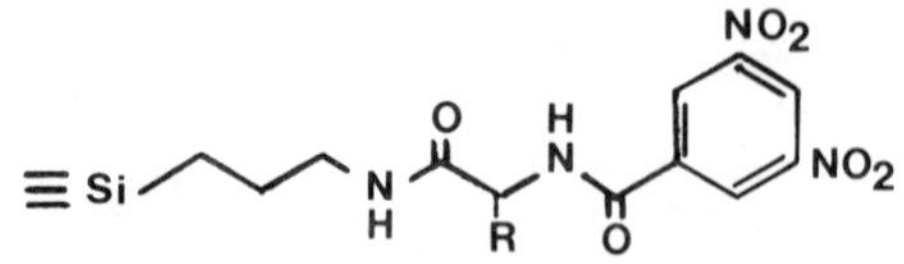

6a: R= Ph; 6b: R= iso-Butyl

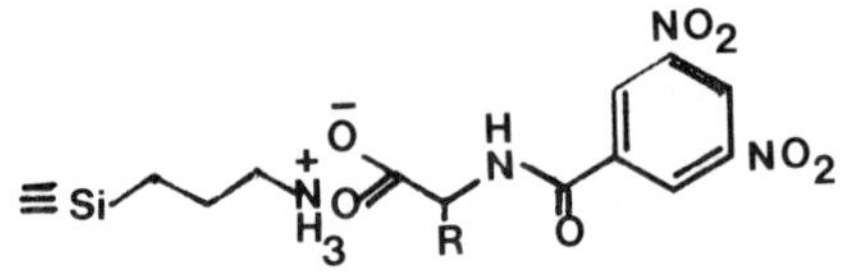

7a: R = Ph; 7b: R= iso–Butyl

Substituted (9-anthryl)trifluoromethylcarbinols [41] bi-β-naphthols [43,72] arylakyl carbinols [42] α-hydroxy acids, hydroxyindanes, hydroxytetralines, hydroxyamines, sulfoxides and hydroxysulfoxides, lactones and lactams could be resolved, if proper π-electron donors were present in the solutes [34,43]. This phase was used successfully for the separation of enantiomers of di- and tetrahydroxy polynuclear hydrocarbons, but it was found that the ionic Pirkle phase 2 became unstable if polar solvents had to be applied [73,74]. Even preparative resolutions became possible. Pirkle and coworkers reported on the preparative separation of the enantiomers of naphthamides with the phenylglycine phase [75]. Lately, a large number of applications of resolutions of enantiomers became known: pyrethroid insecticides [76], norephedrine derivatives [77], ephedrine derivatives [78], amphetamine derivatives [79,80], primary and secondary amides as α-naphthamides [81], antiphlogistics after introduction of suitable π-donor groups [82], and tropic acid derivatives [83]. Benzodiazepinones were resolved analytically and preparatively on (R)-phenylglycine or (S)-pleucine [50]. In addition, epoxide enantiomers of polycyclic aromatic hydrocarbons could be resolved on Pirkle phases 1 and 2 [84].

The importance of this technique for testing the optical purity of compounds of pharmaceutical interest was recently demonstrated in a study in which the performances of the commercially available, covalently and ionically bonded phenylglycine and leucine phases were also studied [85].

In most cases, the elution order of the enantiomers was explained by a rational design of the molecular structure of the diastereomeric complexes which should form between the π-acceptor and π-donor groups. Therefore a report by Kasai and coworkers seems to be of some importance. They studied the resolution of racemic mixtures of acyclic carbinols and their acetates and found that the enantiomers, more strongly retained on the column in these series, differ in absolute configuration. Consequently, the reliability of using elution orders to assign absolute stereochemistry is questionable [45].

Other phases with π-electron acceptors with two or three nitro groups were prepared and evaluated. Trinitrophenylpropylether bonded to silica (8) and 3-(3,5-dinitrobenzamido)propyl silica (9) were synthesized and used for the chromatographic separation of aromatic compounds [86]. Similarly, 2,4-dinitrobenzenesulfonamidopropyl silica (10) was used for the separation of aromatic hydrocarbons, weak nitrogen bases of pyrrol and aniline classes, pyridines, and aza arenes [87].

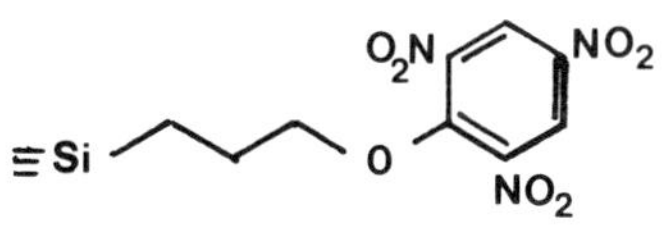

$\equiv$Si O $O_2N$ $NO_2$ $NO_2$

8

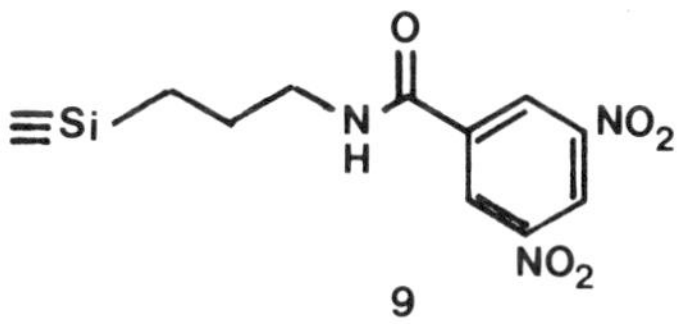

$\equiv$Si N H O $NO_2$ $NO_2$

9

$\equiv$Si N H S O O $NO_2$ $NO_2$

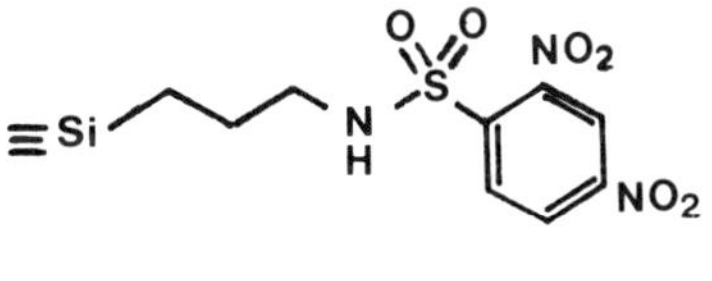

10

A very strong π-electron acceptor widely used in achiral and chiral chemically bonded phases represents the tetranitrofluorenoneimine unit which was linked to silica by a variety of methods. Lochmüller and coworkers directly reacted aminopropyl silica with 2,4,5,7-tetranitrofluorenone and used this packing (11a) in HPLC. The phase was stable under various solvent conditions and afforded very high selectivity for alkyl-substituted PAHs [88]. Along the same lines Lochmüller et al. prepared a series of phases with the fluorenoneimine moiety where the number and positions of the nitro groups were varied.

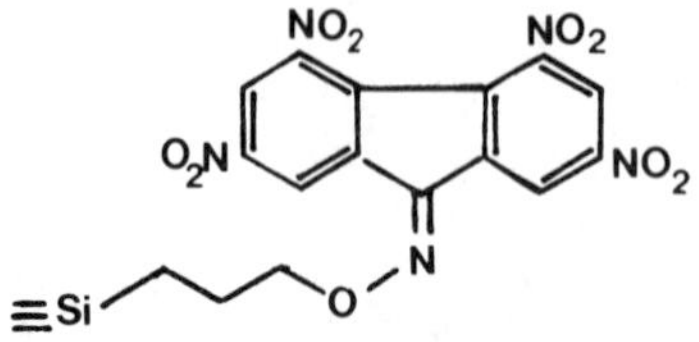

11b

Using three cholanthrene solutes it was found that the capacity factors normalized for fluorenoneimine mass varied linearly with the number of nitro groups in inductively equivalent positions of the fluorene nucleus. The selectivity behavior is less reflective of such a trend and appears for the system studied sterically controlled [89].

In analogy to the work of Lochmüller, Hemetsberger and coworkers prepared phase (11b) for HPLC use and compared the capacity factors with the DAC association constants of *O*-propyltetranitrofluorenoneoxime measured in the same solvent mixtures, *n*-hexane/methylene

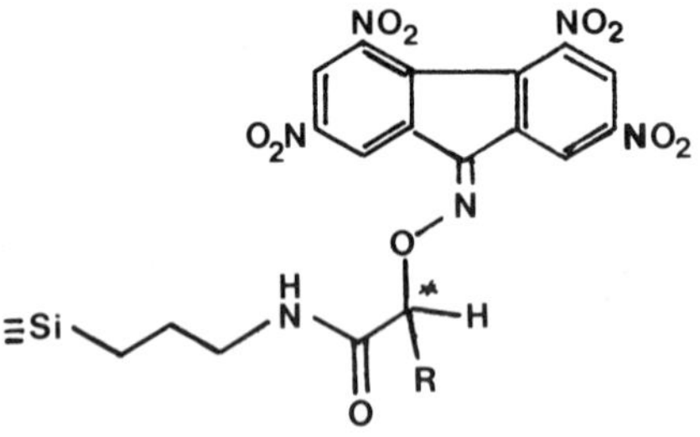

12

chloride, and found a fair correlation between k' and $K^{AD}$. While this correlation is only qualitative in nature, linear plots were obtained for log k' vs. 1/DE (DE = dielectric constant of mobile phase), showing a pronounced decrease of k' with increasing DE [19]. These results led to the investigation of that phase under reversed phase conditions and demonstrated the usefulness of DAC chromatography with polar mobile phases. Some results were presented in Fig. 3 (vide supra) [29]. Recently, a tetranitrofluoreneimidopropyl derivatized packing was prepared by Fetzer, and the retention behavior of 26 large PAHs using *n*-hexane/methylene chloride was studied and compared to a column with nitropropyl silica. Changes in the elution order were dependent on the degree of nonplanarity of the compounds in the solvent system [90].

Separations of helicenes with chemically bonded TAPA linked to aminopropyl silica via an amide bond were performed by Mikes and Gil-Av. The resolution of enantiomers could be achieved on this covalently linked TAPA as well as on those only coated with TAPA. The selectivities were even better but the resolutions turned out to be lower is all cases [14,15,17]. The authors felt that modified supports never pack as efficiently as untreated supports, and they present higher resistance to mass transfer [15].

In addition to the polynitro compounds used as π-electron acceptors in DAC chromatography, a series of other compounds showing more or less acceptor properties were chemically bonded to some sorbents. Hunt et al. prepared phthalimidopropyltrichlorosilane which was bonded to Partisil 5 (13). Since the phthalimido moiety is only a very weak electron acceptor, the results obtained resemble in part the results obtained on a C18 phase [19]. But the usefulness of cyclic aromatic anhydrides for the preparation of chemically bonded stationary phases was pointed out by Holstein and coworkers. They reacted a number of aromatic anhydrides with allylamine which were converted to the trichlorosilanes and linked to silica. Along that route tetrachloro- (14) and tetrabromophthalimidopropyl (15), naphthalene-1,4,5,8-tetracarboxydiimidopropyl (16), pyromellitdiimidopropyl (17), and 3-nitro-1,8-naphthaleneimidopropyl silica (18) were synthesized. Phase (14) was found to be approriate for the separation of alkyl-substituted PAHs and amines [92,93].

For DAC chromatography highly chlorinated aromatic groups, such as pentachlorophenylthio, were investigated by Porath et al. along with some other ligands exhibiting acceptor properties as 4-amino-1,8-naphthamide, 1-chloro-2,4-dinitrobenzene, and 2,3-dichloro-5,6-

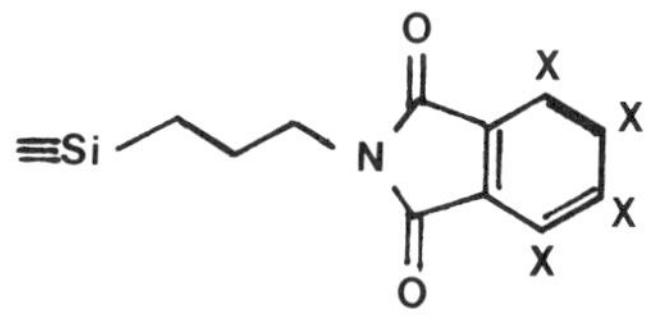

13 X = H;

14 X = Cl;

15 X = Br;

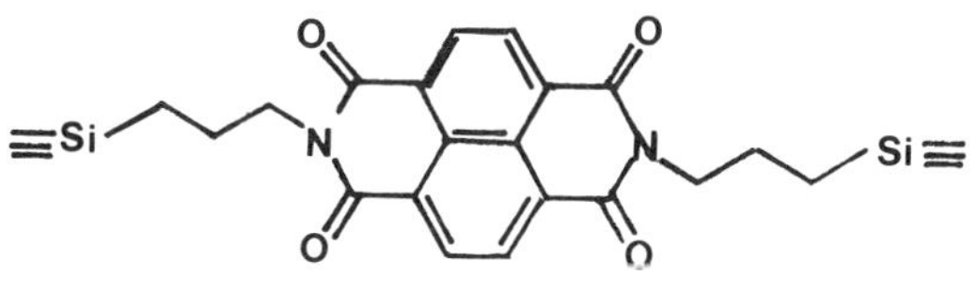

16

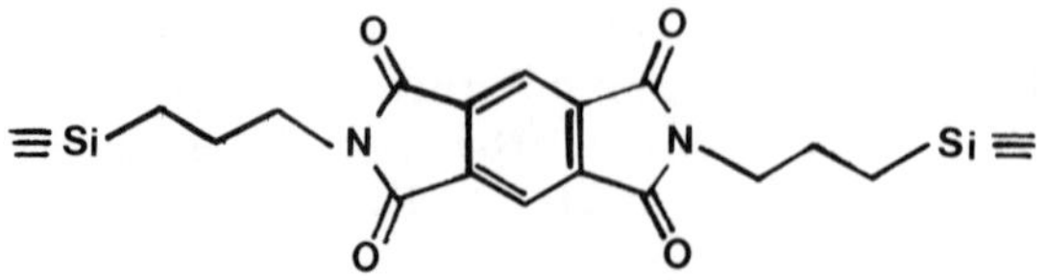

17

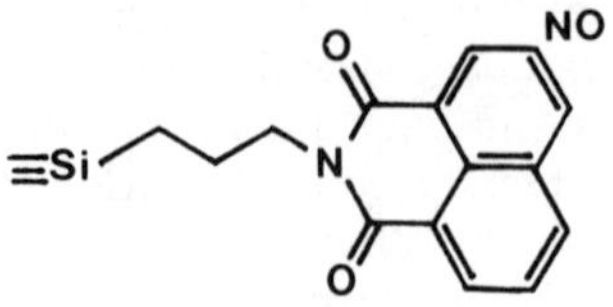

18

dicyano-*p*-benzoquinone. These acceptor ligands as well as the others were bonded to a highly crosslinked dextran. The behavior of amino acids, peptides, vitamins, and a few other substrates studied on these DAC sorbents indicated the potentialities of this type of chromatography [11-13]. Another interesting approach to DAC chromatography represents the preparation of a chemically bonded uracil ligand to silica (19). The tendency of this phase to show distinct hydrogen

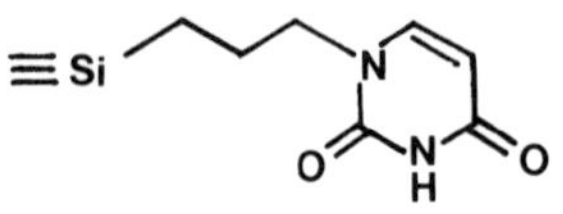

19

bonding beside its tendency to form molecular complexes was investigated by Langer and coworkers, especially for the separation of adenine nucleotides [94,95]. Felix and Bertrand report the preparation of a caffeine-bonded silica [20], which was found to be useful for the sepa-

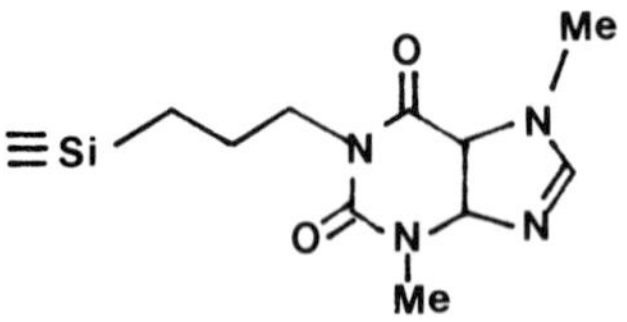

20

rations of PAHs. The synthesis of this phase can be easily achieved by reaction of theobromine with allylhalides, conversion to the trialkoxysilane, and grafting to the surface of silica [96].

The preparation of chemically bonded phases using natural products, which are easily commercially available, as ligands was demonstrated recently by Rosini and coworkers as well. They reacted quinine in the presence of azoisobutyronitrile as radical initiator with mercaptopropyl silica. That way the authors obtained a chiral support (21) which could be used successfully for the resolution of racemic arylalkyl carbinols and binaphthol derivatives [97].

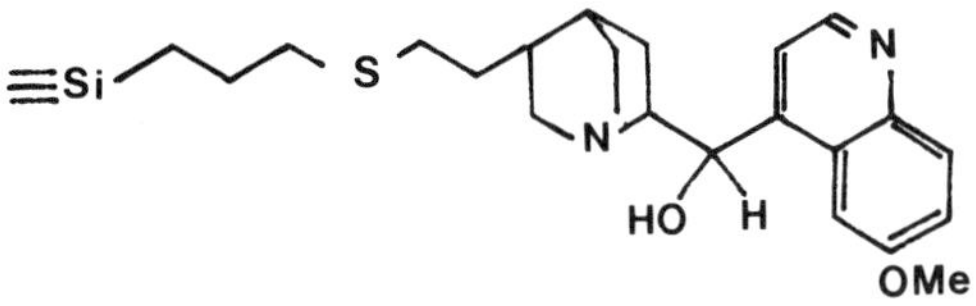

≡Si S N N HO H OMe

21

## CHEMICALLY BONDED ELECTRON DONOR PHASES

In the previous two sections packing materials for various kinds of column DAC chromatography were considered which used electron acceptors either coated or chemically bonded. Generally, the method of running DAC chromatography can be reversed, and $\pi$-electron donors can be used either coated or linked to a sorbent for the separation of mixtures of compounds showing $\pi$-electron acceptor properties. The technique of bare coating of sorbents with $\pi$-electron donors is not expected to be very productive since donor compounds are mostly sufficiently soluble in nonpolar and polar solvents. In TLC there are few examples of layering different aromatic amines for the separation of high explosives like 2,4,6-trinitrotoluene (TNT), 2,4,6-trinitrophenyl-*n*-methylnitramine (tetryl), picrylchloride, and others [98-101]. The authors pointed out that in these days other chromatographic techniques failed to resolve technical mixtures of explosives, even if other sorbent and mobile phases were used.

The simplest $\pi$-electron donor compounds are alkylether of phenols. It was demonstrated by Rose that the phenoxy group is a powerful donor for nitroaromatic compounds, as can be seen by its ability to form solid picrates [102]. Mourey and Siggia synthesized a chemically bonded 3-(phenoxy)propyl silica (22) and compared the chromatographic behavior of this new phase with C18 and phenyl phase under reversed

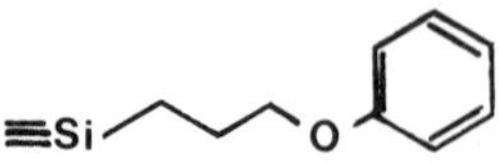

≡Si–(CH₂)₃–O–C₆H₅ ≡Si–(CH₂)₃–O–C₆H₄–C(=O)Me

22

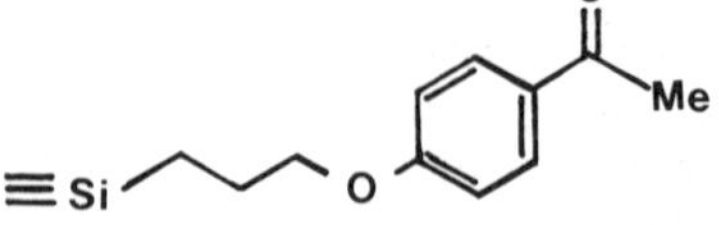

phase conditions [30]. The elution order for nitrobenzenes was found to be the same on octadecyl and phenyl silicas. In contrast, the solutes were eluted in reversed order on the phenylether phase. These findings demonstrate that on octadecyl and phenyl phase the retention is controlled mainly by solvophobic interactions, whereas the retention mechanism on the phenylether phase is altered and at least has a contribution of DAC formation. On the other hand, with nitroanilines and nitroacetanilides the elution order was not reversed in total; the ordering of only some solutes was changed. This clarifies that the phenylether phase does not correlate significantly with the hydrocarbon phases representing a valuable alternative in HPLC. In continuation of the investigation of chemically bonded donor ligands, Siggia and coworkers synthesized 3-(4-acetylphenoxy)propyl silica and studied the retention behavior of primary and secondary aromatic amines by HPLC. They found that this phase was useful for the separation of amines and assumed an attack of the amine group at the carbonyl function in addition to DAC interactions [103].

For further development of new packing materials showing donor properties, two chemically bonded phases are of some importance, although they are only used in TLC. Stetter and Schroeder esterified glucose with 3-(pentamethylphenyl)- (24) and 3-(9-phenanthrylpropionic acid (25) and studied the retention behavior of nitrotoluenes and quinones using a mixture of ether/petrolether/acetone for elution [104]. From the Rf values it is seen that the pentamethylbenzene phase is a much stronger donor than the phenanthryl ligand, showing the importance of alkyl substitution at the π-donor group.

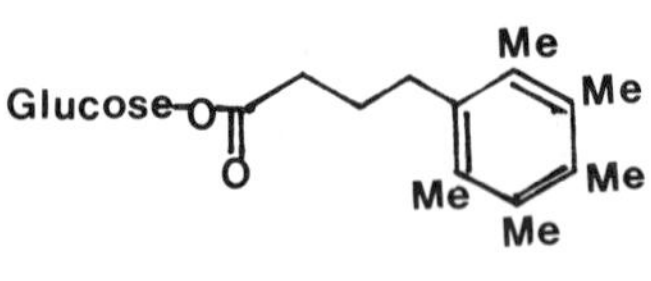

Glucose–O–C(=O)–(CH₂)₂–C₆Me₅ Glucose–O–C(=O)–(CH₂)₂–(9-phenanthryl)

24 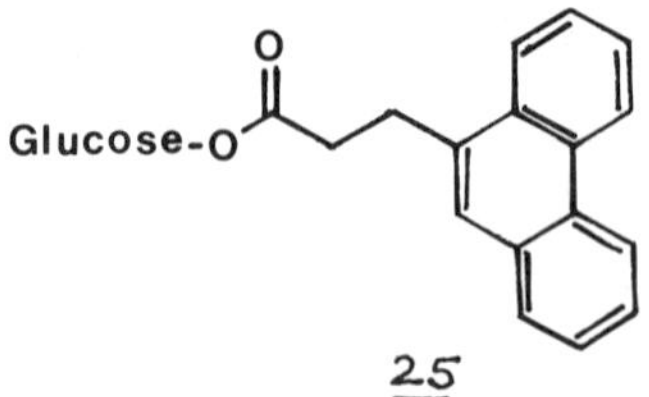 25

Oi and coworkers synthesized two chemically bonded donor phases [(26a) and (26b)] containing the naphthyl moiety by linking (S)-1-(1-naphthyl)ethylamine via an amide bond for resolution of enantiomers

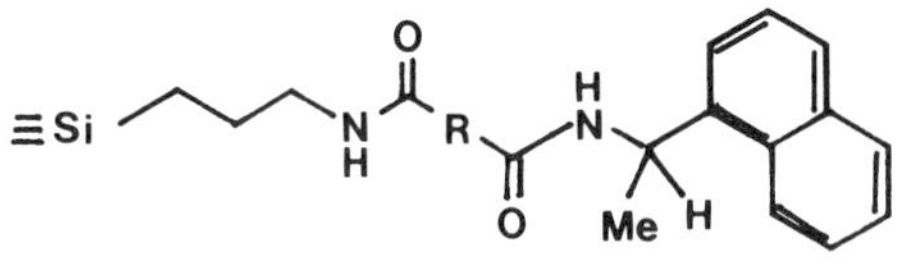

26a: R = -Ph-; 26b: R = $-(CH_2)_2-$

of various alcohols. Since the solutes contained no π-acceptor unit, it became necessary to derivatize with an appropriate reagent. The authors could demonstrate that the 3,5-dinitrophenylurethane prepared by reaction of the alcohols with 3,5-dinitrophenylisocyanate could be resolved efficiently on phase (26a) [105]. In contrast, using 3,5-dinitrobenzoylchloride as derivatizing agent the racemic esters could not be separated, whereas the 3,5-dinitrobenzamide of amines and amino acid esters and the 3,5-dinitroanilide derivatives of racemic carboxylic acids could be resolved. It was suggested that the enantioselectivity of this phase toward the urethane may be the result of a hydrogen-bonding interaction in addition to the DAC interaction [106].

Pirkle and coworkers synthesized chemically bonded chiral stationary phases based on 1-arylalkylamines and found that 1-(6,7-dimethyl-1-naphthyl)isobutylamine silica (27a, 27b) was quite effective for the

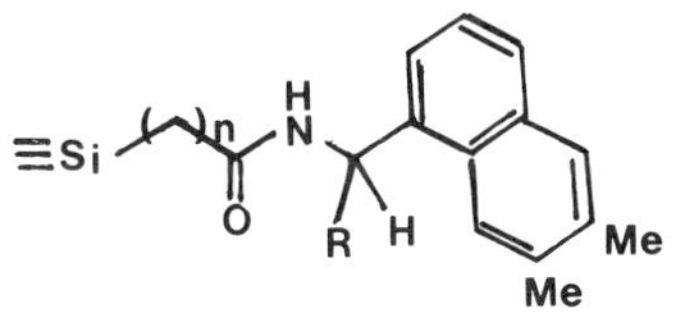

27a: R = iso-$C_4H_9$, n = 10; 27b: n = 3

resolution of *N*-3,5-dinitrobenzoyl derivatives of amino acids, their esters and amides, amino alcohols, and amines. A chromatographic separation factor of 4.73 was found for the derivatized enantiomers of methylphenylalanate. The band shapes were excellent and the resolution value was 16.5. In addition, a number of derivatized alcohols are resolvable on this phase. An analogous stationary phase derived from 1-(-1-naphthyl)ethylamine performed less well for derivatized amino acid esters [107,108]. Two optically active 1-(1-naphthyl)alkylamines were grafted to the *n*-alkyl chain of the spacer by an urea linkage too (28). The urea linkage was found to be reasonable means of connecting a chiral moiety to a silica support and, additionally, seems to exert some control between two competing chiral recognition pro-

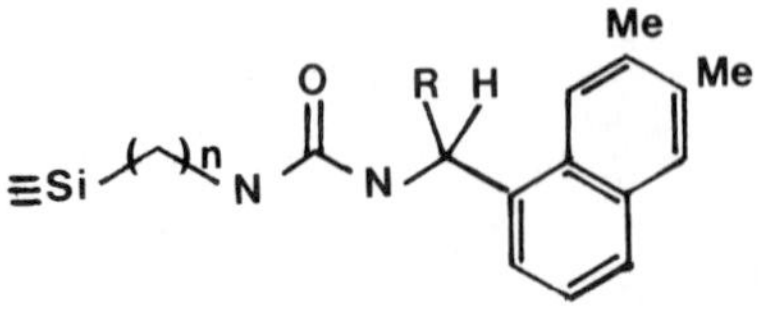

28 : R= iso - Butyl, n = 3, 11

cesses owing to the conformational rigidity and polar nature of the urea linkage [109].

A promising hydantoin-based, chiral stationary phase (29) was prepared and the enantioselectivity was investigated [110]. This phase turned out to be effective again for the resolution of 3,5-dinitrobenzamides derived from amino acids, amino ester, aminoamides, aminophosphonates, amino alcohols, and amines. This packing became valuable for the investigation of various factors influencing the chiral recognition. For example, enhancing the π basicity of the naphthyl

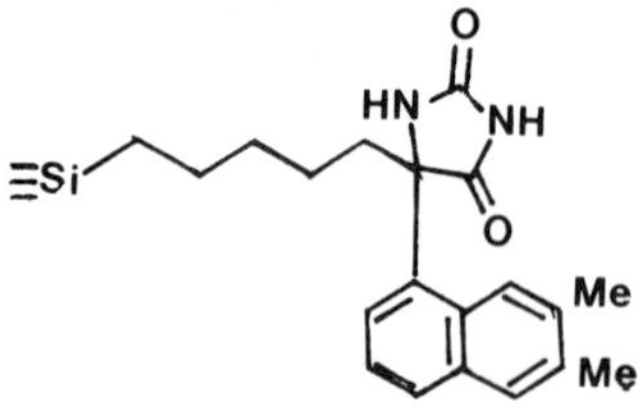

29

system by introducing methyl groups in positions 6 and 7 of the naphthyl moiety significantly improved the performance relative to the stationary phase lacking these methyl groups.

The complexity of interactions which are responsible for chiral recognition was demonstrated by Lloyd [111]. He synthesized three chiral stationary phases—two diastereomeric phases based on (R)-phenylglycyl-(R/S)1-1(1-naphthyl)ethylamide (30, 31) and one enantiomeric phase—based on glycyl-(S)-1-(1-naphthyl)ethylamide (32). 3,5-Dinitrobenzoyl-derivatized amino acid esters were investigated on these phases. It was found that there were only minor differences among these phases if isopropanol/hexane was used as mobile phase. However, marked differences in the chromatographic behavior were found if dichloromethane/hexane was used as eluent. With this later mobile phase the R/R chiral packing did not separate any of the derivatized enantiomeric amino acid esters and the R/S chiral packing

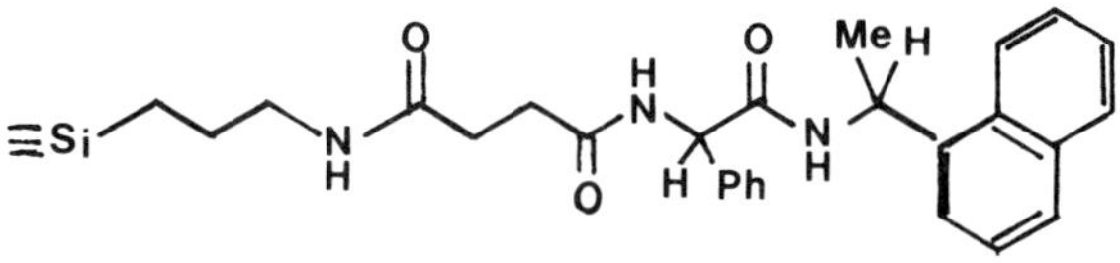

30: (R)/(R); 31: (R)/(S)

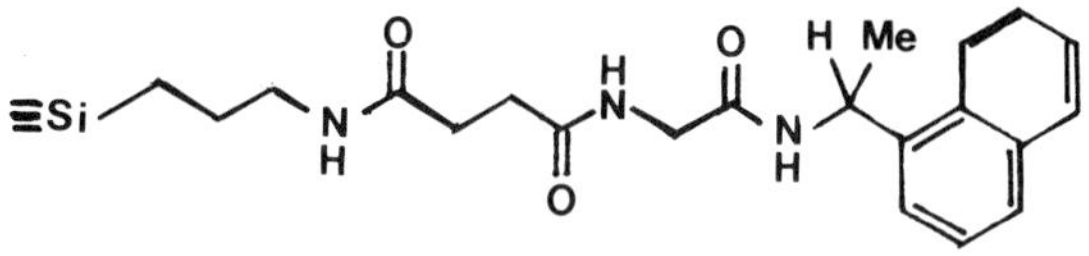

32

gave larger separation factors than the enantiomeric. Additionally, the R/S chiral support gave similar chromatographic separation factors with either mobile phase.

Oi et al. described the preparation of further, new, chiral phases. They bonded (S)-2-(4-chlorophenyl)isovaleric acid and (1R,3R)-*trans*-chrysanthemic acid to aminopropyl silica either directly (33) or as has

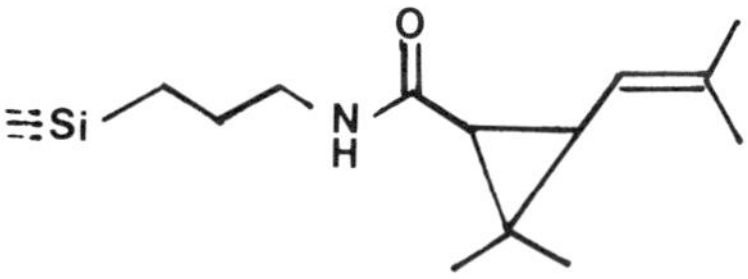

33

been done with the second one via a D-phenylglycine moiety as a connecting link (34a). Phase (34b) was not chemically bonded but ionically linked to aminopropyl silica, and L-valine was used instead of D-phenylglycine. They investigated the enantioselectivities for the separation of dinitroaryl derivatives of amines, amino acids, carboxylic acids, and alcohols under normal phase conditions [112,113]. It was found that phase (33) which contains only one chirality center attached to the carbon atom of the amide group had little enantioselectivity for amine or amino acid derivatives, but carboxylic acid enantiomers were separated to a considerable extent. Phase (34a), which contains two chirality centers attached to the amide bond, showed improved enantioselectivities. On this phase one could separate not only carboxylic acid enantiomers but also amine and amino acid enantiomers. The

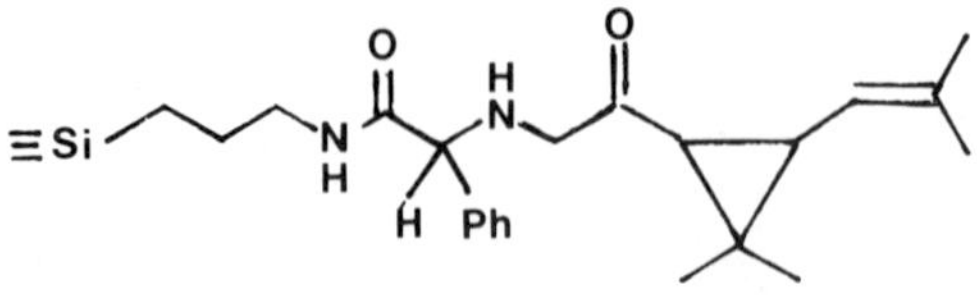

34a

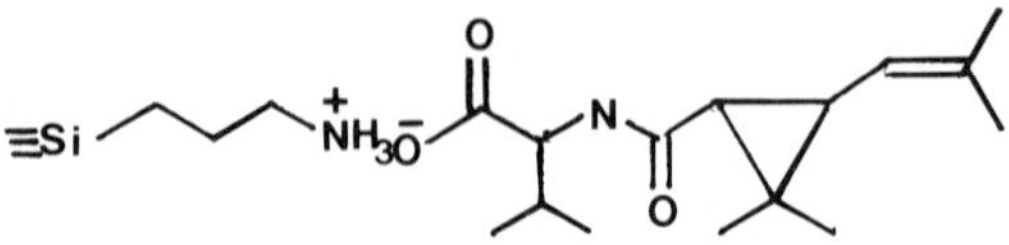

34b

authors emphasize that especially 3,5-dinitrobenzoyl derivatives of amino acids are separated with very high separation factors. For example, the separation factor was 4.0 for *N*-3,5-dinitrobenzoyl-DL-valine-*n*-butylamide on phase (34). It is remarkable that these phases can be used for resolution of enantiomers, and it must be assumed that molecular complexation is due to other than DAC interactions.

Anthracenes are well known to form fairly strong DACs with electron acceptors, and this property was used with 2,2,2-trifluoro-1-(9-anthryl)ethanol for determinations of optical purity by H-NMR [114]. Based on the conception derived by the results obtained in NMR investigations Pirkle and House prepared a chiral, chemically bonded stationary phase using this trifluoromethylanthrylcarbinol ligand to study the chromatographic resolution of suitable compounds showing $\pi$-acceptor properties [41,114]. The anthryl moiety was linked with a 10-chloromethyl group to a mercaptopropyl silica. The authors showed that a large number of racemic-substituted arylalkyl sulfoxides, 3,5-dinitrobenzoyl derivatives of amines, alcohols, thiols, amino alcohols, hydroxy acids, and amino acids could be resolved. Using cyclic solutes it was found that restricted conformations did not necessarily facilitate chiral recognition, since it may aid or even prevent the necessary interactions.

Since chiral recognition works in two directions, the quality of the resolution of enantiomers obtained on this chiral donor packing can be used as a guide to design new, chiral acceptor packings which in turn should separate enantiomers of the trifluoromethylanthrylcarbinol and its analogs. Pirkle used this kind of reasoning successfully to develop new, chiral phases with $\pi$-acceptor properties (vide supra) and termed this a second-generation stationary phase. Results of separation of enantiomers with these second-generation chiral phases in turn will again direct fast and effeciently to new $\pi$-donor phases, which of course can be considered as the third-generation chiral phase [41].

Verzele and van de Velde reported on a chemically bonded anthracene phase. They reacted 9-chloromethylanthracene with the aminopropyl silica and removed residual hydrogen atoms in unreacted amino groups by crosslinking with epichlorohydrin. This phase was compared with octadecyl-, phenyl-, and aminopropyl-bonded silcia gels, and it was found that this phase exhibited specific selectivity for some aromatic compounds [115].

Of the many chemically bonded phases prepared for use in HPLC, there are a few which cannot be classified either as π-acceptor or π-donor type. The separations that are achieved on these packings will be the result of a number of interactions, and probably one of these will be polar in nature. For example, coumarin dimers linked to a polyamide was used for separation of *trans*-stilbeneoxide, mandelamide, benzyl mandeleat, and *trans*-1,2-cyclobutanecarboxamide [52].

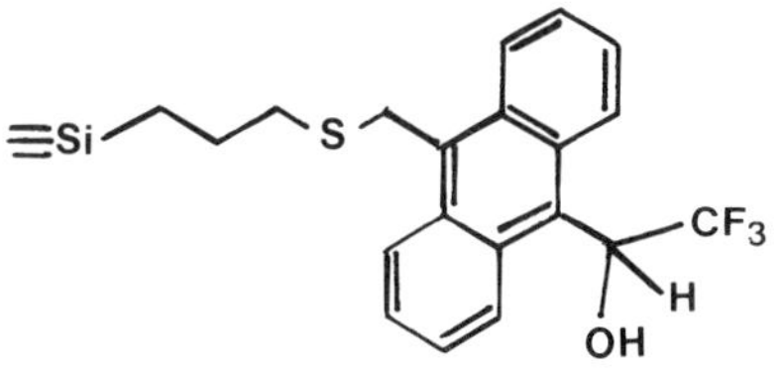

≡Si S $CF_3$ H OH

35

Another chiral ligand which was chemically bonded to silica for separation of racemic mixtures was binaphthyl-2,2'-diylhydrogenphosphate (36). Helicenes could be resolved on this phase demonstrating

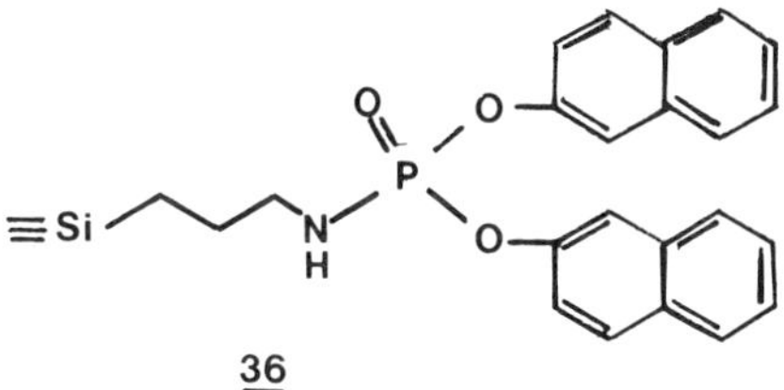

≡Si N H P O O O

36

chiral recognition. The fact that helicenes containing N heteroatoms or electron-donating substituents were resolved better than the parent hydrocarbons indicated that the binaphthyl derivatives must be a π donor or a very weak π acceptor [16,17].

## CONCLUSION

In addition to the interactions controlling selectivity in chromatographic separations, DAC formation introduces a further dimension. Proper choice of π donors and π acceptors, their ability to form either weak or strong complexes, their size, and their geometrical position within a ligand will lead to phases showing altered and adjusted selectivities for many purposes. The aspects of this type of chromatography are numerous. Many new phases will be prepared and investigated in future, particularly for separation of enantiomeric mixtures.

On the basis of current knowledge an exciting theoretical field has opened up. Further study will increase our understanding of the specific interactions governing selectivity and, especially, chiral recognition.

## REFERENCES

1. (a) G. Briegleb, *Elektron-Donator-Acceptor-Komplexe*, Springer-Verlag, Berlin-Göttingen-Heidelberg, 1961; (b) R. Foster, *Organic Charge-Transfer Complexes*, Academic Press, New York, 1969; (c) W. B. Person and R. S. Mulliken, *Molecular Complexes: A Lecture and Reprint Volume*, John Wiley and Sons, New York, 1969; (d) R. Foster (ed.), *Molecular Complexes*, Elek Science, London, 1973; (e) J. Yarwood (ed.), *Spectroscopy and Structure of Molecular Complexes*, Plenum Press, New York, 1973; (f) R. Foster (ed.), *Molecular Association*, Vol. I, 1975, Vol. II, 1979, Academic Press, New York.
2. (a) L. J. Andrews, *Chem. Rev.*, *54*:713 (1954); (b) M. J. S. Dewar, *Bull. Soc. Chim. France*, *18*:C71 (1951); (c) S. P. McGlynn, *Chem. Rev.*, *58*:1113 (1958); (d) S. F. Mason, *Quart. Rev. (London)*, *15*:287 (1961); (e) J. N. Murell, *Quart. Rev. (London)*, *15*:191 (1961); (f) E. M. Kosower, *Progress in Physical Organic Chemistry*, Vol. 3 (S. G. Cohen, A. Streitwieser, Jr., and R. W. Taft, eds.), Interscience, New York, 1965; (g) R. Paetzold, *Z. Chem.*, *15*:377 (1975).
3. F. Feigl, *Spot Test in Organic Analysis*, Elsevier, New York, 1958.
4. (a) K. G. Cunningham, W. Dawson, and F. S. Spring, *J. Chem. Soc.*, 2305 (1951); (b) E. K. Andersen, *Acta Chem. Scand.*, *8*:157 (1954).
5. M. Godlewicz, *Nature*, *164*:1132 (1949).
6. Ref. 1b.
7. P. A. Kollman, *Modern Theoretical Chemistry* (H. F. Schaefer, ed.), Plenum Press, New York.
8. M. J. S. Dewar and A. R. Lepley, *J. Am. Chem. Soc.*, *83*:4560 (1966).
9. K. Morukuma, *Accounts of Chem. Res.*, *10*:294 (1977).
10. L. H. Klemm and D. Reed, *J. Chromatogr.*, *3*:364 (1960).
11. J. Porath and B. Larsson, *J. Chromatogr.*, *155*:47 (1978).

12. J. Porath and K. D. Caldwell, *J. Chromatogr.*, *133*: (1977).
13. J. Porath, *J. Chromatogr.*, *159*:13 (1978).
14. F. Mikes, G. Boshart, and E. Gil-Av, *JCS Chem. Commun.*, 99 (1976).
15. F. Mikes, G. Boshart, and E. Gil-Av, *J. Chromatogr.*, *122*:205 (1976).
16. F. Mikes and G. Boshart, *JCS Chem. Commun.*, 173 (1978).
17. F. Mikes and G. Boshart, *J. Chromatogr.*, *149*:455 (1978).
18. L. Nondek and I. Malek, *J. Chromatogr.*, *155*:187 (1978).
19. H. Hemetsberger, H. Klar, and H. Ricken, *Chromatographia*, *13*:277 (1980).
20. G. Schomburg and B. Vonach, *J. Chromatogr.*, *149*:417 (1978).
21. D. B. Parihar, S. P. Sharma, and K. K. Verma, *J. Chromatogr.*, *31*:120 (1967).
22. D. B. Parihar, S. P. Sharma, and K. K. Verma, *Explosivstoffe*, *12*:281 (1968).
23. D. B. Parihar, O. Prakash, I. Bajaj, R. p. Tripathi, and K. K. Verma, *Microchim. Acta (Wien)*, 393 (1971).
24. C. H. Lochmüller, R. R. Ryall, and C. W. Amoss, *J. Chromatogr.*, *178*:298 (1979).
25. S. A. Matlin, W. I. Lough, and D. G. Bryan, *J. High-Resol. Chromatogr. Chromatogr. Commun.*, *3*:33 (1980).
26. R. G. Harvey and M. Halonen, *J. Chromatogr.*, *25*:294 (1966).
27. L. Nondek, M. Minorik, and I. Malek, *J. Chromatogr.*, *178*:427 (1979).
28. Y. Y. Fialkov and A. Y. Borovikov, *Zhurnal Obshchei Khimii*, *48*:248 (1978).
29. H. Hemetsberger and H. Ricken, *Chromatographia*, *15*:236 (1982).
30. T. H. Mourey and S. Siggia, *Anal. Chem.*, *51*:763 (1979).
31. W. H. Pirkle and C. J. Welch, *J. Org. Chem.*, *49*:138 (1984).
32. W. H. Pirkle and H. M. Hyun, *J. Chromatogr.*, *322*:295 (1985).
33. W. H. Pirkle and H. M. Huynn, *J. Chromatogr.*, *322*:309 (1985).
34. S. Allenmark, L. Nielsen, and W. H. Pirkle, *Acta Chem. Scand.*, *B37*:325 (1983).
35. M. S. Newman, W. B. Lutz, and D. Lednicer, *J. Am. Chem. Soc.*, *77*:3420 (1955).
36. M. S. Newman and D. Lednicer, *J. Am. Chem. Soc.*, *78*:4765 (1956).
37. M. S. Newman and W. B. Lutz, *J. Am. Chem. Soc.*, *78*:2469 (1956).
38. M. S. Newman and H. Junjappa, *J. Org. Chem.*, *36*:2606 (1971).
39. L. H. Klemm, D. Reed, and C. Lind, *J. Org. Chem.*, *22*:739 (1957).
40. C. E. Dalgliesh, *J. Chem. Soc.*, *137*:3940 (1952).
41. W. H. Pirkle, D. W. House, and J. M. Finn, *J. Chromatogr.*, *192*:143 (1980).

42. W. H. Pirkle and J. M. Finn, *J. Org. Chem.*, *46*:2935 (1981).
43. W. H. Pirkle, J. M. Finn, J. L. Schreiner, and B. C. Hamper, *J. Am. Chem. Soc.*, *103*:3964 (1981).
44. I. W. Wainer and T. D. Doyle, *J. Liq. Chromatogr.*, *2*:88 (1984).
45. M. Kasai, C. Froussios, and H. Ziffer, *J. Org. Chem.*, *48*:459 (1983).
46. W. Holstein and H. Hemetsberger, *Chromatographia*, *15*:186 (1982).
47. W. Holstein and H. Hemetsberger, *Chromatographia*, *15*:251 (1982).
48. G. Gübitz, *GIT-Suppl.*, *4*:6-8, 10-11 (1985).
49. D. W. Armstrong, *J. Liq. Chromatogr.*, *7* (*Suppl. 2*):353 (1984).
50. W. H. Pirkle and A. Tsipouras, *J. Chromatogr.*, *291*:291 (1984).
51. E. Sondheimer and I. E. Pollak, *J. Chromatogr.*, *8*:413 (1962).
52. K. Saigo, Y. Chen, N. Yonezawa, K. Tachibana, T. Kanoe, and M. Hasegawa, *Chem. Lett.*, 1891 (1985).
53. J. J. Burger and E. Tomlinson, *Anal. Proc.*, 126 (1982).
54. L. H. Klemm, K. B. Desai, and J. R. Spooner, Jr., *J. Chromatogr.*, *14*:300 (1964).
55. L. H. Klemm, W. Stalick, and D. Bradway, *Tetrahedron*, *20*:1667 (1964).
56. H. Wynberg and K. Lammertsma, *J. Am. Chem. Soc.*, *95*:7913 (1973).
57. F. Mikes and G. Boshart, *J. Chromatogr.*, *149*:455 (1978).
58. F. Mikes, G. Boshart, and E. Gil-Av., *J. Chromatogr.*, *122*:205 (1976).
59. H. Numan, R. Helder, and H. Wynberg, *Recl. Trav. Chim. Pays-Bas*, 211 (1976).
60. M. Haenel and H. A. Staab, *Chem. Ber.*, *106*:2203 (1973).
61. W. Rebafka and H. A. Staab, *Angew. Chem. Int. Ed.*, *12*:776 (1973).
62. A. Krebs, J. Odenthal, and H. Kimmling, *Tetrahedron Lett.*, 4663 (1975).
63. J. T. Ayres and C. K. Mann, *Anal. Chem.*, *36*:2185 (1964).
64. G.-P. Blümer and M. Zander, *Fresenius Z. Anal. Chem.*, *288*:277 (1977).
65. K. Pecka, P. Smidl, M. Vavra, Z. Havel, and S. Hala, *Vys. Sk. Chem.-Technol. Praze, Technol. Paliv*, *D46*:49 (1982).
66. J. S. Thomson and J. W. Reynolds, *Anal. Chem.*, *56*:2434 (1984).
67. P. L. Grizzle and J. S. Thomson, *Anal. Chem.*, *54*:1971 (1982).
68. C. H. Lochmüller and R. R. Ryall, *J. Chromatogr.*, *150*:511 (1978).
69. S. A. Matlin, A. Tito-Lloret, N. J. Lough, D. G. Bryan, T. Browne, and S. Mehani, *J. High-Resol. Chromatogr. Chromatogr. Commun.*, *4*:81 (1981).
70. G. Eppert and I. Schinke, *J. Chromatogr.*, *260*:305 (1983).

71. L. Nondek, *J. High-Resol. Chromatogr. Chromatogr. Commun.*, *8*:302 (1985).
72. W. H. Pirkle and J. L. Schreiner, *J. Org. Chem.*, *46*:4988 (1981).
73. H. B. Weems and S. K. Yang, *Anal. Biochem.*, *125*:156 (1982).
74. S. K. Yang and H. B. Weems, *Anal. Chem.*, *56*:2658 (1984).
75. W. H. Pirkle and J. M. Finn, *J. Org. Chem.*, *47*:4037 (1982).
76. R. A. Chapman, *J. Chromatogr.*, *258*:175 (1983).
77. I. W. Wainer, T. D. Doyle, Z. Hamidzahdeh, and M. Aldridge, *J. Chromatogr.*, *268*:107 (1983).
78. I. W. Wainer, T. D. Doyle, Z. Hamidzahdeh, and M. Aldridge, *J. Chromatogr.*, *261*:123 (1983).
79. I. W. Wainer and T. D. Doyle, *J. Chromatogr.*, *259*:465 (1983).
80. I. W. Wainer, T. D. Doyle, and W. M. Adams, *J. Pharm. Sci.*, *73*:1162 (1984).
81. W. H. Pirkle and C. J. Welch, *J. Org. Chem.*, *49*:138 (1984).
82. J. B. Crowther, T. R. Covey, E. A. Dewey, and J. D. Henion, *Anal. Chem.*, *56*:2921 (1984).
83. I. W. Wainer, T. D. Doyle, and C. D. Breder, *J. Liq. Chromatogr.*, *7*(4):731 (1984).
84. H. B. Weems, M. Mushtag, and S. K. Yang, *Anal. Biochem.*, *148*:328 (1985).
85. Z. Y. Yang, S. Barkan, C. Brunner, J. D. Weber, T. D. Doyle, and I. W. Wainer, *J. Chromatogr.*, *324*:444 (1985).
86. G. Felix and C. Bertrand, *J. High-Resol. Chromatogr. Chromatogr. Commun.*, *7*:714 (1984).
87. L. Nondek and V. Chvalovsky, *J. Chromatogr.*, *312*:303 (1984).
88. C. H. Lochmüller and C. W. Amoss, *J. Chromatogr.*, *108*:85 (1975).
89. C. H. Lochmüller, A. A. Ryall, and C. W. Amoss, *J. Chromatogr.*, *178*:298 (1979).
90. J. C. Fetzer and W. R. Briggs, *J. Chromatogr.*, *346*:81 (1985).
91. C. C. Hunt, P. J. Wild, and N. T. Crosby, *J. Chromatogr.*, *130*:320 (1977).
92. H. Deymann and W. Holstein, *Erdöl Kohle*, *34*:353 (1981).
93. W. Holstein, *Chromatographia*, *14*:468 (1981).
94. J. Langer and Z. Kornetka, *Pol. J. Chem.*, *6*:1303 (1978).
95. K. Szyfter and J. Langer, *J. Chromatogr.*, *175*:189 (1979).
96. G. Felix and C. Bertrand, *J. Chromatogr.*, *319*:432 (1979).
97. C. Rosini, C. Bertucci, D. Pini, P. Altemura, and P. Salvadori, *Tetrahedron Lett.*, 3361 (1985).
98. D. B. Parihar, S. P. Sharma, and K. K. Verma, *Explosivstoffe*, *12*:281 (1968).
99. A. K. Dwivedy, D. B. Parihar, S. P. Sharma, and K. K. Verma, *J. Chromatogr.*, *29*:120 (1967).

100. D. B. Parihar, S. P. Sharma, and K. K. Verma, *J. Chromatogr.*, *29*:258 (1967).
101. D. B. Parihar, S. P. Sharma, and K. K. Verma, *J. Chromatogr.*, *31*:120 (1967).
102. I. Rose, *Molecular Complexes*, Pergamon Press, New York, 1967.
103. D. Y. Pharr, P. C. Uden, and S. Siggia, *J. Chromatogr. Sci.*, *23*:391 (1985).
104. H. Stetter and J. Schroeder, *Angew. Chem. Int. Ed.*, *7*:948 (1968).
105. N. Oi and H. Kitahara, *J. Chromatogr.*, *265*:117 (1983).
106. N. Oi, M. Nagase, and T. Doi, *J. Chromatogr.*, *257*:111 (1983).
107. W. H. Pirkle and M. H. Hyun, *J. Org. Chem.*, *49*:3043 (1984).
108. W. H. Pirkle, M. H. Hyun, and B. Bank, *J. Chromatogr.*, *316*:585 (1984).
109. W. H. Pirkle and M. H. Hyun, *J. Chromatogr.*, *322*:295 (1985).
110. W. H. Pirkle and M. H. Hyun, *J. Chromatogr.*, *322*:309 (1985).
111. M. J. B. Lloyd, *J. Chromatogr.*, *351*:219 (1986).
112. N. Oi, M. Nagase, Y. Indo, and T. Doi, *J. Chromatogr.*, *259*:487 (1983).
113. N. Oi, M. Nagase, Y. Indo, and T. Doi, *J. Chromatogr.*, *265*:111 (1983).
114. W. H. Pirkle and D. W. House, *J. Org. Chem.*, *44*:1957 (1979).
115. M. Verzele and N. Van de Velde, *Chromatographia*, *20*:239 (1985).

# 9

# Packings in Ligand Exchange Chromatography

V. A. Davankov / *Nesmeyanov Institute of Organo-Element Compounds, Academy of Sciences, Moscow, USSR*

In the majority of real chromatographic systems, retention of solutes on the column packing (as well as the separation of solutes in the column) is due to a superposition of several different types of interactions between the solute molecules on the one hand, and the packing material and components of the mobile phase on the other. Nevertheless, it is convenient to identify the major type of interaction which is supposed to be responsible for the solute's resolution in a given chromatographic system and classify modes of column liquid chromatography in accordance with these interaction types. From this viewpoint, we have to place the relatively new technique, ligand exchange chromatography (LEC), in an autonomous position in the series of such well-known methods as ion exchange, size exclusion, charge-transfer chromatography, etc.

Ligand exchange chromatography is a process in which the formation and breaking of labile coordinate bonds to a central metal cation is responsible for the separation of complex-forming solutes. It separates solutes by causing them to change places with other ligands composing the coordination sphere of the metal ion.

## PRINCIPLES AND PECULIARITIES OF LIGAND EXCHANGE CHROMATOGRAPHY

A "ligand" is something "tied on." This term is often used in various branches of chemistry to denote the active graft of a macromolecular structure, as in the case of sorbents for affinity chromatography. In coordination chemistry, which is dealt with in LEC, the ligand is a neutral molecule or an anion capable of donating lone electron pairs into the coordination sphere of the central metal ion. Thereby, a coordinate link is formed between the ligand and the metal ion, and the latter completes its electron shell up to the stable shell structures characteristic of the noble gases. As a matter of fact, only solute molecules which possess heteroatoms (usually, O, N, S) or $\pi$-electronic systems can be separated using the ligand exchange technique. Fortunately, this requirement is met by many classes of organic compounds, both natural and synthetic, low and high molecular ones.

The principles of ligand exchange chromatography are best represented by the simple separation experiments which were described in 1961 by Helfferich [1] in a short report entitled "Ligand Exchange: A Novel Separation Technique" and two subsequent, more detailed papers [2,3]. Helfferich solved the problem of recovering a diamine, 1,3-diamino-2-hydroxypropane, from a dilute aqueous solution (0.001 M) that also contained a 100-fold excess of ammonia. He used a carboxylic cation exchanger, Amberlite IRC-50, that was loaded with blue-green nickel(II)-ammonia complex cations. On passing the solution through a glass column that was packed with the above ligand-exchanging material, he saw a darker colored band forming at the top of the column where the diamine was retained by coordinating to Ni(II) ions. Four monodentate ammonia ligands in the nickel(II) ion coordination sphere were replaced by two bidentate diamine molecules:

$$(RCOO)_2Ni(NH_3)_4 + 2(\text{diamine}) \rightleftarrows (RCOO)_2Ni(\text{Diamine})_2 + 4NH_3$$

where $RCOO^-$ denotes the carboxylic functional ion of the resin. The absorbed diamine was then recovered as a sharp effluent band of high concentration (up to 40% w/v) by passing concentrated aqueous ammonia through the column. Thereby, nickel(II) ions remained in the resin phase, being converted again to the ammonia complex form. It is evident that the reversible exchange of ammonia for diamine ligands in the nickel(II) ion coordination sphere was responsible for the chromatographic process. Note that Ni(II) ions permanently resided in the stationary phase, being bound to the resin-fixed carboxylic anions by ion-coordinative links. No regeneration of the packing material was required for the subsequent sorption-desorption cycle.

However, the column packing in LEC does not necessarily need to display high affinity toward the complexing metal ion. Complexation reactions and ligand exchange may predominantly take place in the mobile phase as well. In this case, the eluent is doped with transition metal ions and sometimes with a special complexing organic compound. Solute molecules which have to be separated in the chromatographic process form complexes with the transition metal ions or mixed-ligand ternary complexes involving the complexing dopant in addition to the metal ion. A typical example for this mode of LEC can be taken from the paper by Gil-Av and coworkers [4]. It is the resolution of amino acid enantiomers on a reversed phase (RP) silica column employing an aqueous eluent that contains $8 \times 10^{-3}$ M of an optically active complex of copper(II) ions with a natural α-amino acid L-proline, $Cu(L\text{-}Pro)_2$. Here ligand exchange takes place in the mobile phase to form ternary species, Cu(L-Pro)(AA), where AA denotes the amino acid enantiomers. These mixed-ligand complexes partition between the mobile phase and the RP packing to result in the separation of various amino acids and enantiomeric pairs of each individual amino acid. Note that the RP stationary phase does not retain the Cu(II) ions other than in the form of complex species formed in the mobile phase.

The above examples represent two extreme modes of LEC: in the experiments of Helfferich the nickel(II) ions were mainly retained on the stationary phase, whereas in the system of Gil-Av et al. the copper(II) ions predominantly resided in the mobile phase. Accordingly, it has been suggested [5] that a distinciton be made between two types of LEC: the chromatography of ligands and that of complexes.

Only ion-exchanging or, still better, complex-forming packing materials can be used for the chromatography of ligands. A series of special ligand exchangers have been synthesized for this mode of LEC that strongly retain transition metal ions. Especially popular are sorbents containing tridentate imino diacetate ligands:

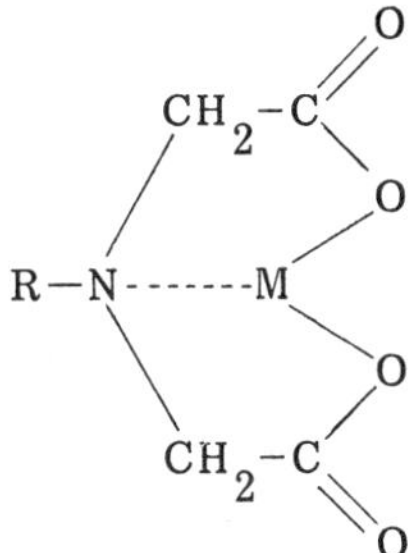


that are covalently bonded to a crosslinked polystyrene matrix (as is the case with chelating Dowex A-1 and Chelex 100 resins), to a crosslinked agarose gel (as is the case with sorbents designed for LEC of enzymes and other natural macromolecules), or to the surface of a porous silica gel (which is more convenient for high-performance liquid chromatography, or HPLC, systems). Chromatography of ligands is applicable to both analytical and preparative scale separations. In the latter case, traces of transition metal ions stripped off the column packing need occasionally be removed from the solutes separated by LEC.

LEC of complexes is mainly reduced to analytical scale separations, since the effluent by necessity contains much less solute resolved than metal ions and complexing eluent dopants. (By the way, even in the case where no special complex-forming compound is added to the eluent and only an inorganic salt is dissolved, one can refer to the chromatographic process as ligand exchange, since the coordination sphere of the metal ion is certainly saturated by solvent molecules such as water, methanol, or acetonitrile. These molecules are typical electron-donating ligands and they have to be displaced from the coordination sites by solute molecules.) The great advantage of LEC of complexes that makes this technique extremely popular is that conventional HPLC packings can be employed immediately.

Between the two extreme modes of LEC, the chromatography of ligands and that of complexes, there is a whole spectrum of intermediate systems where metal ions are located in both the stationary and the mobile phase. Indeed, metal ions should generally partition between the two phases, depending on the concentration and electron-donating power of the ligands present in the stationary and mobile phases, as well as on electrostatic interactions with the matrix-fixed and counterions. Accordingly, specific metal distribution equilibria exist in chromatographic bands that correspond to various resolved solute types. However, they have in common that the total concentration of ligands within the chromatographic band exceeds that of the eluent background. Therefore, each solute band emerging from the column contains a more or less enhanced concentration of transition metal ions. This phenomenon appears to be extremely useful for detecting solutes that do not absorb ultraviolet irradiation. Thus, using ligand-exchanging chromatographic systems which contain copper(II) ions, one can easily analyze aliphatic amino acids, amino alcohols, hydroxy acids, sugars, and many other types of solutes with spectrophotometric detectors and simple flow photometers operating at a 254-nm wavelength.

A typical example for LEC systems between the chromatography of ligands and the chromatography of complexes was published by Karger et al. [6] and should be mentioned. The authors prepared a complex-forming packing by treating silica with an excess of 3-(2-aminoethylamino)propyltrimethoxysilane. When equilibrated with a $10^{-3}$ M Cd(II) salt solution, one-tenth of the surface-bonded diamine ligands converted to cadmium complexes. The packing was found to be very efficient in separations of several sulfa drugs or dipeptides in acetonitrile-water (40:60 v/v) containing $10^{-3}$ M $CdSO_4$ and $5 \times 10^{-2}$ M ammonium acetate. The above solutes also form labile cadmium complexes in solution. As a consequence, it is impossible to discriminate whether the mobile ligands (i.e., solute molecules) coordinate to the matrix-fixed cadmium complexes, or the solute-cadmium complexes coordinate to the bonded diamine ligands. Most probably, both processes occur. Whether chromatography of ligands or of complexes, it is apparent that formation and dissociation of ternary sorption complexes comprising fixed diamine ligands, Cd(II) ions, and solute molecules is responsible for retention and discrimination of the latter, as demanded by the definition of LEC.

The above examples indicate the scope of LEC and show that its potential for separations with high selectivity and high performance using conventional or tailor-made column packings and appropriate metal or metal-chelate additives to the eluent appears unlimited.

At this stage it is useful to formulate some general requirements for ligand-exchanging systems and emphasize important peculiarities, advantages, and drawbacks of this technique.

All the manifold modes of LEC have in common that complexing transition metal ions are present in one or another form in the chromatographic system and that the solutes to be separated enter coordination interactions with these metal ions. An important requirement is that the complexes formed be kinetically labile which means that the ligand exchange in the metal ion coordination sphere is assumed to be sufficiently fast and not diminish the rate of establishment of the interphase equilibrium. Otherwise, the efficiency of the chromatographic columns would drop dramatically. This requirement limits the series of useful transition metal ions to Cu(II), Ni(II), Zn(II), Cd(II), Ag(I), Hg(II) for nitrogen- or sulfur-containing ("soft") electron donors; Ca(II), Mg(II), Mn(II), Fe(III), Al(III), La(III), Tl(III), Ti(IV) for oxygen-containing ("hard") donors; and Ag(I), Hg(II) for $\pi$-donating ligands.

Such typical complex-forming ions as Co(III), Cr(III), and Pt(IV) produce kinetically inert complexes with an extremely slow exchange of ligands. Nevertheless, these complexes have their uses in LEC, but only through "outer-sphere" complexing: ligands of the second, outer shell are exchanged rapidly.

A problem occasionally encountered in LEC is poor chromatographic efficiency. Presumably this is associated with the slow exchange of ligands. In the majority of cases this drawback is best combatted by raising the temperature of the column.

The greatest advantage of LEC is attributed to the unrivaled selectivity of the complex formation process. Inorganic electrolytes and organic compounds missing functions with lone electron pairs do not affect sorption and chromatography of complex-forming ligands. It is ligand exchange, for instance, that made it possible to sorb selectively free $\alpha$-amino acids from seawater on the Cu(II) form of the complexing iminodiacetate resin, Chelex 100 [7]. Individual enzymes and proteins which possess complexing histidine residues exposed to the surface of their macromolecules are selectively sorbed from naturally occurring protein mixtures on highly permeable agarose gels that contain iminodiacetate chelates. This powerful technique, known as "metal chelate affinity chromatography" [8], is especially important for the development of biochemistry and biotechnology. Due to the extreme sensitivity of the coordination process to the sterical structure of ligands, LEC won a spectacular triumph in the field of separation of optical isomers. The possibility was first suggested and proven in 1968 by Davankov and Rogozhin [9,10]. They grafted an optically active amino acid, L-proline, on to a crosslinked polystyrene, saturated the resin with Cu(II) ions, and resolved into enantiomeric pairs racemic mixtures of several amino acids in aqueous ammonia eluents.

The popularity of LEC technique is steadily growing. Especially important for the development of the method have been achievements connected with HPLC of optical isomers; analysis of amino acids, peptides, and amines; fractionation of proteins; analysis of unsaturated hydrocarbons and esters.

A general overview and theoretical evaluation of LEC can be found in review articles [5,11,12]. Special reviews are devoted to the use of LEC in separating optical isomers [13,14], amino acids, peptides, and proteins [15,16], as well as to the application of transition metal complexes in gas and thin-layer chromatography [17,18]. A special monograph on ligand exchange chromatography was published by Davankov et al. [19].

In the following sections, only selected typical chromatographic procedures will be described, with emphasis on the packing materials used.

## LIGAND EXCHANGE WITH NORMAL PHASE SYSTEMS

The striking efficiency of silver nitrate-ethylene glycol stationary phases for the separation of butene isomers using gas-liquid chromatography was first reported as early as 1955 [20]. Argentation of silica packing materials via impregnation with silver nitrate [21] or with ammonia-containing silver nitrate solutions [22] was also found to be very useful for the separation of unsaturated compounds such as fatty acid methyl ester mixtures according to the degree of unsaturation and the geometry of the double bonds in both thin-layer chromatography (see review [18]) and column liquid chromatography [23, 24]. Argentation chromatography, employing the selectivity of the coordination of Ag(I) ions to π-electronic systems, was thus one of the early breakthroughs of LEC into an important field of analysis, still preserving its actuality [25-28]. For the user, the in situ coating of silver onto silica gel in commercially available HPLC columns [29] is recommended. To achieve this a suitable reservoir of about 1/2-in. (o.d.) × 30-in. length is connected with the top of the column, a solution of 0.35 g of $AgNO_3$ in 15 ml of acetonitrile is poured into the reservoir, the remaining void is filled with hexane, and 30 ml of hexane is pumped through the column at a flow rate of 1 ml/min. After equilibration of the column with 30 ml of benzene at a flow rate of 3 ml/min, it is observed that the column acquires the ability to separate isomers of decene, octadecene, and tetradecen-1-ol acetate. The retention of solutes of interest can be adjusted by changing the benzene to hexane ratio in the eluent, benzene being a more competitive component for the interaction with Ag(I). The above procedure appears to produce a sufficient covering of columns of 10 to 25-cm length, since additional treatment of the column with an $AgNO_3$ solution does not result in an increase in retention, selectivity, or resolution. It is worth mentioning that the coating decreases the column efficiency by approximately 50% when comparing the peak width of naphthalene chromatographed in hexane before the coating procedure with the peak width of E or Z isomers of 9-tetradecen-1-01 acetate separated in benzene on the argentated column [29]. Still, this comparison does not simply mean deterioration of the column packing or a slow exchange of ligands around Ag(I) ions.

Silver-loaded aluminosilicate can also serve as a stationary phase for the LEC separation of unsaturated compounds [30].

Kunzru and Frei [31] described a cadmium-impregnated silica that was prepared by first stirring the dried adsorbent in a 15% solution of sodium hydroxyde in methanol for 1 h, followed by thorough washing with methanol by decantation, stirring in a 25% cadmium iodide-methanol solution, and again washing with methanol. The pretreatment of the silica with sodium hydroxide was shown to be essential for any impregnation to occur. Otherwise, such solvents as acetone, dioxane, and methanol easily dissolved the impregnated cadmium from the silica. A series of separations of chloroaniline and toluidine isomers was demonstrated in less than 6 min using a Cd(II)-impregnated packing and hexane-methanol (1-3%) eluents.

A very promising new packing material for LEC, copper(II)-modified silica, has been introduced by Caude and Foucault [32-34]. The in situ coating procedure is applicable to any commercial HPLC columns, and it consists simply of percolating an aqueous solution containing 0.01 M $CuSO_4$ and 1 M $NH_3$ through the column. Equilibrium is reached when about 0.75 mmol of Cu(II) per g of silica is fixed and the effluent of the column acquires blue color. After washing, the analysis of the silica shows a total lack of sulfate anions, which means that a formation of cupric silicate did occur on the silica surface. All columns modified by this procedure gave reproducible results. They were stable from several weeks to several months if they were eluted with a mobile phase of less than 50% water and an ammonia concentration of less than 1 M. Even with such strong basic eluents, the Cu(II) concentration in the effluents never exceeded 0.04 ppm.

Figure 1 shows the rapid decline of the silica solubility in media that are rich in organic components and explains the high stability of

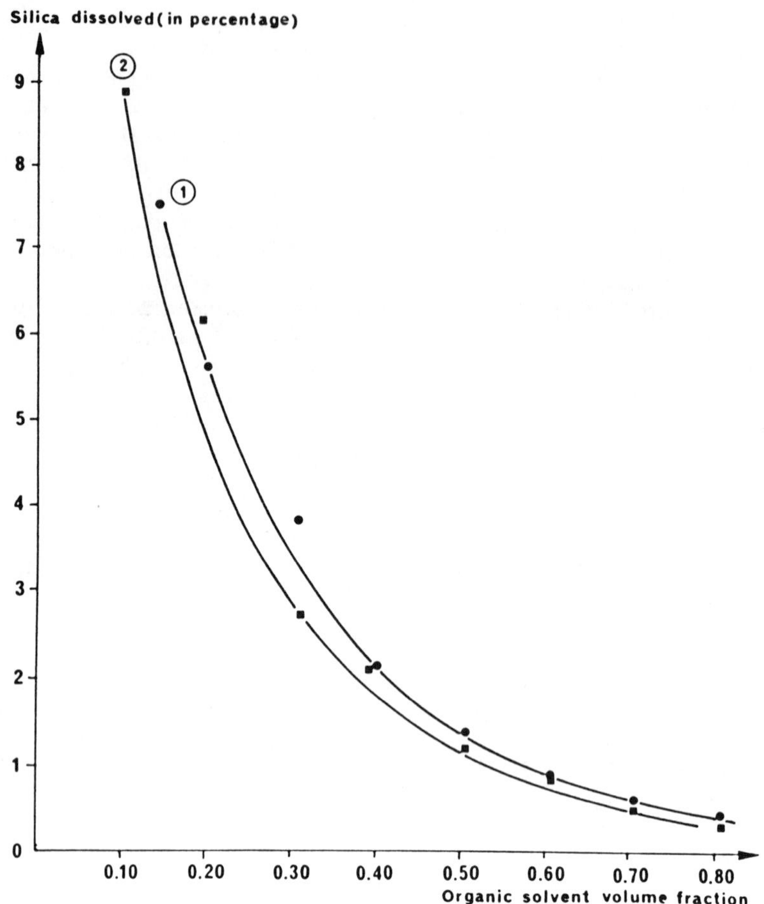


Fig. 1 Solubility to silica (percentage) vs. volume fraction of organic solvent in binary water-organic solvent mixtures. 1, Methanol, 2, acetonitrile. Static method used. Silica, LiChroprep Si 60 (15-25 μm) 200 mg. Liquid phase volume, 20 ml; $NH_3$ 0.5 M. Temperature, 20°C; contact time, 100 hr. (From Ref. 36.)

columns. The lifetime of the columns can be further increased when a guard column of crude particles of copper(II)-modified silica is employed in order to saturate the eluents with silica and copper(II) before they enter the main analytical column. The retention parameters of the column can be additionally stabilized by adding trace amounts (around 1 ppm) of Cu(II) to the eluent [35]. With this precaution, the surface layer of copper(II) silicate remains intact on the column packing. Since there is a stoichiometric ratio of Cu(II) ions to the surface hydroxyl groups, it is evident that copper uptake of the silica under standard conditions is strictly proportional to its specific surface area [37]. In this way, it is even possible to rapidly estimate the degree of covering of the surface with a reversed phase, the influence of endcapping, etc.

With properly packed HPLC solumns, sharp peaks are obtained in LEC on the copper(II)-modified silicas. Caude and Foucault [32-35] used this material for the resolving amino acids and peptides in ammonia-containing mixtures of water and acetonitrile, as shown in Fig. 2. A mixture of eight amino acids was quantitatively analyzed at 235 nm in 15 min with a detection limit of approximately 0.1 nmol. The elution sequence of tripeptides, dipeptides, heavy amino acids, light amino acids, and basic amino acids proves the ligand exchange mechanism of sorption and the influence of the negative charge of the surface on solute retention. The same packing material made it possible to quantitatively analyze a mixture of seven sugars in less than 15 min using acetonitrile-water (75:25 v/v), 1.5 M in ammonia and 0.02 mM in Cu(II), as mobile phase and detecting the effluent at 254 nm [38]. The simplicity of the preparation of the copper(II)-modified silica columns, high selectivity of the ligand exchange procedure, and the ease of photometric detection of various solutes should make this method increasingly popular.

Successful attempts have been made to separate mixtures of 2-alkylpyridines on $\gamma$-alumina-impregnated columns with an ethanolic solution of $CoCl_2 \cdot 6H_2O$ using dioxane-isooctane (1:9 v/v) as mobile phase [39] and to separate phenols on an Fe(III) form of alumina [40].

A promising approach, still scarcely explored, consists of the addition of a metal-organic ligand complex to the eluent in combination with a normal phase HPLC column. Oelrich et al. [41] successfully resolved thyroxine and triiodothyronine (naturally occurring amino acids) on LiChrosorb Si 60 in the presence of 0.2 mM bis(L-prolinato)-copper(II) in a nonpolar eluent consisting of hexane/*n*-propanol/water of 60:37.5:2.5. L enantiomers of the above solutes were observed to retain longer than did D isomers. Kurganov and Davankov [42] resolved 14 different amino acids into enantiomers in acetonitrile-water (9:1) doped with 0.5 mM of $Cu_2(tmpn)_2(OH)_2(ClO_4)_2$ and 5 mM tmpn, where tmpn is an optically active diamine, *N*,*N*,*N'*,*N'*-tetramethyl-(R)-propanediamine. In both systems, the resolving chiral ligands of the eluent, L-proline and (R)-tmpn, form ternary copper complexes with the solutes which otherwise would not resolve into enantiomers. The column packing, being in equilibrium with the mobile phase, certainly contains definite amounts of copper(II) ions and chiral eluent dopants, which means that the packing is dynamically coated with the metal chelates, thus acquiring the unique ability to resolve amino acid enantiomers.

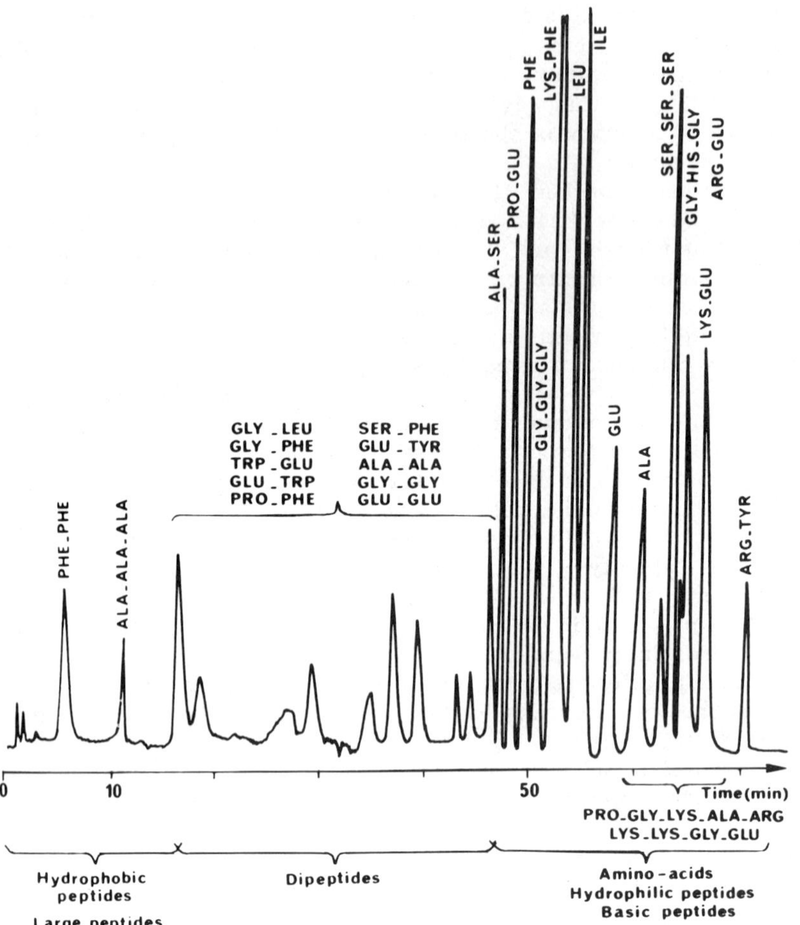


Fig. 2 Composite chromatogram of some peptides and amino acids. Column, 15 × 0.48 cm. Packing, copper(II)-modified LiChrosorb Si 60, dp = 7 μm. Mobile phase gradient from A: water-acetonitrile (10:90)-0.1 M ammonia, 1 ppm of Cu(II) to B: water-acetonitrile (60:40)-0.95 M ammonia, 1 ppm of Cu(II). Flow rate, 2 ml/min. Detection, UV at 254 nm. Sample, 30 μl containing a few μg of the solutes. (From Ref. 35.)

## LIGAND EXCHANGE IN REVERSED PHASE SYSTEMS

The hydrocarbonaceous interphase layer of a reversed phase packing material prevents residual hydroxyl groups from exchanging their protons for transition metal cations or coordinating the latter. Accordingly, RP columns exhibit minimal affinity for complex-forming metal ions. Metal ions have to be added to the eluent in the form of an inorganic salt or as a complex with an organic chelating dopant. Where the latter is sufficiently hydrophobic, metal ions partition on the stationary phase as well.

Of the purely inorganic salts, silver salts (nitrate, perchlorate) alone proved to be useful eluent additives. They are fairly soluble in conventional hydroorganic eluents and do not hydrolyze below a pH of approximately 7.5 [43]. As expected, silver(I) ions strongly influence the retention and separation selectivities of unsaturated compounds and azarenes. By coordination to the aromatic nitrogen atom or C=C double bond, silver ions affect the hydrophobicity and occasionally the conformation of solute molecules. Since the extent of complexation strongly depends on the spatial structure of ligands, reversed phase argentation chromatography is especially useful in separations of isomeric species. The method has been applied to the chromatography of various isomeric open-chain and cyclic unsaturated hydrocarbons [44,45], some vitamins and steroids [46], retinyl esters [47], unsaturated insect pheromones [48], sulfur- [45] and nitrogen-containing [43] aromatic heterocycles.

Reversed phase argentation chromatography represents one of the extreme examples of LEC of complexes. The question may arise whether the exchange of ligands around the Ag(I) ions really plays an important role after the solute-silver(I) complexes have been formed. The answer should be unambiguously positive. In order to give a single and sharp chromatographic peak, each solute species should either be completely coordinated to the metal ion or rapidly exchange its position in the Ag(I) ion coordination sphere with solvent molecules. A complete conversion of solutes into complexes under conditions of extreme dilution is rather improbable. Instead, only a certain portion of the solute interacts in the form of a complex with the stationary phase. Therefore, capacity factors depend on the metal ion concentration in the mobile phase, which can be utilized for estimating the stability constants of the complexes [43]. It is important to emphasize that the kinetics of the ligand exchange contorlling the rate of interconversion of complexed and noncomplexed species determines the peak sharpness in this chromatographic process, as it generally does in all modes of LEC.

Similar to reversed phase argentation chromatography, the addition of copper(II) acetate to a hydroorganic eluent made it possible to effectively resolve mixtures of peptides [49,50], amino acids [51], and *N*-methyl amino acids [52]. Again, the high efficiency and the ease of photometric detection of the solutes are remarkable in these separations. Since the retention of solutes in RP systems in governed by hydrophobic interactions, the elution order of amino acids in copper(II)-containing eluents differs completely from that observed on copper(II)-coated silica.

Reversed phase LEC with metal chelate additives to the eluent opens inexhaustible possibilities. Karger et al. [6] suggested Zn(II) complexes with a hydrophobic diamine, 4-dodecyldiethylenetriamine ($C_{12}$-dien), $CH_3(CH_2)_{11}N(CH_2CH_2NH_2)_2$, and demonstrated fairly good resolutions of mixtures of sulfa drugs, dansyl amino acids, aromatic carboxylic acids, or dipeptides. The common eluent was acetonitrile-water (35:65 v/v), $10^{-3}$ M $ZnSO_4$, 0.025% $C_{12}$-dien, 1% ammonium acetate. Both $C_8$ and $C_{18}$ packings gave equally good results. The above solutes produced labile ternary Zn(II) complexes with $C_{12}$-dien, with the stability strongly varying in each series of solutes.

Selective separations of a series of aromatic amines and pyridine derivatives [53] as well as pyrimidine bases [54] were obtained using a neutral square planar nickel(II)-β-diketonate, bis(2,2,6,6-tetramethylheptane-3,5-dionato)nickel(II), in aqueous methanol in combination with an ODS-2 bonded phase column. The interaction of solutes with the metal complex was shown to be strongly dependent on steric effects.

Particularly impressive and informative results have been obtained in the field of resolution of optical isomers of amino acids, as well as some other classes of chiral compounds, in the presence of chiral metal chelate additives in hydroorganic eluents. Excellent resolution enantioselectivities have been observed for underivatized amino acids, induced by copper(II) complexes with L-proline [4], L-phenylalanine [41,55], *N,N*-dialkyl-L-amino acids [56-59], *N*-benzyl-L-proline [60], L-aspartyl-L-phenylalanine methyl ester (Aspartame) [61], L-aspartyl-*N*-alkylamide [62], *N*-tosyl-L-phenylalanine [63], *N*-tosyl-D-phenylglycine [64], *N,N,N′,N′*-tetramethyl-(R)-propanediamine-1,2 [42]. Optical resolution of 5-dimethylaminonaphthalene-1-sulfonyl (dansyl) derivatives of amino acids was observed to occur in reversed phase systems containing complexes of zinc(II) with L-2-isopropyl-4-octyldiethylenetriamine ($C_3$-$C_8$-diene) [65], nickel(II) complexes with L-prolyl-*N*-octylamide [66], copper(II) complexes with L-proline [67,68], L-arginine [69], *N,N*-dipropyl-L-alanine [70], or L-hystidine methyl ester [68,71].

In several of the above systems, the retention of solutes was observed to rise with the rising extent of formation of ternary complexes with the resolving chiral metal chelate additive. In other systems, however, the increasing concentration of the dopant resulted in diminished retention of solutes. There are also cases where different solutes behave differently. As shown by Davankov et al. [14], there is no contradiction in these results. If the eluent additive is less hydrophobic than the solute of interest, the formation of ternary complexes will diminish the retention of that solute. Vice versa, relatively hydrophobic metal chelates should enhance retention of more polar solutes. Similarly, it has been shown [72] that the interaction with the hydrophobic packing surface mainly determines the relative stability of two diastereomeric ternary complexes comprising the initial metal chelate additive and one or the other enantiomer of a given solute. Contrary to the suggestions of some authors, the enantioselectivity of formation of these ternary structures in solution, i.e., the relative stability of two diastereomeric complexes in the mobile phase, exerts only minor influence on the total selectivity of separation of two enantiomers in the chromatographic column.

There is no doubt that hydrophobic metal chelate additives, in particular those containing long alkyl chains in the organic ligand of the complex, should exhibit marked affinity to the reversed phase column packing. Accordingly, the latter is easily coated with the ligand-exchanging active component of the mobile phase. When the content of the organic component in the eluent is kept below a certain critical level, the adsorbed amount of the modifying ligand sticks permanently to the packing matrix. This simple idea of converting a conventional RP column packing into a highly efficient ligand-exchanging sorbent was first introduced in the work of Davankov, Unger, et al. [73]. The authors suggested a novel chiral phase system composed of an RP silica gel coated with *N*-alkyl-L-hydroxyproline (where alkyl is $n$-$C_7H_{15}$, $n$-$C_{10}H_{21}$, or $n$-$C_{16}H_{33}$) and a hydroorganic eluent doped with copper(II) acetate. The coating of the respective *N*-alkyl-L-Hyp on the reversed phase support was accomplished by first forcing 2 $cm^3$ of a solution of 100 mg *N*-alkyl-L-Hyp in methanol or methanol-water through the prepacked column, followed by washing with 2-4 $cm^3$ concentrated solution of copper(II) acetate in methanol-water (15:85 v/v). In order to achieve a sufficient coating, *N*-alkyl-L-Hyp was dissolved in the following solutions:

$C_7$-L-Hyp in methanol-water 60:40 (v/v)
$C_{10}$-L-Hyp in methanol-water 80:20 (v/v), and
$C_{16}$-L-Hyp in pure methanol

Figure 3 illustrates the resolution of seven racemic amino acids into individual enantiomers using a $C_7$-L-Hyp coated LiChrosorb RP-18 5-μm column (100 × 4.2 mm). The detection limit with a common 254-nm UV detector was estimated to lie around $10^{-10}$ mol, i.e., about $10^{-8}$ g amino acid, with a capacity factor of around 2. The columns were observed to remain stable and did not show any bleeding of the coating material, provided the concentration of the organic component in the eluents was kept under 20-25% (v/v). The following scheme illustrates

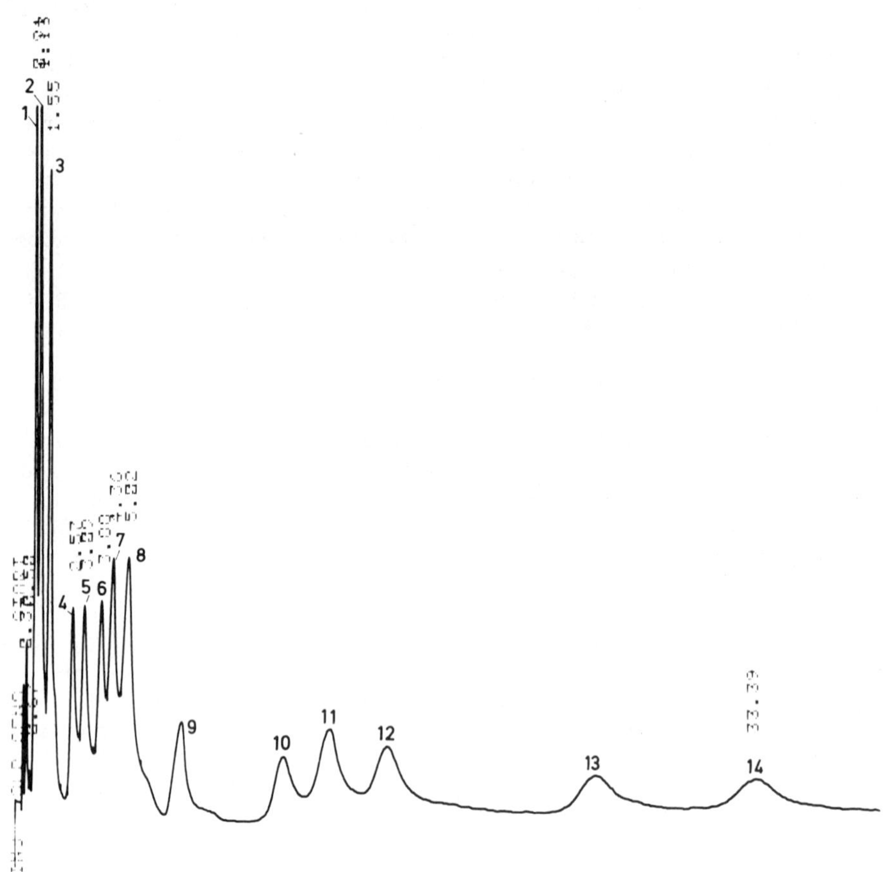


Fig. 3 Separation of seven racemic amino acids on $C_7$-L-hydroxyproline-coated LiChrosorb RP-18, dp = 5 μm. Column, 10 × 0.42 cm. Mobile phase, methanol-water (15:85), $10^{-4}$ M $Cu(AcO)_2$, pH 5.0. Flow rate, 2 ml/min. Column temperature, 293 K. Detection, UV at 254 nm. Elution sequence: 1 L-ala, 2 D-Ala, 3 L-Val, 4 L-Arg, 5 D-Arg, 6 L-Leu, 7 L-Nleu, 8 D-Val, 9 L-phen, 10 D-Leu, 11 L-Trp, 12 D-Nleu, 13 D-Phe, 14 D-Trp. The additional numbers in the chromatogram indicate the retention time in min. (From Ref. 83.)

the suggested molecular structure of the coated interphase layer and the two diastereomeric ternary complexes formed with D and L enantiomers of the amino acid solutes. It is seen that with the *N*-alkyl-Hyp coating, having the L configuration of the asymmetric α-carbon atom, the retention of D solutes appears to be enhanced by an additional hydrophobic interaction of the D-amino acid side group R with the surface alkyl groups of the packing. Indeed, D-amino acids having nonpolar R groups were observed to elute with higher capacity factors than the corresponding L enantiomers [73]. Similar mechanisms of chiral recognition that involve the interaction of diastereomeric ternary complexes with the hydrophobic sorbent surface are also represented by other systems: RP packing, chiral complexing eluent, racemic amino acid solutes [72].

The same scheme appears to include major features of the chiral recognition mechanism of enantiomeric α-hydroxy acids that also resolve on RP packings in eluents containing copper(II) complexes with *N,N,N',N'*-tetramethyl-(R)-propanediamine-1,2 [42], L-phenylalanine [41,74], di-*n*-propyl-L-alanine, or, still better, dimethyl-L-valine [75].

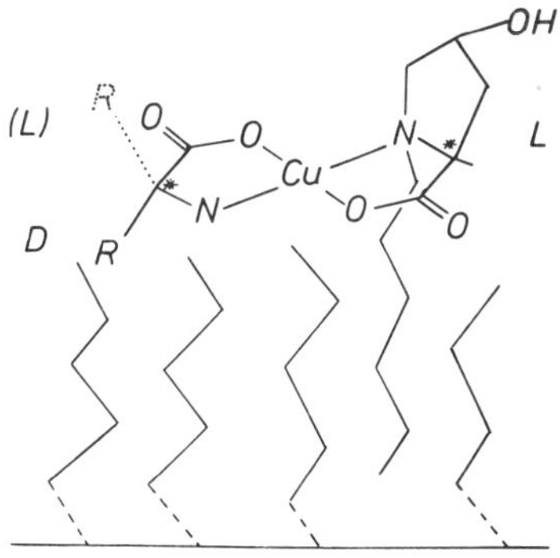

## CHEMICALLY BONDED LIGAND EXCHANGE PHASES

Chemically bonded phases that contain polar functional groups can be used straightaway as ligand-exchanging supports. The addition of organic complexing ligands to the mobile phase is superfluous in this case, although generally it might be useful in adjusting the phase distribution of metal ions and the retention of solutes. Depending on the coordinating ability of matrix-fixed ligands, metal ions can reside predominantly in the stationary or the mobile phase. The latter situation relates to LEC of complexes. A typical example here is the separation of peptide diastereomers on cyanopropyl silica in acetonitrile-0.01 mM copper(II) acetate (30:70 v/v) mixtures [49,50]. Cyanopropyl groups are extremely weak ligands. The γ-aminopropyl bonded phase already adsorbs up to 0.23 μmol $m^{-2}$ of copper(II) ions when treated in situ in the column with a 0.01 M $CuSO_4$ solution in methanol. The ratio of the amino to Cu(II) amounts therefore to 10:1. Six different groups of aromatic amine isomers were separated by a column with such a support using a methanol-cyclohexane mobile phase

[76]. With the cyclohexane content rising, the retention of amines was observed to increase—in contrast to the situation with a metal-free packing—which indicates a significant contribution of the ternary sorption complex formation to the retention of the solute ligands. When polar mobile phases are combined with the amino stationary phase, the eluents should be doped with metal ions in order to compensate for its elution. Thus, it has been found that $10^{-3}$ M cadmium(II) sulfate-acetonitrile-trifluoroethanol (83:15:2 v/v) as mobile phase gave a good separation of amino sugars (glucosamine, galactosamine, mannosamine) on 600-NH amino columns (Alltech). A series of peptides was also separated in $5 \times 10^{-4}$ M cadmium sulfate-acetonitrile-trifluoroethanol (89:10:1 v/v) [77].

Bonding of 3-(2-aminoethylamino)propyltrimethoxysilane to controlled-pore glass [78] or porous silica [6,79] results in ligand exchangers that chelate transition metal ions and retain them much more strongly. In the presence of copper(II) ions, these diamine bonded phases were used for LEC of amino sugars and amino acids [78], and in the presence of cadmium(II) for chromatography of sulfa drugs or dipeptides [6]. The eluent solutions were $10^{-4}$ M Cu(II) in 0.1 M aqueous ammonia (adjusted to pH 9.5 with hydrochloric acid) and $10^{-3}$ M Cd(II) and $5 \times 10^{-2}$ M ammonium acetate in acetonitrile-water (40:60 to 35:65 v/v), respectively. The latter mobile phase system is less aggressive to the bonded diamine phase, which is generally not very stable [79]. More detailed information on the synthesis, metal-binding ability, and hydrolytic stability of a series of chelating bonded phases has been published by Gimpel and Unger [79]. The phases were prepared by reacting a silica of 14-nm pore size with the following silanes (according to two different procedures): 3-aminopropyltriethoxysilane 3-(1-aminoethylamino)propyltriethoxysilane, 3-[2-(2'-aminoethylamino)-ethylamino]propyltriethyoxysilane, *N*-(3-triethoxysilanepropyl)-*N*,*N*-diacetic acid dimethyl ester, *N*-(3-trisodiumsilanolatepropyl)-*N*,*N*-diacetic acid disodium salt, *N*-(3-trisodimsilanolatepropyl)ethelnediamine-*N*,*N'*,*N'*-triacetic acid trisodium salt. The reaction of the triethoxysilane with the silica was carried out under the usual anhydrous conditions. The last two silanolates were coupled to the silica by evaporating the aqueous solutions. Tests on the chemical and chromatographic stability of the packings showed that only silicas with bonded iminodiacetate and the ethylenediaminetriacetate groups remained hydrolytically stable. Eight amino acids could be separated on the last sorbent in an eluent composed of sodium acetate/acetic acid (0.1 M, pH 4.8), $NaClO_4$ (0.3 M), methanol (30%), and $Cu(ClO_4)_2$ ($8 \times 10^{-5}$ M).

Chow and Grushka [80] converted 3-aminopropyl silica into a "dithiocarbamate" bonded phase:

$$-\overset{|}{\underset{|}{Si}}-(CH_2)_3-NH_2 + CS_2 \rightarrow -\overset{|}{\underset{|}{Si}}-(CH_2)_3-NH-C\begin{smallmatrix} \nearrow S \\ \searrow S^- \end{smallmatrix}$$

and a "diketone" bonded phase:

$$-\overset{|}{\underset{|}{Si}}-(CH_2)_3-NH_2 + PhCOCH_2COOC_2H_5 \rightarrow -\overset{|}{\underset{|}{Si}}-(CH_2)_3-NH-\underset{\underset{O}{\|}}{C}CH_2\underset{\underset{O}{\|}}{C}Ph$$

$$+ -\overset{|}{\underset{|}{Si}}-(CH_2)_3-N{=}C(Ph)CH_2\overset{\overset{O}{\|}}{C}OC_2H_5$$

When loaded with copper(II), both packings were found to be highly stable and selective in separating diverse aromatic amines. Optimum concentration of methanol in the mobile phase allows the separation of 12 aromatic amines in less than 13 min. The dithiocarbamate column also proved useful in the separation of diphosphate and triphosphate nucleotides with a Mg(II) gradient in the mobile phase [81]. As expected, the addition of Mg(II) affects the retention of the triphosphate nucleotide more than that of the diphosphate, whereas the nucleoside and monophosphate remain uncomplexed.

More recently [82], an interesting example of efficient LEC consisted of phenol mixtures on an 8-quinolinol silica gel-iron(III) stationary phase of the composition:

$$-\overset{|}{\underset{|}{Si}}-(CH_2)_3-NH-\underset{\underset{O}{\|}}{C}-C_6H_4-N{=}N-C_9H_5N(-O)\cdots Fe^{III}$$

This material has been shown to react with a 1:1 stoichiometry with several metal ions. No bleeding of chelated Fe(III) ions was observed. Therefore, the presence of metal ions in the mobile phase was not necessary. Preferred eluents were water-acetonitrile buffered at pH 4 with acetic acid-sodium acetate. Mixtures of 11 phenols, including Environmental Protection Agency priority pollutant species, were analyzed effectively.

Many prospective chiral ligand exchangers have been synthesized by binding optically active ligands to microparticulate silica [14,19]. Following the first reports on LEC of optical isomers of amino acids [9,10], the majority of authors used L-proline and L-hydroxyproline as a source of chirality and the metal-chelating site of bonded phases.

One approach is to bind the above cyclic amino acids to the silica surface via hydrocarbon spacer groups of different lengths and structures:

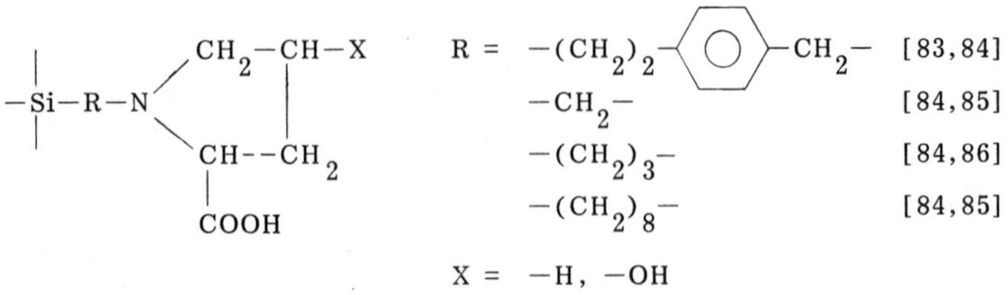

$$-\overset{|}{\underset{|}{Si}}-R-N\begin{matrix} CH_2-CH-X \\ | \quad\quad | \\ CH--CH_2 \\ | \\ COOH \end{matrix}$$

$R = -(CH_2)_2-C_6H_4-CH_2-$ [83,84]

$-CH_2-$ [84,85]

$-(CH_2)_3-$ [84,86]

$-(CH_2)_8-$ [84,85]

$X = -H, -OH$

The starting silylating reagents were 2-(*p*-chloromethylphenyl)ethyl-dimethylchlorosilane, chloromethyltrimethoxysilnae, 3-chloropropyl-triethoxysilane, 8-bromooctyltrichlorosilane, respectively. They were bonded to the silica surface by a simple method, and the halogen atoms were then substituted for L-proline or L-hydroxyproline methyl esters in the presence of sodium iodide. An alternative route is to first react chloroalkyltrialkoxysilane with the amino acid methyl esters and then bond the chiral siloxane obtained to silica. The removal of the protective ester groups proceeds automatically on treating the material with a copper acetate solution. At a concentration of surface functional groups of about 3 $\mu$mole/m$^2$, a deep blue color of the material indicates formation of copper-bis(amino acidato) complexes. All the above chiral bonded phases offer good enantioselectivity in the separation of common amino acids [84-86]. A complex combination of hydrophobic, electrostatic, and complexation interactions with the support governs the retention of amino acid solutes. Depending on the length of the hydrocarbon spacer group, the hydrophobicity of the sorbents and, accordingly, the retention of heavy amino acids varies rather strongly but can be easily adjusted to the desired level by adding acetonitrile to the aqueous eluent, $10^{-5}$-$10^{-4}$ M in copper(II) acetate.

Gubitz et al. [87-91] prepared a chiral bonded phase by treating silica first with 3-glycidoxypropyltrimethoxysilane in boiling benzene and then with sodium L-prolinate in dimethylformamide or methanol at room temperature:

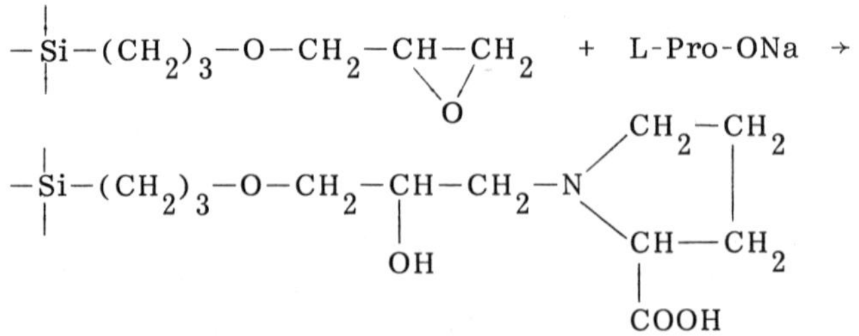

$$-\overset{|}{\underset{|}{Si}}-(CH_2)_3-O-CH_2-\underset{\diagdown O \diagup}{CH-CH_2} \quad + \quad \text{L-Pro-ONa} \rightarrow$$

$$-\overset{|}{\underset{|}{Si}}-(CH_2)_3-O-CH_2-\underset{OH}{\underset{|}{CH}}-CH_2-N\begin{matrix} CH_2-CH_2 \\ | \quad\quad | \\ CH-CH_2 \\ | \\ COOH \end{matrix}$$

This material is now available commercially from Serva, Heidelberg (FRG). Combined with copper(II) ions, the packing allows the resolution of several racemic amino acids, whereas cobalt(II), nickel(II), and zinc (II) give unacceptable results [89]. Various amino acids other than L-proline were also bonded to the 3-glycidoxypropyltrimethoxysilane-treated silica, with the result that cyclic stationary ligands (azetidine carboxylic acid, proline, hydroxyproline, and pipecolic acid) exhibited higher resolving ability than alicyclic matrix-fixed ligands [89,90].

In the third approach [33], the carboxyl function of L-proline was engaged in forming a link to the silica surface:

```
                            CH2—CH2
   |                        |    |
 —Si—(CH2)3—NH—C—CH        CH2
   |           ||  \      /
               O     NH
```

It is convenient to synthesize the packing by treating a 3-aminopropylsilylated silica with the hydroxysuccinimide ester of *N*-*t*-butyloxycarbonyl-L-proline and then hydrolyzing the *t*-BOC protecting group with trifluoroacetic acid [92]. Using the copper(II)-saturated packing and acetonitrile-0.1 M ammonium acetate mixtures of pH 9.0 as eluent, the resolution of dansyl derivatives of all amino acids, except proline, was achieved. In the case of dansyl methionine, cadmium(II) was shown to give acceptable results as well [92]. Dansyl amino acids resolve equally well on the following chiral bonded phases containing *t*-BOC-protected L-proline or L-valine [93]:

```
                          CH2—CH2
   |                      |    |                       |
 —Si—(CH2)3—NH—C—CH      CH2                        —Si—(CH2)3—C—CH—CH(CH3)2
   |           ||   \    /                             |          ||  |
               O      N                                           O   NH
                      |                                               |
                    O=C—O—C(CH3)3                                   O=C—O—C(CH3)3
```

Unprotected amino acids were resolved on the L-histidine-containing phase [94] and hydroxy acids on the L-hydroxyproline pahse [91] belonging to the same series:

```
                                                                          CH2—CH—OH
   |                                                    |                 |    |
 —Si—(CH2)3—NH—C—CH—CH2—C=CH                          —Si—(CH2)3—NH—C—CH    CH2
   |           ||  |     |  |                           |           ||  \   /
               O   NH2   N  NH                                      O    NH
                          \\/
                          CH
```

In all the above systems copper(II) served as complexing metal ions. It is remarkable that the chromatography of hydroxy acids requires pure aqueous eluents, $10^{-4}$ M in $CuSO_4$, whereas in the case of amino acids and dansyl amino acids buffer solutions in hydroorganic media provide better resolution.

Along with the structural parameters of silica, the concentration of the complexing ligands in the chemically bonded phase has been pointed out [95] as an important variable, strongly influencing retention, selectivity, and the efficiency characteristics of the packing. Two types of chiral bonded phases were compared. One was *N*-ω-(dimethylsiloxyl)undecanoyl-L-valine bonded to 5-μm spherical silica. In the presence of Cu(II), it provided resolution of a number of racemic dansyl amino acids [95] as well as salicylaldehyde Schiff bases of primary amino alcohols [96]. In preparations of the second type of sorbent [95], the above chiral valine-containing ligand was cobonded, in varying concentration ratios, with another cholorsilane which can act as a diluent with respect to the ligand exchange process. Thus, a series of inert bonded groups were introduced:

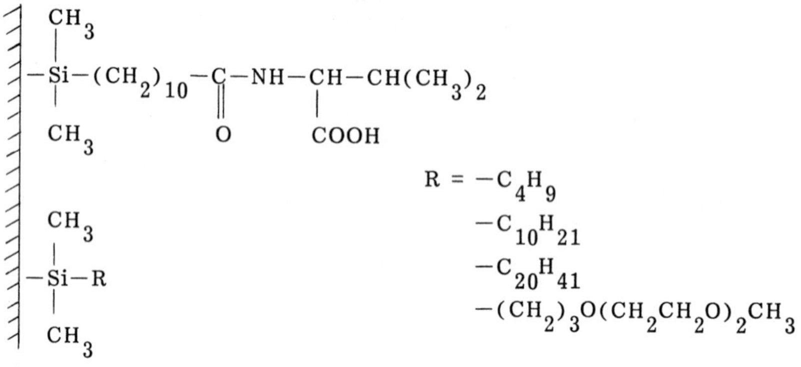

$$-Si(CH_3)_2-(CH_2)_{10}-C(=O)-NH-CH(COOH)-CH(CH_3)_2$$

$$-Si(CH_3)_2-R \qquad R = -C_4H_9,\ -C_{10}H_{21},\ -C_{20}H_{41},\ -(CH_2)_3O(CH_2CH_2O)_2CH_3$$

The diluents alter the environment surrounding the ligand exchange site of the sorbent, thus altering the additional interactions of ternary sorption complexes with the stationary phase. In several cases it was found that the retention and selectivity of dansyl amino acids was much greater on chiral phases diluted with decyl than on nondiluted phases [95].

With the majority of the above-described chiral bonded phases it has been noted that the efficiency of chromatographic columns improves significantly on raising the column temperature. This probably indicates the slow exchange of ligands in the Cu(II) ion coordination sphere contributing greatly to band broadening. In this respect, the hydrolytic stability of chelating bonded phases at elevated temperatures aquires the utmost importance. Generally, bonded phases that were prepared starting from 3-aminopropylsilylated silica seem to be rather labile, especially at pH 7-9. A recent idea of enhancing the hydrolytic stability of silica bonded phases, namely by preparing a polymeric organic interface layer with several links to the matrix, was first carried out by a Russian group [97]. The authors bonded polystyrene chains containing chiral residues of L-proline or L-hydroxyproline to the silica surface. According to one method, a copolymer of styrene with small amounts of methylvinyldimethoxysilane was prepared. It was subjected to chloromethylation in solution, then bonded to the silica, and finally substituted with the chiral amino acid residues. In the second method, a linear chloromethylated polystyrene was reacted with small amounts of 3-aminopropyltriethoxysilane, bonded to silica, and finally converted into the chiral ligand exchanger. The chemical structures of polymeric chiral bonded phases thus prepared can be represented as follows:

```
                                     CH2-CH-X
                                    /      |
                             CH2-N         |           n = 90, X = H
                             |      \      |
                             |       CH--CH2           n = 14, X = OH
                           (C6H4)    |
                             |       COOH
     |                       |
-CH2CH-CH2-CH-CH2-(CH-CH2)n-CH-
           |
           Si-CH3
          /  \
         O    O
        ///////
```

```
                                   |
-CH-CH2-CH-CH2-(CH-CH2)n-CH-
         |        |
       (C6H4)   (C6H4)    CH2-CH-X
         |        |      /      |
         CH2      CH2-N         |                     n = 90, X = OH
         |               \      |
         NH               CH--CH2
         |                |
       (CH2)3             COOH
         |
         Si-OH
        /  \
       O    O
      ///////
```

As expected, the polymeric bonded phases showed good stability up to column temperatures of 75°C. Under optimized conditions, the polymeric chiral bonded phases exhibited an acceptable efficiency in resolving amino acid enantiomers, as shown in Fig. 4. Here, silica-bonded polystyrene chains contained chiral ligands of the type $N^1,N^1$-dibenzyl-1,2-propanediamine coordinated to copper(II) ions:

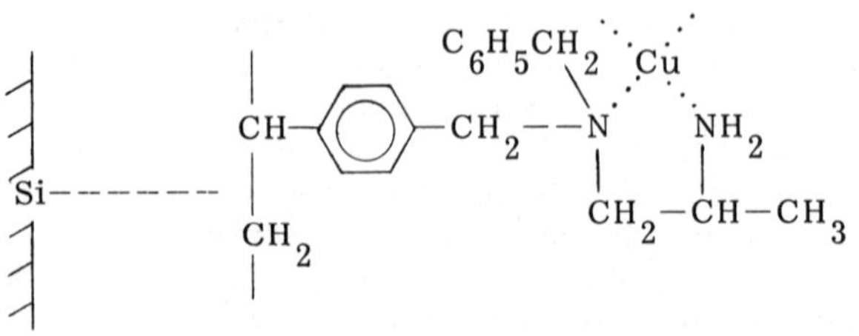

```
                          C6H5CH2   Cu
                                 \  .  .
       |                          \.    .
       CH—⟨○⟩—CH2——N        NH2
       |                           |          |
Si-------                          CH2—CH—CH3
       CH2
       |
```

A second example of bonded phases containing chiral ligands other than the amino acid type was prepared recently [98]: optically active tartaric acid were attached to a 3-aminopropylsilylated support. Amino acids and catecholamines were separated on a column loaded with Cu(II) using methanol-water buffered solutions in the pH range 4.5-7.5.

## LIGAND EXCHANGE CHROMATOGRAPHY ON ION EXCHANGE RESINS

Conventional exchangers saturated with bivalent transition metal ions, copper(II) and nickel(II), were historically the first sorbents in the development of the idea of LEC [1-3].

The most common organic cation exchanger is sulfonated polystyrene, which is easily available in various particle sizes. A corsslinking of 8% gives sufficient rigidity to spherical particles. Since double-charged metal ions cause contraction of beads by pulling the polymer chains together, a crosslinking of 6% and even 4% can be acceptable for LEC as well. The common gel-type resins were observed to give narrower and more symmetric chromatographic peaks in analytical chromatography than the macroporous resins. The latter should be preferred in combination with nonaqueous media that fail to swell gel-type resins. In aqueous eluents, macroporous sulfonated resins retain metal ions less well than do the gel-type resins, but even with gel-type exchangers it is necessary to add Cu(II) and Zn(II) to the mobile phase in order to compensate for the metal loss from the column. Nickel(II) ions are held more strongly. A loss of metals is especially severe in cases where mobile phases contain high concentrations of electrolytes or when complexation with solute molecules diminishes the positive charge of the central metal ion.

The following examples illustrate the variety of classes of organic compounds that were recovered or analyzed using LEC on commercially available sulfonated polystyrene-type resins. When loaded with Fe(III)

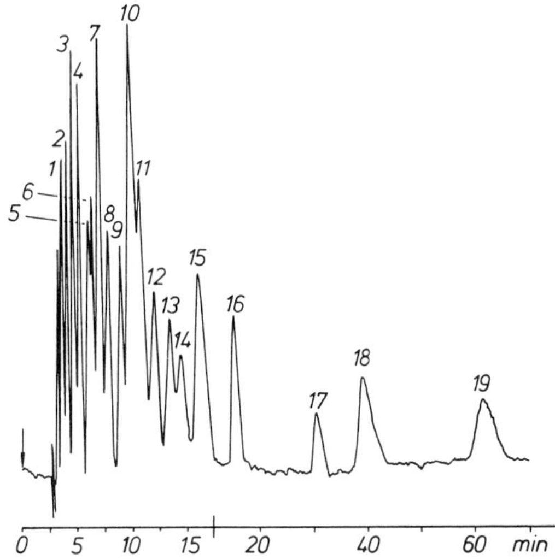


Fig. 4 Separation of 11 racemic amino acids. Column, 25 × 0.42 cm. Packing, *N,N'*-dibenzyl-1,2(R)-propanediamine-containing polystyrene chains bonded to silica gel, dp = 10 μm. Mobile phase, water-acetonitrile (70:30), $10^{-4}$ M $Cu(AcO)_2$, 0.01 M $NH_4AcO$, pH 4.0. Flow rate, 0.5 ml/min. Column temperature, 348 K. Detection, UV at 254 nm. Elution sequence: 1 D-Pro, 2 L-Pro, 3 D-Ala, 4 L-Ala, 5 D-Ser, 6 L-Ser, 7 D-Val, 8 D-Thr, 9 L-Thr, 10 D-Leu + D-Met + D-Ileu, 11 L-Val, 12 D-Tyr, 13 L-Met, 14 L-Leu, 15 L-Tyr + L-Ileu, 16 D-Phe, 17 L-Phe, 18 D-Trp, 19 L-Trp. (From Ref. 19.)

and Al(III) ions, they were used for the chromatography of phenols [99], alcohols and sugars [100], and nucleic acids [101,102]. Fe(III) and Ti(IV) forms of the resins were used by Funasaka's group to separate aromatic acids, amino- and hydroxybenzoic acids, and nitrosophenols [103,104]; they also used Hg(II) for hydroxybenzoic acids [105]. Hg(II) exhibits a high affinity for sulfur-containing compounds [106]. Sulfur-containing components were separated as a group from petroleum fractions by sorption on a macroporous resin carrying Hg(II) ions [107], but once on the resin they could not be desorbed. Cu(II)- and Ni(II)-loaded macroporous resins allow collection of β-diketones [108] and amine bases [109] from petroleum. Cation exchangers loaded with lanthanum(III) ions are selective sorbents for anions of aromatic acids and for dipolar ions like amino acids and trigonelline [110], though poor peak sharpness at room temperature indicates a relatively slow ligand exchange.

Sulfonated polystyrene with nickel(II) ions were preferably employed in the chromatography of aromatic amines and pyridine derivatives [111-113], ethanolamines [114], aziridines [114] that are rapidly hydrolyzed in neutral and acidic aqueous media but proved stable in ammonia solutions, and alkylhydrozine [115] (copper ions catalyze oxidation of these compounds). Copper(II)-loaded resins gave high separation of isomeric hexosamine [116]. Navratil and Walton [117] did an extensive study of aliphatic diamines on three exchangers (sulfonated polystyrene, polyacrylic acids, and carboxymethylcellulose) with three different ions (Cu, Zn, Ni). Varying elution orders were observed for diamines in the above combinations of exchangers with metal ions. A copper(II)-loaded carboxylic resin, Bio-Rex 70, gave the best separation factors, whereas a zinc(II)-loaded sulfonated polystyrene, Aminex A-7, gave the best all-round performance. The eluent was 5.5 M ammonia 0.002 M in Zn(II), the column temperature 55°C. Arikawa [118,119] succeeded in developing a Hitachi ligand exchange amino acid analyzer. The acid and neutral amino acids were eluted at 55°C from a sulfonated polystyrene-type resin with a buffer of pH 4.10 containing $4 \times 10^{-4}$ M zinc acetate and $5.5 \times 10^{-2}$ M sodium acetate. The elution of basic amino acids required $10^{-3}$ M zinc concentration, 0.6 M sodium acetate, and pH 5.5. Hexosamine was also resolved under similar conditions [120].

Chromatography on a sulfonated polystyrene resin saturated with calcium(II) ions is today one of the most popular methods for analyzing carbohydrate mixtures and preparative obtaining individual sugars. With pure water as the eluent, the column temperature was kept at 60-80°C [121]. Retention and resolution of weakly retained hexose can be enhanced by adding alcohol [122] or acetonitrile [123]. A silver(I)-loaded resin was found to give better separations of oligosaccharides [124]. The retention of carbohydrates correlated with the ability of the latter to coordinate metal ions [125].

Silver(I)-loaded macroporous sulfonated polystyrene resins were used in nonaqueous media for the chromatography of unsaturated esters [126,127]; silica-based strongly acidic ion exchangers in the Ag(I) form are alternative packings [127,128].

Phosphorylated polystyrene resins loaded with copper(II) ions behave similarly to sulfonated resins in LEC of amino acids [129], but retain metal ions more strongely.

Stronger retention of copper(II) and nickel(II) is also characteristic of carboxylic-type cation exchangers prepared by copolymerization of acrylic or methacrylic acids with a crosslinking agent. Obviously, carboxy anions retain transition metal ions both by electrostatic interactions and through entering the coordination sphere of the metal. The latter fact explains the frequently reported observations that the coordinating capacity of metal ions in a carboxylic-type resin appears lower than that in a sulfonated polystyrene. Another distinguishing feature of acrylic resins is that they do not undergo hydrophobic interactions with solute molecules and therefore show higher efficiency in chromatography of aromatic solutes than polystyrene-type ion and ligand exchangers. A great disadvantage of acrylic and methacrylic resins is their softness. They are intolerant of high-pressures in chromatographic columns.

Hernandez and Walton [130], studying the chromatography of amphetamine drugs by LEC, found that the acrylic resin Bio-Rex 70 in combination with Cu(II), Ni(II), or Cd(II) gave much sharper bands than a polystyrene resin. Copper(II)-loaded carboxylic resins were used in the chromatography of alkaloids [131], isomeric hexosamines [116], and amino acids [129]. Aqueous or aqueous-alcoholic ammonia solutions were the preferred eluents. Sulfur-containing components from petroleum were sorbed from pentane solutions and eluted by pentane diethyl ether mixtures [132]. Though more weakly, aromatic hydrocarbons were also retained suggesting an interaction with copper-carboxylate complexes in noncoordinating media (pentane).

Exchangers based on cellulose and dextran gels have occasionally been used in LEC, but they release metal ions rather easily. Cellulose, DEAE-, *p*-aminobenzyl-, and phosphate cellulose in combination with antimony, cobalt, mercury, and silver ions were tested in LEC of aliphatic and aromatic amines in organic eluents [133]. Chitosan, a natural glucoseamine polymer obtained from crab shells, retained copper(II) more strongly than did cellulose, and it was used in LEC of amino acids, peptides, and enzymes [134]. Crosslinked dextran gel, Sephadex G-25 [135,136], as well as cellulose-based anion exchangers [137] were employed for group separations of amino acids and peptides in copper(II)-containing alkaline systems.

## POLYMER-BASED CHELATING LIGAND EXCHANGERS

### Chiral Ligand Exchangers

The successful resolution of amino acid enantiomers by Davankov and Rogozhin in 1968 [9,10] attracted the attention of many researchers to ligand exchange chromatography and gave new impact to developing this powerful chromatographic technique, which still maintains its leading positions with respect to selectivity. Numerous publications on LEC of optical isomers were reviewed by Davankov and coworkers [13,14].

Davankov's research group centered much effort on the synthesis of polystyrene-type chiral sorbents containing residues of natural amino acids. Copolymers of styrene with 0.5-5% divinylbenzene were quantitatively chloromethylated into a para position with monochlorodimethylether under mild conditions [138]. Since the alkylation of amino acids by chloromethylpolystryrene was found to require drastic conditions, much more reactive [139] iodomethylated and bromomethylated polystyrenes were synthesized [138,140]. The most convenient way to obtain iodomethylpolystyrene was found to be by boiling the chloromethylated precursor with an acetonic solution of sodium iodide [141]. Finally, conditions were found for an immediate reaction of chloromethylpolystyrene with amino acids in the presence of sodium iodide as catalyst [139,142]. The reaction was carried out in dioxane-methanol mixtures of 6:1 at 50-60°C. However, of all the natural amino acids examined, only proline was found to react with chloromethyl groups via the nitrogen atom resulting in the chiral sorbent with 2.0-2.8 mmol/g of recurring units of the following structure [143]:

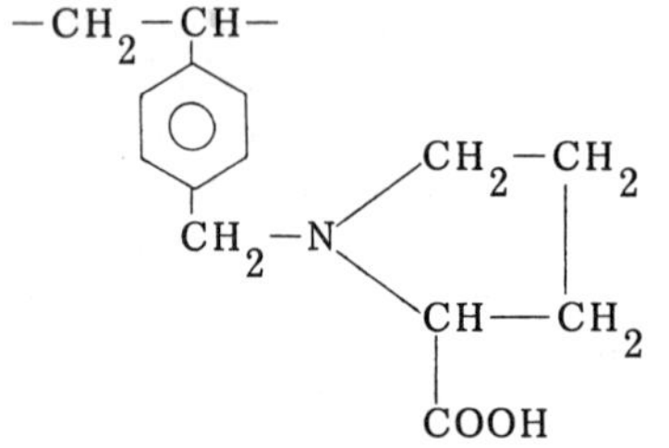

```
—CH2—CH—
      |
    (C6H4)
      |
     CH2—N      CH2—CH2
            \  /       |
             \         |
              CH——————CH2
              |
              COOH
```

Other amino acids reacted with halogen-methylated polystyrene by both their amino and their carboxylic functions. They had to be used in the form of methyl esters. Thus, a general method of obtaining chiral ligand exchangers was developed [144]: reaction of beads of *p*-chloromethylated polystyrene with amino ester hydrochlorides (mole ration 1:1.1) in a mixture of dioxane and methanol (6:1 v/v) in the presence of sodium iodide and sodium bicarbonate (0.3 and 2.5 mol, respectively). The reaction temperature was 50-60°C excluding any noticeable racemization of the chiral amino acid [145].

However, the hydrolysis of the ester function still required prolonged heating of the amino ester-containing polymers with aqueous or aqueous-organic solutions of sodium hydroxide. This obstacle was overcome by introducing "macronet isoporous" polystyrene matrixes that are much more penetrable to any chemical reagent than conventional styrene-divinylbenzene copolymers. One suggestion has been to crosslink the linear polystyrene chains in a dissolved state where the chains are loosely arranged in space. This arrangement is also characteristic of the final three-dimensional network provided a sufficient amount of rigid crosslinks is introduced between the solvated polymer chains [146]. With 4,4'-bischloromethyldiphenyl or *p*-xylilenedichloride as crosslinking bifunctional agents in the Friedel-Crafts reaction with polystyrene, long and rigid cross-bridges were formed:

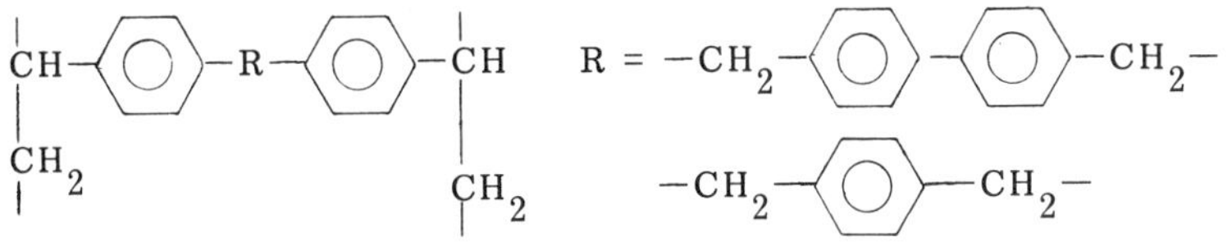

$$\begin{array}{c} | \\ CH-C_6H_4-R-C_6H_4-CH \\ | \qquad\qquad\qquad\qquad | \\ CH_2 \qquad\qquad\quad CH_2 \\ | \qquad\qquad\qquad\qquad | \end{array} \qquad R = -CH_2-C_6H_4-C_6H_4-CH_2- \quad -CH_2-C_6H_4-CH_2-$$

If strictly defined amounts of monochlorodimethylether are employed as crosslinking agents, the link $-R-$ is a $-CH_2-$ group. The macronet isoporous styrene polymers proved to be outstanding supports for the synthesis of various sorbents and exchangers, including chiral ligand exchangers, size exclusion packings and others. More detailed information on the synthesis and properties of macronet isoporous structures were published in review papers [147-149]. For application in LEC, the most important parameters of the macronet isoporous exchangers are enhanced swelling ability [150-152], pressure resistance, exchange capacity, permeability and kinetic properties.

Table 1 presents a list of chiral functional groups that were introduced into macronet isoporous styrene polymers. In the majority of cases, the initial amino acids were used in the form of methyl esters, the ester function being finally hydrolyzed in weakly alkaline or copper(II)-containing media at room temperature. However, with polyfunctional amino acids the introduction of special protecting groups was needed in order to prevent undesired side reactions of the polymer chloromethyl groups with active functions of the amino acid side group. These special synthetic methods cannot be discussed here, but they can be found in the corresponding original publications.

Some amino acid-containing chiral ligand exchangers have been synthesized by other research groups, e.g., sorbents with *N*-carboxymethyl-L-valine [170], L-proline [171], L-histidine [172], L-leucine [173]. However, they were based on a conventional styrene-divinylbenzene matrix exhibiting a low exchange capacity and poor chromatographic efficiency. In this respect, better results were obtained by binding L-phenylalanine and other ligands to chlorosulfonated styrene copolymers [174]. Here an $-SO_2-$ link is formed between the amino group of the ligand and the matrix phenyl ring. Finally, a series of polystyrene-type ligand exchangers was described [175], containing chiral residues of 1-phenylethylamine or 1,2-propanediamine. One of them, bearing $N^1$-benzyl-1,2-propanediamine groups, is worth mentioning as highly selective:

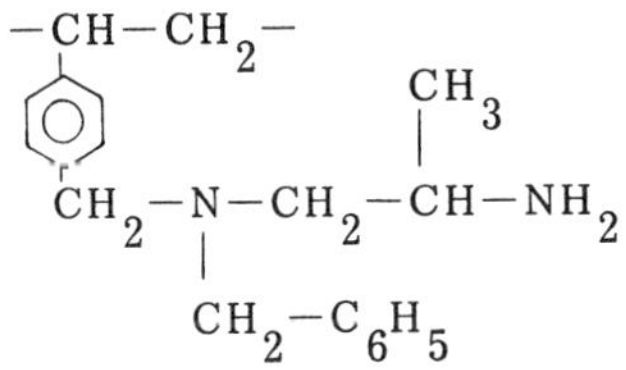

$$\begin{array}{l} -CH-CH_2- \\ \quad | \\ \quad C_6H_4 \qquad\qquad\quad CH_3 \\ \quad | \qquad\qquad\qquad\quad\; | \\ \quad CH_2-N-CH_2-CH-NH_2 \\ \qquad\quad\;\; | \\ \qquad\quad\;\; CH_2-C_6H_5 \end{array}$$

**Table 1** Chiral Chelating Ligands Fixed to the Macronet Isoporous Polystyrene Matrix[a]

| Fixed ligand | Capacity (mmol/g) | Ref. |
|---|---|---|
| Alanine | 2.2 | 144 |
| Valine | 2.3 | 153 |
| Isoleucine | 2.5 | 153 |
| Proline | 2.3 | 154 |
| Serine | 3.0 | 144 |
| Threonine | 2.9 | 144 |
| Tyrosine | 2.2 | 144 |
| Hydroxyproline | 3.0 | 144 |
| allo-Hydroxyproline | 3.2 | 154 |
| Aspartic acid | 2.5 | 155 |
| Pyroglutamic acid | 2.2 | 155 |
| Diaminobutyric acid | 2.2 | 156 |
| 1-Aminobutyrolactam | 2.3 | 156 |
| Lysine | 1.7 | 157 |
| Histidine | 2.0 | 158 |
| Methionine | 2.1 | 159 |
| Methionine-(dl)-sulfoxide | 2.1 | 160 |
| Methionine-(d)- or (1)-sulfoxide | 2.2 | 161 |
| Cysteine | 2.3 | 162 |
| Cysteic acid | 1.2 | 163 |
| *O*-Hydroxyproline | 1.0 | 154 |
| Azetidine carboxylic acid | 2.4 | 164 |
| Phenylalanine | 2.1 | 165 |
| Ar-Phenylalanine | 1.1 | 165 |
| *N*-Carboxymethylvaline | 1.5 | 166 |
| *N*-Carboxymethylaspartic acid | 1.2 | 166 |
| *S*-Carboxymethylcysteine | 1.6 | 167 |

Table 1 (continued)

| Fixed ligand | Capacity (mmol/g) | Ref. |
|---|---|---|
| *S*-(2-Aminoethyl)cysteine | 1.8 | 167 |
| *S,S'*-Ethylenebiscysteine | 1.1 | 167 |
| *N,N'*-Ethylenebismethionine | 0.9 | 167 |
| 1-Aminobenzylphosphonic acid | 2.0 | 168 |
| 1-Amino-1-methylbenzylphosphonic acid | 0.9 | 169 |
| 1-Amino-1-methylpropylphosphonic acid | 0.6 | 169 |
| 1-Amino-1-methylpentylphosphonic acid | 0.8 | 169 |
| 1-Aminobenzylphosphonic acid monoethyl ester | 2.2 | 168 |

[a]Crosslinking degree 6-10%

Among all the above polystyrene-type chiral ligand exchangers synthesized and examined thus far, the best selectivities in resolving amino acid enantiomers were exhibited by cyclic amino acid matrix-fixed ligands, namely proline, hydroxyproline, allo-hydroxyproline, azetidine carboxylic acid [164,176-178], and $N^1$-benzyl-1,2-propanediamine [179]. When taken in the copper(II)-saturated form, these chiral ligand exchangers completely resolve racemates of all common amino acids in aqueous ammonia-containing (0.1-1.5 M, depending on the solute retention) eluents. These sorbents were found to exhibit excellent chemical and configurational stability [180], and they can be used for years without losing their resolving power. High durability and high exchange capacity make the above polystyrene-type chiral sorbents particularly suitable for preparative scale separations of optically active compounds [10,181]. Up to 20 g of DL-proline ($\alpha = 3.95$) or 6 g of DL-threonine ($\alpha = 1.52$) were quantitatively resolved into enantiomers on a column containing 300 g of L-hydroxyproline-incorporating resin [181]. Tritium-containing amino acids were commercially produced in the optically, radiochemically, and chromatographically pure state using the LEC technique [182-184]. Analytical scale separations are feasible as well [179,185,186], since the macronet isoporous structure of the polymeric matrix allows a sufficiently fast mass transport.

It is not possible to present here detailed results on the resolution of optical isomers using the above sorbents (see review papers

[13,14,19]). Worth mentioning are the mechanism of chiral recognition of amino acid enantiomers [18] and enantiomeric resolutions of some hydroxy acids, amino alcohols, aminoamides, and diamines on the L-hydroxyproline-containing resin [18].

Though extremely popular in the synthesis of various resins, crosslinked polystyrene is not the only possible choice for a matrix for chiral ligand exchangers. Lefebvre et al. suggested a highly hydrophilic matrix, crosslinked polyacrylamide, which can be easily grafted with amino acid residues, proline [187], alanine, valine, threonine, phenylglycine, histidine, azetidine carboxylic acid, hydroxyproline, pipecolic acid, or phenylalanine [188]. This was done by treatment of the polyacrylamide beads with formaldehyde and the above amino acids in an aqueous alkaline media at room temperature [189]. The exchange capacity of the obtained resins amounted to up to 3.0 mmol/g. Again, the amino acid residues preserve the ability to strongly coordinate copper(II) ions:

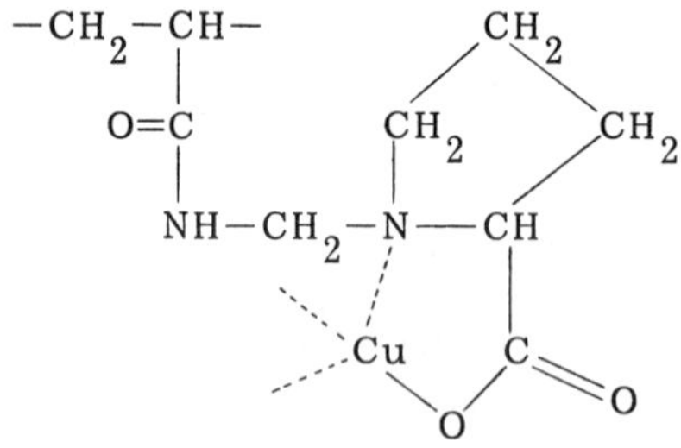

```
 —CH2—CH—                 CH2
      |                  /   \
     O=C              CH2     CH2
      |                |     /
      NH—CH2—N———CH
              :       |
           ·.Cu       C
          ·   \      / \\
               \    /    O
                 O
```

The highest resolving ability was displayed by fixed ligands of cyclic structure, proline, hydroxyproline, pipecolic acid [188]. L-phenylalanine-containing polyacrylamide-type resin showed the widest application range. It resolved racemates of all common amino acids [184]. In order to enhance the mass transport, Boue et al. [189] adsorbed the L-proline-grafted linear polyacrylamide chains onto the surface of porous silica. However, the results indicated the slow ligand exchange to be the efficiency-reducing factor in this type of sorbent.

Finally, copper(II)-chelating chiral resins were prepared by grafting L-proline and L-hydroxyproline [190], L-phenylalanine, L-histidine, and L-tryptophan [19], onto porous hydrophilic polymers of the type Separon [190] and TSK-Gel [191] using epichlorohydrine to activate the hydroxy groups of the polymer. The matrix-fixed ligand has the following structure:

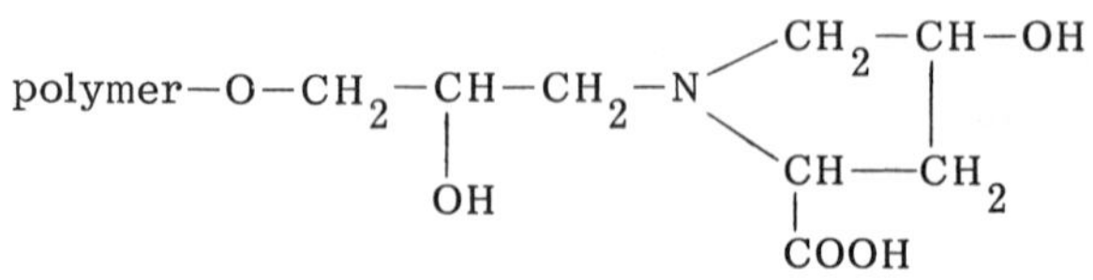

```
                                   CH2—CH—OH
                                  /     |
polymer—O—CH2—CH—CH2—N                  |
              |        \                |
              OH        CH———CH2
                        |
                        COOH
```

The sorbents resolved enantiomers of amino acids with selectivities of about 2, similar to those observed for the polyacrylamide-based resins, but lower than in the case of the polystyrene-type exchangers.

Several other metal ions have been examined in the LEC of amino acid enantiomers with the result that copper(II) generates the highest enantioselectivity.

## Sorbents for Metal Chelate Affinity Chromatography

In 1975, Porath et al. [8] published a paper entitled "Metal Chelate Affinity Chromatography: A New Approach to Protein Fractionation." The new technique utilized the ability of some protein molecules to enter into coordination interactions with metal ions that were immobilized on a polymeric matrix in the form of a complex with a fixed iminodiacetate ligand. As initially suggested and later shown by Sulkowsky [192], histidine or, to a lesser degree, tryptophan residues should be exposed to the surface of the protein molecule in order to produce retention of the latter on a copper(II) iminodiacetate-incorporating stationary phase. The strength of binding correlates positively with the multiplicity of available histidine residues, whereas retention on IDA-Zn(II) and IDA-Co(II) requires the presence of two proximal histidine residues.

Being a typical ligand exchange process, the sorption of proteins on the matrix-fixed metal chelate can be reversed in one of the three following ways: (a) lowering the pH of the aqueous eluent until imidazole groups of the histidine residues become protonated; (b) displacing the protein by imidazole, which is a stronger ligand; (c) destroying the ternary sorption complexes by stripping the metal from the column with histidine, EDTA, or acid. Any of the above types of elution procedures can be performed stepwise or by a gradient.

The main requirements of the sorbents that are intended for use in metal chelate affinity chromatography are a high permeability to large protein molecules and the absence of any hydrophobic and electrostatically charged groups which could interfere with the desired coordination interaction between solute molecules and stationary phase. These requirements are best met by ligand exchangers based on a natural polysaccharide, agarose, crosslinked with epichlorohydrin. Such gels are available under the trade name Sepharose. In a typical preparation experiment [193], Sepharose 6B was treated with epichlorohydrin in 2 M NaOH in the presence of small amounts of $NaBH_4$ at room temperature, and the oxirane-activated agarose thus obtained was subjected to reaciton with disodium iminodiacetate in 2 M $Na_2CO_3$, again in the presence of $NaBH_4$, the suspension being kept at 60°C overnight. Sometimes, longer spacer groups were introduced between the matrix and IDA chelate, for instance [8]:

$$\text{Sepharose}-OCH_2CH(OH)CH_2-O(CH_2)_4O-CH_2CH(OH)CH_2-N(CH_2CO^-)_2Me^{2+}$$

Other hydrophilic supports like CM-Sephadex C-50 (carboxymethylated dextran gel) [194], tris-acryl GF 2000 (prepared by coolymerization of a new acrylic monomer, *N*-acryloyl-2-amino-2-hydroxymethyl-1,3-propanediol, and a bifunctional hydrophilic monomer) [195], and TSK-Gel HW-55 [196] were shown to be useful supports for LEC of enzymes. Even porous silicas with covalently bonded iminodiacetate chelates were recently employed [196,197]. Some possible electrostatic interactions with charged groups of these supports were minimized by using aqueous eluents of high ionic strength (1 M NaCl).

For a long period of time, IDA chelates of Cu(II), Zn(II), and, more rarely, Co(II), Ni(II), and Fe(III) were the most popular bonding sites of sorbents in protein fractionation. Later, tris(carboxymethyl)ethylenediamine-grafted Sepharose (TED-Sepharose 4B) was introduced [193], which was prepared by first reacting the oxirane-activated agarose with ethylenediamine and then with sodium bromoacetate. More recently [198], hydroxamic acid analogs of the above IDA- and TED-Sepharose gels were suggested, with functional groups of the following structure:

$$\text{Sepharose}-OCH_2CH(OH)CH_2-N(CH_2CONHOH)_2$$

$$\text{Sepharose}-OCH_2CH(OH)CH_2-N(CH_2CONHOH)CH_2CH_2N(CH_2CONHOH)_2$$

These chelators should preferably be combined with Fe(III).

In an elegant experiment with tandem columns that were loaded with different combinations of chelating sorbents and metal ions, Porath and Olin [193] showed individual proteins to bind selectively to either one or the other packing, independently of the order in which the columns were connected. This experiment emphasizes the extreme selectivity of the LEC technique in fractionating complex mixtures of natural proteins. Metal chelate affinity chromatography has been used mainly as a preparative, not analytical, separation method thus far (see review papers [15,16,19]). However, it can be of great analytical value in assessing the surface topography of protein molecules [192].

### Miscellaneous Applications of Chelating Resins

Commonly used in many LEC studies is the chelating resin Chelex 100 (Dowex A-1) which has iminodiacetate groups bonded via $-CH_2-$

link on a polystyrene matrix. The resin retains transition metal ions rather strongly, but three positions in the metal ion coordination sphere appear occupied by the resin functional group so that the ligand-binding capacity of the packing is limited compared to the capacities of metal-containing ion exchangers. Further, mass transport in the available resins is slow and their chromatographic performance poor. The most important use of Chelex 100 in LEC is the selective recovery and concentration of trace substances form large volumes of water. Thus, copper(II)-loaded Chelex 100 recovered free amino acids from seawater [7] and from urine [199,200].

When eluted with ammonia solutions, mixtures of amino acids and peptides were fractionated on a copper(II)-Chelex 100 column [129, 199-203]. Acidic and neutral peptides and also acidic amino acids, forming negatively charged copper complexes, eluted first; neutral amino acids and basic peptides forming neutral complexes required 1.5 M ammonia eluents, whereas basic amino acids emerged with 6 M ammonia. Similarly, at pH 8.5-9.5, Ni(II) [204], Zn(II) or Co(II) forms [205] were suitable for the selective isolation of basic amino acids.

Phenols were adsorbed from industrial wastewater by a chelating resin loaded with Fe(III) and eluted with sodium hydroxide [206]. The affinity of Fe(III) for phenols varied depending on the nature of the polymeric support [207].

Aniline, pyridine, and benzylamine were eluted in that order by ammonia from a Ni(II)-Chelex 100 column [208]. Dicarboxylic acids and hydroxy acids were separated on the Zn(II) form [209].

Many purine derivatives and xanthines, including caffeine and theobromine, were clearly separated on copper(II)-loaded Chelex 100 by elution with 1 M ammonia; caffeine was the most strongly retained and could be easily quantitated in beverages [210]. Nucleosides and nucleic acid bases were adsorbed on a copper-loaded Chelex 100 column and successively eluted with ammonia; nucleotides were not adsorbed [210-212]. According to the data obtained on Sepharose-IDA-Cu, the retention increased in the order purines-pyrimidines, and generally bases-nucleosides-nucleotides [213].

## OUTER-SPHERE LIGAND EXCHANGE CHROMATOGRAPHY

Kinetically inert complexes of Co(III), Cr(III), and Rh(III) are unable to quickly exchange their ligands under normal conditions. However, it was known that they loosely bind various anions and other electron-donating ligands in their second, outer coordination sphere. Thus, the idea was formulated [5] to employ ligand exchange in the outer coordination sphere in chromatographic systems. Moreover, surprisingly good resolutions of racemic Co(III) complexes on natural polysaccharide-type sorbents (cellulose, starch, lactose) were ascribed [13,214,215] to enantioselectivity of the outer-sphere ligand exchange: the asymmetrically distributed hydroxy groups of the above carbohydrates were supposed to fit into the external coordination sphere of one enantiomer of the complex more readily than into the second shell of its antipode. Nowadays, outer-sphere ligand exchange chromatography already accumulates its first immediate experimental material.

In 1979, Chow and Grushka [216] developed a new bonded phase, containing a $Co(en)_3^{3+}$ moiety, in order to study the chromatographic behavior of solutes capable of forming outer-sphere associates with octahedral Co(III) complexes. With nucleotides and nucleosides as test solutes, it was found that coordinative interactions with the bonded complexes may strongly contribute to solute retention. This was particularly so with nucleotide triphosphate. In order to achieve an isocratic elution of nucleotide triphosphate in one chromatographic run with weaker retained nucleotide diphosphate and monophosphate, magnesium(II) ions were added ($10^{-3}$ M) to the mobile phase which was 0.037 M $Na_2HPO_4$, pH 6.4. Mg(II) ions affected the retention of triphosphates much more strongly than that of diphosphates but left retention of nucleotide monophosphates unaffected (Fig. 5). The $Co(en)_3^{3+}$ bonded phase was prepared in accordance with the following scheme:

$$-\overset{|}{\underset{|}{Si}}-(CH_2)_3-NH(CH_2)_2NH_2 + [Co(en)_2Cl_2]CL \xrightarrow{\text{pH 7, 70°C, 4 hr}}$$

$$-\overset{|}{\underset{|}{Si}}-(CH_2)_3-Co(en)_3^{3+}(Cl^-)_3$$

The packing had a red color and contained 1.08 $\mu mol/m^2$ of $Co(en)_3^{3+}$ sorption sites.

Another outer-sphere ligand exchanger for the chromatography of some mononucleotides and dinucleotides was prepared [217] from hydrophilic ω-aminobutyril-Sephadex by inserting primary amino groups of the latter into the coordination sphere of a cobalt(III) complex with tetraethylenepentamine, $[Co(tetren)Cl]^{2+}$. Finally, preparation procedures for four substitution-inert cobalt(III) complex bonded phases—$[Co(en)_3]Cl_3$, [Co(edda)(en)]Cl, [Co(dmedda)(en)]Cl, and [Co(deedda)(en)]Cl—were described [218], where en = ethylenediamine, edda = ethylenediamine-*N,N'*-diacetate ion, dmedda = *N,N'*-dimethylethylenediamine-*N,N'*-diacetate ion, and deedda = diethylethylenediamine-*N,N'*-diacetate ion. Similar to the synthesis described by Chow and Grushka, ethylenediamine-containing bonded phases were reacted with cobalt(III) complexes possessing two exchangeable chloride ligands. The phases thus obtained showed specific interactions with glucose, fructose, and sucrose [219], as well as chloro- and nitroanilines [220].

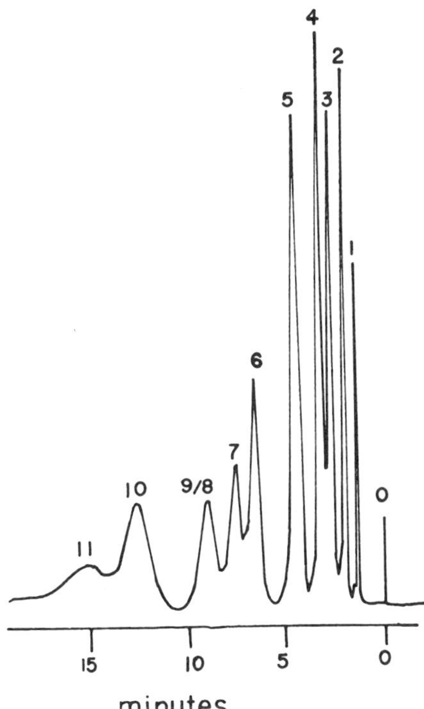


Fig. 5 Separation of some nucleotides and nucleosides on $Co(en)_3^{3+}$ bonded phase column. Column, 20 × 0.31 cm, dp 10 μm. Mobile phase, 0.037 M $Na_2HPO_4 \cdot 7H_2O$ with $10^{-3}$ M $MgSO_4 \cdot 7H_2O$, pH 6.4. Flow rate, 1 ml/min. Detection, UV at 254 nm. Elution sequence: 1 uridine 2 UMP, 3 AMP, 4 GMP, 5 UDP, 6 CDP, 7 ADP, 8 GDP, 9 UTP, 10 ATP, 11 GTP. (From Ref. 216.)

In an early paper Gaal and Inszedy [221] succeeded in the separation of optical isomers of aspartic acid and mandelic acid on a weakly acidic cation exchange column loaded with D-$[Co(en)_3]^{3+}$. This stationary phase may be referred to as a chirally coated packing. Japanese authors [222] have had great success in separating optical isomers of several inert complexes according to the "chiral eluent" technique, using d-tartrate anions or antimony d-tartrate, $[Sb_2(d\text{-}tart)_2]^{2-}$, as eluent additives, and sulfopropylated SP-Sephadex as stationary phase. Numerous examples of chromatographic resolutions of inert racemic complexes on various stationary phases in systems that were modified with optically inert complexes were presented [19]. Extremely informative are the thorough studies of Yoneda (see review [223]) on the stereochemistry of association interactions in the outer coordination sphere of octahedral Co(III) complexes.

## CONCLUSION

With the majority of chromatographic techniques, one operates on the basis of selecting a system composed of a packing material and mobile phase in order to optimize the separation of a given series of solutes. In ligand exchange chromatography, additional components are introduced into the chromatographic system, namely complexing metal ions and frequently resolving ligands. They either reside in the mobile or the stationary phase, or they are distributed between the two phases. This circumstance increases immensely the number of possible variants of designing LEC systems. It allows the use, by one way or another, of any of numerous known packing materials, and both polar and nonpolar mobile phases. Thus, ligand exchange chromatography opens an inexhaustible field for exciting theoretical studies and important practical applications.

## REFERENCES

1. F. G. Helfferich, *Nature, 189*:1001 (1961).
2. F. G. Helfferich, *J. Am. Chem. Soc., 84*:3237 (1962).
3. F. G. Helfferich, *J. Am. Chem. Soc., 84*:3242 (1962).
4. E. Gil-Av, A. Tishbee, and P. E. Hare, *J. Am. Chem. Soc., 102*:5115 (1980).
5. V. A. Davankov and A. V. Semechkin, *J. Chromatogr., 141*:313 (1977).
6. N. H. C. Cooke, R. L. Viavattene, R. Eksteen, W. S. Wong, G. Davies, and B. L. Karger, *J. Chromatogr., 149*:391 (1978).
7. A. Siegel and E. T. Degens, *Science, 151*:1098 (1966).
8. J. Porath, J. Carlsson, I. Olsson, and G. Belfrage, *Nature, 258*:598 (1975).
9. S. V. Rogozhin and V. A. Davankov, *Chem. Abstr., 72*:90875c (1970); Ger. Offen. 1932190 (1970).
10. S. V. Rogozhin and V. A. Davankov, *Dokl. Akad. Nauk SSSR, 192*:1288 (1970); *Chem. Commun.*, 490 (1971).
11. H. F. Walton, in *Ion Exchange and Solvent Extraction* (J. A. Marinsky and Y. Marcus, eds.), Vol. 4, Marcel Dekker, New York, 1973, p. 121.
12. H. F. Walton and J. D. Navratil, in *Recent Developments in Separation Science*, (N. N. Li, ed.), CRC Press, Boca Raton, 1985.
13. V. A. Davankov, in *Advances in Chromatography*, Vol. 18 (J. C. Giddings, E. Grushka, J. Cazes, and P. R. Brown, eds.), Marcel Dekker, New York, 1980, p. 139.

14. V. A. Davankov, A. A. Kurganov, and A. S. Bochkov, in: *Advances in Chromatography*, Vol. 22 (J. C. Giddings, E. Grushka, J. Cazes, and P. R. Brown, eds.), Marcel Dekker, New York, 1983, p. 71.
15. V. A. Davankov, in *Handbook of HPLC for the Separation of Amino Acids, Peptides, and Proteins,* Vol. 1 (W. S. Hancock, ed.), CRC Press, Boca Raton, 1984, p. 393.
16. B. Lönnerdal and C. L. Keen, *J. Appl. Biochem.*, *4*:203 (1983).
17. W. Szczepaniak, J. Nawrocki, and W. Wasiak, *Chromatographia, 12*:484, 559 (1979).
18. O. K. Guha and J. Janak, *J. Chromatogr.*, *68*:325 (1972).
19. V. A. Davankov, J. D. Navratil, and H. F. Walton, *Ligand Exchange Chromatography*, CRC Press, Boca Raton, 1966.
20. B. W. Bradford, D. Harvey, and D. E. Chalkey, *J. Inst. Petrol.*, *41*:80 (1955).
21. E. Dunn and P. Robson, *J. Chromatogr.*, *17*:501 (1965).
22. R. Wood and F. Synder, *J. Am. Oil Chem. Soc.*, *43*:53 (1966).
23. B. de Vries, *Chem. Ind. (London)* 1049 (1962).
24. B. de Vries, *J. Am. Oil Chem. Soc.*, *40*:184 (1963).
25. R. R. Heath, J. H. Tumlinson, R. E. Doolittle, and A. T. Proveaux, *J. Chromatogr. Sci.*, *13*:380 (1975).
26. C. R. Vogt, J. S. Baxter, and T. R. Ryan, *J. Chromatogr.*, *150*:93 (1978).
27. E. Heftmann, G. A. Sanders, and W. F. Haddon, *J. Chromatogr.*, *156*:71 (1978).
28. E. Fuggerth, *J. Chromatogr.*, *169*:469 (1979).
29. R. R. Heath and P. E. Sonnet, *J. Liq. Chromatogr.*, *3*:1129 (1980).
30. S. Lam and E. Grushka, *J. Chromatogr. Sci.*, *15*:234 (1977).
31. D. Kunzru and R. W. Frei, *J. Chromatogr. Sci.*, *12*:191 (1974).
32. M. Caude and A. Foucault, *Anal. Chem.*, *51*:459 (1979).
33. A. Foucault, M. Caude, and L. Oliveros, *J. Chromatogr.*, *185*: 345 (1979).
34. F. Guyon, A. Foucault, and M. Caude, *J. Chromatogr.*, *186*:677 (1979).
35. A. Foucault and R. Rosset, *J. Chromatogr.*, *317*:41 (1984).
36. F. Guyon, L. Chardonnet, M. Caude, and R. Rosset, *Chromatographia, 20*:30 (1985).
37. E. C. Jennings, Jr., VIII Int. Symp. Column Liquid Chromatogr., May 1984, New York, 2p-30.
38. J. L. Leonard, F. Guyon, and P. Fabiani, *Chromatographia, 18*:600 (1984).
39. Yu. I. Chumakov, M. S. Alyabyeva, and B. D. Kabulov, *Chromatographia, 8*:242 (1975).
40. J. P. Rawat, and M. Iqbal, *Chromatographia, 17*:701 (1983).

41. E. Oelrich, H. Preusch, and E. Wilhelm, *J. High Resol. Chromatogr. Chromatogr. Commun.*, *3*:269 (1980).
42. A. A. Kurganov, and V. A. Davankov, *J. Chromatogr.*, *218*: 559 (1981).
43. L. A. D'Avila, H. Colin, and G. Guiochon, *Anal. Chem.*, *55*:1019 (1983).
44. G. Schomburg and K. Zegarski, *J. Chromatogr.*, *114*:174 (1975).
45. B. Vonach and G. Schomburg, *J. Chromatogr.*, *149*:417 (1978).
46. R. J. Tscherne and G. Capitano, *J. Chromatogr.*, *136*:337 (1977).
47. M. G. M. DeRuyter and A. P. DeLeenheer, *Anal. Chem.*, *51*:43 (1979).
48. P. L. Phelan and J. R. Miller, *J. Chromatogr. Sci.*, *19*:13 (1981).
49. C. Hunter, K. Sugden, and J. G. Lloyd-Jones, *J. Liq. Chromatogr.*, *3*:1335 (1980).
50. K. Sugden, C. Hunter, and J. G. Lloyd-Jones, *J. Chromatogr.*, *204*:195 (1981).
51. E. Grushka, S. Levin, and C. Gilon, *J. Chromatogr.*, *235*:401 (1982).
52. E. Grushka, J. Atamna, C. Gilon, and M. Chorly, *J. Chromatogr.*, *281*:125 (1983).
53. C. H. Lochmüller and H. H. Hangac, *J. Chromatogr. Sci.*, *20*: 171 (1982).
54. C. H. Lochmüller, W. B. Hill, R. M. Porter, and H. H. Hangac, *J. Chromatogr. Sci.*, *21*:70 (1983).
55. R. Wernicke, *J. Chromatogr. Sci.*, *23*:39 (1985).
56. S. Weinstein, M. H. Engel, and P. E. Hare, *Analyt. Biochem.*, *121*:370 (1982).
57. S. Weistein and N. Grinberg, *J. Chromatogr.*, *318*:117 (1985).
58. E. Gil-Av and S. Weinstein, in *Handbook of HPLC for the Separation of Amino Acids, Peptides, and Proteins*, Vol. 1 (W. S. Hancock, ed.), CRC Press, Boca Raton, 1984, p. 429.
59. S. Weinstein, *Angew. Chem., Int. Ed. Engl.*, *21*:218 (1982).
60. V. A. Davankov and A. A. Kurganov, *Chromatographia*, *17*:686 (1983).
61. C. Gilon, R. Leshem, Y. Tapuhi, and E. Grushka, *J. Am. Chem. Soc.*, *101*:7612 (1979).
62. C. Gilon, R. Leshem, and E. Grushka, *J. Chromatogr.*, *203*:365 (1981).
63. N. Nimura, T. Suzuki, Y. Kasahara, and T. Kinoshita, *Anal. Chem.*, *53*:1380 (1981).
64. N. Nimura, A. Toyama, and T. Kinoshita, *J. Chromatogr.*, *316*: 547 (1984).
65. W. Lindner, J. N. LePage, G. Davies, D. E. Seitz, and B. L. Karger, *J. Chromatogr.*, *185*:323 (1979).
66. Y. Tapuhi, N. Miller, and B. L. Karger, *J. Chromatogr.*, *205*: 325 (1981).

67. S. K. Lam and F. K. Chow, *J. Liq. Chromatogr.*, *3*:1579 (1980).
68. S. Lam, *J. Chromatogr. Sci.*, *22*:416 (1984).
69. S. Lam, F. Chow, and A. Karmen, *J. Chromatogr.*, *199*:295 (1980).
70. S. Weinstein and S. Weiner, *J. Chromatogr.*, *303*:242 (1984).
71. S. Lam and A. Karmen, *J. Chromatogr.*, *239*:451 (1982).
72. V. A. Davankov and A. A. Kurganov, *Chromatographia*, *17*:686 (1983).
73. V. A. Davankov, A. S. Bochkov, A. A. Kurganov, P. Roumeliotis, and K. K. Unger, *Chromatographia*, *13*:677 (1980).
74. W. Klemisch, A. von Hodenberg, and K. O. Vollmer, *High Resol. Chromatogr. Chromatogr. Commun.*, *4*:535 (1981).
75. I. Benecke, *J. Chromatogr.*, *291*:155 (1984).
76. F. K. Chow and E. Grushka, *Anal. Chem.*, *49*:1756 (1977).
77. V. K. Dua and C. A. Bush, *J. Chromatogr.*, *244*:128 (1982).
78. R. G. Masters and D. E. Leyden, *Anal. Chim. Acta*, *98*:9 (1978).
79. M. Gimpel and K. K. Unger, *Chromatographia*, *16*:117 (1982).
80. F. K. Chow and E. Grushka, *Anal. Chem.*, *50*:1346 (1978).
81. E. Grushka and F. K. Chow, *J. Chromatogr.*, *199*:283 (1980).
82. G. J. Shahwan and J. R. Jezorek, *J. Chromatogr.*, *256*:39 (1983).
83. A. S. Bochkov, Yu. A. Zolotarev, Yu. P. Belov, and V. A. Davankov, Proc. Second Danube Symp. Progress in Chromatography, Carlsbad, Czechoslovakia, 1979, Paper B3.23.
84. P. Roumeliotis, K. K. Unger, A. A. Kurganov, and V. A. Davankov, *Angew. Chem., Int. Ed. Engl.*, *21*:930 (1982).
85. P. Roumeliotis, A. A. Kurganov, and V. A. Davankov, *J. Chromatogr.*, *266*:439 (1983).
86. P. Roumeliotis, K. K. Unger, A. A. Kurganov, and V. A. Davankov., *J. Chromatogr.*, *255*:51 (1983).
87. G. Gübitz, W. Jellenz, G. Löfler, and W. Santi, *J. High Resol. Chromatogr. Chromatogr. Commun.*, *2*:145 (1979).
88. G. Gübitz, W. Jellenz, and W. Santi, *J. Liq. Chromatogr.*, *4*:701 (1981).
89. G. Gübitz, W. Jellenz, and W. Santi, *J. Chromatogr.*, *203*:377 (1981).
90. G. Gübitz, F. Juffman, and W. Jellenz, *Chromatographia*, *16*:103 (1983).
91. G. Gübitz and S. Mihellyes, *Chromatographia*, *19*:257 (1984).
92. W. Lindner, *Naturwiss.*, *67*:354 (1980).
93. H. Engelhardt and S. Kromidas, *Naturwiss.*, *67*:353 (1980).
94. N. Watanabe, *J. Chromatogr.*, *260*:75 (1983).
95. B. Feibush, M. J. Cohen, and B. L. Karger, *J. Chromatogr.*, *282*:3 (1983).
96. L. R. Gelber, B. L. Karger, J. L. Neumeyer, and B. Feibush, *J. Am. Chem. Soc.*, *106*:7729 (1984).
97. A. A. Kurganov, A. B. Tevlin, and V. A. Davankov, *J. Chromatogr.*, *261*:223 (1983).

98. H. G. Kicinski and A. Kettrup, *Fresenius Zeit. Anal. Chem.*, *320*:51 (1985).
99. J. Maslowska and W. Pietek, *J. Chromatogr.*, *201*:293 (1980).
100. V. Shaw and H. F. Walton, *J. Chromatogr.*, *68*:267 (1972).
101. V. Shankar and P. N. Joshi, *J. Chromatogr.*, *90*:99 (1974); *ibid.*, *95*:65 (1974).
102. R. M. Kothari, *J. Chromatogr.*, *64*:85 (1972).
103. W. Funasaka, K. Fujimura, and S. Kuriyama, *Bunseki Kagaku*, *18*:19 (1969); *ibid*, *19*:104 (1970).
104. K. Fujimura, T. Koyama, T. Tanigawa, and W. Funasaka, *J. Chromatogr.*, *85*:101 (1973).
105. W. Funasaka, T. Hanai, K. Fujimura, and T. Ando, *J. Chromatogr.*, *78*:424 (1973).
106. L. A. E. Sluyterman and J. Wijdenes, *Biochem. Biophys. Acta*, *200*:593 (1970).
107. L. R. Snyder, *Anal. Chem.*, *41*:314 (1969).
108. P. V. Webster, J. N. Wilson, and M. C. Franks, *Anal. Chem. Acta*, *38*:193 (1967).
109. P. V. Webster, J. N. Wilson, and M. C. Franks, *J. Inst. Petrol.*, *56* :50 (1970).
110. J. Otto, C. M. DeHernandez, and H. F. Walton, *J. Chromatogr.*, *247*:91 (1982).
111. J. Inczedy, P. Klatsmanyi-Gabor, and L. Erdey, *Acta Chim. Acad. Sci. Hung.*, *69*:137, 265 (1971).
112. O. R. Skorokhod and A. A. Kalinina, *Zh. Fiz. Khim.*, *48*:2830 (1974); *ibid.*, *49*:317 (1975).
113. G. N., Altshuler, E. A. Saveliev, and M. Kh. Achmetov, *Zh. Fiz. Khim.*, *50*:1263 (1976).
114. K. Shimomura, T. J. Hsu, and H. F. Walton, *Anal. Chem.*, *45*:501 (1973).
115. K. Shimomura, L. Dickson, and H. F. Walton, *Anal. Chim. Acta*, *37*:102 (1967).
116. J. D. Navratil, E. Murgia, and H. F. Walton, *Anal. Chem.*, *47*:122 (1975).
117. J. D. Navratil and H. F. Walton, *Anal. Chem.*, *47*:2443 (1975).
118. Y. Arikawa and K. Tochida, *Hitachi Rev.*, *16*:236 (1967); *Chem. Abstr.*, *69*:16008u (1968).
119. Y. Arikawa, Br. Patent 1,173,996 (1967); *Chem. Abstr.*, *72*:45513x (1970).
120. F. W. Wagner and S. L. Shepherd, *Anal. Biochem.*, *41*:314 (1971).
121. L. E. Fitt, W. Hassler, and D. E. Just, *J. Chromatogr.*, *187*:381 (1980).
122. S. J. Angyal, G. S. Bethell, and R. J. Beveridge, *Carbohydrate Res.*, *73*:9 (1979).
123. C. Vidal-Valverde, B. Olmedilla, and C. Martin-Villa, *J. Liq. Chromatogr.*, 7:2003 (1984).

124. H. D. Scobell and K. M. Brobst, *J. Chromatogr.*, *212*:51 (1981).
125. R. W. Goulding, *J. Chromatogr.*, *103*:229 (1975).
126. A. C. Lanser and E. A. Emken, *J. Chromatogr.*, *256*:460 (1981).
127. J. D. Wathen, *J. Chromatogr. Sci.*, *14*:513 (1976).
128. N. W. H. Houx and S. Voerman, *J. Chromatogr.*, *129*:456 (1976).
129. M. Doury-Berthod, C. Poitrenaud, and B. Tremillon, *J. Chromatogr.*, *131*:73 (1977), *ibid.*, *179*:37 (1979).
130. C. M. de Hernandez and H. F. Walton, *Anal. Chem.*, *44*:890 (1972).
131. E. Murgia and H. F. Walton, *J. Chromatogr.*, *104*:417 (1975).
132. J. W. Vogh and J. E. Dooley, *Anal. Chem.*, *47*:816 (1975).
133. R. A. A. Muzzarelli, A. F. Martelli, and O. Tubertini, *Analyst*, *94*:616 (1969).
134. R. A. A. Muzzarelli, F. Tanfani, M. G. Muzzarelli, G. Scarpini, and R. Rocchtti, *Sep. Sci. Technol.*, *13*:869 (1978).
135. R. Gräsbeck and R. Karlssohn, *Acta Chem. Scand.*, *17*:1 (1963).
136. E. Rothenbühler, R. Waibel, and J. Solms, *Anal. Biochem.*, *97*:367 (1979).
137. D. K. J. Tommel, J. F. G. Vliegenthart, T. J. Penders, and J. F. Arens, *Biochem. J.*, *107*:335 (1968), *ibid.*, *99*:48 (1966).
138. S. V. Rogozhin, V. V. Korshak, V. A. Davankov, and L. A. Maslova, *Vysokomol. Soed.*, *8*:1275 (1966).
139. V. A. Davankov, S. V. Rogozhin, V. V. Korshak, and M. P. Tsyurupa, *Isv. Akad. Nauk USSR, Ser. Khim.*, 1612 (1967).
140. S. V. Rogozhin, V. A. Davankov, and V. V. Korshak, *Isv. Akad. Nauk USSR, Ser. Khim.*, 1498 (1966).
141. V. V. Korshak, S. V. Rogozhin, and V. A. Davankov, *Vysokomol. Soed.*, *8*:1686 (1966); *Polym. Sci. USSR*, *8*:1860 (1966).
142. S. V. Rogozhin, V. A. Davankov, and V. V. Korshak, *Vysokomol. Soed.* *10A*:1283 (1968).
143. S. V. Rogozhin, V. A. Davankov, S. G. Vyrbanov, and V. V. Korshak, *Vysokomol. Soed.*, *10A*:1277 (1968).
144. V. A. Davankov, S. V. Rogozhin, and I. I. Piesliakas, *Vysokomol. Soed*, *14B*:276 (1972).
145. S. V. Rogozhin, V. A. Davankov, V. V. Korshak, V. Vesa, and L. A. Belchich, *Isv. Akad. Nauk USSR, Ser. Khim.*, 502 (1971).
146. V. A. Davankov, S. V. Rogozhin, and M. P. Tsyurupa, U.S. Patent 3729457 (1970); *Chem. Abstr.*, *75*:6841v (1971).
147. V. A. Davankov, S. V. Rogozhin, and M. P. Tsyurupa, in *Ion Exchange and Solvent Extraction*, Vol. 7 (J. A. Marinsky and Y. Marcus, eds.), Marcel Dekker, New York, 1977, p. 29.
148. V. A. Davankov and M. P. Tsyurupa, *Angew. Macromol. Chem.*, *91*:127 (1980).
149. G. I. Rosenberg, A. S. Shabaeva, V. S. Moryakov, T. G. Musin, M. P. Tsyurupa, and V. A. Davankov, *Reactive Polym.*, *1*:175 (1983).

150. M. P. Tsyurupa, V. A. Davankov, and S. V. Rogozhin, *J. Polym. Sci.*, (Symp.), *47*:189 (1974).
151. V. A. Davankov, S. V. Rogozhin, and M. P. Tsyurupa, *Angew. Makromol. Chem.*, *32*:145 (1973).
152. V. A. Davankov, M. P. Tsyurupa, and S. V. Rogozhin, *Angew. Makromol. Chem.*, *53*:19 (1976).
153. V. A. Davankov, S. V. Rogozhin, I. I. Piesliakas, and V. S. Vesa, *Vysokomol. Soed.*, *15B*:115 (1973).
154. V. A. Davankov, *Habilitationsschrift*, Ineos, Moscow, 1975.
155. S. V. Rogozhin, I. A. Yamskov, V. A. Davankov, T. F. Kolesova, and V. M. Voevodin, *Vysokomol. Soed.*, *17A*:564 (1975).
156. S. V. Rogozhin, I. A. Yamskov, and V. A. Davankov, *Vysokomol. Soed.*, *17B*:107 (1975).
157. S. V. Rogozhin, I. A. Yamskov, and V. A. Davankov, *Vysokomol. Soed.*, *16B*:849 (1974).
158. S. V. Rogozhin, V. A. Davankov, and I. A. Yamskov, *Isv. Akad. Nauk USSR, Ser. Khim.*, 2325 (1971).
159. V. A. Davankov, S. V. Rogozhin, I. A. Yamskov, and V. P. Kabanov, *Isv. Akad. Nauk USSR, Ser. Khim.*, 2327 (1971).
160. S. V. Rogozhin, V. A. Davankov, I. A. Yamskov, and V. P. Kabanov, *Zh. Obshch. Khim.*, *42*:1614 (1972).
161. B. B. Berezin, I. A. Yamskov, and V. A. Davankov, *J. Chromatogr.*, *261*:301 (1983).
162. S. V. Rogozhin, V. A. Davankov, and I. A. Yamskov, *Vysokomol. Soed.*, *15B*:216 (1973).
163. S. V. Rogozhin, V. A. Davankov, I. A. Yamskov, and V. P. Kabanov, *Vysokomol. Soed.*, *14B*:472 (1972).
164. V. A. Davankov and Yu. A. Zolotarev, *J. Chromatogr.*, *155*: 295 (1978).
165. I. A. Yamskov, B. B. Berezin, V. E. Tikhonov, L. A. Belchich, and V. A. Davankov, *Bioorgan. Khim.*, *4*:1170 (1978).
166. S. V. Rogozhin, I. A. Yamskov, A. S. Pushkin, L. A. Belchich, L. Ya. Zhuchkova, and V. A. Davankov, *Isv. Akad. Nauk USSR, Ser. Khim.*, 2378 (1976).
167. I. A. Yamskov, B. B. Berezin, and V. A. Davankov, *Makromol. Chem.*, *179*:2121 (1978).
168. Yu. P. Belov, S. V. Rogozhin, and V. A. Davankov, *Isv. Akad. Nauk USSR, Ser. Khim.*, 2320 (1973).
169. Yu. P. Belov, V. A. Davankov, and S. V. Rogozhin, *Isv. Akad. Nauk USSR, Ser. Khim.*, 1856 (1977).
170. R. V. Snyder, R. J. Angelichi, and R. B. Meck, *J. Am. Chem. Soc.*, *94*:2660 (1972).
171. J. Jozefonvicz, M. A. Petit, and A. Szubarga, *J. Chromatogr.*, *147*:177 (1978).
172. N. Spassky, M. Riex, J. Quette, M. Quette, and J. Blanchard, *J. Compt. Rend.*, *C287*:589 (1978).

173. E. Tsuchida, H. Nishikawa, and E. Terada, *Eur. Polym. J.*, *12*:611 (1976).
174. V. S. Vesa, *Zh. Obshch. Khim.*, *42*:2780 (1972).
175. A. A. Kurganov, L. Ya. Zhuchkova, and V. A. Davankov, *Makromol. Chem.*, *180*:2101 (1979).
176. V. A. Davankov, S. V. Rogozhin, and A. V. Semechkin, *J. Chromatogr.*, *91*:493 (1974).
177. V. A. Davankov and Yu. A. Zolotarev, *J. Chromatogr.* *155*:285 (1978).
178. V. A. Davankov and Yu. A. Zolotarev, *J. Chromatogr.* *155*:303 (1978).
179. V. A. Davankov and A. A. Kurganov, *Chromatographia, 13*:339 (1980).
180. I. I. Piesliakas, S. V. Rogozhin, and V. A. Davankov, *Zh. Obshch. Khim.*, *44*:468 (1974).
181. V. A. Davankov, Yu. A. Zolotarev, and A. A. Kurganov, *J. Liq. Chromatogr.*, *2*:1191 (1979).
182. Yu. A. Zolotarev, N. F. Myasoedov, V. I. Penkina, O. V. Petrenik, and V. A. Davankov, *J. Chromatogr.*, *207*:63 (1981).
183. N. F. Myasoedov, O. B. Kuznetsova, O. V. Petrenik, V. A. Davankov, and Yu. A. Zolotarev, *J. Labelled Comp. Radiopharm.*, *17*:439 (1979).
184. Yu. A. Zolotarev, N. F. Myasoedov, V. I. Penkina, I. N. Dostovalov, O. V. Petrenik, and V. A. Davankov, *J. Chromatogr.*, *207*:231 (1981).
185. V. A. Davankov, Yu. A. Zolotarev, and A. V. Tevlin, *Bioorgan. Khim.*, *4*:1164 (1978).
186. V. A. Shirokov, V. A. Tsyryapkin, L. V. Nedospasova, A. A. Kurganov, and V. A. Davankov, *Bioorgan. Khim.*, *9*:878 (1983).
187. B. Lefebvre, R. Audebert, and C. Quivoron, *Israel J. Chem.*, *15*:69 (1977).
188. B. Lefebvre, R. Audebert, and C. Quivoron, *J. Liq. Chromatogr.*, *1*:761 (1978).
189. J. Boue, R. Audebert, and C. Quivoron, *J. Chromatogr.*, *204*: 185 (1981).
190. I. A. Yamskov, B. B. Berezin, V. A. Davankov, Yu. A. Zolotarev, I. N. Dostovalov, and N. F. Myasoedov, *J. Chromatogr.*, *217*:539 (1981).
191. N. Watanabe, N. Ohzeki, and E. Niki, *J. Chromatogr.*, *216*:406 (1981).
192. E. Sulkovski, *Trends Biotechnol.*, *3*:1 (1985).
193. J. Porath and B. Olin, *Biochemistry*, *22*:1621 (1983).
194. A. Gozdzicka-Jozefiak and J. Augustyniak, *J. Chromatogr.*, *131*:91 (1977).
195. Y. Moroux, E. Boschetti, and J. M. Egly, *LKB Instrument Journal*, *32*(1):1 (1985).

196. V. P. Varlamov, S. A. Lopatin, and S. V. Rogozhin, *Bioorgan. Khim.*, *10*:927 (1984).
197. L. Fanou-Ayi and M. Vijayalakshmi, in Annales of the N. Y. Academy of Sciences, Biochem. Engeneering III, Venkatasubra-acian, K., Constantinides, A., and Vieth, W. R., Eds., Vol. 413, N. Y., p. 300 (1983).
198. N. Ramadan and J. Porath, *J. Chromatogr.*, *321*:81, 93, 105 (1985).
199. N. R. M. Buist and D. O'Brien, *J. Chromatogr.*, *29*:398 (1967).
200. J. F. Bellinger and N. R. M. Buist, *J. Chromatogr.*, *87*:513 (1973).
201. J. Boisseau and P. Jouan, *J. Chromatogr.*, *54*:231 (1971).
202. J. Boisseau and P. Jouan, *Bull. Soc. Chim. France*, 153 (1973).
203. R. Maurer, *J. Biochem. Biophys. Meth.*, *2*:183 (1980).
204. B. Hemmasi, *J. Chromatogr.*, *104*:367 (1975).
205. B. Hemmasi and E. Bayer, *J. Chromatogr.*, *109*:43 (1975).
206. B. M. Petronio, E. De Caris, and L. Iannuzzi, *Talanta, 29*:691 (1982).
207. B. M. Petronio, A. Lagana, and G. D. Andrea, *Talanta, 31*:357 (1984).
208. S. M. Bak, *Daehan Hwahak Hwoejee, 11*(2):56 (1967); *Chem. Abstr.*, *69*:16030v (1968).
209. P. N. Bedetti, V. Carunchio, and A. Marino, *J. Chromatogr.*, *95*:127 (1974).
210. J. C. Wolford, A. J. Dean, and G. Goldstein, *J. Chromatogr.*, *62*:148 (1971).
211. G. Goldstein, *Anal. Biochem.*, *20*:477 (1967).
212. C. A. Burtis and G. Goldstein, *Anal. Biochem.*, *23*:502 (1968).
213. P. Hubert and J. Porath, *J. Chromatogr.*, *198*:247 (1980), *ibid.*, *206*:164 (1981).
214. S. V. Rogozhin and V. A. Davankov, *Usp. Khim.*, *37*:1327 (1968); *Russ. Chem. Rev.*, *37*:565 (1968).
215. S. V. Rogozhin, V. A. Davankov, and I. I. Piesliakas, in *Chemistry and Technology of High Molecular Compounds*, Vol. 4 (A. M. Sladkov, ed.), VINITI, Moscow, 1973, p. 45.
216. F. K. Chow and E. Grushka, *J. Chromatogr.*, *185*:361 (1979).
217. D. Corradini, M. Sinibaldi, and A. Messina, *J. Chromatogr.*, *235*:273 (1982).
218. C. A. Chang, C. S. Huang, and C. F. Tu, *Anal. Chem.*, *55*: 1390 (1983).
219. C. A. Chang, *Anal. Chem.*, *55*:971 (1983).
220. C. A. Chang, C. F. Tu, and C. S. Huang, *J. Chromatogr. Sci.*, *22*:321 (1984).
221. J. Gaal and J. Inszedy, *Talanta, 23*:78 (1976).
222. Y. Yoshikawa and K. Yamasaki, *Coord. Chem. Rev.*, *28*:205 (1979).
223. H. Yoneda, *J. Chromatogr.*, *313*:59 (1985).

# 10

# Ion Exchangers

Donald J. Pietrzyk / *The University of Iowa, Iowa City, Iowa*

## INTRODUCTION

### History of Ion Exchange

Ion exchangers are by definition insoluble solid substances that contain charge centers and exchangeable counterions of opposite charge. When the ion exchanger is in contact with a solution of an electrolyte its counterions can be exchanged for an equivalent amount of ions of the same charge provided by the electrolyte. If the exchangeable counterion provided by the insoluble solid substance is a cation, it is participating in cation exchange and the solid substance is commonly called a cation exchanger. If it is an anion, anion exchange is taking place and the substance is called an anion exchanger.

A typical cation exchange process is given by

$$LiR(s) + K^{+}Cl^{-}(aq) \rightleftharpoons KR(s) + Li^{+}Cl^{-}(aq) \quad (1)$$

while a typical anion exchange process is given by

$$RCl(s) + K^{+}I^{-}(aq) \rightleftharpoons RI(s) + L^{+}Cl^{-}(aq) \quad (2)$$

where R is the ion exchanger matrix or structural unit containing a fixed-charge site and an equivalent amount of a counterion of opposite charge, (aq) is the aqueous phase, and (s) indicates the solid phase. In both cases the exchange is stoichiometric, i.e., $K^+$ replaces $Li^+$ and $I^-$ replaces $Cl^-$ on an equivalent basis. Furthermore, the exchange can continue until all the $Li^+$ or $Cl^-$ in Eqs. (1) and (2) is replaced by $K^+$ and $I^-$, respectively. If in Eq. (1), for example, the solution electrolyte is $MgCl_2$, the exchange process takes place on an equivalent basis or $2Li^+$ is replaced per $Mg^{2+}$.

In general, ion exchange processes are reversible. Thus, Eqs. (1) and (2) can be reversed by treatment of KR(s) and RI(s) with $Li^+Cl^-(aq)$ and $K^+Cl^-(aq)$, respectively. This means that when the exchanger becomes exhausted, i.e., all the $Li^+$ in Eq. (1) and all the $Cl^-$ in Eq. (2) is replaced, they can be regenerated back to their original forms by treatment with $Li^+Cl^-(aq)$ and $K^+Cl^-(aq)$ solutions, respectively. Regeneration back to their original forms is also possible by treatment with $Li^+Br^-(aq)$ and $Li^+Cl^-(aq)$, respectively, indicating that the exchange in the first case involves only cations and in the second only anions. The ions of opposite charge in the salt solution, assuming no competing equilibria are present, do not participate in the exchange process other than in maintaining electroneutrality. Recognition of the ion exchange process is not a recent occurrence. Thompson in 1850 [1] and Way in 1850 [2] and 1852 [3] carried out the first definitive studies in what they called "base exchange" (cation exchange). In attempts to understand why ammonia was lost from manure they passed ammonium sulfate solutions through columns packed with soil and observed that the effluent from the column contained no ammonium salts but considerable amounts of calcium sulfate. These workers suggested that the exchange process, which they observed, was caused chiefly by the clay fraction of the soil. Furthermore, Way described base exchange (cation exchange) in an essentially correct interpretation of ion exchange, as the conversion of an insoluble aluminosilicic acid from one salt form to another. Interestingly, this work was done prior to the understanding of electrolytic dissociation of salts.

These early studies were of great significance to agricultural chemistry and because of this interest further development of the understanding of the ion exchange process occurred. Subsequently, other workers showed that clays of various kinds, glauconites, zeolites, humic acids, and other naturally occurring minerals and salts were capable of participating in ion exchange.

Ion exchange was eventually recognized as a unit operation along with other separation strategies such as filtration, distillation, and adsorption. At the turn of the century and up to 1935, the type of ion exchangers used were naturally occurring clays, zeolites, minerals, oxide salts, humic acids, and related material. During this time zeolites and claylike materials, which were shown to participate in ion exchange, were synthesized in the laboratory, were well characterized, and replaced many of the naturally occurring materials that were used. In 1935 a major breakthrough occurred in ion exchange technology with the discovery by Adams and Holmes [4] that organic polymeric ion exchangers could be synthesized in the laboratory. This subsequently proved to be a remarkable achievement for several reasons. First, the organic polymeric ion exchangers provided ion exchange properties that were markedly improved over the inorganic mineral and organic naturally occurring type ion exchangers. Second, the synthetic procedures were such that organic polymeric ion exchangers could be prepared with reproducible properties. Third, the development of the synthesis allowed researchers to carefully and systematically alter the properties of the resulting organic polymeric ion exchangers. And fourth, the ability to be able to systematically change the organic polymeric ion exchangers aided the rapid development of the fundamental understanding of ion exchange.

Following the emergence of the organic polymeric ion exchanger, industrial applications flourished. For example, ion exchange strategies were developed for metal recovery from industrial wastes; to separate rare earths; to isolate, concentrate, and separate organic molecules of biochemical and pharmaceutical interest; to isolate and concentrate radioactive ions including man-made elements; as catalysts in various synthetic operations; and, perhaps of most significance, to purify and demineralize water. Even though the analytical separation applications of columns of ion exchangers yielded low-efficiency-type separations relative to todays standards, they were of tremendous importance and remain so even today. Not only are the low-efficiency separations still useful, but they provide the basis for modern, highly efficient column ion exchange strategies and for the optimization of ion exchanger and mobile phase parameters.

In the 1960s column liquid chromatography entered an era of rapid development that yielded vast improvements in column efficiency, resolution of complex mixtures, instrumental control of the separation, and detection of the separated compounds. Column liquid chromatography prior to this period including ion exchange column liquid chromatography, was low in efficiency, and yielded column plate heights that at best approached about 10 mm. At present, columns often are capable of yielding column plate heights better than 0.01 mm. Column chromatography that utilizes highly efficient columns of this type is presently referred to as high-performance liquid chromatography (HPLC). It is not surprising that existing ion exchangers, like other common stationary phases known prior to 1960, were evaluated as stationary phases in HPLC since the nature of the analyte-stationary phase interaction remains essentially the same.

HPLC requires that the column be made from stationary phase microparticles (as particle size decreases, column efficiency increases) that are uniform in size. Interest in small, uniform ion exchange particles had already existed for several years prior to 1960. During this period researchers had begun to realize that small, uniform particles of ion exchangers gave better separations of amino acid mixtures (also for the separation of other charged compounds of biochemical interest) [5] than when using larger ion exchange particles. Thus, small ion exchange particles were available for evaluation. As HPLC developed it appeared that physically stronger and smaller ion exchange particles were needed and research efforts grew toward solving this apparent problem. Early HPLC ion exchange interest focused almost entirely on the separation of organic analyte ions even though classical low-efficiency, column ion exchange liquid chromatography had been shown to be an excellent strategy for the separation of inorganic ions. In 1975 Small and coworkers [6] introduced an ion exchange strategy that permitted the separation and sensitive detection of inorganic analyte ions. Subsequently, this technique, which utilizes highly efficient ion exchange columns, was and still is referred to as ion chromatography. While the major impact of ion chromatography is in inorganic analyte separations and determinations, it is also applicable to organic analyte ion separations [7,8].

In the analytical laboratory applications of ion exchangers and the ion exchange process became and still are a major part of separation science. This chapter introduces the basic principles and properties of ion exchangers and focuses on the emergence of ion exchange as a separation strategy in modern liquid column chromatography and the application and optimization of ion exchangers in separation problems. While the emphasis here is on the ion exchanger and its role in applied separation science, it should be noted that the ion exchange process and its fundamental understanding have been instrumental in the development of many other important areas of science. Some key examples of this are the following:

1. Part of many vital functions in living organisms involve ion exchange.
2. Ion exchange membranes and related phenomena have far-reaching physiological, biophysical, and electrochemical significance.
3. The ion exchange process readily explains certain agricultural, geological, and environmental phenomena related to movement of materials through the soil, water, plants, and environment.
4. The ion exchange process can be taken advantage of in pharmaceutical and agricultural applications for the controlled release of therapeutically useful drugs or agricultural chemicals, respectively.

### The Ion Exchange Process

A key feature of an ion exchanger is that it consists of a framework that is held together by lattice energy or chemical bonds and contains either cation or anion charge centers. The charge at these centers are satisfied by diffuse, freely moving ions (counterions) of opposite charge and can be repalced by other ions of the same charge. In essence, it resembles a porous, insoluble electrolyte and when it is placed in a solution the counterions are free to move into the solution that enters the pores providing electroneutrality is satisfied, i.e., the solution provides simultaneously an equivalent amount of counterion to replace those that are leaving.

Consider a $Li^+$ form cation exchanger in an aqueous solution of KCl as in Eq. (1). The KCl solution permeates the pores of the cation exchanger. This means the pores contain not only the $Li^+$ but also the $K^+$, $Cl^-$, and its solvent. Thus, the electrolyte content of the pores is increased due to the presence of $K^+$ and $Cl^-$. In principle sorption of the KCl has taken place, i.e., the KCl is distributed between the internal pore liquid and the solution external to the pore. $Li^+$ migrates from the exchanger into the solution while $K^+$ migrates from the solution to the exchanger. Eventually, a cation exchange equilibrium is established and the cation exchanger and the solution contain both $Li^+$ and $K^+$. The ratio of $Li^+/K^+$ in the two phases is not necessarily equal and the cation exchange can show a preference for one cation over another. Anion exchange can be described in a similar manner.

The solution as it permeates the pores of the ion exchanger can cause the ion exchanger to swell. Whether swelling is extensive or not depends on the rigidness of the structural framework of the ion exchanger. If the solution contains a solute it also enters the pores with the solvent and consequently undergoes sorption or distribution between the pore liquid and the external solution.

Clearly, the exchanger framework and the coions are essential to maintain electroneutrality. Thus, the ion exchange process is a distribution of counterions between the internal pore liquid and the external solution. Equations (1) and (2), illustrating cation and anion exchange, respectively, are more correctly written as

$$\underline{Li}^+ + K^+ \rightleftharpoons \underline{K}^+ + Li^+ \tag{3}$$

$$\underline{Cl}^- + I^- \rightleftharpoons \underline{I}^- + Cl^- \tag{4}$$

where the underline refers to the interior of the ion exchanger. In Eq. (3) and (4) the concentration ratio of the two counterions in the solid ion exchanger and in the solution are not necessarily the same at equilibrium, indicating that the ion exchanger will show a preference of one counterion over another. Thus, distribution of counterions between the solid ion exchanger pore liquid and the external solution is not purely a statistical one and other factors must influence the distribution. This is one major limitation of viewing the ion exchange process in such a simplified way as done in Eqs. (3) and (4) because these two equilibria showing cation and anion exchange, respectively, do not adequately account for the preference of one counterion over another. A measure of the ion exchanger's preference for one counterion over another is called the ion exchange selectivity. Awareness of ion exchange selectivities, as discussed later, on an absolute or relative basis is one of the major parameters in applying and optimizing the ion exchange strategy for separations because this is the basis for predicting both elution order and eluent strength.

The simple ion exchange model represented by a porous framework containing charge centers and counterions of opposite charge suggests that the kinetics of the counterion exchange within the pore liquid and external solution is a diffusion process. This is the case and the mobilities of the counterions are key factors in determining the rate. Since electrical forces are involved in the exchange process, these must be accounted for in applying simple diffusion rate laws.

Equations (3) and (4), which represent simple cation and anion exchange processes, describe another important characteristic property of ion exchangers, namely, their ion exchange capacity. This constant is a measure of the counterion content of the ion exchanger which also indicates the total number of charge centers within the ion exchanger framework. It is independent of the type of counterion providing the equivalency is taken into account. A shortcoming of the simple model, however, is that some counterion is present in the pore liquid due to sorption of electrolyte. Hence, a correct measure of the ion exchange capacity which reflects the number of framework charge centers must differentiate in terms of quantity between counterions satisfying the framework charge center and those present due to electrolyte sorption.

The simple ion exchange process represented by the simple model shown in Eqs. (3) and (4) representing diffusion and electrostatic attraction suggests that heats of exchange should be small. Additional chemical processes may be part of the overall ion exchange process and contribute to the heat evolved. If these are accounted for, heats of ion exchange are found to be small, often less than 4kJ/eq. [9].

The simple qualitative ion exchange model outlined, while possessing limitations, is still very valuable and useful because it provides the basis for describing the effects of the most important properties of ion exchangers. This is particularly true when applying ion exchangers to separation problems.

## A Quantitative Approach to Ion Exchange

Ion exchangers and the ion exchange process are complicated and their quantitative description in a unifying theory and model is very difficult to accomplish. Quantitative theories and models in the developing field of ion exchange focused on inorganic ion exchangers and the equilibria associated with the ion exchange process. While these initial approaches were valuable and served as the basis for future work, they were in general limited as organic polymeric-type ion exchangers became the ion exchanger of choice. The major reason for this is that early theories and models did not deal with the properties of swelling and electrolyte sorption, and subsequently the effect of these parameters on various aspects of ion exchange equilibria.

It is beyond the scope of this chapter to document and discuss the historical quantitative development of ion exchange. Several other sources focus on this in detail where inorganic ion exchangers [10,11] and organic polymeric ion exchangers [12-16] are considered.

A common underlying view of early researchers, largely because of the prominence of inorganic ion exchangers, was to describe ion exchangers as solid solutions of strong electrolytes or concentrated aqueous solutions of strong electrolytes in which one ionic species is immobile (see, for example, the work of Kielland [17], Vanselow [18], and Bauman and Eichhorn [19]). It wasn't until about 1948, when the work of Gregor first appeared, that swelling factors were taken into account thermodynamically. Gregor [20,21] suggested that the organic polymeric framework (or the ion exchanger matrix) is a network of elastic springs. When the exchanger swells, the framework stretches causing a pressure on the pore liquid. This swelling pressure effects ion exchanger swelling, sorption, and ion exchange equilibria, which are three key parameters that are vital in applying ion exchangers. While this approach was a major step forward, it had limitations because it essentially forcused on the process as being mechanical in that swelling pressure was the major property considered. Subsequent work by Lazare and Gregor [22] modified this model to include the effects of electrostatic interaction. This was done by suggesting that the ion exchanger is a series of parallel planar plates, which carry a uniform electric surface charge, connected by elastic plates. The plates repel one another because of the charge causing the elastic springs to stretch. The pore liquid lies between the plates and has a surplus of charge equal to, but of opposite sign to, the plates charge that is distributed in such a way that it is a minimum halfway between the plates.

In contrast to the macroscopic approach of Gregor, others have attempted to view ion exchange on the molecular scale (see, for example, the work of Katchalsky and coworkers [23-25] and Rice and coworkers [26]). The main starting point of these models, which is based on ideas that were originally applied to linear polyelectrolytes, is to view the organic polymeric matrix as crosslinked chains of interconnected rigid, rodlike segments, each of which contains one inorganic group. The elasticity of the matrix is due to entropy increase rather than the mechanical increase that accompanies chain coiling, which can occur in many ways. They differ in how electrostatic interactions are accounted for, the contribution of ion pairing and its effect, and the effect of the degree of crosslinking.

These models as well as others [10-16] are mathematically complex, often for the simplest cases. While they are important from a fundamental point of view and have certainly encouraged researchers to pursue their experimental verification in a variety of ways, they are difficult to use in a practical, predictive way. The major problem is that it is difficult, and perhaps impossible, to account for all the properties exhibited by ion exchangers in one, unifying, descriptive model.

## PREPARATION OF ION EXCHANGERS

### Synthesis of Organic Polymeric Iion Exchangers

A turning point in ion exchange occurred in 1935 when Adams and Holmes [4] reproted the first study that lead to the synthesis of ion exchangers containing an organic, polymeric matrix and chemically bonded ionogenic groups. Prior to this, ion exchangers were synthetic and naturally occurring inorganic materials and naturally occurring or synthetically modified high molecular weight organic matter containing ionogenic groups. Once the idea of an organic polymeric-type ion exchanger was introduced and advantages were realized, the major effort toward developing new exchangers was in this direction. The key goal is to synthesize a three-dimensional crosslinked matrix which contains ionogenic groups. Two basic approaches were used. One is to polymerize a monomer or comonomers which provides the matrix and then chemically attach the ionogenic groups through appropriate chemical reactions. The second general approach requires synthesis of monomers that contain the ionogenic groups followed by the polymerization step. In general, the first approach is more versatile and is the one usually taken for the synthesis of modern ion exchangers.

The early organic polymeric-based ion exchangers were obtained by condensation polymerization techniques, and the synthetic studies during this time focused on monomers suitable for this polymerization approach. In the late 1940s procedures for the formation of additional polymers using styrene-type monomers were introduced and the focus of the studies quickly turned to these kinds of ion exchange matrices. Styrene-based ion exchangers were quickly shown to be versatile in application and for the most part a styrene-based polymer was the matrix of choice for commercially available ion exchanger. Since about 1960 styrene-based ion exchanger research has focused on (a) refinements in polymerization procedure, (b) development of polymerization procedures which magnified matrix effects, (c) preparation of ion exchangers that would provide uniform, reproducible physical and ion exchange properties, (d) preparation of ion exchangers with a specific type of selectivity, (e) development of ion exchangers that are more compatible with mixed and nonaqueous solvents, and (f) most recently, the development of ion exchangers that are compatible with high-performance liquid chromatographic technology.

As the synthetic technology for the preparation of organic polymeric-based ion exchangers was being developed, the advantages (and the limitations) that this type of ion exchanger offered was realized and synthetic directions focused on emphasizing the advantages and limitations are given below; it should be noted that in general the comparison is between the ion exchange properties provided by organic polymeric ion exchangers and those provided by inorganic ion exchangers.

*Major Advantages of Organic Polymeric-Based Ion Exchangers*

1. Ion exchange capacities are significantly higher, e.g., cation exchanger capacity is often several times larger on organic polymeric cation exchangers than that obtained with many common inorganic cation exchangers even at their most favorable conditions.
2. Ion exchangers are stable throughout the pH range of 1-14 except the anion exchanger whose stability starts to drop off as the environment becomes very basic.
3. Stability is often favorable in oxidizing and reducing solutions.
4. Ion exchange is reversible and reproducible.
5. Ion exchange equivalency is well defined and quantitative.
6. Physical strength is favorable.
7. The synthetic polymerization procedure readily provides well-defined spherical ion exchange beads of appropriate size.
8. Ion exchange kinetics are favorable.
9. Organic polymeric based ion exchangers are virtually insoluble in all common solvents.

*Major Limitations of Organic Polymeric-Based Ion Exchangers*

1. Physical strength is less than that for inorganic ion exchangers unless the matrix is highly crosslinked.
2. Swelling-contracting of the ion exchanger can lead to column channeling.
3. In general, inorganic ion exchangers are more stable than the polymeric ion exchanger toward
   a. radioactive environment.
   b. high temperature.
   c. sudden changes in electrolyte content of its environment.
   d. a large column pressure drop.

Most organic polymeric ion exchangers used in the analytical laboratory are commercial preparations even though the literature provides directions for their synthesis. While many different kinds of organic polymeric ion exchangers have been described in the literature, particularly the patent literature, only a few of these have been subjected to a careful and quantitative determination of their structure and their ion exchange properties. Often suggested structures are based on the monomers used, the polymerization procedure, the expected chemistry, and the assumption that the polymerization is uniform rather than on a careful characterization of the polymer obtained from the reaction. Notable exceptions to this are the commercially available ion exchangers for low-efficiency column liquid chromatography, which are styrene-based and methacrylate-based polymers and modified cellulose and related materials, and those suited to high-efficiency column liquid chromatography (see below). Patents on ion exchanger synthesis including styrene-based polymers continue to be issued and often the differences between the new and old are not major.

This section focuses only on the highlights of the synthesis of ion exchangers suitable for low-efficiency ion exchange column liquid chromatograph (pre-HPLC) and those suitable for high-efficiency ion exchange column liquid chromatography (post-HPLC). Several reviews and reference books focus on the details and provide access to the procedures required for their preparation [12,27-29]. Also, no attempt is made in the following to review in detail the many kinds of inorganic ion exchangers that have been prepared and subsequently evaluated. Their synthesis or preparation from naturally occurring sources are reviewed elsewhere [01,11,30-35].

### *Ionogenic Groups*

A cation ion exchanger has a strong acid or weak acid ionogenic group while an anion exchanger has a strong base or weak base group. Table 1 lists the more common ionogenic groups that have been incorporated into organic polymeric ion exchangers.

Ion exchangers with strong acid and strong base ionogenic groups possess strong electrolytic porperties, are essentially completely dissociated, and are freely used throughout the entire pH range. In contrast, if the ion exchanger has a weak acid or weak base ionogenic group, they provide weak electrolytic properties, are partially dissociated, and are best used in a pH range where the ionogenic group is dissociated. The pH range over which the latter type of ion exchangers are most useful therefore depends on the apparent ionization constant for the group. In Table 1 only $-SO_3H$, $-CO_2H$, $-PO_3H_2$, $-NR_3^+$, and amine-type ion exchangers are commercially available.

Other kinds of groups can be attached to an organic crosslinked polymeric matrix to introduce specific kinds of properties. For example, optically active groups, complexing groups, and oxidizing or reducing groups can be synthetically attached to the matrix; see above for a discussion of these kinds of specific ion exchangers.

### *Polymerization Strategies*

When designing and preparing an organic polymeric matrix the major requirements that must be satisfied are that the polymer (a) be insoluble,

**Table 1** Common Ionogenic Groups in Organic Polymeric Ion Exchangers

| Cation Exchange Group | | |
|---|---|---|
| *Strong acid* | *Weak acid* | |
| $-SO_3^-H^+$ | $-CO_2H$ | $-PHO_2H$ |
| | $C_6H_5-OH$ | $-AsO_3H_2$ |
| | $-PO_3H_2$ | $-SeO_3H$ |
| | $C_6H_5-SH$ | |

| Anion Exchange Group | |
|---|---|
| *Strong base*[a] | *Weak base*[b] |
| $-NR_3^+$ | $-NR_2$ |
| $-PR_3^+$ | $-NHR$ |
| $-SR_2^+$ | $-NH_2$ |

[a]Stored and often used as the halide form rather than the $OH^-$ form because of poor stability of the latter.
[b]The anion exchange site is produced when treated with a strong acid such as HCl to form the hydrochloride salt.

(b) be resistent to oxidation, (c) be physically strong, (d) have porous character, and (e) provide a uniform structure. Two major polymerization routes have been used: condensation polymerization and addition polymerization. In general, a condensation polymerization involves a condensation reaction, usually by an ionic reaction mechanism, occurring between two small polyfunctional monomers by splitting out $H_2O$, $NH_3$, or alcohol. A typical example, which is also one of the earlier developed ion exchangers, is the product from the reaction between phenol and formaldehyde to give the crosslinked polymer shown in Eq. (5). If a *p*-substituted phenol is used, crosslinking does not take

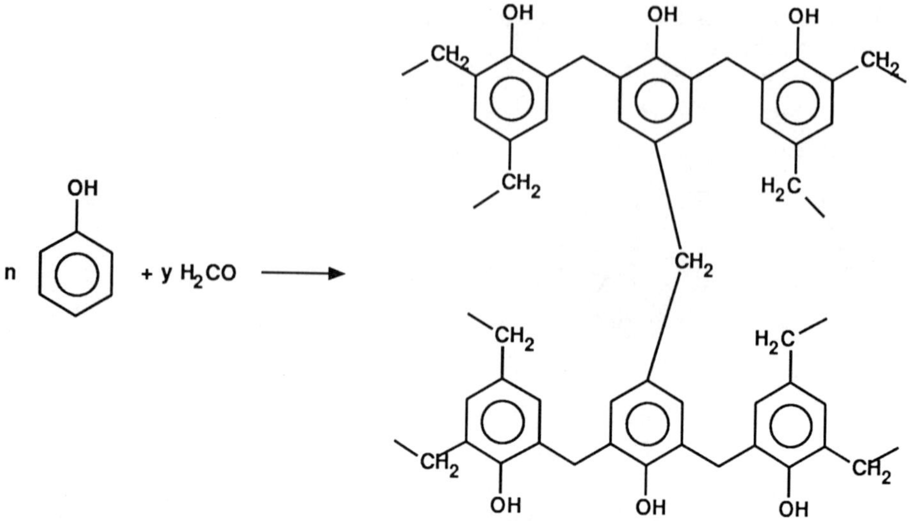

(5)

place and a linear polymer is obtained. Branching and other modified structures can be obtained through the use of mixed phenols and/or substituted phenols. The polymer in Eq. (5) is a weak cation exchanger because of the weak-acid phenolic group. However, it can be converted into a strong-acid cation exchanger by sulfonation; actually, it will be bifunctional since it will contain both strong ($-SO_3H$) and weak (Ar-OH) acid ionogenic groups. A typical example of an anion exchanger prepared by a condensation polymerization strategy is the crosslinked polymer obtained from the reaction of *m*-phenylenediamine and formaldehyde as shown in Eq. (6). This weak base polymer in acid solution

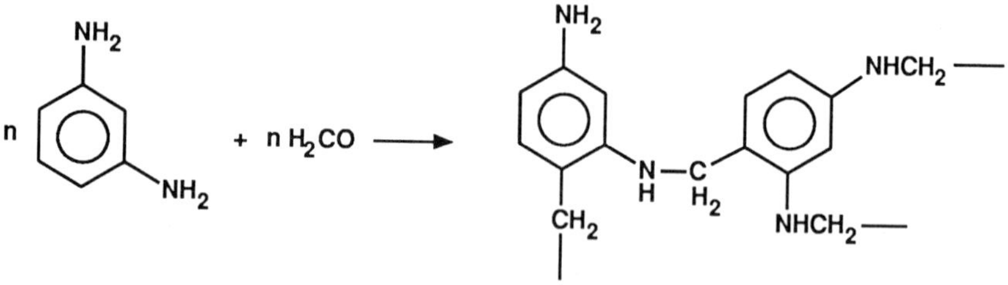

(6)

becomes an anion exchanger. Condensation polymerization, while offering a route to several structurally different types of polymers that have cation or anion exchange properties or can be further modified to yield these properties, is at present more of historical interest than of practical interest, and organic polymeric matrices currently used for ion exchangers are not generally of this type. In general, currently used polymeric ion exchangers are based on an addition polymerization approach.

In general, the major addition polymers which are of ion exchange interest are formed by free radical polymerization of mixtures of olefinic and diolefinic compounds. Since only C-C bonds are present in the resulting polymeric matrix, stability toward hydrolysis, pH, and temperature is markedly improved. One of the more common combinations is the reaction of styrene with divinylbenzene to form the polystyrene-divinylbenzene (PSDB) copolymer as shown in Eq. (7). The

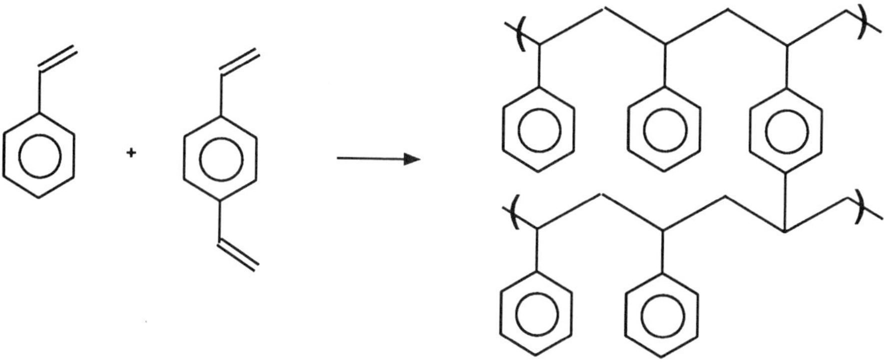

(7)

divinylbenzene provides the crosslinking. The PSDB copolymer and its variations (e.g., crosslinking can be systematically varied by adjustment of the ratio of divinylbenzene to styrene in the reaction mixture) are the polymeric matrix of choice for almost all analytical and large-scale commercial applications of ion exchangers.

Organic polymeric matrices containing ionogenic groups will swell in the presence of polar solvents and electrolytes. Thus, sharp changes in these variables will result in swelling or contracting, and this osmotic shock can cause beads to fracture. Improvements in addition polymers by increasing resistance toward osmotic shock effects can be brought about via synthetic modifications of the suspension polymerization procedure [27] which is generally used to make the gel-type polymer. The gel-type polymer, e.g., a PSDB copolymer, contains 2-16% divinylbenzene which determines the crosslinking. The resulting polymer is not truly porous like a sponge but rather yields its porous-like property because of swelling and expansion between polymer chains. Consequently, its apparent porosity and surface area are dependent on the extent of swelling.

A recent polymeric synthetic modification (27,36-38) in addition polymerization is to use a solvent inert to the polymer but not to the monomers. Refinements of this basic strategy provide what is called a macroporous polymer matrix (also called a porous, isoporous, or macroreticular matrix). By choosing different solvents, a continuous range of porosities can be obtained. The macroporous polymer, which is still derived from styrene and divinylbenzene (other monomers can be used), is highly crosslinked (usually >>16%) and provides a network of microsphere gel beads joined to form a single large bead. Since different crosslinking agents can also be used, the kind of macroporous polymers that can be obtained is increased.

Figure 1 shows electron microscope pictures of typical macroporous PSDB beads. The porosity, which is more permanent-like, can reach pore diameters of over 10 nm while surface areas are often well above 100 $m^2/g$ and are the result of the interstitial spaces between the micro gel beads. In contrast, gel-type PSDB matrices, depending on crosslinking and swelling, have typically apparent porosities of 0.5-5 nm and surface areas less than 0.1 $m^2/g$ (see below for additional discussion of these properties and their effects on the ion exchange process). The macroporous PSDB copolymer will still expand and contract due to the presence of the microsphere gel bead but the degree of swelling is less. Consequently, bead fracture is not as serious as with the gel bead. Furthermore, much of the porosity and surface area is still retained even though the macroporous bead is not in its most favorable swollen condition.

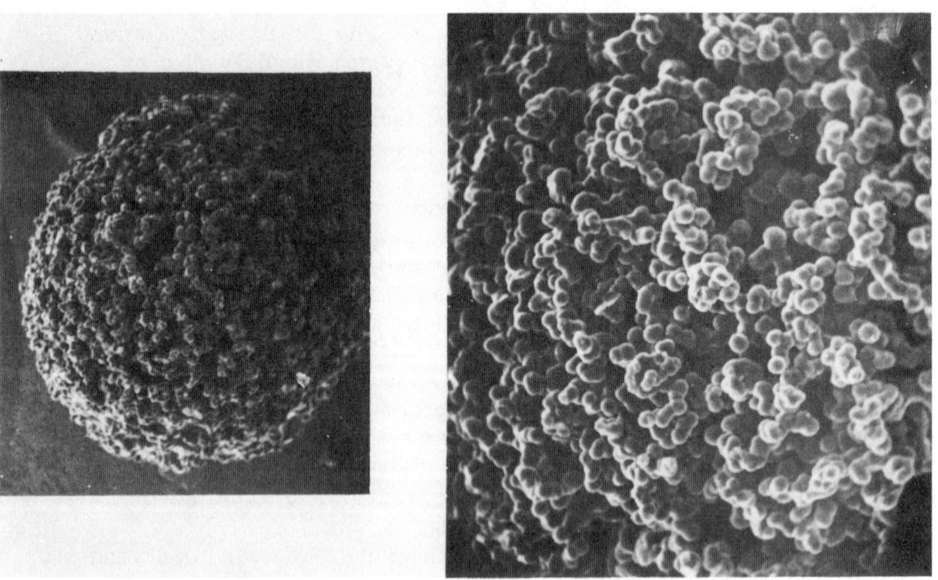

Fig. 1 A scanning electron micrograph of a macroporous quaternary ammonium PSDB anion exchanger Amberlite IRA-938. Left is 100 X magnification and right is 300 X magnification. (Courtesy of Rohm and Haas Company.)

The variables in macroporous polymerization procedure are several fold; a detailed review of the extensive study of these procedures is available [27,36-38]. While improvements have been extensive, problems in polymerization reproducibility from an analytical chromatographic application point of view still exist.

### *Cation Exchangers*

The condensation polymer shown in Eq. (5) provides weak acid cation exchanger(-ArOH ionogenic group) properties. Sulfonation of the polymer or using a sulfonated phenol in the polymerization step yields a bifunctional ion exchanger that contains a strong-acid cation exchange ionogenic group as well as a weak-acid cation exchange ionogenic group.

As previously indicated, PSDB addition polymers are currently the matrix of choice. Thus, a PSDB copolymer is first prepared whereby the crosslinking, spherical character, particle diameter, and internal physical characteristics (gel vs. macroporous) are controlled via the polymerization procedure. The PSDB beads are then sulfonated to yield a strong-acid cation exchanger. Several sulfonation reagents can be used; the more common ones are shown in Eq. (8). In general,

$H_2SO_4 \longrightarrow$

$HSO_3Cl \longrightarrow$

$SO_3 \longrightarrow$

$SO_3^-H^+$ $SO_3^-H^+$

(8)

sulfonation can be done uniformly and will yield cation exchange capacities of about 5 meq/g which corresponds to monosulfonation of the aromatic ring. Contact time and temperature, sulfonation reagent concentrations, organic swelling agents, and the presence of trace polar monomers in the copolymer are variables that are optimized in sulfonation [27]. Since certain sulfonation conditions are also oxidizing ones care must be exercised when pushing the reaction in order to achieve the highest level of sulfonation to avoid polymer charring and degradation.

Several different types of weak-acid cation exchangers containing a carboxyl ionogenic group can be prepared. Condensation polymers derived from dihydroxybenzoic acid and formaldehyde provide this type of exchanger. Addition polymers containing the $-CO_2H$ group are prepared by reaction on the aromatic rings of PSDB; several strategies are shown in Eq. (9). The sulfonium structure can also be converted to a strong-acid cation exchanger by reaction with $Na_2SO_3$ as

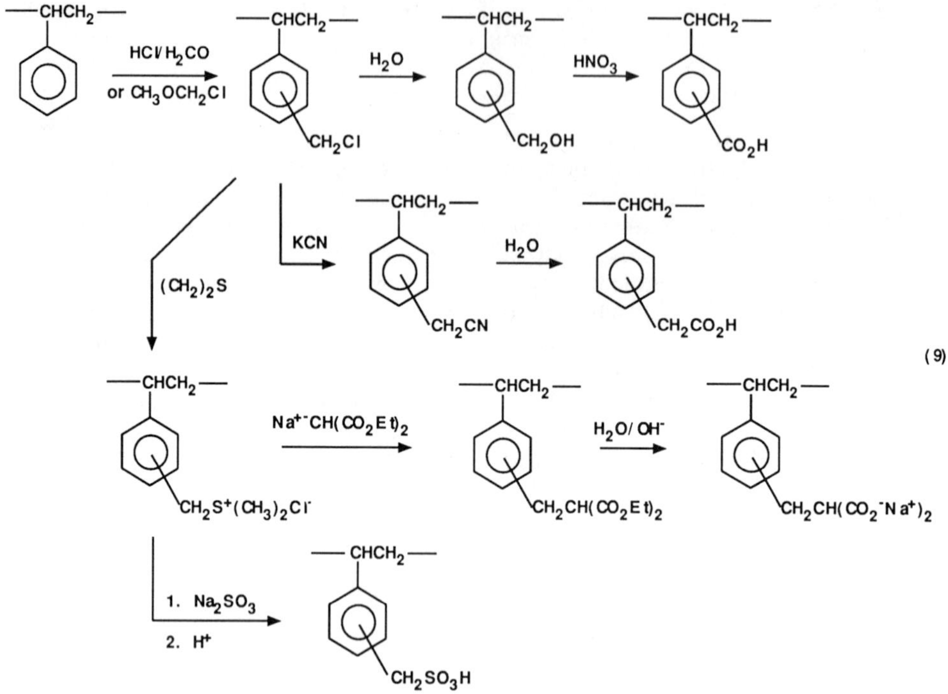

(9)

shown in Eq. (9). In this case the strong-acid exchange site is located on a methylene side chain rather than on the aromatic ring.

Alternatively, acrylic acid, methacrylic acid, their esters, or acrylonitrile can be polymerized with divinylbenzene to form the crosslinked addition copolymer which contains $-CO_2H$ groups. Equation (10) shows the reaction using an acrylic ester followed by hydrolysis. Both gel and macroporous weak-acid cation exchangers can be prepared by this reaction chemistry.

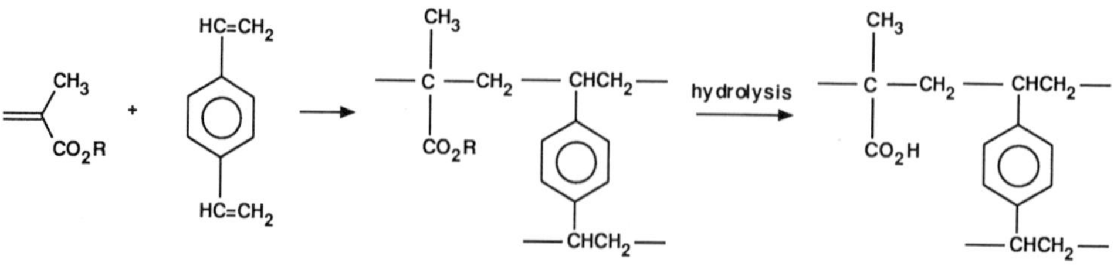

(10)

The aromatic ring system in a PSDB copolymer provides a modestly reactive reaction center for adding groups. Alternately, reactive centers [see Eq. (8)] such as the chloromethyl group and the sulfonium group can be introduced into the polymer and these also provide unique reaction centers. Thus, a phosphonic PSDB cation exchanger can be made by a series of steps of direct reaction with the aromatic ring as shown in Eq. (11) or by reaction of the chloromethylated polymer with triethylphosphite as shown in Eq. (12).

$$-CHCH_2-(C_6H_5) + PCl_3 \xrightarrow{AlCl_3} -CHCH_2-(C_6H_4)PCl_2 \xrightarrow[H_2O]{NaOH} -CHCH_2-(C_6H_4)P(O^-Na^+)_2 \xrightarrow{HNO_3} -CHCH_2-(C_6H_4)PO(OH)_2 \qquad (11)$$

$$-CHCH_2-(C_6H_4)CH_2Cl + P(OEt)_3 \longrightarrow -CHCH_2-(C_6H_4)CH_2PO(OEt)_2 \xrightarrow[H_2O]{NaOH} -CHCH_2-(C_6H_4)CH_2PO(O^-Na^+)_2 \qquad (12)$$

Polyhydroxy high molecular weight compounds can be converted into weak ($-CO_2H$, $-PO_3H_2$) and strong acid ($-SO_3H$) cation exchangers. Crosslinking often is introduced by reaction with epichlorohydrin. Polyhydroxy derivatives such as cellulose, agarose, and other dextrans have been modified this way [39,40]. Improved cellulose-based ion exchangers, in terms of particle size and shape, containing strong-acid and weak-acid exchange sites have also been prepared [41-43].

A bifunctional sulfonated phenolic polymer was made and examined as a cation exchanger [44]. It was suggested that PSDB-type ion exchangers were physically stronger if crosslinked with bis(4-vinylphenyl)-methane than with divinylbenzene [45]. Sulfonated starch was used as a cation exchanger [46]. Improvements in porous methacrylate polymers were described [47] while sulfonated perfluorinated hydrocarbons ground into small particles were evaluated as a cation exchanger [48]. Polyvinylalcohol was crosslinked and cation exchangers containing $-SO_3H$, $-CO_2H$, and $-PO_3H_2$ groups were prepared [49]. In other studies nitration of sulfonated PSDB was suggested to improve retention of alkali metal ions [50].

*Anion Exchangers*

Equation (6) describes a typical condensation polymer that exhibits weak-base anion exchange properties. This is one of the earliest organic polymeric anion exchangers prepared. Subsequently, several others were prepared using other aromatic amines, aminophenols, and aliphatic polyamines [12,27-29]. Condensation reactions between epichlorohydrin and amines is another type of reaction that has been used extensively for the preparation of anion exchangers. The reaction with $NH_3$ is shown in Eq. (13). Although weak-base anion exchangers

$$ClCH_2\overset{O}{CH-CH_2} + NH_3 \longrightarrow \text{—}CH_2CHOHCH_2\text{—}N(CH_2CHOHCH_2\text{—}N(CH_2CHOHCH_2\text{—})_2)\text{—}CH_2CHOHCH_2\text{—}N(CH_2CHOHCH_2\text{—})\text{—} \quad (13)$$

are generally obtained, it is possible to quaternize the amine group to yield a strong-base anion exchanger via postalkylation with an alkyl halide.

Strong-base anion exchangers that were found to be the most successful in commercial and analytical applications are based on addition polymers derived from polystyrene and divinylbenzene. The PSDB copolymer [see Eq. (7)] is quaternized by chloromethylation followed by reaction with a tertiary amine. This is shown in Eq. (14) where

$$\text{—}CHCH_2\text{—}(C_6H_5) \xrightarrow[\text{methylation}]{\text{chloro-}} \text{—}CHCH_2\text{—}(C_6H_4)CH_2Cl \xrightarrow{R_3N} \text{—}CHCH_2\text{—}(C_6H_4)CH_2N^+(CH)_3Cl^- \quad (14)$$

either chlorodimethylether or $H_2CO/HCl$ is used as the chloromethylation reagent. Usually, the tertiary amine $(CH_3)N$, is used, however, the strong-base anion exchanger derived this way shows a selectivity that favors $Cl^-$ over $OH^-$. Other tertiary amines can be used, e.g., anion exchangers derived from dimethylaminoethanol have been prepared and are available commercially. This exchanger provides a more favorable $Cl^-/OH^-$ selectivity, which is of more interest in commercial applications because of regeneration costs than in analytical applications. Treatment of the sulfonium-modified PSDB [see Eq. (9) for its preparation] with a tertiary amine [Eq. (15)] will also yield a strong-base anion exchanger [27].

If the chloromethylated PSDB is heated with $NH_3$, a primary amine, or a secondary amine, a weak-based anion exchanger is produced. These reactions are illustrated in Eq. (16) where the weak-base sites

$$\text{—CHCH}_2\text{—}(C_6H_4)CH_2S^+R_2Cl^- \xrightarrow{R_3N} \text{—CHCH}_2\text{—}(C_6H_4)CH_2N^+R_3Cl^- \qquad (15)$$

$$\text{—CHCH}_2\text{—}(C_6H_4)CH_2Cl \xrightarrow{NH_3} \text{—CHCH}_2\text{—}(C_6H_4)CH_2N^+H_3Cl^-$$

$$\text{—CHCH}_2\text{—}(C_6H_4)CH_2Cl \xrightarrow{RNH_2} \text{—CHCH}_2\text{—}(C_6H_4)CH_2N^+H_2RCl^- \qquad (16)$$

$$\text{—CHCH}_2\text{—}(C_6H_4)CH_2Cl \xrightarrow{R_2NH} \text{—CHCH}_2\text{—}(C_6H_4)CH_2N^+HR_2Cl^-$$

are shown as the hydrochloride salts. Because of the reactivity of the chloromethylated group toward other reagents and the lower stability of the ionogenic group derived from an amine, particularly the quaternary ammonium group in the $OH^-$ form, (see below for a discussion of anion exchanger stability) compared to the $-SO_3H$ group, anion exchange capacity is generally not larger than 3.5-4 meq/g of ion exchanger. The weak-base anion exchangers as shown in Eq. (16) can also be obtained by conversion of the chloromethylated PSDB to the sulfonium PSDB [see Eq. (9)] followed by treatment with $NH_3$, a primary amine, or a secondary amine.

Other types of weak- and strong-base anion exchangers based on addition polymers are well known. Acrylic and methacrylic acid can be polymerized in the presence of polyamines to produce weak-base anion exchangers [12,27] as shown in Eq. (17). Crosslinked polyacrylonitrile can be first converted into a weak-base anion exchanger

$$-CHCH_2- \text{ (with } CO_2H\text{) or } \left[ HC{=}CH_2 \text{ (with } CO_2H\text{)} \right]_n + HN(C_2H_4NH_2)_2 \longrightarrow -CHCH_2-CO-NHC_2H_4NHC_2H_4NH-CO-CH_2CH- \quad (17)$$

and then via quaternization into a strong-base anion exchanger [27]. Other synthetic approaches leading to weak- and strong-base anion

$$-CH(CN)-CH_2- \xrightarrow[NH_3]{Ni,\ H_2} -CH(CH_2NH_2)-CH_2- \xrightarrow{3C_2H_5Br} -CH(CH_2N(C_2H_5)_3\ Br^-)-CH_2- \quad (18)$$

exchangers are based on the preparation of crosslinked polymers in which the vinyl compound used also carries amino or cyano side-chain groups. The former provides a weak-base site while the latter can be converted to a weak-base or strong-base site by reactions similar to those shown in Eq. (18). It is also possible to polymerize vinylpyridine, vinylquinoline, and other amine-containing vinyl compounds with various kinds of difunctional crosslinking agents to yield weak basic ion exchangers.

Polyhydroxy compounds derived from crosslinked cellulose and dextran derivatives can be converted into anion exchangers [39-42]. Typical basic ionogenic groups are diethylaminoethyl, and diethyl-(2-hydroxypropyl)aminoethyl. Polyhydric alcohol was synthetically attached to a polystyrene framework and showed a preference for borate ion [51]. Naturally occurring aminosugar polymers chitin and chitosan were evaluated as ion exchangers [52,53].

Aromatic polyamine polymers were prepared and shown to retain anions [54]. In other studies a polymer was made where the quaternary ammonium group was a bridging group in a macroporous polymeric network [55]. Crosslinked polyvinylalcohol can be chemically modified to contain a quaternary ammonium group and act as an anion exchanger [49].

### *Specialty Exchangers*

Synthetic procedures for the preparation of several types of specialty polymers have been reported. These include (a) amphoteric ion exchangers, (b) complexing polymers, (c) polymers with chiral groups, (d) oxidation-reduction polymers, and (e) ion retardation polymers. Of all of these only complexing polymers and polymers with optically active groups have had significant impact on analytical separations.

*Amphoteric Ion Exchangers* Amphoteric ion exchangers contain both acidic and basic groups attached to the polymeric matrix. While there has been interest in their synthesis, none have been commercially available nor have their dual ion exchange properties and applications been carefully evaluated. Most exchangers with amphoteric properties have been used in the mode that tends to be more dominant. For example, the condensation product between phenol, diethylenetriamine, and formaldehyde yields a crosslinked polymer which contains the acidic phenolic and basic amine group [12,27]; the later is the dominant one. Many of the complexing polymers (see the next section) contain both acidic and basic groups; however, their application is based on coordination rather than ion exchange solely at the cation or anion site. A review covering amphoteric ion exchangers is available elsewhere [56].

*Complexing Polymers* The primary characteristic of a complexing or chelating polymer is that the polymer contains a group in significant number that can participate in coordination with the analyte. Synthetically, the problem is to design a procedure which allows the introduction of this group onto a crosslinked polymer or to use a monomer that contains the group, which is retained in the polymerization procedure. In general, the former approach has been more effective. One of the more successful chelate resins, which is commercially available, is the one derived from the synthetic attachment of an iminodiacetate group to the PSDB copolymer backbone. Synthetically, this can be done by reaction of the iminodiacetate salt with chloromethylated PSDB [see reaction (14) for its preparation] or by reaction with sulfonium PSDB [see reaction (9) for its preparation] as shown in Eq. (19). In analytical applications the iminodiacetic acid chelating exchanger shows a high affinity for polyvalent cations due to coordination similar to aqueous solutions of iminodiacetate salts. For example, this exchanger is effective in stripping many transition metal ions from a brine solution. Table 2 lists the selectivities that a commercially available iminodiacetic acid PSDB exchanger shows toward transition metal ions. The iminodiacetic acid group has also been attached to a macroporous PSDB copolymer [57] and to dextran [58].

Several examples of incorporation of the chelating group within the polymer are the following. Formaldehyde-polyhedric phenol condensation polymers [59] and polymers derived from diallyl phosphate

$$\text{---CHCH}_2\text{---}(\text{C}_6\text{H}_4)\text{CH}_2\text{Cl} \text{ or } \text{---CHCH}_2\text{---}(\text{C}_6\text{H}_4)\text{CH}_2\text{S}^+\text{R}_2\text{Cl}^- + \text{HN(CH}_2\text{CO}_2^-)_2 \longrightarrow \text{---CHCH}_2\text{---}(\text{C}_6\text{H}_4)\text{CH}_2\text{N(CH}_2\text{CO}_2^-)_2 \quad (19)$$

[60] were prepared and evaluated for metal ion retention. Nitrated poly-*m*-phenylene diamine will show increased retention for $K^+$ [61]. Unsaturated crown ethers can be polymerized to yield a crown ether-containing polymer chain [62]. A polymer made from vinylpyridine and acrylonitrile is selective for Au [63] while polyamine-polyurea resins were shown to be selective for heavy metals [64].

Table 2 Selectivity Coefficients for an Iminodiacetic Acid PSDB Chelating Exchanger[a]

| Cation | Log selectivity | Cation | Log selectivity |
|---|---|---|---|
| $Hg^{2+}$ | 4.18 | $Zn^{2+}$ | 1.29 |
| $UO_2^{2+}$ | 2.69 | $Co^{2+}$ | 1.18 |
| $Cu^{2+}$ | 2.65 | $Mn^{2+}$ | 0.69 |
| $VO^{2+}$ | 2.30 | $Ca^{2+}$ | 0.00[b] |
| $Pb^{2+}$ | 2.15 | $Ba^{2+}$ | -0.17 |
| $Ni^{2+}$ | 1.67 | $Sr^{2+}$ | -0.21 |
| $Cd^{2+}$ | 1.41 | $Mg^{2+}$ | -0.32 |

[a] Dowex A-1; Dow Chemical Co.
[b] $Ca^{2+}$ is the standard.
*Source*: From Ref. 13, p. 46, courtesy of Pergamon Press. See also Ref. 68.

Another approach used to modify ion exchangers to promote coordination-type interactions is to use a ligand as a counterion and hold it up on an ion exchanger via ion exchange. For example, mercapto-azobenzenesulfonate is held on an anion exchanger through the sulfonate and is selective for Hg [65] while 8-hydroxyquinoline sulfonate can be held on an anion exchanger through the sulfonate group and retain transition metal ions [66]. Adsorbents selective toward metal ions have also been prepared by impregnating adsorbents such as activated carbon and alumina with precipitating ligands such as arsonic acid derivatives, dimethylglyoxime, and dithiozone [67]. These kinds of modified ion exchange systems are suitable for isolating and concentrating trace inorganic analyte ions.

Chromatographic analytical separation of polyvalent metal ions from one another has been modestly successful using chelating ion exchangers. In many cases kinetics are not ideal for an efficient column operation. Usually the kinetic problem is associated with the coordination. Even though coordination constants between the metal ion and the coordination ionogenic group are favorable and may differ significantly for metal ions, the kinetic factor becomes a limiting one. A major use of the chelating exchanger, however, is for the stripping and concentrating of trace inorganic analyte ions.

A host of chelating resins have been synthesized and tested for analytical separations and/or stripping applications. Virtually all common metal ion complexing groups have been attached to a polymeric matrix. A summary of several of these is given in Table 3.

*Chiral Exchangers* A chiral-modified polymer is one that contains a chiral center synthetically attached to the matrix that often also contains an ionogenic group. For example, an approach which has been studied extensively for the separation of D,L-amino acids is to attach a chiral amino acid to a PSDB copolymer [91-93]. This can be accomplished by reaction of a chiral amino acid with the chloromethylated PSDB [see Eq. (14)] or with the sulfonium PSDB [see Eq. (9)] to give the structure I. While several different chiral amino acid-modified

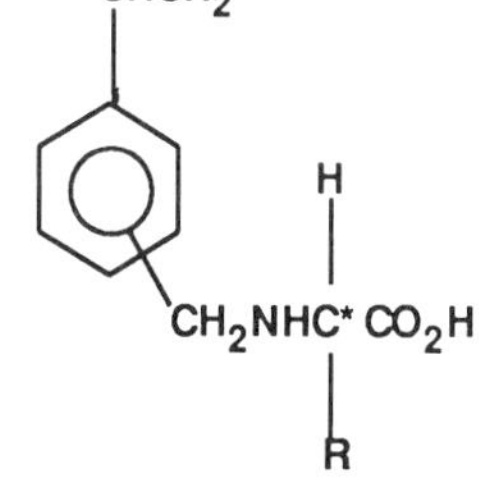

$-CHCH_2-$

H

$CH_2NHC^*CO_2H$

R

polymers have been studied, the one that appears to provide the best resolving power is that which incorporates proline into the PSDB matrix [92,93].

The application of these types of chiral exchangers is a form of ligand exchange (see below). In this form of ligand exchange a metal ion capable of participating in coordination with the chiral ionogenic group is in the mobile phase. Thermodynamic enantioselective effects develop in the formation of a mixed sorption complex between the stationary chiral ligand, the metal ion, and the racemate analyte. Separation arises because of the difference in the two complexes. This strat-

**Table 3** Several Chelating Resins

| Chelating group | Application | Ref. |
|---|---|---|
| Iminodiacetic acid | Metal ions other than alkali metals | 68,69 |
| Nitrilotriacetic acid | Metal ions other than alkali metals | 70 |
| Ethylenediaminetetraacetic acid | Metal ions other than alkali metals | 71,72 |
| Anthranilic acid | Transition metals | 72 |
| Hydroxamic acid | V, Fe, Mo, Ti, Hg, Cu, $UO_2$, Ce | 73 |
| Dimethylglyoxime | Ni | 74,75 |
| 8-Hydroxyquinoline | Alkaline earths, Cu, Co, rare earths | 76,77 |
| Salicylic acid | Fe | 78 |
| Arsonic acid | Cd, Co, Cu, Cr, Fe, Mn, Ni, Pb, Zn | 79 |
| Dithiocarbamate | Metal ions other than alkali metals | 80 |
| Thioglycolate | Transition metals, $UO_2$, Ca | 81 |
| Rescorinal | Transition metals | 76,82 |
| N-substituted hydroxylamine | Fe, Cu, $UO_2$ | 83 |
| Crown ether | Alkali metal and alkaline earth salts of common anions | 84 |

*NOTE*: See Refs. 85-90 for a detailed listing.

egy is not limited to chiral amino acid-modified stationary phases, and other aminocarboxylic and aminophosphonic acid groups can be attached to the PSDB copolymer [92].

*Oxidation-Reduction Polymers* Oxidation reduction polymers are characterized by electron transfer or electron exchange rather than by ion exchange. Two general approaches have been used to prepare oxidation-reduction polymers (see, for example, Ref. 12, 27). In the first a crosslinked copolymer is made form a monomer that is capable of participating in oxidation or reduction. For example, a mercapto group II, a diazonium group III, or a hydroquinone group IV can be introduced into the polymer chain. The second approach is to use a

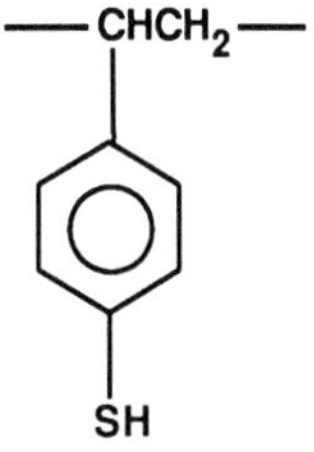

—CHCH₂— (SH) II — —CHCH₂— ($N_2^+Cl^-$) III — —CHCH₂— (HO, OH) IV

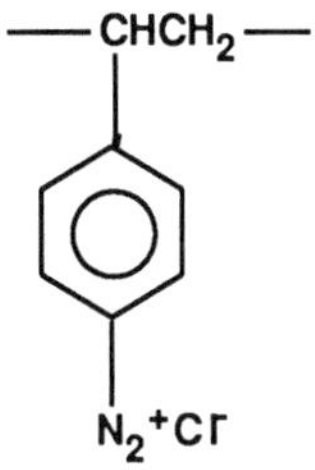

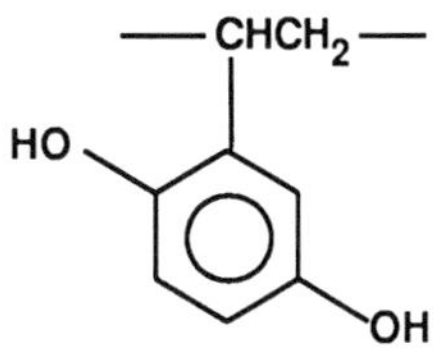

conventional ion exchanger and charge it with a counterion that is capable of participating in oxidation-reduction. For example, a cation exchanger can be charged in an $Fe^{2+}$-$Fe^{3+}$, $Ce^{3+}$-$Ce^{4+}$, or $Ti^{3+}$ form while anion exchangers can be charged in the $HSO_3^-$, $S_2O_3^{2-}$, or $H_2PO_2^-$ form. Commercial and/or analytical applications of oxidation-reduction polymers have not been particularly successful and current interest, although not extensive, is associated more with applications of these polymers in biological studies.

*Ion Retardation Polymers* Ion retardation polymers contain both acid and base groups and show a sharp preference for the retention of strong electrolytes over weak or nonelectrolytes. The most useful type of ion retardation polymer is the so-called snake cage polyelectrolytes. These are conventional cation or anion exchangers on which linear polycations or polyanions, respectively, have been formed by polymerization [12,27]. For example, a strong-base anion exchanger is converted into the acrylate form followed by polymerization of the acrylate anion to produce linear chains intertwined around the polymer matrix. These chains contain carboxyl groups which provide not only an ionogenic group for cations but can also serve as a polycounter-anion for the quaternary ammonium ionogenic group. Although the snake cage-type polymer has both anionic and cationic charge sites, it is different from the amphoteric type ion exchanger because the two types of exchange sites are not attached to the same matrix. Thus, the charge of the poly counteranion and the matrix have more freedom and may neutralize one another. If the latter occurs, mobile counterions (counterion to the matrix charge site and counterion to the snake cage charge site) are not present and electrical neutrality is maintained by the two types of fixed ionogenic groups.

### Inorganic Ion Exchangers

As indicated earlier, the first ion exchangers studied and used were inorganic oxides and salts, naturally occurring zeolites, and synthetic compounds of similar composition. When the organic polymeric ion exchangers were introduced, interest in inorganic ion exchangers diminished. However, once it was realized that the polymeric ion exchangers suffered from swelling-contraction limitations, were not as physically strong or thermally stable, and were of limited stability in relation to certain radioactivities, interest in inorganic ion exchangers was renewed. This interest has continued even though the major limitations of inorganic ion exchangers relative to the polymeric type have never been overcome. That is, most inorganic ion exchangers undergo slower ion exchange reactions, their ion exchange capacities often do not approach those obtainable with the polymeric-based ion exchangers, and they are very dependent on the conditions of the separation. While inorganic ion exchangers have not shared the success of polymeric ion exchangers in analytical separations, it should be noted that almost all studies that focus on the use of inorganic ion exchange in column chromatography have, even at present, employed relatively large particles of a broad range of sizes generally that are irregular in shape. This type of particle, coupled with slower ion exchange rates, is not likely to provide efficient column separations that are presently common in HPLC strategies. Since liquid chromatographic detectors have undergone tremendous improvements in detection limits, the need for large ion exchange capacities to handle the large analyte loading required for its deteciton is no longer a major problem in analytical applications. Thus, it appears that favorable HPLC ion exchange separations should be possible with many inorganic ion exchangers providing spherical microparticles of uniform size distribution are used. Recent studies with silica and alumina where aqueous mobile phases were used demonstrated that efficient and useful ion exchange separations of both organic and inorganic analyte ions are possible.

A large number of inorganic substances have been shown to possess ion exchange properties [10,11,30-35]. In general, any hydrated, insoluble inorganic salt or oxide has the potential to provide ion exchange characteristics. These can be divided into the following useful types:

1. Hydrous oxides
2. Multivalent metal acidic salts
3. Heteropolyacid salts
4. Insoluble ferrocyanide salts
5. Aluminosilicate salts

### *Inorganic Ion Exchanger; Ion Exchange Process*

An exact, single, unifying description and quantitative model that satisfies all ion exchange properties exhibited by inorganic ion exchangers is not possible. The major reason for this is that inorganic ion exchangers are very complex in their surface properties. Furthermore, these properties are very dependent on the method of preparation, conditioning, temperature, and chemical and solvent environment. Consider as an example zirconium phosphate gels. Their ion exchange properties were first reported in 1956 [94,95]. Subsequently, different crystalline forms, phases, single crystals, and products of intermediate crystalline stages were prepared and shown to have different ion exchange properties. In general, this is a characteristic of most inorganic oxides and salts that exhibit ion exchange properties.

Solid oxides in aqueous suspension are generally electrically charged. Usually electrophoresis experiments are used to determine this charge. Two apparently indistinguishable mechanisms have been suggested to account for this surface charge phenomena [96-101]. One involves amphoteric dissociation of surface MOH groups while the second is derived from the adsorption of metal hydroxy complexes due to hydrolysis of the oxide and/or amphoteric dissociation of MOH groups. Both mechanisms explain quantitatively a pH-dependent surface charge and the existence of a pH that results in zero net charge. This latter point is called the isoelectric point pH. It represents both the pH at which an immersed solid surface has zero net charge and the pH resulting in equivalent concentrations of positive and negative complexes in a system of multihydroxy complexes. The isoelectric point pH is dependent on the type of oxide, its pretreatment, and its aqueous environment, namely, the type and concentration of cations and anions present. Because of this dependence the isoelectric point pH for a given oxide can cover a broad range of values. Typical isoelectric pH values for several common oxides are listed in Table 4. At a pH value below the isoelectric pH the oxide is an anion exchanger while at large pH it is a cation exchanger.

Most insoluble metal oxides will exhibit cationic and/or anionic exchange properties. A simple description of the ion exchange assuming a hydrated oxide surface (in reality, the surface of the oxide or salt is very complex and not necessarily uniform in surface properties) is illustrated in Eqs. (20) and (21):

**Table 4** Isoelectric pH Values for Several Inorganic Oxide Ion Exchangers

| Oxide | Isoelectric pH value |
|---|---|
| Silica | 2.2 |
| Alumina | 7.5 |
| Titanium oxide | 4.7 |
| Zirconium oxide | 6.7 |
| Tungstic oxide | 0.5 |
| Lead oxide | 9.8 |
| Ferric oxide | 6.5 |
| Manganese oxide | 4.5 |
| Thorium oxide | 9.0 |
| Stannic oxide | 4.7 |

See Ref. 100.

$$M\text{-}O^-H^+_{(s)} + X^+ \rightleftharpoons M\text{-}O^-X^+_{(s)} + H^+ \tag{20}$$

$$M\text{-}OH_2^+C^-_{(s)} + X^- \rightleftharpoons M\text{-}OH_2^+X^-_{(s)} + C^- \tag{21}$$

where M is the central ion of the oxide and X is the analyte ion. The exchange in Eq. (20), where the metal oxide acts as a cation exchanger, is favored as the pH is increased. The exchange in Eq. (21), where the metal oxide acts as an anion exchanger, is favored as the pH is decreased. Many oxides are capable of exhibiting both properties because of favorable isoelectric pH values. Since the exchange sites are derived from weak-acid or weak-base groups, pH is a very important parameter in determining the availability of the ion exchange site.

A simplified representation of ion exchange with an insoluble inorganic salt, which most often involves cation exchange, is illustrated in Eq. (22).

$$H^{\pm}\text{salt}_{(s)} + M^+ \rightleftharpoons M^{\pm}\text{salt}_{(s)} + H^+ \tag{22}$$

Depending on the conditioning and stiochiometry of the salt, the cation exchange may involve several steps. For example, for zirconium phosphate as the inorganic cation exchanger the exchange involves two steps, or

$$Zr(HPO_4)_{2(s)} + Na^+ \rightleftharpoons ZrNaH(PO_4)_{2(s)} + H^+ \quad (23)$$

$$ZrHaH(PO_4)_{2(s)} + Na^+ \rightleftharpoons Zr(NaPO_4)_{2(s)} + H^+ \quad (24)$$

In Eqs. (23) and (24) no attempt is made to show the quantitative nature of the hydration of the zirconium phosphate, a factor which is also important in determining the cation exchange property.

A key structural property of the more useful inorganic salts that exhibit favorable ion exchange properties is that they have a defined crystallinity. Often these are of the type $M^{4+}(HX^{5+}O_4)_2 \cdot YH_2O$. Their structures are layered and they are capable of exhibiting ion sieve properties [10,30,31]. Each layer consists of a plane of the tetravalent M sandwiched between tetrahedral $(HXO_4)^{2-}$ groups. Modestly strong covalent bonds are within the layers while those between the adjacent layers are weak van der Waals-type interactions. The layers are thus able to move relative to one another as protons are replaced by other cations or if the number of water molecules changes. Knowledge of the details of the exact crystal structure, then, is invaluable for a quantitative understanding of the ion exchange process and its reversibility.

In ion exchange, whether using an inorganic oxide or a salt, the process requires the ingoing and outgoing analyte ions and counterions to be attached to the charged oxide or salt matrix purely by electrostatic attractions. The process should be reversible and obey the mass action law. In general, most oxides and salts follow this and are in agreement with the simplified ion exchange interactions shown in Eqs. (20)-(21) and (22). However, the retention of polyvalent ions on many inorganic oxides and salts appears to involve more than electrostatic attraction. While the process involves ion exchange in that a stoichiometric amount of counterion provided by the charged oxide or salt is replaced by the polyvalent ion, replacement of the polyvalent ion by counterion is not easily accomplished. Thus, it appears that additional interactions are causing retention of the analyte ion [10,30-33], namely, complexation and hydrolytic or precipitation between the analyte ion and the atoms within the oxide or salt structure. For example, the retention order of inorganic anions on $Al_2O_3$, $La_2O_3$, $Bi_2O_3$, ZnO, and PbO follows the solubilities of the respective salts formed between the metal of the oxide and the anion [102]. Some interactions, e.g., $F^-$ on alumina, are so strong that from a chromatographic point of view they are irreversible. As a general and qualitative guideline if the analyte ion is known to form a very stable complex or insoluble salt with the oxide or salt components, departure from a true ion exchange process should be anticipated. The more stable the complex or insoluble the product of the interaction, the more irreversible is the ion exchange.

*Inorganic Ion Exchanger Preparation*

A variety of strategies are used to prepare oxides and salts that have ion exchange properties. Even for a given oxide or salt many procedures are often available because individual oxides and salts possessing ion exchange properties are not single compounds. The type of procedure used will influence surface area, porosity, crystallinity, stiochiometry, hydration, as well as particle size and shape. The latter two are significant relative to column use of the oxides or salts in that these factors will influence efficiency, mobile phase flow properties, and rates of exchange while the first three not only influence these properties but also play a dominant role in determining the ion exchange behavior exhibited by the oxide or salt. Thus, it is not unusual to observe differing ion exchange behaviors for given types of ion exchange oxides or salts because different synthetic procedures were used for their preparation. The need to carefully characterize a given oxide or salt for a given preparation is therefore a requirement.

A major characteristic of ion exchange oxides and salts is that they are insoluble in water and all procedures, if not based on naturally occurring materials, should involve combinations of reagents and an environment that will produce the insoluble oxide or salt in a reproducible, crystalline form. The ratios and concentration of the compounds, rates and type of addition, temperature of reaction, time of reaction, presence of additives, nature of solvent and its composition, pH control, and presence of ligands are major experimental factors that are adjusted and, depending on these adjustments, yield the oxide or salt usually of a specific form. The procedure is often but not always reproducible. Many different kinds of oxides and salts that possess ion exchange properties have been prepared. Many of these procedures are reviewed elsewhere [10,11,30-35].

Typically, for the preparation of many oxides, aqueous solutions of metal salts and $NH_3$ or alkali hydroxide are combined while for the preparation of inorganic ion exchange salts, metal oxides or salts are combined or refluxed with acids that provide the precipitating anion. For example, zirconium phosphate can be prepared by combination of $Zr^{4+}$ and $PO_4^{3-}$, by refluxing gel zirconium phosphate with concentrated $H_3PO_4$, or by precipitation from an HF solution of a zirconium salt by the addition of $H_3PO_4$; the details of these procedures are provided elsewhere [33]. The specific ion exchange properties exhibited by the zirconium phosphate will depend on the procedure and amounts of reagents used. Table 5 lists the different kinds of crystalline zirconium phosphates that have been prepared. Structural characteriza-

**Table 5** Crystalline Zirconium Phosphates that Exhibit Ion Exchange

| Formula | Proposed designation | Interlayer spacing (Å) | Method of preparation |
|---|---|---|---|
| $Zr(HPO_4)_2 \cdot H_2O$ | α-ZrP | 7.56 | Reflux ZrP gel in 10-15 M phosphoric acid |
| $Zr(HPO_4)_2$ | β-ZrP | 9.4 | Dry γ-ZrP |
| $Zr(HPO_2)_2 \cdot 2H_2O$ | γ-ZrP | 12.2 | Reflux ZrP gel in acid solution of sodium phosphate |
| $Zr(HPO_4)_2 \cdot 1/2H_2O$ | δ-ZrP[a] | 7.13 | Reflux ZrP gel in enriched phosphoric acid |
| $Zr(HPO_4)_2$ | ε-ZrP[a] | 5.59 | Reflux ZrP gel in enriched phosphoric acid |
| $Zr(HPO_4)_2$ | ζ-ZrP | 7.41 | Heat α-ZrP 100-200°C |
| $Zr(HPO_4)_2$ | η-ZrP | 7.37 | Heat α-ZrP 250-300°C |
| $Zr(HPO_4)_2 \cdot 8H_2O$ | θ-ZrP | 10.4 | Wash $Zr(NaPO_4)(HPO_4) \cdot 5H_2O$ with acid |
| Not established | i-ZrP | 7.73 | Wash $Zr(NaPO_4)_2$, phase H, with acid |
| Not established | k-ZrP | 6.11 | Dissolve α-ZrP in HF, boil, and add $SiO_2$ |
| Not established | $PO_4$/Zr-2.2[a] | | $ZrOCl_2 \cdot 8H_2O$ + fused $H_3PO_4$ at 100-120°C |
| $Zr(HPO_4)(H_2PO_4)_2$ | α-ZrTP | | $ZrOCl_2$(Anhyd) + fused $H_3PO_4$ at 85°C |
| $Zr(HPO_4)(H_2PO_4)_2$ | β-ZrTP | | $ZrOCl_2 \cdot 8H_2O$ + fused $H_3PO_4$ at 70°C |

[a]Ion exchange behavior has not yet been firmly established.

*Source*: From Ref. 33 courtesy of Marcel Dekker, Inc. See original reference for literature references.

tion studies of the zirconium phosphates or the complete establishment of their ion exchange properties while extensive is not complete nor should it be assumed that this list of zirconium phosphate structures is complete.

Most inorganic oxides and salts that exhibit ion exchange characteristics are not available commercially in a chromatographic quality form, i.e., spherical microparticles in a uniform size distribution are not available. Two notable exceptions are silica and alumina. Both of these have a long chromatographic history of preparation, their understanding, and application. No attempt will be made to review their preparation procedure (see Chap. 6). Commercially, they are available, particularly silica, in chromatographic grades of known, reproducible porosities, surface area, particle size, and particle shape.

Table 6 lists several metal oxides that have been studied. Of these silica and alumina have received the most attention. Typical ion exchange selectivities, which must be treated qualitatively since these are very dependent on the oxide preparation, environment, and pretreatment, are listed for silica, alumina, and several other metal oxides that appear to have a wide range of ion exchange applicability. Most oxides because of favorable isoelectric pH values can exhibit both anion and cation exchange properties. This is also illustrated in Table 6.

Table 7 lists many of the inorganic salts that have been shown to possess ion exchange properties, most of them cation exchangers. No attempt is made to list ion exchange selectivity orders because of the large dependence on the inorganic salts preparation, the environment, and its pretreatment.

It should not be assumed that ion exchange selectivities are the same as those observed when using PSDB strong-acid or base ion exchangers (see below). Many reversals are noted when comparing selectivities found for inorganic oxide and salt ion exchangers to selectivities for organic polymeric-type ion exchangers.

### Ion Exchangers for HPLC

An effective stationary phase in HPLC must possess a set of physical properties, in addition to properties which influence analyte-stationary phase retention, that leads to a favorable flow of mobile phase and produces narrow, well-defined chromatographic peaks. The closer these properties are to the optimum, the more efficient the column operation becomes and the better its ability to resolve mixtures of analytes assuming other factors that influence analyte-stationary phase interactions are favorable. Table 8 summarizes the key stationary phase physical properties that must be satisfied for an efficient column operation. Since the ion exchange process is the same whether carried out by a classical, low-pressure, low-column-efficiency approach or

**Table 6** Inorganic Oxides that Exhibit Ion Exchange Properties

| Oxide | Selectivity | Ref. |
|---|---|---|
| | Anion exchangers | |
| $Al_2O_3$ | $F^- > SO_4^{2-} > Cr_2O_7^{2-}$, $HCO_3^- >$ Benzoate $> ClO_3^- > BrO_3^- > Cl^- > NO_3^- > Br^- > ClO_3^- > SCN^- > I^- > ClO_4^- > C_2H_3O_2^-$ | 103 |
| $Fe_2O_3$ | $OH^-$, $AsO_4^{3-}$, $S^{2-}$, $B_4O_7^{3-}$ $PO_4^{3-}$, $F^-$, $CO_3^{2-} >$ $Fe(CN)_6^{4-}$, $CrO_4^{2-}$, $SO_3^{2-}$, $SO_4^{2-}$, $IO_4^- > S_2O_3^{2-} >$ $Fe(CN)_6^{3-} > BrO_3^- > SCN^- > NO_2^-$, $NO_3^- > Cl^-$, $ClO_3^- >$ $Br^- > I^-$, $ClO_4^-$ | 104 |
| $SnO_2$ | $PO_4^{3-} \gg C_2O_4^{3-} \gg SO_4^{2-} > Cr_2O_7^{2-} > Fe(CN)_6^{4-} >$ $Fe(CN)_6^{3-} \gg Cl^- > MnO_4^- > Br^- > I^-$ | 105 |
| ZnO | $S^{2-} > Fe(CN)_6^{4-} > PO_4^{3-} > C_2O_4^{2-} > SO_3^{2-} > S_2O_3^{2-} >$ $F^- > SO_4^{2-} > Fe(CN)_6^{3-} > I^- > CNS^- > Cl^- > NO_3^-$ | 106 |
| $Bi_2O_3$ | $S^{2-} > CrO_4^{2-} > C_2O_4^{2-} > PO_4^{2-} > I^- > Fe(CN)_6^{4-} >$ $Fe(CN)_6^{3-}$ | 106 |
| PbO | $S^{-2} > CrO_4^{2-} > Fe(CN)_6^{4-} > PO_4^{3-} > Cl^- > S_2O_3^- >$ $SO_3^{2-} >> I^- > SO_4^{2-} > Fe(CN)_6^{3-}$ | 106 |
| $La_2O_3$ | $PO_4^{3-} > C_2O_4^{2-} > CrO_4^{2-} > MnO_4 > I^- > S^{2-} >$ $Fe(CN)_6^{4-} > SO_4^{2} > Fe(CN)_6^{3-} > Cl^-$ | 106 |

Table 6 (continued)

| Oxide | Selectivity | Ref. |
|---|---|---|
| ZrO | $PO_4^{3-} > Cl^- > Br^- > I^-$ | 107 |
| | Cation exchangers | |
| $SiO_2$ | Fe(III), Hg(II) > Al > $UO_2$(II) > Cu > Ag, Zn > Cd > Co > Ni; Nb, Zr > U(IV), Pu(IV) > $UO_2$(II) > Gd > Ba > Ca > Mg > Cs > Rb > K > Na | 108, 109 |
| $Al_2O_3$ | Th, Al, U(IV) > Zr, Ce(IV) > Cr > Fe(III), Ce(III) > Ti > Hg(II) > $UO_2$(II) > Pb > Cu > Ag > Zn > Co, Fe(II) > Ni > Tl(I) > Mn > Ba > Sr > Ca > Mg > Cs > Rb > K > Na | 110, 111 |
| $SnO_2$ | Cr, Al > Cu > $UO_2$(II), Zn > Co, Fe(II) > Ni > Mn > alkaline earths > Na > Cs | 112 |
| $MnO_2$ | Cu > Co > Zn > Ni >> Mg alkaline earths > alkali metals | 30 |
| $TiO_2$ | Cu > Ni > Co; Na > Rb > Cs | 10 |
| $Sb_2O_5$ | Na > Rb > Cs > K > Li; Sr > Ca > Ba > Ra | 113 |
| ZrO | Zn > Cu > Fe > Ni > Co; Be > Sr > Ba > Ca > Mg; Li > Rb, Cs > Na > K | 114 |

The selectivity orders listed are dependent on the conditions, the preparation of the oxide, and the treatment of the oxide. Differences in selectivity are common as these variables are altered.
*Source*: Taken in part from Ref. 30, courtesy of Elsevier Science Publishers. See also Refs. 10, 11, 31-34.

Table 7 Widely Studied Inorganic Salts that Exhibit Ion Exchange Properties

| Type | |
|---|---|
| Phosphate | $Zr^{4+}$, $UO_2^{2+}$, $Ti^{4+}$, $Ti^{4+}$, $Sn^{4+}$, $Hf^{4+}$, $Ce^{4+}$, $Th^{4+}$, $Sb^{5+}$, $Cr^{3+}$, $R.E.^{3+}$ |
| Arsenate | $Zr^{4+}$, $Ti^{4+}$, $Sn^{4+}$, $Ce^{4+}$, $Th^{4+}$ |
| Tungstate | $Zr^{4+}$, $Ti^{4+}$, $Sn^{4+}$ |
| Heteropoly complexes | Mo, W (as phosphate, silicate, arsonate, germanate salts) |

A wide range of salts exhibit ion exchange properties (usually cation exchange). These salts contain multivalent cations such as Zr, Hf, Th, Ti, Ce, Sn, Al, Fe, Cr, U, and anions such as phosphate, arsenate, antimonate, vanadate, mobydate, tungstate, tellurate, silicate, oxalate. See Refs. 10, 13, 31, 33.

Table 8 Key Stationary Phase Physical Properties

1. Uniform particle size
2. Microsize particles
3. Rigid particles
4. Physical strength to resist fracture due to column pressure drop
5. Spherical particles
6. Porous properties that provide access to retention sites
7. Chemical stability toward a wide range of mobile phase conditions

by a modern, high-pressure, high-column-efficiency approach, the key to the development of ion exchange in HPLC is largely due to the establishment of procedures for the synthesis of ion exchangers that satisfy the properties listed in Table 8.

The properties listed in Table 8 deal with column performance and not with analyte-stationary phase interactions. Since the ion exchange process is the same, an ion exchanger suitable for HPLC applications, just like its counterpart in low-efficiency chromatography, must provide the following basic components: (a) It should possess an ionogenic group. (b) The ionogenic groups within the ion exchanger must be accessible. (c) A favorable number (ion exchange capacity) of the ionogenic groups should be present. (d) A preference of one analyte over another (ion exchange selectivity) should take place. (e) The ion exchange should be reversible and follow mass action effects. (f) The kinetics of the exchange (rate of exchange) should be rapid enough to permit a column operation.

Ion exchangers suitable for HPLC applications are of four basic types:

1. Organic polymeric ion exchanger
2. Bonded phase ion exchanger
3. Pellicular ion exchanger
4. Inorganic oxide ion exchanger

Figure 2 indicates their basic physical form. A macroporous structure is achieved by the network of microspheres that are jointed together or by the channels that are present in the matrix. Since the ionogenic groups are attached to the matrix, they are accessible because of the porosity. This leads to a favorable effect on mass transfer and column efficiency. The pellicular-type ion exchanger differs from the others in that the macroporous network, which contains the ionogenic group, is at the surface. Since its thickness over the inert core is about 1-3 μm rather than throughout the bead, mass transfer is even more favorable. The useful result, however, is partially offset by the subsequent decrease in the number of ionogenic groups (ion exchange capacity) present and the difficulty in preparing the pellicular form in a <10-μm size and a narrow reproducible size range.

As interest in ion exchange separations using a HPLC strategy grew and examples of these basic types of ion exchangers were introduced and/or evaluated, improvements in their physical properties were achieved. That is, techniques were developed or improved which produced spherical particles, microsize particles, particles of narrow size range, and favorably packed columns. Consequently, ion exchange has been rapidly transformed from a low column inlet pressure technique of low column efficiency to a high-pressure technique that yields a high column efficiency. On the average, column efficiencies in modern

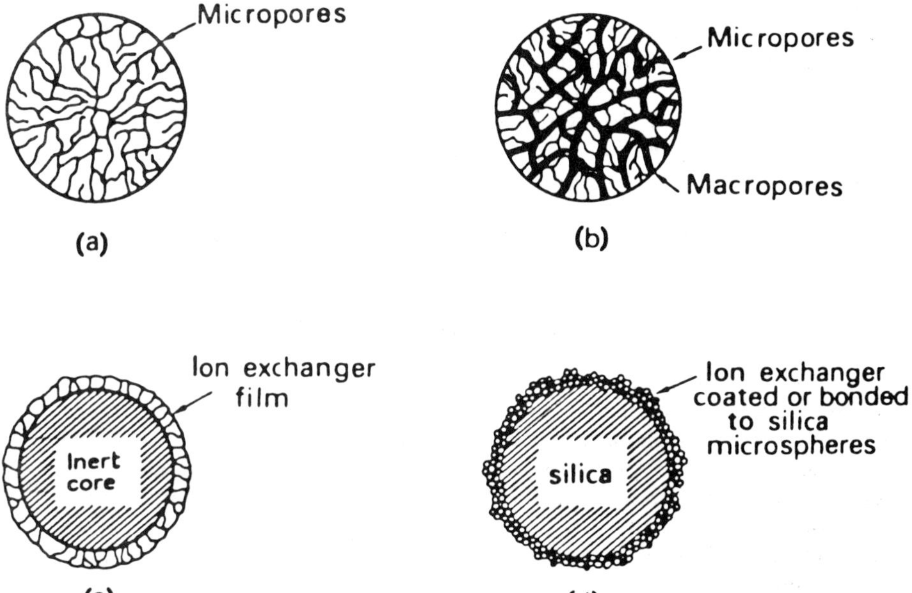


Fig. 2 Types of ion exchangers. (a) Microporous or gel organic polymeric ion exchanger; (b) macroporous organic polymeric ion exchanger; (c) pellicular ion exchanger; (d) bonded phase ion exchanger.

ion exchange column chromatography may not always approach those obtained in reversed and normal phase HPLC, for reasons that will become apparent. Nevertheless, the modern ion exchange column is still vastly improved over the low-efficiency ion exchange column of the past.

*Organic Polymeric Ion Exchanger*

Historically, ion exchange chromatography was recognized as a powerful strategy for the separation of amino acids, proteins, nucleic acids, nucleosides, nucleotides, and many other compounds of biochemical interest. Because of this strong interest, particularly in amino acid separations, organic polymeric ion exchangers were the first type of stationary phase to be used under modern liquid chromatographic conditions [5]. For example, in the early 1960s when automated amino acid ion exchange strategies were being developed, the effect of ion exchanger particle size on resolution was recognized and much effort was directed toward the synthesis of small, uniform-size organic polymeric ion exchangers for this kind of application. Essentially, the ion exchangers were conventional-type ion exchangers, i.e., they were sulfonated PSDB for cation exchange application and quaternized PSDB for anion exchange applications. Modified synthetic procedures and sizing operations were used to obtain exchangers in the 50-, 25-, and eventually 10-μm size. These small, conventional gel ion exchanger particles satisfied several of the requirements listed in Table 8. However, in the early days of HPLC they did not satisfy all of them because of the tendency at that time to use excessive column inlet pressures, which was generally due to excessive column length and the lack of uniform stationary phase particle size. Since the ion exchange particles were gel-type polymers, they still had disadvantages. (a) They underwent considerable swelling and contraction depending on the mobile phase changes. (b) Because of swelling-contraction bead fracture was often a problem. (c) The swelling-contraction influenced mass transfer in analyte-stationary phase interactions. (d) The softness of the polymer often lead to tightening of the packed column resulting in an increased column pressure drop and subsequent interference in mobile phase flow. The bonded phase type of ion exchanger did not suffer from these problems, and as modern liquid chromatography was developing attention quickly turned to these types of ion exchangers, particularly for applications in the separation of organic-type analytes.

The totally polymeric-type ion exchanger was not completely forgotten, and as column packing procedures and synthetic procedures improved interest in the polymeric type ion exchanger was renewed. Key improvements were the ability to produce spherical, microparticle (<10-μm) ion exchangers of a very narrow particle size range. Cross-linking was often increased which provided better physical stability and the high level of ion exchange capacity found with the conventional polymeric type ion exchanger could still be retained.

Another improvement in the polymeric ion exchanger was to use a macroporous PSDB matrix for the cation or anion ion exchanger (see above). Several of the problems associated with microparticle organic polymer gel-type ion exchangers can be minimized by using the macroporous PSDB copolymer matrix. Their chromatographic performance in terms of column efficiency, while not matching the bonded phase-type ion exchanger, is still favorable. Furthermore, the PSDB-type ion exchanger, whether a gel or macroporous type, offers the distinct advantage of being chemically stable over a wide range of mobile phase conditions (for example, pH 1-13) and for this reason it is particularly valuable in the separation of inorganic analyte ions by ion chromatographic and related strategies.

Many different types of polymeric-based ion exchangers were synthesized during the period that HPLC was being developed as a chromatographic strategy (see above). While interest was often present to produce smaller, more uniform particles, major success, particularly from a commercial availability point of view, has been achieved primarily with the PSDB copolymer-based ion exchanger. One other major improvement was the refinement of polymerization strategies that lead to a macroporous-type polymer matrix [27,36,37].

When ion chromatography was introduced in 1975 [6], interest in ion exchangers for the ion chromatographic separation of inorganic ions grew rapidly. Accompanying this development was the preparation of ion exchangers that would be suited to this strategy and would provide a favorable high column efficiency. The novel approach of Small et al. [6] requires two columns: first the separator column and second the stripper, more often called the suppressor column. Both are ion exchangers and in the case of the work of Small et al. both were PSDB-based ion exchangers. If anions are to be separated, the separator column is an anion exchanger and the suppressor column is a cation exchanger. For separation of cations the reverse applies. In both bases the role of the suppressor column is to remove mobile phase electrolyte so that the analyte ions can be more easily detected by a conductivity detector. Thus, to avoid a high frequency of regeneration the suppressor columns were conventional ion exchangers of high ion exchange capacity and small particle size to minimize band broadening. The separator columns were sulfonated PSDB cation exchanger for cation exchange and quaternary ammonium PSDB anion exchanger for anion separations. In the initial experiments [6] a 2% crosslinked PSDB copolymer (180-325 mesh) was sulfonated to yield a low-capacity cation exchanger (about 0.02 mdq/g). The anion exchanger was prepared by powdering a high-capacity, 8% crosslinked quaternary amine PSDB anion exchanger. The powder (0.5-2 μm) was agglomerated onto the surface of a low-sulfonated PSDB cation exchanger by passing a slurry of the powder through a column of the cation exchanger. The anion exchanger thus coats the cation exchanger, which serves as a core material. In both cases a superficial-type ion exchanger, which should provide favorable mass transfer characteristics, was obtained.

The experiments of Small and coworkers [6] not only demonstrated enormous potential but also clearly established that ion exchange could be readily carried out with favorable column efficiency using low-capacity ion exchangers. Following this work, low-capacity PSDB ($-SO_3H$) cation and ($-CH_2N\overset{+}{R}_3X^-$) anion exchangers suitable for the high-efficiency, two-column ion chromatographic technique became commercially available.

Previous studies on the sulfonation of PSDB [115-118] had shown that a superficial sulfonated cation exchanger could be prepared. The key in the procedure is the type of sulfonation reagent used and its concentration [see Eq. (8)], temperature, solvent, and time of sulfonation. Low-capacity anion exchangers were prepared by controlling the stoichiometry of the chloromethylation step [see Eq. (14)]. Subsequently, an alternate low-capacity anion exchanger was prepared by taking a surface-sulfonated PSDB copolymer and coating this with a thin monolayer (0.1-0.5 μm thickness) of latex which is then converted into an anion exchange material [119]. Other PSDB surface-modified strong-acid cation and strong-base anion exchangers have been prepared and are commercially available [120].

Macroporous spherical copolymers of 2-hydroxyethylmethacrylate that were modified to contain diethylaminoethyl, $-CO_2H$, $-PO_3H_2$, and $-SO_3H$ groups were synthesized, are commercially available, and are suitable for protein separation [121] while a macroporous spherical cellulose was prepared and converted into a $-CO_2H$ and *t*-amine weak-acid and base ion exchanger, respectively [43]. Phenyl Kel F was converted into a strong-acid action exchanger by sulfonation and into a strong-base anion exchanger by chloromethylation followed by reaction with trimethylamine [122].

Subsequent workers [123,124] demonstrated that suppressor columns could be eliminated and highly efficient separations still achieved on low-capacity ion exchangers (from 10 to 200 μeq/g). In general PSDB copolymers, usually of the macroporous variety, are sulfonated [123-125] for cation exchangers or are quaternized via chloromethylation and $R_3N$ reaction [126-128] for anion exchange.

Although many different kinds of polymeric organic ion exchangers have been prepared, most do not meet all the requirements listed in Table 8, particularly if being commercially available is also added to the list. Of all of these, the major ones that most closely satisfy the requirements and which are also widely available commercially are the cation and anion exchangers based on the PSDB copolymer. A summary of the PSDB ion exchanger and their properties that were commercially available during the early developing period of HPLC is available elsewhere [129-133].

### *Bonded Phase Ion Exchanger*

Introduction of the bonded stationary phase to HPLC was a major turning point in the development of HPLC. This type of stationary phase is widely used and accounts for a large majority of HPLC applications. The bonded phase has the physical strength of an adsorbent and the resolving power associated with a partitioning system provided by a stationary phase coated with a liquid layer. Since the bonded phase involves chemical bonding of the phase layer to the stationary phase, it is not lost during elution with mobile phase.

Bonded phases are made by chemically bonding hydrocarbon groups to a microsilica particle. If polar groups are contained on the hydrocarbon, then the bonded phase exhibits normal phase chromatographic properties while if they are absent it exhibits reversed phase chromatographic properties. If the hydrocarbon group contains an ionogenic group or one is chemically attached to it in a subsequent reaction, then the bonded phase is an ion exchanger.

The chemistry of chemically bonding a group onto silica can be accomplished in several ways [134-136]. One approach is based on conversion of the silanol by reaction with an alcohol to form a silicate ester as shown in Eq. (25). The silanol group will readily undergo

$$\equiv \text{Si-OH} + \text{ROH} \longrightarrow \equiv \text{Si-OR} \qquad (25)$$

reactions with organochlorosilanes [see Eq. (26)]. If an organodichlorosilane is used, a terminal -OH group is obtained which can then

$$\equiv \text{SiOH} + \text{R}_3\text{SiCl} \longrightarrow \equiv \text{Si-O-SiR}_3 \qquad (26)$$

be reacted with another type of organochlorosilane. The silanol can be converted into the chloride, e.g., by reaction with thionyl chloride, and the resulting chloride is treated with a Grignard organolithium reagent to give the product as shown in Eq. (27). The silane chemistry

$$\equiv \text{SiOH} + \text{SOCl}_2 \longrightarrow \equiv \text{Si-Cl} + \text{LiR} \longrightarrow \equiv \text{Si-R} \qquad (27)$$

illustrated in Eq. (26) and the Grignard chemistry in Eq. (27) offer the better and more versatile synthetic routes to the bonded phase ion exchangers.

If the hydrocarbon chain contains a terminal amine group (e.g., V), it is a normal bonded phase which can also exhibit weak-base anion

$$\equiv \text{Si(CH}_2)_n\text{NH}_2 \qquad\qquad \equiv \text{Si(CH}_2)_n\overset{+}{\text{N}}\text{R}_3\text{X}^-$$

V VI

exchange properties. Conversion of V to a quaternary ammonium group by alkylation gives VI, which gives a strong-base cation exchanger. If aromatic groups are attached, then these can be sulfonated to yield a strong-acid cation exchanger [137-139]; similarly, aliphatic side chains be sulfonated [140]. Another approach that was used to obtain the strong-acid cation exchanger was to prepare a vinylated silica VII which was then polymerized with styrene to give VIII which can then be sulfonated to give the cation exchanger IX [141].

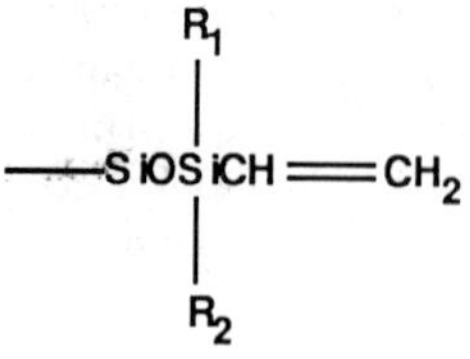

$R_1$

—SiOSiCH$=CH_2$

$R_2$

VII

$R_1$—Si—$R_2$

OSi—

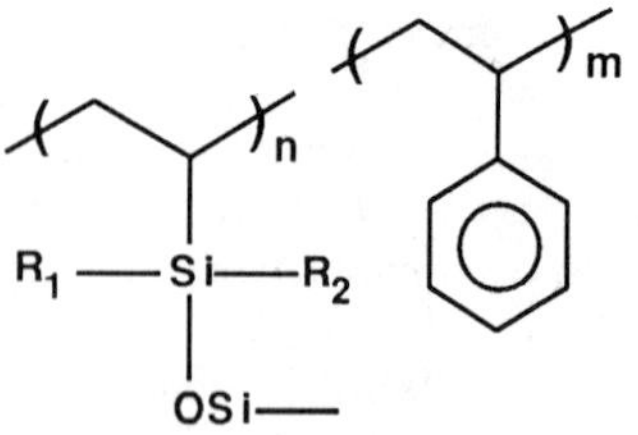

VIII

These workers also prepared a strong-base anion exchanger by chloromethylation of VIII and reaction with a *t*-amine to yield the quaternary ammonium derivative X. Weak-base anion exchangers were prepared

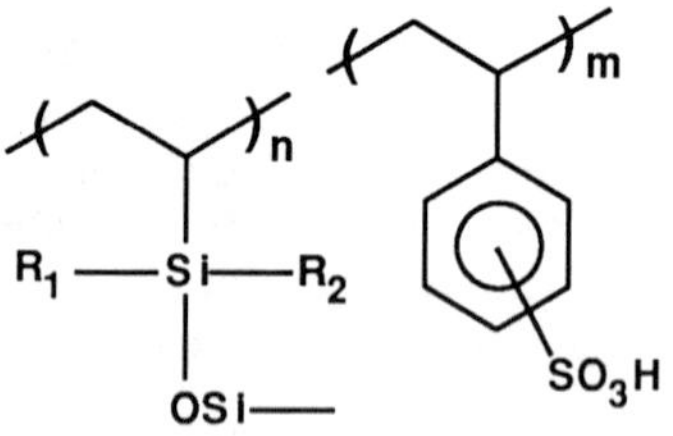

$R_1$—Si—$R_2$

OSi—

$SO_3H$

IX

$R_1$—Si—$R_2$

OSi—

$CH_2N^+R_3X^-$

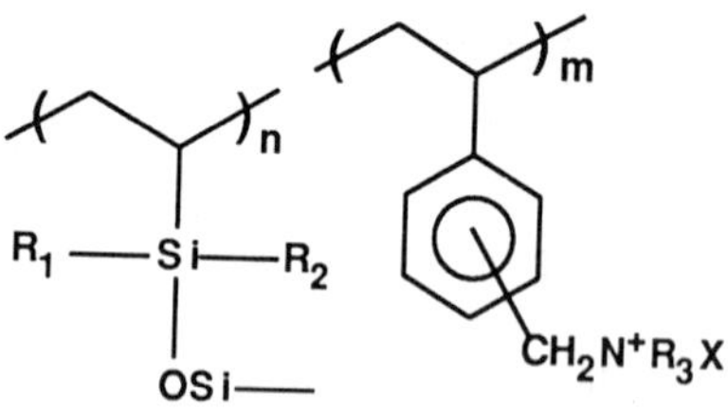

X

by attaching a diethylaminopropyl group to the silica and by attaching an amine epoxy derivative silica. A major difference in those ion exchangers over most others is that their ion exchange capacities were much higher than typical bonded phase ion exchangers.

A strong-base quaternary ammonium ion exchanger was prepared by first adding the phenyl group to silica via silane chemistry, chloromethylating the ring, and then adding *N,N,N*-dimethylthanolamine [137]. Alternatively, the chloromethylphenyl group is introduced in the silane chemistry. In this case chlorodimethyl [4-(4-chloromethyl)] phenylbutylsilane was combined with silica and then treated with $Me_3N$ [142]. If the chloromethyl group is treated with $CN^-$ the nitrile is obtained which can then be hydrolyzed to give a weak-acid ($-CO_2H$) cation exchanger [138]. A strong-acid cation exchanger [143] was also obtained by sulfonation of the product from the reaction of chlorodimethylphenylbutylsilane with silica. Alkaloid, nucleotides, nucleosides, and organic base retention was determined on these ion exchangers.

Reaction of silica (a) with 2-phenylethyltrichlorosilane followed by sulfonation with chlorosulfonic acid, (b) with 3-chloropropyltrichlorosilane followed by reaction with benzyldimethylamine, and (c) with 3-chloropropylsilane followed by reaction with 3-alanine gave strong-acid, strong-base, and weak-acid ion exchangers, respectively [144]. These ion exchangers were used to separate nucleosides, alkoloids, and weak-acid pharmaceuticals. In a slightly different approach, a strong-acid cation exchanger was made by treating silica with mercaptopropyltrimethyloxysilane to give the mercaptan XI, which was then oxidized to the sulfonic acid; this exchanger was used to separate phenothiazines [145].

$$-OSi(OCH_3)_2-CH_2CH_2CH_2SH \xrightarrow[H^+]{KMnO_4} -O-Si(OCH_3)(OCH)-CH_2CH_2CH_2SO_3H \quad (28)$$

XI XII

The effects of post- and presulfonation in the preparation of a strong-acid cation exchanger were considered [145]. In one case chlorodimethyl(3-phenylpropyl)silane was combined with silica and then sulfonated with chlorosulfonic acid. In the second case, the silane derivative was first sulfonated with chlorosulfonic acid and then combined with silica. Presulfonation produced a bonded phase of higher coverage and exchange capacities while postsulfonation produced exchangers that retained a more hydrophobic property due to lower exchange capacity. The presulfonation ion exchanger was also more reproducible from batch to batch probably because postsulfonation may cause additional side reactions.

A bifunctional weak-base anion exchanger was shown to be useful for nucleoside, nucleotide, purine, and pyrimidine separations [147] while treatment of silica with 3-aminopropyltriethyloxysilane gave a weak-base anion exchanger which was used to separate carbohydrates [148]. A protein-compatible ion exchange bonded phase was prepared by reaciton of silica with glycidoxypropyltrimethoxysilane to yield XIII, which can be converted to the cation exchanger XIV or the anion exchanger XV [149]. A weak-base anion exchanger was prepared by reaction of silica (different pore diameters and surface areas) with $\gamma$-amino(alkyl)trimethoxysilane and it was suggested that pore diameter and surface area rather than alkyl chain was more important in determining the retention of adenosine mono- and diphosphate [150].

Weak-base groups were immobilized on glass by reaction with silanol groups and used to concentrate inorganic anions [151]. Complexing groups such as 8-hydroxquinoline [152,153] were bound to silica.

$$\text{—SiOSi(CH}_2)_3\text{OCH}_2\text{CH—CH}_2 \text{ (epoxide, XIII)} \xrightarrow{Na_2SO_3} \text{—SiOSi(CH}_2)_3\text{OCH}_2\text{CH(OH)CH}_2\text{SO}_3\text{H (XIV)}$$

$$\text{XIII} \xrightarrow{HN(Et)_2} \text{—SiOSi(CH}_2)_3\text{OCH}_2\text{CH(OH)CH}_2\text{N(Et)}_2 \text{ (XV)} \quad (29)$$

Bonded phase ion exchangers containing a chiral center were prepared by reaction of triethoxypropylaminosilane [157] or 3-glycidoxypropyltrimethoxysilane [158] with silica and then with L-proline. When charged in the $Cu^{2+}$ form they can be used to separate DL-amino acids. In a novel application of ion exchange, quaternary ammonium groups were attached to glass silanols using silane chemistry. In these studies the glass beads varied from 4 to 2500 nm in pore size and the quaternized ammonium glass bead was used for the exclusion chromatographic separation of water-soluble polymers [159].

It should be noted that the amine-type bonded phase is also a normal stationary phase. In many cases these were prepared for this kind of application rather than for applications as a weak-base anion exchanger.

Since silica is the base material for the bonded phase ion exchanger, its properties both chemical and physical must be considered as part of the stationary phase. No attempt will be made to discuss this here and the reader is referred to Chap. 6. In general, silicas used commercially for bonded phase ion exchangers are the same as those used for normal and reversed bonded stationary phases. That is, they are amorphous, porous (typical surface areas from about 200 to 800 $m^2/g$ and average pore diameters of 5-25 nm), microparticles of narrow range in size distribution typically 3, 5, and 10 μm. The silica-based bonded phase ion exchangers have a limited, useful pH range of about pH 2-8. Silica solubility, particularly in the basic direction, and bond breaking within the bonded phase increases at pH conditions beyond this pH range. Because of this limitation, the PSDB-based ion exchanger, which is pH-stable from 1 to 13, is often preferred in inorganic analyte ion separations.

Silica-based bonded phase ion exchangers meeting the requirements listed in Table 8 are readily available commercially, reproducible, and widely used. While the ionogenic group is identified, the chemistry used to add the group or the exact structure of the carbon portion of the bonded phase is often not readily available.

*Pellicular Ion Exchanger*

A pellicular ion exchanger is formed by depositing a matrix onto a surface of an inert core that is microparticle in size. The ionogenic groups are then attached to the matrix. It is similar to a superficial PSDB-type ion exchanger [115-118] in that the exchange sites are on the surface. However, unlike the PSDB ion exchanger, the pellicular ion exchanger core is usually inert, nonporous, and not accessible to the mobile phase. The inertness means that the pellicular ion exchanger is essentially independent of the physical and chemical composition of the core. For example, in the PSDB ion exchanger crosslinking will influence bead swelling and contracting. This is not the case in the pellicular ion exchanger. Since the core is rigid it retains a high level of physical strength. The number of ion exchange sites is low because the coating is thin. However, the thin coating of the pellicular ion exchanger means that mass transfer, which is essentially at the surface since the core is usually not accessible, is very favorable. Horvath and coworkers [160] introduced pellicular ion exchangers to HPLC using glass as the core material, polystyrene-divinylbenzene as the polymer coating, and sulfonation [see Eq. (8)] to yield the strong-acid cation exchanger and chloromethylation and reaction with trimethylamine [see Eq. (14)] to yield the strong-base anion exchanger. Following this, Kirkland [161,162] developed a similar type of pellicular ion exchanger except that a silica (Zipax) core was used. A fluorocarbon containing a $-SO_3H$ group was used to produce the cation exchanger with a cation exchange capacity of 3.5 μeq/g while the anion exchanger was based on a quaternary ammonium group and had an anion exchange capacity of 12 μeq/g. These coverings were a thin porous shell that surrounded the siliceous core. A scanning electron micrograph of the anion exchanger is shown in Fig. 3.

The PSDB copolymer was the most widely used coating partly because it could easily be converted into a cation and anion exchanger by sulfonation and chloromethylation/*t*-amine reaction, respectively. Since the silica or other core material provides physical strength, extensive crosslinking is not required for this purpose and the amount can be varied over a wide range. Similarly, thickness of the layer can be varied. If the layer thickness is increased, the number of exchange sites on the layer can be increased. However, if the layer is too thick, then the advantage of surface mass transfer is reduced.

A key part of the preparation of a pellicular ion exchanger is the coating procedure since a uniform thin coat on a microparticle is desired [160-163]. Details of the commercial procedures for coating are not readily available. Since a PSDB coating is usually used, the chemistry of adding ionogenic gorups is well established. Core materials are frequently silica or glass. When the former is used, mobile phase pH is generally restricted to a range of about 2-8. Aqueous mobile

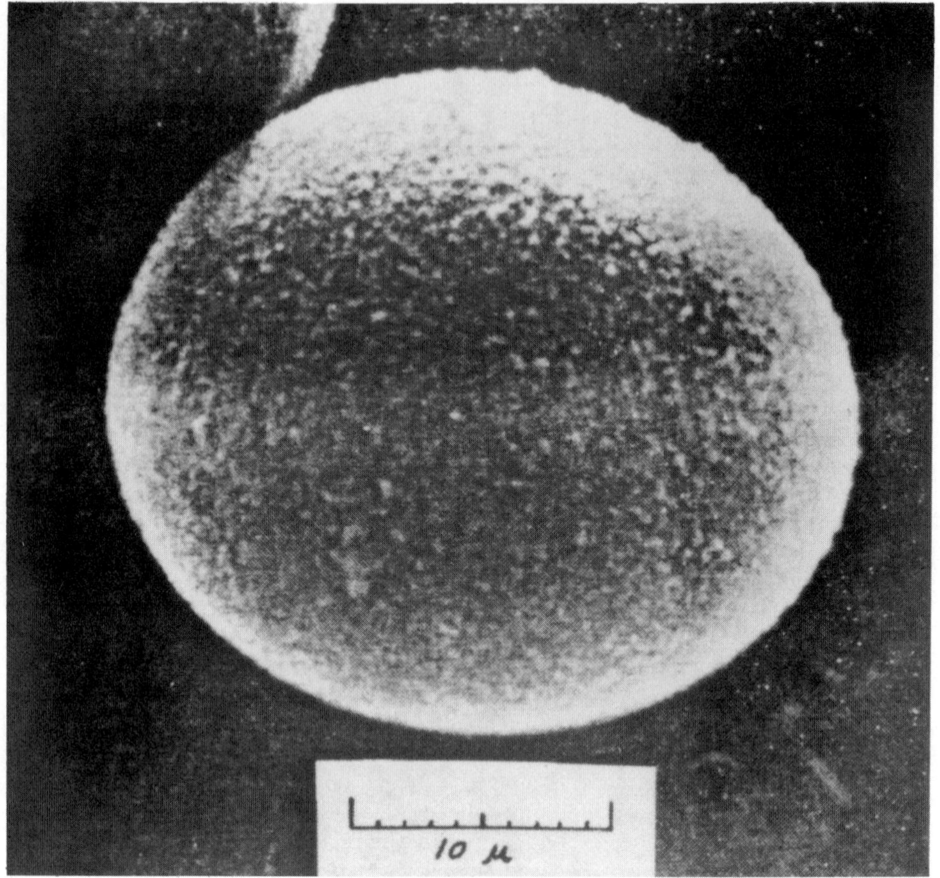


Fig. 3 A Steroscan scanning electron micrograph of a Zipax anion exchange particle. (From Ref. 161, courtesy of Preston Publication Inc.)

phases are generally the best and the tolerance of the pellicular ion exchanger toward the type and amount of organic solvents present will depend on the polymer coating.

Other special kinds of pellicular ion exchangers have been prepared. For example, short-chain polyglycine peptides were bonded to a glass core pellicular resin-coated bead and used to separate amino acids, phenols, and aromatic amines [164], and crosslinked diethylaminoethyl-modified dextran was coated on glass beads and used to separate gangliosides [165].

A polyethyleneimine covering of silica, which exhibits weak-base ion exchange properties, shows good promise as a stationary phase for the separation of proteins, oligonucleotides, and related derivatives [166-167]. The polyethyleneimine is so strongly held onto the silica that crosslinking can also be initiated to improve the polymer coating. A polyethyleneimine-coated PSDB copolymer core, which can also be used for proteins, is commercially available [168]. Another pellicular weak acid-base ion exchanger which shows promise for protein separations was made by first preparing aminopropylsilica by silane chemistry. This was then combined with polysuccinimide which, when hydrolyzed, gives a polyaspartic acid coating on silica. Poly-2-aminoethylaspartamide silica was also prepared and used for protein separations [169].

Glass beads of limited porosity covered with chemically attached groups, silica with certain kinds of bonded phases, silicas with extensive bonded phase coverage, and modified latex-covered ion exchangers described in the previous two sections exhibit characteristics of a pellicular-type ion exchanger. The distinction between bonded phase and pellicular is not always a clear-cut one.

The use of pellicular ion exchangers has diminished, particularly as a stationary phase for the column separation of organic analyte ions. Several reasons have contributed to this change. (a) Pellicular ion exchangers have a low ion exchange capacity. (b) They have a tendency toward a larger particle size. (c) The pH limits of silica as a core are limited. (d) Use of reversed phase stationary phases for the separation of many organic analyte ions provides an excellent strategy. (e) Use of a pairing ion as a mobile phase additive for the separation of organic analyte ions on reversed stationary phases is widely applicable. (f) A marked improvement in the availability of highly efficient, reproducible organic polymeric-type ion exchangers and bonded phase ion exchangers has taken place. Surveys of the commercial availability of pellicular ion exchangers and their applications during their peak use are available elsewhere [129-133].

### *Inorganic Oxide Ion Exchanger*

While the ion exchange properties of inorganic oxides and salts have been known and studied for a long time (see below), essentially no direct effort has been made commercially to develop these materials as stationary phase ion exchangers for HPLC applications. However, considerable effort has gone into the development of alumina and silica as normal phase adsorbents for HPLC applications or, as in the case of silica, as precursors to bonded reversed phases. Thus, these materials are readily available in a microparticle size that is spherical and uniform in size distribution suitable for HPLC application. The fact that these oxides are prepared for normal phase chromatography means that they must first be thoroughly hydrated by equilibration with an aqueous mobile phase prior to their use as ion exchangers. This is an important consideration because suppliers that presently supply ready-packed alumina and silica columns do so for applications toward normal phase chromatography, i.e., they are supplied in an inert solvent. The minimal data on the use of silica and alumina in HPLC applications as ion exchangers (see below) suggest that they are very efficient, reproducible, very selective, and applicable to the separation of both organic and inorganic analyte ions providing their pH limits and counterion compatibility are acocunted for. Silica is useful only as a cation exchanger whereas alumina exhibits both anion and cation exchange properties with the former being more versatile in application.

## PROPERTIES OF ION EXCHANGERS

### Organic Polymeric Ion Exchangers

#### *Ion Exchanger Equilibria-Selectivity*

Consider an ion exchanger charged in a counterion A form that is placed in a solution of an electrolyte that provides counterion B. Ion exchange takes place, which is reversible and stoichiometric, and reaches equilibrium such that the exchanger and the solution contain both counterions. For cation exchange, it can be written as

$$\underline{RA} + B^+ \rightleftharpoons \underline{RB} + A^+ \tag{30}$$

while for anion exchange the equilibrium is

$$\underline{RA} + B^- \rightleftharpoons \underline{RB} + A^- \tag{31}$$

where R is the exchange matrix containing a fixed anionic or cationic site, respectively.

Ion exchange equilibrium can be expressed by the ion exchange isotherm, the selectivity coefficient, or the distribution coefficient or capacity factor. The ion exchange isotherm describes the ion exchanger ionic composition as a function of the experimental conditions. In practice, the isotherm is obtained by determining the amount of the given counterions in solution or the counterions on the ion exchanger after equilibrium is reached. Unlike the isotherm, which describes the distribution over a range of counterion concentration, the selectivity coefficient and distribution coefficient or capacity factor apply to a given point on the isotherm.

Most often the competition of the two counterions for the fixed ionogenic site is not the same and one counterion is preferred over the other. This selectivity is often called the selectivity coefficient; for Eq. (30) it is given by

$$K_{A^+}^{B^+} = \frac{[\underline{RB}][A^+]}{[\underline{RA}][B^+]} \tag{32}$$

while for eq. 31 it is

$$K_{A^-}^{B^-} = \frac{[\underline{RB}][A^-]}{[\underline{RA}][B^-]} \tag{33}$$

where $K_{A^+}^{B^+}$ and $K_{A^-}^{B^-}$ are cation and anion molar selectivity coefficients if concentrations are in molar units. When evaluating selectivity it is more convenient to define ion concentrations within the ion exchanger as gram equivalents per kilogram of dry exchanger. If this is done the ionic form of the dry ion exchanger must be specified. For polymeric anion exchangers the $Cl^-$ form is usually the standard used while for a polymeric cation exchanger the $H^+$ form is used. Also, selectivity equilibria, such as those shown in Eqs. (30) adn (31), and their respective selectivity coefficients, shown in Eqs. (32) and (33), are written so that the selectivity coefficient is greater than 1.

A very useful way to deal with equilibrium which is of practical significance is to describe equilibrium as a distribution coefficient. The distribution coefficient expresses any point on the ion exchange isotherm since it is expressed as

$$K_D = \frac{\text{amount of A retained by the ion exchanger}}{\text{amount of A in solution}} \tag{34}$$

where A is the analyte counterion and $K_D$ the distribution coefficient. The $K_D$ value is dependent on the experimental conditions and is most often determined by a batch equilibration procedure. If the $K_D$ value is for dilute concentrations of the analyte ion, its location on the ion exchange isotherm will often be on the linear portion of the isotherm and therefore will be independent of analyte ion concentration.

Concentration units for $K_D$ are dependent on its application. In classical ion exchange where $K_D$ data are very useful for predicting elution order and subsequent separations, it is common to define $K_D$ in the following manner:

$$K_D = \frac{\text{volume of solution, ml}}{\text{weight of ion exchanger, g}} \times \frac{\text{amount analyte ion retained}}{\text{amount analyte ion in solution}} \tag{35}$$

which reflects a batch determination of $K_D$ where a solution of known volume and concentration of analyte ion whose exchange is being deter mined is equilibrated with a known weight of ion exchanger. For a more fundamental and quantitative treatment of distribution coefficients and their significance, activity in the solution and in the ion exchanger must be taken into account [12].

In HPLC applications of ion exchangers the equilibrium position in terms of a selectivity is more conveniently expressed as the ratio of capacity factors for the two analyte ions that are being compared. Capacity factor k' is given by

$$k' = \frac{V - Vo}{Vo} \tag{36}$$

where V and Vo are retention volumes for the analyte ion and a measure of the column interstitial volume (see, for example, Ref. 170 for a discussion of k' in HPLC), and can be readily determined from a column experiment in contrast to the batch equilibration procedures of the past [12]. The k', which can be calculated from the chromatogram obtained in the column experiment, like the $K_D$ represents a single point on the ion exchange isotherm. Thus, if k' is determined for a nonlinear portion of the ion exchange isotherm, its value will be dependent not only on the conditions of the experiment but also on counter-ion analyte concentration.

Many factors will influence ion exchange selectivity [12,13,16]. In general, for a given pair of analyte ions and a typical sulfonated strong-acid or quaternary ammonium strong-base PSDB-type cation and anion exchanger, the ion exchanger will tend to show a preference for

1. The counterion with the larger polarizability
2. The counterion with the smaller solvated volume
3. The counterion with the higher oxidation state
4. The counterion which has the larger interaction with the ionogenic group
5. The counterion which participates the least in association with mobile phase components
6. The counterion that contributes the least to ion exchanger swelling

When comparing organic polymeric ion exchangers or experimental conditions, ion exchange selectivity is generally favored by an increase in crosslinking, a decrease in analyte solution concentration, a decrease in temperature, and interactions favoring coordination or other association phenomena between the analyte ion and the ion exchanger site or matrix.

Because of the many variables and the competing aspects of the variables, exceptions to the above general guidelines exist. On an absolute scale, therefore, ion exchange selectivities are not universal and for a quantitiative listing they must be carefully determined for a given ion exchanger under a given set of conditions.

Selectivity for common inorganic monovalent and divalent analyte cations on a gel-type sulfonated strong-acid PSDB-type cation exchanger, in general, follows the order shown in Eqs. (37) and (38), respectively:

$$Ba^{2+} > Pb^{2+} > Sr^{2+} > Ca^{2+} > Ni^{2+} > Cu^{2+} > Co^{2+} > Zn^{2+} > Mg^{2+} \quad (37)$$

$$Tl^{+} > Ag^{+} > Cs^{+} > Rb^{+} > K^{+} > NH_4^{+} > Na^{+} > Li^{+} \quad (38)$$

while for a gel-type quaternary ammonium strong-base PSDB-type anion exchanger selectivity, the selectivity order for common divalent and monovalent anions is

$$\text{citrate}^{2-} > SO_4^{2-} > C_2O_4^{2-} > I^{-} > NO_3^{-} > CrO_4^{2-} > Br^{-} > SCN^{-} > Cl^{-} > \text{formate}^{-} > \text{acetate}^{-} > F^{-} \quad (39)$$

A quantitative measure of the selectivity can be obtained by an experimental determination of the selectivity coefficient. Since a comparison between two counterion analytes is being made, a standard must be used. For cation exchange the comparison is to either $H^+$ or $Li^+$ where its selectivity is defined as 1 while for anion exchange $OH^-$ or $Cl^-$ is usually the reference anion. Table 9 lists quantitative selectivities for several metal ions on a sulfonated PSDB copolymer (Dowex 50) as a function of percent crosslinking [171]. Form these data it can be seen that selectivity undergoes a modest change through the mono- and divalent metal series. The effects that the exchanger itself can impose on selectivity is indicated by comparing selectivities for different crosslinking which determines the swelling properties of the exchanger. As crosslinking increases, selectivity increases significantly. Table 10 list selectivities for several common inorganic and organic analyte anions using $Cl^-$ as the reference on a quaternary alkylammonium PSDB strong-base (Dowex 1) anion exchanger [172]. Comparing selectivity change between inorganic cations and anions indicates that the change is greater for the latter. However, since many other factors are involved, including ion exchanger structure, the difference as evident in the comparison of Tables 9 and 10 is not always this large.

The cation and anion selectivity orders listed in Eqs. (37)-(39) are not followed for all ionogenic groups. For example, for methacrylate cation exchangers the order is $Li^+ > Na^+ > K^+$ and the difference increases as the crosslinking increases [173]. A similar order is ob-

**Table 9** Selectivity Coefficients for Inorganic Cations Relative to $Li^+$ on a Sulfonated Polystyrene-Divinylbenzene Ion Exchanger

| | Selectivity coefficient percent crosslinking | | | | Selectivity coefficient percent crosslinking | | |
|---|---|---|---|---|---|---|---|
| Cation | 4 | 8 | 16 | Cation | 4 | 8 | 16 |
| $Li^+$ | 1.00 | 1.00 | 1.00 | $Mg^{2+}$ | 2.95 | 3.29 | 3.51 |
| $H^+$ | 1.32 | 1.27 | 1.47 | $Zn^{2+}$ | 3.13 | 3.49 | 3.78 |
| $Na^+$ | 1.58 | 1.98 | 2.37 | $Co^{2+}$ | 3.23 | 3.74 | 3.81 |
| $NH_4^+$ | 1.90 | 2.55 | 3.34 | $Cu^{2+}$ | 3.29 | 3.85 | 4.46 |
| $K^+$ | 2.27 | 2.90 | 4.50 | $Cd^{2+}$ | 3.37 | 3.88 | 4.95 |
| $Rb^+$ | 2.46 | 3.16 | 4.62 | $Ni^{2+}$ | 3.45 | 3.93 | 4.06 |
| $Cs^+$ | 2.67 | 3.25 | 4.66 | $Ca^{2+}$ | 4.15 | 5.16 | 7.27 |
| $Ag^+$ | 4.73 | 8.51 | 22.9 | $Sr^{2+}$ | 4.70 | 6.51 | 10.1 |
| $Te^+$ | 6.71 | 12.4 | 28.5 | $Ba^{2+}$ | 7.47 | 11.5 | 20.8 |

*Source*: From Ref. 12, p. 168, courtesy of McGraw-Hill. See also Ref. 171.

**Table 10** Selectivity Coefficients for Inorganic Anions Relative to $Cl^-$ on a Quaternary Ammonium Polystyrene-Divinylbenzene Anion Exchanger

| Anion | Selectivity coefficient | Anion | Selectivity coefficient |
|---|---|---|---|
| $I^-$ | 8.7 | $Cl^-$ | 1.00 |
| $ArO^-$ | 5.2 | $HCO_3^-$ | 0.32 |
| $HSO_4^-$ | 4.1 | $H_2PO_4^-$ | 0.25 |
| $NO_3^-$ | 3.8 | $HCO_2^-$ | 0.22 |
| $Br^-$ | 2.8 | $C_2H_3O_2^-$ | 0.17 |
| $CN^-$ | 1.6 | $H_2NCH_2CO_2^-$ | 0.10 |
| $HSO_3^-$ | 1.3 | $F^-$ | 0.09 |
| $NO_2^-$ | 1.2 | $OH^-$ | 0.09 |

*Source*: From Ref. 172 courtesy of New York Academy of Sciences.

served for phosphonic acid-type cation exchangers [174]. If the quaternary ammonium group in the strong-base anion exchanger is changed from $-CH_2N^+(CH_3)X^-$ to $-CH_2N^+(CH_3)_2(CH_2OH)X^-$, some alteration in anion selectivity is observed; the major change, however, is that the difference in selectivity is sharply reduced.

Selectivity can be altered by changing the structure within the ionogenic group. For example, this effect was studied by altering the R groups in the quaternary ammonium ionogenic groups in an anion exchanger [175]. A series of alkylated, hydroxylalkylammonium, and N-ring-containing quaternary ammonium PSDB anion exchangers were prepared and Table 11 summarizes selected selectivity data for several of these strong-base anion exchangers. As the ionogenic $R_4N^+$ group structure is altered, selectivity changes little for some ions and significantly for others, particularly for large polarizable anions. These differences, which tend to be small, would not lead to a major change in resolution if these kinds of anion exchangers were used in a low-column-efficiency separation. However, for a high-column-efficiency separation the differences were significant and improved resolution can be obtained by optimizing the $R_4N^+$ ionogenic group structure relative to the separation that is being attempted.

**Table 11** Inorganic Anion Selectivity on Anion Exchangers as a Function of $R_4N^+$ Structure

| | Anion selectivity on anion exchangers[a] | | | |
|---|---|---|---|---|
| Anion | TMA | TBA | TEA | DMBA |
| $Cl^-$ | 1.0 | 1.0 | 1.0 | 1.0 |
| $F^-$ | 0.66 | 0.71 | 0.64 | 0.72 |
| $Br^-$ | 1.20 | 1.34 | 1.30 | 1.22 |
| $I^-$ | 2.51 | 3.82 | 2.95 | 2.62 |
| $NO_2^-$ | 0.82 | 0.90 | 0.79 | 0.86 |
| $NO_3^-$ | 1.30 | 1.54 | 1.38 | 0.30 |
| $C_2H_3O_2^-$ | 0.25 | 0.25 | 0.14 | 0.26 |

[a]Chloromethylated PSDB was treated with trimethylamine (TMA), triethanolamine (TEA), and *N,N*-dimethylbenzylamine (DMBA) to give the quaternary ammonium anion exchangers with capacity of about 0.027 meq/g.
*Source*: From Ref. 175.

Chelating ion exchangers will show a high selectivity for those analyte ions that coordinate with the chelating ionogenic group. Because formation constants for the coordination are now involved, selectivity can be large and specific. Selectivity data for several metal ions that coordinate with iminodiacetic acid are listed in Table 2 for a commercially available PSDB copolymer which contains the iminodiacetic acid ionogenic group. When compared to the sulfonated PSDB cation exchanger (see Table 9) it can be seen that the coordination strongly enhances selectivity.

Since many of the ions in Eqs. (37)-(39) are derived from weak bases or acids, respectively, the order listed is also dependent on the pH used in the equilibration. Similarly, cations, particularly the transition metal ions, are readily coordinated and their location in the series is significantly influenced by the type of anions and/or the presence of other complexing ligands in the solution.

In general, data for the quantitative determination of ion exchange selectivities of organic analyte ions are not readily available. Although not shown in Table 10, selectivities for tetralkylammonium cations are greater than those for $NH_4^+$ and increase as the alkyl chain increases. Similarly, alkylsulfonate selectivities are large and increase as alkyl chain length increases.

Side chain contributions in the retention of organic analyte ions on ion exchangers can be extensive. Factors such as side chain size, steric properties, hydrophobicity, and electron-withdrawing and donating capabilities will influence retention through their effects on ionization of the organic analyte as well as interaction with the ion exchanger. In general, a relative selectivity order can be obtained by determining the elution order for a series of organic analyte ions for a given elution condition. This approach is practical since it gives immediate information about the potential of the separation.

The nature of organic analyte ions and their compatibility with an organic polymeric matrix suggests that non-ion-exchange interactions are a possibility at the matrix surface. Indeed, this can occur even with ion exchangers that have high ion exchange capacities. A measurement of elution order under these circumstances yields, then, not relative ion exchange selectivity but a selectivity which represents a contribution of the contributing interactions accounting for retention. As the ion exchange capacity (the number of ionogenic groups) decreases, the availability of polymeric matrix surface increases. For low-capacity ion exchangers (<200 μeq/g) (e.g., the type used in ion chromatography, which have a macroporous, high-surface-area, polymeric matrix), retention of an organic analyte ion is due to both ion exchange and surface adsorption [176]. As ion exchange capacity increases, ion exchange becomes dominant, whereas for a decrease adsorption becomes dominant. Furthermore, by manipulating the mobile phase conditions it is possible to favor one interaction over the other. This has practical implications since this permits retention order to be sharply altered depending on how the organic analytes structure favors ion exchange and adsorption [177,178].

Even though inorganic analyte ion selectivities, which were usually obtained from a batch equilibration procedure, and organic analyte ion relative selectivities, which were usually obtained from a column experiment, have been historically determined on large-particle-size ion exchangers, these data in a qualitative sense are readily applied to current microparticle ion exchangers of similar structure and ionogenic group. If selectivities are to be determined on the latter ion exchangers, then analyte ion concentrations must be reduced since these ion exchangers have very low ion exchange capacities. When this is done the determination of the selectivity is on the linear portion of the ion exchange isotherm. With currently available sensitive column effluent detectors this does not generally represent a problem, and relative elution order selectivities are easily and rapidly obtainable with columns packed with microparticle ion exchangers.

### *Sorption of Solutes*

In addition to a reversible, stoichiometric ion exchange, ion exchangers have the ability to take up strong, weak, and nonelectrolyte-type solutes from a solution. The kind and number of ionogenic groups present will influence the sorption but are not directly involved in the stoichiometry of the sorption as is the case when ion exchange occurs. Sorption can be an equilibrium one and, in general, is a reversible process. Often by switching to the pure solvent the sorbed solute is removed from the ion exchanger. Sorption of weak and nonelectrolytes is not the same as the sorption of strong electrolytes. The latter, which is unique to ionic sorbents, is the result of the electrostatic influence of the ionogenic group and its counterion leading to a Donnan-type sorption equilibrium.

Strong electrolytes are sorbed to differing degrees. Low ion exchange capacity, low crosslinking in the ion exchanger, high electrolyte concentration in the solution, high counterion oxidation state, and low coion oxidation state favor electrolyte sorption. Secondary equilibria, such as interactions favoring association or complex formation between the coion, the counterion, the ionogenic group, or the matrix, will disturb the Donnan equilibrium and lead to enhanced electrolyte sorption.

Weak and nonelectrolytes are sorbed by ion exchangers because of several kinds of interactions that may occur. In the absence of interactions solute concentration in the ion exchanger interior would be equal to the solute concentration in the exterior solution. Salting-out phenomena, interactions between the weak or nonelectrolyte solute and the counterion, sieving action, and solute interactions with the polymeric matrix of the ion exchanger can all contribute to the sorption.

Sorption of weak and nonelectrolytes by interaction with the polymeric matrix of the ion exchanger can cover a wide range of interaction energies. On the weak side are London interactions. Of intermediate strength are dipole-dipole interactions whereas strong interactions are favored by $\pi$-electron and charge-transfer interactions. Weak electrolytes in general are sorbed much like nonelectrolytes and are not affected significantly by Donnan equilibrium. The influence of the polymeric matrix in the ion exchanger on solute sorption can be extensive. For example, the macroporous PSDB copolymer without exchange sites has been shown to be a very useful reversed phase adsorbent for the retention of nonelectrolytes and weak electrolytes [179,180].

### *Ion Exchange Kinetics*

The understanding of ion exchange kinetics, particularly from a quantitative viewpoint, is not at the level of the understanding of ion exchange equilibria. However, much progress has been made in both fundamental understanding and mathematical treatment. The discussion here is only qualitative; a more quantitative discussion can be found elsewhere [12-14,181,182].

Consider the system where the spherical ion exchanger beads charged in counterion Y form are placed in a well-stirred solution of electrolyte providing counterion Z. Counterion Y diffuses out of the bead into the solution and counterion Z diffuses from the solution into the bead as equilibrium is approached. The ion exchange process, according to Boyd and workers [180], who were the first to quantitatively treat ion exchange kinetics, is made up of five steps:

1. Counterion Z must reach the ion exchanger bead surface. Even though the solution is stirred, it must diffuse through a thin stationary film whose thickness depends on the rate of stirring that surrounds the ion exchanger bead.
2. Shortly after the exchange has started Y counterions at the bead surface are exchanged and Z counterions must diffuse into the bead to reach other sites.
3. The exchange of Y counterions for Z counterions takes place.
4. The Y counterions replaced by Z counterions in the interior of the bead must diffuse to its surface.
5. The Y counterions after reaching the bead surface must diffuse through the thin stationary film on the bead surface into the solution.

Since electroneutrality must be maintained, the first and last step must occur simultaneously, i.e., diffusion of every counterion Z through the stationary phase toward the bead interior must be matched by diffusion of counterion Y from the bead through the stationary film into the solution. Similarly, diffusion within the bead interior of counterion Z to the sites and Y from the site must be equal. Thus, the number of steps is reduced to film diffusion, bead diffusion, and chemical exchange, and only the diffusion process becomes rate controlling. For dilute solutions film diffusion is the slow step while for more concentrated solutions diffusion within the bead is the slow step [181]. At intermediate concentration (about 0.01 to 0.1 M for the bulk solution electrolyte), the rate may be affected by both steps.

The coion in the bulk solution has little effect on the kinetics and rate of exchange. Other factors, such as ion exchanger bead particle size, bead swelling, temperature, solvent type of counterions, agitation, counterion mobilities, and solution sitrring, will affect the rates. Decreasing the bead size, which favors more rapid exchange, will affect both film and bead diffusion with the effect being greater for the latter. This effect is particularly evident from a kinetic and, most importantly, from a practical, applied point of view when considering the currently used microparticle ion exchanger. Increased bead swelling (favored by low crosslinking) and elevated temperature favor a more rapid ion exchange. Solvent polarity will effect diffusion and bead swelling; as the solvent or mixed-solvent polarity decreases, bead swelling decreases leading to reduced exchange rates. The type of counterions (hydrated size and oxidation state) affects their diffusion ability while increased agitation decreases film thickness. Also, it should be noted that the general features of ion exchange kinetics can be extended to the kinetics of electrolyte and nonelectrolyte sorption and to bead swelling.

### *Ion Exchange Capacity*

A quantitative measure of the total ion exchange capabilities of an ion exchanger is its ion exchange capacity. Ion exchange capacity is defined as the number of counterion equivalents in a specific amount of the ion exchanger. For sulfonated and quaternized ammonium PSDB, strong-acid- and strong-base-type cation and anion exchangers of average (about 8%) crosslinking, the theoretical weight ion exchange capacity approaches >5.0 meq/g. This assumes that there is one ionogenic group per benzene ring. Since crosslinking also provides aromatic rings, they can also contain ionogenic groups and consequently changes in crosslinking amounts do not necessarily grossly affect the ion exchange capacity. In practice, maximum ion exchange capacities approach 5 meq/g for the sulfonated PSDB ion exchanger and 3-4 meq/g for the quaternized ammonium PSDB anion exchanger. The fact that the anion exchanger frequently has less than the theoretical amount is due to synthesis [see Eq. (14)], the reactivity of the chloromethylated group toward other reagents, and the modest stability of the quaternary ammonium group. Ion exchangers with lower weight ion exchange capacity (see above) are obtained by controlling the procedure for introducing the ionogenic group. Weight ion exchange capacities for other organic polymeric ion exchangers can cover a wide range, assuming a high-capacity ion exchanger is the goal, but typically are in the 1-5 meq/g range. For example, a crosslinked polyacrylic acid weak-acid cation exchanger can provide exchange capacities often exceeding 10 meq/g.

Several factors will influence the actual number of counterions taken up by the ion exchanger and a capacity that expresses the total number of ionogenic groups present is not always a true indication of the useful ion exchange capacity. Thus, an effective or apparent ion exchange capacity is used. This is more useful from a practical viewpoint since it expresses the ion exchange capacity for a given ion exchanger under a given set of conditions.

One factor affecting ion exchange capacity associated with the ion exchanger itself is its ability to swell when placed in the solvent. Swelling favors easier penetration by diffusion and excess to interior ionogenic groups. If not swollen the interior groups are not completely available. Counterion size, if large enough, can hinder excess to interior ionogenic groups. If the ionogenic group is a weak acid or base, dissociation of this group will strongly influence its capacity. Major solution factors influencing capacity are those that influence swelling, provide favorable diffusion, and favor dissociation of the ionogenic group.

Ion exchangers are also capable of taking up strong, weak, and nonelectrolytes. This uptake, while dependent on the number and type of ionogenic groups, is not a measure of the number of ionogenic groups in the exchanger. This kind of uptake, particularly for weak and nonelectrolytes, is more dependent on the experimental conditions that are employed.

Weak-acid and base ion exchangers have capacities that are pH-dependent. The high selectivity for $H^+$ on a weak-acid ion exchanger and $OH^-$ on a weak-base ion exchanger, which is due to association, is responsible for the pH dependence.

Ion exchange capacity is frequently determined by taking a weighed amount of the cation exchanger in the $H^+$ form, replacing the $H^+$ with an alkali metal countercation, and titrating the liberated $H^+$ with standard base. Anion exchangers in the $OH^-$ or $Cl^-$ form are treated similarly using an appropriate counteranion to liberate the $OH^-$ or $Cl^-$, which is then titrated with standard acid or $Ag^+$, respectively [12,13]. For low-capacity ion exchangers which are currently used in HPLC and ion chromatography, acid-base titrations are more difficult to carry out because the amount of $H^+$ or $OH^-$ liberated per gram of low-capacity ion exchanger (<200 μeq/g) is very small and often large quantities of the ion exchanger are not available. In this case replacement of $K^+$ or $Na^+$ form the cation exchanger by an appropriate countercation and atomic absorption or flame emission analysis can be done. Alternatively, a microconductance measurement of replaced $H^+$ by NaOH can be done [176]. For the anion exchanger a micro $Ag^+$-$Cl^-$ potentiometric titration is suitable.

### Ion Exchanger Surface Area and Porosity

Organic polymeric ion exchanger surface area, which is largely due to bead interior and not outer-sphere surface and porosity, is directly related to crosslinking and the extent of bead swelling. For the PSDB ion exchanger the type of process used in the polymerization of the matrix is a major factor. If the resulting matrix is microreticular (a gel polymer), it is not truly porous, i.e., its porous-like property is evident only when the polymer is swollen. Thus, if the imbibed solvent, e.g., water, is removed, the matrix contracts, the porous-like property disappears, the surface area on a relative basis becomes insignificant, and exchange sites are no longer available. As crosslinking increases the dependence of the porous-like character and surface area on swelling-contracting decreases. In contrast, for the macroporous PSDB matrix (see Fig. 1) a true porosity is maintained, and even though the bead-swelling solvent (such as water) is removed, porosity, surface area, and subsequently the availability of interior exchange sites are still present. This is illustrated in Fig. 4 where retention of the weak-base *p*-nitroaniline was determined in ethanol as a function of equilibration time on a microporous or gel PSDB sulfonated cation exchanger (D50 × 8 and IR-120) with 8% crosslinking and a macroporous PSDB sulfonated cation exchanger (A-15 and A-15c, which is a small particle A-15) in the H form. The retention, measured as a batch distribution coefficient (see above), is for an acid-base interaction or

$$RSO_3^-H^+ + O_2N\text{-}C_6H_4\text{-}NH_2 \rightleftharpoons O_2N\text{-}C_6H_4\text{-}NH_3^+(RSO_3^-) \qquad (40)$$

In these experiments the ion exchangers are reproducibly dry and contain only trace levels of water; thus they are in their contracted state. As shown in Fig. 4, the uptake of the base by the microporous ion exchanger is much slower in ethanol and reaches a lower equilibrium

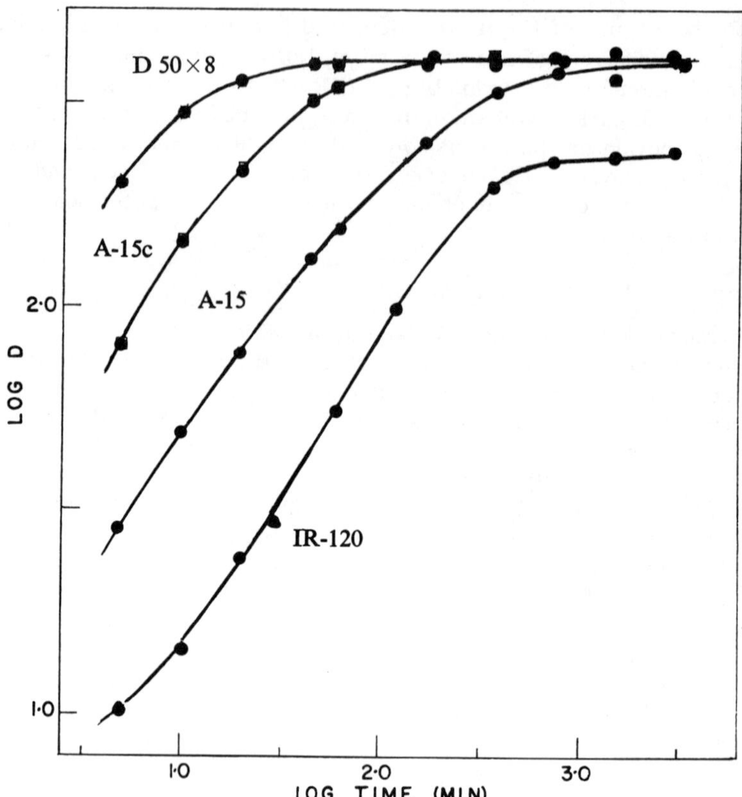


Fig. 4 Sorption rate of *p*-nitroaniline from ethanol on a reproducibly dry strong-acid PSDB cation exchanger in the H form. From Ref. 183, courtesy of Pergamon Press.)

level of retention. If benzene is used, the base is not even retained on the dry microporous D50 × 8 cation exchanger because the ion exchanger will not swell in benzene while it is taken up by the dry macroporous A-15c cation exchanger [183], which is consistent with its macroporous structure. If the D50 × 8 is preswollen with water before equilibration, then the base is retained.

Typical porosities and surface areas for gel and macroporous PSDB sulfonated H-form cation exchangers are listed in Table 12. Values for gel and macroporous unsulfonated PSDBcopolymer matrices are also included. The variable data for the microporous or gel exchangers are the result of the large dependence of porous-like character and surface area on bead swelling. Although not shown, PSDB anion ex-

Table 12 Surface Area and Porosities for Gel and Macroporous Cation Exchangers and PSDB Copolymers

| Ion exchanger | Surface area $m^2/g$ | Average pore diameter Å |
|---|---|---|
| | Gel matrix | |
| Dowex 50 × 8[a] | <0.1 | 5-50 |
| Amberlite IR-120[b] | <0.1 | 5-50 |
| Polystyrene-Divinylbenzene[a,e] | <0.1 | 5-50 |
| | Macroporous matrix | |
| Amberlyst, 15[b,f] | 42 | 290 |
| XAD-1[b,d,e] | 100 | 205 |
| XAD-2[b,d,e] | 300 | 90 |
| PLRP-S[c,d,e] | 550 | 100 |
| PRP-1[d,e] | 415 | 75 |

[a]Dow Chemical Company.
[b]Rohm and Haas Chemical Co.
[c]Polymer Laboratories.
[d]Hamilton Co.
[d]A PSDB adsorbent.
[e]Quaternary ammonium macroporous anion exchangers of similar surface areas and porosities are available.
*Note*: See Also Refs. 36-48, 184-186.

changers would have similar porosities and surface areas. Since macroporous PSDB polymers of differing porosity and surface areas (XAD, PRP-1, and PLRP-S in Table 12) can be synthesized, it is possible to prepare cation and anion exchangers that differ in surface area and porosity. This property is of practical value in organic analyte ion separations and in inorganic analyte ion separations using low-capacity, high-efficiency ion exchange columns.

### *Ion Exchanger Stability*

Generally, organic polymeric ion exchangers, particularly those containing a PSDB copolymeric matrix, have a favorable stability toward a wide variety of conditions. The PSDB ion exchangers, however, are not totally inert and they can undergo changes in their properties under certain conditions. The least stable is the anion PSDB exchanger, particularly the quaternized ammonium-type exchanger when charged in the $OH^-$ form.

The PSDB ion exchangers are insoluble in water and all common organic solvents. Only in the absence of crosslinking (or very low crosslinking) is solubility a concern. The three major changes that can occur in the PSDB ion exchanger which will utlimately affect their column performance in analytical liquid column chromatography are

1. Bead fracture
2. Loss of ion exchange capacity due to loss of the ionogenic group
3. C-C bond breaking at the crosslink

Bead fracture most often is the result of the sudden shock the exchanger experiences when conditions are abruptly changed that causes the exchanger to swell or contract extensively. As ion exchanger crosslinking increases, the tendency toward bead fracture decreases. The PSDB ion exchanger, while possessing a rigid, three-dimensional network, still exhibits softness relative to high pressure. Thus, in column applications, which is very common in analytical liquid chromatography, excessive pressure drop across the column can produce bead fracture, particularly if swelling-contracting is also occurring at the high inlet pressures. By using increased crosslinking or a macroporous PSDB matrix rather than a gel one (see above), the effects of pressure drop can be reduced. In column operation bead fracture is undesirable because (a) it changes the mobile phase flow through the column, (b) it can create channeling in the column, (c) it can plug the column or screens holding the exchanger bed, and (d) for low-capacity ion exchangers, the column ion exchange capacity can decrease sharply on a percent basis due to fractured particles passing out of the column. All of these factors result in poorer column performance by affecting peak shape and/or retention time.

In applications of ion exchangers in analytical chromatography loss of ionogenic group is more of a problem than C-C breaking. The latter generally requires a drastic condition or one which is usually not encountered in typical analytical ion exchange column chromatography. Loss of ionogenic groups and C-C bond breaking for PSDB-type ion exchangers, which usually occurs first at the crosslinks, will occur when the ion exchanger is exposed to high temperature. Decomposition patterns for ion exchangers containing other types of organic polymeric matrix will depend on the matrix. High-temperature organic polymer matrices can be prepared; however, loss of attached ionogenic groups will still take place. Decomposition temperatures are not part of a typical column operation but might be approached if heat-drying techniques are used to remove water or other solvents present in the ion exchanger interior. Since water solvation energies are appreciable, a high temperature is required to remove this water. Infrared, mass spectrometry, and nuclear magnetic resonance techniques have been extensively used to determine ion exchanger decomposition products [187]. Often these techniques are combined with thermogravimitry and related thermostrategies to monitor decomposition as a function of temperature [187].

The loss of ionogenic group through thermal decomposition depends on the ionogenic group and its counterion. Figure 5 summarizes thermal stability data for PSDB ion exchangers containing the $-CH_2NR_3^+$, $-SO_3^-$, and $PO_3^{-2}$ ionogenic groups. The strong- and weak-acid cation exchangers are thermally more stable than the quaternized ammonium strong-base anion exchanger. The former at elevated temperature involves a combined desulfonation (more readily for the H form over an M form), C-C breaking, and oxidation depending on the availability of oxygen. The anion exchanger, particularly when charged in the OH form, decomposes at the quaternary ammonium site before the PSDB matrix is affected as temperature increases. Depending on the R groups in the quaternary ammonium site, and whether water is present or not, amine, olefin, and alcohol decomposition products are formed.

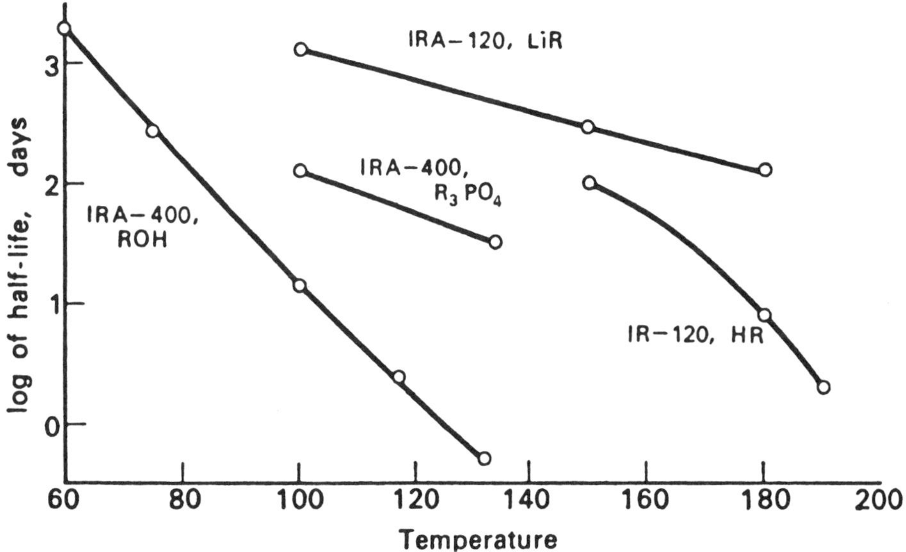


Fig. 5 Thermal stability of organic polymeric cation and anion exchangers. From Ref. 13, courtesy of Pergamon Press.

Because the thermal stability of the quaternary ammonium ionogenic group is low, drying of these ion exchangers must be done carefully. Usually a combined thermal (<60°C) and vacuum procedure will reduce water levels. If the anion exchanger is in the OH form, loss of quaternary ammonium group will occur in exchanger-drying procedures.

Almost all analytical column liquid chromatographic applications of ion exchangers do not require the use of solvent-free ion exchangers. Storage and packing of columns is done with water-preswollen (solvated) ion exchangers. Dry ion exchangers have favorable desiccant properties and the sulfonated cation exchanger can be used as a desiccant [187].

In general, ion exchangers are not decomposed by strong acid or base except when combined with elevated temperature. Strong oxidizing agents, particularly when combined with heat, will cause ion exchanger decomposition. Reactions leading to loss of the ionogenic site and C-C breaking will occur. Decomposition is also caused by radiation.

### *Ion Exchanger Swelling*

Organic ion exchangers will sorb the solvent in which they are placed. This causes the ion exchanger, particularly the organic polymeric ion exchanger, to swell or expand in volume. If the solvent is removed the polymeric ion exchanger will contract. Swelling-contracting is an equilibrium property and depends on the solvent, the ion exchanger polymeric matrix and its degree of crosslinking, the ionogenic group, its counterion and their solvation, and the presence of electrolytes in the solvent.

When ion exchangers swell they do so only to a limited degree, and when equilibrium is reached swelling ceases. Swelling equilibrium results from a balancing of opposing forces. On the one hand, solvent surrounds the ionic and polar constituents of the polymeric ion exchanger causing the matrix to stretch. The structure and properties of the matrix resist this process. At equilibrium the elastic forces of the matrix balance the tendency toward dissolution due to solvation.

Organic polymeric ion exchanger swelling is favored by

1. Polar solvents
2. High ion exchange capacity
3. Low crosslinking in the polymer matrix
4. Low counterion oxidation state
5. Strong solvation of the fixed ionogenic groups
6. Large and strongly solvated counterions
7. High dissociation between ionogenic group and counterion
8. Low concentration of electrolyte in the external solution

The swelling of organic polymeric ion exchangers is of considerable practical importance in column operation. Since swelling can be large, ion exchanger particles packed in glass columns may cause the glass column to burst when the exchanger swells. In HPLC applications, where ion exchange particles are packed into stainless steel columns, swelling will not burst the column but it can cause the packed bed to tighten. Consequently, this will affect the flow of the mobile phase through the column, produce excessive column inlet pressure, and cause ion exchange bead fracture. The reverse process, or ion exchange bead contraction, can be equally undesirable in column operation. In this case when the mobile phase is sharply changed to produce shrinking of the ion exchanger particles, column channeling, particularly at the column walls, can occur. This also has a significant effect on how the mobile phase flows through the column.

### *Ion Exchangers in Organic Solvents*

The properties of ion exchangers described in the previous sections will often differ when the exchangers are placed in water-organic solvent mixtures or nonaqueous solvent [187-189]. Furthermore these differences will vary depending on whether the exchanger is microporous (gel) or macroporous, and whether the organic polymeric ion exchanger is high or low capacity. Since low-capacity ion exchangers can participate in adsorption [176-178], particularly if the ion exchanger polymer matrix is a macroporous PSDB, the role of the effect of solvent on adsorption must also be taken into account.

In general, selectivities for the exchange of a given pair of ions are greater in an organic solvent of a lower dielectric constant than water. For water-organic systems the selectivities may be intermediate, larger, or smaller than selectivities for water. Often reversal in selectivities is observed.

The dielectric constant of the solvents involved appears to be the most important factor affecting exchange efficiency. Other factors (such as solvation of the ionogenic group, hydrogen bonding between the ionogenic group and solvent, coordination properties of the solvent, and interactions between the solvent and the polymer backbone) can be of equal, if not greater, importance under certain circumstances. Clearly, exchanger swelling will be strongly dependent on the nature of the solvent.

A practical consequence of the enhanced or altered selectivity is that separation factors are often improved. Thus, certain separations become possible, whereas others are improved or more readily completed by utilizing an organic solvent or mixed solvent in place of water in the eluting mixture.

Rates of exchange are generally slower in mixed or nonaqueous solvents; as the polarity fo the solvent and the swelling of the exchanger decrease, the exchange rate decreases. However, there are few detailed kinetic studies employing this kind of media. The effect of the solvent on exchange rate will also be influenced by whether the exchanger is microporous or macroporous. In general, particle diffusion is the rate-determining step for exchange of electrolytes in mixed or nonaqueous solvents. Since weak electrolytes and nonelectrolytes can be retained by exchangers by processes other than ion exchange, the kinetic phenomena and factors which influence it become more complex.

Ion exchangers are capable of taking up anhydrous solvents as well as mixed solvents. For mixtures of water and common water-miscible organic solvents, water is often preferred. This preference generally holds up to about 80% by volume organic solvent after which the interior solvent composition approaches that of the bulk composition. The nature of the charged form of the exchanger, whether it is an anion or cation exchanger, the extent of swelling (amount of crosslinking), whether the exchanger is microporous or macroporous, the ratio of organic solvent to water, and the polarity of the organic solvent used in the solvent mixture are significant factors that influence the solvent preference. (It is possible to observe a preference of the organic solvent over water for certain combinations of these parameters.) Because of the differences in the bulk vs. interior solvent composition, the role of partitioning must be considered in ion exchange separation applications, particularly when weak and nonelectrolytes are being separated.

Uptake of anhydrous solvents generally decreases with a decrease in polarity. Also, the type of anhydrous solvents taken up and the extent of uptake are markedly different for the macroporous exchanger when comapred to the microporous exchanger.

The microporous exchangers will swell to the greatest extent in the more polar solvents. In nonpolar solvents this type of exchanger will remain collapsed, provided that the exchanger is free of water. If the exchanger is preswollen in water and then added to the nonpolar solvent, it will remain swollen because of the exchanger's preference for water.

In general, ion exchange capacities for strong electrolyte counterions will remain the same even in mixed or nonaqueous solvents. Capacity for weak and nonelectrolyte retention can vary substantially since molecular retention is often involved. Furthermore, the type of solvent mixture, its polarity, and the charged form of the exchanger will have a significant effect on the capacity exhibited by the exchanger toward the weak electrolyte and nonelectrolyte.

Because of the rigid, permanent-like porosity of the macroporous ion exchanger, it has many practical advantages over the gel-type ion exchanger, particularly in the presence of organic solvents. For example, the macroporous ion exchanger tends to be physically and chemically more stable, will function in completely nonpolar solvents even if the exchanger is not preswollen, becomes organically fouled up in continuous column operation at a much slower rate, will take up solvent (and swell) much differently from the gel-type exchanger, and will be permeable to large organic ions and molecules in nonaqueous as well as aqueous media.

## Bonded Phase Ion Exchangers

### *Ion Exchanger Equilibria-Selectivity*

Detailed isotherm data, selectivity coefficient data, or distribution coefficient data determined by batch equilibration procedures as a function of the ion exchange parameters, are not readily available for bonded phase ion exchangers. Most studies with bonded phase ion exchangers deal with a dynamic determination of selectivity and equilibrium, i.e., capacity factor measurements are made as a function of the ion exchange parameters using a column technique. There are several reasons for this: (a) Bonded phase ion exchangers are expensive relative to organic polymeric ion exchangers for the quantities needed for batch equilibration. (b) Bonded phase ion exchangers are microparticle in size and their separation from an equilibrated solution is more difficult. (c) Since the ionogenic groups on the bonded phase ion exchanger ($-SO_3H$, $-CO_2H$, $-PO_3H_2$ for cation exchange and $-CH_2N^+(R)_3X$, $CH_2NH_2$ for anion exchange) are the same as on the organic polymeric ion exchanger, a major change in equilibria-selectivity is not likely. (d) The column experiment using a high-performance strategy is fast, reliable, and the information obtained this way is readily applied to the prediction of resolution of complex mixtures.

Because of the similarity in ionogenic groups, Eqs. (30) and (31) describing cation and anion exchange and Eqs. (32) and (33) which define cation and anion molar selectivity coefficients for cation and anion exchange, respectively, apply to the bonded phase ion exchangers.

A relative ion exchange selectivity on bonded phase ion exchangers is obtained by comparing capacity factors [see Eq. (36)] for given analyte ions calculated from chromatograms for the analyte ions that are determined as a function of mobile parameters affecting ion exchange. Table 13 compares retention times for a series of inorganic analyte anions obtained on a quaternary ammonium bonded phase silica anion exchanger to a quaternary ammonium low-capacity PSDB anion exchanger. These elution orders are in good agreement and are consistent with the selectivity order found for a high-capacity quaternary ammonium PSDB ion exchanger [see Eq. (39)].

Matrix effects on selectivity on bonded phase ion exchangers are different than those for organic polymeric ion exchangers. Bonded

**Table 13** Comparison of Retention Times for Common Inorganic Anions on a Low-Capacity Quaternary Ammonium PSDB Anion Exchanger to a Quaternary Ammonium Bonded Phase Anion Exchanger

| | Retention time (min) | |
|---|---|---|
| Anion | PSDB anion exchanger[a] | Bonded phase anion exchanger[b] |
| $F^-$ | 2.90 | |
| $CH_3COO^-$ | 2.90[c] | 2.10 |
| $H_2PO_4^-$ | 3.24[c] | 3.15 |
| $CN^-$ | | 3.40 |
| $BrO_3^-$ | | 3.60 |
| $HCO_3^-$ | 3.30[c] | |
| $Cl^-$ | 2.96 | 3.82 |
| $NO_2^-$ | 3.10 | 4.40 |
| $Br^-$ | 3.38 | 5.00 |
| $ClO_3^-$ | | 5.05 |
| $NO_3^-$ | 3.40 | 5.72 |
| $I^-$ | 4.42 | 11.5 |
| $CrO_4^{2-}$ | 6.70 | |
| $SCN^-$ | 7.04 | 22.4 |
| $ClO_4^-$ | 8.46 | 27.0 |
| $SO_4^{2-}$ | 9.04 | 13.6 |
| $SO_3^{2-}$ | 9.14 | 13.6 |
| $C_2O_4^{2-}$ | 9.60 | 14.6 |
| $S_2O_3^{2-}$ | 14.4 | 21.6 |

[a] A quaternary ammonium PSDB anion exchanger with a 0.007 meq/g capacity, 500 × 3.0 mm, a mobile phase of 0.1 mM phthalate at pH = 6.25 at 2.0 ml/min.
[b] Vydac 302 quaternary ammonium silica bonded phase anion exchanger, 250 × 4.6 mm, a mobile phase of 5.0 mM phthalate at pH = 4.0 at 2.0 ml/min.
[c] Appears as a negative peak.
*Source*: From Ref. 190, courtesy of Elsevier Science Publishers.

phase ion exchangers do not swell or contract; thus, this parameter is not a factor. While organic polymeric ion exchangers, particularly low-capacity ion exchangers derived form macroporous, high-surface-area PSDB [176,178], will exhibit adsorption in addition to ion exchange, the bonded phase ion exchanger silica matrix is capable of providing a cation type of ion exchange site. This site is due to residual silanol groups on the silica surface and is a problem primarily in cation exchange. In the course of preparing the bonded phase not all silanols are consumed (see above) as the bonded phase is chemically attached. The remaining silanols, which are low in number, can participate in cation exchange since the silanol is a weak-acid ionogenic group. The silanols, due to cation exchange, have been shown to affect the chromatography of amines [191] and metal cations [192] even though the number of silanols remaining on the silica surface in a bonded phase is low. The number present is dependent on the bonding chemistry and the strategies that are used to block free silanols. Consequently, ion exchange capacity due to free silanols is likely to be variable for different manufactured bonded phase silica-based ion exchangers. Figure 6 shows that cation exchanger capacity due to free silanol presence on the silica surface is sufficient to yield resolution of cation mixtures.

### *Other Properties of Bonded Phase Ion Exchangers*

Bonded phase ion exchangers have been prepared commercially to meet the requirements (see Table 8) that a stationary phase should meet in order to provide optimum results in HPLC applications. Experience gained in the preparation of silica for reversed and normal bonded stationary phases has contributed to the rapid development of bonded phase ion exchangers.

In general, ion exchange on bonded phase ion exchangers is rapid. Two major factors contribute to this rapidity: (a) Bonded phase ion exchangers are microparticles of uniform size distribution with modestly high surface area. (b) Bonded phase ion exchangers have a tendency to provide the ionogenic group which is part of the bonded layer at the surface. Thus, diffusion throughout a macroporous solid stationary phase is lessened and mass transfer is increased. The quality of the silica in terms of its porosity, surface area, and method of preparation as well as the amount of bounded phase coverage is a critical factor in determining surface vs. bead interior diffusion. In general, a porous silica is used in the preparation.

The major drawback of the silica-based bonded phase ion exchanger is that its stability is limited to a pH range of 2-9. Out of this range both silica dissolution and/or hydrolysis of the bonded layer occurs. This also places a limit on the pH range of the mobile phase that can be used in the separation. For example, this becomes a disadvantage

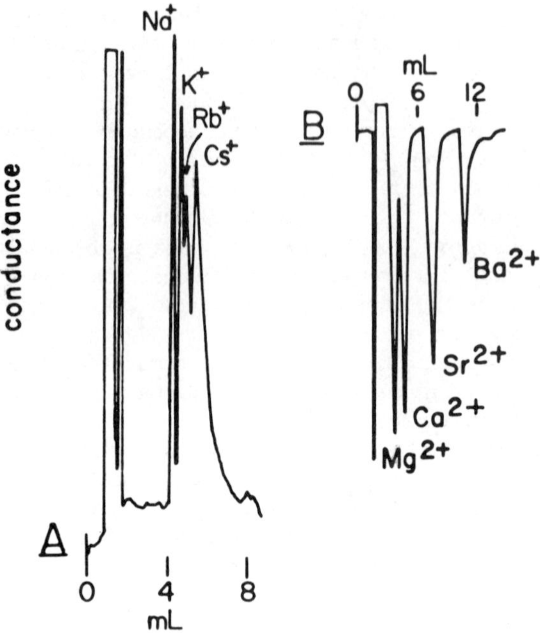


Fig. 6 Separation of alkali metals (A) and alkaline earths (B) on a Silica C-8 bonded phase. Column: Zorbax C-8, 150 × 4.6 mm; mobile phase: (A) 3:7 MeOH-$H_2$), 2.5 × $10^{-3}$ M LiCl, (B) 3:7 MeOH-$H_2O$, 5 × $10^{-4}$ Na citrate (pH - 7.0); conditions: 1.0 ml/min, conductivity. From Ref. 192, courtesy of Marcel Dekker.

in the separation of weakly acidic or basic-type analytes since strong-base and strong-acid mobile phases, respectively, are required to convert the analytes to their dissociated forms. In general, the $SO_3H$ bonded phase is more stable with respect to loss of the ionogenic group than the quaternary ammonium group. However, since a basic condition above pH = 8 is not used, loss of the quaternary ammonium group when charged in the OH form is not as serious a problem as in the case of organic polymeric anion exchangers, where a basic pH is frequently used. The temperature limit is about 80°C but rarely is the bonded phase ion exchanger used at this temperature.

Bonded phase ion exchangers are low-capacity ion exchangers. Typical ion exchange capacities are in the order of 100-300 μeq/g.

### Pellicular Ion Exchanger

The ionogenic groups in a pellicular ion exchanger are the same as those used in organic polymeric and bonded phase ion exchangers. Thus, the selectivity order would be expected to be the same.

The properties of a pellicular ion exchanger are determined by the structure and crosslinking of the polymer coating, the uniformity of the coating, its thickness, contribution of the core material, and the particle size. Since the coating is a thin layer, the ion exchange capacity msut be small even if all the theoretically available positions on the polymer coating are used. Exchange capacities most often are less than 100 μeq/g and can approach as little as 5 μeq/g. The fact that ion exchange capacities are low requires that the analyte ion sample concentration also be low, so that the column chromatography can be carried out on the linear portion of the ion exchange isotherm. If this is done, then a column overload is avoided.

The core material makes up as much as 98% of the particle. Thus, its physical properties of density, size (assuming a thin coating), and column packing techniques are determined by the core material. Typically, the coating layer to which the ionogenic groups are chemically bonded is about 1% of the particle radius.

A PSDB copolymer coating containing the ionogenic group appears to be stable under normal column chromatographic conditions. Conditions that produce swelling or contracting do not permanently change the layer. On the other hand, mechanical factors such as rubbing of adjacent beads due to stirring, excessive compression, or vibration will damage the layer through chipping and pealing, particularly if the layer is lowly crosslinked and/or is swollen.

The major advantage that a pellicular ion exchanger should offer is a favorable column efficiency due to a rapid mass transfer. The latter should occur because ion exchange is essentially a surface exchange and diffusion into bead interior is minimal. A detailed discussion of the factors influencing efficiency is provided elsewhere [163].

Table 14 compares column efficiencies obtained on a sulfonated PSDB layer glass core pellicular cation exchanger to sulfonated gel-type PSDB microparticle high-capacity cation exchangers for the separation of nucleosides. The totally polymeric cation exchanger provides the better efficiency in terms of plate height and a modestly faster separation time in terms of plates generated per second. The third factor, however, favors the pellicular cation exchanger and indicates that the inlet pressure required to achieve the plate level is much higher on the totally polymeric cation exchanger. Two factors contribute to what appears to be a poorer performance by the pellicular ion exchanger. First, the particle size of the pellicular ion exchanger is 5-10 times larger than the totally polymeric cation exchangers. A smaller uniform particle range would provide better efficiency. Second,

**Table 14** Comparison of Organic Polymeric and Pellicular Cation Exchangers for the Separation of Nucleosides

| Cation exchanger | Particle size (μm) | Plate height (mm) | Plates/sec | Plates/atm |
|---|---|---|---|---|
| Conventional[a] | 3-7 | 0.29 | 6.2 | 2.9 |
| Conventional[b] | 7-14 | 0.74 | 4.6 | 1.0 |
| Pellicular | 44-53 | 1.94 | 3.1 | 22.3 |

[a]DC-X cation exchanger, Durram Chemical Co.
[b]VC-10 cation exchanger, Sondell Scientific Instrumetns.
*Source*: From Ref. 163, courtesy of Marcel Dekker, Inc.

the pellicular ion exchanger is a low-capacity cation exchanger while the totally polymeric cation exchangers are high capacity and analyte loadings are well below the overload limits for the latter cation exchangers.

Several potential advantages are offered by pellicular ion exchangers due to their unique structure and properties [163]:

1. Columns can be dry-packed. This property also permits easy packing of a long, narrow-diameter column.
2. Columns of pellicular ion exchangers withstand high column inlet pressure. If a glass core pellicular ion exchanger is used, column compression is low and permeability remains high even though inlet pressure is increased.
3. Pellicular ion exchangers provide good stability and if a glass core is used they are useful throughout the entire pH range.
4. Since the solid core provides physical strength, crosslinking in the polymer coating can be varied to alter its swelling and contracting properties based on their influence on retention rather than on physical strength.
5. Since exchange capacity is low, eluents of low electrolyte concentration can be used even in cases where ion exchange is very favorable.

Even though the glass core pellicular ion exchangers offer several useful advantages, they are not widely used. It should be noted, however, that low-capacity, macroporous, totally polymeric ion exchangers that have their ionogenic groups on the surface and bonded phase ion exchangers containing a bonded phase layer chemically attached to macroporous silica possess many of the useful features of the pellicular ion exchanger.

## Inorganic Salt and Oxide Ion Exchangers

### *Ion Exchange Equilibrium–Selectivity*

The concept of ion exchange selectivity at equilibrium applies to inorganic ion exchangers providing an ion exchange process is present. While a quantitative selectivity determination based on the establishment of an ion exchange isotherm for competing ions is desired, most selectivities on inorganic ion exchangers are not obtained this way. The approaches most used are based on a batch equilibration method in which the two competing analyte ions are allowed to reach equilibrium with the inorganic ion exchanger or an establishment of an elution order obtained by a column elution procedure. Alternatively, elution on a thin layer of the inorganic oxide or salt ion exchanger or cellulose impregnated with the inorganic ion exchanger can be used to establish an elution order. Relative selectivities of these types have been obtained for many inorganic oxide and salt ion exchangers.

Crosslinking-swelling effects are not significant in inorganic ion exchangers and therefore have no influence on analyte ion selectivity on inorganic ion exchangers. The major factors are (a) analyte ion size, (b) steric effects and lattice forces, particularly when dealing with zeolite structures or similar-type ion exchangers, (c) interactions favoring association between the analyte ion and the inorganic ion exchanger site and matrix, and (d) association phenomena, such as complex formation, between the analyte ion and the mobile phase components. Because of these contributions, particularly the first two, selectivity on inorganic ion exchangers is often different than the selectivity found for typical strong-acid sulfonated PSDB and strong-base quaternary ammonium PSDB cation and anion exchangers, respectively. Furthermore, these factors are so significant that selectivities can differ in magnitude as well as in order from one kind of inorganic ion exchanger to another.

Table 6 lists cation and anion selectivities for several oxide ion exchangers. For oxides that exhibit cation selectivity, the general order is:

$$\text{ter- and polyvalent cations} > \text{transition metal ions} > \text{alkaline earth ions} > \text{alkali metal ions} \quad (41)$$

Within the families listed in Eq. (41) the order may differ from one inorganic ion exchanger oxide to another. While inorganic salts show a high selectively for multivalent cations, particularly the ter- and polyvalent cations, crossover between the groups in the selectivity order shown in Eq. (41) is common and depends on the ion exchanger salt.

The preference for multivalent ions is high for all types of ion exchangers. However, the preference shown by inorganic oxides and salts, particularly the former, is often unusually high and is likely the result of a contribution of hydrolysis to the ion exchange process. In some cases the affinity is so high that retention is still obtained even when the pH is well below the isoelectric pH value where cation exchange sites should no longer be present.

Often the selectivity order for the alkaline earths and the alkali metals is similar to the order found on typical sulfonated PSDB-type cation exchanger, i.e., a decreasing selectivity is found for an increasing size of the hydrated cation. There are exceptions and observed reversals in selectivity are suggested to be due to a high polarisability of anionic groupings in the oxide. For example, $ZrO_2$, $SnO_2$, and $TiO_2$ exhibit this property and because of this they prefer the larger, more polarizable hydrated metal ion. Another reason appears to be steric effects. This is the case with $Sb_2O_5$, which has a zeolite-type structure.

Inorganic anion selectivities on inorganic oxide ion exchangers are indicated in Table 6. A general qualitative trend is not indicated and many reversals in anion selectivity are apparent when comparing the different inorganic oxides. The only general trend appears to be that selectivity rises sharply as anionic charge increases. Furthermore, retention of polyvalent anions can be so strong that even strong eluents and a pH well above the isoelectric pH where anion exchange sites should no longer be present will not desorb the retained polyvalent anion. The fact that there are reversals and irreversible sorption (mostly polyvalent anions) suggests that processes other than anion exchange contribute to anion sorption. These are likely hydrolysis and complexation effects.

Selectivities for organic analyte ions, particularly those of more complex structures, due to ion exchange are more difficult to identify. Furthermore, studies of this type are not extensive. Inorganic oxides and salts are adsorbents and they readily exhibit adsorptive properties toward organic analyte ions. Since the inorganic oxides and salts are used as ion exchangers in aqueous solution, their adsorptive properties are sharply diminished. However, retention order of organic analyte ions often reflects a contribution of adsorption, ion exchange, and sieving if the organic analyte ion is large. Another limiting factor that affects ion exchange of organic analyte ions on inorganic oxide and salt ion exchangers is that the pH required to ionize the organic analyte weak acid or base is often not compatible with the pH stability range of the inorganic ion exchanger.

### *Ion Exchange Kinetics*

Quantitative kinetic studies of ion exchange on inorganic salts and oxides are few. In general, there are many similarities between the inorganic ion exchanger and organic polymeric ion exchangers, particularly when focusing on general features. Qualitatively, both particle diffusion-controlled and film diffusion-controlled processes are indicated [10,30,32,193] depending on the oxide or salt. The kinetics are often dependent on environment, analyte ion, inorganic ion exchange pretreatment, and pH, particularly in the case of the inorganic oxide ion exchanger. For example, for alumina [193] between pH 6 and 9, rates of anion or cation exchange are directly proportional to the concentration of the ion being exchanged while at a higher or lower pH value a two-step exchange is indicated. The first is due to a more rapid exchange at surface sites on the alumina while the other is due to a slower exchange at sites within the alumina.

While quantitative details of ion exchange rates are lacking, enough qualitative information is available to indicate that exchange rate is often rapid enough to use several of the inorganic ion exchangers in a dynamic condition as encountered in column chromatography. Because most inorganic oxide and salt ion exchangers are not available in microparticles of uniform size, this factor is also an important one in influencing exchange rates in column experiments. The fact that tailing is often found when using inorganic ion exchangers in column chromatography suggests that factors other than rates of exchange and particle size are involved. Irrevisible retention, complexation, and hydrolytic tendencies also contribute to the tailing.

### *Ion Exchange Capacity*

Inorganic oxides and salts do not possess a clearly defined ion exchange capacity as is the case with organic polymeric and bonded phase ion exchangers. Several factors contribute to this. First, inorganic ion exchangers are weak-acid or base ion exchangers and are poorly defined in terms of an ionogenic group structure. Thus, the capacity is not only pH-dependent but is also dependent on many other environmental factors. These include the ion being investigated; its coion; ion exchanger pretreatment such as aging, heat treatment, and preparation procedure; degree of crystallinity; and surface area. Therefore, in general, ion exchange capacity for a given inorganic ion exchanger is a measure of the maximum uptake of a given ion under a given set of experimental conditions.

A few inorganic salt ion exchangers will provide ion exchange capacities similar to or larger than the high-capacity organic polymeric ion exchangers while inorganic oxides in general will provide maximum capacities of about 2-3 meq/g at a favorable pH assuming monovalent ions are used for this determination. If more specific interactions are present, i.e., interactions other than electrostatic ones, then exchange capacity for that ion can be large. This would be the case when polyvalent ions (high selectivity due to the contribution of hydrolysis and/or irreversible sorption) are used for a capacity determination. A high capacity can also be indicated if complexation is a contributing factor.

Since inorganic ion exchangers are weak-acid or weak-base ion exchangers, capacity changes with pH. Several inorganic oxides are also able to provide both cation and anion exchange properties because of a favorable isoelectric pH value. Figure 7 shows how ion exchange capacity for alumina, which is an oxide that exhibits anion and cation exchange properties, changes with pH. A pH titration of the type shown in Fig. 7 is a common experimental technique used for the determination of inorganic ion exchanger capacity. As the pH decreases alumina becomes an anion exchanger, its anion exchange capacity increases, and the capacity approaches a maximum value of about 2.0 meq/g. As the pH increases alumina becomes a cation exchanger, its cation exchange capacity increases, and the capacity approaches a maximum value of 2.0 meq/g. Similar capacity titration curves are reported for many other oxides [195].

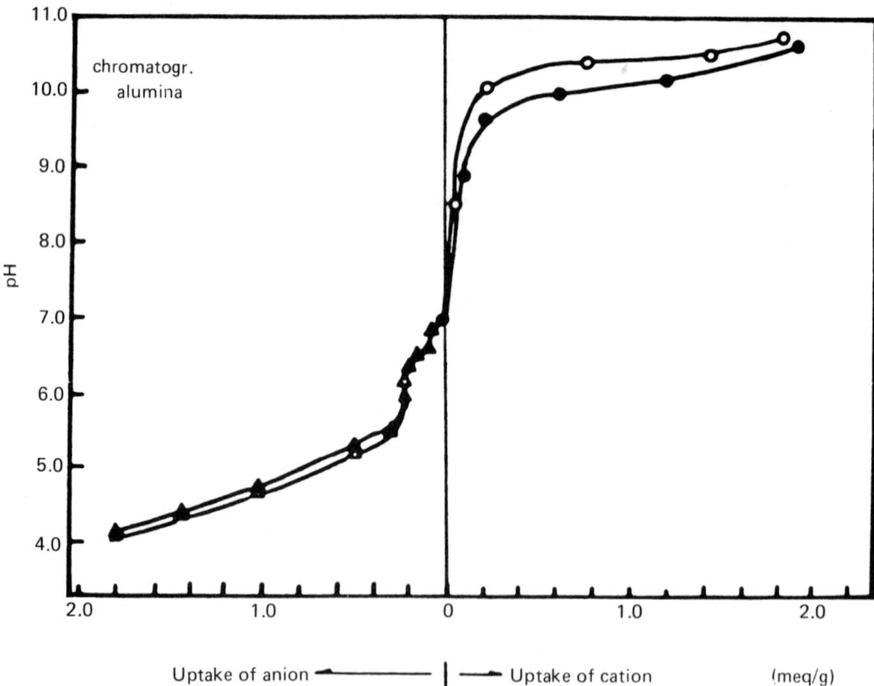


Fig. 7 Titration curve of Brockman alumina with a strong-acid and base titrant. Conditions: (○) 0.02 M LiCl; (●) 0.02 M KOH; (▲) 0.02 M HCl; (△) 0.02 M $HNO_3$. (From Ref. 10, courtesy of CRC Press, Inc. See also Refs. 194 and 195.)

### *Other Properties of Inorganic Ion Exchangers*

Inorganic ion exchangers are more truly porous than organic polymeric ion exchangers because the former have much more rigid structures and are not subject to swelling and contracting like the later ion exchangers. However, the dependence on hydration, crystalline form, and structure stoichiometry is greater in determining surface area and porosity. Table 15 shows how the surface area, S, of three metal oxides changes as a function of aging on heating. As can be seen from these data, methods of preparation and subsequent treatment are crucial in determining surface area and porosity. If reproducibility is required, then it is necessary to carefully control the preparation procedure, the pretreatment, and the conditioning of the inorganic ion exchanger.

Inorganic ion exchangers have a rigid framework type of structure but are still able to take up solvent within the interstitial volume. This occurs with little change in the external volume of the solid material.

The uptake of solvent depends on the internal volume within the inorganic exchanger and the type of counterion present since it also consumes a portion of the internal volume. While solvation energies are a contributing factor, often the more important factor is interstitial size because of the rigid framework. Thus, the uptake of solvent becomes largely dependent on steric requirements. This is particularly true in the case of zeolite-type structures where zeolite pore size determines the type of solvent sorbed based on solvent molecule dimensions. For other crystalline-type inorganic ion exchangers that have larger structures, swelling takes place anisotropically. In these cases solvent is sorbed and occupies space between the lattice layers. As solvent is take up, interlayer distance increases depending on the crystalline structure, the spacking and layer distances, the counterion form, and the solvent. For water the distance can be large enough that the inorganic ion exchanger takes on a gel-type structure.

Useful inorganic ion exchangers are insoluble in aqueous solution and possess physical strength because of their highly rigid crossnetwork. When packed in columns they are able to withstand high column inlet pressures. In general, inorganic ion exchangers are not affected by radiation while their reactivity to oxidizing and reducing agents depends on both the exchanger and the oxidizing and reducing agents.

A major factor that influences inorganic ion exchanger stability is acidity and basicity. In general, inorganic ion exchangers have a pH range over which they are insoluble and retain ion exchange properties. This useful range may be small and is dependent on the inorganic ion exchanger. For example, silica stability is limited to the pH range of 2-8 while the useful pH range of alumina is 2-12.

Table 15 Surface Areas for Several Metal Oxide Gels

| Aluminum hydroxide gel Hydrolysis of $Al(C_4H_9O)_3$ | | Ferric oxide gel $FeCl_3 + NH_3(aq)$ | | Silica gel Hydrolysis of $Si(OC_2H_3)_4$ | | Ceric oxide gel $Ce(NO_2)_3 + NH_3(aq) + H_2O_3$ | |
|---|---|---|---|---|---|---|---|
| $S(m^2g^{-1})$ | Aged at R.T. (hr) | $S(m^2g^{-1})$ | Heating temp. (°C) | $S(m^2g^{-1})$ | Treatment | $S(m^2g^{-1})$ | Heating temp. (°C) |
| 391 | 0 | 390 | 50 | 557 | Without | 92.9 | 35 |
| 415 | 2 | 294 | 100 | 388.0 | Hydrothermal in open system | 88.8 | 100 |
| 251 | 24 | 217 | 150 | 320.4 | Hydrothermal in open system | 70.2 | 300 |
| 37 | 2352 | 186 | 300 | 203.4 | Hydrothermal in open system | 63.1 | 500 |
| 438 | 20 (98% EtOH) | 105 | 400 | 139 | In an autoclave | 52.4 | 600 |
| | | 66 | 500 | 61.8 | In an autoclave | 38.6 | 700 |
| | | 11 | 800 | 34.6 | In an autoclave | 7.9 | 1000 |
| | | | | 237-727 | Commercial | | |

*Source*: From Ref. 10, p. 165, courtesy of CRC Press, Inc.

Most inorganic ion exchangers are hydrated and this hydration plays a major role in their ability to exhibit ion exchange properties. This is particularly true for inorganic oxide-type ion exchangers. Heating the oxide to high enough temperature that the water is lost will change the ion exchange properties including ion exchange capacity. If the oxide or salt is to be used as an ion exchanger in a column application with aqueous mobile phases, excessive heating or drying of the oxide or salt is not a requirement. Frequently, oxides or salts should be dried or heated under a controlled temperature and humidity environment to produce a hydrated ion exchanger of reproducible water activity. In some cases in the preparation of the oxides or salts they are obtained as gelatinous, amorphous, or chalklike particles and are difficult to pack into a column. Modest heating at lower temperatures often produces a free flowing powder. Since heating also affects hydration level and consequently ion exchange capacity, an alternate procedure that is often used with oxides to produce particles rather than the gelatinous material is to coprecipitate the desired oxide with a second oxide.

## APPLICATIONS OF ION EXCHANGERS

### Ion Exchangers as Stationary Phases

Organic polymeric, bonded phase, and pellicular ion exchangers are mainly used in columns. All three can also be used in thin-layer or in an impregnated cellulose sheet form. Inorganic ion exchangers, while used in columns, are equally used in thin layers or as impregnated cellulose sheets. Only column applications of ion exchangers are considered in this chapter. In general, the strategies used to separate analyte ions on ion exchange columns are essentially the same as those used in sheet methods. The major difference in terms of the chromatography is that with sheet methods additional interaction due to the binder must be taken into account. Thus, in sheet methods ion exchange is often accompanied by adsorption and partition effects, particularly when dealing with organic analytes.

Historically, the fundamental properties exhibited by polymeric and inorganic ion exchangers have been studied for the most part by using batch equilibration techniques. This approach allows equilibrium to be reached. Column chromatography, on the other hand, is a continuous process and equilibrium can only be approached. Thus a fundamental theory must take this into account. While the conclusions from the batch studies are appropriate to apply to ion exchange column chromatography, they are not exact. This fundamental problem can be addressed by coupling the ion exchange process with continuous-flow plate theory and random walk theory, which are attempts to account for the continuous processes that occur in column chromatography.

Although there are limitations in these concepts, particularly the plate theory, they satisfactorily deal with the effect of stationary and mobile phase parameters on selectivity, efficiency, capacity, and subsequently resolution. With appropriate approximations, plate theory can be used to qualitatively predict column design to achieve a given resolution in ion exchange column chromatography [196]. A detailed discussion of these two approaches and their limitations are available elsewhere [12,13,170,197].

High-capacity organic polymeric ion exchangers are generally used in modestly wide-diameter (0.5-2 cm) columns of modest length for analytical chromatography and in wide diameters from 2 cm to diameters measured in feet for preparative and industrial applications. Since particles, which are available in spherical or irregular shape, are usually large and of a broad size range, a low column efficiency is obtained. Typical sizes and ranges commercially available are 38-75 μm, 75-130 μm, 150-300 μm, and larger. As interest in improving column efficiency grew, both particle size and range were reduced and these improved particles are now commercially available. In general, dry packing, but not in a glass column because of swelling, or slurry pouring is sufficient to pack the larger ion exchanger beads into columns. Table 16 lists representative examples of high-capacity, low-efficiency organic polymeric ion exchangers which are commercially available. Cation exchanger capacity approaches 5 meq/g while anion exchangers are typically 3-4 meq/g. Crosslinking varies from 2 to 16% (gel type). Usually, the exchangers are supplied in the H or Cl form, respectively.

In HPLC, where high column efficiency is desired in order to obtain a high resolution, microparticles with a narrow size range must be used. Low-capacity organic polymeric bonded phase and pellicular ion exchangers which satisfy the requirements are available. In general, these are used in small-diameter columns of 4.6 mm or less at lengths of 5-25 cm. Slurry packing at high pressure must be used in order to obtain optimum column efficiency.

In general, macroporous low-capacity PSDB ion exchangers, bonded phase ion exchangers, and pellicular ion exchangers are rigid and are able to withstand high column inlet pressures required for slurry packing and modestly high column inlet pressures that are required in normal everyday application. For the gel-type PSDB ion exchanger even at high crosslinking high inlet pressure can be a problem. Since these are relatively soft polymers they can be compressed and if this is excessive bead fracture is likely to occur.

Table 17 lists representative polymeric-type anion and cation exchangers. In general, the areas of applications are those recommended by the manufacturer. In many cases exchange capacities are approximate. If this quantity needs to be known accurately, the user should be prepared to make this determination.

Table 18 lists representative bonded phase and pellicular ion exchangers that are commercially available. In general, for comparable particle size the bonded phase ion exchanger is capable of delivering a higher column efficiency than the polymeric ion exchanger. Even though this difference exists, the bonded phase ion exchanger has not replaced the low-capacity PSDB ion exchanger. The latter (a) has a longer lifetime, (b) is less likely to get fouled up by analyte sample matrix, (c) is stable throughout the entire pH range unlike the silica bonded phase ion exchanger, which is limited to pH 2-8, (d) is less expensive, and (e) can be readily prepared to have different, useful levels of low ion exchange capacity. While the silica bonded phase ion exchangers can be used for the separation of inorganic analyte ions, their applications are more often focused on the separation of organic analyte ions.

## Ion Exchange Strategies in Analytical Column Liquid Chromatography

Ion exchangers are versatile stationary phases for analytical separations using column liquid chromatographic techniques. They can be applied to many kinds of separation problems. Typical kinds of samples handled by ion exchange are

1. Inorganic and organic analyte anions and cations
2. Ionic analytes from nonionic analytes
3. Large and small molecular weight analyte ions
4. Closely related and unrelated analyte ions
5. Nonelectrolytes
6. Analyte ion quantities at trace levels
7. Concentration and/or stripping of analyte ions at ultratrace levels
8. Preparative analyte quantities, often in multigrams

Ion exchangers are useful as part of other analytical and synthetic strategies. For example, they can be used to deionize water, prepare specific kinds of electrolytes, purify electrolytes, remove interfering anions or cations, dissolve insoluble salts, concentrate trace analytes, act at catalysts, and remove electrolytes from synthetic mixtures. Many of these operations have industrial significance and can be performed as commercially useful levels from both an economic and an applied point of view. Inorganic ion exchangers for the most part have the potential to be used similarly. Their versatility in these kinds of applications, however, is more limited because of their specific type of ion exchange-determining properties.

This section focuses on the analytical separation strategies for the application of ion exchangers as stationary phases in column liquid

**Table 16** Commercially Available High-Capacity Organic Polymeric Ion Exchangers

| Type and exchange groups | Bio-Rad Analytical Grade Ion Exchange Resins | Dow Chem. Company "Dowex" | Rohm & Haas Co. "Amberlite" | Ionac Chemical Company |
|---|---|---|---|---|
| Type 1, Strongly basic polystyrene gel-type resins $\phi$-$CH_2N^+(CH_3)_3Cl^-$ | AG 1-X2<br>AG 1-X4<br>AG 1-X8 | 1-X2<br>I-X4<br>I-X8 (SBR)<br>SBR-P<br>11 | IRA 401<br>IRA 402<br>IRA 400<br>IRA 420<br>IRA 430<br>IRA-67,IRN 78 | A-540<br>A-548<br>ASB-1<br>A-440<br>A-546<br>A-935 |
| Type 1, strongly basic nonpolystyrene gel-type resins $\phi$-$CH_2N^+(CH_3)_3Cl^-$ | | | IRA 458 | |
| Type 2, strongly basic polystyrene gel-type resins $\phi$-$CH_2N^+(CH_3)_2(C_2G_4OH)Cl^-$ | AG 2-X8 | 2-X4<br>2-X8(SAR) | IRA 410 | A-550<br>ASB-2 |
| Strongly basic, macroporous resins, type 1 $\phi$-$CH_2N^+(CH_3)_3Cl^-$ | AG MP-1 | MSA-1 | IRA 900<br>IRA 904<br>IRA 938<br>IRA 958<br>Amberlyst A-26,A-27 | AFP-100<br>A 641<br><br>PA 318,PA 320 |
| Strongly basic, macroporous resins, type 2 $\phi$-$CH_2N^+(CH_3)_2$ $(C_2H_4OH)Cl^-$ | | MSA-2 | IRA 910 | A-642<br>A-651 |
| Intermediate base resins $R$-$N^+(CH_3)_2Cl^-$ and $R$-$N^+(CH_3)_2$ $(C_2H_4OH)Cl^-$ | Bio-Rex 5 | | IRA 47 | A-305<br>MS-170 |
| Weakly basic gel-type polystyrene, phenolic or polyamine $\phi$-$CH_2N^+(R)_2Cl^-$ | AG 3-X4A | WGR<br>WGR-2 | IRA 45<br>IR 48<br>IRA 47<br>IRA 68<br>IRA 60, IRP 58 | A-375<br>A-260 |

| Permutit Company (England) | Mitsubishi Chem. Co. "Diaion" | Bayer "Lowatit"<br>Permutit, Inc. Germany | Fisher "Rexyn"<br>Merck | Akzo Chem. Co. "Imac"<br>Montedision "Kastel" | Wallen Dye Factories "Wofatit" |
|---|---|---|---|---|---|
| Zerolit FF<br>(lightly crosslinked)<br>Zerolit FF<br>Permutit S-1 | SA11A,SA11B<br>SA10A,SA108<br>SA100 | M 500<br>M 5020<br>M 5080<br>ESP | 201<br>111 | S 5-40<br>S 5-50<br>A 500 | |
| | | MN | | | L 165<br>L 150 |
| S-2 | SA21A,SA21B<br>SA20A,SA20B | M 600<br>ES | | S 5-42<br>A 300 | |
| PA 306,PA 308<br>PA 310,PA 312 | PA 304 | MP 5080 | | A 500 P | |
| | PA 404<br>PA 406,PA 408<br>PA 410,PA 412<br>PA 414,PA 416<br>PA 418,PA 420 | | | A 300 P | |
| F | | | 208<br>205 | | |
| G | WA 10<br>WA 11 | MIH<br>E | 207<br>203<br>206<br>11 | A13, 17, 19<br>A 27<br>A 101 | MD |

Table 16 (continued)

| Type and exchange groups | Bio-Rad Analytical Grade Ion Exchange Resins | Dow Chem. Company "Dowex" | Rohm & Haas Co. "Amberlite" | Ionac Chemical Company |
|---|---|---|---|---|
| Macroporous intermediate and weak-base exchangers $\phi\text{-}CH_2N^+(R)_2Cl^-$ | | MWA-1 | IRA-35<br>IRA-93<br>IRA-94<br>IRA-99<br>Amberlyst A-21 | MG-1<br>A-328<br>AFP 329 |
| Strongly acidic polystyrene, gel-type $\phi\text{-}SO_3^-H^+$ | AG 50W-X2<br>AG 50W-X4<br>AG 50W-X8<br><br>AG 50W-X10<br>AG 50W-X12<br>AG 50W-X16 | 50W-X2<br>50W-X4<br>50W-X8<br>HCRW-2<br>50W-X10<br>HCR,HCR-S | IR-116,IR-118<br>IR-120,IRN-77<br>IRN-218<br>IRN-163<br>IRN-169<br>IR-122<br>IR-124,IR-130<br>IR-140,IR-169 | C-298<br>C-249, CF<br>C-240, 242<br>C-250, 251<br>C-253, 255<br>C-256, 257<br>C-256, 299 |
| Strongly acidic nonpolystyrene, $\phi\text{-}SO_3^-H^+$ | | | | |
| Macroporous strong-acid cation exchangers $\phi\text{-}SO_3^-H^+$ | AG MP-50 | MSC-1 | 200<br>252<br>Amberlyst 15 | CFP 110<br>CFS |
| Weak-acid cation exchange resins $R\text{-}COO^-Na^+$ | Bio-Rex 70 | CCR-2<br>MWC-1 | IRC-84<br>IRC-50,CG-50<br>DP-1,IRC-72<br>IRP-64 | CC<br>CCN |
| Weakly acidic chelating resin $\phi\text{-}CH_2N(CH_2COO^-H^+)_2$ | Chelex 100 | A-1 | | |
| Mixed-bed resins<br>$\phi\text{-}SO_3^-H^+$ & $\phi\text{-}CH_2N^+(CH_3)_2OH^-$ | AG 501-X8 | | IRN-150<br>MB-1 | NM-60,NM-40<br>NM-65,M-747<br>MI-747,NM-42 |
| $\phi\text{-}SO_3H^+$ & $\phi\text{-}CH_2N^+(CH_3)_3OH^-$ indicator dye | AG 501-X8(D) | | MB-3 | |

| Permutit Company (England) | Mitsubishi Chem. Co. "Diaion" | Bayer "Lowatit" / Permutit, Inc. Germany | Fisher "Rexyn" / Merck | Akzo Chem. Co. "Imac" / Montedision "Kastel" | Wallen Dye Factories "Wofatit" |
|---|---|---|---|---|---|
| | WA 20<br>WA 21<br>WA 30 | MP 7080 | | A 20 | |
| Zeocarb 225 (X4)<br>Zeocarb 225 | SK 102, SK 103<br>SK 104, SK 106<br>SK 1A, SK 1B<br>SK 110, SK 112<br>SK 116 | PN<br>S 100<br>S 1080<br>RS[b] | 101<br>1 | C-22<br>C-12<br>C-300 | KPS 200 |
| Zeocarb 215 | | | | | F.P. |
| | PK 204, PK 208<br>PK 212, PK 216<br>PK 220, PK 224<br>PK 228 | | | C 8P<br>C 16P | |
| Zeocarb 226 | WK 10<br>WK 11 | CNO<br>C<br>CP 3050 | 102<br>IV | C 101<br>Z-5 | CN<br>CP 300 |
| | CR 10 | | | | |
| | SMN-1 | | 300<br>V<br><br>I-300 | | |

This is a cross-reference of commercially available high-capacity ion exchangers. They are not identical in composition or physical properties. In many applications they can be used interchangeably, particularly if the mobile phase conditions are adjusted accordingly. Also, some ion exchangers listed are no longer available.

*Source*: Reprinted courtesy of Bio-Rad Laboratories, Chemical Division.

**Table 17** Commercially Available High-Performance Organic Polymeric Ion Exchangers

| Name | Ionogenic group | Polymer | Particle size (μm) | Capacity (μeq/g) | Specific application | Company |
|---|---|---|---|---|---|---|
| | | | *Anion Exchanger* | | | |
| Aminex HPX-72S | $-NR_3^+SO_4^{2-}/2$ | PSDB | 11 | 1400 | Organic weak bases | BioRad |
| Aminex HPX-720 | $-NR_3^+OH^-$ | PSDB | 11 | 1400 | Organic strong bases | BioRad |
| Aminex A-14 to A-29 | $-NR_3^+$ | PSDB | 5, 11, 15, 18, 20 | 1200, 1400 | General, nucleic acids | BioRad |
| Anion PW | $-NR_3^+$ | Polyacrylate | 10 | 30 | Ion chromatography | BioRad |
| MA7P | -R-N=NH | Methacrylate | 7 | | Peptides | BioRad |
| Polypore A | $-NR_3^+$ | PSDB | 10 | | Desalting | Brownlee Labs |
| HPIC-AS4A | $-NR_3^+$ | PSDB-latex | 15 | 20 | Ion chromatography | Dionex |
| HPIC-AS5 | $-NR_3^+$ | PSDB-latex | 15 | | Ion chromatography | Dionex |
| HPIC-AS6 | $-NR_3^+$ | PSDB-latex | 10 | 80 | Ion chromatography | Dionex |
| PRP-X100 | $-NR_3^+$ | PSDB | 10 | 200 | Ion chromatography | Hamilton |
| Ion-100 | $-NR_3^+$ | PSDB | | | Ion chromatography | Interaction |
| Ion-110 | $-NR_3^+$ | PSDB | 10 | | Ion chromatography | Interaction |
| Mono Q HR 5/5 | $-CH_2N(CH_2)_3^+$ | Styrene | 9.8 | 320 | Proteins | Pharmacia |
| Mono P HR 5/20 | Amine groups | Styrene | 9.8 | 170 | Proteins | Pharmacia |
| TSK DEAE-5PW | $-NEt_2$ | PSDB | 10 | >60 | Proteins, nuclectides, enzymes, polypeptides | Toyo Soda |
| IC PAK | $-NR_4^+$ | Methacrylate | 10 | 30 | Ion chromatography | Waters |
| Protein-PAK DEAE | $-(CH_2)_2N(Et)_2$ | Methacrylate | 10 | | Proteins | Waters |
| | | | *Cation Exchanger* | | | |
| Aminex HPX-87C | $-SO_3^-\ {}_2Ca^{2+}/2$ | PSDB | 9 | 1700 | Monosaccharides | BioRad |
| Aminex HPX-87P | $-SO_3^-\ {}_2Pb^{2+}/2$ | PSDB | 9 | 1700 | Pentose sugars | BioRad |

| | | | | | | |
|---|---|---|---|---|---|---|
| Aminex HPX-87H | $-SO_3^-H^+$ | PSDB | 9 | 1700 | Carbohydrates, organic acids | BioRad |
| Aminex-42C | $-SO_3^-\ _2Ca^{2+}/2$ | PSDB | 25 | 1200 | Oligosaccharides | BioRad |
| Aminex-42A | $-SO_3^-Ag^+$ | PSDB | 25 | 1200 | Oligosaccharides | BioRad |
| Aminex-65A | $-SO_3^-Ag^+$ | PSDB | 17.5 | 1700 | Oligosaccharides | BioRad |
| Aminex A-4 to A-9 | $-SO_3^-$ | PSDB | 5,9,11,13,17.5,20 | 1700,200 | General, amino acids | BioRad |
| Aminex Q-150S, -15S | $-SO_3^-$ | PSDB | 22,28 | 1700 | Carbohydrates | BioRad |
| Aminex 50WX4 | $-SO_3^-$ | PSDB | 25,32.5 | 1200 | Oligosaccharides | BioRad |
| Cation | $-SO_3^-$ | PSDB | 10 | 12 | Ion chromatography | BioRad |
| MA7C | $-CH_2CO_2^-$ | Methacrylate | 7 | | Proteins | BioRad |
| Polypore H | $-SO_3^-Ca^{2+}/2$ | PSDB | 10 | | Organic acids | Brownlee Labs |
| Polypore CA | $-SO_3^-Ca^{2+}/2$ | PSDB | 10 | | Carbohydrates | Brownlee Labs |
| Polypore PB | $-SO_3^-Pb^{2+}/2$ | PSDB | 10 | | Carbohydrates | Brownlee Labs |
| HPIC - CS3 | $-SO_3^-$ | PSDB-latex | 10 | | Ion chromatography | Dionex |
| HPIC - CS2 | $-SO_3^-$ | PSDB | 15 | 20 | Ion chromatography | Dionex |
| HPICE - AS1 | $-SO_3^-$ | PSDB | 7 | 2000 | Ion chromatography | Dionex |
| HPICE - AS5 | $-SO_3^-$ | PSDB | 8 | | Ion chromatography | Dionex |
| PRP-X200 | $-SO_3^-$ | PSDB | 10 | 35 | Ion chromatography | Hamilton |
| HC-40 | $-SO_3^-Ca^{2+}/2$ | PSDB | 10,15 | 5000 | Carbohydrates | Hamilton |
| HC-75 | $-SO_3^-Ca^{2+}/2$ | PSDB | 10,15 | 5000 | Carbohydrates | Hamilton |
| IC Ion 200 | $-SO_3^-$ | PSDB | 5 | 100 | Ion chromatography, mono- and divalent cations | Interaction |
| IC Ion 210 | $-SO_3^-$ | PSDB | | 500 | Ion chromatography, transition metals | Interaction |
| AA 511 | $-SO_3^-Na^+$ | PSDB | 6 | | Amino acids | Interaction |

Table 17 (continued)

| Name | Ionogenic group | Polymer | Particle size (μm) | Capacity (μeq/g) | Specific application | Company |
|---|---|---|---|---|---|---|
| AA 503 | $-SO_3^-Li^+$ | PSDB | 6 | | Amino acids | Interaction |
| AA 911 | $-SO_3^-Na^+$ | PSDB | 9 | | Amino acids | Interaction |
| ARH-801 | $-SO_3^-$ | PSDB | | | Organic acids | Interaction |
| Ion 300 | $-SO_3^-$ | PSDB | | | Organic acids | Interaction |
| ARH-601 | $-SO_3^-$ | PSDB | | | Organic acids | Interaction |
| CHO-611 | $-SO_3^-$ | PSDB | | | Carbohydrates | Interaction |
| CHO-620 | $-SO_3^-Ca^{2+}/2$ | PSDB | | | Carbohydrates | Interaction |
| CHO-682 | $-SO_3^-Pb^{2+}/2$ | PSDB | | | Carbohydrates | Interaction |
| BiO-10,20 | $-SO_3^-$ | PSDB | | | Peptides | Interaction |
| Mono S | $-CH_2SO_3^-$ | Styrene | 9.8 | 150 | Proteins | Pharmacia |
| TSK SP-5PW | $-SO_3^-$ | PSDB | 10 | >60 | Proteins, nucleotides, enzymes, polypeptides | Toya Soda |
| IC PAK-TM | $-SO_3^-$ | PSDB | 10 | 400 | Ion chromatography | Waters |
| IC PAK | $-SO_3H$ | PSDB | 10 | 12 | Ion chromatography | Waters |
| Sugar PAK 1 | $-SO_3^-Ca^{2+}/2$ | PSDB | | | Carbohydrates | Waters |
| Protein-PAK SP | $-CH_2CH_2CH_2SO_3^-$ | Methacrylate | 10 | | Proteins | Waters |
| Cation | $-SO_3^-$ | PSDB | 8 | 40 | Ion chromatography | Wescan |
| Ion Ex | $-SO_3^-$ | PSDB | 13 | 5000 | Organic acids | Wescan |

**Table 18** Commercially Available High-Performance Bonded and Pellicular Ion Exchangers

| Name | Ionogenic group | Silica pore size (Å) | Particle size (μm) | Capacity (μeq/g) | Specific applications | Company |
|---|---|---|---|---|---|---|
| | | | *Anion Exchangers* | | | |
| Bakerbond 7100 | $-(CH_2)_xN(R_3)^+$ | 120 | 5 | 400 | Nucleotides, organic acids, general | J. T. Baker |
| Bakerbond 7099 | $-(CH_2)_3NH_2$ | 120 | 5 | 700 | Carbohydrates, general | J. T. Baker |
| Bakerbond PEI | $-(CH_2CH_2NH)_x$ | 300 | 5, 15, 40 | -0.15 g protein/g | Proteins | J. T. Baker |
| Anion SW | $NR_3^+$ | 175 | 5 | 400 | Ion chromatography | BioRad |
| Amino 5S | $-NH_2$ | 80 | 5 | | Disaccharides | BioRad |
| Aquapore AX-300 | $-(CH_2CH_2NH)X$ | 300 | 7, 5, 20 | 300 | Small organic anions, proteins | Brownlee Labs |
| Aquapore Amino | $-(CH_2)_3NH_2$ | 300 | 5, 7 | | Normal phase, strong acids | Brownlee Labs |
| Zipax | $-NR_3^+$ | Pellicular | 30 | | General, organic acids, nucleotides | Du Pont |
| Zorbax SAX | $-NR_3^+$ | 80 | 5 | | General, organic acids, nucleotides | Du Pont |
| Zorbax Bio Series SAX | $-N(CH_3)_3^+$ | 300 | 7 | | General, organic acids, nucleotides | Du Pont |
| Zorbax $NH_2$ | $-NH_2$ | 80 | 5 | | Normal phase, carbohydrates | Du Pont |
| Zorbax Bio Series WAX | $-NH_2$ | 300 | 7 | | Nucleotides | Du Pont |
| SAX | $-(CH_2)_xNR_3^+$ | 60, 100, 300 | 5, 10 | 800-2000 | General, organic acid | ES Industries |
| RP SAX | $-(CH_2)_xNR_3^+$ butyl, end-capped | 60, 100, 300 | 5, 10 | 800-2000 | General, organic acids, reversed phase effects | ES Industries |
| M-WAX | $-(CH_2)_xNH_2$ | 60, 100, 300 | 5, 10 | | Nucleotides, general, carbohydrates, antibiotics | ES Industries |
| D-WAX | $-(CH_2)_xNHR$ | 60, 100, 300 | 5, 10 | | Nucleotides, general, carbohydrates, antibiotics | ES Industries |
| T-WAX | $-(CH_2)_xNR_2$ | 60, 100, 300 | 5, 10 | | Nucleotides, general, carbohydrates, antibiotics | ES Industries |

Table 18 (continued)

| Name | Ionogenic group | Silica pore size (Å) | Particle size (μm) | Capacity (μeq/g) | Specific applications | Company |
|---|---|---|---|---|---|---|
| Nucleosil SB | $-N(CH_3)_3^+$ | 100 | 5, 10 | 1000 | General, organic acids | Macherey-Nagel, Co. |
| Nucleosil $NH_2$ | $-N(CH_2)_xNH_2$ | 100, 120 | 5, 10 | | Normal phase | Macherey-Nagel, Co. |
| Nucleosil $N(CH_2)_2$ | $-N(CH_3)_2$ | 100 | 5, 10 | | Normal phase | Macherey-Nagel, Co. |
| LiChrosorb $NH_2$ | $-(CH_2)_xNH_2$ | 250 | 5, 10 | | Normal phase, carbohydrates peptides | E. Merck, EM Science |
| LiChrospher $NH_2$ | $-(CH_2)_xNH_2$ | 180 | 3, 5, 10 | | Normal phase, carbohydrates peptides | E. Merck, EM Science |
| LiChrospher SAX | $-(CH_2)_xNR_3^+$ | 180 | 5, 10 | | General, organic acids | E. Merck, EM Science |
| Spherisorb SAX | $-(CH_2)_3N(CH_3)_3$ | 220 | 5, 10 | 400 | General organic acids | Phase Separations, Inc. |
| Spherisorb $NH_2$ | $-(CH_2)_3NH_2$ | 220 | 3, 5, 10 | 600 | Normal phase, carbohydrates nucleotides, hydroxylated compounds | Phase Separations, Inc. |
| PolyMethyl A | Polypeptide | 300, 2000 | 5, 7, 15-20 | | Proteins, peptides, hydrophobic interaction (pH 7.0) | PolyLC |
| PolyEthyl A | Polypeptide | 300, 1000 | 5, 7, 15-20 | | Proteins, peptides, hydrophobic interaction (pH 7.0) | PolyLC |
| PolyPropyl A | Polypeptide | 300, 1000 | 5, 7, 15-20 | | Proteins, peptides, hydrophobic interaction (pH 7.0) | PolyLC |
| PolyButyl A | Polypeptide | 300, 1000 | 5, 7, 15-20 | | Proteins, peptides, hydrophobic interaction (pH 7.0) | PolyLC |
| APS-Hypersil | $-NH_2$ | 120 | 3, 5, 10 | | Normal phase, carbohydrates | Shandon |
| Sotaphase SAX | $-NR_3^+$ | | 5 | | General, nucleotides, proteins, organic organic acids | Sota Chromatography |
| Sotaphase QA | $-NR_3^+$ | 100-4000 | 7 | | General, nucleotides, proteins, organic organic acids | Sota Chromatography |
| Sotaphase Amine | $-NH_2$ | | 3, 5 | | Normal phase, carbohydrates | Sota Chromatography |
| Sotaphase DE | $-(CH_2)_2N(Et)_2$ | 100-4000 | 7 | | Normal phase, carbohydrates, proteins | Sota Chromatography |

| | | | | | | |
|---|---|---|---|---|---|---|
| SynChropak 0300 | $-NR_3^+$ | 30 | 6.5 | | Proteins | SynChrom |
| SynChropak AX100 to AX1000 | $-NR_2$ | 100, 200, 300 500, 1000 | 5, 6.5, 55 | | Proteins | SynChrom |
| TSK IEX-260A SIL | $-N(CH_3)_2(CH_2C_6H_5)^+$ | | 12 | >500 | Nucleic acids, general | Toya Soda |
| TSK IEX-540DEAE SIL | $-N(Et)_2$ | | 5, 10 | >200 | General, proteins, nucleotides, peptides | Toya Soda |
| TSK gel DEAE-2SW | $-NEt_2$ | 125 | 5 | >300 | Proteins, nucleotides, enzymes, polypeptides | Toya Soda |
| TSK gel DEAE-2SW | $-NEt_2$ | 150 | 10 | >300 | Proteins, nucleotides, enzymes, polypeptides | Toya Soda |
| MicroPak SAX | $-NR_3^+$ | | 10 | | General nucleotides, organic acids | Varian |
| MicroPak AX | $-NH_2$ | | 5, 10 | | Normal phase, nucleotides, polycarboxylic acids, peptides, carbohydrates | Varian |
| Vydac 300IC405 | $-(CH_2)_xNR_3^+$ | | 5 | | Ion chromatography | Vydac |
| Vydac 301TP104 | $-(CH_2)_xNR_3^+$ | | 10 | | General nucleotides, organic acids | Vydac |
| Vydac 302IC4.6 | $-(CH_2)_xNH_3^+$ | | 10 | | Ion chromatography | Vydac |
| Vydac 301 SC | glass-silica-$(CH_2)NR_3^+$ | Pellicular | 30 | | General nucleotides, organic acids | Vydac |
| Vydac 601 PSC | glass-silica-$(CH_2)$-$NH_2$ | Pellicular | 30 | | Normal phase, general | Vydac |
| Bondapak Amine | $-NH_2$ | | 10 | 300-400 | Normal phase, polar compounds, organic acids | Waters |
| Anion | $-NR_3^+$ | | 10 | 200 | Ion chromatography | Wescan |
| Partisil SAX | $-NR_3^+$ | 85 | 5, 10 | | General, nucleotides, organic acids | Whatman |
| Partisphere SAX | $-NR_3^+$ | 120 | 5 | 200 | General, nucleotides, organic acids | Whatman |
| Partisil PAC | -NRH/-CN | 85 | 5, 10 | | Normal phase, bifunctional | Whatman |
| Partisphere PAC | -NRH/-CN | 120 | 5 | | Normal phase, bifunctional | Whatman |
| Partisphere WAX | $-(CH_2)_2N(Et)_2$ | 120 | 5 | 200 | Normal phase | Whatman |

Table 18 (continued)

| Name | Ionogenic group | Silica pore size(Å) | Particle size (μm) | Capacity (μeq/g) | Specific applications | Company |
|---|---|---|---|---|---|---|
| | | | *Cation Exchangers* | | | |
| Bakerbond 7101 | $-(CH_2)_3SO_3^-$ | 120 | 5 | 300 | General, organic bases | J. T. Baker |
| Bakerbond 7108 | $-(CH_2)_3CO_2H$ | 120 | 5 | 200 | General, organic bases | J. T. Baker |
| Bakerbond CBX | $-CH_2CH_2CO_2H$ | 300 | 5, 15, 40 | | Proteius | J. T. Baker |
| Aquapore CX-300 | Polyaspartic acid | 300 | 7, 10, 20 | | Proteins | Brownlee Labs |
| Zipax 300 SCX | $-SO_3^-$ | Pellicular | 30 | | Organic bases | Du Pont |
| Zorbax SCX | $-SO_3^-$ | 300 | 5 | | Organic bases | Du Pont |
| Zorbax Bio Series SCX | $-SO_3^-$ | 300 | 7 | | Organic bases, proteins | Du Pont |
| Zorbax Bio Series WCX | $-CO_2H$ | 300 | 7 | | Peptides | Du Pont |
| P-SCX | $-ArSO_3^-$ | 60, 100, 300 | 5, 10 | 800-2000 | General, organic bases | ES Industries |
| A-SCX | $-(CH_2)_xSO_3^-$ | 60, 100, 300 | 5, 10 | 800-2000 | General, organic bases | ES Industries |
| RP-SCX | $-ArSO_3^-$, butyl end-capped | 60, 100, 300 | 5, 10 | 800-2000 | General, organic bases reversed phase effects | ES Industries |
| P-WCX | $-ArCO_2H$ | 60, 100, 300 | 5, 10 | | Nucleotides | ES Industries |
| A-WCX | $-(CH_2)_xCO_2H$ | 60, 100, 300 | 5, 10 | | Nucleotides | ES Industries |
| Nucleosil-SA | $-SO_3^-$ | 100 | 5, 10 | 1000 | General, organic bases | Machery-Nagel, Co. |
| LiChrosorb | $-SO_3^-$ | | 10 | | General, organic bases | E. Merck; EM Science |
| Perisorb KAT | $-SO_3^-$ | | | | General, organic bases | E. Merck; EM Science |

| | | | | | | |
|---|---|---|---|---|---|---|
| PolyCATA | Polyaspartic acid | 300, 100 | 5, 7, 15-20 | | Proteins | PolyLC |
| Sotaphase QA | $-(CH_2)_xCO_2H$ | 100-4000 | 7 | | General, proteins | Sota Chromatography |
| SynChropak S300 | $-SO_3^-$ | 300 | 6.5 | | Proteins | SynChrom |
| SynChropak CM300 | $-CO_2H$ | 300 | 6.5 | | Proteins | SynChrom |
| TSK CM-2-SW | $-CO_2^-$ | 125 | 5 | >300 | Proteins, enzymes, nucleotides, polypeptides | Toya Soda |
| TSK CM-3-SW | $-CO_2^-$ | 250 | 10 | >300 | Proteins, enzymes, nucleotides, polypeptides | Toya Soda |
| Vydac 400 IC405 | $-SO_3^-$ | 300 | 5 | | Ion chromatography | Vydac |
| Vydac 401 TP104 | $-SO_3^-$ | 300 | 10 | | General, organic bases, nucleotides | Vydac |
| Vyda 401 SC | glass-silica-$(CH_2)-SO_3^-$ | Pellicular | 30 | | General | Vydac |
| Partisil SCX | $-ArSO_3^-$ | 85 | 5, 10 | | General, organic bases, nucleotides, amino acids | Whatman |
| Partisphere SCX | $-ArSO_3^-$ | 120 | 5 | 200 | General, organic bases, nucleotides | Whatman |
| Partisphere WCX | $-CO_2H$ | 120 | 5 | 100 | General, proteins, organic bases, nucleotides | Whatman |

chromatography. Discussions of the aforementioned other types of applications are available elsewhere [12,13]. It should be noted that these strategy concepts have their origin in low-efficiency, high-capacity column ion exchange techniques. However, these concepts are not limited to only this kind of column technology and with appropriate column and instrument modifications they are equally applicable to high-efficiency ion exchange column techniques (i.e., HPLC). Furthermore, with this vast improvement in column efficiency offered by an HPLC approach, mobile phase manipulation according to these separation strategies becomes even more powerful and versatile in achieving rapid resolution of the mixture.

### *Inorganic Analyte Ions; Basis for Ion Exchange Separation*

Ion exchange selectivity [12,13,16] is the major basis for the ion exchange separation of simple inorganic analyte ions. Furthermore the selectivity, in general terms, is predictable in terms of three major factors. These are:

1. Analyte ion hydrated radii
2. Analyte ion oxidation state
3. Analyte ion secondary equilibria

If all other factors are equal, analyte ion retention correlates with its size. As the hydrated radius increases retention increases. This is illustrated in Fig. 8, where the halide ions are separated on a high-capacity strong-base anion exchanger according to their size. Similarly, the elution order for alkali metal cations and alkaline earth cations on a strong-acid cation exchanger follows size and has the order $Li^+$, $Na^+$, $K^+$, $Rb^+$, $Cs^+$ and $Mg^{2+}$, $Ca^{2+}$, $Sr^{2+}$, $Ba^{2+}$, $Ra^{2+}$, respectively. Examination of the ion exchange selectivities listed in Tables 9 and 10, although not complete, are consistent, in general, with this size-retention relationship. If the ionogenic group structure is altered, which is more easily accomplished with the $-CH_2N(R_3)^+$ group than with the $-SO_3^-$ group, selectivities (see Table 11) are also altered.

In general, if all other factors are equal, inorganic analyte ion retention increases as its oxidation state increases. Thus, for example, $Ca^{2+}$ is more retained than $Na^+$ on a cation exchanger and $SO_4^{2-}$ is more retained than $Cl^-$ on an anion exchanger.

The effects of selectivity and oxidation state can be used to adjust mobile phase eluting power. Thus, a stronger eluent is favored by using a mobile phase counterion that has a higher ion exchange selectivity or one that has a higher oxidation state value.

A major strategy in liquid column chromatographic separations is to introduce secondary equilibria by manipulation of mobile phase

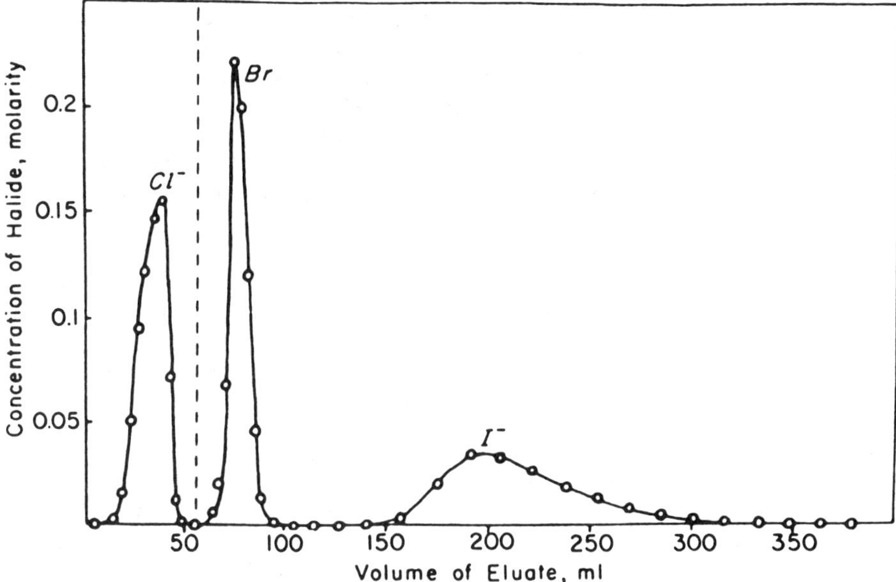


Fig. 8 Anion exchange separation of the halides. Column: Dowex 1 × 10, 100-200 mesh, 6.7 × 3.4 $cm^2$; mobile phase: 55 ml of 0.50 M $NaNO_3$ then 2.0 M $NaNO_3$. (From Ref. 198, courtesy of American Society for Testing Materials.)

additives, pH, solvent composition, and the addition of ligands [199]. This strategy is used in adsorption, partition, as well as in ion exchange. In ion exchange the two major types of secondary equilibrium that are taken advantage of are pH and coordination.

If an inorganic analyte ion has weak-acid or base characteristics, its charge will be determined by the mobile phase pH. A pH that favors dissociation increases retention. Thus, $PO_4^{3-}$ as a trivalent anion is highly retained while $H_2PO_4^-$ as a monovalent anion at a lower pH is much less retained.

The most versatile strategy in low-efficiency, high-capacity ion exchange separations to alter inorganic analyte ion retention is to use a ligand in the mobile phase and take advantage of coordination that can occur between the analyte ion and the ligand. When anion exchange selectivities are compared (see Table 9), it is observed that a considerable difference exists between even many adjacent anions, i.e., separation factors (ratio of ion exchange selectivities) are favorable. In contrast, comparison of cation exchange selectivities (see Table 10), indicates that differences are small and separations of complex mixtures will only be feasible providing the cation exchange column has the capability of delivering a high level of efficiency. Adding a ligand to the mobile phase introduces additional coordination equilibria which ultimately influences metal ion retention.

Consider an aqueous HCl solution of metal ion $M^{n+}$. Depending on the metal ion, its coordination number, and the $Cl^-$ concentration, the equilibrium concentration of $M^{2+}$ and the metal chloro complexes present will depend on the stability constants for the formation of the complexes. By controlling the $Cl^-$ concentration the metal ion can be converted from a cation, to a neutral species, to an anion complex, and thus separations should be feasible on either a cation or an anion exchanger. If a cation exchanger is used, increasing the mobile phase $Cl^-$ concentration causes the metal ions to elute because of the coordination. If an anion exchanger is used, the metal ions should be retained at a high $Cl^-$ mobile phase concentration and be eluted as the $Cl^-$ concentration is decreased. A factor which is key in determining metal ion elution order and resolution therefore becomes the stability constant for the metal-chloro complexes.

The separation of metal ions on a strong-base anion exchanger from a $Cl^-$ mobile phase has been studied extensively and this strategy has proven to be very useful in many practical situations. In the classic study by Kraus and Nelson [200,201], retention of the elements on an anion exchanger was determined as a function of mobile phase HCl concentration up to 12 M. Table 19 is a brief summary of these data. It indicates the HCl concentration that can be used to elute the metal ion while retaining those listed below on the anion exchanger; for a more complete listing of retention data, see Ref. 13, p. 154.

Figure 9 shows the separation of the transition metal ions accomplished by a consecutive decrease in mobile phase HCl concentration. Also, extensively studied was the use of $F^-$ as a mobile phase ligand. Fluoride ion is particularly useful because it forms stable complexes with metal ions that are difficult to keep in solution because of hydrolysis.

Although the separation shown in Fig. 9, which is typical, is low in column efficiency, separations of this type are still of considerable practical importance. Furthermore, the strategy can be applied to many kinds of complex metal ion mixtures. The strategy is not limited to $Cl^-$ and $F^-$, and all kinds of inorganic and organic ligands have been applied to metal ion separations [13,202,203]. Although the details may differ, the basic principle is the same. Since stability constants for complex formation do not necessarily have to be large, the range of useful ligands is great. Furthermore, the coordination itself or the lack of coordination can introduce specificity into the separation. For example, $NH_3$, $Br^-$, $I^-$, $NO_3^-$, $SO_4^{2-}$, $CN^-$, $SCN^-$, $CO_3^{2-}$, and $PO_4^{3-}$ are other common inorganic ligands that have been studied extensively. Typical organic ligands that have been used successfully are citrate, tartrate, iminodiacetic acid, ethylenediaminetetraacetic

**Table 19** Elution Order for Selected Metal Ions on a Strong-Base Anion Exchanger as a Function of HCl Concentration

| MHCl | Metal ion |
|---|---|
| 12 | Rare earths, $Ni^{2+}$, $Al^{3+}$ $Mn^{2+}$ (slightly adsorbed) |
| 9.5 | $Ti^{4+}$ |
| 7.5 | $Zr^{4+}$ |
| 7.0 | $Fe^{2+}$ |
| 4.5 | $Co^{2+}$ |
| 3.0 | $Cu^{2+}$ |
| 1.0 | $Fe^{3+}$ |
| 0.02 | $Zn^{2+}$ |
| 0.001 | $Cd^{2+}$ |

See also Ref. 13, p. 154, and Ref. 200.

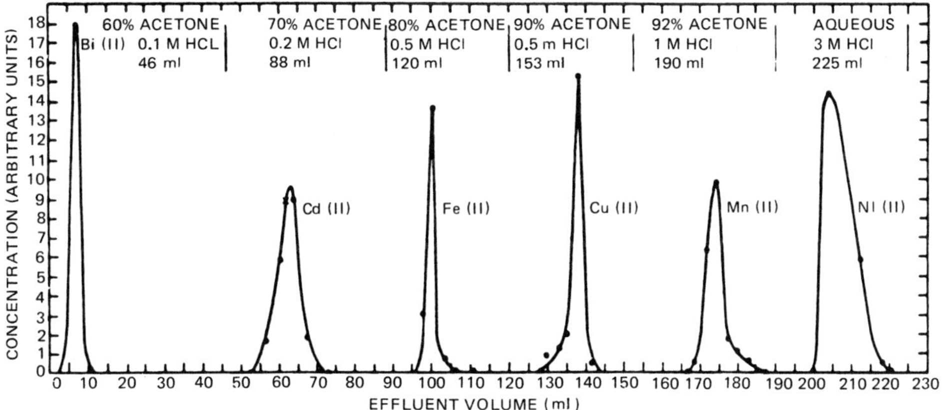


Fig. 9 Anion exchange separation of transition metal ions on an anion exchanger from an HCl mobile phase. Column: Dowex 1 × 8, 200-230 mesh, 20 × 0.03 $cm^2$. (From Ref. 201, courtesy of the *Journal of the American Chemical Society*.)

acid (EDTA), α-hydroxybutyric acid (αHBA), and polyamines. Particularly noteworthy is that several of these ligands, particularly citrate, EDTA, and αHBA, are useful not only for successful analytical separations but also for the commercial separation of rare earths, related elements, and transuranium elements. Although many organic ligands are known, not all are chromatographically used with ion exchange stationary phases. One major reason is that rates of complex formation-dissociation are not always favorable for the ion exchange column separation.

Adding an organic solvent to the mobile phase that also contains a ligand affects both the ion exchange and the formation constants for the analyte-ligand complexes. In general, formation constants are sharply increased thus allowing mobile phase eluent strength to be reduced. With the increase in mobile phase variables due to the effect of the type and concentration of organic solvent in addition to ligand effects, the options in optimizing mobile phase composition are increased. Figure 10 illustrates a separation of several transition metal ions on a strong-base, high-capacity, PSDB anion exchanger using HCl-organic solvent-water combinations. Compared to Fig. 9, where an aqueous HCl eluent is used, metal ions are eluted at much lower HCl concentrations. Increasing the organic solvent concentration (95% ETOH) requires an even lower HCl concentration for a successful separation [204]. Figure 11 shows the separation of transition metal ions on a sulfonated, high-capacity, PSDB cation exchanger using an acetone-HCl-$H_2O$ mobile phase [205]. In this case a modest increase in HCl and acetone, which favors formation of the chloro complexes, causes the metal ions to be eluted from the cation exchanger.

These examples illustrate how selectivity and coordination can be used advantageously in inorganic analyte ion separations. Many other useful examples based on these kinds of strategies utilizing low-efficiency, high-capacity ion exchangers have been developed. The scope of these separations are reviewed in detail elsewhere [12,13, 88,202,203].

In general, inorganic analyte separation strategies on inorganic ion exchangers are also based on selectivity which is influenced by analyte radius and oxidation state and secondary equilibria. Inorganic ion exchangers, however, can differ significantly in chemical structure and this difference can result in sharply altered ion exchange selectivities. This is illustrated by comparing selectivities in Table 5 for inorganic oxide ion exchangers to selectivities for PSDB ion exchangers in Tables 9 and 10.

Application of secondary equilibria can be used effectively to manipulate mobile phase strength with inorganic ion exchangers. However, this application is not as wide in scope as with polymeric ion exchangers. The major reason for this is due to the differences in ion exchanger chemical stability. Any alteration in mobile phase pH,

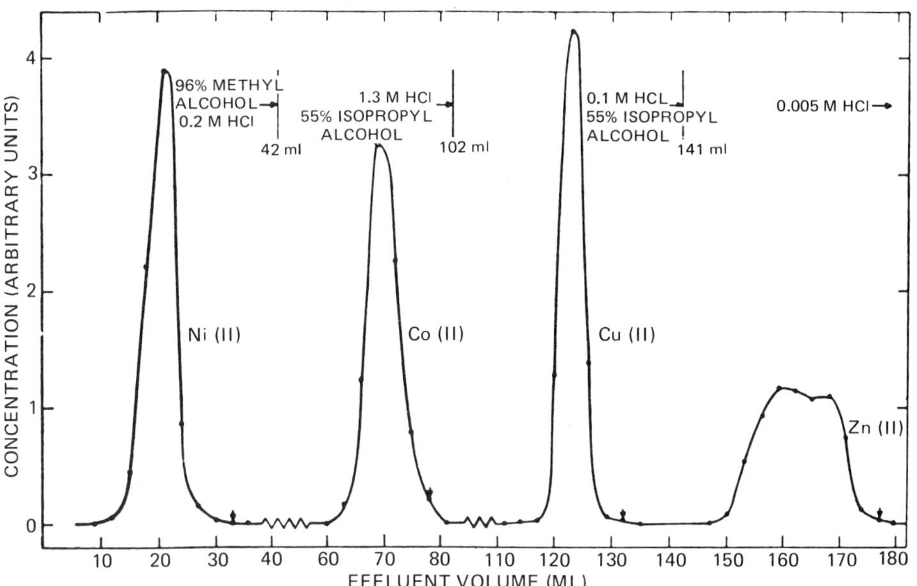


Fig. 10 Separation of transition metals on an anion exchanger using an organic solvent-water-HCl mobile phase. Column: Dowex 1 × 8, 100-200 mesh, 6 × 2.2 cm. (From Ref. 204, courtesy of Pergamon Press.)

which will influence dissociation of weak-acid and base analytes, must be within the range of inorganic ion exchanger pH stability. If ligands are used, their choice must take into account the possibility of complexaction between the ligand and the inorganic ion exchanger. Depending on the stability the inorganic ion exchanger can be permanently changed, dissolved, or reduced in ion exchange capacity.

Most applications of inorganic ion exchangers in inorganic analyte ion separations involve low-efficiency columns or sheet methods. These are reviewed (see Table 16) in detail elsewhere [10,11,13,30-33,88].

*Organic Analytes; Basis for Ion Exchange Separations*

Strategies for the separation of organic molecules on ion exchangers are based on (a) ion exchange selectivity, (b) ligand exchange, and (c) salting out, ion exclusion, and related phenomena. The first involves an ion exchange process and thus the organic analyte must

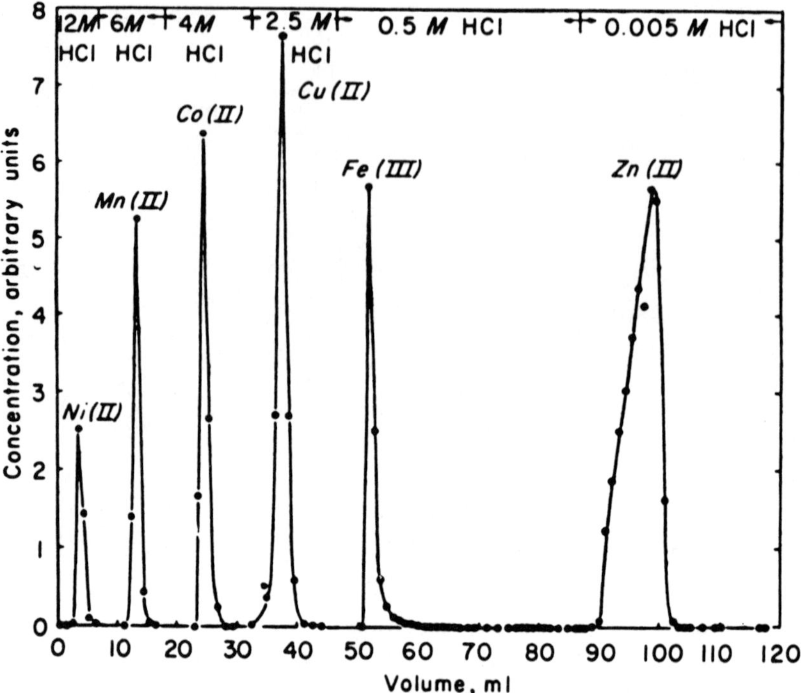


Fig. 11 Separation of transition metals on a cation exchanger using an acetone-water-HCl mobile phase. Column: Dowex 50 × 8, 100-200 mesh, 12.5 × 1.2 cm. (From Ref. 205, courtesy of *Analytical Chemistry*.)

possess electrolytic properties. Ligand exchange, while involving the exchange process, requires that the organic analyte have the capabilities to coordinate with the ionogenic group's counterion. Salting out, ion exclusion, and related phenomena do not involve the ion exchange process and consequently the type of ionogenic group or its charged form are not critical factors in the chromatography. This strategy is best applied to the separation of organic weak and nonelectrolytes.

While many organic molecules have acidic or basic properties and are separable by ion exchange, the major impact of ion exchange, historically, has been in the separation of organic molecules of biochemical interest. This is still true except that high column efficiency type ion exchangers are also available for the separations.

While the ion exchange separation of organic strong and weak electrolytes is based on selectivity, selectivity coefficient data are generally not available. Furthermore, selectivity orders are difficult to predict because the ion exchange process can be accompanied by adsorption (organic analyte-ion exchanger matrix interactions), partition (organic analyte-ion exchanger interior solvent interactions), and sieving action (organic analyte size-porosity size exclusion properties of the ion exchanger).

Position and type of substituents within the organic analyte will affect the retention process by influencing the analyte ionization constant, which affects ion exchange, and the matrix interactions, which affect adsorption. For example, aromatic-containing organic analyte ion retention is significantly greater than nonaromatic organic analyte ion retention on a PSDB ion exchanger due to matrix effects providing ionization for the analytes being compared is similar.

Mobile phase pH is a significant parameter in controlling organic analyte retention particularly for those analytes that are weak acids or bases. By altering the mobile phase pH organic bases can be converted into cations for cation exchange and anions for anion exchange. Since acid and base strengths of organic analytes can differ, a careful adjustment of the pH relative to the ionization constants for the organic analytes can lead to a favorable separation.

Organic solvents are frequently included in the mobile phase partly for solubility reasons and partly because their presence will influence the retention of organic analytes on ion exchangers. Organic solvents, if used in high concentration, will also alter ionization of the organic analytes and buffer salts used in the elution and this effect can be taken advantage of in designing suitable mobile phase conditions.

Figure 12 illustrates the separation of a chlorinated phenol mixture on a quaternary ammonium PSDB strong-base anion exchanger using a MeOH-$H_2O$-increasing $HC_2H_3O_2$ gradient. At low $HC_2H_3O_2$ concentration the stronger phenols are more dissociated and retention on the anion exchanger is high. Increasing the $HC_2H_3O_2$ concentration suppresses ionization; thus the phenols elute off the column according to $K_a$ values with the weakest eluting first. A similar strategy can be used for the separation of other substituted phenols and benzoic acid derivatives. Aliphatic carboxylic acids are retained by anion exchangers but to a lesser degree because of the absence of a matrix effect on retention. A pH gradient or a fixed pH of about 6, depending on the aliphatic mono- and dicarboxylic acids in the sample, can be used. Because of low retention mobile phase counteranion concentration should be modest. For the separation of sulfonic acid derivatives, which are strong electrolytes, the mobile phase eluting power must be increased or the ion exchanger's strength decreased. The former is achieved by increasing mobile phase electrolyte concentration while the latter is achieved by using a weak-base anion exchanger. As with other aromatic acids, aromatic sulfonic acids in general are more retained than alkyl sulfonic acids due to the matrix effect if a

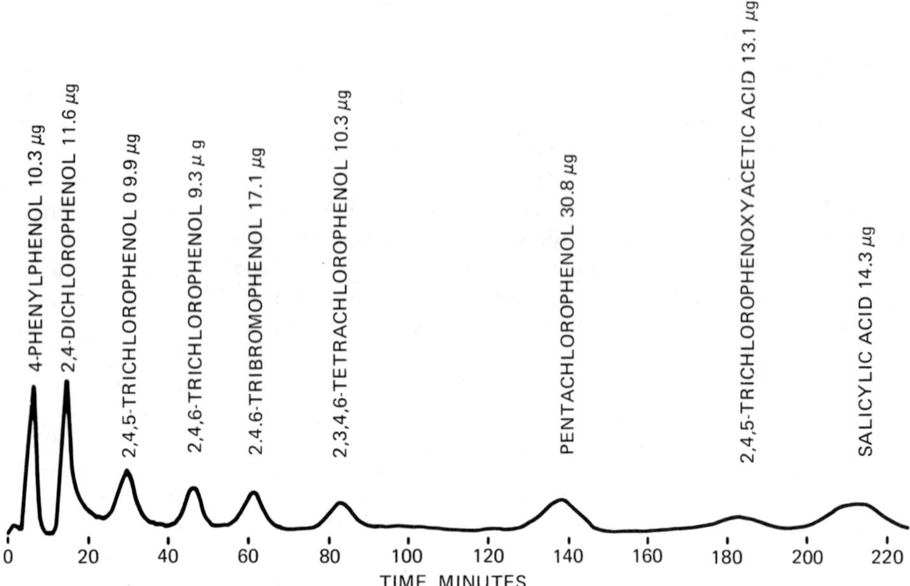


Fig. 12 Anion exchange separation of phenol derivatives. Column: AG 1 × 2, 200-400 mesh, 2.8 × 500 mm; mobile phase; 50% acetic acid-methanol continuous gradient, 100 ml of methanol in the mixing flask; conditions: 1 ml/min, 285 nm. (From Ref. 206, courtesy of CRC Press.)

PSDB-based anion exchanger is used. In general, retention will increase as the number of rings increases or as the alkyl chain increases to the point where sieving becomes a factor.

The separation of organic bases including amines, pyridines, alkaloids, purines, pyrimidines, nucleic acid derivatives, nucleosides, and related compounds can be separated on cation exchangers. A fixed pH favoring ionization and/or type and concentration of counter-cation in the mobile phase are manipulated to achieve the separation. Since many of the pharmaceutical and biochemical compounds listed are amphoteric, they can also be separated on anion exchangers; whether an anion or cation exchanger is used depends on the ionization constants and the complexity of the mixture. Like the organic acids, aromatic amines and bases are more retained than nonaromatic bases of similar ionization constants. Quaternary ammonium salts are strong electrolytes and their separation requires a stronger eluent mixture. If an aromatic group is present in the $R_4N^+$ compound retention will increase; also retention increases as the size of the R group increases.

Strong-acid PSDB cation exchangers will show a high enough affinity for other polar aromatic compounds due to the analyte $\pi$-electron interactions with the PSDB matrix that they can be separated on the exchanger. In essence the cation exchanger is acting as a reversed phase adsorbent and the $-SO_3H$ ionogenic group is modifying the hydrophobic character of the matrix. If an acrylic cation exchanger is used, retention is sharply reduced because this exchanger is free of aromatic $\pi$ electrons. For organic analytes that are also weakly acidic the retention is the highest when it is associated and neutral, and lowest when it is dissociated and anionic. Adding nonaqueous solvents to the mobile phase decreases retention. This type of strategy can be used for separating phenols, benzoic acids, aromatic amides, purines, analgesic drugs, and other polar aromatic-containing organic compounds [207].

In some cases organic molecules, which are nonelectrolytes, can be converted into electrolytes and then separated on ion exchangers. For example, aldehydes and certain ketones will undergo a reaction with bisulfite to form bisulfite addition products. These are anionic products and thus the aldehydes and ketones are separated as anions on an anion exchanger. Similarly, anionic borate esters are formed between borate anion and diols and these can be separated on anion exchangers. This strategy has been applied to the separation of glycols, glycerols, sugar phosphates, sugar alcohols, and mono- and disaccharides. Since the borate and bisulfite reactions are equilibrium reactions, mobile phase condition adjustments must reflect this property.

One of the more successful applications of ion exchange in organic separations is the use of a cation exchanger to separate the amino acids. The approach is to take advantage of the amphoteric property of amino acids and the fact that the neutral form at intermediate pH is in fact highly charged since the amino acids are zwitterions. Because of the importance of this separation, which also permits the accurate determination of individual amino acids, all facets of the separation and detection have been studied extensively [5]. The elution of the amino acids on a sulfonated cation exchanger is based on control of $Na^+$ concentration, pH (citrate buffer), and temperature. The elution, wich usually involves a pH gradient from about pH 3.2 to 4.3, causes the amino acids to gradually shift from a more positive species to a less positive but still highly charged zwitterion form. During this mobile phase change the amino acids are eluting off in the order acidic side-chain amino acids, neutral amino acids, and then basic amino acids. Resolution within these families takes place because of slight difference in ionization constants and hydrophobicity of the side-chain groups.

The importance of ion exchanger particle size on amino acid resolution was recognized, and small, crushed, and eventually small spherical cation exchanger particles were used for amino acid separation prior to the emergence of HPLC. It is not surprising that a major application of ion exchangers in HPLC is for the separation of amino acids.

Amino acids can also be separated on an anion exchanger. In this case a basic mobile phase pH would be used to convert the amino acids to anions. For complex mixtures a gradient from a basic to a less basic pH which shifts the amino acid toward the zwitterionic form could be used. The amino acid elution order would be different on the anion exchanger vs. the cation exchanger because of side-chain effects. Because of the different elution order, anion exchange might be preferred for simplier mixtures. However, for complex mixtures the cation exchange approach is preferred because of better reproducibility and stability of the cation exchanger and amino acid analytes. Also, the type of detection used will often favor a separation on the acidic side rather than on the basic side.

Peptides and proteins are similar to amino acids in that they have terminal amine and carboxyl groups but are more complex because of the large number of side-chain groups due to the individual amino acid subunits. Thus, their retention and subsequent elution from ion exchangers is more complex but is still based on conditions that produce cations, zwitterions, or anions. Usually, pH and electrolyte (buffer) concentration are the major mobile phase variables that are optimized. The more complex the peptide or the longer the peptide chain, in general, the higher the retention on the ion exchanger and the stronger the eluent must be to elute the peptide in a reasonable time period.

Sugars and sugar-related derivatives can be separated on ion exchangers as a result of partitioning, size, and anion exchange [208]. In partitioning either a cation or an anion exchanger can be used. The separation is based on the distribution of the sugar between the ion exchanger interior and exterior solvent mixtures. The ionic form of the ion exchanger influences the interior solvent composition and therefore influences sugar partitioning. Figure 13 illustrates a separation of mono- and disaccharides on a strong-acid PSDB cation exchanger. Similar separations can be accomplished by partitioning on an anion exchanger. Figure 14 illustrates the separation of a series of oligomeric acids and in this case anion exchange is a major contributing factor. If the sugars become too large, then an exclusion limit is reached.

In ligand exchange chromatography a cation exchanger is charged with cations of a metal that is capable of forming stable complexes [209]. The cation electrostatically satisfies the cation exchanger but

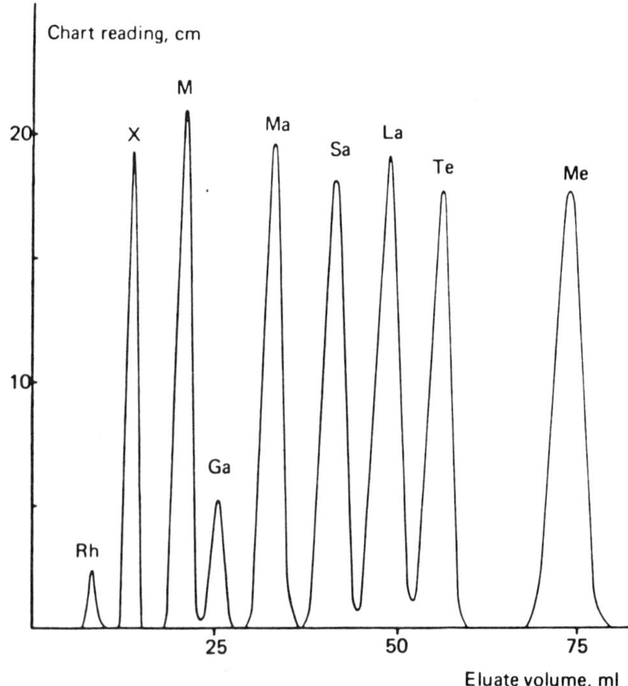


Fig. 13 Separation of mono- and disaccharides on a cation exchanger. Mobile phase: 85% ethanol at 75°C; column: 2.6 × 1225 mm, Dowex 50 × 8, Li form, 14-17 μm; 5.2 cm/min. Rh, rhamnose (2 μg); X, xylose (20 μg); M, mannose (70 μg); Ga, galactose (10 μg); Ma, maltose (100 μg); Sa, saccharose (100 μg); La, lactose (100 μg); Te, trehalose (100 μg); Me, melibiose (100 μg). (From Ref. 208, courtesy of Marcel Dekker, Inc.)

still retains free coordination sites which are able to interact with organic analytes that also have ligand properties. Usually, the cation form is $Cu^{2+}$, $Zn^{2+}$, $Cd^{2+}$, or $Ni^{2+}$. In the separation the metal ions stay on the column and the ligands are coordinated to the metal ion. Elution is obtained by using another ligand in the mobile phase which competes with the analyte ligand for the metal ion. For example, for a $Cu^{2+}$ form cation exchanger, $NH_3$ in the mobile phase will compete with the ligand X for the coordination site according to the equilibrium shown in Eq. (42).

$$Cu(NH_3)^{2+}_{4(resin)} + nX_{(solution)} \rightleftharpoons CuXn^{2+}_{(resin)} + nNH_{3(solution)} \tag{42}$$

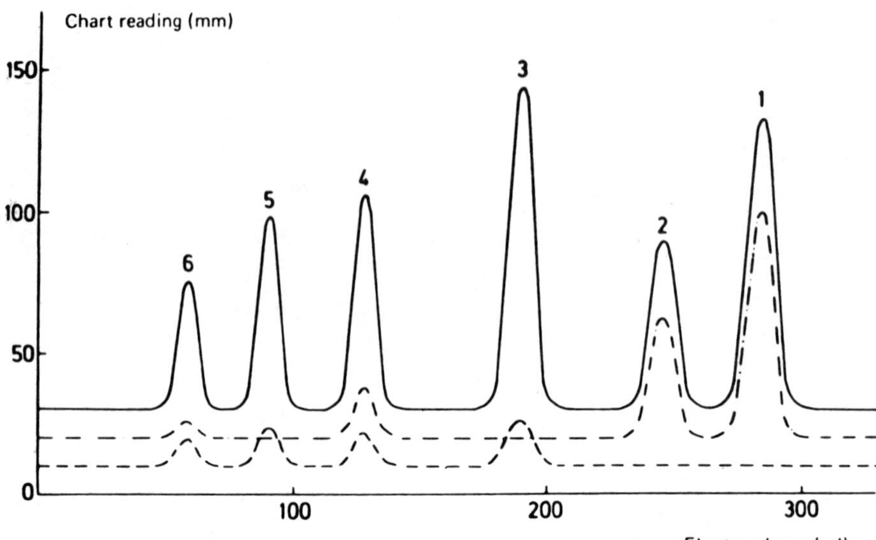


Fig. 14 Separation of oligomeric acids on an anion exchanger. Mobile phase: 0.02 $NaC_2H_3O_2$ (pH = 5.9) at 30°C; column: 4 × 670 mm Dowex 1 × 8, acetate, 13-18 μm; 8.5 cm/min; 1, xylonic (1.0 mg); 2, gluconic (0.6 mg); 3, xylobionic (1.0 mg); 4, cellobionic (0.6 mg); 5, xylotrionic (0.5 mg); 6, cellotrionic acid (0.3 mg). Analysis channels: solid line, chromic acid; dashed line, carbazole; dash-dot line, periodate-formaldehyde. (From Ref. 14, courtesy of Marcel Dekker, Inc.)

The ligand that forms the weakest complex is often eluted first but not always. While formation constants for the M ligand are significant, other factors also contribute to retention. These are ligand-ion exchanger matrix interactions and ligand-ion exchanger interior solvent interactions. For these reasons formation constant order is not always a reliable indication of analyte ligand elution order.

Ligand exchange has been applied to the separation of amines, amino acids, diamines, hydrazines, purines, and pyrimidine bases, other N- and S-containing compounds, and olefins (using an $Ag^+$ form cation exchanger). While many useful separations are possible, this strategy often suffers from three major limitations: (a) Kinetics of coordination are involved and the rate of formation-dissociation is not always favorable. (b) A basic mobile phase is usually used and this can lead to oxidation of many N-containing analytes. (c) The metal ion can be slowly removed from the column; this effect is minimized by including the metal ion in the mobile phase which establishes an equilibrium loading of the metal ion on the column.

Salting out, ion exclusion, and related phenomena-type chromatography on ion exchangers are special cases of partition chromatography in which the ion exchanger interior solvent serves as the stationary phase. Since the ion exchange process is absent, the charged form of the ion exchanger is important only to the extent that it influences the solvent interior and is of a form that is compatible with the mobile phase electrolyte so that ion exchange will not occur. Salting out and related phenomena are applied to nonelectrolyte sorptions and separation [12,13]. By using a modestly high salt concentration in an aqueous or a mixed solvent, aldehydes, ketones, alcohols, diols, ethers, amines, and acids can be separated. Elution times are long and separated peaks are broad, thus limiting this approach relative to HPLC strategies.

Ion exclusion is readily applied to the sorption and separation of organic acids by suppressing their ionization and using a cation exchanger to take advantage of Donnan equilibrium [12,13]. Strong acids elute first followed by weak acids according to their ionization constants. If the eluent also contains a modest concentration of strong acid, dissociation is suppressed and selectivity can be improved. This strategy can be applied to the separation of simple alkyl carboxylic acid, benzoic acid, other aromatic acids, and nucleic acid mixtures.

## HPLC Ion Exchange Separations

In the previous section several kinds of ion exchange separation strategies for inorganic and organic analyte ions were outlined. Their development occurred primarily with ion exchange columns that yielded low column efficiency, and consequently long analysis times were required for the separation. In addition, many complex mixtures were often impossible to separate or only partial resolution was obtained. Even though selectivity differences were measurable for given mobile phase conditions, resolution was not always obtained because of broad bands due to the low column efficiency.

With the emergence of HPLC technology and the development of high-efficiency ion exchange columns, the scope and power of ion exchange separations has changed. In principle the basic strategies of manipulating the mobile-stationary phase parameters are essentially the same. The difference is that with the modern ion exchange column and instrumentation it is possible to take full advantage of slight differences in selectivity and realize resolution in the separation. Furthermore, this can be done rapidly, often in minutes vs. hours as is the case when using high-capacity, low-efficiency ion exchangers, with samples even at trace levels. As improvements in ion exchange columns and HPLC instrumentation have taken place, and awareness of how to use ion exchange principles advantageously in HPLC has increased, separation of more and more complex mixtures has become more and more routine.

The use of ion exchangers in HPLC has had major impact in two main areas. One is the separation and determination of inorganic analyte ions and the second is the separation and determination of compounds of biochemical interest. Particularly striking is the application to inorganic separation and analysis. Other kinds of stationary phases, such as normal and reversed phase adsorbents, and bonded phases were successfully developed and applied to organic separations. Quantitative inorganic separations, on the other hand, lagged far behind in development and for the most part molecular and atomic spectroscopic methods dominated inorganic analysis. This changed in 1975 when ion chromatography was introduced [6]. This approach demonstrated that high-efficiency ion exchange separations of inorganic analyte ions was possible.

The following two sections focus on ion exchanger applications in the HPLC separation of inorganic and organic analytes. Applications are numerous and extend to all areas of the chemical, biological, and health sciences and industries. No attempt is made to survey these applications but rather to focus on selected examples which illustrate the scope of ion exchangers in HPLC.

### *Ion Chromatography*

Ion chromatography applications in inorganic separations can be conveniently divided into two major types. These are:

1. Suppressed anion and cation exchange chromatography
2. Single-column anion and cation exchange chromatography

Both are ion exchange separation strategies that depend on the differences in ion selectivity for separation of analyte anions on an anion exchanger and analyte cations on a cation exchanger. Optimization of the mobile phase involves choosing a counterion that enhances the selectivity difference and is at the same time compatible with the detection system.

*Suppressed Ion Exchange Chromatography* Suppressed ion exchange chromatography was introduced by Small and coworkers [6]. This approach, which was described previously, employs a separator column and a suppression column. In the initial studies of Small and coworkers, an anion exchanger (separator column) was used to separate the analyte anions followed by a $H^+$ form cation exchanger (suppression column) which exchanges mobile phase cations for $H^+$. Since the eluent is a strong base or salt of a weak acid, the suppressor column converts it to $H_2O$ or HA, thus sharply reducing the mobile phase background conductance so that a conductivity detector can be used to detect the separated anions. For analyte cation separations the reverse is used, i.e., the separator column is a cation exchanger, the suppressor column is a hydroxide form anion exchanger, and the eluent is a strong acid or a salt of a weak base.

Two other types of suppression units have been devised for anion separations. One, introduced in 1981, is a continuously regenerating fiber suppressor [210]. Subsequently, a micromembrane suppressor was introduced [211]. The overall goal of the fiber and membrane suppressors, which are cation suppressors, is the same as that of the cation exchanger suppressor in that they substitute $H^+$ for mobile phase cations. The advantages of the membrane suppressor over the fiber suppressor is that dead volume is reduced, which improves detection limit, and it increases suppressor capacity, thus permitting stronger eluents to be used.

Figure 15 shows a typical separation of inorganic anions using an anion exchange separator column and a micromembrane suppressor. If a cation exchanger is used as the suppressor column, separation is still obtained but peak widths are larger because of peak broadening that will occur in the supporessor column. Figure 16 shows that detection limits for the membrane suppressor is increased over that of the fiber suppressor. Furthermore, since the capacity of the membrane suppressor is increased, a stronger eluent can be used. This allows the elution of anions with large anion exchange selectivities, such as those separated in Fig. 16, to be eluted rapidly without loss in resolution. Favorable separation of polyphosphates are possible [211].

In general, eluents that have been used successfully for inorganic analyte anion separations are $OH^-$ and weak-acid anions such as borate, carbonate, bicarbonate, glycine, tyrosine, silicate, *p*-cyanophenate, benzoate, gluconate, and phthalate. All of these except the last three can be used with the membrane suppressor [211]. Figure 17 shows the selectivity that can be obtained by using a tyrosine eluent with the membrane suppressor.

Inorganic cations are readily separated and detected by conductivity using a cation exchange separator column and either an anion exchange suppression column or an anion exchange membrane suppressor. Figure 18 illustrates the separation of alkaline earths using the two column system. Since a relatively strong eluent countercation is used (a divalent cation), the alkali metal ions appear as an early unresolved peak. However, these, including $NH_4^+$, can be readily resolved by using a dilute mineral acid (0.001 to 0.01 M) eluent. Besides mineral acids and phenylenediamine salts, other useful eluents for cation suppression ion exchange chromatography are hydrochloride salts of histidine, lysine, triethanolamine, and hydroxylamine.

*Single-Column Ion Exchange Chromatography* Fritz and others [126,212,213] showed that a suppression system could be omitted if the inorganic anions and cations are separated on a low-capacity anion

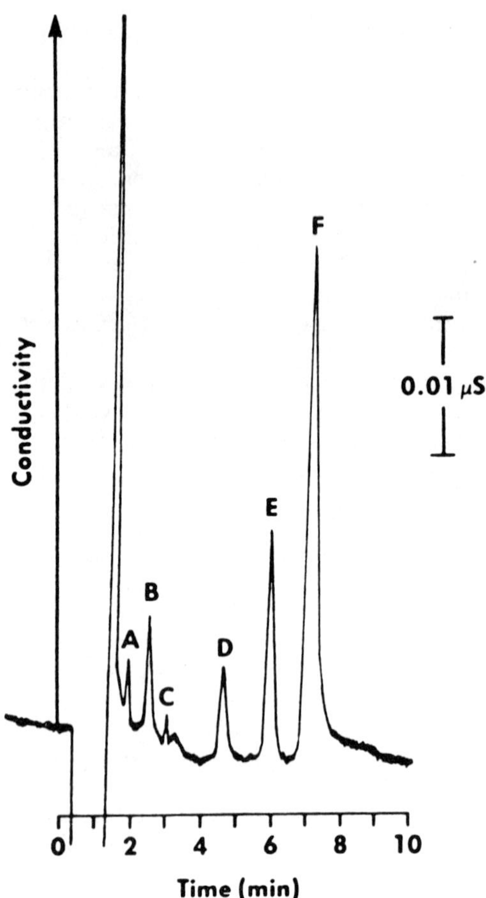


Fig. 15 Anion exchange separation of inorganic anions using a micromembrane suppressor. Column: 25 cm × 4 mm HPIC-AS4; mobile phase: 2.8 mM NaHCO, + 2.2 mM $Na_2CO_3$; flow rate: 2 ml/min; detection: conductometric; injection volume: 100 μl; Temperature: ambient. Peaks: A = $Cl^-$ (3 ppb), B = $NO_2^-$ (10 ppb), C = $HPO_4^{2-}$ (10 ppb), D = $Br^-$ (15 ppb), E = $NO_3^-$ (20 ppb), F = $SO_4^{2-}$ (40 ppb). (From reference 211, courtesy of Aster Publishing.)

or cation exchanger (<100 μeq/g), respectively. Since the exchange capacity is low, eluents of low ionic strength are sufficient to elute the analytes and still allow conductivity detection. Furthermore, by using salts of weak acids for anion elution and salts of weak bases for cation elution, mobile phase background conductivity is further reduced. Cations can also be detected by postcolumn reaction with a

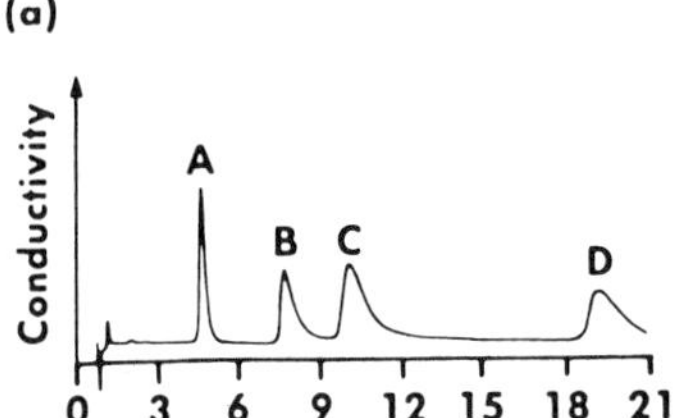


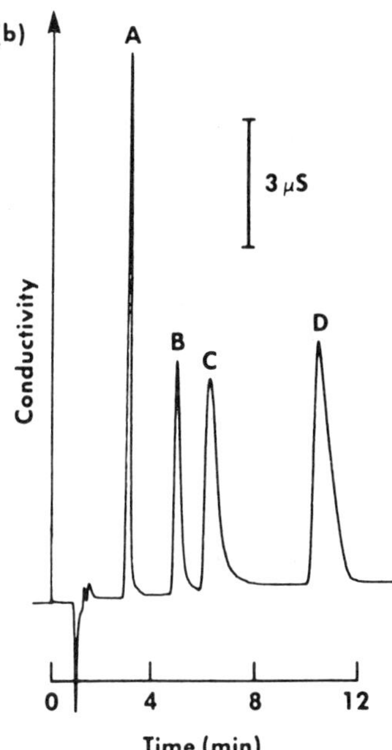


Fig. 16 Separation of oxy-metal anions by anion exchange using (a) fiber suppression and (b) micromembrane suppression. Column: 25 cm × 4 mm HPIC-AS5; mobile phase: (a) 2.8 mM $Na_2CO_3$ + 2.2 mM $Na_2CO_3$ and (b) 10 mM NaOH + 5 mM $Na_2CO_3$; flow rate: 2 ml/min; injection volume: 50 μl; Temperature: ambient. Peaks: A = $SO_4^{2-}$ [(a) 25 ppm and (b) 50 ppm], B = $WO_4^{2-}$ [(a) 25 ppm and (b) 50 ppm], C = $MoO_4^{2-}$ [(a) 25 ppm and (b) 50 ppm], D = $CrO_4^{2-}$ [(a) 25 ppm and (b) 50 ppm]. (From Ref. 211, courtesy of Aster Publishing.)

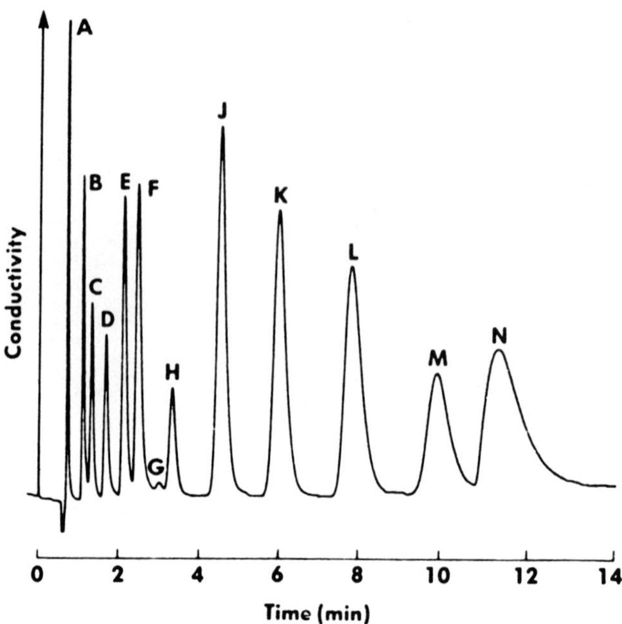


Fig. 17 Anion exchange separation of inorganic anions using a tyrosine eluent and a micromembrane suppressor. Column: 25 cm × 4 mm HPIC-AS4A; mobile phase: 1 mM tyrosine; flow rate: 2 ml/min; injection volume: 50 μl; temperature: ambient. Peaks: A = fluoride, B = chloride, C = nitrate, D = benzoate, E = bromide, F = nitrate, G = carbonate, H = selenite, I = sulfate, K = selenate, L = phosphate, M = phthalate, N = iodine. (From Ref. 211, courtesy of LC Magazine Aster Publishing.)

complexing agent and detected by absorbance. Thus, the choice of the eluent for cations does not have to be tied to conductivity detector compatibility.

The ion exchangers initially prepared by Fritz and coworkers [123, 124,126,213] were of only modest column efficiency. Subsequently, much effort has gone into the preparation of low-capacity, high-efficiency, spherical, uniform size distribution anion and cation PSDB-based ion exchangers. Presently, ion exchange columns for single-column ion exchange separations are capable of yielding 16,000-20,000 plates/m depending on the manufacturer, the test sample, and the conditions.

Figure 19 shows a single-column separation of inorganic anions while transition cations are separated in Fig. 20. The anions are detected by conductivity while the cations are detected by absorbance following postcolumn reaction with 4-(2-pyridylazo)resorcinal (PAR).

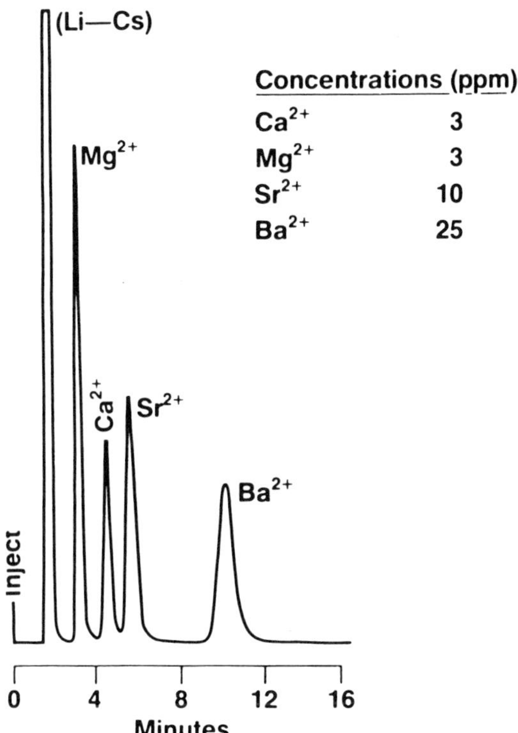


Fig. 18 Separation of alkaline earths by cation exchange using anion exchanger suppression. Eluent: 2.5 mM *m*-phenylenediamine dihydrochloride, 2.5 mM $HNO_3$; flow rate: 3.5 ml/min; separator column: 6 × 250 mm packed with surface-sulfonated 35- to 55-μm cation exchanger; suppressor column: 4 × 250 mm packed with 200-400 mesh Dowex 1 × 10; injection volume: 100 μl; conductivity. (From Ref. 7, courtesy of Preston Publications.)

Stronger eluents allow the separation of multivalent anions while a weaker eluent in terms of both ionic strength and countercation can be used for the successful separation of alkali metals; a more dilute ethylenediammonium eluent than that used in Fig. 20 permits the resolution of the alkaline earths.

Low-capacity silica bonded phase ion exchangers can be used for single-column ion chromatography. However, the choice of eluents is limited because silica-based ion exchangers are stable only through the pH range of approximately 2-8 thus requiring eluents to be within this pH range. Column efficiencies tend to be modestly higher for the silica-based ion exchanger vs. the low-capacity PSDB ion exchanger.

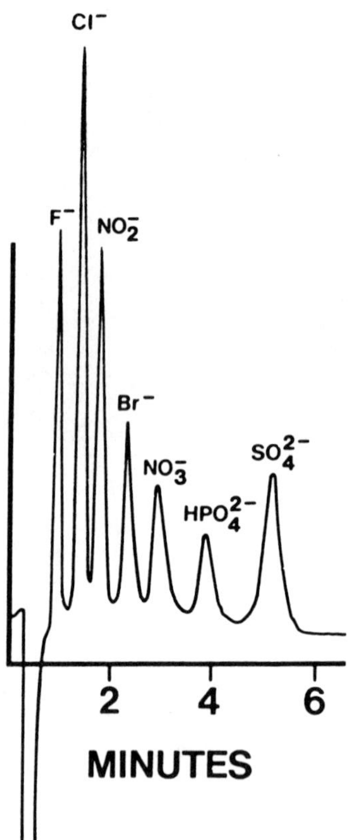


Fig. 19 Single-column separation of inorganic anions on an anion exchanger. Column: PRP-X100, 150 × 4.1 mm; mobile phase: 4 mM *p*-hydroxybenzoic acid at pH 8.6; sample: 100 μl injection, 20 ppm of each anion; conditions: 3 ml/min, conductivity. (From Ref. 128, courtesy of Preston Publications.)

However, in most applications this difference is not a major one. Since the ionogenic groups on the bonded phase are also $-SO_3H$ and $-CH_2NR_3^+X^-$, ion selectivities are similar to those observed for the low-capacity PSDB ion exchangers. Consequently, the separations are similar to those shown previously with the major difference being the pH limitation of the eluent and its subsequent effect on the choice of elution conditions.

Recently, it was shown that silica and alumina were suitable low-capacity, high-column-efficiency stationary phase ion exchangers for single-column ion exchange separations. Silica, because of its isoelec-

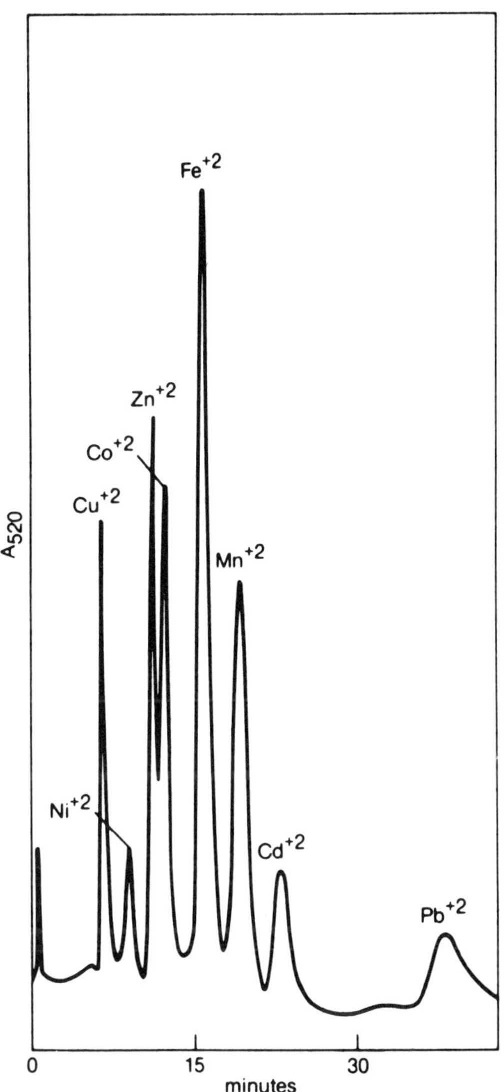


Fig. 20 Separation of transition metals by single-column cation exchange. Column: ION 210, 3 × 100 mm; eluent: mM ethylenediamine plus 10 mM citric acid. Eluent A run for 3 min, eluent B introduced for remainder of analysis; flow rate, 1.0 ml/min; pressure, 135 atm; detection, postcolumn addition of PAR reagent detection at 520 nm. (From Ref. 214, courtesy of International Scientific Communications, Inc.)

tric pH value and limited pH range of stability, is useful only as a cation exchanger [109], while alumina, because of its isoelectric pH value, can be used as both an anion and a cation exchanger [103]. Figure 21 shows the separation of alkali metals and alkaline earths on silica while Fig. 22 shows inorganic anion separations on alumina. The alumina column delivers 40,000 to 70,000 plates/m, which is a distinct increase over the low-capacity PSDB anion exchangers. Also, of practical importance is the fact that the anion selectivity on alumina is not identical to the selectivity on a $-CH_2\overset{+}{N}(CH_3)_3X^-$ PSDB anion exchanger. For example, the elution order in Fig. 22 is $I^-$, $Br^-$, $Cl^-$ ($F^-$ is so strongly held that it can not be eluted), which is exactly the opposite for the quaternary ammonium PSDB anion exchanger.

Ion chromatography has had a major impact on quantitative analysis and is applicable to many kinds of complex analytical problems. The area has been and will continue to be reviewed frequently because of this impact; only recent reviews are cited here [7,8,88,89,133,190, 214-217].

### *Organic Ion Exchange Chromatography*

Both high-efficiency, low-capacity PSDB ion exchangers and bonded phase ion exchangers are widely used in organic HPLC separations.

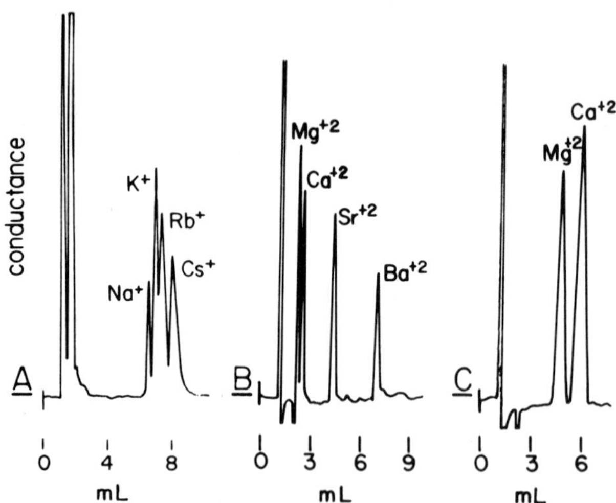


Fig. 21 Separation of alkali metals (A) and alkaline earths (B, C) on a silica column. Column: 150 × 4.6 mm Zorbax Sil; mobile phase: (A) aqueous 2.5 mM LiCl, (B) 1:4 MeOH-$H_2O$ 3.0 mM Na citrate, pH = 7.41, and (C) same as B except 1.0 mM Na citrate; conditions: 1 ml/min, conductivity. (From Ref. 109, courtesy of *Analytical Chemistry*.)

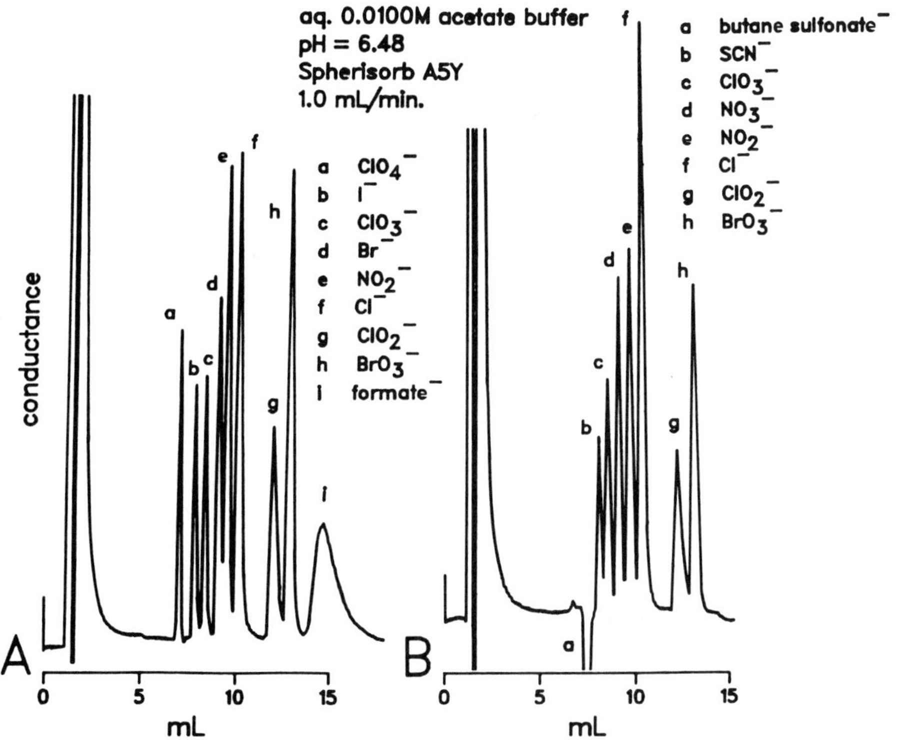


Fig. 22 Separation of inorganic anions on an alumina column. Column: 150 × 4.1mm Spherisorb A5Y; mobile phase: aqueous 10 mM Na acetate buffer, pH = 6.48; conditions: 1 ml/min, conductivity. (From Ref. 103, courtesy of *Analytical Chemistry*.)

For comparable particle sizes the bonded phase ion exchanger provides a higher column efficiency but a more limited useful pH range because it is silica-based.

The strategies for separating organic acids and bases on high-efficiency anion and cation exchangers are essentially the same as those used on high-capacity, low-efficiency ion exchange columns. One major difference that affects selectivity, however, is that matrix effects on the bonded phase ion exchanger is not comparable to the matrix effect provided by the PSDB-type ion exchanger due to interaction between aromatic analytes and the π electrons of the polymer matrix.

While many kinds of organic acids and bases can be conveniently separated on HPLC ion exchangers, the ion exchange approach is not always the preferred strategy. Most of these compounds can also be separated by reversed and normal phase chromatography and usually one of these approaches (particularly reversed phase) is taken. The fact that the organic analyte may be acidic or basic adds to the power of reversed and normal phase chromatography because the effect of pH on retention complements the other mobile phase variables that are part of reversed and normal phase chromatography. Thus, retention and resolution are possible with the analytes in their dissociated or undissociated form while manipulating the other mobile phase variables. Furthermore, organic analyte acids and bases are also readily separated by ion pair chromatographic strategies.

Major areas where HPLC-suitable ion exchangers are in greatest use are (a) ion chromatographic separation of simple, nonaromatic organic acids and bases, (b) ion exclusion strategies for weak acids and certain nonelectrolytes, (c) separation of nucleosides, nucleotides, and related derivatives, (d) ligand exchange strategy for the separation of compounds of biochemical interest including D,L amino acids, (e) amino acid separations, (f) separation of peptides and proteins, and (g) separation of compounds of biochemical interest that contain weak acid and base properties. No attempt is made to review these assorted applications; however, several selected examples are briefly discussed below which emphasize the chromatographic strategy and scope of the separation.

Table 20 lists the retention times for amines (for comparison alkali metals are also included) using a single-column cation exchanger and a dilute $HNO_3$ eluent. A suppression column strategy can also be used. Because the columns provide high efficiency, baseline separations are obtained for most adjacent pairs. Tetraalkylammonium salts are not included but their elution should be expected to be greater than $NH_4^+$ retention and should increase with increasing chain size until exclusion factors become important.

Simple alkyl and aromatic carboxylic acids can be separated in a similar manor using either a single-column or a suppression strategy except that an anion exchanger is used for the separation. A better strategy in terms of resolution and efficiency for the separation of these acids is based on ion exclusion. Table 21 lists the retention order relative to $HC_2H_3O_2$, which is designated as the standard and assigned the value 1.00, for the elution of organic acids on a high-capacity gel PSDB sulfonated cation exchanger. Using a low-capacity, high-efficiency cation exchanger and a dilute $H_2SO_4$ eluent provides an elution consistent with Table 21. Since peak shapes are narrow due to high efficiency, many adjacent pairs are resolvable even though the selectivity difference in Table 21 appears to be small. For example, as shown in Fig. 23, a complex mixture of simple carboxylic acids is virtually baseline resolved in less than 14 min.

Carbohydrates, oligosaccharides, and derivatives can be separated on high-efficiency cation exchangers based primarily on partitioning

**Table 20** Retention Times for Amines and Alkali Alkali Metals on a Low-Capacity, Single-Column Cation Exchanger

| Compound | Adjusted retention time (min) |
|---|---|
| Ammonium | 4.25 |
| Methylamine | 3.11 |
| Dimethylamine | 4.64 |
| Diethylamine | 7.36[a] |
| Triethylamine | 9.02[a] |
| Ethanolamine | 2.91 |
| *n*-Butylamine | 17.95[a] |
| Hydroxylamine | 3.66 |
| *N*-Methylhydroxylamine | 4.92 |
| Hydrazine* | 5.97 |
| Cyclohexylamine | 27.20 |
| Pyridine | 11.25[a] |
| Lithium | 1.77 |
| Sodium | 2.68 |
| Potassium | 5.87 |
| Rubidium | 6.93 |
| Cesium | 8.74 |

[a]Peaks tall.
*Source*: From Ref. 8, courtesy of Dr. Alfred Huthig Verlag.

effects. Usually, a $Ca^{2+}$ form cation exchanger provides the better resolution. Another approach to separate carbohydrates and oligosaccharides is to use a single-column strategy that employs a low-capacity, high-efficiency anion exchanger. At a high mobile phase pH these analytes can be separated; as the mobile phase NaOH concentration increases, their retention decreases [218].

Amino acids have been successfully separated on polymeric ion exchangers for over 30 years [5,88,89]. Cation exchangers are used

**Table 21** Retention Data for Organic Acids Relative to Acetic Acid by Ion Exclusion Chromatography on an Organic Polymeric Strong-Acid Cation Exchanger

| Acid | Ratio acid/ acetic acid | Acid | Ratio acid/ acetic acid |
|---|---|---|---|
| Sulfuric | 0.57 | Fumaric | 1.0 |
| Toluenesulfonic | 0.57 | Glutaric | 1.0 |
| Sulfurous | 0.58 | Chloroacetic | 1.0 |
| 5-Sulfosalicylic | 0.58 | Acetic | 1.00 |
| Sulfamic | 0.58 | Levulinic | 1.0 |
| Hydrochloric | 0.59 | Nadic | 1.0 |
| Acetylenedicarboxylic | 0.59 | L-Pyroglutamic | 1.13 |
| Trichloroacetic | 0.60 | Methylenebismercapto-acetic | 1.15 |
| Mucic | 0.60 | Propionic | 1.17 |
| L-Cysteic | 0.61 | Tetrahydrophthalic | 1.21 |
| Maleic | 0.61-0.71 | Acrylic | 1.23 |
| Oxalic | 0.62 | Carbonic | 1.26 |
| Phosphoric | 0.63 | Isobutyric | 1.32 |
| Citric | 0.64 | Butyric | 1.45 |
| Nitroform | 0.67 | Mandelic | 1.49 |
| Itaconic | 0.70 | Pivalic | 1.49 |
| Pyruvic | 0.71 | α-Hydroxybutyric | 1.57 |
| Malonic | 0.72 | Methacrylic | 1.63 |
| α-Ketobutyric | 0.74 | Isovaleric | 1.66 |
| Glyceric | 0.75 | *tert*-Butylacetic | 1.67 |
| Boric | 0.75-0.79 | Crotonic | 1.95 |
| α-Ketovaleric | 0.79 | Valeric | 2.09 |
| Cyanuric | 0.80 | Furoic | 2.09 |
| Mercaptosuccinic | 0.82 | Cyclohexanecarboxylic | 3.26 |
| Succinic | 0.82 | 2,4-Dihydroxybenzoic | 3.80 |
| Glycolic | 0.82 | *p*-Hydroxybenzoic | 4.46 |
| Lactic | 0.84 | Hydrocinnamic | 5.40 |
| Formic | 0.91 | Benzohydroxamic | 5.95 |
| Adipic | 1.0 | | |

*Source*: From Ref. 8, courtesy of Dr. Alfred Huthig Verlag.

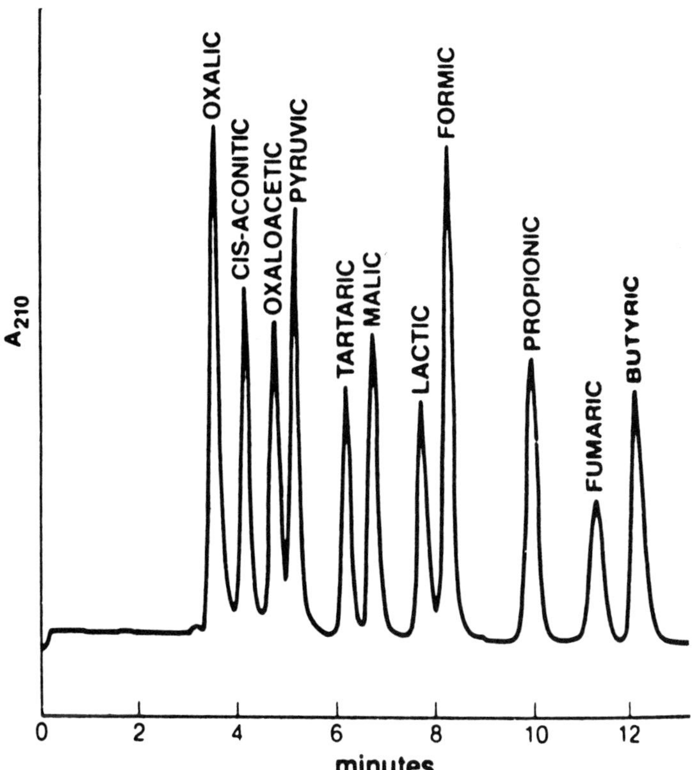


Fig. 23 Separation of organic acids by ion exclusion chromatography on a cation exchanger. Column: ORH-801, 0.65 × 30 cm; mobile phase: 0.01 N sulfuric acid; flow rate, 0.8 ml/min; temperature 35°C; detection: 210 nm. (From Ref. 219, courtesy of Preston Publications.)

with acidic mobile phases to separate amino acids as cations while anion exchangers are used with basic mobile phases for amino acid separations as anions. For complex mixtures, pH gradients are the most effective.

While silica based bonded phase cation exchangers can be used bonded phase anion exchangers have limited applications because of the pH limitation of the silica. High efficiency polymeric cation and anion exchangers do not suffer from this limitation and are widely used for the separation of complex mixtures of free amino acids. In general, most post column derivitization reactions are applicable for detection of the separated amino acids. Figure 24 illustrates a typical separation of a complex mixture of amino acids on a highly efficient polymeric cation exchanger (0.46 cm × 12.0 cm) using a gradient in-

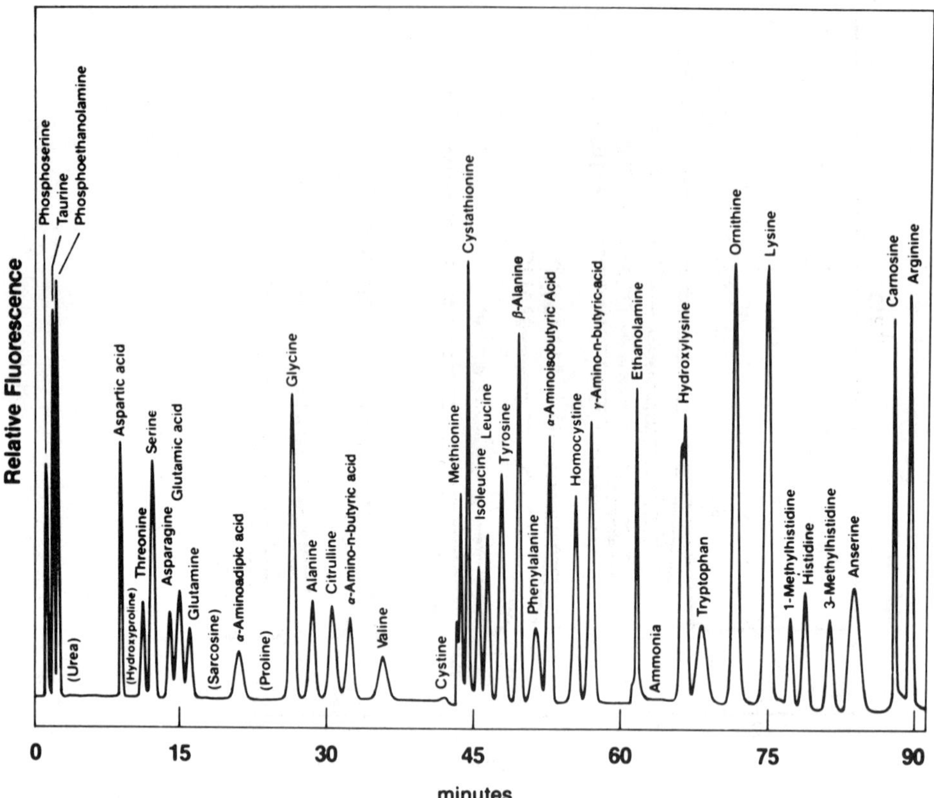


Fig. 24 Separation of amino acids on a high efficiency polymeric cation exchanger. (From Ref. 219, courtesy of Preston Publications.)

volving five discrete buffered solutions containing lithium salts. Each amino acid was 2.5 nmol. Detection was by fluorescence based on a post column reaction between each amino acid and o-phthalaldehyde (OPA). The compounds in parenthesis were present in the sample but were not detected by OPA because hypochlorite was not added to the column effluent. Because of interactions with the polymeric matrix the amino acid elution order on a polymeric cation exchanger will differ from the order found on a silica based bonded phase cation exchanger. Also, since certain amino acids contain acidic side chains or basic side chains, elution order on a cation exchanger will differ from that obtained on an anion exchanger.

D,L-Amino acids can be separated by a ligand exchange-related strategy. In this case a chiral eluent that coordinates with the metal ion is included in the mobile phase along with the metal ion. For D,L-amino acid separation L- or D-proline is the chiral agent; $Cu^{2+}$ is the coordinating metal ion, and a cation exchanger is used. Figure 25 shows the separation of D,L-amino acids. If the L-proline is used the L-amino acid for each D,L pair appears before its D enantiomer. If the D form is used the D enantiomer is first. When D,L-proline is used, amino acids are still separated but not as their enantiomers as shown in Fig. 25.

Nucleic acid derivatives can be readily separated on ion exchangers. Because of higher efficiency the bonded phase ion exchangers are preferred; however, the pH range offered by the low-capacity PSDB ion exchanger can be an advantage when separating certain mixtures. In addition to higher efficiency, the bonded phase ion exchanger is equilibrated more rapidly particularly if a gradient is used, and it can be used to separate in one separation members from each of the base group, nucleoside group, and nucleotide group. Figure 26 shows a separation of a complex mixture of nucleotides on a quaternary ammonium strong-base bonded phase anion exchanger. By including more KCl in the mobile phase (chromatogram a in Fig. 26), the elution time for resolution of the mixture is almost halved.

Both bonded phase and organic polymeric ion exchangers are useful for peptide separations. The latter can offer several advantages: (a) If it has a very large pore size its size exclusion limit is increased. (b) A higher exchange capacity is possible which is useful for preparative chromatography. (c) Being stable throughout the entire pH range allows a basic mobile phase pH to be used. Figure 27 shows a separation of a series of peptides on a sulfonated PSDB cation exchanger which has 1000-Å pore size and a capacity for peptides of 1-5 mg. Although not shown, this ion exchanger can also be used for protein separation and purification.

High-efficiency anion and cation exchangers are widely used in both analytical and preparative chromatography of proteins. Since proteins contain both acidic and basic residues, they exhibit amphoteric properties. Thus, in general, cation exchange would be used at an acidic pH where the net charge of the protein is negative. Because of protein size, interest in retaining activity, and isolation and purification goals, much effort in recent years has gone into the development of ion exchangers that are suited to protein chromatography. In fact, protein retention involves more than the simple protein charge-surface charge type of interaction described above [222]. Not only are the charge characteristics of the surrounding medium important but it also appears that only a fraction of the protein surface is involved, which results in a charge asymmetry effect. Also the type of displacing salt used in the elution alters the nature of the retention.

Table 22 lists different kinds of proteins that have been purified by anion and cation exchangers. Figure 28 shows a typical separation. In this case *E. coli* RNA is separated on a weak-base, *t*-amine, PSDB anion exchanger. This exchanger, which has a 1000-Å porosity, allows a favorable resolution for even large protein fragments.

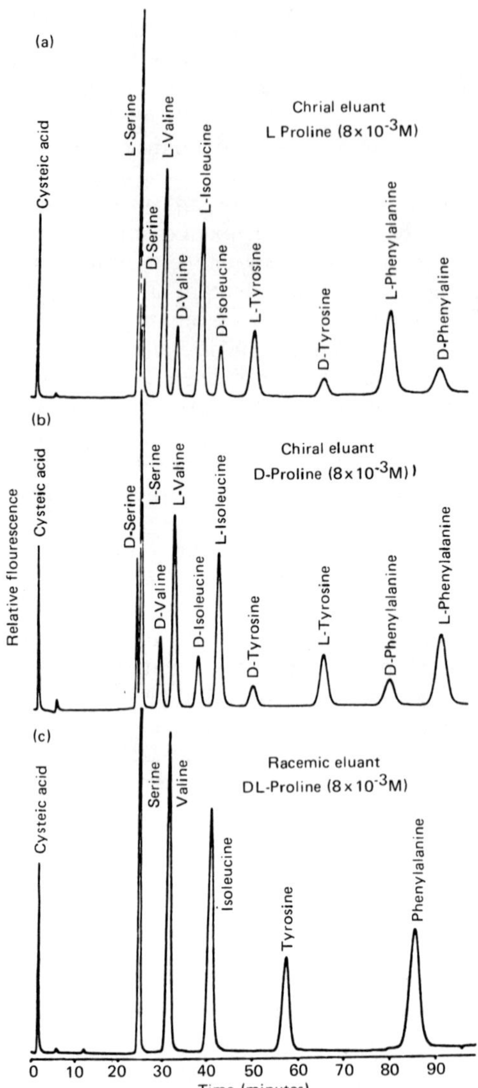


**Fig. 25** Effect of the chiral eluent on the separation of D,L-amino acids. Column: 120 mm × 2 mm packed with cation exchanger DC 4a resin; mobile phase: sodium acetate buffer (0.05 M, pH 5.5) containing $4 \times 10^{-3}$ M $CuSO_4$ and $8 \times 10^{-3}$ M of (a) L-proline, (b) D-proline (c) D,L-proline; temp.: 75°C. (From Ref. 220, courtesy of Aster Publishing.)

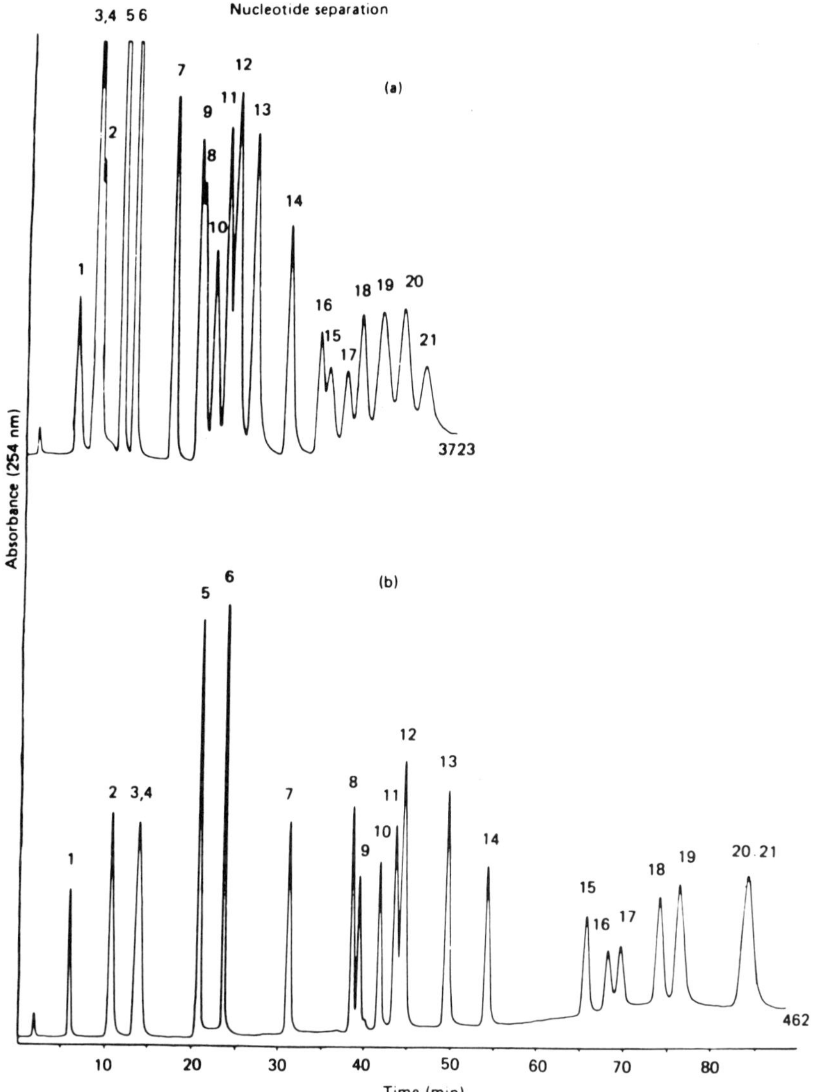


Fig. 26 Nucleotide separation on a bonded phase ion exchanger. Column: mobile phase: (a) conditions: low concentration eluent, 200 mM $KH_2PO_4$, pH 4.0; high concentration eluent, 200 mM $KH_2PO_4$ and 250 mM KCl, pH 4.5; gradient, from 0 to 100% in 45 min; flow rate, 1.5 ml/min. (b) conditions: low concentration eluent, 7 mM $KH_2PO_4$ and 7 mM KCl, pH 4.0; high concentration eluent, 250 mM $KH_2PO_4$ and 50 mM KCl, pH 5.0; gradient, from 0 to 100% in 35 min; flow rate 2.0 ml/min. Both: ambient temperature, detection at 254 nm. Key: 1 = CMP; 2 = AMP; 3 = TMP; 4 = UMP; 5 = IMP; 6 = GMP; 7 = XMP; 8 = TDP; 9 = UDP; 10 = CDP; 11 = IDP; 12 = ADP; 13 = GDP; 14 = XDP; 15 = UTP; 16 = TTP; 17 = CTP; 18 = ITP; 19 = ATP; 20 = GTP; 21 = XTP. (From Ref. 221, courtesy of Elsevier Scientific Publishing Co.)

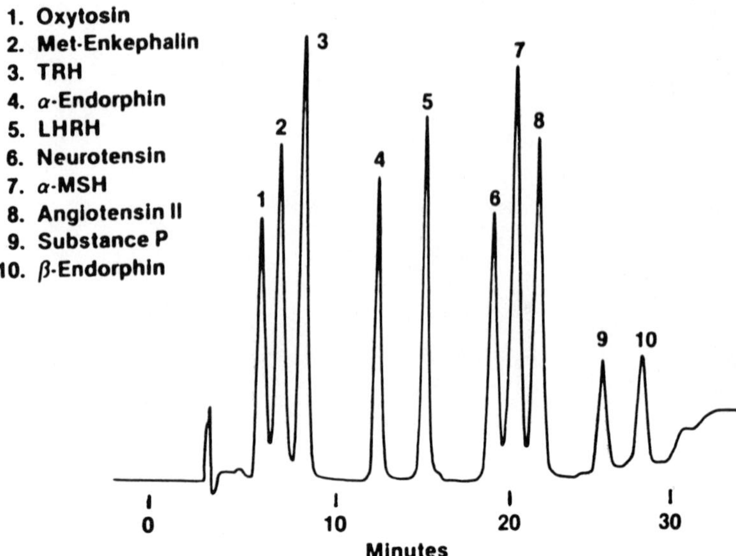


Fig. 27 Separation of peptides on a strong-acid cation exchanger. Column: Bio-Gel TSK SP-5PW, 75 × 7.5 mm; Sample: 2 μg of each peptide in 50 μl except LHRH (1 μg); Mobile phase: 30 ml linear gradient from A to B. (A) 0.02 M phosphate buffer of pH 3/$CH_3CN$ (70:30), (B) 0.5 M phosphate buffer of pH 3/$CH_3CN$ (70:30); Conditions: 1 ml/min, 20°C, 220 nm absorbance. (Reprinted courtesy of Bio-Rad Chemical Division.)

Table 22 Proteins Purified by High-Performance Ion Exchange Chromatography

| Protein | Column | Type of ion exchanger |
|---|---|---|
| Lactate dehydrogenase isoenzymes | DEAE-glycophase[d] | WAX[a] |
| | SynChropak AX300[e] | WAX |
| Creatine kinase isoenzymes | DEAE-glycophase | WAX |
| | SynChropak AX300 | WAX |
| Alkaline phosphatases | DEAE-glycophase | WAX |
| Hexokinase isoenzymes | SynChropak AX300 | WAX |
| Arylsulfatase isoenzymes | DEAE-glycophase | WAX |
| Hemoglobins | SynChropak AX300 | WAX |
| | DEAE-glycophase | WAX |
| | IEX 545 DEAE[f] | |
| | IEX 535 CM[f] | WCX[b] |
| | Bio-Rex 70[g] | WCX |

Table 22 (continued)

| Protein | Column | Type of ion exchanger |
|---|---|---|
| Cytochrome c | CM-polyamide | WCX |
| Lysozyme | CM-polyamide | WCX |
| Myoglobin | CM-polyamide | WCX |
| | IEX 535 CM | WCX |
| Soybean trypsin inhibitor | CM-glycophase[d] | WCX |
| Interferon | Partisil SCX[h] | SCX[c] |
| Lipoxygenase | SynChropak AX300 | WAX |
| Trypsin | DEAE-glycophase | WAX |
| Chymotrypsinogen | SP-glycophase | SCX |
| | IEX 535 CM | WCX |
| Immunoglobulin G | SynChropak AX300 | WAX |
| Ovalbumin | SynChropak AX300 | WAX |
| | IEX 545 DEAE | WAX |
| Albumin | SynChropak AX300 | WAX |
| | DEAE-glycophase | WAX |
| | IEX 545 DEAE | WAX |
| Apolipoproteins | SynChropak AX300 | WAX |
| Adenylsuccinate synthetase | SynChropak AX300 | WAX |
| Insulin | Partisil SCX | SCX |
| | IEX 535 CM | WCX |
| β-Lactoglobulin | Partisil SCX | SCX |
| Carbonic anhydrase | Partisil SCX | SCX |
| Monoamine oxidase | SynChropak AX300 | WAX |

[a]WAX designates weak anion exchanger.
[b]WCX designates weak cation exchanger.
[c]SCX indicates strong cation exchanger.
[d]DEAE and CM-glycophase are products of Pierce Chemical Company, Rockford, Illinois.
[e]SynChropak AX300 is produced by SynChrom, Linden, Indiana.
[f]IEX 535 CM and 545 DEAE are the products of Toya Soda Corporation, Yamaguchi, Japan.
[g]Bio-Rex 70 is supplied by Bio-Rad, San Francisco, California.
[h]Partisil SCX is manufactured by Whatman, Clifton, New Jersey.
*Source*: From Ref. 223, courtesy of Academic Press, Inc. See original reference for literature references.

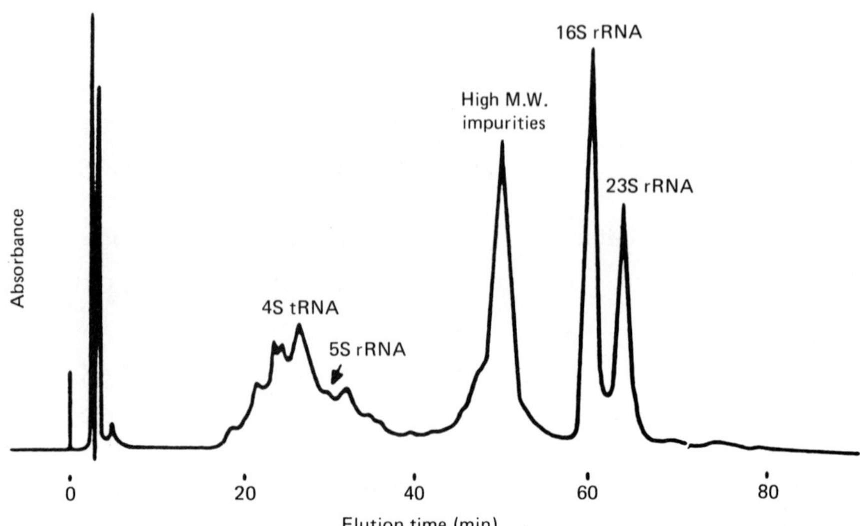


**Fig. 28** Separation of *E. coli* RNA on a weak-base anion exchanger. Column: Bio-Gel TSK DEAE-5-PW, 75 × 7.5 mm; Sample: 0.1 ml of a 1:5 ml diluted solution; Mobile phase: 300 min linear gradient from 0.1 M Tris-HCl buffer, pH 7.6 containing 0.3 M NaCl to 0.1 M Tris-HCl buffer, pH 7.6 containing 1 M NaCl; Conditions: 1 ml/min, 25°C, 260 mm absorbance. (Reprinted courtesy of Bio-Rad Chemical Division.)

## REFERENCES


1. H. S. Thompson, *J. Roy. Agr. Soc. Engl.*, *11*:68 (1850).
2. J. T. Way, *J. Roy. Agr. Soc. Engl.*, *11*:313 (1850).
3. J. T. Way, *J. Roy. Agr. Soc. Engl.*, *13*:123 (1852).
4. B. A. Adams and E. L. Holmes, *J. Soc. Chem. Ind. (London)*, *54*:1T (1935).
5. S. Blackman, *Amino Acid Determination*, 2nd ed., Marcel Dekker, New York, 1978.
6. H. Small, T. S. Stevens, and W. C. Bauman, *Anal. Chem.*, *47*:1801 (1975).
7. C. A. Pohl and E. L. Johnson, *J. Chromatogr. Sci.*, *18*:442 (1980).
8. J. S. Fritz, D. T. Gjerde, and C. Pohlandt, *Ion Chromatography*, Hüthig, Heidelberg, 1982.
9. Y. Marcus and A. S. Kertes, *Ion Exchange and Solvent Extraction of Metal Complexes*, Wiley-Interscience, New York, 1969, p. 283.
10. A. Clearfield, *Inorganic Ion Exchange Materials*, CRC Press, Boca Raton, 1982.

11. C. B. Amphlett, *Inorganic Ion Exchangers*, Elsevier, Amsterdam, 1964.
12. F. Helfferich, *Ion Exchange*, McGraw-Hill, New York, 1962.
13. W. Rieman III and H. F. Walton, *Ion Exchange in Analytical Chemistry*, Pergamon Press, Oxford, 1970.
14. F. Helfferich, *Ion Exchange*, Vol. 1 (J. A. Marinsky, ed.), Marcel Dekker, New York, 1966, p. 65.
15. J. Marinsky, *Ion Exchange*, Vol. 1 (J. A. Marinsky, ed.), Marcel Dekker, New York, 1966, p. 353.
16. D. Reichenberg, *Ion Exchange*, Vol. 1 (J. A. Marinsky, ed.), Marcel Dekker, New York, 1966, p. 227.
17. J. Kielland, *J. Soc. Chem. Ind. (London)*, *54*:232T (1935).
18. A. P. Vanselow, *J. Am. Chem. Soc.*, *54*:1307 (1932).
19. W. C. Bauman and J. Eichhorn, *J. Am. Chem. Soc.*, *69*:2830 (1947).
20. H. P. Gregor, *J. Am. Chem. Soc.*, *70*:1293 (1948).
21. H. P. Gregor, *J. Am. Chem. Soc.*, *73*:642 (1951).
22. L. Lazare, B. R. Sundheim, and H. P. Gregor, *J. Phys. Chem.*, *60*:641 (1956).
23. A. Katchalsky and S. Lifson, *J. Polym. Sci.*, *11*:409 (1953).
24. A. Katchalsky, *Progr. Biophys.*, *4*:1 (1954).
25. I. Michaeli and A. Katchalsky, *J. Polym. Sci.*, *23*:683 (1957).
26. S. A. Rice and M. Nagasawa, *Polyelectrolyte Solutions*, Academic Press, New York, 1961, p. 461.
27. R. M. Wheaton and M. J. Hatch, *Ion Exchange*, Vol. 2 (J. A. Marinsky, ed.), Marcel Dekker, New York, 1969, p. 191.
28. G. Nickless and G. R. Marshall, *Chromatog. Rev.*, *6*:154 (1964).
29. A. O. Jakubovic, *Chem. Prod.*, *23*:510 (1960).
30. M. J. Fuller, *Chromatogr. Rev.*, *14*:45 (1971).
31. V. Vesely and V. Pekarek, *Talanta*, *19*:219 (1972).
32. I. S. C. Churms, *S. Afr. Ind. Chemist*, *19*:148 (1965).
33. A. Clearfield, G. H. Nancollas, and R. H. Blessing, *Ion Exchange and Solvent Extraction*, Vol. 5 (J. Marinsky and Y. Marcus, eds.), Marcel Dekker, New York, 1973, p. 1.
34. R. K. Iler, *The Chemistry of Silica*, John Wiley and Sons, New York, 1979.
35. E. A. Materova, F. A. Belinskaya, E. A. Militsina, and P. A. Skabichevskii, *Ionnyi Obmen, Leningr. Gos. Univ.*: 3 (1965).
36. J. R. Millar, D. G. Smith, W. E. Marr, and T. R. E. Kressman, *J. Chem. Soc.*, 218 (1963).
37. R. Kunin, E. F. Meitzner, J. A. Oline, S. A. Fisher, and N. W. Frisch, *Ind. Eng. Chem. Prod. Res. Dev.*, *1*:140 (1962).
38. V. A. Davankov, S. V. Rogozhin, M. P. Tsyuripa, *Advances in Ion Exchange and Solvent Extraction*, Vol. 7 (J. Marinsky and Y. Marcus, eds.), Marcel Dekker, New York, 1978, p. 29.
39. C. S. Knight, *Advances in Chromatography*, Vol. 4 (J. C. Gid-

dings, E. Grushka, J. Cazes, P. R. Brown, eds.), Marcel Dekker, New York, 1967, p. 61.
40. J. Peska, J. Stamberg, and J. Hradil, *Angew. Macromol. Chem.*, *53*:73 (1976).
41. J. N. Done, *J. Chromatogr.*, *125*:43 (1976).
42. J. S. Ayers, M. J. Peterson, B. E. Sheerin, and G. S. Bethell, *J. Chromatogr.*, *294*:195 (1984).
43. Y. Motozato and C. Hirayama, *J. Chromatogr.*, *298*:499 (1984).
44. F. Wolfe and U. Schallert, *Z. Chem. (Leipzig)*, *12*:184 (1972).
45. C. Schwachula and D. Lukas, *J. Chromatogr.*, *102*:123 (1974).
46. J. Pastyr and L. Kuniak, *Cellul. Chem. Technol.*, *7*:715 (1973).
47. H. Galina and B. N. Kolarz, *J. Appl. Polym. Sci.*, *24*:901 (1979).
48. H. L. Yeager and A. Steck, *Anal. Chem.*, *51*:862 (1979).
49. C. Hirayama, K. Matsumoto, and Y. Motozato, *Nippon Kagaku Kaishi*, 998 (1976).
50. R. Ratner, J. Itzchaki, and H. D. Kohn, *J. Appl. Chem.*, *18*: 48 (1968).
51. J. Deson and R. Rosset, *Bull. Soc. Chim. Fr.*, 4307 (1968).
52. R. A. A. Muzzarelli and B. Spalla, *J. Radioanal. Chem.*, *10*:27 (1972).
53. R. A. A. Muzzarelli adn R. Rocchetti, *Anal. Chim. Acta*, *70*:293 (1974).
54. E. E. Ergozhin, B. A. Zhubanov, V. N. Prusova, and S. R. Rafikov, *Izv Akad. Nauk SSSR, Ser. Khim.*, 972 (1972).
55. I. Hashida, O. Tanizawa, and M. Nishimura, *Nippon Kagaku Kaishi*, 1987 (1974).
56. B. Kapparov, L. N. Prodius, and E. E. Ergozhin, *Tr. Inst. Khim. Nauk Akad. Nauk Kaz. SSR*, *58*:70 (1983).
57. R. Hering, *Z. Chem. (Leipzig)*, *12*:345 (1972).
58. G. V. Myasoedova, S. B. Sarvin, and N. I. Uryanskaya, *Zh. Anal. Kim.*, *26*:1820 (1971).
59. R. Juroda, K. Ishida, and T. Kiriyama, *Anal. Chem.*, *40*:1502 (1968).
60. N. R. Piper, *Anal. Chim. Acta*, *42*:423 (1968).
61. J. S. Coleman and G. L. Gilbert, *J. Chromatogr.*, *34*:289 (1968).
62. M. Shirai, A. Ueda, and M. Tanaka, *Makromol. Chem.*, *186*:493 (1985).
63. F. Wolf, R. Hauptmann, and D. Warnecke, *Z. Anal. Chem.*, *238*: 432 (1968).
64. J. Dingman, Jr., S. Siggia, C. Barton, and K. B. Hiscock, *Anal. Chem.*, *44*:1351 (1972).
65. M. Chikuma, M. Nakayama, T. Tanaka, and H. Tanaka, *Talanta*, *26*:911 (1979).
66. K. S. Lee, W. Lee, and D. W. Lee, *Anal. Chem.*, *50*:255 (1978).
67. A. M. Gurvich and T. B. Gapon, *Zh. Anal. Khim.*, *27*:933 (1972).
68. R. Rosset, *Bull. Soc. Chim. Fr.*, 59 (1966).

69. J. P. Riley and G. Skirrow, *Chemical Oceanography*, Vol. 3, Academic Press, New York, 1975.
70. R. Hering, *J. Prakt. Chem.*, *14*:285 (1961).
71. G. Kuehn, E. Hoyer, and R. Hering, *Z. Chem.*, *4*:262 (1964).
72. H. P. Gregor, M. Taifer, L. Citarel, and E. I. Becker, *Ind. Eng. Chem.*, *44*:2834 (1952).
73. G. Petrie, D. Locke, and C. E. Meloan, *Anal. Chem.*, *37*:919 (1965).
74. V. A. Kyachks, *Dokl. Adak, Nauk SSSR*, *81*:235 (1951).
75. J. N. King and J. S. Fritz, *J. Chromatogr.*, *153*:507 (1978).
76. L. D. Pennington and M. B. Williams, *Ind. Eng. Chem.*, *51*:759 (1959).
77. J. R. Jezorek and H. Freiser, *Anal. Chem.*, *51*:759 (1959).
78. R. C. DeGeiso, L. G. Donaruma, and E. A. Tomic, *Anal. Chem.*, *34*:845 (1962).
79. J. S. Fritz and E. M. Myers, *Talanta*, *23*:590 (1976).
80. R. M. Barnes and J. S. Genna, *Anal. Chem.*, *51*:1065 (1979).
81. P. Burba and K. H. Lieser, *Fresenius Z. Anal. Chem.*, *298*:373 (1979).
82. A. Sugii, N. Ogawa, and N. Hashiyuma, *Talanta*, *26*:970 (1979).
83. F. Vernon and T. W. Kyffin, *Anal. Chim. Acta*, *94*:317 (1977).
84. E. Blasius, K. P. Janzen, W. Adrian, G. Klautke, R. Lorscheider, P. G. Maurer, V. B. Nguyen, T. Nguyen Tien, G. S. Scholten, and J. Stockemer, *Fresenius Z. Anal. Chem.*, *284*:337 (1977).
85. G. Nickless and G. R. Marshall, *Chromatogr. Rev.*, *6*:154 (1964).
86. N. Guivetchi, *J. Rech. Centre Natl. Rech. Sci., Lab Bellevue (Paris)*, *14*:73 (1963).
87. J. R. Millar, *Chem. Ind. (London)*, 606 (1957).
88. H. F. Walton, *Anal. Chem.*, *52*:15R (1980). (See also previous reviews in even years.)
89. R. E. Majors, H. G. Barth, and C. H. Lochmüller, *Anal. Chem.*, *56*:300R (1984). (See also previous reviews in even years.)
90. S. K. Sahni and J. Reedijk, *Coord. Chem. Rev.*, *59*:1 (1984).
91. V. A. Davankov and S. Rogozhin, *J. Chromatogr.*, *60*:280 (1971).
92. V. A. Davankov, *Advances in Chromatography*, Vol. 18 (J. Giddings, E. Grushka, J. Cazes, P. Brown, eds.), Marcel Dekker, New York, 1980, p. 139.
93. V. A. Davankov, Yu. A. Zolotarev, and A. A. Kurganov, *J. Liq. Chromatogr.*, *2*:1191 (1979).
94. K. A. Kraus and H. O. Phillips, *J. Am. Chem. Soc.*, *78*:644 (1956).
95. C. B. Amphlett, L. A. McDonald, and M. J. Redman, *Chem. Ind. (London)*, 1314 (1956).
96. K. K. Unger, *Porous Silica* (J. Chromatogr. Lib., Vol. 16), Elsevier, Amsterdam, 1979.
97. D. J. O'Connor and A. S. Buchanan, *Trans. Faraday Soc.*, *52*:397 (1956).

98. G. A. Parks and D. L. de Bruyn, *J. Phys. Chem.*, *66*:967 (1962).
99. J. A. Yopps and D. W. Fuerstenau, *J. Colloid Sci.*, *19*:61 (1964).
100. G. A. Parks, *Chem. Revs.*, *65*:177 (1965).
101. J. Lyklema, *Croat. Chim. Acta*, *43*:249 (1971).
102. E. Hayck and H. Schimann, *Monatsh. Chem.*, *88*:686 (1957).
103. G. L. Schmitt and D. J. Pietrzyk, *Anal. Chem.*, *57*:2247 (1985).
104. A. Lewandowski and S. Tustanowski, *Chem. Anal. (Warsaw)*, *14*:77 (1969).
105. J. D. Donaldson and M. J. Fuller, *J. Inorg. Nucl. Chem.*, *32*:1703 (1970).
106. E. Hayek and H. Schimann, *Monatsh. Chem.*, *88*:686 (1957).
107. S. Tustanowski, *J. Chromatogr.*, *31*:268 (1978).
108. S. Ahrland, I. Grenthe, and B. Noren, *Acta Chem. Scand.*, *14*:1059 (1960).
109. R. L. Smith and D. J. Pietrzyk, *Anal. Chem.*, *56*:610 (1984).
110. F. H. Pollard and J. F. W. McOmie, *Chromatographic Methods of Inorganic Analysis*, Butterworths, London, 1953, p. 7.
111. G. M. Schwab and A. N. Ghosh, *Angew. Chem.*, *53*:39 (1940).
112. A. Lewandowski and S. Idzikowski, *Chem. Anal. (Warsaw)*, *11*:611 (1966).
113. M. Abe, *Bull. Chem. Soc. Jap.*, *42*:2683 (1969).
114. W. J. Maeck, M. E. Kussy, and J. E. Rein, *Anal. Chem.*, *35*:2086 (1963).
115. H. Small, *J. Inorg. Nucl. Chem.*, *18*:232 (1961).
116. L. C. Hansen and T. W. Gilbert, *J. Chromatogr. Sci.*, *12*:464 (1974).
117. J. R. Parrish, *Nature*, *207*:402 (1965).
118. M. Skafi and K. H. Lieser, *Z. Anal. Chem.*, *250*:306 (1970).
119. T. S. Stevens and M. A. Langhorst, *Anal. Chem.*, *54*:950 (1982).
120. C. A. Pohl and S. C. Papanu, Eur. Patent Appl. EP 134,099; U.S. Patent Appl. 522,827, Aug. 12, 1983; *Chem. Abstr.*, *103*:54859a (1985).
121. O. Mikes, P. Strop, Z. Hostomska, M. Smrz, S. Slovakova, and J. Coupek, *J. Chromatogr.*, *301*:93 (1985).
122. R. W. Siergiej and N. D. Danielson, *J. Chromatogr. Sci.*, *21*:362 (1983).
123. J. S. Fritz and J. N. Story, *Anal. Chem.*, *46*:825 (1974).
124. J. S. Fritz and J. N. Story, *J. Chromatogr.*, *90*:267 (1974).
125. T. Okada and T. Kuwamoto, *Anal. Chem.*, *55*:1001 (1983).
126. D. T. Gjerde, G. Schmuckler, and J. S. Fritz, *J. Chromatogr.*, *187*:35 (1980).
127. R. E. Barron and J. S. Fritz, *Reactive Polym.*, *1*:215 (1983).
128. D. P. Lee, *J. Chromatogr.*, *22*:327 (1984).
129. R. E. Majors, *Am. Lab.*, *4*(5):37 (1972).
130. R. E. Majors, *Am. Lab.*, *7*(10):13 (1975).
131. R. E. Majors, *J. Chromatogr. Sci.*, *15*:334 (1977).

132. R. E. Majors, *J. Chromatogr. Sci.*, *18*:488 (1980).
133. R. Wood, L. Cummings, and T. Jupille, *J. Chromatogr. Sci.*, *18*:551 (1980).
134. A. Pryde, *J. Chromatogr. Sci.*, *12*:486 (1974).
135. R. E. Majors and M. J. Hopper, *J. Chromatogr. Sci.*, *12*:767 (1974).
136. R. E. Majors, *High Performance Liquid Chromatography: Advances and Perspectives*, Vol. 1 (C. Horvath, ed.), Academic Press, Orlando, 1980, p. 76.
137 K. Unger and D. Nyamah, *Chromatographia*, *7*:63 (1974).
138. R. A. Barford, L. T. Olszewski, D. H. Suanders, P. Magidam, and H. L. Rothbart, *J. Chromatogr. Sci.*, *12*:555 (1974).
139. D. H. Saunders, R. A. Barford, P. Magidman, L. T. Olszewski, and H. L. Rothbart, *Anal. Chem.*, *46*:834 (1974).
140. N. Weigand, I. Sebastian, and I. Halasz, *J. Chromatogr.*, *102*: 325 (1974).
141. M. Caude and R. Rosset, *J. Chromatogr. Sci.*, *15*:405 (1977).
142. P. A. Asmus, C. E. Low, and M. Novotny, *J. Chromatogr.*, *119*:25 (1976).
143. P. A. Asmus, C. E. Low, and M. Novotny, *J. Chromatogr.*, *123*:109 (1976).
144. G. B. Cox, C. R. Loscombe, M. J. Slucutt, K. Sudgen, and J. A. Upfield, *J. Chromatogr.*, *117*:269 (1976).
145. B. B. Wheals, *J. Chromatogr.*, *177*:263 (1979).
146. J. B. Crowther, P. Griffiths, S. D. Faziv, J. Magram, and R. A. Hartwick, *J. Chromatogr. Sci.*, *22*:221 (1984).
147. E. H. Edelson, J. G. Lawless, C. T. Wehr, and S. R. Abbott, *J. Chromatogr.*, *174*:409 (1979).
148. R. Schwarzenbach, *J. Chromatogr.*, *117*:206 (1976).
149. A. K. Roy, A. Burgum, and S. Roy, *J. Chromatogr. Sci.*, *22*: 84 (1984).
150. M. Okamoto and F. Yamada, *J. High Resol. Chromatogr. Chromatogr. Commun.*, *5*:443 (1982).
151. D. E. Leyden, G. H. Luttrell, W. K. Nonidez, and D. B. Werho, *Anal. Chem.*, *48*:67 (1976).
152. K. F. Sugawara, H. H. Weetall, and G. D. Schucker, *Anal. Chem.*, *46*:489 (1974).
153. M. M. Guedes de Mota, F. G. Romer, and B. Griepink, *Fresenius Z. Anal. Chem.*, *287*:19 (1977).
154. D. E. Leyden, G. H. Luttrell, A. E. Sloan, and N. J. De Angelis, *Anal. Chim. Acta*, *84*:97 (1976).
155. E. Blasius, K. P. Janzen, H. Luxenburger, V. B. Nguyen, H. Klotz, and J. Stockemer, *J. Chromatogr.*, *167*:307 (1978).
156. E. Blasius, K. P. Janzen, and J. Zendu, *Fresenius Z. Anal. Chem.*, *320*:435 (1985).
157. A. Foucault, M. Caude, and L. Olivers, *J. Chromatogr.*, *185*: 345 (1979).

158. G. Guebitz, W. Jellenz, and W. Santi, *J. Liq. Chromatogr.*, *4*:701 (1981).
159. C. P. Talley and L. M. Bowman, *Anal. Chem.*, *51*:2239 (1979).
160. C. Horvath, B. Preiss, and S. R. Lipsky, *Anal. Chem.*, *39*: 1422 (1967).
161. J. J. Kirkland, *J. Chromatogr. Sci.*, *8*:72 (1970).
162. J. J. Kirkland, U.S. Patent 3,488,922 (1970).
163. C. Horvath, *Ion Exchange and Solvent Extraction*, Vol. 5 (J. Marinsky and Y. Marcus, eds.), Marcel Dekker, New York, 1973, p. 207.
164. E. Grushka and R. P. W. Scott, *Anal. Chem.*, *45*:1626 (1973).
165. P. Freedman, O. Nilsson, J. L. Tayot, and L. Svennerholm, *Biochim. Biophys. Acta*, *618*:42 (1980).
166. J. Alpert and F. E. Regnier, *J. Chromatogr.*, *185*:375 (1979).
167. J. D. Pierson and F. E. Regnier, *J. Chromatogr.*, *255*:137 (1983).
168. Y. Kato, K. Nakamura, and T. Hashimoto, *J. Chromatogr.*, *266*: 385 (1983).
169. A. J. Alpert, *J. Chromatogr.*, *266*:23 (1983).
170. L. R. Snyder and J. J. Kirkland, *Introduction to Modern Liquid Chromatography*, 2nd ed., Wiley-Interscience, New York, 1979.
171. O. D. Bonner and L. L. Smith, *J. Phys. Chem.*, *61*:326 (1957).
172. S. Peterson, *Ann. N.Y. Acad. Sci.*, *57*:144 (1953).
173. H. P. Gregor, M. J. Hamilton, R. J. Oza, F. Bernstein, *J. Phys. Chem.*, *60*:263 (1956).
174. J. I. Bregman and Y. Murata, *J. Am. Chem. Soc.*, *74*:1867 (1952).
175. R. E. Barron and J. S. Fritz, *J. Chromatogr.*, *284*:13 (1984); see also *316*:201 (1984).
176. R. A. Hux and F. F. Cantwell, *Anal. Chem.*, *56*:1258 (1984).
177. A. S. Khan and F. F. Cantwell, *Talanta*, *32*:901 (1985).
178. D. J. Pietrzyk, Z. Iskandarani, and G. L. Schmitt, *J. Liq. Chromatogr.* *9*:2633 (1986).
179. Z. Iskandarani and D. J. Pietrzyk, *Anal. Chem.*, *53*:489 (1981).
180. D. P. Lee, *J. Chromatogr. Sci.*, *20*:203 (1982).
181. G. E. Boyd, A. W. Adamson, and L. S. Myers, Jr., *J. Am. Chem. Soc.*, *69*:2836 (1947).
182. V. A. Kuzminykh and V. P. Meleshko, *Zh. Fiz. Khim.*, *54*:973 (1980).
183. D. J. Pietrzyk, *Talanta*, *13*:209 (1966).
184. K. A. Kun and R. Kunin, *Polym. Lett.*, *2*:587 (1964).
185. B. Chu and D. M. Tan Creti, *J. Phys. Chem.*, *71*:1943 (1967).
186. D. J. Pietrzyk, *Talanta*, *16*:169 (1969).
187. D. J. Pietrzyk, *CRC Critical Reviews in Analytical Chemistry*, Vol. 6, 1976, p. 131.
188. Y. Marcus, *Ion Exchange and Solvent Extraction*, Vol 4 (J.

Marinsky and Y. Marcus, eds.), Marcel Dekker, New York, 1973, p. 1.
189. G. J. Moody and J. D. R. Thomas, *Analyst, 93*:557 (1968).
190. P. R. Haddad and L. A. Heckenberg, *J. Chromatogr., 300*:357 (1984).
191. B. A. Bidlingmeyer, J. K. Del Rios, and J. Korpi, *Anal. Chem., 54*:442 (1982).
192. R. L. Smith, Z. Iskandarani, and D. J. Pietrzyk, *J. Liq. Chromatogr., 7*:1935 (1984).
193. S. C. Churms, *J. S. Afr. Chem. Inst., 19*:108 (1966).
194. S. C. Churms, *J. S. Afr. Chem. Inst., 19*:98 (1966).
195. M. Abe and T. Ito, *J. Chem. Soc. Jap., 86*:817 (1965).
196. F. W. Cornish, *Analyst, 83*:634 (1958).
197. J. C. Giddings, *Dynamics of Chromatography*, Marcel Dekker, New York, 1965.
198. R. Kunin, *Symposium on Ion Exchange and Chromatography in Analytical Chemistry*, American Society for Testing Materials, Publication No. 195, Philadelphia, 1958, p. 3.
199. B. L. Karger, J. N. LePage, and N. Tanaka, *High Performance Liquid Chromatography Advances and Perspectives*, Vol. 1 (C. Horvath, ed.), Academic Press, New York, 1980, p. 113.
200. K. A. Kraus and F. Nelson, *Proc. 1st Int. Conf. on Peaceful Uses of Atomic Energy, 7*:113 (1955).
201. K. A. Kraus and G. E. Moore, *J. Am. Chem. Soc., 75*:1460 (1953).
202. J. Inczedy, *Analytical Applications of Ion Exchangers*, Pergamon Press, Oxford, 1966.
203. J. Korkisch, *Modern Methods for the Separation of Rare Metal Ions*, Pergamon Press, Oxford, 1969.
204. J. S. Fritz and D. J. Pietrzyk, *Talanta, 8*:143 (1961).
205. J. S. Fritz and T. A. Rettig, *Anal. Chem., 34*:1562 (1962).
206. N. E. Skelly and R. H. Stehl, *Recent Development in Separation Science*, Vol. 1 (N. N. Li, ed.), CRC Press, Cleveland, 1972, p. 141.
207. H. F. Walton, G. A. Eicemann, and J. L. Otto, *J. Chromatogr., 180*:145 (1979).
208. O. Samuelson, *Advances in Chromatography*, Vol. 16 (J. C. Giddings, E. Grushka, J. Cazes, and P. R. Brown, eds.), Marcel Dekker, New York, 1978, p. 113.
209. V. A. Davankov and A. V. Semechkin, *J. Chromatogr. Rev., 141*:313 (1977).
210. T. S. Stevens, J. C. Davis, and J. H. Small, *Anal. Chem., 53*:1488 (1981).
211. J. Stillian, *LC Magazine*, 3(9):802 (1985).
212. J. S. Fritz, D. T. Gjerde, and R. M. Becker, *Anal. Chem., 52*:1519 (1980).

213. D. T. Gjerde, J. S. Fritz, and G. Schmuckler, *J. Chromatogr.*, *186*:509 (1979).
214. J. Benson, *Am. Lab.*, *17*(6):30 (1985).
215. F. C. Smith, Jr. and R. C. Chang, *The Practice of Ion Chromatography*, Wiley-Interscience, New York, 1983.
216. R. A. Wetzel, C. Pohl, and J. M. Riviello, *Chem. Anal. (N.Y.)*, *78*:355 (1985).
217. J. S. Fritz, *LC Magazine*, *2*(6):446 (1984).
218. R. D. Rocklin and C. A. Pohl, *J. Liq. Chromatogr.*, *6*:1577 (1983).
219. J. R. Benson and D. J. Woo, *J. Chromatogr. Sci.*, *22*:386 (1984).
220. W. Lindner and C. Pettersson, *Liquid Chromatogrpahy in Pharmaceutical Development* (I. W. Wainer, ed.), Aster, Springfield, Oregon, 1985, p. 117.
221. M. McKeag and P. R. Brown, *J. Chromatogr.*, *152*:253 (1978).
222. W. Kopaciewicz, M. A. Rounds, J. Fausnaugh, and F. E. Regnier, *J. Liq. Chromatogr.*, *266*:3 (1983).
223. F. E. Regnier, *Anal. Biochem.*, *126*:1 (1982).

# 11

# Packings and Stationary Phases for Ion Pair Chromatography

**B.-A. Persson and Per-Olof Lagerström** / *AB Hässle, Mölndal, Sweden*

## PRINCIPLES OF ION PAIR CHROMATOGRAPHY

The ion pair extraction concept can be applied to liquid chromatography either in liquid-liquid or liquid-solid mode in normal or reversed phase systems.

In ion pair extraction two ions of opposite charge are extracted as an ion pair to an organic phase:

$$HB^+_{aq} + X^-_{aq} = HBX_{org} \qquad (1)$$

where $HB^+$ may illustrate a cation such as a protonated amino compound and $X^-$ the anion of an acid. The equilibrium may be characterized by an extraction constant:

$$K_{ex} = \frac{(HBX)_{org}}{(HB^+)_{aq} \cdot (X^-)_{aq}} \qquad (2)$$

The magnitude of the extraction constant depends on the hydrophobicity of the two ions forming the ion pair, the kind of interaction forces between the ions, and the nature of the organic phase, e.g., its polarity and hydrogen-accepting and donating properties [1].

The ion pair extraction is influenced by side reactions, the extent of which depends on the properties of the ionic species, the organic phase, the concentration range studied, and other agents present in the system. Adduct formation is another kind of side reaction. For a simple ion pair extraction the distribution ratio for an ammonium compound $HB^+$ is given by

$$D_{HBX} = \frac{(HBX)_{org}}{(HB^+)_{aq}} = K_{ex} \cdot (X^-)_{aq} \tag{3}$$

The distribution ratio will depend on the magnitude of the extraction constant and on the concentration of the counterion $X^-$. In chromatographic systems based on liquid-liquid distribution of ion pairs it is important that $D_{HBX}$ has a constant value irrespective of the solute concentration. Side reactions must then be controlled and the counterion concentration be high enough to ensure constant $D_{HBX}$. This has to be considered in the choice of composition of mobile and stationary phase [2]. In normal phase liquid-liquid systems the capacity factor is given by

$$k' = \frac{1}{D_{HBX}} \cdot \frac{V_s}{V_m} \tag{4}$$

The ion pair is migrating into the organic mobile phase, the counterion $X^-$ being supplied by the aqueous stationary phase [3]. The magnitude of $D_{HBX}$ most often has a value of 1 and lower to give appropriate retention. According to Eqs. (3) and (4), increasing $K_{ex}$ and $(X^-)_{aq}$ will decrease the capacity factor. In liquid-liquid systems for reversed phase chromatography the capacity factor is expressed by

$$k' = D_{HBX} \cdot \frac{V_s}{V_m} \tag{5}$$

The counterion is present in the aqueous mobile phase and the solute ion is retained as ion pair by distribution into the organic stationary phase [4]. Usually conditions are chosen for $D_{HBX} = 1$ and higher.

If the retention is due to adsorption of the solute as ion pair to a hydrophobic solid phase as in bonded phase chromatography, a similar expression can be devised:

$$HB^+_m + X^-_m + A_s = HBXA_s \tag{6}$$

where $A_s$ is a binding site of the surface. This equilibrium can be expressed by a binding constant $K_{HBX}$ and the capacity factor given by

$$k' = K^{o} \cdot q \cdot K_{HBX} \cdot (X^{-})_{m} \qquad (7)$$

where $K^{o}$ is the capacity of the adsorbent and q the phase volume ratio. Increased hydrophobicity and concentration of the counterion increases the capacity factor [5]. The presence and adsorption of ionic eluent components, coions, with the same charge as the solute ion will decrease the capacity factor by competing for the binding sites of the hydrophobic surface. This can be written as

$$k' = K^{o} \cdot q \cdot K_{HBX} \frac{(X^{-})_{m}}{1 + K_{QX} \cdot (Q^{+})_{m}(X^{-})_{m}} \qquad (8)$$

The concentration of the coion $Q^{+}$ will obviously influence the capacity factor and together with the counterion concentration determine the retention of the solute ion pair. Linear correlations are obtained in limited regions only. The situation may be even more complicated if the hydrophobic surface of the solid phase cannot be assumed to be homogeneous but binding sites of different properties have to be included in a retention model [6].

Different retention models have been used in ion pair liquid chromatography on bonded phases which also is reflected by expressions like ion-ion interaction and solvent-generated or dynamic ion exchange chromatography. Under certain experimental conditions one model is more plausible but in other cases it is hard to judge how to best characterize the system.

## NORMAL PHASE CHROMATOGRAPHY

The first applications on ion pair technique in liquid chromatography were in the normal phase mode, i.e., with a mobile organic phase. Nowadays reversed or bonded phase chromatography is the dominant technique using ion pair distribution. However, for selectivity or sample workup reasons the normal phase mode should be recognized as a valuable alternative.

### Liquid Adsorption

There are only a few examples of systems in normal phase mode where retention due to ion pair adsorption has been utilized. For the separation of amines silica has been used as stationary phase [7-12] with aqueous perchloric acid [7-11] or dodecyl sulfate [12] as ion-pairing agent in methanol (Fig. 1). The bulk organic solvent has in most

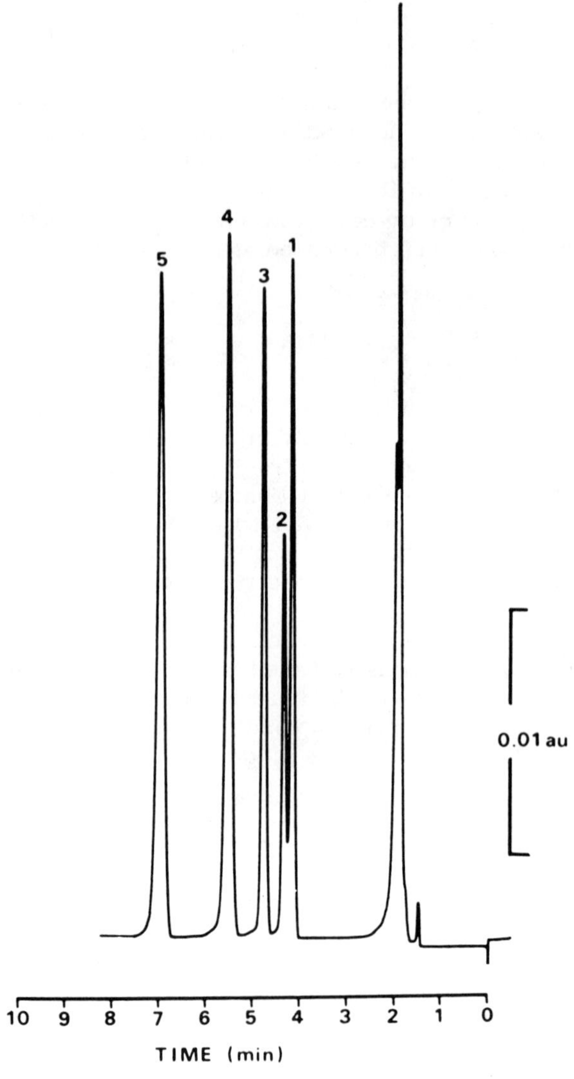


Fig. 1 Chromatography of tocainide (1), mexiletine (2), monoethylglycinexylidide (3), R17251 (internal standard) (4), and lignocaine (5) on a 250-mm column packed with Spherisorb 5 silica. Eluent: methanol-containing perchloric acid (0.02%, v/v; 1.85 mM). (From Ref. 11.)

instances been either dichloromethane or diethylether or a mixture of these. The polar components of the mobile phase, the ion-pairing agent and water-methanol, are adsorbed onto the silica, 0.3 ml/g being the estimate of Mellström and Braithwaite [8]. This corresponds to 1/5-1/3 of the volume of a precoated stationary aqueous phase. Increased alcohol concentration will decrease retention and so will increased counterion concentration [7,8,11]. Lagerström and Persson [7] found that the capacity factor for divalent amines increased with the concentration of perchloric acid in the mobile phase, the amines being transformed to divalent cations.

Whether changes in the ion pair retention are due to effects in the stationary phase imposed by the acid adsorbed from the eluent or to solvation effects in the mobile phase has not been discussed very much. Eriksson and coworkers [9] found that shortest retention was obtained by perchloric acid followed by nitric and hydrochloric acids, while organic acids like trichloroacetic and methanesulfonic gave stronger retention. This indicates that the elution is a combination of suppression of acidic groups on the silica surface and solvation of the ion pairs in the mobile organic phase. Silica was also used as stationary phase for separation of basic drugs with methanol as mobile phase and NaBr and $NaClO_4$ as ion-pairing agents [13]. The additives promoted the elution of sharp and symmetrical peaks despite the lack of pH control.

In the systems discussed so far the choice of stationary solid phase is not critical to performance. However, for separation of enantiomers of amines by ion pair adsorption Pettersson and coworkers [14-17] showed that the properties of the adsorbent are decisive for the outcome. Mobile phases of low polarity containing a chiral counterion were used, the formed diastereomeric ion pairs without identical distribution and adsorption properties [16,17] (Fig. 2). LiChrosorb DIOL as adsorbent was preferred for giving peak symmetry superior to other solid phases tested, such as silica of different surface area, nitrile, nitro, and C2 phases. The chiral counterions, camphor-10-sulfonic acid for resolution of amines [14,15] and alprenolol [15] or quinine [16,17] for resolution of racemic acids, in the mobile phases are adsorbed onto the solid phase, the resolution being strongly dependent on the interaction between the solutes and phase components.

## Liquid-Liquid Chromatography

In the beginning ion pair distribution in liquid chromatography comprised normal phase systems, where the counterion in an aqueous solution was coated onto a solid phase. Schill and coworkers [18] used purified Celite 545, diatomaceous earth, as support for stationary aqueous phases of inorganic acids. Good correlation was obtained between retention volumes found and those calculated from batch extraction

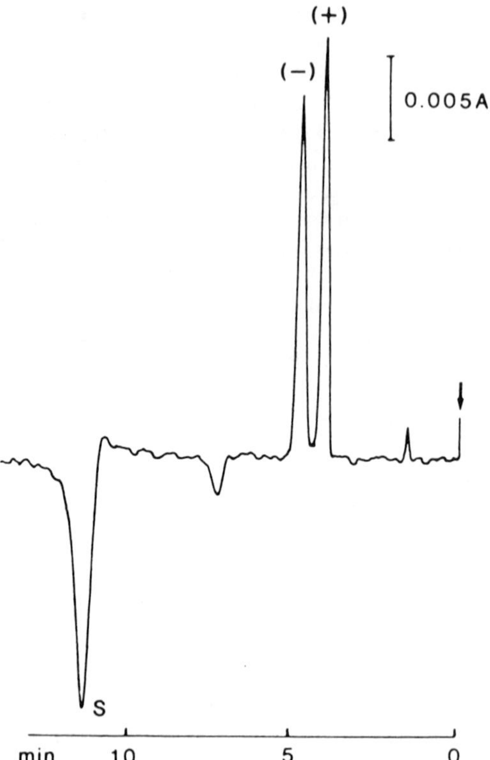


Fig. 2 Resolution of (±)-*N*-tert-butoxycarbonylphenylalanine. Solid phase: LiChrosorb DIOL. Mobile phase: $3.5 \times 10^{-4}$ M quinidine and $3.5 \times 10^{-4}$ M acetic acid in dichloromethane-1-pentanol (99:1). (From Ref. 16.)

studies. Chloroform was the mobile phase giving symmetrical peaks, except for promazine tailing severely. Porous ethanolyzed cellulose was introduced as support for the separation of tricyclic amines as chloride ion pairs [19] with cyclohexane plus pentanol as mobile phase. Cellulose was also used for separation of acetylcholine [20,21] and choline [22] wtih picrate as detector-sensitive counterion in the stationary aqueous phase. Eksborg and Schill [3] compared cellulose, diatomaceous earth, and Porasil D (silica) of medium particle size in a thorough investigation of ion pair partition chromatography of organic ammonium compounds as picrate ion pairs. It was shown that the degree of loading with stationary phase influenced column efficiency significantly.

By the access of microparticulate silica as packing material and support for the stationary phase, much improved peak sharpness can be obtained. Persson and Karger [23] used perchloric acid as stationary aqueous phase for the separation of biogenic amines and tetrabutylammonium at neutral pH for acidic metabolites. The stationary phases were applied by an in situ coating technique onto the silica support, which despite its high surface area was not found to contribute to the retention of the biogenic amines. The same type of stationary and solid phases was used for thyroid hormones and sulfa drugs [24], and in a following study [25] the influence of pH and counterion content in the stationary phase was examined including a comparison with static distribution data. Knox and Jurand [26] used aqueous perchlorate on silica for separation of tricyclic amines and also compared the selectivity and efficiency of this liquid-liquid system with a liquid-solid system with acetic acid in the mobile phase. Mixtures of aqueous perchloric acid and sodium perchlorate as stationary phases on silica with butanol and dichloromethane in the mobile phase gave decent chromatographic separations [27-29]. For separation of diastereoisomeric pairs of *N*-propylajmaline such a system was superior to separations based on other liquid chromatographic principles [30]. Liquid-liquid chromatographic systems have to be carefully thermostatted and equilibrated. This was demonstrated by Lagerström [31] in the separation of carboxylic acids as ion paris with tetrabutylammonium. Silica gel supports were compared with ethanolyzed cellulose as regards retention of solutes and peak symmetry. Silica gel of different surface area has been used as microparticulate support in the liquid-liquid systems presented so far. Counterions such as perchlorate and tetrabutylammonium in the aqueous stationary phase often create little problem as regards chromatographic performance. However, counterions with high detectability in the photometric detector through condensed, chromophore structures will as a rule make additional considerations necessary. Picrate was used as counterion [32] and different supports (silica, unmodified or silanized, alumina, and kieselguhr) were tested. The last one was preferable but the systems were generally of low stability and performance.

Crommen et al. [33] used LiChrospher SI 100 as support for an aqueous stationary phase containing naphthalene-2-sulfonate as counterion. Amino acids, dipeptides, and alkylamines were separated and made detectable in the UV detector (Fig. 3). Naphthalene sulfonate was purified by repeated extraction and used in concentrations of 0.01 and 0.1 M. The lower practical limit for the counterion content in the stationary phase is often 0.01 M if good stability and sample capacity are to be obtained. The aqueous stationary phase was applied onto the silica support by either injection or pumping technique, after which recycling with mobile phase equilibrated with stationary phase was started. Particularly with detector active counterions, very care-

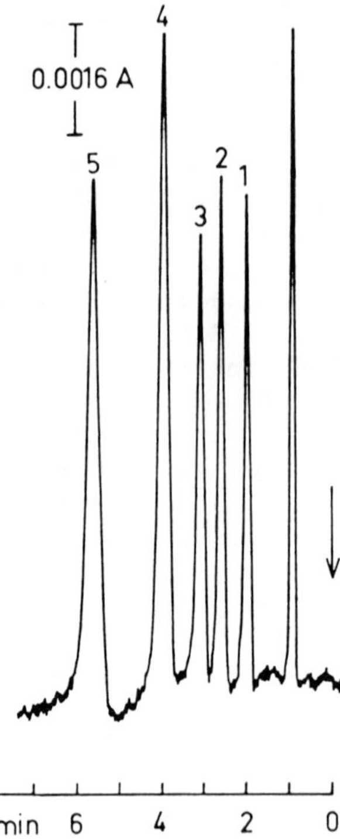


Fig. 3 Separation of dipeptides. Mobile phase: chloroform-1-pentanol (9:1). Stationary phase: naphthalene-2-sulfonate, 0.01 M, pH 2.3. Support: LiChrospher SI-100 (10 μm). Samples: 1 = leucylleucine (74 ng); 2 - phenylalanylvaline (104 ng); 3 = valylphenylalanine (106 ng); 4 = leucylvaline (170 ng); 5 = methionylvaline (214 ng). (From Ref. 33.)

ful thermostatting is necessary as is the choice of eluents with low background extraction of counterion. At equilibrium the content of stationary aqueous phase was measured to 1.0 ml/g of support, which corresponds to a porosity of 0.38, implying that the pores are completely filled with stationary phase. The correlation was good between found and calculated capacity factors, the most hydrophobic amines being an exception. The authors concluded that for these compounds ion pair distribution was not the predominant retention factor. Good efficiency and peak symmetry were as a rule obtained and the systems were reported to be quite stable. Crommen [34] showed that liquid-liquid distribution has a dominating influence on the retention at maximum loading of stationary phase and that adsorption effects increase at lower loading. He also demonstrated [35] that lower surface area silica was advantageous for the separation of quaternary ammonium ions as ion pairs with naphthalene sulfonate.

LiChrosorb DIOL was introduced as solid phase for a stationary phase containing dimethylprotriptyline as counterion [36]. This tricyclic quaternary ammonium ion forms detectable ion pairs with carboxylates and alkyl sulfates. The conditions used with very low loading of stationary phase and undersaturated (90%) mobile phases gave retention characteristics and response factors, indicating that adsorption of the ion pairs to the solid phase largely governed the retention. The border between one mode of distribution to the other is not very distinct. However, liquid-liquid ion pair systems obviously require a high degree of loading of stationary aqueous phase on a solid support with low interaction and a mobile phase completely saturated with the stationary phase [2].

## REVERSED PHASE CHROMATOGRAPHY

Ion pair adsorption to a lipophilic solid phase is by far the most common mode in reversed and normal phase liquid chromatography of ionized compounds. Only a limited number of separation systems employing ion pair liquid-liquid chromatography on bonded phases have appeared.

### Liquid Adsorption

#### *Ion Pair Retention*

Different retention models have been proposed during the last 10 years. Generally the nature of the alkyl bonded phase is of minor importance for the retention mechanism compared with the effect of the ion-pairing agents. Apart from the ion pair adsorption model mentioned above [6], "dynamic ion exchange" [37] or "solvent-generated dynamic ion exchange" [38], "hetaeric chromatography" [39], "ion interaction" [40], "dynamic complex exchange" [41], and "ion exchange desolvation" [42,43] have been used. Horvath and coworkers [44,45] used the solvophobic theory to support the hypothesis of the formation of a stoichiometric complex between the ionic components in the mobile phase and binding of the neutral complex to the stationary phase. Retention is in most instances explained by ion exchange mechanisms, i.e., only the ion-pairing agent bound to the stationary phase will interact with the ionic solutes. This is accomplished by the following events: the counterion binds strongly to the stationary phase, added salt decreases retention of solute and increases binding of the counterion to the stationary phase, and organic solvents such as propanol decrease counterion binding as well as the retention of the solutes. In the model of dynamic complex exchange [41] the most important assumptions are that counterions in the mobile phase form ion pairs with the ionic solute which migrate to the surface where these solutes are exchanged between the counterion of the ion pair and the counterions covering the solid stationary phase.

The ion exchange desolvation theory [42,43] suggests that the counterion is adsorbed from an eluent to an extent governed by its hydrophobicity. The oppositely charged solute interacts electrostatically with the adsorbed counterion. Retention is largely a result of desolvation of counterion and neutralization of coulombic charges. The degree of occupation of the surface by adsorbed counterion determines the surface area available for desolvation of the solute.

In the ion interaction mode [40,46] a layer of lipophilic ions is adsorbed onto the nonpolar surface. A primary ion layer and an oppositely charged coion layer are formed on the surface in this electrical double-layer model. If the counterion concentration in the mobile phase is increased, the amount of adsorbed counterion and charge on the surface is also increased. Transfer of solute ions through the double layer is a function of electrostatic and van der Waals forces. The retention results from this coulombic attraction and from an additional lipophilic "sorption" onto the nonpolar surface. A pair of ions, not necessarily an ion pair, has been adsorbed onto the stationary phase.

A similar theory was proposed by Cantwell [47], who attributes the retention to ion exchange of the solute for a coion of the same charge in the diffuse part of the electrical double layer combined with adsorption of the solute on the surface. Smith [48], Rudzinski [49], and coworkers used the same approach and compared alkyl-bonded silica and polystyrene-divinylbenzene (PRP-1). Hydrophobic binding dominates for both materials but the latter one has no residual silanol groups for cationic exchange.

Knox and Laird [50] demonstrated a nearly parabolic dependence of retention on counterion concentration which was referred to formation of micelles into which the solute can be distributed. Another explanation for the decrease in k' at higher counterion concentration includes the decreasing hydrophobic surface available for desolvation in the ion exchange process, but also the simultaneously increasing coion concentration [42,43]. Under certain experimental conditions one retention model is more plausible but in other cases it is hard to judge how to best characterize the effects in the stationary phase.

### *Counterion in the Stationary Phase*

The extent to which the counterion is adsorbed onto the stationary phase plays a key role in most retention models. The shape of the adsorption isotherm provides information on the nature of the adsorption mechanism. According to Giles and coworkers [51], there are four main classes of isotherms (Fig. 4), characterized by their initial slope, the L type (Langmuir) and H type being frequently recognized in the adsorption of counterions to alkyl-bonded silica. The H-type isotherm can be considered as a special case of the L-type, the solute-substrate affinity being very high as for alkyl sulfonates, sulfates, or tetra-alkylammonium counterions with increasing chain length [42].

Knox and Hartwick [52] calculated the maximum attainable surface concentration of alkyl sulfates on ODS-Hypersil and found that the surface coverages approached those of the chemically alkyl-bonded phases (Fig. 5). They also found that to a first approximation retention was a linear function of the charge density on the surface of the packing material, which arises from adsorption of the counterion, and that it was more or less independent of the chain length of the counterion at any surface concentration (octyl, decyl, and dodecyl sulfate).

The addition of an organic modifier, such as methanol or acetonitrile, to the eluent influences the adsorption of the counterion to the stationary phase. Hung and Taylor [43] reported that the shape of the isotherms was drastically altered, from H type with pure water via L type with 30% acetonitrile to almost C type with 60% acetonitrile.

Tilly-Melin and coworkers [5,6] examined the adsorption of tetra-alkylammonium ions as ion pairs onto LiChrosorb RP-18 and μBondapak $C_{18}$ and found that the data fitted well with the Langmuir type of

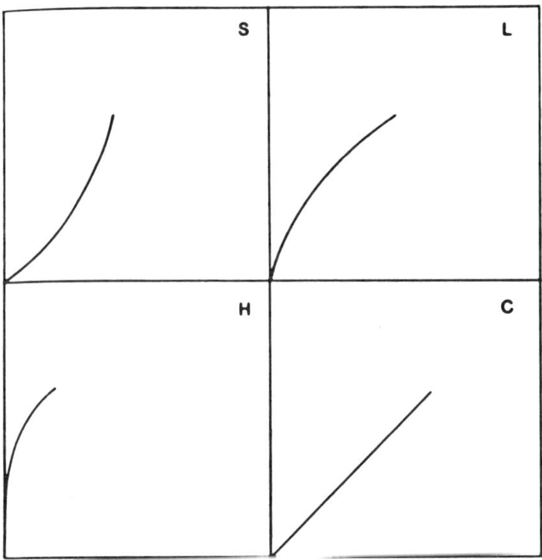


Fig. 4 Shapes of adsorption isotherms according to the Giles classification. (From Ref. 51.)

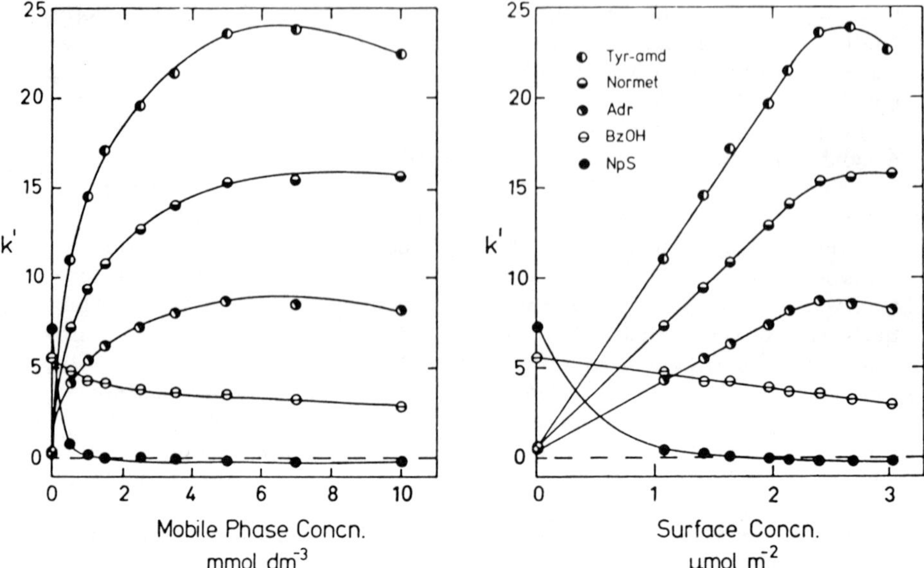


Fig. 5 Dependence of k' for various solutes upon concentrations of decyl sulfate in standard eluent. Left: k' vs. $C_m$; right: k' vs. $C_s$. Tyr-amd = tyrosine amide, Normet = normetadrenaline, Adr = adrenaline, BzOH = benzyl alcohol, NpS = naphthalene-2-sulfonate. Pairing ion: decyl sulfate. (From Ref. 52.)

equation. Because of a slight but significant decrease of k' for uncharged acids at increasing concentration of tetrabutylammonium, they postulated the existence of two kinds of adsorption sites with different tendencies to bind the ion pairs [5]. Using tertiary amines with small amino substituents as solutes, a change in pH of the eluent from 3 to 6 affected the binding of the ion pair to one of the sites only moderately, while the binding to the other one increased substantially. They concluded that the other site had weakly acidic properties and was not easily accessible to cations with bulky substituents such as tetrabutylammonium, which was only slightly affected. It seems likely that nonderivatized silanol groups on the solid phase are involved in the adsorption. Jansson and coworkers [53] and Sokolowski and Wahlund [54] used the same theory of two binding sites accounting for effects observed with dimethyloctylamine and tetraalkylammonium ion pairs (Fig. 6).

Bartha and Vigh [55] studied adsorption isotherms of symmetrical tetraalkylammonium ion pairs on LiChrosorb RP-18 at constant ionic strength and found that the data fitted poorly to the Langmuir or

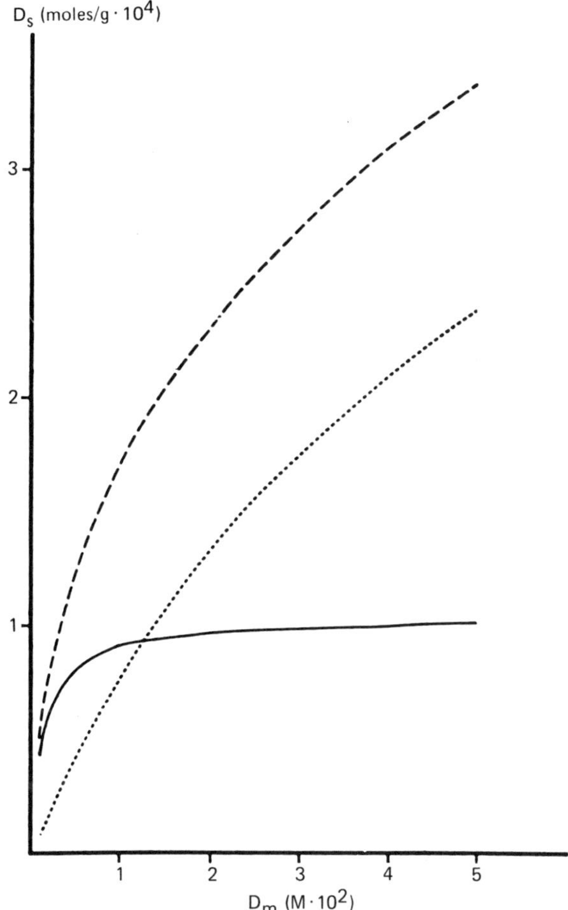


Fig. 6 Adsorption isotherms for dimethyloctylamine (DMOA) calculated from estimated constants. Mobile phase: DMOA, 0.0345 M 1-pentanol and 0.05 M KBr in aqueous phosphate buffer (pH 2.2). (----) Total adsorption of DMOA on the solid phase; (••••) adsorption of DMOA on site $A^x$; (——) adsorption of DMOA on site A. (From Ref. 53.)

Freundlich isotherms at higher mobile phase concentrations of counter-ion.

Ion pair systems for the reversed phase separation of ammonium compounds seem rather straightforward. However, with respect to peak shape the chromatographic performance is in many instances not as good as anticipated [46,54,56-58]. The problem was referred to free nonreacted silanol groups and efforts have been made by "endcapping" to produce bonded phase material with minimal residual silanols. In recent times special columns for separation of amines have been introduced but, even though progress has been made, precautions are still required.

Sokolowski and Wahlund [54] and Persson and coworkers [58] studied the retention behavior of amines on a couple of different ODS-packing materials and found large variations in asymmetry factor and column efficiency. All materials improved substantially after addition of suitable modifiers, which is a means to control minor variations in the properties of the packing material and make separations reproducible from one batch to the other (Fig. 7).

Efficient separation of ammonium compounds on bonded phases by ion pair chromatography most often requires the presence of a modifier in the aqueous mobile phase [46,54,56-61]. Neutral modifiers such as acetonitrile, butanol, or other organic solvents are tools to regulate retention but are in general not able to improve peak shape. Ionic ammonium modifiers are then preferred, and di- and trimethylsubstituted amines and ammonium ions such as *N,N*-dimethyloctylamine and *N,N,N*-trimethyloctylammonium have been found most efficient [54,56-58]. The molecular geometry of the masking agent is of great importance as discussed by Bij and coworkers [62] and Sokolowski and Wahlund [54] (Fig. 8). The structural demands on the amine modifier are to some extent dependent on the properties of the solutes and the

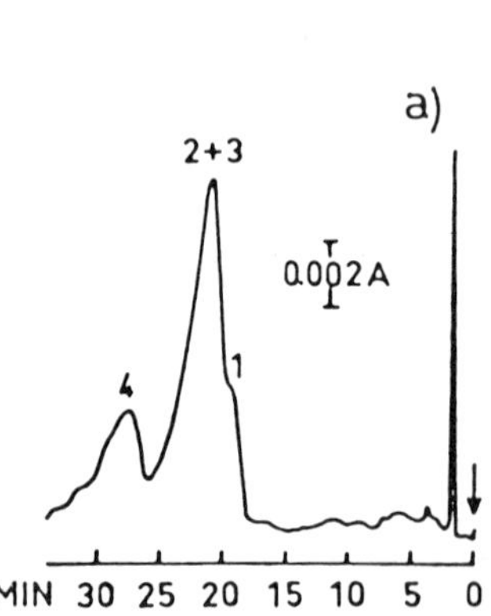


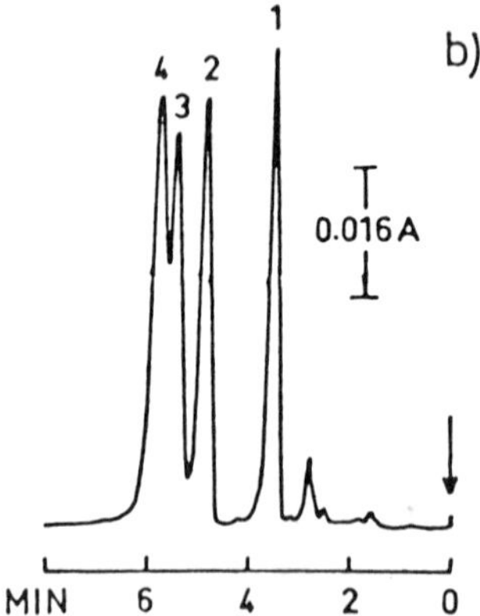


Fig. 7 Comparison of peak shapes in the absence and presence of DMOA in the eluent. Solid phase: ODS-Hypersil, 5 μm. Peaks: 1 = *N*-methylimipramine; 2 = imipramine; 3 = desipramine; 4 = trimipramine. (a) Eluent: 1:1 methanol-phosphate buffer (pH 2.84). (b) Eluent: DMOA 0.05 M in 1:1 methanol-phosphate buffer (pH 2.84). (From Ref. 54.)

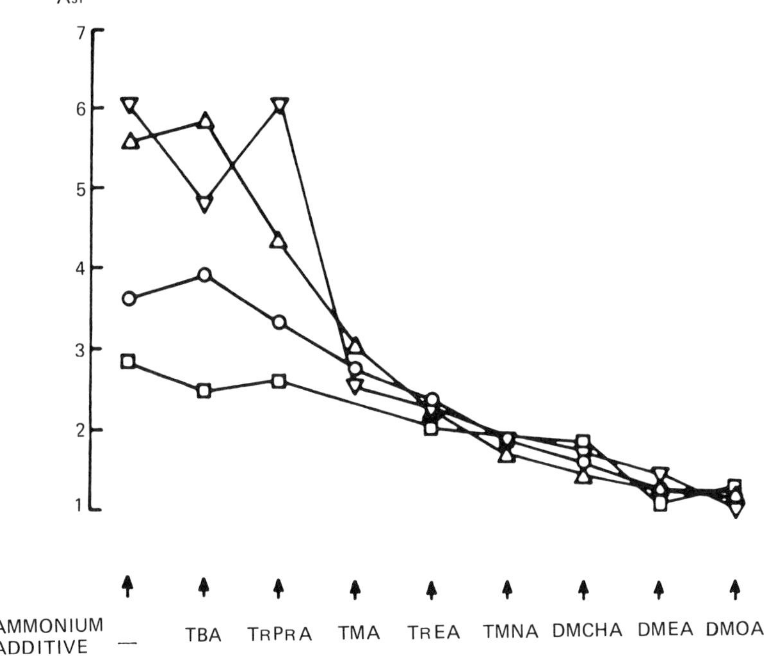


Fig. 8 Peak asymmetry of ammonium compounds in the presence of different ammonium additives in the eluent. Additives: TBA = tetrabutylammonium; TrPrA = tripropylammonium; TMA = tetramethylammonium; TrEA = triethylammonium; TMNA = trimethylnonylammonium; DMCHA = dimethylcyclohexylammonium; DMEA = dimethylethylammonium; DMOA = dimethyloctylammonium. Eluent: 0.05 M additive in 1:1 methanol-phosphate buffer pH 2.0-3.3 (exceptions: TMA and TMNA 0.03 M). Solid phase: LiChrosorb RP-8, 5 μm. Samples: (▽) imipramine; (○) desipramine; (□) propranolol. (From Ref. 54.)

bonded phase packing used. Gill et al. [59] and Persson et al. [58] found that more hydrophobic solutes showed more tailing. The asymmetry factor generally decreased as the hydrophobic character of the amine additive increased [54,59]. Asmus and Freed [63] showed that hydrophilic 2-phenylethylamines such as catecholamines gave good peak shapes on ODS-silica with eluents containing only inorganic buffers. Hung and coworkers [61] found with mobile phases containing dodecyl sulfate as counterion that amine modifiers did not improve column efficiency and peak shape for tricyclic amines but only decreased their retention. This can probably be explained by dodecyl sulfate having already neutralized active sites on the packing material. Residual silanol groups may contribute to the presence of adsorption sites of different activity on the surface of the bonded phase. Some of the sites with high activity are obviously neutralized by an amine modifier having high affinity for those sites in particular [53,58-60,62,64]. Secondary and tertiary amine modifiers exert greater hydrogen-bonding characteristics than primary and quaternary amines and may have greater hydrogen-bonding interaction at the site [60]. Besides the influence on chromatographic performance, increased concentration of amine modifier will decrease retention of the amine solutes by the competition for binding sites on the solid phase. The adsorption of amine modifier onto the bonded phase will increase with increased hydrophobicity and concentration of the counterion. The isotherm will approach a maximum level, where the retention of the solutes will correlate with an ion exchange mechanism [65]. Melander and Horvath [66] and Sorel and Hulshoff [67] thoroughly discussed the retention mechanism on bonded phase chromatography in recent reviews.

Bonded phase systems with mobile phases containing both ion-pairing agents, neutral and ionic modifiers will offer a number of possibilities to regulate the retention characteristics of the stationary phase. In this connection it may be appropriate to note that the composition of the sample injected may influence the chromatographic performance. This is not specific for ion pair chromatography but distortion of peaks or ghost peaks [68,69] may appear whenever the sample solution is not compatible with the mobile phase. The disturbances are caused by a migrating zone with a deviating concentration of one of the mobile phase components. However, such effects can be utilized in the so-called UV visualization liquid chromatography, where the mobile phase contains a UV-absorbing component, the UV trace of which is monitored [70,71]. Nilsson and Westerlund [72] explored such phenomena in order to obtain very narrow, compressed peaks of selected compounds and improve the detection limit and selectivity. The amine solutes were injected together with a high concentration of an organic sulfate or sulfonate. The acidic mobile phase contained a tertiary amine as a cationic modifier. The injection of the organic anion gave rise to a migrating zone with a deficiency of this cationic mobile phase component. A sufficiently high concentration of the injected anion gave a totally depleted zone and an extremely compressed peak was observed for the analyte coeluting with this zone (Fig. 9).

Kraak and Huber [73] early combined silica as solid phase and aqueous mobile phases for separation of sulfonic and carboxylic acids as ion pairs. Tri-*n*-octylamine was coated onto the silica and aqueous perchloric acid was the eluent, with the low solubility of the amine perchlorate making the system quite stable. Later on silica was rediscovered as solid phase for ion pair separations with aqueous eluents, primarily as an alternative to alkyl bonded phases devoid the problems of low stability and reproducibility. Crommen [74] found that solute

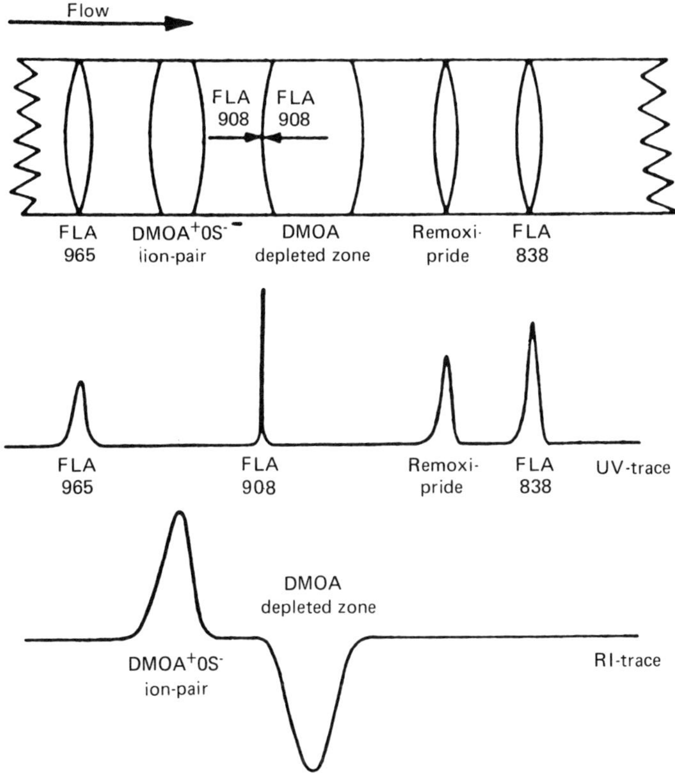


Fig. 9 Schematic representation of the peak compression effect. (From Ref. 72.)

retention was related to the surface area of the silica, the retention order being similar to that of alkyl bonded phases. An increasing degree of substitution of the nitrogen atom gave increased retention for ammonium compounds, quaternary ammonium ions being the most strongly retained. Solute retention was unchanged between pH 2 and 5 but increased for cationic compounds at above pH 5 due to increasing ionization of the silanol groups on the silica surface. Svendsen and Greibrokk [75] used a similar separation system but for the hydrophilic amines studied they found no ion pair effect by the counterion used. Bidlingmeyer and coworkers [76] suggested that retention of organic amines on silica using reversed phase eluents is dependent on electrostatic and adsorption forces. A decreased degree of bonding of the silica improved the chromatographic behavior of lipophilic amines significantly, nonmodified adsorbents being preferable. The interesting conclusion was that bad peak symmetry for amines in bonded phase chromatography is not caused by residual silanol groups, provided those are freely accessible at the surface. Hansen et al. [77-81] used dynamically modified silica as stationary phase for separations generally performed on bonded phases (Fig. 10). Cetyltrimethylammonium (CTMA) has a strong affinity to ionized silanol groups and is adsorbed onto bare silica from an aqueous mobile phase at pH 7-8. With a CTMA concentration of 2.5 mM and an eluent containing 50% of methanol, 0.5 mmol of CTMA per gram of silica is adsorbed constituting a monolayer of surfactant. Ion exchange plays a minor role as retention mechanism but the stationary CTMA-coated phase behaves as a regular bonded phase. Cations and anions are retained by reversed phase mechanism, anions as ion pairs with CTMA [77-79]. With the mobile phase used about 10 mg of silica per liter is dissolved, which is why a guard column has to be inserted [75]. Only long-chain

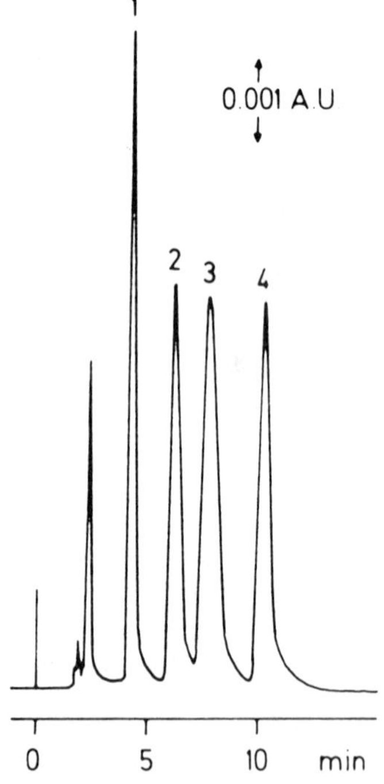


**Fig. 10** Separation of aliphatic amines and a quaternary ammonium compound. Mobile phase: methanol-water-0.2 M potassium phosphate (pH 7.7) (60:15:25) containing 0.20 g/liter of CTMA. Peaks: 1 = didesmethylimipramine; 2 = desmethylimipramine; 3 = imipramine; 4 = *N*-methylimipramine ion. (From Ref. 77.)

quaternary ammonium ions were found to adsorb onto silica in appreciable amounts, and two symmetrical compounds, tetrabutyl- and tetrapentylammonium, only to a minor extent [79]. These separation systems with CTMA seem durable and very reproducible as was the aim with the approach taken.

Dynamically coated silica was also used by Jansson and coworkers [58,82], however, with aqueous mobile phases of pH about 2. They used a hydrophobic ammonium modifier, dimethyl- or trimethyloctylammonium, to improve the chromatographic behavior of the amine solutes. By introducing a hydrophobic counterion, an octyl sulfate, in the mobile phase the ammonium modifier is adsorbed as ion pair onto the silica. At high content of lipophilic ammonium modifier and counterion there is a drastically increased adsorption of these species as ion pairs on the solid phase, causing a change in the retention properties of the stationary phase. A retention pattern similar to that of alkyl bonded phases is obtained while at low concentration other effects, such as hydrogen-bonding influence, change the retention order. The maximum amount of adsorbed amine modifier was estimated at about 1 mmol/g of silica, while the monolayer capacity was about 0.5 mmol/g. Generally, the systems employing silica adsorbent and aqueous mobile phases have demonstrated sharp chromatographic peaks and good stability. However, they may be a complement rather than an alternative to alkyl bonded phases.

## Liquid-Liquid Chromatography

Ion pair liquid-liquid chromatography in reversed phase mode has been utilized only to a small extent. The major reason for this is the access of hydrophobic packing materials for aqueous mobile phases. By immobilizing the lipophilic moiety as a bonded phase, precautions of a technical nature in liquid-liquid systems can be avoided. However, it must be emphasized that the bonded phase cannot be regarded as a liquid phase attached onto the surface.

Wahlund made the pioneering work with liquid-liquid systems. In the first paper [83], he and Gröningsson used silicon-treated cellulose as support coated with 50% (v/w) pentanol and hexanol as liquid stationary phases for the separation of hydrophobic amines as ion pairs with inorganic anions. A good correlation was reported between calculated and found retention volumes. A low-efficiency system but with dioctylphosphate in chloroform as stationary phase was used by Eksborg and coworkers [84] for ion pair separation of aminophenols.

Wahlund and coworkers [85] showed ion pair separations of different acidic compounds with pentanol as stationary phase on LiChrosorb RP-2 as the support. The pentanol was either applied by in situ technique [86] or by dynamic coating [87,88], with both methods giving about the same degree of loading. Efforts to increase the content of stationary phase gave low efficiency [86], while 0.5 ml of pentanol per g of support obtained by spontaneous coating gave good stability. Rapid recoating could be obtained owing to the high content of pentanol (2.5%) in the saturated mobile aqueous phase. It is emphasized that careful thermostatting is necessary for good stability but precolumn is not required. Tetrabutylammonium was present as counterion in the mobile aqueous phase for the separation of carboxylates, sulfonates, sulfonamides, and barbiturates, butyronitrile being tested as well as liquid stationary phase [87]. $C_2$, $C_8$, and $C_{18}$ packings were compared as support for pentanol, which was dynamically coated [4]. About the same extent of coating, 0.6 ml/g, was obtained, although saturation was much more rapidly reached for the $C_2$ phase than for the other two. It was estimated that the pores of RP-2 were filled to a degree of 70% with pentanol, RP-8 and RP-18 to 80%. For RP-8 5 mol of pentanol corresponded to each mol of octylsilane.

It was shown that the quotient of volume stationary phase over mobile phase ($V_s/V_m$) increased linearly during the coating process approaching a constant level. When the relative saturation of the mobile phase was lowered slightly, to 95%, a strong decrease in $V_s/V_m$ was obtained [4] which was continued for 75% saturation. From these studies Wahlund and Beijersten concluded that the retention of acids as ion pairs was mainly due to distribution to the stationary pentanol phase, the adsorption to the support being negligible. In contrast, the support was strongly interacting with the hydrophobic acids but not with hydrophilic ones when distributed in acidic form. The retention mechanism for this kind of system was discussed thoroughly in a following paper [88] where the adsorption isotherm for pentanol was evaluated. A model was suggested implying that the retention was a combination of adsorption to the solid phase and partition to the bulk phase of pentanol.

Systems with pentanol as stationary phase were also used for ion pair separation of hydrophobic amines. Wahlund and Sokolowski [54, 56] tried to explain the strong tailing of the peaks by ion pair dissociation effects in the pentanol phase but finally stated that the disturbances were due to an additional retention mechanism, interaction with the bonded phase support. They introduced long-chain amines and quaternary ammonium ions as additives in the mobile phase, which substantially improved peak symmetry and column efficiency by modifying the surface of the support.

A strong protonacceptor, tri-*n*-butylphosphate, has been used as stationary liquid phase on a hydrophobic support [89-92], forming a system with good efficiency and stability and with interesting selectivity (Fig. 11). The stationary phase is applied in situ by injection of 20-µl aliquots of TBP until overloading while mobile phase saturated with TBP is pumped through. Supersaturation of the system may occur but can be avoided if certain precautions are taken [91]. After

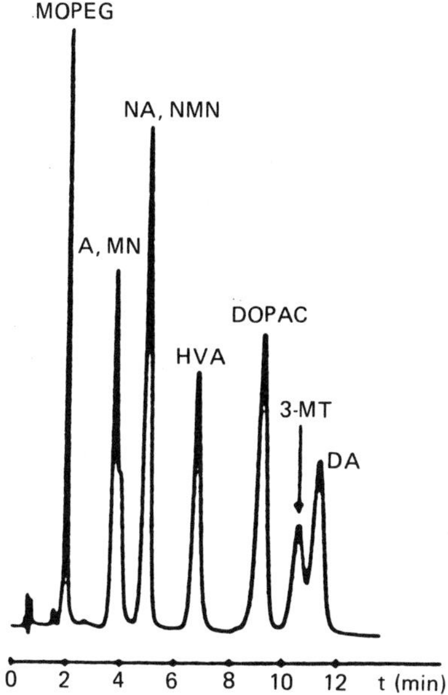


Fig. 11 Separation of the nine compounds under optimized conditions with respect to the perchlorate concentration (0.20 M) and the pH (4.9) of the mobile phase. Stationary phase: tri-*n*-butyl phosphate coated on Nucleosil $C_8$, 5 μm. (From Ref. 92.)

loading with stationary phase the pores of the packing material are almost completely filled with TBP according to measurements made [89]. LiChrosorb RP8, Polygosil $C_{18}$, and Nucleosil $C_8$ have been used as solid phases. TBP is considered to form adducts with the ion pairs and also has strong solvating ability for perchloric acid present in the mobile aqueous phase. The perchlorate concentration and addition of methanol to the mobile phase were used to regulate retention for separation of different biogenic amines and their metabolites.

## REFERENCES

1. G. Schill, *Ion Exchange and Solvent Extraction*, Vol. 6, (J. A. Marinsky and Y. Marcus, eds.), Marcel Dekker, New York, 1974, p. 1.
2. B. Fransson, K-G. Wahlund, I. M. Johansson, and G. Schill, *J. Chromatogr.*, *125*:327 (1976).
3. S. Eksborg and G. Schill, *Anal. Chem.*, *45*:2092 (1973).
4. K-G. Wahlund and I. Beijersten, *J. Chromatogr.*, *149*:313 (1978).

5. A. Tilly-Melin, Y. Askemark, K-G. Wahlund, and G. Schill, *Anal. Chem.*, *51*:976 (1979).
6. A. Tilly-Melin, M. Ljungcrantz, and G. Schill, *J. Chromatogr.*, *185*:225 (1979).
7. P-O. Lagerström and B-A. Persson, *J. Chromatogr.*, *149*:331 (1978).
8. B. Mellström and R. Braithwaite, *J. Chromatogr.*, *157*:379 (1978).
9. B-M. Eriksson, B-A. Persson, and M. Lindberg, *J. Chromatogr.*, *185*:575 (1979).
10. R. J. Flanagan, G. C. A. Storey, and D. W. Holt, *J. Chromatogr.*, *187*:391 (1980).
11. R. J. Flanagan, G. C. A. Storey, R. K. Bhamra, and I. Jane. *J. Chromatogr.*, *247*:15 (1982).
12. P. Haefelfinger, *J. Chromatogr. Sci.*, *17*:345 (1979).
13. J. E. Greving, H. Bouman, J. H. G. Jonkman, H. G. M. Westenberg, and R. A. de Zeeuw, *J. Chromatogr.*, *186*:683 (1979).
14. C. Pettersson and G. Schill, *J. Chromatogr.*, *204*:179 (1981).
15. C. Pettersson and G. Schill, *Chromatographia*, *16*:192 (1982).
16. C. Pettersson and K. No, *J. Chromatogr.*, *282*:671 (1983).
17. C. Pettersson, *J. Chromatogr.*, *316*:553 (1984).
18. G. Schill, R. Modin, and B-A. Persson, *Acta Pharm. Suecica*, *2*:119 (1965).
19. B-A. Persson, *Acta Pharm. Suecica*, *5*:343 (1968).
20. S. Eksborg and B-A. Persson, *Acta Pharm. Suecica*, *8*:205 (1971).
21. B. Ulin, K. Gustavii, and B-A. Persson, *J. Pharm. Pharmacol.*, *28*:672 (1976).
22. S. Eksborg and B-A. Persson, *Acta Pharm. Suecica*, *8*:605 (1971).
23. B-A. Persson and B. L. Karger, *J. Chromatogr. Sci.*, *12*:521 (1974).
24. B. L. Karger, S. C. Su, S. Marchese, and B-A. Persson, *J. Chromatogr. Sci.*, *12*:678 (1974).
25. S. C.Su, A. V. Hartkopf, and B. L. Karger, *J. Chromatogr.*, *119*:523 (1976).
26. J. H. Knox and J. Jurand, *J. Chromatogr.*, *103*:311 (1975).
27. B-A. Persson and P-O. Lagerström, *J. Chromatogr.*, *122*:305 (1976).
28. D. Westerlund, L. B. Nilsson, and Y. Jaksch, *J. Liq. Chromatogr.*, *2*:373 (1979).
29. D. Westerlund, L. B. Nilsson, and Y. Jaksch, *J. Chromatogr.*, *211*:181 (1981).
30. I. Grundevik and B-A. Persson, *J. Liq. Chromatogr.*, *5*:141 (1982).
31. P-O. Lagerström, *Acta Pharm. Suecica*, *13*:213 (1976).
32. W. Santi, J. M. Huen, and R. W. Frei, *J. Chromatogr.*, *115*:423 (1975).
33. J. Crommen, B. Fransson, and G. Schill, *J. Chromatogr.*, *142*:283 (1977).

34. J. Crommen, *Acta Pharm. Suecica, 16*:111 (1979).
35. J. Crommen, *J. Chromatogr., 193*:225 (1980).
36. L. Hackzell, M. Denkert, and G. Schill, *Acta Pharm. Suecica, 18*:271 (1981).
37. C. P. Terweij-Groen, S. Heemstra, and J. C. Kraak, *J. Chromatogr., 161*:69 (1978).
38. J. C. Kraak, K. M. Jonker, and J. F. K. Huber, *J. Chromatogr., 142*:671 (1977).
39. Cs. Horvath, W. R. Melander, and I. Molnar, *J. Chromatogr., 125*:129 (1976).
40. B. A. Bidlingmeyer, S. N. Deming, W. P. Price, Jr., B. Sachok, and M. Petrusek, *J. Chromatogr., 186*:419 (1979).
41. W. R. Melander and Cs. Horvath, *J. Chromatogr., 201*:211 (1980).
42. C. T. Hung and R. B. Taylor, *J. Chromatogr., 202*:333 (1980).
43. C. T. Hung and R. B. Taylor, *J. Chromatogr., 209*:175 (1981).
44. Cs. Horvath, W. R. Melander, I. Molnar, and P. Molnar, *Anal. Chem., 49*:2295 (1977).
45. W. R. Melander, K. Kalghatgi, and Cs. Horvath, *J. Chromatogr., 201*:201 (1980).
46. B. A. Bidlingmeyer, *J. Chromatogr. Sci., 18*:525 (1980).
47. F. F. Cantwell, *J. Pharm. Biomed. Anal., 2*:153 (1984).
48. R. L. Smith, Z. Iskandarani, and D. J. Pietrzyk, *J. Liq. Chromatogr., 7*:1935 (1984).
49. W. E. Rudzinski, D. Bennett, and W. Garcia, *J. Liq. Chromatogr., 5*:1295 (1982).
50. J. H. Knox and G. R. Laird, *J. Chromatogr., 122*:17 (1976).
51. C. H. Giles, D. Smith, and A. Huitson, *J. Colloid Interface Sci., 47*:755 (1974).
52. J. H. Knox and R. A. Hartwick, *J. Chromatogr., 204*:3 (1981).
53. S-O. Jansson, I. Andersson, and B-A. Persson, *J. Chromatogr., 203*:93 (1981).
54. A. Sokolowski and K-G. Wahlund, *J. Chromatogr., 189*:299 (1980).
55. A. Bartha and Gy. Vigh, *J. Chromatogr., 260*:337 (1983).
56. K-G. Wahlund and A. Sokolowski, *J. Chromatogr., 151*:299 (1978).
57. W. R. Melander, J. Stoveken, and Cs. Horvath, *J. Chromatogr., 199*:35 (1980).
58. B-A. Persson, S-O. Jansson, M-L. Johansson, and P-O. Lagerström, *J. Chromatogr., 316*:291 (1984).
59. R. Gill, S. P. Alexander, and A. C. Moffat, *J. Chromatogr., 247*:39 (1982).
60. J. S. Kiel, S. L. Morgan, and R. K. Abramson, *J. Chromatogr., 320*:313 (1985).
61. C. T. Hung, R. B. Taylor, and N. Paterson, *J. Chromatogr., 240*:61 (1982).
62. K. E. Bij, Cs. Horvath, W. R. Melander, and A. Nahum, *J. Chromatogr., 203*:65 (1981).

63. P. A. Asmus and C. R. Freed, *J. Chromatogr.*, *169*:303 (1979).
64. A. P. Goldberg, E. Nowakowska, P. E. Antle, and L. R. Snyder, *J. Chromatogr.*, *316*:241 (1984).
65. S-O. Jansson, *J. Liq. Chromatogr.*, *5*:677 (1982).
66. W. R. Melander and Cs. Horvath, *Ion-Pair Chromatography* (M. T. W. Hearn, ed.), Marcel Dekker, New York, 1985, p. 27.
67. R. H. A. Sorel and A. Hulshoff, *Adv. Chromatogr.*, *21*:87 (1983).
68. D. Berek, T. Bleha, and Z. Pevna, *J. Chromatogr. Sci.*, *14*:560 (1976).
69. J. J. Stranahan and S. N. Deming, *Anal. Chem.*, *54*:1540 (1982).
70. M. Denkert, L. Hackzell, G. Schill, and E. Sjögren, *J. Chromatogr.*, *218*:31 (1981).
71. F. V. Warren, Jr. and B. A. Bidlingmeyer, *Anal. Chem.*, *56*:487 (1984).
72. L. B. Nilsson and D. Westerlund, *Anal. Chem.*, *57*:1835 (1985).
73. J. C. Kraak and J. F. K. Huber, *J. Chromatogr.*, *102*:333 (1974).
74. J. Crommen, *J. Chromatogr.*, *186*:705 (1979).
75. H. Svendsen and T. Greibrokk, *J. Chromatogr.*, *212*:153 (1981).
76. B. A. Bidlingmeyer, J. K. del Rios, and J. Korpi, *Anal. Chem.*, *54*:442 (1982).
77. S. H. Hansen, *J. Chromatogr.*, *209*:203 (1981).
78. S. H. Hansen, P. Helboe, M. Thomsen, and U. Lund, *J. Chromatogr.*, *210*:453 (1981).
79. S. H. Hansen, P. Helboe, and U. Lund, *J. Chromatogr.*, *240*:319 (1982).
80. S. H. Hansen, P. Helboe, and U. Lund, *J. Chromatogr.*, *270*:77 (1983).
81. S. H. Hansen and P. Helboe, *J. Chromatogr.*, *285*:53 (1984).
82. S-O. Jansson, I. Andersson, and M-L. Johansson, *J. Chromatogr.*, *245*:45 (1982).
83. K-G. Wahlund and K. Gröningsson, *Acta Pharm. Suecica*, *7*:615 (1970).
84. S. Eksborg, P-O. Lagerström, R. Modin, and G. Schill, *J. Chromatogr.*, *83*:99 (1973).
85. K-G. Wahlund, *J. Chromatogr.*, *115*:411 (1975).
86. K-G. Wahlund and U. Lund, *J. Chromatogr.*, *122*:269 (1976).
87. I. M. Johansson and K-G. Wahlund, *Acta Pharm. Suecica*, *14*:459 (1977).
88. K-G. Wahlund and I. Beijersten, *Anal. Chem.*, *54*:128 (1982).
89. H. J. L. Janssen, U. R. Tjaden, H. J. de Jong, and K-G. Wahlund, *J. Chromatogr.*, *202*:223 (1980).
90. J. de Jong, U. R. Tjaden, W. van't Hof, and C. F. M. van Valkenburg, *J. Chromatogr.*, *282*:443 (1983).

91. J. de Jong, J. P. Schouten, R. G. Muusze, and U. R. Tjaden, *J. Chromatogr.*, *319*:23 (1985).
92. J. de Jong, C. F. M. van Valkenburg, and U. R. Tjaden, *J. Chromatogr.*, *322*:43 (1985).

# 12

# Packings in Affinity Chromatography

Jan-Christer Janson and Tore Kristiansen / *Pharmacy LKB Biotechnology AB, Uppsala, Sweden*

## INTRODUCTION

Originally, the term *affinity chromatography* was used to describe adsorption phenomena based on biological pair formations of the type listed in Table 1 [1]. Today, however, the term is often used as a collective description of a large family of adsorption chromatography methods, all of which utilize more or less specific interactions between biological macromolecules in solution and covalently attached ligand molecules on a solid phase. In addition to classical biospecific affinity chromatography the methods are hydrophobic interaction chromatography, charge-transfer affinity chromatography, immobilized metal affinity chromatography, dye ligand affinity chromatography, immobilized affinity chromotography, immunoaffinity chromotography (immunosorption), and covalent chromatography (chemisorption).

Modern affinity chromatography was initiated by the now classical paper by Cuatrecasas et al. [1] on the purification of α-chymotrypsin and staphylococcal nuclease, and most of the early work was concerned with affinity adsorbents based on immobilized enzyme inhibitors such as cofactor analogs [2]. During the early years after 1970 the number of affinity chromatography papers grew at an exponential rate that did not abate until around 1975. A factor contributing to this rapid spread of the technique was the simultaneous commercial availability of suitable supports and the development of simple methods for covalent attachment of affinity ligands.

The design of an affinity chromatographic experiment depends on the answers to a number of critical questions. The most important are as follows:

**Table 1** Examples of Biospecific Pair Formations Utilized in Affinity Chromatography

| | |
|---|---|
| Enzyme | Substrate, substrate analog<br>Cofactor, cofactor analog<br>Inhibitor |
| Hormone | Carrier protein<br>Receptor |
| Glycoprotein | Lectin |
| Antibody | Antigen<br>Hapten |
| Nucleic acid | Complementary polynucleotide<br>Polynucleotide-binding protein |

1. Which is the most suitable ligand?
2. Are packings carrying this ligand commercially available?
3. If not, is information on the synthesis of adsorbents based on this ligand known in the literature?
4. Again if not, to what support should the ligand be immobilized to produce an affinity adsorbent for a particular biomolecule?
5. Which chemistry is most appropriate for the covalent attachment of the ligand?
6. The question of whether or not to use a spacer (leash) should be addressed.
7. Which displacement (elution) agents should be investigated?
8. What are the proper operating conditions, e.g., temperature, buffers, flow rate, and column size?

The aim of this chapter is to provide the reader with an up-to-date review of currently available supports, preactivated supports, spacer supports, and ready-to-use affinity chromatography column packings. In addition, there will be brief discussions of the most widely used coupling chemistries, the choice of ligands, the use of spacers, and the problem of leakage. The goal is not to provide a range of experimental methodologies used in this field of chromatography. For such information the reader is referred to a number of excellent monographs and handbooks published over the last few years [3-10].

## CHOICE OF SUPPORT

The final aim of an affinity separation is to isolate a maximum of material of the highest purity in a minimum of time. With respect to the packing, this means finding the optimum combination of binding capacity, specificity, and flow rate.

In response to this goal, a wide variety of supports for affinity chromatography have been described in the literature [3-10], ranging from inorganics like glass and silica, synthetic polymers such as vinyl and acrylate polymers and polyacrylamide, to supports derived from natural sources such as collagen, dextran, cellulose, and agarose. With respect to function, they are all compromises, since an ideal carrier material for all applications of affinity chromatography would have to meet a formidable number of demands, some of which are contradictory both in theory and practical design.

In addition to being hydrophilic, an ideal support should be chemically and physically stable in order to tolerate a wide variation in temperature, pH, and ionic strength, as well as prolonged exposure to chaotropic agents, urea, guanidine hydrochloride, and detergents. Still, it must be reactive enough to permit derivatization by a variety of chemical routes.

Binding capacity is directly related to the surface available for substitution with ligand in a given volume of adsorbent. In order to obtain an acceptable surface-to-volume ratio for particles down to about 5 μm diameter, they must be porous. Pore geometry should be optimized for unrestricted access of proteins to ligands which may themselves be immobilized proteins, e.g., antibodies, enzymes, or lectins. Since most biomolecules are in the range 2-30 nm, materials with peaks of pore size distribution roughly five times larger, i.e., between 10 and 150 nm, will cover the majority of applications.

The matrix should be sufficiently rigid to allow the production of particles down to 10 μm or less and their use at linear flow rates as high as 5 cm $min^{-1}$. A combination of hydrophilicity with chemical and physical inertness is best achieved by design of media giving rise to alcohol hydroxyls as the solvent-contacting phase. This can be obtained either by direct use of polymer matrices containing hydroxyl groups or by subsequent hydroxylation of polymers with an initial lack of alcohol hydroxyls. The first alternative has been the predominant approach for traditional, low-pressure applications with average particle diameters around 100 μm, whereas the second alternative is prevalent for medium- and high-pressure media.

With the firm establishment of affinity chromatography as a major separation technique across the life sciences as well as in commercial biotechnology, it is natural to briefly consider carrier selection in relation to user category. Some factors of importance in selecting a support, either the basic material with or without preactivation, or with a ligand attached, include:

Scale of operation
Concentration of wanted substance in starting material

Required yield of purified material per cycle
Required purity of isolated material
Acceptable cycle time
Service life of packing
Overall process economy

The relative weight of these factors depends on the context of each particular application. We may consider three principal types of experimental setups:

1. Process scale affinity chromatography
2. Intermediate (traditional) laboratory scale affinity chromatography
3. High-performance liquid affinity chromatography (HPLAC)

### Process Scale Affinity Chromatography

In this category the final economic balance of the entire process tends to govern the choice of packing, with the cost/performance ratio being more important than its purely chromatographic qualities. Further, a support which appears unpromising judged by its general characteristics and reputation can sometimes be chemically and physically customized to a particular purification scheme and become the primary choice among several others with inherently superior properties. For instance, modern agarose-based separation media specially adapted to fast, large-scale chromatography, e.g., Sepharose Fast Flow, are greatly superior to cellulose on all counts except price. On the other hand, cellulose is comparatively cheap even when processed into bead form, is in constant supply, and can be efficiently derivatized. Therefore it should not be dismissed offhand, but may still have merit as an affinity packing, e.g., to remove unwanted material in an early stage of purification.

Affinity chromatography has now matured into a firmly established separation technique backed by wide application in large-scale procedures [11]. Accordingly, the major manufacturers of separation equipment have gathered considerable expertise in the handling of specific biomolecules and substance classes, and the design of large-scale separation schemes. Since the choice and tailoring of the affinity packing are governed by the overall economy of the planned operation, the pool of information accumulated by manufacturers may help the user to find the proper balance between all process parameters in a particular application and avoid overemphasizing the purely chromatographic properties of the packing.

### Intermediate Laboratory Scale

During the almost 70 years since Starkenstein observed that amylase binds to insoluble starch (for a historical account, see Ref. 12), a wide range of materials have been introduced and promoted as affinity packings (Tables 2-11). With time and experience, beaded agarose has pulled far ahead of other base materials and is at present the primary alternative when an affinity separation is contemplated. Although not the ideal affinity packings, it is generally recognized as the best compromise so far with respect to mechanical and chemical properties

**Table 2** Commercially Available Preactivated Affinity Chromatography Packings

| Activation method | Matrix | Manufacturer[a] |
|---|---|---|
| *Low-pressure media (dp ∿ 100 μm)* | | |
| CNBr | Sepharose 4B | A |
| CNBr | Sepharose 6MB | A |
| Tresyl[b] | Sepharose 4B | A |
| Bisoxirane | Sepharose 6B | A,H |
| *N*-hydroxysuccinimide | Sepharose 4B | A,H |
| *N*-hydroxysuccinimide | Agarose 6% | B |
| 1,1'-Carbonyldiimidazole | Agarose 6% | D |
| Glutaraldehyde | Ultrogel AcA22 | C |
| 2-Pyridyldisulfide | Sepharose 4B | A |
| Bromoacetyl | Cellulose | M |
| *Medium-pressure media (dp ∿ 50 μm)* | | |
| Oxirane | Eupergite C | J |
| Bisoxirane | Fractogel TSK HW-65F | E |
| 1,1'-carbonyldiimidazole | Fractogel TSK HW-65F | D,E |
| 1,1'-carbonyldiimidazole | Trisacryl GF-2000 | D |
| 2-Fluoro-1-methylpyridinium toluene-4-sulfonate | Fractogel TSK HW-75F | K |
| *High-pressure media (dp ∿ 10 μm)* | | |
| Tresyl[b] | Silica (500 Å) | D |
| Oxirane | Silica (NG) | L |

[a]See Appendix 1 for a list of manufacturers.
[b]2,2,2-Trifluoroetanesulfonylchloride.

**Table 3** Commercially Available Affinity Chromatography Packings Based on Controlled-Pore Glass

| Ligand | Particle diameter (μm) | Pore diameter (Å) | Manufacturer[a] |
|---|---|---|---|
| Aminoaryl | 125-177 | 500 | D |
| Aminopropyl | " | 500 | D |
| Carboxyl (10-Å spacer) | " | 500 | D |
| Long-chain alkylamine | " | 500 | D |
| *N*-hydroxysuccinimide (10-Å spacer) | " | 500 | D |
| Diazonium borofluoride | " | 500 | D |
| Thiol (10-Å spacer) | " | 500 | D |
| Carbonyl diimidazole | 74-125 | 200 | D |

[a]See Appendix 1 for a list of manufacturers.

**Table 4** Commercially Available Ready-to-Use Affinity Chromatography Packings Based on Medium-Pressure Matrices (dp ∿ 50 μm)

| Ligand | Matrix | Manufacturer[a] |
|---|---|---|
| Iminodiacetic acid | Fractogel TSK HW-65F | D |
| Tris(carboxymethyl) ethylenediamine | " | D |
| Butyl | Fractogel TSK HW-75F | E |
| Aminoglyceryl ether | " | E |
| Phenyl borate | " | E |
| Cibacron blue F3G-A | Fractogel TSK HW-65F | D,E[b] |
| | Trisacryl GF-2000 | B |
| Oligo $(dT)_8$ | " | B |
| Protein A | " | D |
| Protein A | Fractogel TSK HW-65F | E |
| Heparin | " | D |
| Gelatin | Fractogel TSK HW-75F | E |

[a]See Appendix 1 for a list of manufacturers.
[b]Red, green, brown, and orange dye adsorbents are also available from this manufacturer.

**Table 5** Widely Used Commercially Available Ready-to-Use Affinity Chromatography Packings Based on Low-Pressure Matrices (dp ∿ 100 μm)

| Ligand | Matrix | Manufacturer[a] |
|---|---|---|
| Cibacron blue F3G-A | Agarose | A, B, C, D, N |
| Heparin | " | A, B, C, D |
| Protein A | " | A, B, C, D |
| Con A | " | A, B, C, M |
| Gelatin | " | A, B, D |
| Procion red HE3B | " | A, B |
| Protein G | " | A |

[a]See Appendix 1 for a list of manufacturers.

**Table 6** Commercially Available Affinity Packings for the Separation and Purification of Nucleic Acids

| Packing | Application | Manufacturer[a] |
|---|---|---|
| Oligo(dT) cellulose | mRNA | A,H |
| Oligo(dA) cellulose | Poly(U) sequences | A |
| Oligo(dC) cellulose | Poly(I) sequences | A |
| Oligo(dG) cellulose | Poly(C) sequences | A |
| Oligo(dI) cellulose | Poly(C) sequences | A |
| Oligo(dT) Trisacryl M | mRNA | C |
| Oligo(dT) Ultrogel A4R | mRNA | C |
| Poly(A)-Sepharose 4B | Poly(A)-binding RNA | A,H |
| Poly(U)-Sepharose 4B | mRNA | A,H |
| DNA affinity adsorbent | (A-T)-specific | O |
| | (G-C)-specific | O |
| | Supercoiled DNA | O |
| Acriflavine-Ultrogel A4R | DNA and RNA fragments | C |

[a]See Appendix 1 for a list of manufacturers.

**Table 7** Commercially Available Agarose-Based Spacer Packings (Non-preactivated) for Use with Carbodiimide Coupling Reagents

| Spacer ligand | Connector | Spacer length (atoms) | Manufacturer[a] |
|---|---|---|---|
| Ethylamine | Isourea | 5 | A |
| Ethylamine | Amide | 6 | B,C |
| Butylamine | Isourea | 7 | A |
| Hexanoic acid | Isourea | 8 | A |
| Hexylamine | Isourea | 9 | A |
| Diaminodipropylamine | Alkylamine | 9 | D |
| Hexylamine | Amide | 10 | C |
| Ethylaminesuccinamide | Amide | 10 | B |
| Hexanoic acid | Alkylamine | 10 | A |
| Hexylamine | Alkylamine | 11 | A |
| Octylamine | Isourea | 11 | A |
| Decylamine | Isourea | 13 | A |
| Polyacrylic hydrazide | (NG) | Polymer | M,H |
| Poly-L-lysine | (NG) | Polymer | M,H |
| Succinylated poly-L-lysine-poly-D,L-alanine | (NG) | Polymer | M |

[a]See Appendix 1 for a list of manufacturers.
*Note*: A range of agarose-based alkylamines with nonspecified connector groups are available from manufacturers H and M.

balanced against availability, convenience in handling, price, and overall performance. A long list of alternative chemistries have been worked out and documented in detail for coupling amino, carboxyl, hydroxyl, and sulfhydryl functions to native agarose in the laboratory [3-10]. Several manufacturers offer preactivated agarose to which ligands can be added and coupled directly (Table 2). As stated in the following brief descriptions of other base materials for affinity packings, there may be more suitable supports than agarose for specific applications. For instance, cellulose remains a useful alternative, particularly in the field of nucleic acid separation, as reviewed in Ref. 8.

**Table 8** Commercially Available Packings for Low- and Medium-Pressure Hydrophobic Interaction Chromatography (HIC)

| Ligand | Connector | Matrix | Manufacturer[a] |
|---|---|---|---|
| Ethyl | Isourea | Sepharose 4B | A |
| Butyl | Isourea | Sepharose 4B | A |
| Butyl | Ether | Fractogel TSK HW-65 | E |
| Hexyl | Isourea | Sepharose 4B | A |
| Phenyl | Ether | Sepharose CL-4B | A |
| Phenyl | Ether | Sepharose Fast Flow | A |
| Octyl | Isourea | Sepharose 4B | A |
| Octyl | Ether | Sepharose CL-4B | A |
| Decyl | Isourea | Sepharose 4B | A |

[a]See Appendix 1 for a list of manufacturers.
*Note*: A wide selection of HIC matrices with nonspecified connectors based on nonspecified agaroses are available from manufacturers H and M, in kits as well as individually.

### *High-Performance Liquid Affinity Chromatography (HPLAC)*

For several years high-performance liquid chromatography (HPLC) has steadily been penetrating the life sciences, propelled by a multitude of specially designed separation media for ion exchange, reversed phase, and size exclusion chromatography of biomolecules. It was therefore natural to search for advantages in adapting affinity methods as well to an HPLC environment in the bioscience laboratory.

Technically, HPLC instrumentation holds an obvious attraction. A typical column affinity separation invovles four stages: adsorption, washing, displacement, and reequilibration. It is a complex procedure to optimize them. A highly reproducible solvent delivery system with extensive programming and documentation facilities is a great help in handling not only dedicated HPLC columns, but also conventional, medium-scale, and user-packed affinity columns or to establish conditions for a large-scale method. However, in the true HPLC mode additional and severe demands are made on affinity packings. Compared to conventional chromatography, an essential advantage of HPLC methodology is the substantial increase in speed. One design principle working toward this goal is to minimize the bead size of the porous packing

**Table 9** Commercially Available Packings Containing Immobilized Lectins

| Lectin | Specificity | Matrix | Manufacturer[a] |
|---|---|---|---|
| Lentil | α-D-mannosyl | Sepharose 4B | A |
| | | Ultrogel AcA22 | C |
| Wheat germ | *N*-acetyl-β-D-glukosaminyl | Sepharose 4B | A |
| | | Sepharose 6MB | A |
| | | Ultrogel AcA22 | C |
| *Helix pomatia* | *N*-acetyl-β-D-galactosaminyl | Sepharose 6MB | A |
| | | Ultrogel AcA22 | C |
| Soybean | *N*-acetyl-D-galactosaminyl | Sepharose 4B | A |
| Peanut | Galactose-β(1-3)-*N*-acetyl-D-galactosaminyl | Sepharose 4B | A |
| | | Ultrogel AcA22 | C |
| Pea | D-glucosyl, D-mannosyl | Sepharose 4B | A |
| Castor bean | D-galactosyl, *N*-acetyl-D-galactosaminyl | Sepharose 4B | A |
| *Bandereia simplicifolia* | β-D-galactosyl | Sepharose 4B | A |

[a]See Appendix 1 for a list of manufacturers.

*Note*: A range of lectin-containing adsorbents based on nonspecified agarose matrices are available from manufacturer H.

**Table 10** Commercially Available Packings for Immobilized Metal Ion Affinity Chromatography (IMAC)

| Ligand | Connector | Matrix | Manufacturer[a] |
|---|---|---|---|
| *Low-pressure packings (dp ∿ 100 μm)* | | | |
| Iminodiacetic acid | Ether | Sepharose 6B | A |
| | | Sepharose Fast Flow | A |
| | | Agarose 4% | D |
| Tris(carboxymethyl) ethylenediamine | Ether | Agarose 4% | D |
| *Medium-pressure packings (dp ∿ 50 μm)* | | | |
| Iminodiacetic acid | Ether | Fractogel TSK HW-65F | D |
| Tris(carboxymethyl) ethylenediamine | Ether | Fractogel TSK HW-65F | D |
| *High-pressure packings (dp ∿ 10 μm)* | | | |
| Iminodiacetic acid | Ether silane | Silica (100-Å) | G |
| | (NG) | TSK gel G5000PW | E |
| | Ether | Superose 12 | A |

[a]See Appendix 1 for a list of manufacturers.

**Table 11** Commercially Available Ready-to-Use HPLAC Column Packings

| | Particle diameter (μm) | Pore diameter (Å) | Manufacturer[a] |
|---|---|---|---|
| *Silica-based* | | | |
| SelectiSphere-10: | | | |
| Boronate | 10 | 100 | D |
| Concanavalin A | 10 | 500 | D |
| Cibacron blue F3G-A | 10 | 500 | D |
| Protein A | 10 | 500 | D |
| Si = Polyol: | | | |
| Butyl (HIC) | 5,10 | 300,500 | G |
| Cibacron blue F3G-A | 5 | 300 | G |
| Iminodiacetic acid | 3,5,10 | 100 | G |
| Bakerbond HI-Propyl | 5 | 300 | I |
| *Organic polymer-based* | | | |
| TSK gel-5PW: | | | |
| Phenyl | 10 | 1000 | F |
| Ether | 10 | 1000 | F |
| Chelate | 10 | 1000 | F |
| *Agarose-based* | | | |
| Superose: | | | |
| Phenyl | 10 | (12% agarose) | A |
| Alkyl (neopentyl) | 10 | (12% agarose) | A |
| Protein A | 10 | (12% agarose) | A |
| Chelating | 10 | (12% agarose) | A |

[a]See Appendix 1 for a list of manufacturers.

material. This involves minimizing the diffusion pathlength in the inner, stagnant volume of the liquid phase, where molecular transport to and from adsorption sites takes place. The diffusion is considered nonrestricted when the ratio between the molecular stokes radius and particle pore radius is equal to or smaller than 0.2. With decreasing bead size in a porous bead population, the fraction of adsorption sites facing the inner bead volume will be reduced. Consequently, sorption events in this stationary volume will largely be transferred to areas exposed to the interstitial flow and benefit from facilitated sorption kinetics.

A further step is to do away with the inner volume altogether by making solid beads but to reduce their size still more to maintain an acceptable surface-to-volume ratio [13]. Inevitably, bead size reduction will increase flow resistance, which must be compensated by higher pressure for a given column geometry. Agarose beads, hardly surpassed for low-pressure affinity chromatography, need extensive chemical modification in order to withstand the pressures generated in the 5- to 15-μm-diameter range. The increase in rigidity obtained after crosslinking by generally available methods still does not permit flow rates comparable to those routinely obtained on silica-based HPLC media. However, with Superose, an extensively modified agarose of substantially increased rigidity, the gap has narrowed. Porous silica does possess the necessary mechanical stability and is therefore at present the dominating alternative base material for HPLAC packings.

## SUPPORT MATERIALS

### Agarose

The agarose gel structure is an open three-dimensional network of fibers composed of spontaneously aggregated galactan helices (Fig. 1) [14]. The pore-volume-to-matrix-volume ratio is extremely high, with the pore diameters depending solely on the polymer concentration of the agarose solution used during the gel formation step [15]. In a 4% agarose gel (such as Sepharose 4B, the most widely used affinity chromatography matrix) the average pore diameter is around 30 nm [16]. Being a natural product, agarose is subject to considerable variation in quality. The main features affecting its matrix function in affinity chromatography are its content of charged and nonpolar groups, respectively. The main distinction between agarose and its original source, agar, is their content of sulfate ester groups. A common agarose specification limits the sulfate content to 0.3% (i.e., maximum 0.1% S). The presence of carboxyl groups is due to cyclic pyruvate esters. This is normally no problem, since producing an agarose practically devoid of bound pyruvate is a question of choosing the right algal source for the agar. Much less attention has been paid to the effect of nonpolar groups in agarose. It is a well-known fact that there is considerable variation in the content of methoxy groups in agarose preparations, affecting both the content of derivatizable hydroxyls and the level of nonspecific adsorption of agarose products intended for affinity chromatography. The main double disadvantage of agarose gels as matrices in low-pressure affinity chromatography, the limited physical and chemical stability, has largely been overcome by chemical modification procedures. Thus Porath et al. [17], by crosslinking with a variety of bifunctional reagents, could improve both thermal and chemical stability of agarose beads significantly. By further improvement of the crosslinking chemistry, agarose gels with considerably increased rigidity have been developed [18,19], allowing a reduc-

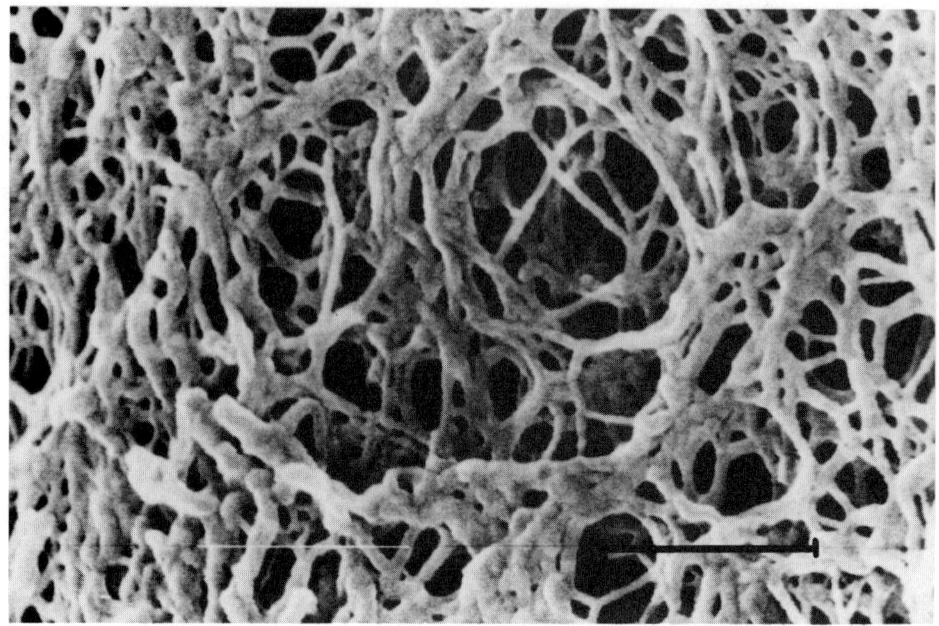

Fig. 1 Scanning electron micrograph of a 2% agarose gel after quick-freezing and lyophilization at -70°C. The black bar to the right (and the thin white bar to the left) represents 500 nm. (Preparation and photo: A. Medin, Biomedical Center, Uppsala, Sweden.)

tion in particle diameter down to the HPLC range [20]. Agarose gels can be activated for coupling of affinity ligands by a variety of chemistries, the most popular one being the CNBr activation method. Some of the most widely used will be briefly discussed later in the chapter.

## Cellulose

One of the oldest affinity chromatography matrices [24,25], cellulose has not yet received widespread acceptance as an affinity matrix for protein chromatography. This was originally the case because of unsatisfactory physical and chemical characteristics such as fibrous structure, lack of porosity, matrix leakage, susceptibility to microbial degradation, and nonspecific adsorption. In recent years, however, a new generation of spherically shaped, macroporous, and crosslinked cellulose preparations has developed [26,27], which could become an alternative to agarose especially in process scale applications. One area in which cellulose has become the dominating support is in the affinity chromatography of oligonucleotides and nucleic acids. A comprehensive treatment of this area was recently published by Schott [8].

### Silica

Its unsurpassed combination of high rigidity and macroporosity has made silica the number one support for high-performance liquid affinity chromatography [28-31]. Most packings are derived from epoxide- or diol-bonded silicas [32], which seem to satisfactorily suppress nonspecific adsorption by masking most of the silanols, but chloro-silanized silicas have been used as well [33]. The major weakness of silica-based packings—their water solubility at pH values above approximately 7.5—can be partly compensated for by zirconium oxide surface stabilization [34,35].

### Porous Glass

Besides silica, controlled-pore glass is the most important inorganic matrix used for the manufacture of affinity chromatography packings [36]. In spite of its obvious advantages, many of which are shared with silica, and the commercial availability of several preactivated derivatives (Table 3), porous glass has not received general acceptance as a matrix in affinity chromatography. The main reason is probably a fear of low chemical stability and presence of nonspecific adsorption of the protein species. An excellent, balanced discussion of the advantages and disadvantages of controlled-pore glass is given in Scouten's book [5]. In a recent report, DuVal et al. [37] discuss the use of thionyl chloride-activated succinamidopropyl-bonded, controlled-pore glass as a covalent immobilization matrix.

### Synthetic Organic Polymers

#### *Polyacrylamide Gels*

The main advantages of polyacrylamide as a matrix in affinity chromatography are (a) low nonspecific adsorption of proteins and (b) high concentration of functional groups for derivatization. There are also a wide range of chemical derivatization procedures described in the literature [21], the most widely used being based on activation with hydrazine [22], and glutaraldehyde [23]. The most serious disadvantages of polyacrylamide, which have restricted its use in affinity chromatography, are (a) small-pore diameters which are often further reduced during activation; (b) elastic deformation under pressure, which will limit flow rate and column size; (c) limited chemical stability, especially in alkaline media, which give rise to carboxyl groups and therefore the risk of nonspecific adsorption effects. A glutaraldehyde activated polyacrylamide-agarose composite gel, Ultrogel AcA22, is commercially available (Table 2).

### *Other Polymers*

From a production technical and economical point of view, the ideal situation would be to make an adequately porous and rigid base matrix lacking nonspecific adsorption and with high physical and chemical stability from simple and readily available bulk monomers. Several attempts have been made in this direction, the most well known probably being the hydroxyalkyl methacrylate gel Spheron/Separon [38,39] containing neutral hydroxyl groups which can be activated by a variety of chemistries [40]. Other examples are the Trisacryl and Fractogel matrices which are commercially available as preactivated and ready-to-use affinity packings (Tables 2 and 4). Eupergite is an epoxy group containing polymer material being marketed for the immobilization of enzymes and as a preactivated matrix for affinity chromatography (Table 2).

### Quartz Fibers

An alternative to spherical packings, 0.5-μm-diameter nonporous quartz fibers bonded with mercaptopropyltrimethoxysilane and tresyl chloride-activated dextran, was recently reported [41]. Before derivatization, the fibers were cut in a homogenizer to an average length of 50 μm with a surface area of approximately 2 $m^2 g^{-1}$. The fibers were derivatized and used packed in columns. The advantages of quartz fiber packings are the low-pressure drops and the favorable adsorption-desorption kinetics. Possible disadvantages are not addressed in the report but should be the same as for silica and porous glass, i.e., low chemical stability and high nonspecific adsorption.

## COUPLING CHEMISTRY

The covalent attachment of an affinity ligand to the desired matrix is usually performed in two steps. In the first step the matrix is subjected to an often harsh reaciton which gives rise to the introduction of reactive groups. In the second step the ligand is coupled to the activated matrix under often comparatively mild conditions..

Today, there are a wide variety of preactivated matrices for ligand immobilization commercially available (for "lazy" people as Meir Wilchek [42] once expressed it). Disregarding their often high prices, preactivated matrices represent a fast and convenient means for anyone to make their own affinity adsorbents without much chemistry involved.

In Table 2 are listed the preactivated matrices currently commercially available. However, often very little experience is needed to also manage the activation step. Volume 34 of *Methods in Enzymology* is entirely devoted to affinity techniques and contains detailed information on the immobilization of a wide variety of ligands to several different matrices. In this review we shall restrict ourselves to a short discussion of the pros and cons of the most popular coupling methods used in the preparation of affinity chromatography packings.

## Cyanogen Bromide Activation

CNBr reacts with hydroxyl group-containing polymers resulting in the formation of very reactive cyanate esters. This coupling method was originally developed by Porath and coworkers [43] for the immobilization of proteins to crosslinked dextran gels but was introduced for the covalent attachment of low molecular weight affinity ligands to agarose by Cuatrecasas et al. [1]. In recent years, the elucidation of the activation mechanism and further development of the coupling technique to significantly increased efficiency is the accomplishment of Wilchek and coworkers. (See Ref. 44 for review and detailed references.)

*Activation*

$$\text{Polymer-OH} + \text{OH}^- \longrightarrow \text{polymer-O}^- \xrightarrow{+\text{ CNBr}} \text{polymer-OCN} + \text{Br}^-$$

*Coupling*

$$\text{Polymer-OCN} + \text{ligand-NH}_2 \longrightarrow \text{polymer-O-}\overset{\text{NH}}{\overset{\|}{\text{C}}}\text{-NH-ligand}$$

The main advantages of the CNBr method are its simplicity, speed, reproducibility, and the mild conditions during both activation and coupling. The coupling yield compares favorably with other methods. The main disadvantages are the noxious character of CNBr which requires the use of a fume hood, the fact that the immobilized ligand becomes positively charged at neutral pH because of the N-substituted isourea connecting group, and, finally, the instability of the same group in the presence of nucleophilic reagents. In spite of these disadvantages, the CNBr method has been applied successfully for almost 20 years and is still the most widely used coupling method for agarose matrices.

### Epoxy Activation

By reacting hydroxyl group containing polymer matrices with bisoxiranes such as 1,4-butanediol diglycidyl ether, Sundberg and Porath [45] obtained epoxy-activated spacer gels which were shown to react with a wide range of functional groups in affinity ligands and generate very stable linkages. Thus not only primary amino groups but also thiols and alcohol groups could participate in coupling reactions. The main advantage of epoxy activated media is the high stability of the linkages involved (ether linkage to the matrix and alkyl, thioether and ether linkages to primary amine, thiol, and alcohol groups, respectively), which makes them ideally suited for the immobilization of small affinity ligands. The main disadvantages are the slow coupling reaction often requiring 24 hr or more, relatively high pH, and the sometimes need for elevated temperature during coupling.

*Activation*

$$\text{Polymer-OH} + \underbrace{CH_2\text{-}CH}_{\text{O}}\text{-}CH_2\text{-O-}(CH_2)_4\text{-O-}CH_2\text{-}\underbrace{CH\text{-}CH_2}_{\text{O}} \longrightarrow$$

$$\text{Polymer-O-}CH_2\text{-}\overset{\text{OH}}{\overset{|}{CH}}\text{-}CH_2\text{-O-}(CH_2)_4\text{-O-}CH_2\text{-}\underbrace{CH\text{-}CH_2}_{\text{O}}$$

*Coupling*

$$\text{Polymer-O-}CH_2\text{-}\overset{\text{OH}}{\overset{|}{CH}}\text{-}CH_2\text{-O-}(CH_2)_4\text{-O-}CH_2\underbrace{CH\text{-}CH_2}_{\text{O}} + \text{ligand-}NH_2 \longrightarrow$$

$$\text{Polymer-O-}CH_2\overset{\text{OH}}{\overset{|}{CH}}\text{-}CH_2\text{O-}(CH_2)_4\text{O-}CH_2\text{-}\overset{\text{OH}}{\overset{|}{CH}}\text{-}CH_2\text{-NH-ligand}$$

### Tresyl Activation

The use of organic sulfonyl chlorides, such as *p*-toluenesulfonyl chloride (tosyl chloride) and 2,2,2-trifluoroethanesulfonyl chloride (tresyl chloride), for the activation of hydroxyl group-containing gel media has been studied by Nilsson and Mosbach and was recently reviewed by these authors [46] and by Scouten et al. [47]. Tresylate esters are approximately 100-fold more reactive than the tosylates and will react readily with thiols and primary amino groups giving 75-100% protein coupling yield after 1 hr at pH 7.5 in the cold [46]. The main advantages of the tresyl activation method are the same as those for the CNBr method. In addition, very stable alkyl linkages are obtained between the ligand and the matrix. The activated matrix is also very stable. It will not lose coupling capacity even after several weeks in aqueous media. The main disadvantage is the high cost of the tresyl chloride. Tresyl-activated amtrices are commercially available (Table 3).

*Activation*

$$\text{Polymer-OH} + CF_3\text{-}CH_2\text{-}SO_2Cl \longrightarrow \text{polymer-O-}SO_2CH_2CF_3$$

*Coupling*

$$\text{Polymer-O-}SO_2CH_2CF_3 + \text{ligand-}NH_2 \longrightarrow \text{polymer-NH-ligand} + HSO_3CH_2CF_3$$

### 1,1'-Carbonyldiimidazole Activation

This method was developed by Bethell et al. [48] for hydroxyl group-containing polymer gel materials and possesses coupling characteristics for primary amine-containing compounds similar to that of CNBr-activated matrices. One advantage, however, is the formation of a noncharged urethane linkage. Also the reagent is less poisonous than CNBr. As a disadvantage should be mentioned the need for nonaqueous solvents during the activation step calling for a dehydration of the gel material with a concomitant risk for decrease in matrix porosity. Another disadvantage is the instability of the bond in alkaline buffers (pH > 10). Carbonyldiimidazole-activated matrices are commercially available (Table 2).

*Activation*

$$\text{Polymer-OH} + \text{Im-}\underset{\underset{O}{\|}}{C}\text{-Im} \longrightarrow \text{polymer-O-}\underset{\underset{O}{\|}}{C}\text{-Im}$$

*Coupling*

$$\text{Polymer-O-}\underset{\underset{O}{\|}}{C}\text{-Im} + \text{ligand-}NH_2 \longrightarrow \text{polymer-O-}\underset{\underset{O}{\|}}{C}\text{-NH-ligand}$$

### Divinylsulfone Activation

Developed by Porath and Sundberg [49], this method shares many characteristics with the bisoxirane activation. It can be used for the covalent attachment of compounds carrying amino, alcohol, and phenol groups through noncharged linkages simultaneously introducing a five-atom spacer arm. The coupling reaction proceeds at lower pH and at lower temperature than with epoxy-activated matrices. The divinylsulfone coupling method has been used successfully for the immobilization of carbohydrates [50]. During the activation of agarose a cross-linking side reaction usually occurs resulting in significantly higher rigidity of the final product [51]. The most serious drawbacks of this method are the high toxicity of the reagent and the instability of the coupled ligand in alakline medium (the amino group link above about pH 8 and the hydroxyl group link above pH 9-10).

*Activation*

$$\text{Polymer-OH} + CH_2=CH\text{-}\overset{\overset{\displaystyle O}{\|}}{\underset{\underset{\displaystyle O}{\|}}{S}}\text{-}CH=CH_2 \longrightarrow$$

$$\text{polymer-O-}CH_2\text{-}CH_2\text{-}\overset{\overset{\displaystyle O}{\|}}{\underset{\underset{\displaystyle O}{\|}}{S}}\text{-}CH=CH_2$$

*Coupling*

$$\text{Polymer-O-}CH_2\text{-}CH_2\text{-}\overset{\overset{\displaystyle O}{\|}}{\underset{\underset{\displaystyle O}{\|}}{S}}\text{-}CH=CH_2 + \text{ligand-}NH_2 \longrightarrow$$

$$\longrightarrow \text{polymer-O-}CH_2\text{-}CH_2\text{-}\overset{\overset{\displaystyle O}{\|}}{\underset{\underset{\displaystyle O}{\|}}{S}}\text{-}CH_2\text{-NH-ligand}$$

## LIGANDS AND SPACERS

### Ligands for Enzymes

The first modern affinity chromatography experiments were performed with enzymes and this protein category is still the largest application area for the technique. Affinity chromatography packings for enzymes can be grouped into two categories. The first involves ligands with high specificity for individual enzymes which are often based on immobilized substrates or inhibitors that are modified substrates. The second category is formed by ligands able to select between groups of enzymes and which are often based on immobilized cofactors or inhibitors obtained by modifying cofactors or cofactor analogs. These so-called general ligands will thus show group specificity during the adsorption stage whereas separation and purification of the individual enzymes within the group will take place by specific elution using either gradient elution with free cofactor or ternary complex formation with cofactor and substrate. Most of the developments in this field were made on model systems based on immobilized nucleotide coenzymes by Mosbach et al. [2,52] in the early 1970s. In a recent review, Wilchek et al. [44] list the conditions for affinity chromatography of 293 enzymes. Of these, 99 were purified on immobilized nucleotide coenzymes and analogs, 141 on immobilized substrates and substrate analogs, 19 on immobilized Con A, 4 on heparin, and 3 on calmodulin. The most frequently used nucleotide ligands were AMP (30), ADP (16), ATP (15), NADP (14), and NAD (13). A wide variety of ready to use nucleotide-containing packings are available from manufacturers A and H (Appendix 1).

Heparin is a sulfated glycosaminoglycan with anticoagulant properties which after immobilization to an appropriate carrier has been shown to selectively bind several different enzymes and other proteins [53,54]. Attached to Sepharose CL-6B, heparin is used for the industrial isolation of antithrombin III from human plasma [55]. Immobilized heparin is commercially available from several manufacturers (Table 5).

Calmodulin is an intracellular, low molecular weight, $Ca^{2+}$-binding protein which regulates the activities of several different enzymes. Calmodulin immobilized to Sepharose 4B has been shown to bind many of these enzymes [56] as well as human interferon $\alpha$ [57] in a $Ca^{2+}$-dependent manner. Calmodulin-containing packings are commercially available from manufacturers A and B (Appendix 1).

## Other Ligands

### *IgG-Binding Ligands*

Protein A of *Staphylococcus aureus* [58] and protein G of *Streptococcus* [59] are cell surface proteins which bind to the Fc region of human and several animal IgG. The main difference between the two proteins is found in the binding of human subclass IgG3 and goat IgG. In both cases protein G shows strong binding and protein A no binding. Ever since its introduction in 1975, protein A-Sepharose CL-4B has been a simple and convenient tool for the isolation of highly purified IgG from normal and hyperimmune sera. In recent years it has become the method of choice for the concentration and purification of appropriate IgG subclasses from hybridoma cell culture supernatants [60]. Through its broader binding specificity, immobilized protein G is likely to gain widespread popularity. Packings containing immobilized proteins A and G are commercially available (Table 5).

### *Boronic Acid Ligands*

In 1970, Weith et al. [61] used phenylboronic acid (PBA) immobilized to cellulose for the separation of vicinal *cis*-diol-containing sugars, nucleosides, and nucleotides. PBA coupled to a variety of other carriers has been used for the isolation of nucleic acids and many different kinds of proteins such as enzymes, interferon, and IgG, as well as separation of glycosylated from nonglycosylated hemoglobin. The area has been reviewed by Dean et al. [62]. PBA-containing packings are available from manufacturers B, D, E, H, and N (Appendix 1).

### *Carbohydrate Ligands*

Packings containing immobilized monosaccharides, oligosaccharides, and analogs of these are useful as affinity adsorbents for sugar-specific enzymes and lectins [63,64]. Several ready-to-use carbohydrate-containing packings are available from manufacturers A, D, and H (Appendix 1).

### *Nucleic Acid-Binding Ligands*

In modern molecular biology research there are a number of instances in which affinity chromatography packings are used for the separation and purification of DNA as well as RNA species. The oldest and most frequently used packing is oligo-(dT)-cellulose, which separates most mRNAs containing 3'-terminal poly(A) segments from total cellular RNA. In addition to oligo-(dT)-cellulose there are a number of other packings available for the purification of messenger RNA. These are listed in Table 6. In the same table are also listed a number of other affinity packings together with their use and origin of manufacture. In 1978 Bünemann and Müller [65] introduced a new affinity medium for base pair-specific separation of double-stranded DNA based on crosslinked *N,N'*-bisacrylamide to which certain intercalating dyes are covalently attached. Thus immobilized A/T-specific malachite green and G/C-specific phenyl neutral red can be used for the separation of DNA molecules with molecular weights up to $25 \times 10^6$ differing in their base pair compositions by not more than 2-3%.

### *Pyrogen-Binding Ligands*

The most important pyrogen, the lipopolysaccharide (LPS) endotoxin, a cell wall constituent of gram-negative bacteria, can be reduced below the detection limit of the Limulus amoebocyte lysate test by adsorption to the lipophilic, cyclic peptide antibiotic polymyxin B coupled to CNBr-activated Sepharose 4B [66]. In a recent review paper, Sofer [67] discusses the use of various adsorbents for pyrogen removal in bio-product recovery processes.

### *Protease-Binding Ligands*

Proteolytic degradation, causing microheterogeneity and the possible formation of neodeterminant immunogenic structures, is one of the major problems in biotechnology downstream processing. The inhibition or quick removal of proteases is therefore of considerable interest. Affinity adsorbents based on immobilized protease inhibitors have primarily been used for the isolation of serin proteases. Examples of such inhibitors are soybean trypsin inhibitor, aprotinin (Trasylol), pepstatin, D-tryptophan methyl ester, glycyl-L-tyrosylazobenzylsuccinic acid, L-alanyl-L-alanyl-L-alanine, and *para*-aminobenzamidine. Several inhibitor affinity packings are available from manufacturers D and H, benzamidine-Sepharose 6B from manufacturer A (Appendix 1). Thiol proteinases can be isolated by mixed disulfide formation with 2-pyridyldisulfide-activated thiol group-containing packings (Table 2). The antibiotic cyclopeptide bacitracin covalently attached to CNBr-activated Sepharose 4B has been shown to efficiently bind serine, aspartyl, and metalloproteinases from various sources [68]. The application of crude extracts from homogenized baker's yeast and *E. coli* to the HIC adsorbents octyl- and phenyl-Sepharose CL-4B has been shown to considerably reduce their content of proteolytic activity [69].

### *Virus-Binding Ligands*

Virus contamination is a serious threat to people working with fluids and extracts of human origin. An affinity adsorbent based on octanoic acid hydrazide-derivatized Sepharose 4B has been successfully used for the removal of hepatitis B virus and non-A non-B hepatitis virus from human plasma protein fractions [70,71].

## The Use of a Spacer (Leash)

The spacer concept is as old as the affinity chromatography technique itself. Thus Cuatrecasas et al. [1] and Steers et al. [72] clearly demonstrated the significance of attaching low molecular weight ligands used in low-affinity systems (dissociation constant $K_D$ of $10^{-3}$-$10^{-5}$) at some distance from the matrix itself, thereby dramatically increasing its functional concentration and availability to binding. Later, however, it was pointed out by Er-el et al. [73] and by O'Carra [74] that there is a considerable risk of introducing new, nonspecific adsorption centers by using long-chain hydrophobic spacers, a discovery which in fact led to the development of hydrophobic interaction chromatography. Regardless of this, most spacer-derivatized packings used, and most of those commercially available, are based on alkyl derivatives. However, there are examples where increased hydrophilicity has been achieved through the introduction of several amide groups, e.g., succinamide or glycine oligomers [75].

Most spacer-containing packings are substituted with either amino or carboxyl groups to allow the subsequent synthesis of desired affinity adsorbents using water-soluble carbodiimide coupling reagents such as 1-cyclohexyl-3-(2-morpholinoethyl)carbodiimide-*p*-toluenemethosulfonate (CMC) or 1-ethyl-3-(3-dimethylaminopropyl)carbodiimide hydrochloride (EDC). This requires the presence of either primary amino groups of carboxyl groups in the ligand which are not involved in the biospecific binding process. Ligands containing hydroxyl and thiol groups can be coupled to gels carrying oxirane-activated spacer groups. A wide variety of spacer packings containing 5-13 atoms are commercially available (Table 7).

### Ligand Leakage

The most popular coupling method is still the one based on CNBr activation. Originally developed for the immobilization of proteins [43], it has been widely used for the covalent attachment of low molecular weight ligands as well. Such monovalently coupled ligands are connected by a N-substituted isourea bond which is not completely stable, especially in the presence of nucleophilic reagents. Thus one should avoid storage and operating conditions involving high pH and/or amino group-containing buffers or else one will experience a small but constant leakage of ligand from the matrix [76-78]. Leakage of multipoint attached macromolecules such as large proteins is less likely to occur even in the presence of low concentrations of nucleophilic reagents. The most probably cause of leakage in this case is leakage of the matrix itself, and often as mechanical leakage due to abrasion damage [79]. By using crosslinked agarose as a matrix, the risk of matrix leakage is significantly reduced. For the immobilization of low molecular weight ligands by single-point attachment, stable covalent linkages are obtained using, for example, oxirane (epoxide)- or tresyl-activated agarose.

### Polymer Spacers

The spacer concept pinpoints the importance of high degrees of substitution in affinity chromatography using low molecular weight ligands. One way to achieve this, and also to reduce the ligand leakage, is to use polymer spacers, as was suggested by Wilchek et al. [80]. The principle is to introduce multiple-point attached hydrophilic polypeptides such as poly(L-lysine)-poly(D,L-alanine)-poly(L-lysine) or poly-acryl-hydrazide. The latter polymer possesses the advantage of being free of ion exchange groups. It can also be multipoint attached to periodate oxidized agarose, thus avoiding the drawbacks of the CNBr method [44].

## SPECIAL-AFFINITY PACKINGS

### Hydrophobic Interaction Chromatography (HIC)

Soon after the spacer concept was introduced as an essential element in the design of affinity chromatography adsorbents [1,81], reports on anomalous behavior of spacer-containing ligands started to appear [72,74,82]. The binding observed could be ascribed to the hydrophobic spacer rather than the ligand. Several authors independently realized the potential of this new separation principle [83-87] and hydrophobic interaction chromatography was soon recognized as a powerful tool in protein separation. Pure hydrophobic interaction can only be achieved using adsorbents lacking nonhydrophobic binding sites [88,89] and is characterized by an increase in binding capacity by increasing ionic strength as well as temperature.

Few proteins will bind to neutral HIC adsorbents at low ionic strength. This is why salts like NaCl, $Na_2SO_4$, or $(NH_4)_2SO_4$ are added to a concentration of up to several moles per liter in order to get adequate binding. Desorption is best achieved by gradually decreasing the ionic strength of the eluent, sometimes in combination with a superimposed gradual increase in the concentration of a less polar solute, e.g., ethylene glycol [90].

The main difference between HIC and RPC is in the degree of substitution of the alkyl or aryl groups. In HIC, values in the range 40-100 mmol/mol galactose are typical and give a reasonable compromise between binding capacity and recovery of biologically active proteins. The binding capacity of HIC adsorbents is also strongly dependent on the structure of the immobilized hydrophobic groups. Shaltiel [91] in a recent review discusses the experience of using kits containing agarose gels substituted with alkyl chains ranging from $C_1$ to $C_{12}$. HIC packings for low-, medium-, and high-pressure systems are commercially available (Tables 8 and 11).

### Immobilized Lectin Affinity Chromatography

Lectins are proteins which bind to carbohydrates and carbohydrate-containing proteins [92]. By immobilizing lectins to suitable supports, affinity adsorbents which can be used for the isolation of glycoproteins are obtained [93]. Binding selectivity is achieved by choosing a lectin with appropriate specificity, e.g., toward $\alpha$-D-mannosyl (concanavalin A and lentil), β-D-galactosyl (castor bean and *Crotalaria*), *N*-acetyl-β-D-glucosaminyl (wheat germ), and *N*-acetyl-β-D-galactosaminyl (*Helix pomatia*). Displacement of adsorbed glycoproteins is accomplished by applying a solution of the corresponding soluble saccharide. Lectin affinity chromatography is one of the mildest existing protein purification methods and has become widely used. Ready-to-use lectin gels are commercially available (Table 9).

### Immunoaffinity Chromatography

The introduction of hybridoma technology for the production of monoclonal antibodies significantly increased the range and importance of immunoaffinity chromatography. Large quantities of defined antibodies with homogeneous binding sites can now be produced for the preparation of highly specific immunosorbents. Monoclonal antibodies have several advantages over conventional, polyclonal antibodies for affinity purification. Thus, by definition monoclonals exhibit single-epitope attachment to their antigens which gives rise to higher and better defined specificities, low to moderate affinities (association constants of $K_A = 10^6$-$10^9$), more favorable desorption kinetics, and at least 10-fold higher binding capacities. The low affinities require less drastic elution conditions (decreasing pH to 3-4 is often sufficient), preserves the antigen, and increases the life span of the adsorbent, often to several hundred cycles. A comprehensive treatment of immunoaffinity chromatography using monoclonal antibodies is found in the second edition of Gooding's book [60]. Monoclonal antibodies against human interferons α, β, and γ, human interleukin-2, and bovine serum albumin, coupled to CNBr-activated Sepharose CL-4B are commercially available from Celltech Ltd., Slough, Berkshire, England.

### Dye Ligand Affinity Chromatography

Reactive triazine dye Cibacron blue F3G-A covalently attached to high molecular weight dextran, introduced as a void volume marker [94] for gel filtration columns in the middle of the 1960s, was found to interact with yeast phosphofructokinase [95] and pyruvate kinase [96] causing anomalous elution behavior. This discovery led to the development of adsorbents based on dyes covalently attached directly to various carriers, predominantly agarose. A wide variety of different dyes have been tested for adsorption of a range of enzymes; still the blue dye (Cibacron blue F3G-A) appears to be the most useful. The dye behaves like an analog of ADP-ribose and will thus show affinity to enzymes which are dependent of nucleotide cofactors (e.g., ATP and NAD). NADP-dependent enzymes seem to bind more strongly to Procion Red HE-3B. Low concentrations of metal ions belonging to the first series transition elements promote the binding of proteins to immobilized triazine dye ligands [97,98]. The dye adsorbents have become very popular, particularly the blue gel, because of their relatively low cost, high binding capacities, and very wide selectivities covering hundreds of enzymes [99] and other proteins (e.g., human serum albumin [100] and human interferons [101,102]). By chemical modification of the dye aromatic ring system it is possible to alter the binding specificity and change the dye-protein affinity by several orders of magnitude [103]. A strategy for enzyme isolation using dye ligand affinity chromatography was recently reported [104]. Ready-to-use dye ligand affinity adsorbents are commercially available based on low-, medium-, and high-pressure media (Tables 4, 5, and 11).

### Immobilized Metal Affinity Chromatography (IMAC)

Introduced by Porath et al. [105] in the mid-1970s, immobilized metal affinity chromatography offered a new dimension in protein separation. Recent studies [106] have shown that the main separation discriminasor appears to be the number of surface-exposed histidine residues. The first step in the synthesis of an IMAC adsorbent is the covalent immobilization of metal chelate complex-forming groups such as iminodiacetic acid (IDA) and triscarboxymethylethylenediamine (TED) to a suitable support. Before use these will have to be saturated with the appropriate metal ion (most often $Me^{2+}$, and this is the second step in the preparation of an IMAC adsorbent.

The most useful metal ions are to be found among the first series transition metals, the most frequently used being $Cu^{2+}$ and $Zn^{2+}$, and to some extent $Co^{2+}$ and $Ni^{2+}$ [106]. IDA complexes of these metal ions will bind exposed imidazole and thiol groups on the surface of proteins. Phosphoproteins have been isolated with IDA-$Fe^{3+}$ gels [107] and serum proteins have been separated using a TED-$Tl^{3+}$ adsorbent [108]. $Zn^{2+}$ chelate gels have been used for the purification of human interferon β [109]. IMAC packings for low-, medium-, and high-pressure systems are commercially available (Table 10).

### High-Performance Liquid Affinity Chromatography (HPLAC)

A natural development, the combination of affinity chromatography and high-performance liquid chromatography, was first reported by Ohlson et al. [110]. As with low- and medium-pressure affinity chromatography systems, the matrices for HPLAC should have the same basic chracteristics as those of the corresponding size exclusion chromatography media [111]. In addition, they should be amenable to derivatization by a variety of chemistries and preferably withstand harsh conditions for regeneration and maintenance (to the extent that these conditions are compatible with the attached ligands).

As in other branches of HPLC, the most popular matrix in HPLAC is silica, mainly because of its unsurpassed rigidity combined with a wide variety of adequate porosities. The main drawbacks of silica, its instability at high pH and tendency to nonspecific adsorption, have spurred attempts to develop HPLAC media based on other materials, notably synthetic organic polymers and agarose. In Table 11 are listed the preactivated and ready-to-use HPLAC column packings available as of 1986.

### Covalent Chromatography

By immobilizing the thiol group-containing tripeptide glutathion to CNBr-activated agarose and reacting the free thiol groups with 2,2'-dipyridyldisulfide, Brocklehurst et al. [112] obtained a mixed disulfide agarose derivative which reacted spontaneously and quantitatively with thiol group-containing proteins (such as papain) to give a new mixed disulfide in which the protein was covalently attached to the agarose matrix. Following the rinsing out of contaminating, non-thiol-containing proteins in the sample, the immobilized protein was eluted by the addition of low concentrations of thiol group-containing solutes such as cystein or mercaptoethanol. In a newer development, a more stable ether linkage was used for the attachment of an activated thiol-containing hydroxypropyl group (2-pyridyldisulfidehydroxypropyl-ether) to Sepharose 6B [113]. A comprehensive review of covalent chromatography in biochemistry and biotechnology was recently published [114]. 2-Pyridyldisulfide-activated thiol packings are commercially available (Table 2).

## OPERATING CONDITIONS IN AFFINITY CHROMOTOGRAPHY

The resolution obtained in any chromatographic column is the result of a combination of the selectivity and the efficiency of the column packing material. In affinity chromatography, the selectivities and binding equilibria normally are so favorable that we often face a highly specific type of all-or-nothing situation for the adsorption and desorption. The kinetics of the adsorption reactions are generally also extremely favorable, allowing very high linear flow rates during sample application even on columns packed with low-pressure media (dp $\sim$ 100 μm). A rate of 5 cm $min^{-1}$ is normally no problem even in cases with macromolecular ligands such as in immunoaffinity applications. Even if the flow rates should be reduced dramatically during the desorption phase to avoid excessive dilution of the eluted protein (at least a 20-fold decrease in linear flow rate is recommended), this does not significantly increase the total chromatographic cycle time. In affinity chromatography the desorption peak volumes are normally very small compared to the sample volumes. In other words, preparative affinity chromatography is generally regarded as a comparatively fast separation technique and this is probably the reason why biospecific high-performance liquid chromatography has not developed as rapidly as one would expect since its introduction in 1978 [110], at least in comparison with the corresponding development in high-performance size exclusion, reversed phase, and ion exchange chromatography.

## CONCLUSIONS AND PROSPECTS

Although a carefully controlled combination of separation principles, such as, for example, size exclusion and ionic charge, can sometimes be effective, a generally successful principle in the development and use of separation media and procedures is to apply one single separation principle at a time and try to exclude the interference from other mechanisms in order to make it easier to interpret and assess the results.

Contrary to that principle, most bioaffinity systems which we try to mimic and exploit for a given separation are extremely complex sets of molecular properties developed by nature for very specific functions, involving a finely tuned interaction of properties like conformation, polarity, and charge. Whenever these systems are removed from their in vivo environment to form the basis of a separation scheme, an additional set of chemical and physical parameters is introduced, related to the affinity support itself as well as general operating conditions like pH, temperature, and flow rate.

In a technical sense every affinity separation system is therefore a deliberate tangle of separation principles and mechanisms, further confounded by a large set of chromatographic parameters. Despite brilliant efforts to put areas like, for example, affinity sorption kinetics and the leakage problem on firmer theoretical ground, interest and progress in the theories of affinity separation has been moderate.

However, an enormous amount of empirical knowledge has accumulated and much of it has already been converted into detailed advice that is useful to both the novice and the expert. Bioaffinity separation may still appear to the general life scientist with a separation problem as a mixture of folklore and scientific rigor, but the spectacular success often obtained after quite simple trial and error within a given frame of advice makes it easier to endure even considerable vagueness of the underlying theory.

The wide selection of preactivated and even ligand affinity supports now available tends to reduce the need for the end user to modify the basic support material or to manipulate more or less hazardous activating reagents, as well as spacer molecules when applicable. From a purely scientific point of view, this situation is not wholly satisfactory since commercially sensitive details about the composition, manufacture, and sometimes properties of the separation tools are not generally available. However, in compensation these supports are extensively characterized with respect to operational parameters, which are highly reproducible. Different outcomes of allegedly similar separation procedures can therefore with greater confidence be ascribed to sample variation rather than more or less untraceable differences in supports or other methodology, which is of obvious importance, e.g., to control laboratories.

Even with the wide selection of affinity separation materials now available, there is room for improvement in both basic supports and ready-made affinity adsorbents. In the biosciences there is a general need to be able to work with ever smaller samples. As the separation scale is reduced, the ratio between nonspecifically adsorbing surface and the affinity surface proper throughout the sample contact area becomes more and more critical, and must therefore be minimized. Accordingly, a particular challenge is to increase the specific capacity after immobilization of the ligand, particularly a macromolecule prone to multipoint attachment. This involves orientation of the native or modified ligand in a uniform and sterically favorable way before coupling to the activated support, which must allow a strong activation and a high ligand substitution without losing its favorable mechanical and physical properties.

Another major task is to arrive at a generally applicable model of adsorption-desorption kinetics, which even casual affinity chromatographers can grasp and turn into practical use to optimize their separation parameters.

## APPENDIX

### Manufacturers of Affinity Chromatography Packings

A = Pharmacia LKB Biotechnology AB, Uppsala, Sweden

B = BioRad, Chemical Division, Richmond, CA, USA

C = Rèactifs IBF, Villeneuve-la-Garenne, France

D = Pierce Chemical Company, Rockford, IL, USA

E = E. Merck, Darmstadt, FRG

F = Toyo Soda Manufacturing Co., Ltd., Tonda, Yamaguchi, Japan

G = Serva Feinbiochemica GmbH and Co., Heidelberg, FRG

H = Sigma Chemical Company, St. Louis, MO, USA

I = J. T. Baker Chemical Co., Phillipsburg, NJ, USA

J = Röhm Pharma GmbH, Weiterstadt, FRG

K = BioProbe International Inc., Tustin, CA, USA

L = Beckman Instruments Inc., Berkeley, CA, USA

M = ICN Biochemicals Inc., Naperville, IL, USA

N = Amicon, Danvers, MA, USA

O = Boehringer Mannheim GMBH, Mannheim, FRG

## REFERENCES

1. P. Cuatrecasas, M. Wilchek, and C. B. Anfinsen, *Proc. Natl. Acad. Sci. USA*, *61*:636 (1968).
2. K. Mosbach, H. Guilford, P-O. Larsson, R. Ohlsson, and M. Scott, *Biochem. J.*, *125*:20 (1971).
3. W. B. Jacoby and M. Wilchek (eds.), *Meth. Enzymol.*, *34*:1-810 (1974).
4. C. R. Lowe, *An Introduction to Affinity Chromatography*, Elsevier, Amsterdam, 1979.
5. W. H. Scouten, *Affinity Chromatography*, John Wiley and Sons, New York, 1981.
6. T. J. C. Gribnau, J. Visser, and H. J. F. Nivard (eds.), *Affinity Chromatography and Biological Recognition*, Academic Press, Orlando, 1983.
7. I. M. Chaiken, M. Wilchek, and I. Parikh (eds.), *Affinity Chromatography and Biological Recognition*, Academic Press, Orlando, 1983.
8. H. Schott, *Affinity Chromatography: Template Chromatography of Nucleic Acids and Proteins*, Marcel Dekker, New York, 1984.
9. P. D. G. Dean, W. S. Johnson, and F. A. Middle (eds.), *Affinity Chromatography: A Practical Approach*, IRL Press, Oxford, 1985.
10. J. Turkova, I. M. Chaiken, and M. T. W. Hearn (eds.), Proceedings of the 6th International Symposium on bioaffinity chromatography and related techniques, *J. Chromatogr.*, *376*:1-451 (1986).
11. J.-C. Janson, *Trends Biotechnol.*, *2*:31 (1984).
12. I. M. Hais, *J. Chromatogr.*, *373*:265 (1986).
13. K. K. Unger, G. Jilge, R. Janzen, H. Giesche, and J. N. Kinkel, *Chromatographia*, *22*:379 (1986).
14. S. Arnott, A. Fulmer, W. E. Scott, I. C. M. Dea, R. Moorhouse, and D. A. Rees, *J. Mol. Biol.*, *90*:269 (1974).
15. G. K. Ackers and R. L. Steere, *Biochim. Biophys. Acta*, *59*:137 (1962).

16. A. Amsterdam, Z. Er-el, and S. Shaltiel, *Arch. Biochem. Biophys.*, *17*:673 (1975).
17. J. Porath, J.-C. Janson, and T. Låås, *J. Chromatogr.*, *60*:167 (1971).
18. Sepharose Fast Flow, Pharmacia AB, Uppsala, Sweden.
19. Superose, Pharmacia AB, Uppsala, Sweden.
20. T. Andersson, M. Carlsson, L. Hagel, P-Å Pernemalm, and J.-C. Janson, *J. Chromatogr.*, *326*:33 (1984).
21. J. K. Inman and H. M. Dintzis, *Biochemistry*, *8*:4075 (1969).
22. J. K. Inman, *Meth. Enzymol.*, *34*:30 (1974).
23. P. D. Weston and S. Avrameas, *Biochem. Biophys. Res. Commun.* *45*:1574 (1971).
24. D. H. Campbell, E. Luescher, and L. S. Lerman, *Proc. Natl. Acad. Sci., USA*, *37*:575 (1951).
25. L. S. Lerman, *Proc. Natl. Acad. Sci., USA*, *39*:232 (1953).
26. J. Stamberg, J. Peska, H. Dautzenberg, and B. Philipp, in *Affinity Chromatography and Related Techniques* (T. C. J. Gribnau, J. Visser, and R. J. F. Nivard, eds.), Elsevier, Amsterdam, 1982, p. 131.
27. Y. Motozato and C. Hirayama, *J. Chromatogr.*, *298*:499 (1984).
28. S. Ohlson, L. Hansson, P.-O. Larsson, and K. Mosbach, *FEBS Lett.*, *93*:5 (1978).
29. P.-O. Larsson, M. Glad, L. Hansson, M.-O. Månsson, S. Ohlson, and K. Mosbach, in *Advances in Chromatography*, Vol. 21 (J. C. Giddings et al., eds.), Marcel Dekker, New York, 1983, p. 41.
30. D. A. P. Small, T. Atkinson, and C. R. Lowe, *J. Chromatogr.*, *266*:151 (1983).
31. Y. D. Clonis, K. Jones, and C. R. Lowe, *J. Chromatogr.*, *363*:31 (1986).
32. F. E. Regnier and R. Noel, *J. Chromatogr. Sci.*, *14*:316 (1976).
33. E. Hagemeier, K.-S. Boos, E. Schlimme, K. Lechtenbörger, and A. Kettrup, *J. Chromatogr.*, *268*:291 (1983).
34. R. W. Stout and J. J. DeStefano, *J. Chromatogr.*, *326*:63 (1985).
35. R. W. Stout, S. I. Sivakoff, R. D. Ricker, H. C. Palmer, M. A. Jackson, and T. J. Odiorne, *J. Chromatogr.*, *352*:381 (1986).
36. H. H. Weetall and A. M. Filbert, *Methods Enzymol.*, *34*:59 (1974).
37. G. DuVal, H. E. Swaisgood, and H. R. Horton, *J. Appl. Biochem.*, *6*:240 (1984).
38. J. Coupec, M. Kriváková, and S. Pokorny, *J. Polym. Sci. Polym. Symp.*, *42*:182 (1973).
39. Spheron is a product of Lachema, Brno and Separon is a product of Laboratory Instrument Works, Prague, Czechoslovakia.
40. J. Turková, *Affinity Chromatography*, John Wiley and Sons, New York, 1981.
41. P. Wikström and P.-O. Larsson, *J. Chromatogr.*, *388*:123 (1987).
42. M. Wilchek, *J. Chromatogr.*, *157*:458 (1978).

43. R. Axén, J. Porath, and S. Ernback, *Nature, 214*:1302 (1967).
44. M. Wilchek, T. Miron, and J. Kohn, *Meth. Enzymol., 104*:3 (1984).
45. L. Sundberg and J. Porath, *J. Chromatogr., 90*:87 (1974).
46. K. Nilsson and K. Mosbach, *Meth. Enzymol., 104*:56 (1984).
47. W. H. Scouten, W. van den Tweel, D. Delhaes, H. Kranenberg, and M. Dekker, *Meth. Enzymol., 376*:289 (1986).
48. G. S. Bethell, J. Ayers, W. S. Hancock, and M. T. W. Hearn, *J. Biol. Chem., 254*:2572 (1979).
49. J. Porath and L. Sundberg, *Nature, 238*:261 (1972).
50. B. Ersson, K. Aspberg, and J. Porath, *Biochim. Biophys. Acta, 310*:446 (1973).
51. J. Porath, T. Låås, and J.-C. Janson, *J. Chromatogr., 103*:49 (1975).
52. K. Mosbach, *Meth. Enzymol., 34*:229 (1974).
53. A. A. Farooqui, *J. Chromatogr., 184*:335 (1980).
54. E. Ber, G. Muszynska, E. Tarantowicz-Marek, and G. Dobrowolska, in *Affinity Chromatography and Biological Recognition* (I. M. Chaiken, M. Wilchek, and I. Parikj, eds.), Academic Press, Orlando, 1983, p. 455.
55. R. Eketorp, in *Affinity Chromatography and Related Techniques* (T. C. J. Gribnau, J. Visser, and R. J. F. Nivard, eds.), Elsevier, Amsterdam, 1982, p. 263.
56. C. B. Klee, D. L. Newton, and M. Krinks, in *Affinity Chromatography and Biological Recognition* (I. M. Chaiken, M. Wilchek, and I. Parikj, eds.), Academic Press, Orlando, 1983, p. 55.
57. T. A. Myöhänen, L. Kågedal, J. Savin, and G. V. Alm, in *Affinity Chromatography and Biological Recognition* (I. M. Chaiken, M. Wilchek, and I. Parikj, eds.), Academic Press, Orlando, 1983, p. 469.
58. H. Hjelm, K. Hjelm, and J. Sjöquist, *FEBS Lett., 28*:73 (1972).
59. L. Björk and G. Kronvall, *J. Immunol., 133*:969 (1984).
60. J. W. Goding, *Monoclonal Antibodies: Principles and Practice*, Academic Press, London, 1986, p. 122.
61. H. L. Weith, J. L. Wiebers, and P. T. Gilham, *Biochemistry, 9*:4396 (1970).
62. P. D. G. Dean, F. A. Middle, C. Longstaff, A. Bannister, and J. J. Dembinski, in *Affinity Chromatography and Biological Recognition* (I. M. Chaiken, M. Wilchek, and I. Parikj, eds.), Academic Press, Orlando, 1983, p. 433.
63. P. Vretblad, *Biochim. Biophys. Acta, 434*:169 (1976).
64. R. Uy and F. Wold, *Anal. Biochem., 81*:98 (1977).
65. H. Bünemann and W. Müller, *Nucl. ACids Res.,* 5:1059 (1978).
66. A. C. Issekutz, *J. Immunol. Meth., 61*:276 (1983).
67. G. Sofer, *Biotechnology, 2*:1035 (1984).
68. V. M. Stepanov and G. N. Rudenskaya, *J. Appl. Biochem.,* 5:420 (1983).

69. P. Hedman and J. G. Gustafsson, *Develop. Biol. Standard, 59*: 31 (1985), S. Karger, Basel.
70. M. Einarsson, L. Kaplan, E. Nordenfelt, and E. Miller, *J. Virol. Meth., 3*:213 (1981).
71. M. Einarsson and B. Flehmig, *J. Virol. Meth., 8*:233 (1984).
72. E. J. M. Steers, P. Cuatrecases, and H. B. Pollard, *J. Biol. Chem., 246*:196 (1971).
73. Z. Er-el, Y. Zaidenzaig, and S. Shaltiel, *Biochem. Biophys. Res. Commun., 49*:383 (1972).
74. P. O'Carra, in *Industrial Aspects of Biochemistry* (B. Spencer, ed.), North Holland, Amsterdam, 1974, p. 107.
75. C. R. Lowe and P. D. G. Dean, *Affinity Chromatography*, John Wiley and Sons, London, 1974, p. 218.
76. G. I. Tesser, H.-U. Fisch, and R. Schwyzer, *Helv. Chim. Acta, 57*:1718 (1974).
77. M. Wilchek, T. Oka, and Y. J. Topper, *Proc. Natl. Acad. Sci., USA, 72*:1055 (1975).
78. M. Wilchek, in *Enzyme Engineering*, Vol. 3 (E. K. Pye and H. H. Weetyall, eds.), Plenum Press, New York, 1978, pp. 283-289.
79. R. Axén, J. Carlsson, J-C. Janson and J. Porath, *Enzymologia, 41*:359 (1971).
80. M. Wilchek and T. Miron, *Meth. Enzymol., 34*:72 (1974).
81. P. Cuatrecasas, *J. Biol. Chem., 245*:3059 (1970).
82. P. O'Carra, S. Barry, and T. Griffin, *FEBS Lett., 43*:169 (1974).
83. R. J. Yon, *Biochem. J., 126*:765 (1972).
84. S. Shaltiel and Z. Er-el, *Proc. Natl. Acad. Sci., USA, 70*:778 (1973).
85. B. H. J. Hofstee and N. F. Otillio, *Biochem. Biophys. Res. Commun., 50*:751 (1973).
86. B. H. J. Hofstee, *Anal. Biochem., 52*:430 (1973).
87. S. Hjertén, *J. Chromatogr., 87*:325 (1973).
88. J. Porath, L. Sundberg, N. Fornstedt, and I. Olsson, *Nature, 245*:465 (1973).
89. S. Hjertén, J. Rosengren, and S. Påhlman, *J. Chromatogr., 101*: 281 (1974).
90. J.-C. Janson and T. Låås, in *Chromatography of Synthetic and Biological Polymers*, Vol. 2 (R. Epton, ed.), Ellis Horwood, Chichester, 1978, p. 60.
91. S. Shaltiel, *Meth. Enzymol., 104*:69 (1984).
92. H. Lis and N. Sharon, *Ann. Rev. Biochem., 42*:541 (1973).
93. T. Kristiansen, *Meth. Enzymol., 34*:331 (1974).
94. Blue Dextran 2000, Pharmacia AB, Uppsala, Sweden.
95. G. Kopperschläger, G. Freyer, W. Diezel, and E. Hofmann, *FEBS Lett., 1*:137 (1968).
96. R. Haeckel, B. Hess, W. Lauterborn, and K. H. Wuster, *Hoppe-Seyler's Z. Physiol. Chem., 349*:669 (1968).

97. P. Hughes, R. F. Sherwood, and C. R. Lowe, *Biochim. Biophys. Acta, 700*:90 (1982).
98. P. Hughes, R. F. Sherwood, and C. R. Lowe, *Eur. J. Biochem., 144*:135 (1984).
99. G. Kopperschläger, H. J. Böhme, and E. Hofmann, in *Advances in Biochemical Engineering*, Vol. 25 (A. Fiechter, ed.), 1982, p. 101.
100. M. J. Harvey, in *Methods in Plasma Protein Fractionation* (J. M. Curling, ed.), Academic Press, New York, 1980, p. 189.
101. W. J. Jankowski, W. von Münchhausen, E. Sulkowski, and W. A. Carter, *Biochemistry, 15*:5182 (1976).
102. S. Pestka, *Arch. Biochem. Biophys., 221*:1 (1983).
103. C. R. Lowe, S. J. Burton, J. C. Pearson, Y. D. Clonis, and V. Stead, *J. Chromatogr., 376*:121 (1986).
104. R. K. Scopes, *J. Chromatogr., 376*:131 (1986).
105. J. Porath, J. Carlsson, I. Olsson, and G. Belfrage, *Nature, 258*:598 (1975).
106. E. Sulkowski, *Trends Biotechnol., 3*:1 (1985).
107. L. Andersson and J. Porath, *Anal. Biochem., 154*:250 (1986).
108. J. Porath, B. Olin, and B. Granstrand, *Arch. Biochem. Biophys., 225*:543 (1983).
109. J. W. Heine, J. Van Damme, M. De Ley, A. Billau, and P. DeSomer, *J. Gen. Virol., 54*:47 (1981).
110. S. Ohlsson, L. Hansson, P.-O. Larsson, and K. Mosbach, *FEBS Lett., 93*:5 (1978).
111. K. Unger, *Meth. Enzymol., 104*:154 (1984).
112. K. Brocklehurst, J. Carlsson, M. P. J. Kiestan, and E. M. Crook, *Biochem. J., 133*:573 (1973).
113. Thiopropyl-Sepharose 6B, Pharmacia AB, Uppsala, Sweden.
114. K. Brocklehurst, J. Carlsson, and M. P. J. Kierstan, *Top. Enzyme Ferment. Biotechnol., 10*:146 (1985).

# 13

# Theory and Design of Chiral Stationary Phases for the Direct Chromatographic Separation of Enantiomers

William H. Pirkle and Thomas C. Pochapsky / *School of Chemical Sciences, University of Illinois, Urbana, Illinois*

That one may speak of *designing* chiral stationary phases (CSPs) for chromatographically separating enantiomers is an indication of how far the field has developed in the last two decades. The acceptance of the technique as the preferred method of monitoring enantiomeric purity has spurred the development of new CSPs and enlivened debate regarding the molecular basis for enantiomer separation on CSPs. At the same time, the answers to these mechanistic questions are pertinent to more general questions regarding how selective complexation ("molecular recognition") occurs and what forces mediate such complexation. In this aspect CSP design has importance beyond the practical benefits that the technology provides and will make major contributions to the understanding of selective complexation. One might justly say that all chromatographic separations are mediated by selective transient interactions between analyte and stationary phase. However, CSPs which differ only subtly in structure may still demonstrate a large disparity in ability to separate the same enantiomeric pairs, indicating that it is the *complementarity* of interactions between stationary phase and analyte which is important for enantioselectivity and makes CSPs such sensitive tools for studying molecular recognition.

The authors of this chapter are in the enviable position of having had access to the chapters by V. Davankov and H. Hemetsberger as we made our selection of topics. Both of these chapters and other current reviews [1,2] describe in detail recent progress in chromatographic separation of enantiomers, information which would be redundant here. It is our aim instead to consider general principles which apply to CSP design and give examples of their application.

Before proceeding further it is appropriate to define what constitutes a "designed" CSP. Most CSPs, especially the commercially available ones, incorporate some structure which is obtained or easily derived from the natural chiral pool (i.e., amino acids, carbohydrates, etc.). As such, it is the rational elaboration of the natural material which constitutes "design." This elaboration can be fairly simple, e.g., the acylation of various polysaccharides in order to improve their chromatographic properties, or it can be relatively complex, involving several synthetic steps to reach the CSP from the natural chiral material. Alternatively, a completely synthetic CSP precursor, designed to achieve a specific type of separation, may be prepared. In either case, the objective is the same, and the same considerations apply. It is these considerations which we will attempt to define.

## THEORETICAL BASIS FOR ENANTIOSELECTIVITY

The property of *chirality*, or *dysymmetry*, is shared by objects which are not superimposable on their mirror image. Chiral objects are not bisected by a mirror plane. Mirror image nonsuperimposable chiral objects are termed *enantiomers*. Although this definition is adequate for chirality in immutable objects, molecular chirality, by virtue of the dynamic nature of molecules, must be further clarified. Molecules that are chiral due to differential substitution of a tetrahedral or higher order center (*stereogenic* or *chiral* center) are *asymmetrical*, with $C_1$ symmetry. A second type of chirality is observed in molecules which have $C_n$ symmetry ($n > 1$) but are able to populate conformers which are chiral. If interconversion between such conformers is fast, these conformations will average over time. However, if the barrier to interconversion between the enantiomeric conformers is sufficiently large with respect to kT,* such enantiomers can be isolated. Chirality in binaphthyl systems is of this sort, as is that in various helical polymers [3].

When a chiral compound is generated in an achiral environment, it is generated as a racemate, an equimolar mixture of both enantiomers. This is because in an achiral environment, enantiomers are energetically degenerate and interact in identical fashion with the environment. It is only in a chiral environment that enantiomers become nondegenerate and may be formed with some stereochemical bias. By the same token, enantiomers may only be differentiated from each other in a chiral environment, and it is by providing such an environment that a CSP operates.

---

*Where k is the Boltzmann constant and T the Kelvin temperature.

The operation of a CSP involves the formation of a transient *diastereomeric* complex (or adsorbate). In this context, *diastereomers* are chiral molecules containing two or more stereogenic (chiral) centers and having the same chemical composition and bond connectivity. However, they differ in stereochemistry about one or more of the chiral centers. Provided two *stereoisomers* are not enantiomers of one another, they are, in principle, chemically distinguishable in an achiral environment. A simple example of diastereoisomerism is that of two threaded bolts which are mirror images of each other, and hence enantiomers. In an achiral environment, such as a box, they are identical, and behave in identical fashion to achiral stimuli. When one attempts to thread these bolts into identical proper sized nuts (diastereomer formation), it is obvious that the two bolts are not equivalent.

Generation of diastereomers from enantiomers takes several forms. One may generate isolable, long-lived diastereomers by chemical derivitization of the enantiomers with a chiral reagent. These may be separated by achiral means. Alternatively, transient interaction between each of two enantiomers and some chiral entity can generate short-lived diastereomeric complexes. These complexes, usually not isolable, may be sufficiently energetically nondegenerate to be used to differentiate and/or separate enantiomers. In chromatographic systems, the chiral agent is either added to the mobile phase or is incorporated into the stationary phase. In the former case, large energetic nondegeneracy is not rigorously essential for enantiomer separation to occur. In the latter case, nondegeneracy is essential to enantiomer separation. All CSPs must form such transient diastereomeric complexes to be successful.

Diastereomeric complexation takes place as a consequence of one or more attractive interactions between the species involved. In a more passive alternative manner, inclusion of enantiomers into chiral cavities, usually in a polymeric network, need not necessarily involve attractive interactions between cavity and the included species. Never-the less, we consider such a situation to be operationally equivalent to complex formation.

Why are diastereomeric complexes energetically nondegenerate? A priori, they differ in symmetry and are members of different point groups. The electronic, vibrational, and rotational energy level ordering of the complex is defined by the symmetries of the various molecular orbitals, which are those allowed by the irreducible representations of the point group to which the complex belongs [4]. Since the molecular orbitals cannot have symmetry elements which are foreign to the point group of the complex, the diastereomeric complexes must be nondegenerate and energetically nonequivalent, except by coincidence.

It is instructive to look at the symmetry differences between two diastereomeric complexes generated by interaction of the enantiomers of one compound with a single enantiomer of a second compound in terms of a "plane of pseudosymmetry" which bisects the axis along which the two molecules approach each other to form the complexes, as shown in Fig. 1. If the complexes to be considered are formed from the enantiomers of species X(ABCD) and one enantiomer of a chiral complexing agent X(A'B'C'D'), the "meso" complex 1 has a plane of "pseudosymmetry" which is not defined for complex 2. This plane is uniquely defined, like all planes, by three nonaligned points. Any three points in the plane may be chosen, but it is most convenient to choose the midpoints of the three interatomic vectors A-A', B-B', and C-C'. The same three vector midpoints in complex 2 will not define the same plane as in complex 1, (i.e., the plane in complex 2 does not bisect the axis between the two chiral centers X, and is not a plane of "pseudosymmetry"). Our consideration of the three simultaneous interactions of A with A', B with B', and C with C' (or D with D') leads to some conclusions regarding the interactions of chiral molecules: although the line defined by any two points in either complex is not unique (it can be duplicated in either complex if the molecules are free to tumble with respect to each other), the planes defined above are unique, and differ between the two complexes. It is only by consideration of the plane defined by three simultaneous points of interaction between the two molecules that nonequivalence between the two complexes is discernible.

The presence of such unique planes is the trivial result of interaction between two chiral species. They occur no matter what the relative orientation of the two molecules in the complex. If this difference in symmetry was *all* that was required for enantiomer separation, any chiral species would suffice to differentiate between the enantiomers of any other species. This is obviously not the case. The practical problem is of course the *degree* of perturbation of the energy levels of both species which results from the interactions between the two species. Recent calculations, treating the interactions between two chiral tetrahedral species in terms of sums of interactions between monopoles, dipoles, and higher order multipoles, have shown that energy differences between diastereomeric complexes 1 and 2 are not the result of atom-atom or dipole-dipole interaction (interaction of tetraherdon edges) but rather are the result of six-center forces occurring simultaneously between triplets of atoms or functionality in both species [5]. This agrees intuitively with the symmetry arguments above. Hence, the greater the degree of interaction occurring between the three sets of atoms A-A', B-B', and C-C' *simultaneously*, the greater the degree of energy difference between complexes 1 and 2. This is not to imply that all the interactions A-A', B-B', and C-C' must be strong interactions, or even attractive. If, for example, two of the interactions are attractive, they may force steric interaction at a third point. In fact, if complexation is passive, as is diffusion

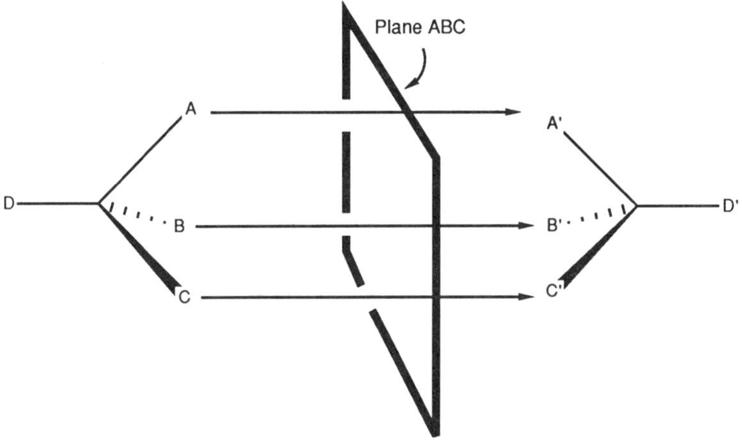


"Meso" complex 1

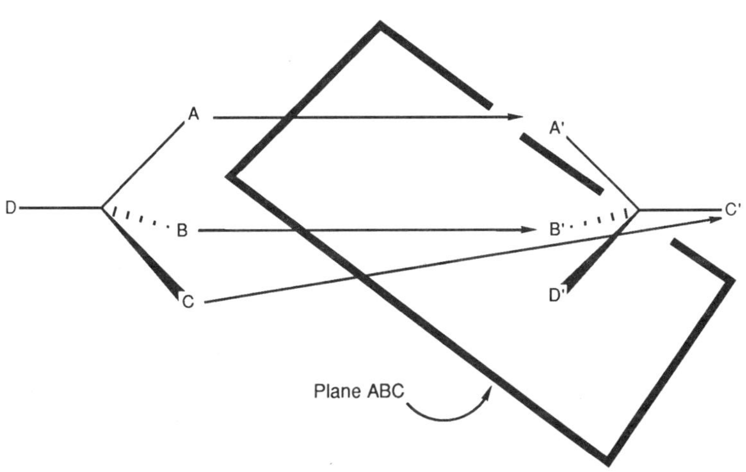


"Homochiral" Complex 2

Fig. 1 Three-point requirement for enantioselective interactions. Any combination of one-point (i.e., AA', BB', AB', etc.) or two-point (AA'BB', AB'BA', etc.) interactions may be duplicated by either complex, assuming complete rotational freedom of both species in the complexes. Midpoints of vectors defined by three *simultaneous* inter actions AA'BB'CC' together define planes ABC which are nonidentical in the two complexes. Six-center interaction is thus essential for enantioselectivity.

into a chiral cavity, none of the interactions need be attractive. If, however, there is strong interaction between only one pair of sites, the chance of significant differential perturbation is small, since the six-center forces will not be significant.

It is important to note that although Fig. 1 represents the interacting substituents of the two species as atoms appended to a tetrahedral stereogenic center, the interacting sites need not be so appended. They must, however, have some preferred orientation with respect to the chiral center, or else either enantiomer will be able to undergo all the given interactions simultaneously and little nonequivalence will be observed.

The first and second requirements for designing an effective CSP follow from these observations. Enantioselectivity is the result of differential interactions between a chiral entity and the enantiomers of (typically) a second substance. These interactions result in diastereomeric complexes which necessarily differ in symmetry and which may be energetically nonequivalent. The degree of energetic nonequivalence is determined exclusively by higher order (six-center or greater) interactions between the two species, and it is therefore desirable to design the CSP to interact with its client compounds at as many points as possible in order to maximize these higher order interactions. Furthermore, there must be at least some stereochemical dependence of these interactions, i.e., there must be some preferred orientation of the interacting functionality with respect to the chiral centers of both species.

The aforementioned requirements, though necessary, are still insufficient. For example, if interactions A-A' and B-B' are similar in nature and of the same relative orientation (A and B both being hydrogen bond donors and A' and B' both being acceptors, for example), one can imagine relative orientations of components of diastereomeric complexes which are nearly identical (see Fig. 2). This leads to a third requirement for successful CSP design, i.e., the multiple interactions available to the analyte molecules must be as mutually exclusive as possible, so as to prevent a given interaction from occurring at multiple sites in the diastereomeric complexes.

These design requirements present somewhat of a dilemma in that a CSP must be essentially tailor-made to suit a particular type of analyte for maximum enantioselectivity. This is because both the type of interactions available to the CSP and analyte, as well as the spatial arrangement of the interaction sites, will determine the effectiveness of the CSP in separating the enantiomers of a given analyte. This also means that optimization for one analyte may lead to only marginal separations, or none at all, with other analytes. A good example of this situation is presented by ligand exchange chromatography (LEC). LEC was applied by Davankov and Rogozhin in 1971 [6], and was the first chromatographic technique for enantiomer separation for which

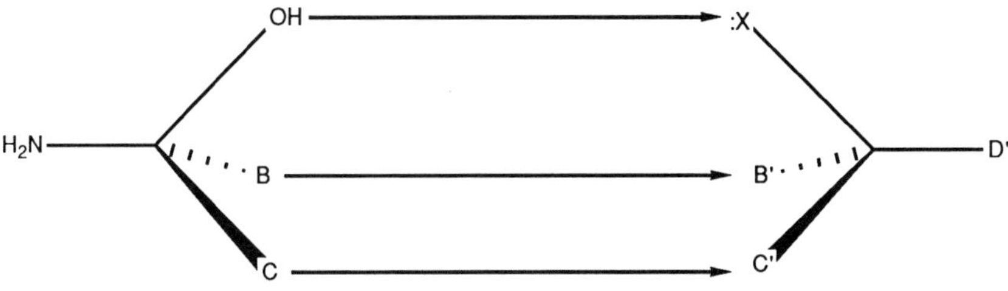


Complex A

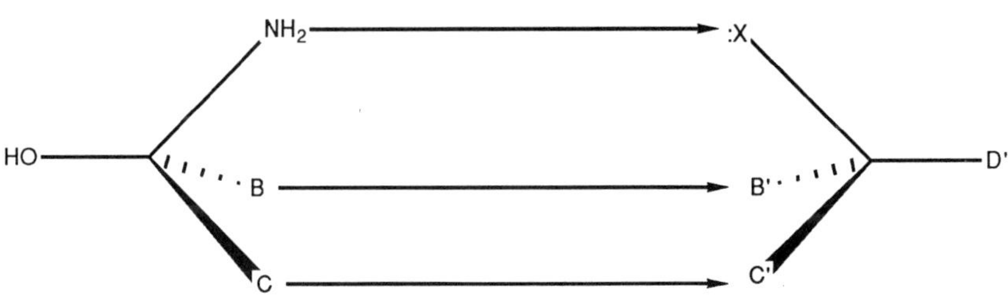


Complex B

**Fig. 2** Nonexclusivity as a source of problems in designed enantioselectivity. Although complexes A and B are diastereomeric, they do not differ greatly in energy, since both the amino and hydroxyl hydrogens of the left-hand partner of each complex are capable of hydrogen-bonding to group X of the right-hand partner.

a testable mechanism was proposed. Based on the formation of diastereomeric divalent metal ion complexes involving two amino acids and one or more solvent molecules, LEC has evolved to become the method of choice for the rapid chromatographic analysis of underivatized α-amino acid enantiomers. This field has often been reviewed [7,8] and a chapter for this book is devoted to recent developments in LEC. Because the formation of the diastereomeric complexes in LEC involves chelation of the metal ion by bidentate ligands, usually α-amino acids, this type of separation is restricted to chelating species having the appropriate spacing between the chelating functional groups. LEC works much less well for β-amino acids [9], indicating that either the spacing of the amino and carboxylate groups of β-amino acids is not appropriate for effective chelation, or that there is insufficient steric rigidity in the molecule to bring about large free-energy differences in the diastereomeric complexes.

To design broad-range CSPs, one must strike some middle ground, where some selectivity is sacrificed for more general applicability. Several avenues for obtaining wide-spectrum applicability from a CSP are available. One approach is to design the CSP so that the specific sites of interaction required for separation on that CSP are easily incorporated by simple derivitization into a wide range of prospective analytes. Alternatively, the CSP itself may be designed to be capable of more than one mode of enantioselectivity by incorporating functionality which is capable of a number of different interactions. This is clearly the case with many CSPs containing amide functionality. Primary and secondary amides are capable of either donating or accepting hydrogen bonds or undergoing dipolar interactions, and are also capable of π-dipole stacking. In several cases, amide-containing CSPs have been found to have multiple modes of enantioselectivity (vide infra).

The problem of appropriate spatial arrangement of the interaction sites on the CSP is also approached in a spirit of compromise. Ideally, two completely rigid systems, perfectly complementary, would give the most exclusive enantioselectivity. However, some flexibility seems to be desirable in a broad-spectrum CSP. This allows it to conform to the steric requirements of a variety of analytes. Of course, care must be taken that there still remains some stereochemical "communication" between the stereogenic center and the sites of interaction, as described above.

In summary, the design elements required for an effective CSP are:

1. At least three points of interaction between CSP and analyte, at least two of which should be attractive (except for intercalative mechanisms)

2. Some orientational bias of the interaction sites with respect to the stereogenic centers
3. A degree of mutual exclusivity among the available interactions.

The remainder of this chapter will examine these principles as they apply to various existing CSPs.

## IMPLICATIONS OF MECHANISMS OF ENANTIOMER SEPARATION ON THE DESIGN OF CSPs

It was very early recognized that in principle one might separate enantiomers by enantioselective adsorption onto some chiral nonracemic stationary phase. Adams and coworkers used a variety of chiral materials such as wool and cellulose to impart optical activity into solutions of racemic dyes by preferential adsorption of one enantiomer [10]. Similar reports are found scattered through the literature of the past half-century [11]. However, not until chromatographic science had itself been developed sufficiently were serious attempts made to prepare designed CSPs. Because of the rapid development of gas chromatography with respect to liquid techniques, some of the earliest successes were in this area, and it is some of the CSPs designed for gas chromatography (GC) use which will be discussed first.

As with any GC stationary phase, two requirements are paramount for GC CSP: thermal stability and nonvolatility. Thermal stability in this case implies stereochemical stability as well. Long-term stability of a chiral GC phase requires a high-activation barrier to racemization. Derivatives of a variety of α-amino esters, dipeptides, and diamides have been found to fulfill these requirements, and the work of Gil-Av, Feibush, and Weinstein, among others, has lead to a number of GC CSPs derived from these compounds. The first successful GC CSP was *N*-trifluoroacetyl-L-isoleucine lauryl ester, *3*, which was used to separate the enantiomers of *N*-trifluoroacetyl-α-amino esters [12]. Similar phases, prepared from *N*-trifluoroacetyl dipeptides, also proved successful [13,14]. Other GC CSPs have been prepared from α-naphthylethylamine lauramide [15], diamides of α-amino acids [16], and tripeptides [17].

Despite the fact that GC CSPs have been studied since the mid-1960s, less is known about their mode of action than that of their liquid chromatography (LC) counterparts.

Any chromatographic separation involves the partitioning of the analyte between stationary and mobile phases. Assuming a fast-exchange situation, this partitioning reaches equilibrium for all regions of the stationary phase, and the separation of two analytes is controlled by the partition coefficients $Kp_i$. Since the $Kp_i$ are equilibrium constants, they are related to the free energies of partition by Eq. (1).

$$\Delta G_{partition} = -RT \ln Kp_i \tag{1}$$

The separation of the two analytes, measured as the ratio of their corrected retention times $\alpha$ (which is also the ratio of the $Kp_i$), is a measure of the difference in free energies of adsorption of the analyte on the stationary support $\Delta\Delta G$, as indicated by Eq. (2).

$$\Delta\Delta G_{partition} = -RT \ln \alpha \tag{2}$$

$\Delta\Delta G$ can be broken down to two terms, $\Delta\Delta H$ and $T\Delta\Delta S$. The enthalpy term $\Delta\Delta H$ reflects the relative binding energies of the diastereomeric complexes and is closely related to observable mechanistic considerations. The entropy term, on the other hand, reflects the change in order of both the stationary and mobile phases upon complexation, and is more difficult to explain mechanistically in any detail. In ambient temperature LC, the $\Delta\Delta H$ term is expected to be most significant, since for enantiomers the ordering of an achiral solvent with respect to enantiomers should be the same. Differences in entropy would almost completely reflect the relative freedom of motion which the analyte enantiomers have in the complex and the extent to which the complexes are solvated. Solution differences seem minimal at ambient temperature except for a few somewhat exceptional cases. One such case, noted by Davankov, results from the two diastereomeric copper complexes involving different numbers of bound species, with one diastereomeric complex incorporating a water molecule and the other not. In this case, the $\Delta\Delta S$ term was dominant at ambient or near-ambient temperatures. For the vast majority of enantiomer separations, the $\Delta\Delta S$ term will not be the controlling term in the free-energy relationship *near ambient temperature*. As temperature is increased, however, even a relatively small entropy term will become increasingly important, and the small free-energy differences typically seen for diastereomeric complexes will become even smaller. This means that GC separation factors are typically quite small. Because of the high number of theoretical plates of most modern capillary GC columns, this is not a serious drawback. But because of the difficulty in separating enthalpic from entropic contributions, proposed mechanisms of interaction become harder to test, and designing CSPs for GC applications is difficult owing to a lack of theoretical models.

Nevertheless, attempts to discern the mechanisms of enantioselectivity on GC CSPs have been made. Preliminary speculation concerning the mechanism of separation of *N*-trifluoroacetyl amino esters on CSP *3* centered around hydrogen bonding between trifluoroacetyl

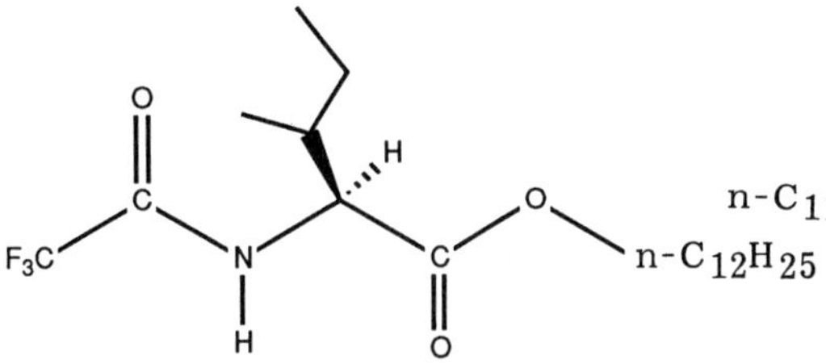

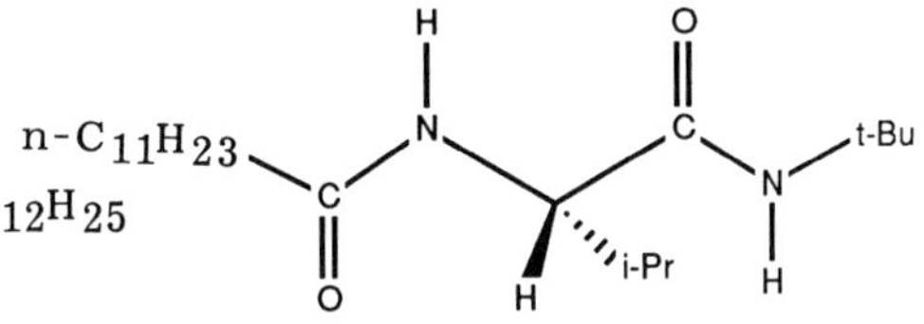

**3** **4**

groups, although no formal model relating configuration to elution order has been proposed [18]. Similarly, for *N*-trifluoroacetyl dipeptide CSPs, no attempt has been made to correlate the structure of the diastereomeric adsorbates with elution order, although it has been speculated that hydrogen bonding is important [19]. A somewhat more exacting proposal has been made concerning the nature of enantioselectivity on diamide GC CSPs such as *4*. It has been observed that polyamides are capable of associating in β-sheetlike structures, which can then intercalate a diamide analyte, as in Fig. 3 [20]. Parallel intercalation will give rise to preferential retention of the homochiral enantiomer of an amino acid diamide (same configuration as the CSP) while antiparallel intercalation would result in the retention of the heterochiral enantiomer. Since the homochiral enantiomer is observed to be the most retained experimentally, the parallel intercalation mechanism is thought to be thermodynamically favored [21].

Other GC CSPs give evidence that cooperative effects may be important to enantioselectivity. X-ray studies of *N*-acetyl-α-naphthylethylamine indicate that a highly ordered environment is maintained by a network of hydrogen bonds in this crystalline substance. By assuming that the molten lauramide has a residue of similar order, it was suggested that the analyte enantiomer which best intercalates into this postulated network will be most retained on the *N*-lauroylnapthyl-

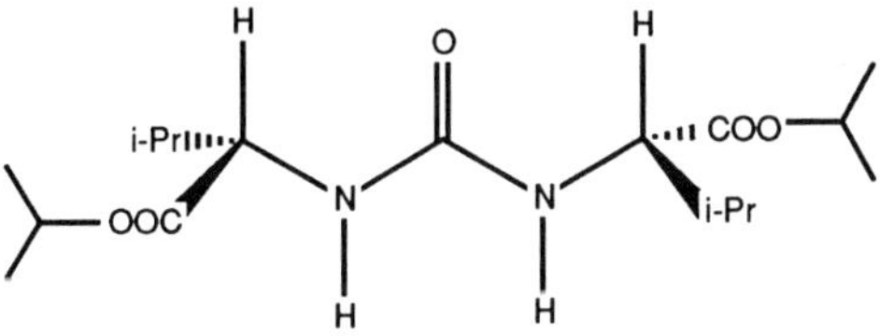

**5**

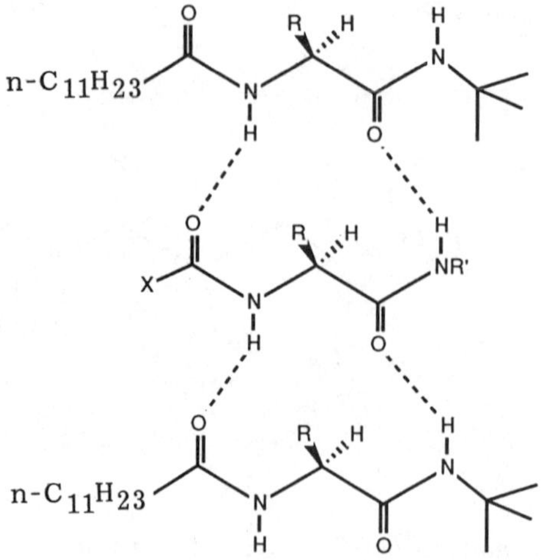

Parallel- S retained in S

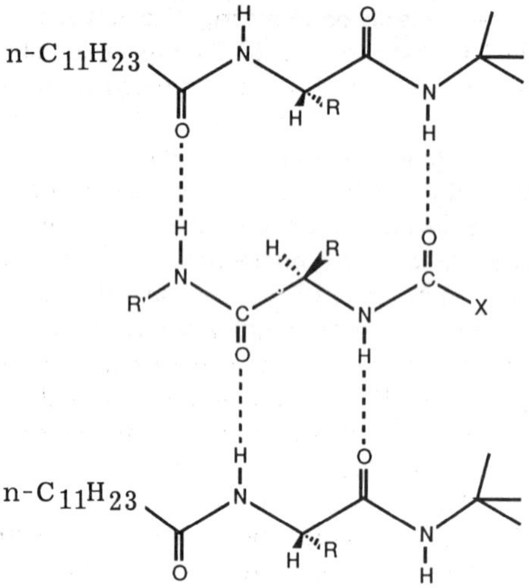

Antiparallel - R retained in S

Fig. 3 Intercalation of analyte into the proposed "β-sheet" structure of GC CSP *4*. Parallel intercalation leads to $C_5$-$C_7$ hydrogen-bonding patterns, preferred in the solid phase by the CSP. This mechanism is thought to favor intercalation of the homochiral analyte enantiomer (same configuration as the CSP). Antiparallel intercalation ($C_5$-$C_5$ and $C_5$-$C_7$) will favor intercalation of the heterochiral enantiomer (opposite configuration to the CSP).

ethylamine CSP. Studies by Lochmuller and Souter indicate that the separation of chiral amides on the GC CSP *5* is very dependent on the long-range order of the stationary phase [23]. Separation is in general much poorer above the melting point of the CSP (i.e., when the CSP is an isotropic melt) than in the smectic mesophase [24]. Indeed, separations were further improved using the CSP in the solid state, verifying the importance of long-range order in the CSP to chiral recognition [25].

These observations seem to preclude simple 1:1 interaction models for GC separation of enantiomers such as those which have been proposed to account for a number of LC enantiomer separations, and reenforce initial suspicions that entropy considerations are relatively more important in GC than in LC. Nevertheless, related LC CSPs have been developed for which 1:1 interaction models have been proposed, and it is to these systems that we now turn. These CSPs provide separations of similar analytes under LC conditions as their GC counterparts [26-28]. They are of interest not so much because of their practicality but because they were for some time at the focal point of a discussion regarding their mechanism of operation. Portions of the debate illustrate nicely some of the concerns regarding elucidation of the mechanism(s) of CSP action. Hara, who synthesized a number of these phases, provided much of the experimental data regarding their mode of action. He initially proposed a dual hydrogen-bonding mechanism, entailing head-to-tail dimers, as shown in Fig. 4. Pirkle suggested that the two-point mechanism proposed was insufficient to account for enantiomer separation since either of the enantiomers was capable of undergoing such interactions [29]. He proposed an alternative mechanism, based on dipole stacking of the π systems of the acylamino acids, which provides the requisite multipoint interaction. Stereoselectivity of the dipole-stacking interaction comes from the preferred conformation of the acylamino acids as shown and the fact that the interaction is between the planes defined by the three-atom π systems, not between single points. Hara presented data indicating that increasing the steric bulk of either the acyl group of the stationary phase or the analyte decreases the separation factors of enantiomers of α-acylamino amides [30]. This seems more consistent with a face-to-face interaction between π systems as opposed to the edge-to-edge model, which tends to place sterically bulky R groups distant from each other in the complex.

At this point, it is tempting to use the above example to illustrate another fairly general theme in CSP design. The π stacking of amide dipoles proposed by Pirkle to rationalize the observed separations on CSP *6* is an example of a common motif in CSP design, i.e., the use as an interaction site of some π or aromatic functionality which is conformationally restrained with respect to the chiral center. Because π-π and π dipole stacking interactions are highly directional and multi-

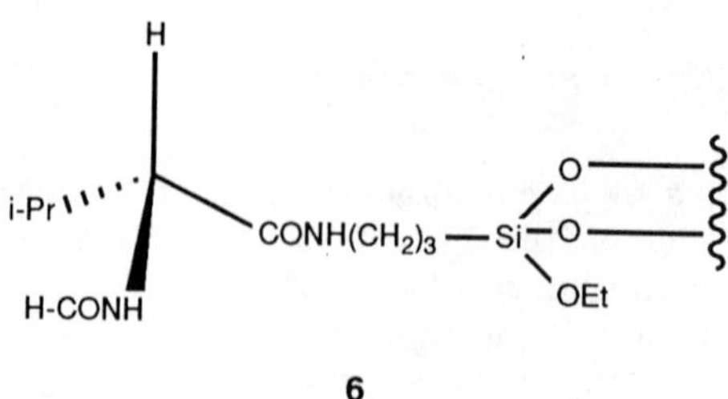


6

Hydrogen-bonding model
(2-point)

Dipole-stacking model (4-point)

Fig. 4 Enantioselectivity models for CSP *6*. The initially proposed hydrogen-bonding model provides only two points of interaction, insufficient for enantioselectivity. The dipole-stacking model provides the requisite number of interactions and also explains the observed elution orders of analyte enantiomers.

point, they provide a good deal of stereoselectivity if incorporated appropriately into a CSP. The nature of such interactions could be considered in the following fashion.

To completely define the relative positions of two circular planar objects, only two vectors need to be defined—one joining the centers and one joining a point on each circumference. Figure 5 demonstrates this relationship. Describing the interactions between two aromatic groups in this fashion, a $\pi$-$\pi$ interaction between aromatics can, in principle, act as two of the three necessary points of interaction required for enantioselectivity. Figure 5 also indicates how enantioselectivity can occur between two conformationally rigid species using a $\pi$-$\pi$ interaction and one other attractive interaction to provide the requisite three points of simultaneous contact. It is implicit in this model that the $\pi$ functionality be conformationally restrained with respect to the chiral center. Attachment of an aromatic group, for example, through a flexible methylene chain to the chiral center is experimentally not very effective in terms of chiral recognition [31]. The use of $\pi$-$\pi$ interactions for increased selectivity in molecular recognition is not unique to CSPs, of course. Many other designed molecular recognition systems make effective use of such interactions as well [32,33].

Two early designed CSPs which incorporated aromatic functionality include the first $\pi$ donor-acceptor-type CSP, designed by Mikes and Boshart [34]. Mikes' CSP 7, which incorporates nitrofluorenyl derivatives of hydroxypropionic acid, is capable of separating the enantiomers of a variety of helicenes. Despite relatively low separation factors, the high chromatographic efficiency of this bonded phase allowed baseline separation of the helicene enantiomers.

Pirkle and House also prepared a bonded phase CSP, *8*, which incorporated $\pi$ functionality [35]. This CSP is based on a chiral solvating agent previously designed for nuclear magnetic resonance (NMR)

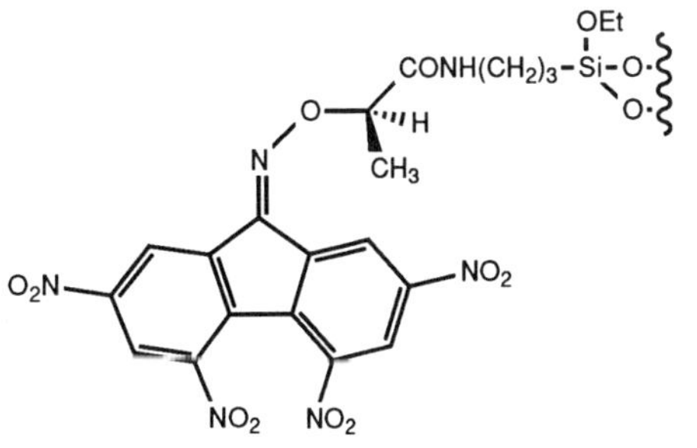

7

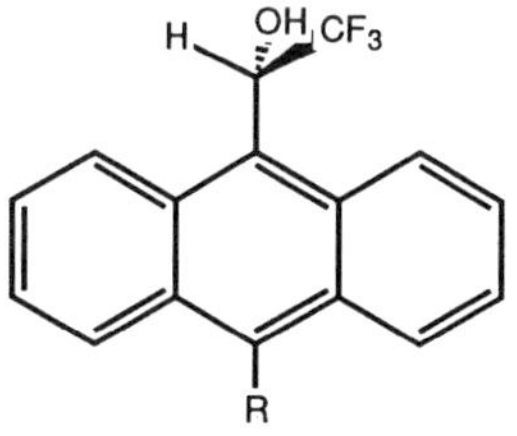

8

a, R = H

b, R = $CH_2SCH_2CH_2Si(OEt)O_2$---

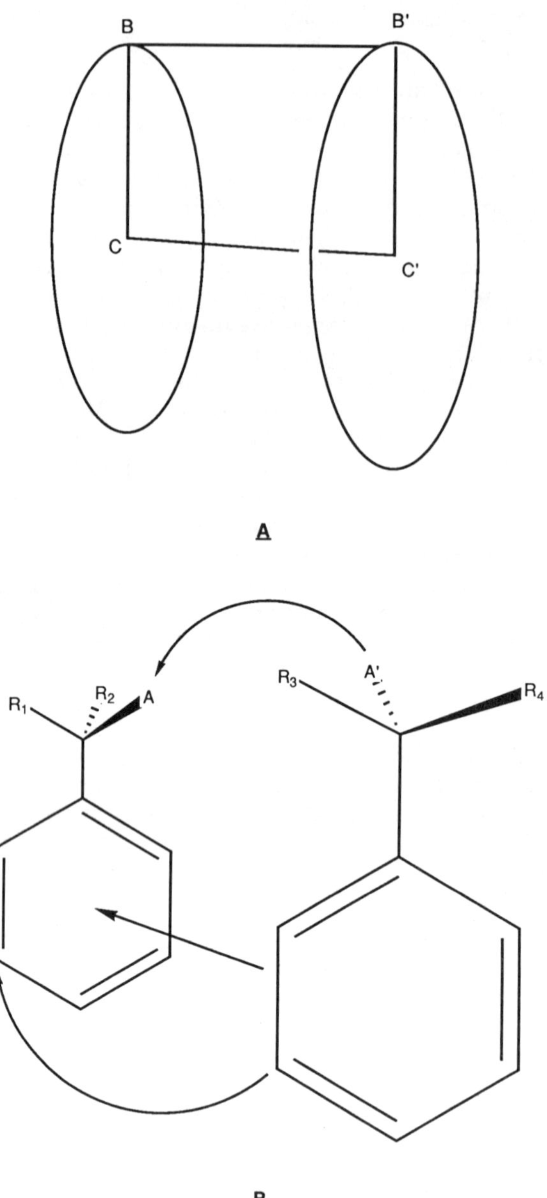


Fig. 5 Multipoint interaction between two planar circular objects. Interaction between the objects are completely defined by two vectors, one joining their centers and the other joining a point on either edge (A). A π-π interaction between two aromatic rings can provide in this fashion two out of the three requisite interactions necessary for chiral recognition (B).

analysis of enantiomer mixtures (through the induction of chemical shift nonequivalence). As more will be said of the relation between NMR and CSP design later, it suffices to note here that from solution NMR experiments it became obvious that in some cases there were significant free-energy differences between the diastereomeric complexes formed with the solvating agent [36]. It was also evident that these free-energy differences might be exploited to effect a separation of enantiomers for suitable analytes if the solvating agent could be immobilized on a chromatographic support. Several different methods of appending the solvating agent to the stationary phase were used to generate a series of broad-spectrum CSPs for the separation of π-acceptor-substituted enantiomers [37,38]. Based on NMR and UV/VIS studies, it was observed that one of the important interactions between the CSP and analyte was a π donor-acceptor interaction, with two hydrogen-bonding interactions implicated as well.

The enantiomers of a series of *N*-3,5-dinitrobenzoyl derivatives of α-amino acid esters and amides are readily separated on type *8* CSPs. CSPs *9a* and *9b* derived from two of these amides, separate the enantiomers of a variety of π-donor-substituted racemates [39]. In general, if a CSP derived from A separates the enantiomers of B, then a CSP derived from B should separate the enantiomers of A. A series of CSPs related by this concept have been developed by the Pirkle group and by others [40-42]. In general, all of these CSPs contain either donor or acceptor π functionality and dipolar or hydrogen-bonding groups.

The concept of reciprocity is expected to be valid for those systems where but a single selector interacts with an analyte enantiomer

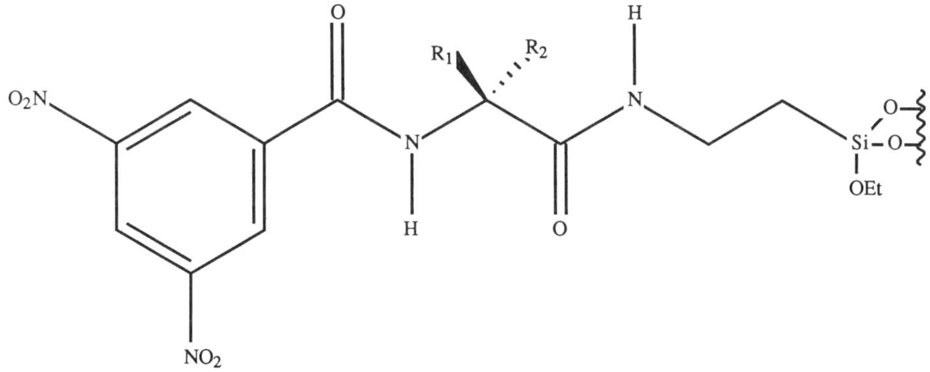

**9**

a, R1 = i-Bu, R2 = H

b, R1 = H, R2 = Ph

at any instant and neither neighboring strands of bonded phase nor the underlying support affect the energetics of complexation. This ideal situation is never actually attained, for one finds that variation in strand spacing, tether length, mode of binding, and nature of the underlying support do affect the extent and sometimes even the sense of chiral recognition afforded by a CSP. If one wishes to optimize the performance of a CSP, an understanding not only of how these variables affect performance but *why* is essential.

## ROLE OF STATIONARY SUPPORT IN CSP ACTION

There are instances where the nature of the stationary support greatly influences the observed enantioselectivity afforded by a CSP. It has been observed in LEC that the elution order of amino acids is determined by the nature of the solid support. On CSPs derived from L-proline utilizing a nonpolar support, the amino acid enantiomer most strongly retained is that which can place its side chain close to the stationary support while maintaining chelation of the metal ion [43]. If the stationary support is capable of providing an axial ligand for the metal ion, however, the amino acid enantiomer which places its side chain away from the stationary support, so as not to interfere with axial ligation during chelation, will be most retained [44].

This degree of dependence is unusual, although it is not uncommon for the proximity of the stationary support to be a factor in determining which of several mechanisms of enantioselectivity will predominate for a given type of analyte. Papers by Pirkle and Hyun address this situation [45-47]. It was noted that for several homologous series of analytes, enantiomer elution order from several CSPs, as well as the degree of separation, is dependent on the chain length of the alkyl substituents of the analyte as well as the length of the arm by which the CSP is connected to the stationary support (microparticulate silica). Specifically, for series of *N*-(3,5-dinitrobenzoyl)-α-arylalkylamines *10*, the enantiomers show a decrease in separation factor on CSP *11* as the length of the alkyl substituent increases, with no separation occurring for the enantiomers of the $C_8$ amide [45]. Further increase in chain length results first in the inversion of elution order with a subsequent increase in separation factors. To explain this observation, two competing mechanisms with opposite senses of enantioselectivity were proposed. The mechanisms are based on evidence obtained from the study of various analytes and are shown in Fig. 6. The (R) enantiomers intercalate their alkyl substituents between strands of bonded phase. This process is energetically favored on CSP *11* for those analytes having short alkyl chains. The (S) enantiomers are retained by a different combination of interactions which does not require intercalation of the alkyl substituent. As the

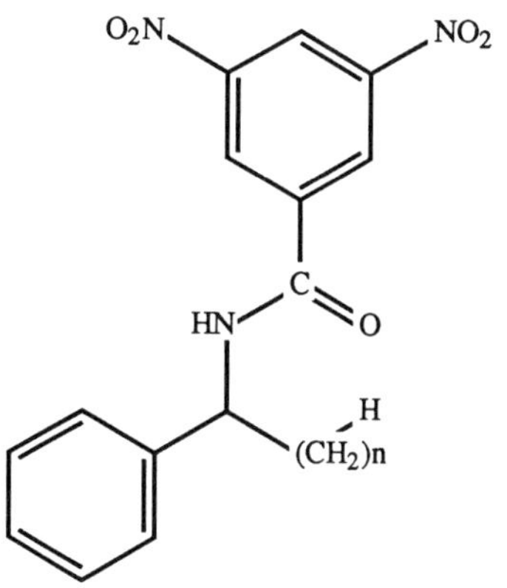

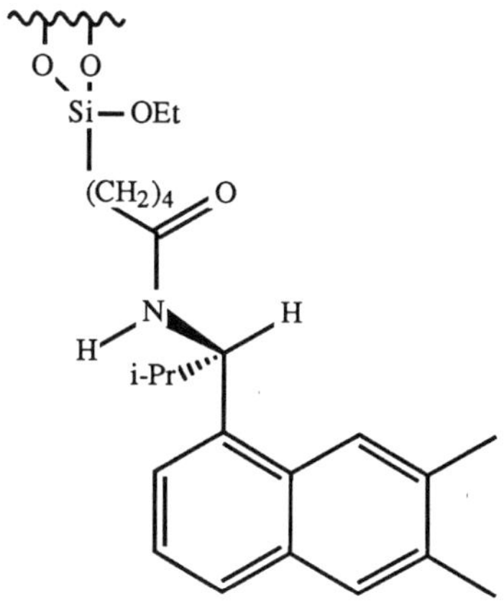

**10** **11**

alkyl substituents are increased in length, the (R) enantiomers elute progressively sooner, and eventually before the nonintercalating (S) enantiomers. Shortening the length of the connecting arm, lessening the average interstrand distance, or packing the space between strands with alkyl silanes makes retention of the (R) enantiomers relatively more difficult. A change of the orientation of the chiral moiety of the CSP with respect to the support changes the relative energetics of the two mechanisms and has the anticipated effects on elution order and selectivity.

A final observation is appropriate at this point. The preceding example demonstrates that the observed separations of enantiomers (or any analyte, in general) are not the result of a single mode of interaction but occur as a consequence of the weighted time-average of all modes of interaction between the CSP and analyte. For enantiomer, any interaction or set of interactions which both enantiomers undergo equally will not contribute to enantiomer separation but will attenuate the observed level of enantioselectivity resulting from those sets of interactions which do distinguish between the enantiomers. That is, the greater the fraction of time the analyte enantiomers spend engaging in such nonstereoselective interactions, the less enantioselectivity (i.e., the magnitude of the separation factor) which will be observed. Thus, not only must a CSP contain ample sites for those interactions which contribute to enantioselectivity, it should as nearly as possible contain no sites which lead to retention without enantioselectivity. For example, it has been noted that end capping (removal of underivatized silanol sites on bonded phase particles by treatment with a silanating agent) reduces retention and increases the level of

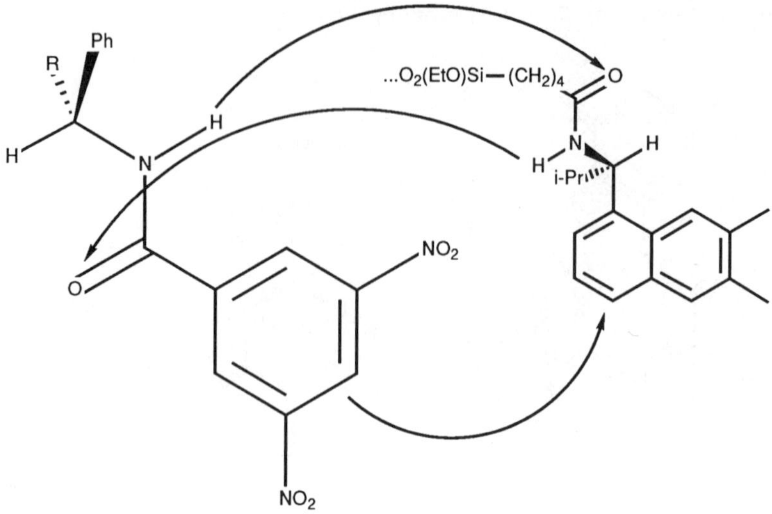

Dipole-stacking model (intercalative)

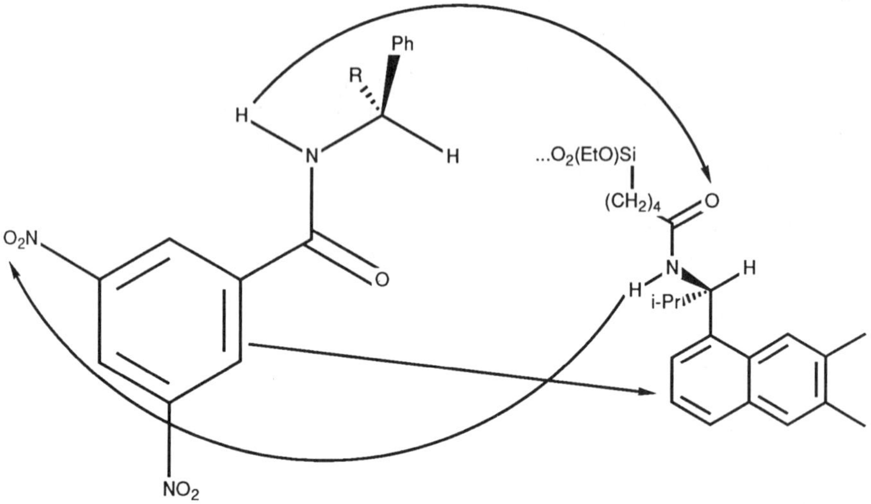

Hydrogen-bonding model (non-intercalative)

**Fig. 6** Dipole-stacking (intercalative) vs. hydrogen-bonding (non-intercalative) models for retention of α-arylalkyl amine dinitrobenzamides of type *10* on CSP *11*. As the length of the alkyl chain R on *10* increases, steric interactions between R and the underlying support tend to disfavor the dipole-stacking mechanism, while the noninter-

enantioselectivity, in some instances very significantly. In one case at least, end capping resulted in the doubling of observed separation factors [48].

## MAGNETIC RESONANCE AS A GUIDE TO CSP DEVELOPMENT

Most models invoked to explain observed enantiomer separations on LC CSPs are based on interactions which occur in solution. Implicit here is the assumption that the chiral portion of the bonded phase is fully solvated by the mobile phase and that a true phase boundary effectively does not exist between the mobile and stationary phases. There are obvious exceptions to this generality, a few of which were noted in the previous section, where the nature of the underlying support is critical to the extent and sense of separation. Nevertheless, it is often convenient to study the mechanism of enantiomer separation on CSPs by observing analogous solution systems spectroscopically. If the nature of enantioselectivity on a CSP is not dependent on the nature of the underlying support, spectroscopic studies of analogous solubilized systems can be of great value in determining the nature of the interaction between CSP and analyte. Of all spectroscopic techniques, nuclear magnetic resonance (NMR) is by far the most valuable for such studies and has been used for in-depth study of a variety of CSP-analyte systems.

An account of the design of CSP *8* illustrates quite nicely the general relationship between NMR and chromatographic techniques. In a historic experiment, the titration of a mixture of a racemic sulfoxide *12* and the enantiomerically pure chiral solvating agent *8a* with an achiral lanthanide shift reagent was followed by NMR. Different degrees of complexation between the lanthanide and the sulfoxide enantiomers were observed, indicating that the diastereomeric complexes between the sulfoxide enantiomers and the solvating agent are of different stabilities. It was expected that the diastereomeric complexes would also have different partition coefficients between a stationary

---

calative hydrogen-bonding model is unaffected. Since the two mechanisms have opposite senses of enantioselectivity, the degree of enantioselectivity diminishes, the elution order of the enantiomers gradually being inverted. That is, the (R) enantiomer is most retained by the dipole-stacking mechanism, which is initially dominant ($R < C_8$), whereas the (S) enantiomer is most retained for $R > C_8$ (hydrogen-bonding mechanism dominant).

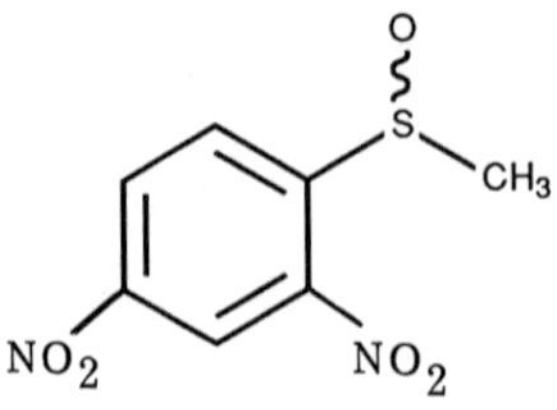

O
S
$CH_3$
$NO_2$ $NO_2$

**12**

and mobile phase and hence prove to be separable chromatographically. This was indeed the case, and chromatography of the racemic sulfoxide over silica in the presence of the chiral solvating agent *8a* resulted in enantiomeric enrichment of sulfoxide *12* in the first (and last) fractions [49]. CSP *8b* was developed directly as a result of these experiments.

The chemical shift nonequivalence between enantiomers induced by chiral solvating agent *8a* has been described in terms of a two-point interaction model [35]. At first, this may seem in violation of the three-point interaction requirement for enantioselectivity. However, the magnetic shielding tensors of the complexes, which describe the shielding a nucleus will experience at a given set of spatial coordinates, are the resultant of all the electronic and nuclear contributions to the magnetic enviornment of the system. Since the diastereomeric complexes are nonequivalent, nuclei in the two complexes will have the same chemical shift only accidentally. The degree of interaction between the two molecules in the complex will determine how much time one chiral species will spend in the vicinity of the second chiral species, and which groups will be most affected by the presence of the other species. Since the kinetics of complexation of the solvating agent and analyte are typically fast with respect to the proton chemical shift time scale, the observed nonequivalence is a time average of chemical shifts for the various allowable orientations of the complex as well as the chemical shifts of the free species. It is logical that the more time the two species spend complexed, the greater the effect exerted by one species on the magnetic environment of the other. In turn, the more time the complex spends in one conformation, the more the chemical shift changes will reflect that particular conformation. Both of these factors are controlled by the number and strength of the interactions between the two molecules in the complex.

Since the same factors important for enantiomer separation on a CSP are also important for chemical shift nonequivalence, the two techniques are complementary: both stem from the formation of short-lived multimolecular complexes. Whereas the NMR technique does not require the diastereomers to be energetically different for spectral nonequivalence to be observed, chromatographic separations do require some difference in the free energies of adsorption.

A number of other designed CSPs have benefitted from intensive NMR investigations of analogous soluble systems. The chiral crown ether systems developed by Cram and coworkers were the subject of a number of NMR studies, from which the development of CSPs proceeded [50,51]. In this case, the initial studies were aimed at examining the partition of racemic amino acids between two phases using the chiral crown ether as a selective complexing agent. It was later shown that an immobilized chiral crown ether, *13*, is effective as a CSP for resolving chiral amino acids and amines (Fig. 7) [33].

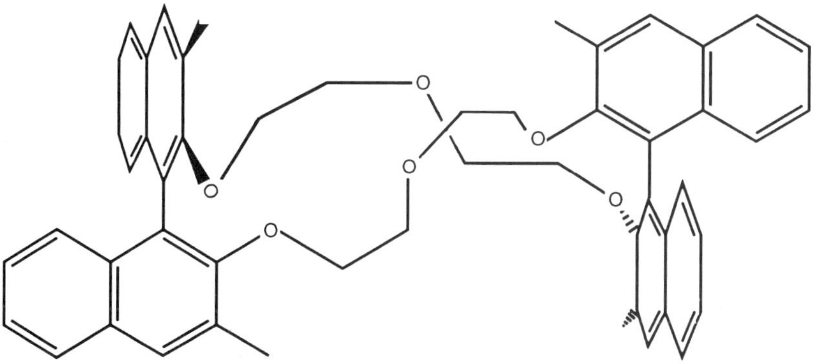

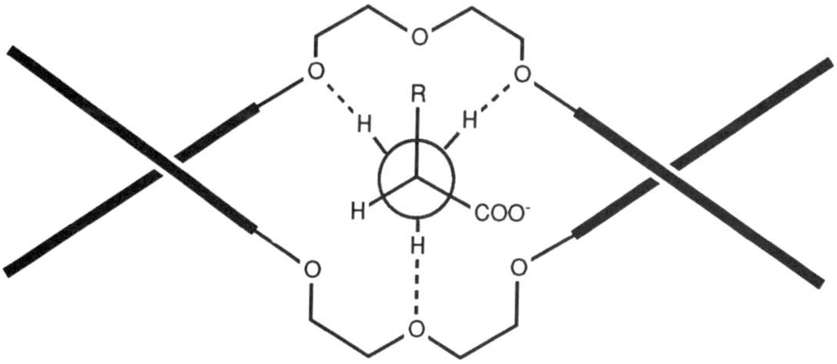

Fig. 7 Enantioselective binding of amino acid by a chiral crown ether *13*. Only one enantiomer of the amino acid is able to maintain complexation between its ammonium ion and the crown ether while avoiding steric interactions between its side chain and the naphthyl ring of *13*.

More recently, a series of solution NMR studies, including intermolecular nuclear Overhauser effect (NOE) experiments, was used to determine the nature of enantioselective complexation between type *14* *N*-aryl amino acid derivatives and *N*-(3,5-dinitrobenzoyl)leucine amide *15*. This system has proven to be well suited for study of its mechanism(s) of enantioselectivity. The final section of this chapter will be devoted to describing this system in terms of the principles previously discussed.

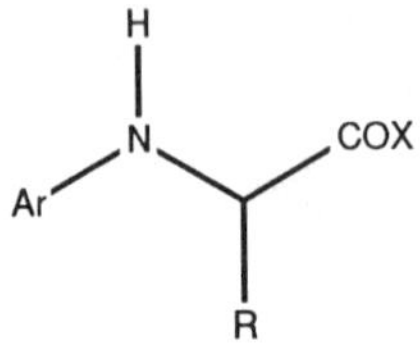

**14** 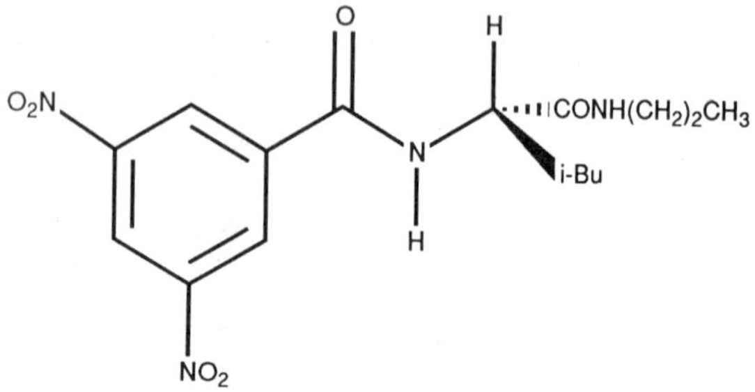 **15**

## A PRACTICAL EXAMPLE: CSPs DERIVED FROM *N*-ARYL AMINO ACIDS

$\pi$-Donor-type CSPs derived from *N*-arylamino acids, *14*, covalently linked to a microparticulate silica support were developed recently at Urbana [53]. These CSPs show a high degree of enantioselectivity for analytes incorporating $\pi$-acceptor functionality, with separation factors of greater than 100 being noted in some cases [54]. These large separation factors are unprecedented for CSPs operating by noncovalent mechanisms, and the mode of action of these phases has been studied in detail. These CSPs resulted from an idea as to what would constitute a successful CSP for the separation of $\pi$-acceptor-substituted species, followed by experimental work aimed at optimizing the basic design. "Optimizing" is, of course, a relative term when applied to CSPs: as noted earlier, optimization of a CSP for the separation of the enantiomers of one type of compound may render the CSP of marginal use for other analytes. Recent work in fact indicates that commercially available *N*-arylamino acid CSP presently marketed (CSP *16*) can be improved on in terms of enantioselectivity [55]. Still, CSP *16* in many ways represents the benefits and drawbacks inherent in a designed CSP and provides an excellent example of how the considerations discussed above are practically applied.

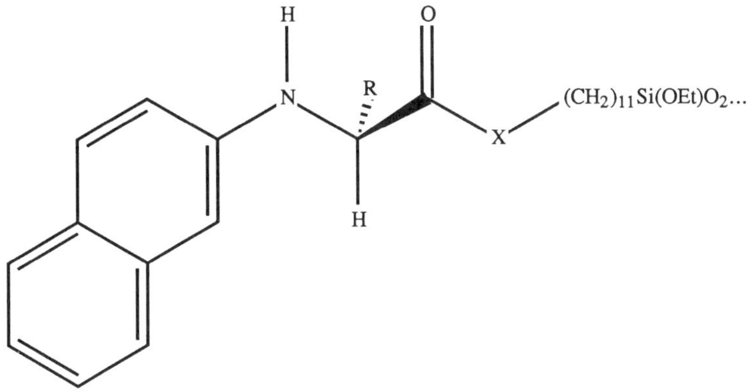

$(CH_2)_{11}Si(OEt)O_2...$

**16**

a, X = NH
b, X = O

The logic of reciprocity discussed above has led to a wide variety of π-donor- and π-acceptor-based CSPs. Separation of *N*-(3,5-dinitrobenzoyl)amino acid derivatives on the initial fluoroalcohol CSPs (e.g., CSP *8b*) led to the very general and widely accepted CSPs *9a* and *9b*. These CSPs were soon commercialized and have been demonstrated to provide facile enantiomer separations for a wide variety of analytes. The popularity of these CSPs may be attributed not only to their generality but also to the ease with which they are prepared using readily available nonracemic precursors. Availability is the key to utility. Hence it was desired that ease of preparation be extended to reciprocal π-donor-type phases. Several reciprocal π-donor-type phases previously prepared, most notably those of Oi and Pirkle, have been described. Despite their scope, few of these have been commercialized, primarily because of the difficulty of preparation. In most instances, the amines from which they are derived are not commercially available. If, however, an enantiomerically pure α-amino acid could be used as a precursor for a π-donor phase, the availability problem would be overcome. The most obvious means of preparing a π-donor CSP from enantiomerically pure amino acids, i.e., by acylation of the free amine with a π-donor-bearing functionality, suffers from a number of drawbacks, most of which are apparent when the previously discussed guidelines are considered. If, for example, the amino acid is *N*-naphthoylated and attached to the silica support through an amide linkage (CSP *17*) in a fashion analogous to that of CSPs *9a* and *9b*, there is now little exclusivity among sites available for interaction with the analyte. In CSPs *9a* and *9b*, the amide proton of the dinitrobenzamide is by far more acidic than the amide proton of the

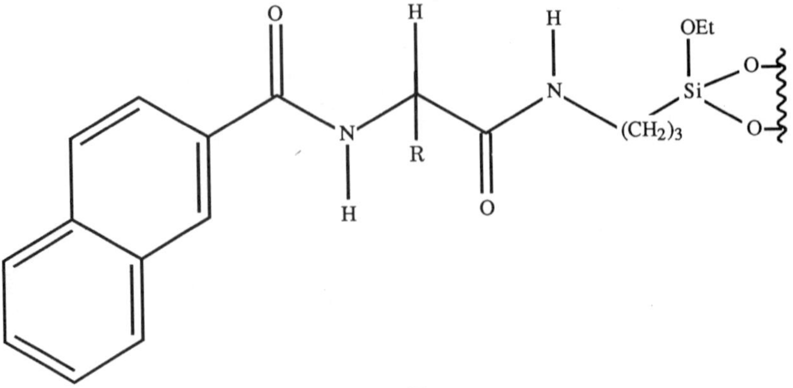

17

C-terminal linking amide as evidenced by its downfield chemical shift. This is due to the electron-withdrawing effects of the dinitrobenzoyl ring. With no such withdrawing effect from the naphthyl ring of *17*, the acidities of the two amide protons are comparable, and hence, our guideline which demands some degree of mutual exclusivity among the available interactions is violated. Furthermore, the electron withdrawing carbonyl substituent attenuates the π-donor character of the naphthyl ring, making this site less useful for donor-acceptor interactions. It seems, then, that CSP *17*, though it may work to some degree, is not really a viable design for a π-donor amino acid-based CSP. To avoid attenuation of π-donor ability, one might acylate the amino group of the α-amino acid with α-(1-naphthyl)acetyl chloride. However, this does not alleviate the site ambiguity which arises from two N-H hydrogen-bonding sites in the same molecule. Furthermore, the π donor loses some of its vital steric connectivity to the chiral center owing to relatively free rotation about the methylene bridge between the carbonyl oxygen and the naphthyl ring [31]. This conformational flexibility between the interaction sites is undesirable, for it will reduce the extent of enantioselectivity.

Still another possibility is to directly bond the amino nitrogen to a π-basic aromatic system. Aromatic amines are good π donors and should interact with π acceptors such as dinitrobenzoyl groups. Furthermore, the delocalization of the nitrogen lone pair into the aromatic ring should provide a degree of conformational restraint with respect to the stereogenic center. The absence of a carbonyl group at the NH site reduces the interactions potentially available, thus causing the phase to be mechanistically simpler and more amenable to optimization. Indeed, CSPs *16a* and *16b*, prepared from *N*-(2-naphthyl)-alanine, show a high degree of enantioselectivity for a variety of π-acceptor-substituted analytes [56]. If, as in CSP *16a*, a C-terminal amide is used to bind the *N*-arylamino acid to a stationary support, there will still be some site ambiguity presented by two available acidic hydrogens, one being the aromatic amine N-H and the other being the amide N-H. This ambiguity is avoided by using an ester rather than an amide linkage for bonding the chiral selector to the stationary support. The superiority of the ester linkage is demonstrated by the larger separation factors afforded by CSP *16b* as compared to those seen on CSP *16a*.

Another variable to be considered in this system is the nature of the substituent appended to the chiral center of the CSP. Although this substituent is not expected to be directly involved in the interaction between CSP and analyte, the choice is nevertheless important as the substituent will control approach to the "backside" of the molecule and will also influence the conformational rigidity of the system. Such indirect effects are quite important, and to obtain a CSP of broad scope one must avoid rigid systems which might afford a very high degree of enantioselectivity for a small number of analytes and seek instead a somewhat more flexible system which affords an attenuated degree of chiral recognition but suffices for a large clientele of analytes. *N*-(2-Naphthyl)alanine was initially the amino acid of choice. Althought the methyl substituent provides less steric bulk at the $\alpha$ carbon than others might, it provides an adequate degree of conformational bias and affords efficient enantioselectivity of a variety of acceptor analytes [48].

Although further optimization of type *16* CSPs is under way, it is clear that, by several logical choices, a CSP showing a great deal of selectivity for a large clientele of $\pi$-acceptor-substituted analytes has been designed from first principles. From a body of chromatographic and spectroscopic data, including intermolecular nuclear Overhauser effects (NOEs) observed in soluble complexes of the homochiral (S-S) pairing of *14n* and *15*, a mechanism accounting for the high degree of enantioselectivity afforded by CSP *16* has been proposed. This model is shown in Fig. 8, along with some of the observed NOEs. The importance of the $\pi$ donor-acceptor interaction is manifest, controlling as it does the allowed relative orientations of the two molecules in the complex, thereby effectively precluding other interactions which might dilute the observed enantioselectivity. As opposed to CSP *11*, this mechanism, or variations thereof, seems to be the only one operant to a significant degree on CSP *16*. This typically limits the clientele of CSP *16* to compounds in which the chiral center is adjacent to the sites of interaction with the CSP, and in which the appropriate $\pi$-acceptor and hydrogen-bonding sites are either present initially or introduccable by derivatization.

These considerations should not be taken for shortcomings, for CSP *16* is performing as it was designed to do. The use of but a single chiral recognition mechanism coupled with a paucity of sites for nonspecific retention make this CSP extremely selective for its client ana-

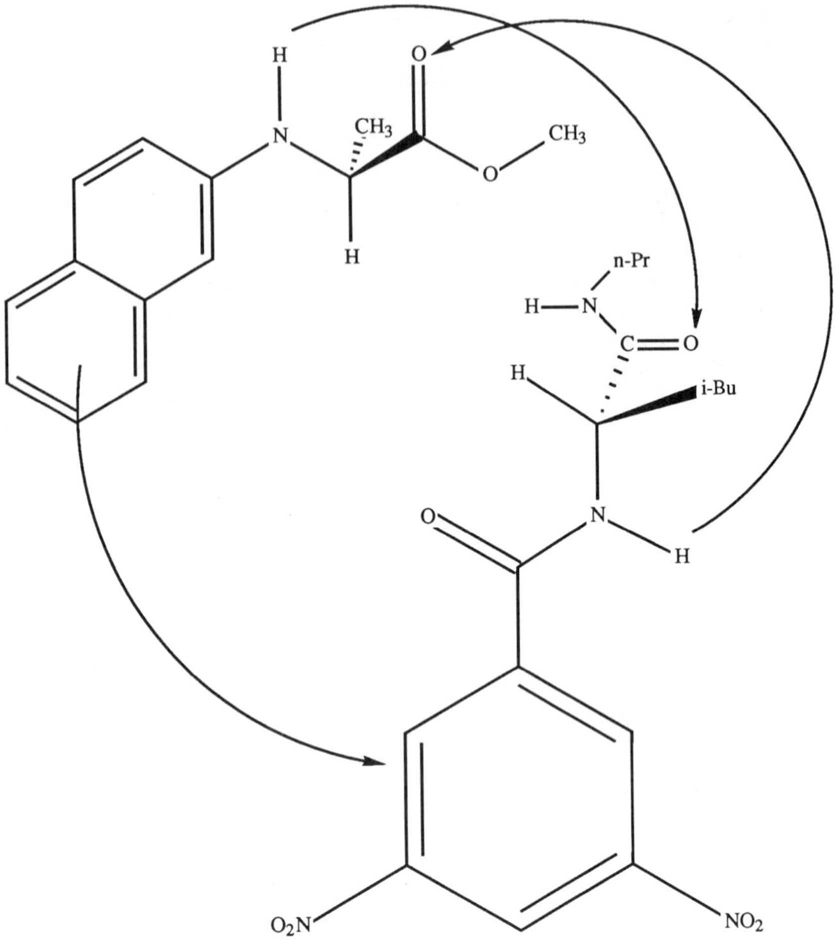

Fig. 8 Energetically favored homochiral complex between type *14* *N*-(2-naphthyl)alanine methyl ester and *N*-(3,5-dinitrobenzoyl)leucine-*n*-propylamide *15*. The three interactions involved in the complexation are a hydrogen bond between the dinitrobenzamide N-*H* of *15* and the carbonyl oxygen of *14*, a second hydrogen bond between the amino N-*H* of *14* and the C-terminal amide carbonyl oxygen of *15*, and efficient donor-acceptor complexation between the naphthyl ring of *14* and the dinitrobenzoyl ring of *15*.

lytes. The mechanistic simplicity makes it possible to relate enantiomer elution orders to absolute configuration.

## CONCLUSION

Recent developments in CSP technology are the result of several factors. Improvements in HPLC equipment and ancillary techniques have made enantiomer separation feasible despite the typically modest free-energy differences observed between diastereomeric complexes. The use of NMR to directly observe the interactions involved in the formation of the diastereomeric complexes has in a number of cases clarified the nature of chiral recognition and suggested avenues for CSP design. Perhaps most importantly, the development of logical design strategies of CSPs has made serendipity a less important factor in CSP development.

Future CSPs will most likely fall into one of two categories. The broad-spectrum CSP, which is capable of separating the enantiomers of a variety of compounds by introduction of appropriate functionality, will certainly be developed further. As scope and selectivity continue to improve, such CSPs become attractive for preparative separations. For commercial scale resolution of a particular racemate, single-client CSPs with a high degree of selectivity are likely to be developed. The prototype for these CSPs are affinity and bound antibody stationary phases, these being extremely selective for a given hapten. Equally intriguing is the use of template-imprinted polymers as selective CSPs, such as the phases designed by Wulff et al. [56,57]. The CSP of Wulff et al. consists of a methylmethacrylate-ethylenedimethacrylate copolymer originally templated around stryrenyl borate esters of d-4-nitrophenylmannopyranoside, *18*. After polymerization and hydrolytic removal of the pyranosyl moiety, the remaining chiral cavities are of a size and shape suitable for reincorporation of the template, the d enantiomer of nitrophenyl mannopyranoside. Passage of a racemic mixture

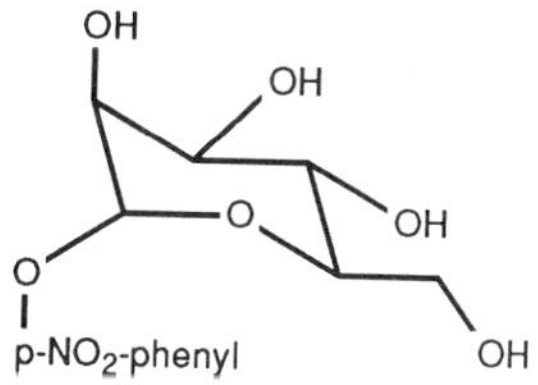

18

of *18* over this polymeric CSP results in the preferred retention of the d enantiomer. Further developments along these lines are likely, and may result in a variety of highly selective, specific CSPs which can be used repetitively for large-scale enantiomer separation.

CSP design is a field only now giving a hint of its full potential. As the principles governing efficient CSP design are elucidated, the expansion of this technique to as yet unexplored areas is inevitable. CSP design is not only practical; it has important theoretical implications as well, since the forces which govern CSP-analyte interactions are the same as those governing all molecular recognition systems, including drug-receptor and enzyme-substrate interactions. As such, discoveries in the (relatively) controllable realm of CSP design will increase understanding of how specific multimolecular complexes are formed.

## ACKNOWLEDGMENT

This work was supported by grants from the National Science Foundation and from Eli Lilly and Company.

## REFERENCES

1. W. H. Pirkle and T. C. Pochapsky, *Advances in Chromatography*, Vol. 27 (J. C. Giddings, E. Grushka, and P. R. Brown, eds.), Marcel Dekker, New York, 1987, Chap. 3.
2. J. M. Finn, *Chromatographic Chiral Separations*, in *Chromatographic Science*, Vol. 40 (M. Zeif and L. J. Crane, eds.), Marcel Dekker, New York, 1987, Chap. 3.
3. For a thorough review of stereochemistry and chirality, see B. Testa, *Principles of Organic Stereochemistry*, Marcel Dekker, New York, 1979, Chap. 1.
4. For a review of symmetry and its relation to molecular orbital theory, see S. F. A. Kettle, *Symmetry and Structure*, John Wiley and Sons, New York, 1985.
5. L. Salem, X. Chapuisat, G. Segal, P. C. Hiberty, C. Minot, C. Leforrestier, and P. Sautet, *J. Am. Chem. Soc. 109*:2887-2894 (1987).
6. V. A. Davankov and S. V. Rogozhin, *J. Chem. Soc.*, 490 (1971).
7. V. A. Davankov, *Advances in Chromatography*, Vol. 18 (J. C. Giddings, eds.), Marcel Dekker, New York, 1980, Chap. 4.
8. V. A. Davankov, A. A. Kurganov, and A. S. Bochkov, *Advances in Chromatography*, Vol. 22 (J. C. Giddings, eds.), Marcel Dekker, New York, 1983, Chap. 3.

9. O. Griffeth, personal communication.
10. R. Adams and A. W. Ingersoll, *J. Am. Chem. Soc. 44*:2930-2937 (1922).
11. H. K. Ihrig and C. W. Porter, *J. Am. Chem. Soc. 45*:1990-1993 (1923).
12. E. Gil-Av, B. Feibush, and R. Charles-Sigler, *Tetrahedron Lett.*, 1009 (1966).
14. W. Parr and P. Y. Howard, *Anal. Chem., 45*:711 (1973).
15. S. Weinstein, B. Feibush, and E. Gil-Av, *J. Chromatogr., 126*: 97 (1976).
16. B. Feibush, *J. Chem. Soc., Chem. Commun.*, 544 (1971).
17. E. Gil-Av, *Chromatogr. Library, 32*:131 (1985).
18. S. Nakaparksin, P. Birrell, E. Gil-Av., and J. Oro, *J. Chromatogr. Sci., 8*:177 (1970).
19. B. Feibush and E. Gil-Av, *Tetrahedron, 26*:1361 (1970).
20. E. Gil-Av and B. Feibush, in *Peptides, 1974*, Proc. 13th European Peptide Symposium, Jerusalem (Y. Wolman, ed.), p. 279.
21. B. Feibush, A. Balen, B. Altman, and E. Gil-Av, *J. Chem. Soc., Perkin Trans. II*, 1230 (1979).
22. S. Weinstein, L. Leiserowitz, and E. Gil-Av, *J. Am. Chem. Soc. 102*:2768 (1980).
23. C. H. Lochmuller and R. W. Souter, *J. Chromatogr., 87*:243 (1973).
24. C. H. Lochmuller and R. W. Souter, *J. Chromatogr., 88*:42 (1974).
25. J. A. Corbin and L. B. Rogers, *Anal. Chem., 42*:974 (1970).
26. A. Dobashi, K. Oka, and S. Hara, *J. Am. Chem. Soc. 102*:7122 (1980) and refs. therein.
27. S. Hara and A. Dobashi, *J. Liq. Chromatogr., 2*:883 (1979).
28. N. Oi and H. Kitahara, *J. Chromatogr., 285*:198 (1984).
29. W. H. Pirkle, *Tetrahedron Lett., 24*:5707 (1983).
30. S. Hara and A. Dobashi, *J. Chromatogr., 186*:543 (1979).
31. H. Berndt and G. Kurger, *J. Chromatogr., 348*:275-279 (1985).
32. J. Rebek and D. Nemeth, *J. Am. Chem. Soc. 108*:5637 (1986).
33. G. D. Y. Sogah and D. J. Cram, *J. Am. Chem. Soc. 98*:3038 (1979).
34. F. Mikes and G. Boshart, *J. Chem. Soc., Chem. Commun.*, 173 (1978).
35. For a review, see W. H. Pirkle and D. J. Hoover, in *Topics in Stereochemistry*, Vol. 13 (S. Wilen and E. Eliel, eds.), John Wiley and Sons, New York, 1982.
36. W. H. Pirkle and D. L. Sikkenga, *J. Org. Chem., 40*:3430 (1975).
37. W. H. Pirkle and D. House, *J. Org. Chem., 44*:1957 (1979).
38. J. M. Finn, Ph.D. thesis, University of Illinoise, 1981.
39. W. H. Pirkle, J. M. Finn, J. L. Schreiner, and B. C. Hamper, *J. Am. Chem. Soc. 103*:3964 (1981).

40. N. Oi and H. Kitahara, *J. Chromatogr.*, *265*:117 (1983).
41. N. Oi, M. Nagase, Y. Inda, and T. Doi, *J. Chromatogr.*, *265*:111 (1983).
42. W. H. Pirkle and M. H. Hyun, *J. Org. Chem.*, *49*:3034 (1984).
43. B. Lefebvre, R. Audebert, and C. Quivoron, *Israeli J. Chem.*, *15*:69 (1977).
44. D. Charmot, R. Audebert, and C. Quivoron, *J. Liq. Chromatogr.* *8*:1753 (1985).
45. W. H. Pirkle, M. H. Hyun, and B. Bank, *J. Chromatogr.*, *316*: 585 (1984).
46. W. H. Pirkle and M. H. Hyun, *J. Chromatogr.*, *322*:295 (1984).
47. W. H. Pirkle and M. H. Hyun, *J. Chromatogr.*, *328*:1 (1985).
48. T. C. Pochapsky, Ph.D. thesis, University of Illinois, 1986.
49. W. H. Pirkle and D. L. Sikkenga, *J. Org. Chem.*, *42*:370 (1977).
50. G. Dotsevi, G. D. Y. Sogah, and D. J. Cram, *J. Am. Chem. Soc.* *97*:1295 (1975).
51. G. D. Y. Sogah and D. J. Cram, *J. Am. Chem. Soc.* *101*:3035 (1979).
52. W. H. Pirkle and T. C. Pochapsky, *J. Am. Chem. Soc.*, *109*:5975 (1987).
53. W. H. Pirkle and T. C. Pochapsky, *J. Am. Chem. Soc.*, *108*:352 (1986).
54. W. H. Pirkle and T. C. Pochapsky, *J. Chromatogr.*, *369*:175 (1986).
55. W. H. Pirkle and K. Deming, unpublished results.
56. G. Wulff, R. Grobe-Einsler, W. Vesper, and A. Sarhan, *Makromol. Chem.*, *178*:2817 (1977).
57. G. Wulff, R. Kemmerer, J. Vietmeier, and H.-G. Poll, *Nouv. J. Chim.*, *16*:681 (1982).

# Index

B

C

H



I

M

P

T



W

Z